Data Science and Machine Learning

Mathematical and Statistical Methods

Chapman & Hall/CRC Machine Learning & Pattern Recognition

Introduction to Machine Learning with Applications in Information Security
Mark Stamp

A First Course in Machine Learning
Simon Rogers, Mark Girolami

Statistical Reinforcement Learning: Modern Machine Learning Approaches
Masashi Sugiyama

Sparse Modeling: Theory, Algorithms, and Applications
Irina Rish, Genady Grabarnik

Computational Trust Models and Machine Learning
Xin Liu, Anwitaman Datta, Ee-Peng Lim

Regularization, Optimization, Kernels, and Support Vector Machines
Johan A.K. Suykens, Marco Signoretto, Andreas Argyriou

Machine Learning: An Algorithmic Perspective, Second Edition
Stephen Marsland

Bayesian Programming
Pierre Bessiere, Emmanuel Mazer, Juan Manuel Ahuactzin, Kamel Mekhnacha

Multilinear Subspace Learning: Dimensionality Reduction of Multidimensional Data
Haiping Lu, Konstantinos N. Plataniotis, Anastasios Venetsanopoulos

Data Science and Machine Learning: Mathematical and Statistical Methods
Dirk P. Kroese, Zdravko I. Botev, Thomas Taimre, Radislav Vaisman

For more information on this series please visit: https://www.crcpress.com/Chapman--HallCRC-Machine-Learning--Pattern-Recognition/book-series/erie

To my wife and daughters: Lesley, Elise, and Jessica

— DPK

To Sarah, Sofia, and my parents

— ZIB

To my grandparents: Arno, Harry, Juta, and Maila

— TT

To Valerie

— RV

CONTENTS

Preface xiii

Notation xvii

1 Importing, Summarizing, and Visualizing Data 1
1.1 Introduction 1
1.2 Structuring Features According to Type 3
1.3 Summary Tables 6
1.4 Summary Statistics 7
1.5 Visualizing Data 8
1.5.1 Plotting Qualitative Variables 9
1.5.2 Plotting Quantitative Variables 9
1.5.3 Data Visualization in a Bivariate Setting 12
Exercises 15

2 Statistical Learning 19
2.1 Introduction 19
2.2 Supervised and Unsupervised Learning 20
2.3 Training and Test Loss 23
2.4 Tradeoffs in Statistical Learning 31
2.5 Estimating Risk 35
2.5.1 In-Sample Risk 35
2.5.2 Cross-Validation 37
2.6 Modeling Data 40
2.7 Multivariate Normal Models 44
2.8 Normal Linear Models 46
2.9 Bayesian Learning 47
Exercises 58

3 Monte Carlo Methods 67
3.1 Introduction 67
3.2 Monte Carlo Sampling 68
3.2.1 Generating Random Numbers 68
3.2.2 Simulating Random Variables 69
3.2.3 Simulating Random Vectors and Processes 74
3.2.4 Resampling 76
3.2.5 Markov Chain Monte Carlo 78
3.3 Monte Carlo Estimation 85

3.3.1 Crude Monte Carlo . . . 85
3.3.2 Bootstrap Method . . . 88
3.3.3 Variance Reduction . . . 92
3.4 Monte Carlo for Optimization . . . 96
3.4.1 Simulated Annealing . . . 96
3.4.2 Cross-Entropy Method . . . 100
3.4.3 Splitting for Optimization . . . 103
3.4.4 Noisy Optimization . . . 105
Exercises . . . 113

4 Unsupervised Learning **121**
4.1 Introduction . . . 121
4.2 Risk and Loss in Unsupervised Learning . . . 122
4.3 Expectation–Maximization (EM) Algorithm . . . 128
4.4 Empirical Distribution and Density Estimation . . . 131
4.5 Clustering via Mixture Models . . . 135
4.5.1 Mixture Models . . . 135
4.5.2 EM Algorithm for Mixture Models . . . 137
4.6 Clustering via Vector Quantization . . . 142
4.6.1 K-Means . . . 144
4.6.2 Clustering via Continuous Multiextremal Optimization . . . 146
4.7 Hierarchical Clustering . . . 147
4.8 Principal Component Analysis (PCA) . . . 153
4.8.1 Motivation: Principal Axes of an Ellipsoid . . . 153
4.8.2 PCA and Singular Value Decomposition (SVD) . . . 155
Exercises . . . 160

5 Regression **167**
5.1 Introduction . . . 167
5.2 Linear Regression . . . 169
5.3 Analysis via Linear Models . . . 171
5.3.1 Parameter Estimation . . . 171
5.3.2 Model Selection and Prediction . . . 172
5.3.3 Cross-Validation and Predictive Residual Sum of Squares . . . 173
5.3.4 In-Sample Risk and Akaike Information Criterion . . . 175
5.3.5 Categorical Features . . . 177
5.3.6 Nested Models . . . 180
5.3.7 Coefficient of Determination . . . 181
5.4 Inference for Normal Linear Models . . . 182
5.4.1 Comparing Two Normal Linear Models . . . 183
5.4.2 Confidence and Prediction Intervals . . . 186
5.5 Nonlinear Regression Models . . . 188
5.6 Linear Models in Python . . . 191
5.6.1 Modeling . . . 191
5.6.2 Analysis . . . 193
5.6.3 Analysis of Variance (ANOVA) . . . 195

5.6.4 Confidence and Prediction Intervals . . . 198
5.6.5 Model Validation . . . 198
5.6.6 Variable Selection . . . 199
5.7 Generalized Linear Models . . . 204
Exercises . . . 207

6 Regularization and Kernel Methods **215**
6.1 Introduction . . . 215
6.2 Regularization . . . 216
6.3 Reproducing Kernel Hilbert Spaces . . . 222
6.4 Construction of Reproducing Kernels . . . 225
6.4.1 Reproducing Kernels via Feature Mapping . . . 225
6.4.2 Kernels from Characteristic Functions . . . 225
6.4.3 Reproducing Kernels Using Orthonormal Features . . . 227
6.4.4 Kernels from Kernels . . . 229
6.5 Representer Theorem . . . 231
6.6 Smoothing Cubic Splines . . . 235
6.7 Gaussian Process Regression . . . 239
6.8 Kernel PCA . . . 243
Exercises . . . 246

7 Classification **253**
7.1 Introduction . . . 253
7.2 Classification Metrics . . . 255
7.3 Classification via Bayes' Rule . . . 259
7.4 Linear and Quadratic Discriminant Analysis . . . 261
7.5 Logistic Regression and Softmax Classification . . . 268
7.6 K-Nearest Neighbors Classification . . . 270
7.7 Support Vector Machine . . . 271
7.8 Classification with Scikit-Learn . . . 279
Exercises . . . 281

8 Decision Trees and Ensemble Methods **289**
8.1 Introduction . . . 289
8.2 Top-Down Construction of Decision Trees . . . 291
8.2.1 Regional Prediction Functions . . . 292
8.2.2 Splitting Rules . . . 293
8.2.3 Termination Criterion . . . 294
8.2.4 Basic Implementation . . . 296
8.3 Additional Considerations . . . 300
8.3.1 Binary Versus Non-Binary Trees . . . 300
8.3.2 Data Preprocessing . . . 300
8.3.3 Alternative Splitting Rules . . . 300
8.3.4 Categorical Variables . . . 301
8.3.5 Missing Values . . . 301
8.4 Controlling the Tree Shape . . . 302
8.4.1 Cost-Complexity Pruning . . . 305

8.4.2 Advantages and Limitations of Decision Trees . . . 306
8.5 Bootstrap Aggregation . . . 307
8.6 Random Forests . . . 311
8.7 Boosting . . . 315
Exercises . . . 323

9 Deep Learning **325**
9.1 Introduction . . . 325
9.2 Feed-Forward Neural Networks . . . 328
9.3 Back-Propagation . . . 332
9.4 Methods for Training . . . 336
9.4.1 Steepest Descent . . . 336
9.4.2 Levenberg–Marquardt Method . . . 337
9.4.3 Limited-Memory BFGS Method . . . 338
9.4.4 Adaptive Gradient Methods . . . 340
9.5 Examples in Python . . . 342
9.5.1 Simple Polynomial Regression . . . 342
9.5.2 Image Classification . . . 346
Exercises . . . 350

A Linear Algebra and Functional Analysis **357**
A.1 Vector Spaces, Bases, and Matrices . . . 357
A.2 Inner Product . . . 362
A.3 Complex Vectors and Matrices . . . 363
A.4 Orthogonal Projections . . . 364
A.5 Eigenvalues and Eigenvectors . . . 365
A.5.1 Left- and Right-Eigenvectors . . . 366
A.6 Matrix Decompositions . . . 370
A.6.1 (P)LU Decomposition . . . 370
A.6.2 Woodbury Identity . . . 372
A.6.3 Cholesky Decomposition . . . 375
A.6.4 QR Decomposition and the Gram–Schmidt Procedure . . . 377
A.6.5 Singular Value Decomposition . . . 378
A.6.6 Solving Structured Matrix Equations . . . 381
A.7 Functional Analysis . . . 386
A.8 Fourier Transforms . . . 392
A.8.1 Discrete Fourier Transform . . . 394
A.8.2 Fast Fourier Transform . . . 396

B Multivariate Differentiation and Optimization **399**
B.1 Multivariate Differentiation . . . 399
B.1.1 Taylor Expansion . . . 402
B.1.2 Chain Rule . . . 402
B.2 Optimization Theory . . . 404
B.2.1 Convexity and Optimization . . . 405
B.2.2 Lagrangian Method . . . 408
B.2.3 Duality . . . 409

B.3 Numerical Root-Finding and Minimization 410
B.3.1 Newton-Like Methods 411
B.3.2 Quasi-Newton Methods 413
B.3.3 Normal Approximation Method 415
B.3.4 Nonlinear Least Squares 416
B.4 Constrained Minimization via Penalty Functions 417

C Probability and Statistics **423**
C.1 Random Experiments and Probability Spaces 423
C.2 Random Variables and Probability Distributions 424
C.3 Expectation 428
C.4 Joint Distributions 429
C.5 Conditioning and Independence 430
C.5.1 Conditional Probability 430
C.5.2 Independence 430
C.5.3 Expectation and Covariance 431
C.5.4 Conditional Density and Conditional Expectation 433
C.6 Functions of Random Variables 433
C.7 Multivariate Normal Distribution 436
C.8 Convergence of Random Variables 441
C.9 Law of Large Numbers and Central Limit Theorem 447
C.10 Markov Chains 453
C.11 Statistics 455
C.12 Estimation 456
C.12.1 Method of Moments 457
C.12.2 Maximum Likelihood Method 458
C.13 Confidence Intervals 459
C.14 Hypothesis Testing 460

D Python Primer **465**
D.1 Getting Started 465
D.2 Python Objects 467
D.3 Types and Operators 468
D.4 Functions and Methods 470
D.5 Modules 471
D.6 Flow Control 473
D.7 Iteration 474
D.8 Classes 475
D.9 Files 477
D.10 NumPy 480
D.10.1 Creating and Shaping Arrays 480
D.10.2 Slicing 482
D.10.3 Array Operations 482
D.10.4 Random Numbers 484
D.11 Matplotlib 485
D.11.1 Creating a Basic Plot 485

D.12 Pandas 487
D.12.1 Series and DataFrame 487
D.12.2 Manipulating Data Frames 489
D.12.3 Extracting Information 490
D.12.4 Plotting 492
D.13 Scikit-learn 492
D.13.1 Partitioning the Data 493
D.13.2 Standardization 493
D.13.3 Fitting and Prediction 494
D.13.4 Testing the Model 494
D.14 System Calls, URL Access, and Speed-Up 495

Bibliography **497**

Index **505**

PREFACE

In our present world of automation, cloud computing, algorithms, artificial intelligence, and big data, few topics are as relevant as *data science* and *machine learning*. Their recent popularity lies not only in their applicability to real-life questions, but also in their natural blending of many different disciplines, including mathematics, statistics, computer science, engineering, science, and finance.

To someone starting to learn these topics, the multitude of computational techniques and mathematical ideas may seem overwhelming. Some may be satisfied with only learning how to use off-the-shelf recipes to apply to practical situations. But what if the assumptions of the black-box recipe are violated? Can we still trust the results? How should the algorithm be adapted? To be able to truly understand data science and machine learning it is important to appreciate the underlying mathematics and statistics, as well as the resulting algorithms.

The purpose of this book is to provide an accessible, yet comprehensive, account of data science and machine learning. It is intended for anyone interested in gaining a better understanding of the mathematics and statistics that underpin the rich variety of ideas and machine learning algorithms in data science. Our viewpoint is that computer languages come and go, but the underlying key ideas and algorithms will remain forever and will form the basis for future developments.

Before we turn to a description of the topics in this book, we would like to say a few words about its philosophy. This book resulted from various courses in data science and machine learning at the Universities of Queensland and New South Wales, Australia. When we taught these courses, we noticed that students were eager to learn not only how to apply algorithms but also to understand how these algorithms actually work. However, many existing textbooks assumed either too much background knowledge (e.g., measure theory and functional analysis) or too little (everything is a black box), and the information overload from often disjointed and contradictory internet sources made it more difficult for students to gradually build up their knowledge and understanding. We therefore wanted to write a book about data science and machine learning that can be read as a linear story, with a substantial "backstory" in the appendices. The main narrative starts very simply and builds up gradually to quite an advanced level. The backstory contains all the necessary background, as well as additional information, from linear algebra and functional analysis (Appendix A), multivariate differentiation and optimization (Appendix B), and probability and statistics (Appendix C). Moreover, to make the abstract ideas come alive, we believe it is important that the reader sees actual implementations of the algorithms, directly translated from the theory. After some deliberation we have chosen Python as our programming language. It is freely available and has been adopted as the programming language of choice for many practitioners in data science and machine learning. It has many useful packages for data manipulation (often ported from R) and has been designed to be easy to program. A gentle introduction to Python is given in Appendix D.

To keep the book manageable in size we had to be selective in our choice of topics. Important ideas and connections between various concepts are highlighted via *keywords* and page references (indicated by a ☞) in the margin. Key definitions and theorems are highlighted in boxes. Whenever feasible we provide proofs of theorems. Finally, we place great importance on *notation*. It is often the case that once a consistent and concise system of notation is in place, seemingly difficult ideas suddenly become obvious. We use different fonts to distinguish between different types of objects. Vectors are denoted by letters in boldface italics, $\boldsymbol{x}, \boldsymbol{X}$, and matrices by uppercase letters in boldface roman font, $\mathbf{A}, \mathbf{K}$. We also distinguish between random vectors and their values by using upper and lower case letters, e.g., $\boldsymbol{X}$ (random vector) and $\boldsymbol{x}$ (its value or outcome). Sets are usually denoted by calligraphic letters $\mathcal{G}, \mathcal{H}$. The symbols for probability and expectation are $\mathbb{P}$ and $\mathbb{E}$, respectively. Distributions are indicated by sans serif font, as in Bin and Gamma; exceptions are the ubiquitous notations $\mathcal{N}$ and $\mathcal{U}$ for the normal and uniform distributions. A summary of the most important symbols and abbreviations is given on Pages xvii–xxi.

KEYWORDS

☞ xvii

Data science provides the language and techniques necessary for understanding and dealing with data. It involves the design, collection, analysis, and interpretation of numerical data, with the aim of extracting patterns and other useful information. Machine learning, which is closely related to data science, deals with the design of algorithms and computer resources to learn from data. The organization of the book follows roughly the typical steps in a data science project: Gathering data to gain information about a research question; cleaning, summarization, and visualization of the data; modeling and analysis of the data; translating decisions about the model into decisions and predictions about the research question. As this is a mathematics and statistics oriented book, most emphasis will be on modeling and analysis.

We start in Chapter 1 with the reading, structuring, summarization, and visualization of data using the data manipulation package pandas in Python. Although the material covered in this chapter requires no mathematical knowledge, it forms an obvious starting point for data science: to better understand the nature of the available data. In Chapter 2, we introduce the main ingredients of *statistical learning*. We distinguish between *supervised* and *unsupervised* learning techniques, and discuss how we can assess the predictive performance of (un)supervised learning methods. An important part of statistical learning is the *modeling* of data. We introduce various useful models in data science including linear, multivariate Gaussian, and Bayesian models. Many algorithms in machine learning and data science make use of Monte Carlo techniques, which is the topic of Chapter 3. Monte Carlo can be used for simulation, estimation, and optimization. Chapter 4 is concerned with unsupervised learning, where we discuss techniques such as density estimation, clustering, and principal component analysis. We then turn our attention to supervised learning in Chapter 5, and explain the ideas behind a broad class of regression models. Therein, we also describe how Python's `statsmodels` package can be used to define and analyze linear models. Chapter 6 builds upon the previous regression chapter by developing the powerful concepts of kernel methods and regularization, which allow the fundamental ideas of Chapter 5 to be expanded in an elegant way, using the theory of reproducing kernel Hilbert spaces. In Chapter 7, we proceed with the classification task, which also belongs to the supervised learning framework, and consider various methods for classification, including Bayes classification, linear and quadratic discriminant analysis, K-nearest neighbors, and support vector machines. In Chapter 8 we consider versatile methods for regression and classification that make use of tree structures. Finally, in Chapter 9, we consider the workings of neural networks and deep learning, and show that these learning algorithms have a simple mathematical interpretation. An extensive range of exercises is provided at the end of each chapter.

> Python code and data sets for each chapter can be downloaded from the GitHub site:
> `https://github.com/DSML-book`

Acknowledgments

Some of the Python code for Chapters 1 and 5 was adapted from [73]. We thank Benoit Liquet for making this available, and Lauren Jones for translating the R code into Python.

We thank all who through their comments, feedback, and suggestions have contributed to this book, including Qibin Duan, Luke Taylor, Rémi Mouzayek, Harry Goodman, Bryce Stansfield, Ryan Tongs, Dillon Steyl, Bill Rudd, Nan Ye, Christian Hirsch, Chris van der Heide, Sarat Moka, Aapeli Vuorinen, Joshua Ross, Giang Nguyen, and the anonymous referees. David Grubbs deserves a special accollade for his professionalism and attention to detail in his role as Editor for this book.

The book was test-run during the 2019 *Summer School of the Australian Mathematical Sciences Institute*. More than 80 bright upper-undergraduate (Honours) students used the book for the course *Mathematical Methods for Machine Learning*, taught by Zdravko Botev. We are grateful for the valuable feedback that they provided.

Our special thanks go out to Robert Salomone, Liam Berry, Robin Carrick, and Sam Daley, who commented in great detail on earlier versions of the entire book and wrote and improved our Python code. Their enthusiasm, perceptiveness, and kind assistance have been invaluable.

Of course, none of this work would have been possible without the loving support, patience, and encouragement from our families, and we thank them with all our hearts.

This book was financially supported by the Australian Research Council *Centre of Excellence for Mathematical & Statistical Frontiers*, under grant number CE140100049.

Dirk Kroese, Zdravko Botev,
Thomas Taimre, and Radislav Vaisman
Brisbane and Sydney

NOTATION

We could, of course, use any notation we want; do not laugh at notations; invent them, they are powerful. In fact, mathematics is, to a large extent, invention of better notations.

Richard P. Feynman

We have tried to use a notation system that is, in order of importance, simple, descriptive, consistent, and compatible with historical choices. Achieving all of these goals all of the time would be impossible, but we hope that our notation helps to quickly recognize the type or "flavor" of certain mathematical objects (vectors, matrices, random vectors, probability measures, etc.) and clarify intricate ideas.

We make use of various typographical aids, and it will be beneficial for the reader to be aware of some of these.

• Boldface font is used to indicate composite objects, such as column vectors $\boldsymbol{x} = [x_1, \ldots, x_n]^\top$ and matrices $\mathbf{X} = [x_{ij}]$. Note also the difference between the upright bold font for matrices and the slanted bold font for vectors.

• Random variables are generally specified with upper case roman letters X, Y, Z and their outcomes with lower case letters x, y, z. Random vectors are thus denoted in upper case slanted bold font: $\boldsymbol{X} = [X_1, \ldots, X_n]^\top$.

• Sets of vectors are generally written in calligraphic font, such as $\mathcal{X}$, but the set of real numbers uses the common blackboard bold font $\mathbb{R}$. Expectation and probability also use the latter font.

• Probability distributions use a sans serif font, such as Bin and Gamma. Exceptions to this rule are the "standard" notations $\mathcal{N}$ and $\mathcal{U}$ for the normal and uniform distributions.

• We often omit brackets when it is clear what the argument is of a function or operator. For example, we prefer $\mathbb{E}X^2$ to $\mathbb{E}[X^2]$.

• We employ color to emphasize that certain words refer to a `dataset`, `function`, or `package` in Python. All code is written in `typewriter` font. To be compatible with past notation choices, we introduced a special blue symbol $\mathbf{X}$ for the model (design) matrix of a linear model.

• Important notation such as $\mathcal{T}$, g, g^* is often defined in a mnemonic way, such as $\mathcal{T}$ for "training", g for "guess", g^* for the "star" (that is, optimal) guess, and ℓ for "loss".

• We will occasionally use a Bayesian notation convention in which the *same* symbol is used to denote different (conditional) probability densities. In particular, instead of writing $f_X(x)$ and $f_{X \mid Y}(x \mid y)$ for the probability density function (pdf) of X and the conditional pdf of X given Y, we simply write $f(x)$ and $f(x \mid y)$. This particular style of notation can be of great descriptive value, despite its apparent ambiguity.

General font/notation rules

x	scalar
$\boldsymbol{x}$	vector
$\boldsymbol{X}$	random vector
$\mathbf{X}$	matrix
$\mathcal{X}$	set
$\widehat{x}$	estimate or approximation
x^*	optimal
$\overline{x}$	average

Common mathematical symbols

$\forall$	for all
$\exists$	there exists
$\propto$	is proportional to
$\perp$	is perpendicular to
$\sim$	is distributed as
$\stackrel{\text{iid}}{\sim}$, $\sim_{\text{iid}}$	are independent and identically distributed as
$\stackrel{\text{approx.}}{\sim}$	is approximately distributed as
∇f	gradient of f
$\nabla^2 f$	Hessian of f
$f \in C^p$	f has continuous derivatives of order p
$\approx$	is approximately
$\simeq$	is asymptotically
$\ll$	is much smaller than
$\oplus$	direct sum

$\odot$	elementwise product
$\cap$	intersection
$\cup$	union
$:=$, $=:$	is defined as
$\xrightarrow{\text{a.s.}}$	converges almost surely to
$\xrightarrow{\text{d}}$	converges in distribution to
$\xrightarrow{\mathbb{P}}$	converges in probability to
$\xrightarrow{L_p}$	converges in L_p-norm to
$\lVert \cdot \rVert$	Euclidean norm
$\lceil x \rceil$	smallest integer larger than x
$\lfloor x \rfloor$	largest integer smaller than x
x_+	$\max\{x, 0\}$

Matrix/vector notation

$\mathbf{A}^\top$, $\boldsymbol{x}^\top$	transpose of matrix $\mathbf{A}$ or vector $\boldsymbol{x}$
$\mathbf{A}^{-1}$	inverse of matrix $\mathbf{A}$
$\mathbf{A}^{+}$	pseudo-inverse of matrix $\mathbf{A}$
$\mathbf{A}^{-\top}$	inverse of matrix $\mathbf{A}^\top$ or transpose of $\mathbf{A}^{-1}$
$\mathbf{A} \succ 0$	matrix $\mathbf{A}$ is positive definite
$\mathbf{A} \succeq 0$	matrix $\mathbf{A}$ is positive semidefinite
$\dim(\boldsymbol{x})$	dimension of vector $\boldsymbol{x}$
$\det(\mathbf{A})$	determinant of matrix $\mathbf{A}$
$\lvert\mathbf{A}\rvert$	absolute value of the determinant of matrix $\mathbf{A}$
$\mathrm{tr}(\mathbf{A})$	trace of matrix $\mathbf{A}$

Reserved letters and words

$\mathbb{C}$	set of complex numbers
d	differential symbol
$\mathbb{E}$	expectation
e	the number $2.71828\ldots$
f	probability density (discrete or continuous)
g	prediction function
$\mathbb{1}\{A\}$ or $\mathbb{1}_A$	indicator function of set A
i	the square root of -1
ℓ	risk: expected loss

Loss	loss function
ln	(natural) logarithm
$\mathbb{N}$	set of natural numbers $\{0, 1, \dots\}$
$\mathcal{O}$	big-O order symbol: $f(x) = \mathcal{O}(g(x))$ if $\lvert f(x)\rvert \leqslant \alpha g(x)$ for some constant α as $x \to a$
o	little-o order symbol: $f(x) = o(g(x))$ if $f(x)/g(x) \to 0$ as $x \to a$
$\mathbb{P}$	probability measure
π	the number 3.14159 ...
$\mathbb{R}$	set of real numbers (one-dimensional Euclidean space)
$\mathbb{R}^n$	n-dimensional Euclidean space
$\mathbb{R}_+$	positive real line: $[0, \infty)$
τ	deterministic training set
$\mathcal{T}$	random training set
$\mathbf{X}$	model (design) matrix
$\mathbb{Z}$	set of integers $\{\dots, -1, 0, 1, \dots\}$

Probability distributions

Ber	Bernoulli
Beta	beta
Bin	binomial
Exp	exponential
Geom	geometric
Gamma	gamma
F	Fisher–Snedecor F
$\mathcal{N}$	normal or Gaussian
Pareto	Pareto
Poi	Poisson
t	Student's t
$\mathcal{U}$	uniform

Abbreviations and acronyms

cdf	cumulative distribution function
CMC	crude Monte Carlo
CE	cross-entropy
EM	expectation–maximization
GP	Gaussian process
KDE	Kernel density estimate/estimator

KL	Kullback–Leibler
KKT	Karush–Kuhn–Tucker
iid	independent and identically distributed
MAP	maximum *a posteriori*
MCMC	Markov chain Monte Carlo
MLE	maximum likelihood estimator/estimate
OOB	out-of-bag
PCA	principal component analysis
pdf	probability density function (discrete or continuous)
SVD	singular value decomposition

CHAPTER 1

IMPORTING, SUMMARIZING, AND VISUALIZING DATA

This chapter describes where to find useful data sets, how to load them into Python, and how to (re)structure the data. We also discuss various ways in which the data can be summarized via tables and figures. Which type of plots and numerical summaries are appropriate depends on the type of the variable(s) in play. Readers unfamiliar with Python are advised to read Appendix D first.

1.1 Introduction

Data comes in many shapes and forms, but can generally be thought of as being the result of some random experiment — an experiment whose outcome cannot be determined in advance, but whose workings are still subject to analysis. Data from a random experiment are often stored in a table or spreadsheet. A statistical convention is to denote variables — often called *features* — as *columns* and the individual items (or units) as *rows*. It is useful to think of three types of columns in such a spreadsheet: FEATURES

1. The first column is usually an identifier or index column, where each unit/row is given a unique name or ID.

2. Certain columns (features) can correspond to the design of the experiment, specifying, for example, to which experimental group the unit belongs. Often the entries in these columns are *deterministic*; that is, they stay the same if the experiment were to be repeated.

3. Other columns represent the observed measurements of the experiment. Usually, these measurements exhibit *variability*; that is, they would change if the experiment were to be repeated.

There are many data sets available from the Internet and in software packages. A well-known repository of data sets is the Machine Learning Repository maintained by the University of California at Irvine (UCI), found at `https://archive.ics.uci.edu/`.

These data sets are typically stored in a CSV (comma separated values) format, which can be easily read into Python. For example, to access the abalone data set from this website with Python, download the file to your working directory, import the pandas package via

```
import pandas as pd
```

and read in the data as follows:

```
abalone = pd.read_csv('abalone.data',header = None)
```

It is important to add `header = None`, as this lets Python know that the first line of the CSV does not contain the names of the features, as it assumes so by default. The data set was originally used to predict the age of abalone from physical measurements, such as shell weight and diameter.

Another useful repository of over 1000 data sets from various packages in the R programming language, collected by Vincent Arel-Bundock, can be found at:

https://vincentarelbundock.github.io/Rdatasets/datasets.html.

For example, to read Fisher's famous iris data set from R's `datasets` package into Python, type:

```
urlprefix = 'https://vincentarelbundock.github.io/Rdatasets/csv/'
dataname = 'datasets/iris.csv'
iris = pd.read_csv(urlprefix + dataname)
```

The iris data set contains four physical measurements (sepal/petal length/width) on 50 specimens (each) of 3 species of iris: setosa, versicolor, and virginica. Note that in this case the headers are included. The output of **read_csv** is a `DataFrame` object, which is pandas's implementation of a spreadsheet; see Section D.12.1. The `DataFrame` method **head** gives the first few rows of the `DataFrame`, including the feature names. The number of rows can be passed as an argument and is 5 by default. For the iris `DataFrame`, we have:

☞ 487

```
iris.head()
```

```
   Unnamed: 0  Sepal.Length    ...     Petal.Width  Species
0           1           5.1    ...             0.2   setosa
1           2           4.9    ...             0.2   setosa
2           3           4.7    ...             0.2   setosa
3           4           4.6    ...             0.2   setosa
4           5           5.0    ...             0.2   setosa

[5 rows x 6 columns]
```

The names of the features can be obtained via the `columns` attribute of the `DataFrame` object, as in `iris.columns`. Note that the first column is a duplicate index column, whose name (assigned by pandas) is `'Unnamed: 0'`. We can drop this column and reassign the `iris` object as follows:

```
iris = iris.drop('Unnamed: 0',1)
```

The data for each feature (corresponding to its specific name) can be accessed by using Python's *slicing* notation []. For example, the object iris['Sepal.Length'] contains the 150 sepal lengths.

The first three rows of the abalone data set from the UCI repository can be found as follows:

```
abalone.head(3)
```

```
   0      1      2      3       4       5       6      7   8
0  M  0.455  0.365  0.095  0.5140  0.2245  0.1010  0.150  15
1  M  0.350  0.265  0.090  0.2255  0.0995  0.0485  0.070   7
2  F  0.530  0.420  0.135  0.6770  0.2565  0.1415  0.210   9
```

Here, the missing headers have been assigned according to the order of the natural numbers. The names should correspond to Sex, Length, Diameter, Height, Whole weight, Shucked weight, Viscera weight, Shell weight, and Rings, as described in the file with the name abalone.names on the UCI website. We can manually add the names of the features to the DataFrame by reassigning the columns attribute, as in:

```
abalone.columns = ['Sex', 'Length', 'Diameter', 'Height',
'Whole weight','Shucked weight', 'Viscera weight', 'Shell weight',
'Rings']
```

1.2 Structuring Features According to Type

We can generally classify features as either quantitative or qualitative. *Quantitative* features possess "numerical quantity", such as height, age, number of births, etc., and can either be *continuous* or *discrete*. Continuous quantitative features take values in a continuous range of possible values, such as height, voltage, or crop yield; such features capture the idea that measurements can always be made more precisely. Discrete quantitative features have a countable number of possibilities, such as a count. QUANTITATIVE

In contrast, *qualitative* features do not have a numerical meaning, but their possible values can be divided into a fixed number of categories, such as {M,F} for gender or {blue, black, brown, green} for eye color. For this reason such features are also called *categorical*. A simple rule of thumb is: if it does not make sense to average the data, it is categorical. For example, it does not make sense to average eye colors. Of course it is still possible to represent categorical data with numbers, such as 1 = blue, 2 = black, 3 = brown, but such numbers carry no quantitative meaning. Categorical features are often called *factors*. QUALITATIVE CATEGORICAL FACTORS

When manipulating, summarizing, and displaying data, it is important to correctly specify the type of the variables (features). We illustrate this using the nutrition_elderly data set from [73], which contains the results of a study involving nutritional measurements of thirteen features (columns) for 226 elderly individuals (rows). The data set can be obtained from:

http://www.biostatisticien.eu/springeR/nutrition_elderly.xls.

Excel files can be read directly into pandas via the **read_excel** method:

```
xls = 'http://www.biostatisticien.eu/springeR/nutrition_elderly.xls'
nutri = pd.read_excel(xls)
```

This creates a `DataFrame` object `nutri`. The first three rows are as follows:

```
pd.set_option('display.max_columns', 8) # to fit display
nutri.head(3)
```

```
   gender  situation  tea ...    cooked_fruit_veg  chocol  fat
0       2          1    0 ...                   4       5    6
1       2          1    1 ...                   5       1    4
2       2          1    0 ...                   2       5    4

[3 rows x 13 columns]
```

You can check the type (or structure) of the variables via the **info** method of `nutri`.

```
nutri.info()
```

```
 <class 'pandas.core.frame.DataFrame'>
RangeIndex: 226 entries, 0 to 225
Data columns (total 13 columns):
gender              226 non-null int64
situation           226 non-null int64
tea                 226 non-null int64
coffee              226 non-null int64
height              226 non-null int64
weight              226 non-null int64
age                 226 non-null int64
meat                226 non-null int64
fish                226 non-null int64
raw_fruit           226 non-null int64
cooked_fruit_veg    226 non-null int64
chocol              226 non-null int64
fat                 226 non-null int64
dtypes: int64(13)
memory usage: 23.0 KB
```

All 13 features in `nutri` are (at the moment) interpreted by Python as *quantitative* variables, indeed as integers, simply because they have been entered as whole numbers. The *meaning* of these numbers becomes clear when we consider the description of the features, given in Table 1.2. Table 1.1 shows how the variable types should be classified.

Table 1.1: The feature types for the data frame `nutri`.

Qualitative	`gender`, `situation`, `fat`
	`meat`, `fish`, `raw_fruit`, `cooked_fruit_veg`, `chocol`
Discrete quantitative	`tea`, `coffee`
Continuous quantitative	`height`, `weight`, `age`

Note that the categories of the qualitative features in the second row of Table 1.1, `meat`, …, `chocol` have a natural order. Such qualitative features are sometimes called *ordinal*, in

Table 1.2: Description of the variables in the nutritional study [73].

Feature	**Description**	**Unit** or **Coding**
`gender`	Gender	1=Male; 2=Female
`situation`	Family status	1=Single 2=Living with spouse 3=Living with family 4=Living with someone else
`tea`	Daily consumption of tea	Number of cups
`coffee`	Daily consumption of coffee	Number of cups
`height`	Height	cm
`weight`	Weight (actually: mass)	kg
`age`	Age at date of interview	Years
`meat`	Consumption of meat	0=Never 1=Less than once a week 2=Once a week 3=2–3 times a week 4=4–6 times a week 5=Every day
`fish`	Consumption of fish	As in `meat`
`raw_fruit`	Consumption of raw fruits	As in `meat`
`cooked_fruit_veg`	Consumption of cooked fruits and vegetables	As in `meat`
`chocol`	Consumption of chocolate	As in `meat`
`fat`	Type of fat used for cooking	1=Butter 2=Margarine 3=Peanut oil 4=Sunflower oil 5=Olive oil 6=Mix of vegetable oils (e.g., Isio4) 7=Colza oil 8=Duck or goose fat

contrast to qualitative features without order, which are called *nominal*. We will not make such a distinction in this book.

We can modify the Python value and type for each categorical feature, using the **`replace`** and **`astype`** methods. For categorical features, such as `gender`, we can replace the value 1 with `'Male'` and 2 with `'Female'`, and change the type to `'category'` as follows.

```
DICT = {1:'Male', 2:'Female'} # dictionary specifies replacement
nutri['gender'] = nutri['gender'].replace(DICT).astype('category')
```

The structure of the other categorical-type features can be changed in a similar way. Continuous features such as `height` should have type `float`:

```
nutri['height'] = nutri['height'].astype(float)
```

We can repeat this for the other variables (see Exercise 2) and save this modified data frame as a CSV file, by using the pandas method `to_csv`.

```
nutri.to_csv('nutri.csv',index=False)
```

1.3 Summary Tables

It is often useful to summarize a large spreadsheet of data in a more condensed form. A table of counts or a table of frequencies makes it easier to gain insight into the underlying distribution of a variable, especially if the data are qualitative. Such tables can be obtained with the methods `describe` and `value_counts`.

As a first example, we load the nutri `DataFrame`, which we restructured and saved (see previous section) as `'nutri.csv'`, and then construct a summary for the feature (column) `'fat'`.

```
nutri = pd.read_csv('nutri.csv')
nutri['fat'].describe()
```

```
count            226
unique             8
top        sunflower
freq              68
Name: fat, dtype: object
```

We see that there are 8 different types of fat used and that sunflower has the highest count, with 68 out of 226 individuals using this type of cooking fat. The method `value_counts` gives the counts for the different fat types.

```
nutri['fat'].value_counts()
```

```
sunflower    68
peanut       48
olive        40
margarine    27
Isio4        23
butter       15
duck          4
colza         1
Name: fat, dtype: int64
```

Column labels are also attributes of a `DataFrame`, and `nutri.fat`, for example, is exactly the same object as `nutri['fat']`.

It is also possible to use **crosstab** to *cross tabulate* between two or more variables, giving a *contingency table*: CROSS TABULATE

```
pd.crosstab(nutri.gender, nutri.situation)
```

```
situation  Couple  Family  Single
gender
Female         56       7      78
Male           63       2      20
```

We see, for example, that the proportion of single men is substantially smaller than the proportion of single women in the data set of elderly people. To add row and column totals to a table, use `margins=True`.

```
pd.crosstab(nutri.gender, nutri.situation, margins=True)
```

```
situation  Couple  Family  Single  All
gender
Female         56       7      78  141
Male           63       2      20   85
All           119       9      98  226
```

1.4 Summary Statistics

In the following, $\boldsymbol{x} = [x_1, \ldots, x_n]^\top$ is a column vector of n numbers. For our `nutri` data, the vector $\boldsymbol{x}$ could, for example, correspond to the heights of the $n = 226$ individuals.

The *sample mean* of $\boldsymbol{x}$, denoted by $\overline{x}$, is simply the average of the data values: SAMPLE MEAN

$$\overline{x} = \frac{1}{n}\sum_{i=1}^{n} x_i.$$

Using the **mean** method in Python for the `nutri` data, we have, for instance:

```
nutri['height'].mean()
```

```
163.96017699115043
```

The *p-sample quantile* $(0 < p < 1)$ of $\boldsymbol{x}$ is a value x such that at least a fraction p of the data is less than or equal to x and at least a fraction $1 - p$ of the data is greater than or equal to x. The *sample median* is the sample 0.5-quantile. The p-sample quantile is also called the $100 \times p$ *percentile*. The 25, 50, and 75 sample percentiles are called the first, second, and third *quartiles* of the data. For the `nutri` data they are obtained as follows. SAMPLE QUANTILE SAMPLE MEDIAN QUARTILES

```
nutri['height'].quantile(q=[0.25,0.5,0.75])
```

```
0.25    157.0
0.50    163.0
0.75    170.0
```

The sample mean and median give information about the *location* of the data, while the distance between sample quantiles (say the 0.1 and 0.9 quantiles) gives some indication of the *dispersion* (spread) of the data. Other measures for dispersion are the *sample range*, $\max_i x_i - \min_i x_i$, the *sample variance*

SAMPLE RANGE

SAMPLE VARIANCE

$$s^2 = \frac{1}{n-1}\sum_{i=1}^{n}(x_i - \overline{x})^2, \tag{1.1}$$

and the *sample standard deviation* $s = \sqrt{s^2}$. For the nutri data, the range (in cm) is:

SAMPLE STANDARD DEVIATION

☞ 457

```
nutri['height'].max() - nutri['height'].min()
```

```
48.0
```

The variance (in cm^2) is:

```
round(nutri['height'].var(), 2)  # round to two decimal places
```

```
81.06
```

And the standard deviation can be found via:

```
round(nutri['height'].std(), 2)
```

```
9.0
```

We already encountered the **describe** method in the previous section for summarizing qualitative features, via the most frequent count and the number of unique elements. When applied to a *quantitative* feature, it returns instead the minimum, maximum, mean, and the three quartiles. For example, the 'height' feature in the nutri data has the following summary statistics.

```
nutri['height'].describe()
```

```
count    226.000000
mean     163.960177
std        9.003368
min      140.000000
25\%      157.000000
50\%      163.000000
75\%      170.000000
max      188.000000
Name: height, dtype: float64
```

1.5 Visualizing Data

In this section we describe various methods for visualizing data. The main point we would like to make is that the way in which variables are visualized should always be adapted to the variable types; for example, qualitative data should be plotted differently from quantitative data.

For the rest of this section, it is assumed that matplotlib.pyplot, pandas, and numpy, have been imported in the Python code as follows.

```
import matplotlib.pyplot as plt
import pandas as pd
import numpy as np
```

1.5.1 Plotting Qualitative Variables

Suppose we wish to display graphically how many elderly people are living by themselves, as a couple, with family, or other. Recall that the data are given in the situation column of our nutri data. Assuming that we already *restructured the data*, as in Section 1.2, we can make a *barplot* of the number of people in each category via the plt.bar function of the standard matplotlib plotting library. The inputs are the x-axis positions, heights, and widths of each bar respectively. ☞ 3

BARPLOT

```
width = 0.35 # the width of the bars
x = [0, 0.8, 1.6] # the bar positions on x-axis
situation_counts=nutri['situation'].value_counts()
plt.bar(x, situation_counts, width, edgecolor = 'black')
plt.xticks(x, situation_counts.index)
plt.show()
```

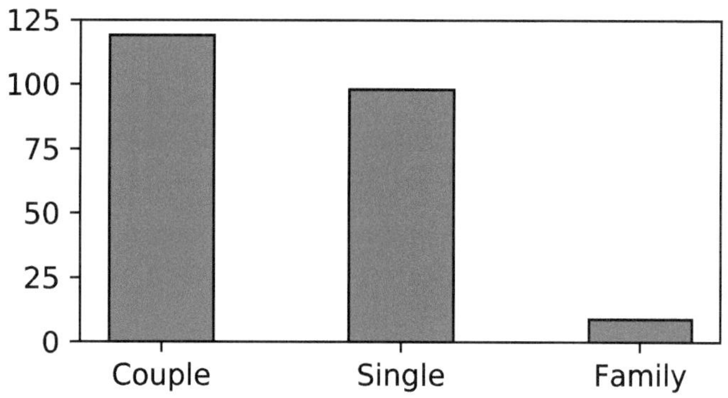

Figure 1.1: Barplot for the qualitative variable 'situation'.

1.5.2 Plotting Quantitative Variables

We now present a few useful methods for visualizing quantitative data, again using the nutri data set. We will first focus on continuous features (e.g., 'age') and then add some specific graphs related to discrete features (e.g., 'tea'). The aim is to describe the variability present in a single feature. This typically involves a central tendency, where observations tend to gather around, with fewer observations further away. The main aspects of the distribution are the *location* (or center) of the variability, the *spread* of the variability (how far the values extend from the center), and the *shape* of the variability; e.g., whether or not values are spread symmetrically on either side of the center.

1.5.2.1 Boxplot

BOXPLOT

A *boxplot* can be viewed as a graphical representation of the five-number summary of the data consisting of the minimum, maximum, and the first, second, and third quartiles. Figure 1.2 gives a boxplot for the `'age'` feature of the `nutri` data.

```
plt.boxplot(nutri['age'],widths=width,vert=False)
plt.xlabel('age')
plt.show()
```

The `widths` parameter determines the width of the boxplot, which is by default plotted vertically. Setting `vert=False` plots the boxplot horizontally, as in Figure 1.2.

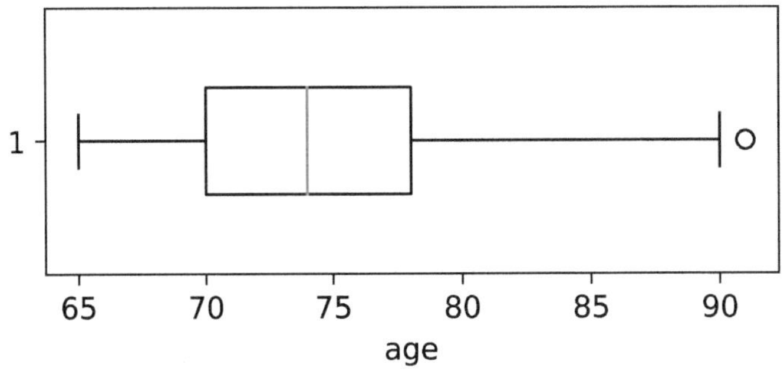

Figure 1.2: Boxplot for `'age'`.

The box is drawn from the first quartile (Q_1) to the third quartile (Q_3). The vertical line inside the box signifies the location of the median. So-called "whiskers" extend to either side of the box. The size of the box is called the *interquartile range*: IQR = $Q_3 - Q_1$. The left whisker extends to the largest of (a) the minimum of the data and (b) $Q_1 - 1.5\,\text{IQR}$. Similarly, the right whisker extends to the smallest of (a) the maximum of the data and (b) $Q_3 + 1.5\,\text{IQR}$. Any data point outside the whiskers is indicated by a small hollow dot, indicating a suspicious or deviant point (outlier). Note that a boxplot may also be used for discrete quantitative features.

1.5.2.2 Histogram

HISTOGRAM

A *histogram* is a common graphical representation of the distribution of a quantitative feature. We start by breaking the range of the values into a number of *bins* or *classes*. We tally the counts of the values falling in each bin and then make the plot by drawing rectangles whose bases are the bin intervals and whose heights are the counts. In Python we can use the function **`plt.hist`**. For example, Figure 1.3 shows a histogram of the 226 ages in `nutri`, constructed via the following Python code.

```
weights = np.ones_like(nutri.age)/nutri.age.count()
plt.hist(nutri.age,bins=9,weights=weights,facecolor='cyan',
         edgecolor='black', linewidth=1)
plt.xlabel('age')
plt.ylabel('Proportion of Total')
plt.show()
```

Here 9 bins were used. Rather than using raw counts (the default), the vertical axis here gives the percentage in each class, defined by $\frac{\text{count}}{\text{total}}$. This is achieved by choosing the "weights" parameter to be equal to the vector with entries 1/266, with length 226. Various plotting parameters have also been changed.

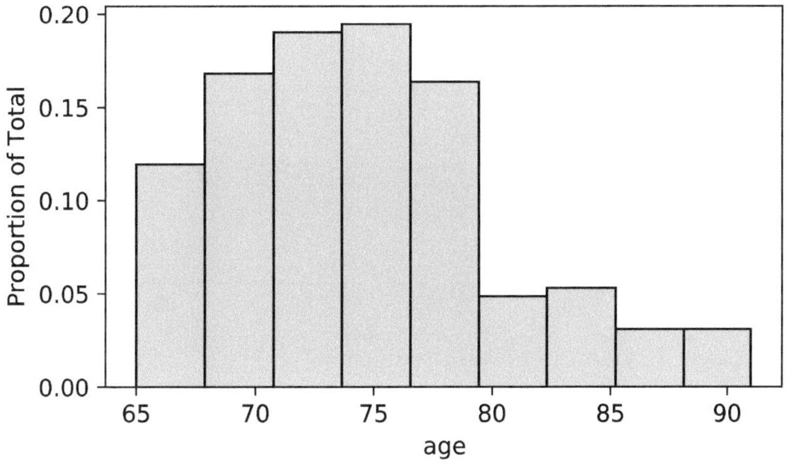

Figure 1.3: Histogram of `'age'`.

Histograms can also be used for discrete features, although it may be necessary to explicitly specify the bins and placement of the ticks on the axes.

1.5.2.3 Empirical Cumulative Distribution Function

The *empirical cumulative distribution function*, denoted by F_n, is a step function which jumps an amount k/n at observation values, where k is the number of tied observations at that value. For observations $x_1, \dots, x_n$, $F_n(x)$ is the fraction of observations less than or equal to x, i.e.,

EMPIRICAL CUMULATIVE DISTRIBUTION FUNCTION

$$F_n(x) = \frac{\text{number of } x_i \leqslant x}{n} = \frac{1}{n} \sum_{i=1}^{n} \mathbb{1}\{x_i \leqslant x\}, \tag{1.2}$$

where $\mathbb{1}$ denotes the *indicator* function; that is, $\mathbb{1}\{x_i \leqslant x\}$ is equal to 1 when $x_i \leqslant x$ and 0 otherwise. To produce a plot of the empirical cumulative distribution function we can use the `plt.step` function. The result for the age data is shown in Figure 1.4. The empirical cumulative distribution function for a discrete quantitative variable is obtained in the same way.

INDICATOR

```
x = np.sort(nutri.age)
y = np.linspace(0,1,len(nutri.age))
plt.xlabel('age')
plt.ylabel('Fn(x)')
plt.step(x,y)
plt.xlim(x.min(),x.max())
plt.show()
```

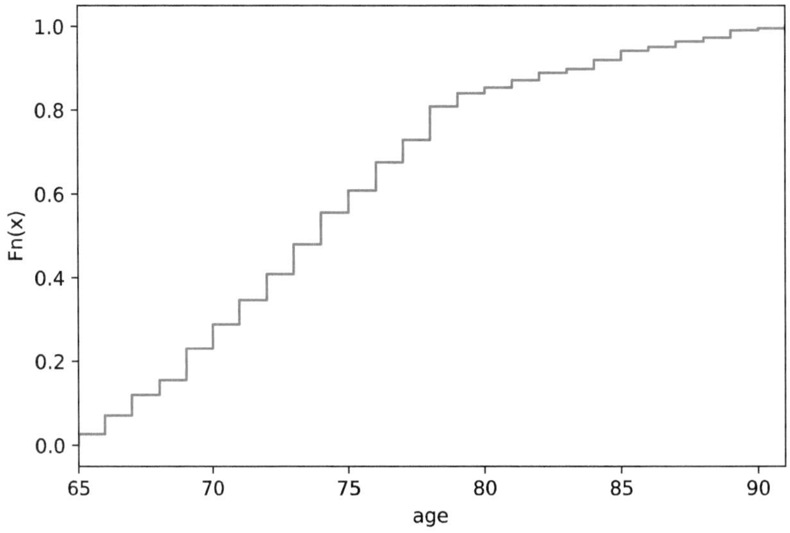

Figure 1.4: Plot of the empirical distribution function for the continuous quantitative feature `'age'`.

1.5.3 Data Visualization in a Bivariate Setting

In this section, we present a few useful visual aids to explore relationships between two features. The graphical representation will depend on the type of the two features.

1.5.3.1 Two-way Plots for Two Categorical Variables

Comparing barplots for two categorical variables involves introducing subplots to the figure. Figure 1.5 visualizes the contingency table of Section 1.3, which cross-tabulates the family status (situation) with the gender of the elderly people. It simply shows two barplots next to each other in the same figure.

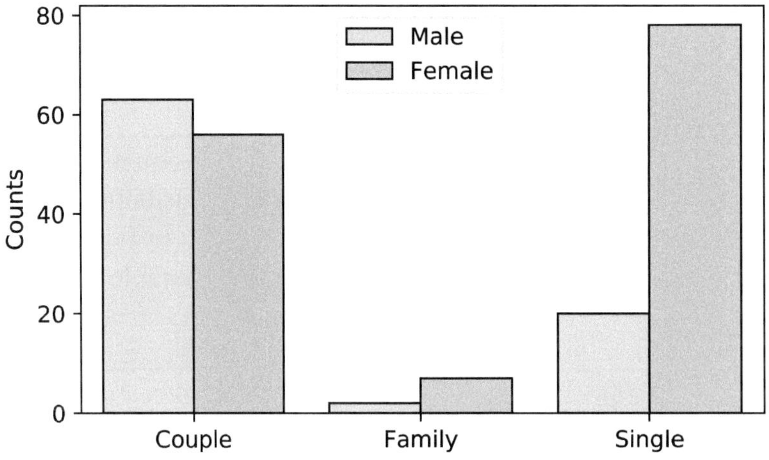

Figure 1.5: Barplot for two categorical variables.

The figure was made using the seaborn package, which was specifically designed to simplify statistical visualization tasks.

```
import seaborn as sns
sns.countplot(x='situation', hue = 'gender', data=nutri,
     hue_order = ['Male', 'Female'], palette = ['SkyBlue','Pink'],
      saturation = 1, edgecolor='black')
plt.legend(loc='upper center')
plt.xlabel('')
plt.ylabel('Counts')
plt.show()
```

1.5.3.2 Plots for Two Quantitative Variables

We can visualize patterns between two quantitative features using a *scatterplot*. This can be done with `plt.scatter`. The following code produces a scatterplot of `'weight'` against `'height'` for the nutri data. SCATTERPLOT

```
plt.scatter(nutri.height, nutri.weight, s=12, marker='o')
plt.xlabel('height')
plt.ylabel('weight')
plt.show()
```

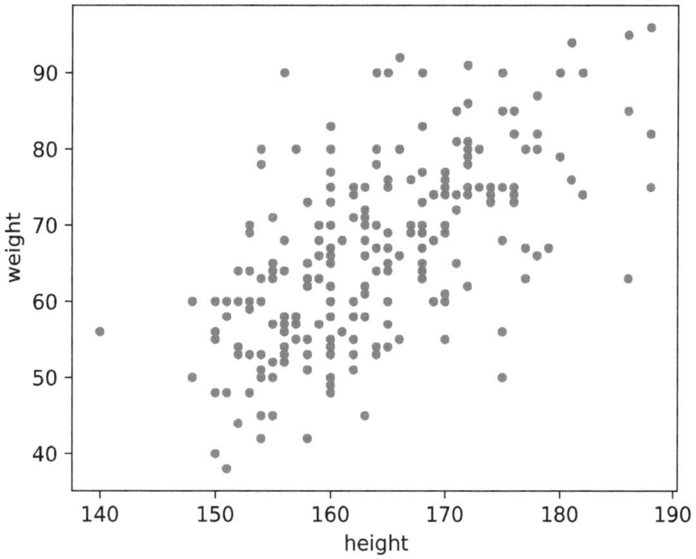

Figure 1.6: Scatterplot of `'weight'` against `'height'`.

The next Python code illustrates that it is possible to produce highly sophisticated scatter plots, such as in Figure 1.7. The figure shows the birth weights (mass) of babies whose mothers smoked (blue triangles) or not (red circles). In addition, straight lines were fitted to the two groups, suggesting that birth weight decreases with age when the mother smokes, but increases when the mother does not smoke! The question is whether these trends are statistically significant or due to chance. We will revisit this data set later on in the book. ☞ 199

```
urlprefix = 'https://vincentarelbundock.github.io/Rdatasets/csv/'
dataname = 'MASS/birthwt.csv'
bwt = pd.read_csv(urlprefix + dataname)
bwt = bwt.drop('Unnamed: 0',1)   #drop unnamed column
styles = {0: ['o','red'], 1: ['^','blue']}
for k in styles:
    grp = bwt[bwt.smoke==k]
    m,b = np.polyfit(grp.age, grp.bwt, 1) # fit a straight line
    plt.scatter(grp.age, grp.bwt, c=styles[k][1], s=15, linewidth=0,
        marker = styles[k][0])
    plt.plot(grp.age, m*grp.age + b, '-', color=styles[k][1])

plt.xlabel('age')
plt.ylabel('birth weight (g)')
plt.legend(['non-smokers','smokers'],prop={'size':8},
            loc=(0.5,0.8))
plt.show()
```

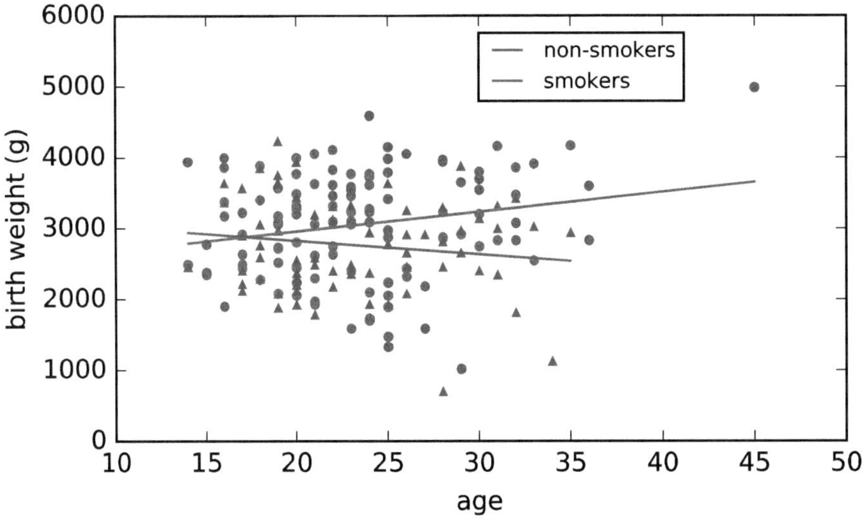

Figure 1.7: Birth weight against age for smoking and non-smoking mothers.

1.5.3.3 Plots for One Qualitative and One Quantitative Variable

In this setting, it is interesting to draw boxplots of the quantitative feature for each level of the categorical feature. Assuming the variables are structured correctly, the function **plt.boxplot** can be used to produce Figure 1.8, using the following code:

```
males = nutri[nutri.gender == 'Male']
females = nutri[nutri.gender == 'Female']
plt.boxplot([males.coffee,females.coffee],notch=True,widths
    =(0.5,0.5))
plt.xlabel('gender')
plt.ylabel('coffee')
plt.xticks([1,2],['Male','Female'])
plt.show()
```

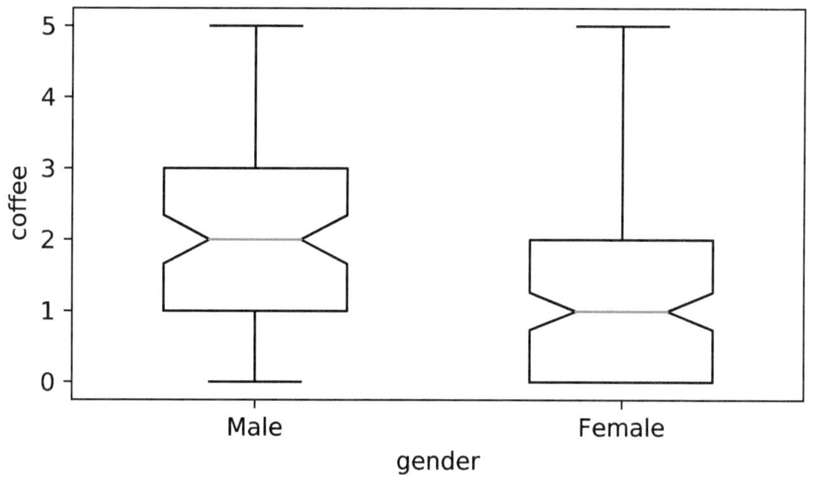

Figure 1.8: Boxplots of a quantitative feature `'coffee'` as a function of the levels of a categorical feature `'gender'`. Note that we used a different, "notched", style boxplot this time.

Further Reading

The focus in this book is on the mathematical and statistical analysis of data, and for the rest of the book we assume that the data is available in a suitable form for analysis. However, a large part of practical data science involves the *cleaning* of data; that is, putting it into a form that is amenable to analysis with standard software packages. Standard Python modules such as numpy and pandas can be used to reformat rows, rename columns, remove faulty outliers, merge rows, and so on. McKinney, the creator of pandas, gives many practical case studies in [84]. Effective data visualization techniques are beautifully illustrated in [65].

Exercises

Before you attempt these exercises, make sure you have up-to-date versions of the relevant Python packages, specifically matplotlib, pandas, and seaborn. An easy way to ensure this is to update packages via the Anaconda Navigator, as explained in Appendix D.

1. Visit the UCI Repository `https://archive.ics.uci.edu/`. Read the description of the data and download the Mushroom data set `agaricus-lepiota.data`. Using pandas, read the data into a `DataFrame` called `mushroom`, via `read_csv`.

   (a) How many features are in this data set?

   (b) What are the initial names and types of the features?

   (c) Rename the first feature (index 0) to `'edibility'` and the sixth feature (index 5) to `'odor'` [Hint: the column names in pandas are immutable; so individual columns cannot be modified directly. However it is possible to assign the entire column names list via `mushroom.columns = newcols`.]

(d) The 6th column lists the various odors of the mushrooms: encoded as `'a'`, `'c'`, ... Replace these with the names `'almond'`, `'creosote'`, etc. (categories corresponding to each letter can be found on the website). Also replace the `'edibility'` categories `'e'` and `'p'` with `'edible'` and `'poisonous'`.

(e) Make a contingency table cross-tabulating `'edibility'` and `'odor'`.

(f) Which mushroom odors should be avoided, when gathering mushrooms for consumption?

(g) What proportion of odorless mushroom samples were safe to eat?

2. Change the type and value of variables in the `nutri` data set according to Table 1.2 and save the data as a CSV file. The modified data should have eight categorical features, three floats, and two integer features.

3. It frequently happens that a table with data needs to be restructured before the data can be analyzed using standard statistical software. As an example, consider the test scores in Table 1.3 of 5 students before and after specialized tuition.

Table 1.3: Student scores.

Student	Before	After
1	75	85
2	30	50
3	100	100
4	50	52
5	60	65

This is not in the standard format described in Section 1.1. In particular, the student scores are divided over two columns, whereas the standard format requires that they are collected in one column, e.g., labelled `'Score'`. Reformat (by hand) the table in standard format, using three features:

- `'Score'`, taking continuous values,
- `'Time'`, taking values `'Before'` and `'After'`,
- `'Student'`, taking values from 1 to 5.

Useful methods for reshaping tables in `pandas` are **`melt`**, **`stack`**, and **`unstack`**.

4. Create a similar barplot as in Figure 1.5, but now plot the corresponding *proportions* of males and females in each of the three situation categories. That is, the heights of the bars should sum up to 1 for both barplots with the same `'gender'` value. [Hint: `seaborn` does not have this functionality built in, instead you need to first create a contingency table and use `matplotlib.pyplot` to produce the figure.]

☞ 2
5. The `iris` data set, mentioned in Section 1.1, contains various features, including `'Petal.Length'` and `'Sepal.Length'`, of three species of iris: setosa, versicolor, and virginica.

(a) Load the data set into a pandas DataFrame object.

(b) Using matplotlib.pyplot, produce boxplots of 'Petal.Length' for each the three species, in one figure.

(c) Make a histogram with 20 bins for 'Petal.Length'.

(d) Produce a similar scatterplot for 'Sepal.Length' against 'Petal.Length' to that of the left plot in Figure 1.9. Note that the points should be colored according to the 'Species' feature as per the legend in the right plot of the figure.

(e) Using the **kdeplot** method of the seaborn package, reproduce the right plot of Figure 1.9, where kernel density plots for 'Petal.Length' are given. ☞ 131

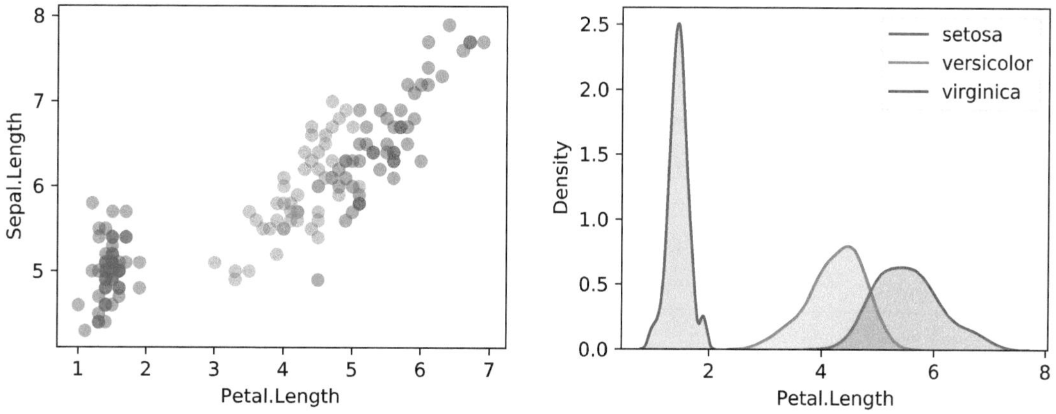

Figure 1.9: Left: scatterplot of 'Sepal.Length' against 'Petal.Length'. Right: kernel density estimates of 'Petal.Length' for the three species of iris.

6. Import the data set EuStockMarkets from the same website as the iris data set above. The data set contains the daily closing prices of four European stock indices during the 1990s, for 260 working days per year.

(a) Create a vector of times (working days) for the stock prices, between 1991.496 and 1998.646 with increments of 1/260.

(b) Reproduce Figure 1.10. [Hint: Use a dictionary to map column names (stock indices) to colors.]

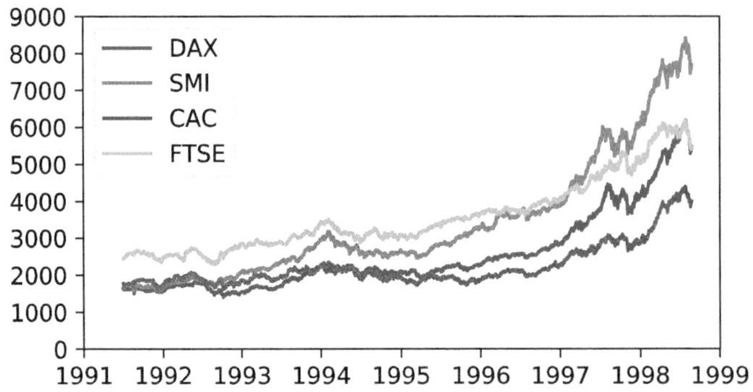

Figure 1.10: Closing stock indices for various European stock markets.

7. Consider the KASANDR data set from the UCI Machine Learning Repository, which can be downloaded from

https://archive.ics.uci.edu/ml/machine-learning-databases/00385/de.tar.bz2.

This archive file has a size of 900Mb, so it may take a while to download. Uncompressing the file (e.g., via 7-Zip) yields a directory de containing two large CSV files: `test_de.csv` and `train_de.csv`, with sizes 372Mb and 3Gb, respectively. Such large data files can still be processed efficiently in `pandas`, provided there is enough memory. The files contain records of user information from Kelkoo web logs in Germany as well as meta-data on users, offers, and merchants. The data sets have 7 attributes and 1919561 and 15844717 rows, respectively. The data sets are anonymized via hex strings.

(a) Load `train_de.csv` into a `pandas` `DataFrame` object `de`, using

```
read_csv('train_de.csv', delimiter = '\t').
```

If not enough memory is available, load `test_de.csv` instead. Note that entries are separated here by tabs, not commas. Time how long it takes for the file to load, using the `time` package. (It took 38 seconds for `train_de.csv` to load on one of our computers.)

(b) How many unique users and merchants are in this data set?

8. Visualizing data involving more than two features requires careful design, which is often more of an art than a science.

(a) Go to Vincent Arel-Bundocks's website (URL given in Section 1.1) and read the `Orange` data set into a `pandas` `DataFrame object` called `orange`. Remove its first (unnamed) column.

(b) The data set contains the circumferences of 5 orange trees at various stages in their development. Find the names of the features.

(c) In Python, import `seaborn` and visualize the growth curves (circumference against age) of the trees, using the `regplot` and `FacetGrid` methods.

CHAPTER 2

STATISTICAL LEARNING

The purpose of this chapter is to introduce the reader to some common concepts and themes in statistical learning. We discuss the difference between supervised and unsupervised learning, and how we can assess the predictive performance of supervised learning. We also examine the central role that the linear and Gaussian properties play in the modeling of data. We conclude with a section on Bayesian learning. The required probability and statistics background is given in Appendix C.

2.1 Introduction

Although structuring and visualizing data are important aspects of data science, the main challenge lies in the mathematical analysis of the data. When the goal is to interpret the model and quantify the uncertainty in the data, this analysis is usually referred to as *statistical learning*. In contrast, when the emphasis is on making predictions using large-scale data, then it is common to speak about *machine learning* or *data mining*.

STATISTICAL LEARNING

MACHINE LEARNING

DATA MINING

There are two major goals for modeling data: 1) to accurately predict some future quantity of interest, given some observed data, and 2) to discover unusual or interesting patterns in the data. To achieve these goals, one must rely on knowledge from three important pillars of the mathematical sciences.

Function approximation. Building a mathematical model for data usually means understanding how one data variable depends on another data variable. The most natural way to represent the relationship between variables is via a mathematical function or map. We usually assume that this mathematical function is not completely known, but can be approximated well given enough computing power and data. Thus, data scientists have to understand how best to approximate and represent functions using the least amount of computer processing and memory.

Optimization. Given a class of mathematical models, we wish to find the best possible model in that class. This requires some kind of efficient search or optimization procedure. The optimization step can be viewed as a process of fitting or calibrating a function to observed data. This step usually requires knowledge of optimization algorithms and efficient computer coding or programming.

Probability and Statistics. In general, the data used to fit the model is viewed as a realization of a random process or numerical vector, whose probability law determines the accuracy with which we can predict future observations. Thus, in order to quantify the uncertainty inherent in making predictions about the future, and the sources of error in the model, data scientists need a firm grasp of probability theory and statistical inference.

2.2 Supervised and Unsupervised Learning

FEATURE
RESPONSE

Given an input or *feature* vector $\boldsymbol{x}$, one of the main goals of machine learning is to predict an output or *response* variable y. For example, $\boldsymbol{x}$ could be a digitized signature and y a binary variable that indicates whether the signature is genuine or false. Another example is where $\boldsymbol{x}$ represents the weight and smoking habits of an expecting mother and y the birth weight of the baby. The data science attempt at this prediction is encoded in a mathematical function g, called the *prediction function*, which takes as an input $\boldsymbol{x}$ and outputs a guess $g(\boldsymbol{x})$ for y (denoted by $\widehat{y}$, for example). In a sense, g encompasses all the information about the relationship between the variables $\boldsymbol{x}$ and y, excluding the effects of chance and randomness in nature.

PREDICTION FUNCTION

REGRESSION

In *regression* problems, the response variable y can take any real value. In contrast, when y can only lie in a finite set, say $y \in \{0, \dots, c-1\}$, then predicting y is conceptually the same as classifying the input $\boldsymbol{x}$ into one of c categories, and so prediction becomes a *classification* problem.

CLASSIFICATION

LOSS FUNCTION

We can measure the accuracy of a prediction $\widehat{y}$ with respect to a given response y by using some *loss function* $\text{Loss}(y, \widehat{y})$. In a regression setting the usual choice is the squared-error loss $(y-\widehat{y})^2$. In the case of classification, the zero–one (also written 0–1) loss function $\text{Loss}(y, \widehat{y}) = \mathbb{1}\{y \neq \widehat{y}\}$ is often used, which incurs a loss of 1 whenever the predicted class $\widehat{y}$ is not equal to the class y. Later on in this book, we will encounter various other useful loss functions, such as the cross-entropy and hinge loss functions (see, e.g., Chapter 7).

> The word *error* is often used as a measure of distance between a "true" object y and some approximation $\widehat{y}$ thereof. If y is real-valued, the absolute error $|y - \widehat{y}|$ and the squared error $(y-\widehat{y})^2$ are both well-established error concepts, as are the norm $\|\boldsymbol{y}-\widehat{\boldsymbol{y}}\|$ and squared norm $\|\boldsymbol{y}-\widehat{\boldsymbol{y}}\|^2$ for vectors. The squared error $(y-\widehat{y})^2$ is just one example of a loss function.

It is unlikely that any mathematical function g will be able to make accurate predictions for all possible pairs $(\boldsymbol{x}, y)$ one may encounter in Nature. One reason for this is that, even with the same input $\boldsymbol{x}$, the output y may be different, depending on chance circumstances or randomness. For this reason, we adopt a probabilistic approach and assume that each pair $(\boldsymbol{x}, y)$ is the outcome of a random pair $(\boldsymbol{X}, Y)$ that has some joint probability density $f(\boldsymbol{x}, y)$. We then assess the predictive performance via the expected loss, usually called the *risk*, for g:

RISK

$$\ell(g) = \mathbb{E}\,\text{Loss}(Y, g(\boldsymbol{X})). \tag{2.1}$$

For example, in the classification case with zero–one loss function the risk is equal to the probability of incorrect classification: $\ell(g) = \mathbb{P}[Y \neq g(\boldsymbol{X})]$. In this context, the prediction function g is called a *classifier*. Given the distribution of $(\boldsymbol{X}, Y)$ and any loss function, we can in principle find the best possible $g^* := \operatorname{argmin}_g \mathbb{E}\,\text{Loss}(Y, g(\boldsymbol{X}))$ that yields the smallest risk $\ell^* := \ell(g^*)$. We will see in Chapter 7 that in the classification case with $y \in \{0, \ldots, c-1\}$ and $\ell(g) = \mathbb{P}[Y \neq g(\boldsymbol{X})]$, we have

CLASSIFIER

☞ 253

$$g^*(\boldsymbol{x}) = \underset{y \in \{0,\ldots,c-1\}}{\operatorname{argmax}}\, f(y \mid \boldsymbol{x}),$$

where $f(y \mid \boldsymbol{x}) = \mathbb{P}[Y = y \mid \boldsymbol{X} = \boldsymbol{x}]$ is the conditional probability of $Y = y$ given $\boldsymbol{X} = \boldsymbol{x}$. As already mentioned, for regression the most widely-used loss function is the squared-error loss. In this setting, the optimal prediction function g^* is often called the *regression function*. The following theorem specifies its exact form.

REGRESSION FUNCTION

Theorem 2.1: Optimal Prediction Function for Squared-Error Loss

For the squared-error loss $\text{Loss}(y, \widehat{y}) = (y - \widehat{y})^2$, the optimal prediction function g^* is equal to the conditional expectation of Y given $\boldsymbol{X} = \boldsymbol{x}$:

$$g^*(\boldsymbol{x}) = \mathbb{E}[Y \mid \boldsymbol{X} = \boldsymbol{x}].$$

Proof: Let $g^*(\boldsymbol{x}) = \mathbb{E}[Y \mid \boldsymbol{X} = \boldsymbol{x}]$. For any function g, the squared-error risk satisfies

$$\begin{aligned}
\mathbb{E}(Y - g(\boldsymbol{X}))^2 &= \mathbb{E}[(Y - g^*(\boldsymbol{X}) + g^*(\boldsymbol{X}) - g(\boldsymbol{X}))^2] \\
&= \mathbb{E}(Y - g^*(\boldsymbol{X}))^2 + 2\mathbb{E}[(Y - g^*(\boldsymbol{X}))(g^*(\boldsymbol{X}) - g(\boldsymbol{X}))] + \mathbb{E}(g^*(\boldsymbol{X}) - g(\boldsymbol{X}))^2 \\
&\geqslant \mathbb{E}(Y - g^*(\boldsymbol{X}))^2 + 2\mathbb{E}[(Y - g^*(\boldsymbol{X}))(g^*(\boldsymbol{X}) - g(\boldsymbol{X}))] \\
&= \mathbb{E}(Y - g^*(\boldsymbol{X}))^2 + 2\mathbb{E}\,\{(g^*(\boldsymbol{X}) - g(\boldsymbol{X}))\mathbb{E}[Y - g^*(\boldsymbol{X}) \mid \boldsymbol{X}]\}\,.
\end{aligned}$$

In the last equation we used the tower property. By the definition of the conditional expectation, we have $\mathbb{E}[Y - g^*(\boldsymbol{X}) \mid \boldsymbol{X}] = 0$. It follows that $\mathbb{E}(Y - g(\boldsymbol{X}))^2 \geqslant \mathbb{E}(Y - g^*(\boldsymbol{X}))^2$, showing that g^* yields the smallest squared-error risk. □

☞ 433

One consequence of Theorem 2.1 is that, conditional on $\boldsymbol{X} = \boldsymbol{x}$, the (random) response Y can be written as

$$Y = g^*(\boldsymbol{x}) + \varepsilon(\boldsymbol{x}), \tag{2.2}$$

where $\varepsilon(\boldsymbol{x})$ can be viewed as the random deviation of the response from its conditional mean at $\boldsymbol{x}$. This random deviation satisfies $\mathbb{E}\,\varepsilon(\boldsymbol{x}) = 0$. Further, the conditional variance of the response Y at $\boldsymbol{x}$ can be written as $\mathbb{V}\text{ar}\,\varepsilon(\boldsymbol{x}) = v^2(\boldsymbol{x})$ for some unknown positive function v. Note that, in general, the probability distribution of $\varepsilon(\boldsymbol{x})$ is unspecified.

Since, the optimal prediction function g^* depends on the typically unknown joint distribution of $(\boldsymbol{X}, Y)$, it is not available in practice. Instead, all that we have available is a finite number of (usually) independent realizations from the joint density $f(\boldsymbol{x}, y)$. We denote this sample by $\mathcal{T} = \{(\boldsymbol{X}_1, Y_1), \ldots, (\boldsymbol{X}_n, Y_n)\}$ and call it the *training set* ($\mathcal{T}$ is a mnemonic for training) with n examples. It will be important to distinguish between a random training set $\mathcal{T}$ and its (deterministic) outcome $\{(\boldsymbol{x}_1, y_1), \ldots, (\boldsymbol{x}_n, y_n)\}$. We will use the notation τ for the latter. We will also add the subscript n in τ_n when we wish to emphasize the size of the training set.

TRAINING SET

Our goal is thus to "learn" the unknown g^* using the n examples in the training set $\mathcal{T}$. Let us denote by $g_{\mathcal{T}}$ the best (by some criterion) approximation for g^* that we can construct

LEARNER

from $\mathcal{T}$. Note that $g_{\mathcal{T}}$ is a random function. A particular outcome is denoted by g_{τ}. It is often useful to think of a teacher–learner metaphor, whereby the function $g_{\mathcal{T}}$ is a *learner* who learns the unknown functional relationship $g^* : \boldsymbol{x} \mapsto y$ from the training data $\mathcal{T}$. We can imagine a "teacher" who provides n examples of the true relationship between the output Y_i and the input $\boldsymbol{X}_i$ for $i = 1, \ldots, n$, and thus "trains" the learner $g_{\mathcal{T}}$ to predict the output of a new input $\boldsymbol{X}$, for which the correct output Y is not provided by the teacher (is unknown).

SUPERVISED LEARNING

The above setting is called *supervised learning*, because one tries to learn the functional relationship between the feature vector $\boldsymbol{x}$ and response y in the presence of a teacher who provides n examples. It is common to speak of "explaining" or predicting y on the basis of $\boldsymbol{x}$, where $\boldsymbol{x}$ is a vector of *explanatory variables*.

EXPLANATORY VARIABLES

An example of supervised learning is email spam detection. The goal is to train the learner $g_{\mathcal{T}}$ to accurately predict whether any future email, as represented by the feature vector $\boldsymbol{x}$, is spam or not. The training data consists of the feature vectors of a number of different email examples as well as the corresponding labels (spam or not spam). For instance, a feature vector could consist of the number of times sales-pitch words like "free", "sale", or "miss out" occur within a given email.

As seen from the above discussion, most questions of interest in supervised learning can be answered if we know the conditional pdf $f(y \,|\, \boldsymbol{x})$, because we can then in principle work out the function value $g^*(\boldsymbol{x})$.

UNSUPERVISED LEARNING

In contrast, *unsupervised learning* makes no distinction between response and explanatory variables, and the objective is simply to learn the structure of the unknown distribution of the data. In other words, we need to learn $f(\boldsymbol{x})$. In this case the guess $g(\boldsymbol{x})$ is an approximation of $f(\boldsymbol{x})$ and the risk is of the form

$$\ell(g) = \mathbb{E}\,\mathrm{Loss}(f(\boldsymbol{X}), g(\boldsymbol{X})).$$

An example of unsupervised learning is when we wish to analyze the purchasing behaviors of the customers of a grocery shop that has a total of, say, a hundred items on sale. A feature vector here could be a binary vector $\boldsymbol{x} \in \{0, 1\}^{100}$ representing the items bought by a customer on a visit to the shop (a 1 in the k-th position if a customer bought item $k \in \{1, \ldots, 100\}$ and a 0 otherwise). Based on a training set $\tau = \{\boldsymbol{x}_1, \ldots, \boldsymbol{x}_n\}$, we wish to find any interesting or unusual purchasing patterns. In general, it is difficult to know if an unsupervised learner is doing a good job, because there is no teacher to provide examples of accurate predictions.

The main methodologies for unsupervised learning include *clustering*, *principal component analysis*, and *kernel density estimation*, which will be discussed in Chapter 4.

☞ 121

In the next three sections we will focus on supervised learning. The main supervised learning methodologies are *regression* and *classification*, to be discussed in detail in Chapters 5 and 7. More advanced supervised learning techniques, including *reproducing kernel Hilbert spaces*, *tree methods*, and *deep learning*, will be discussed in Chapters 6, 8, and 9.

☞ 167
☞ 253

2.3 Training and Test Loss

Given an arbitrary prediction function g, it is typically not possible to compute its risk $\ell(g)$ in (2.1). However, using the training sample $\mathcal{T}$, we can approximate $\ell(g)$ via the empirical (sample average) risk

$$\ell_{\mathcal{T}}(g) = \frac{1}{n}\sum_{i=1}^{n} \mathrm{Loss}(Y_i, g(\boldsymbol{X}_i)), \tag{2.3}$$

which we call the *training loss*. The training loss is thus an unbiased estimator of the risk (the expected loss) for a prediction function g, based on the training data. TRAINING LOSS

To approximate the optimal prediction function g^* (the minimizer of the risk $\ell(g)$) we first select a suitable collection of approximating functions $\mathcal{G}$ and then take our *learner* to be the function in $\mathcal{G}$ that minimizes the training loss; that is,

$$g_{\mathcal{T}}^{\mathcal{G}} = \underset{g \in \mathcal{G}}{\mathrm{argmin}}\, \ell_{\mathcal{T}}(g). \tag{2.4}$$

For example, the simplest and most useful $\mathcal{G}$ is the set of *linear* functions of $\boldsymbol{x}$; that is, the set of all functions $g : \boldsymbol{x} \mapsto \boldsymbol{\beta}^\top \boldsymbol{x}$ for some real-valued vector $\boldsymbol{\beta}$.

We suppress the superscript $\mathcal{G}$ when it is clear which function class is used. Note that minimizing the training loss over all possible functions g (rather than over all $g \in \mathcal{G}$) does not lead to a meaningful optimization problem, as any function g for which $g(\boldsymbol{X}_i) = Y_i$ for all i gives minimal training loss. In particular, for a squared-error loss, the training loss will be 0. Unfortunately, such functions have a poor ability to predict new (that is, independent from $\mathcal{T}$) pairs of data. This poor generalization performance is called *overfitting*. OVERFITTING

⚠ By choosing g a function that predicts the training data exactly (and is, for example, 0 otherwise), the squared-error training loss is zero. Minimizing the training loss is not the ultimate goal!

The prediction accuracy of new pairs of data is measured by the *generalization risk* of the learner. For a *fixed* training set τ it is defined as GENERALIZATION RISK

$$\ell(g_{\tau}^{\mathcal{G}}) = \mathbb{E}\, \mathrm{Loss}(Y, g_{\tau}^{\mathcal{G}}(\boldsymbol{X})), \tag{2.5}$$

where $(\boldsymbol{X}, Y)$ is distributed according to $f(\boldsymbol{x}, y)$. In the discrete case the generalization risk is therefore: $\ell(g_{\tau}^{\mathcal{G}}) = \sum_{\boldsymbol{x},y} \mathrm{Loss}(y, g_{\tau}^{\mathcal{G}}(\boldsymbol{x})) f(\boldsymbol{x}, y)$ (replace the sum with an integral for the continuous case). The situation is illustrated in Figure 2.1, where the distribution of $(\boldsymbol{X}, Y)$ is indicated by the red dots. The training set (points in the shaded regions) determines a fixed prediction function shown as a straight line. Three possible outcomes of $(\boldsymbol{X}, Y)$ are shown (black dots). The amount of loss for each point is shown as the length of the dashed lines. The generalization risk is the average loss over all possible pairs $(\boldsymbol{x}, y)$, weighted by the corresponding $f(\boldsymbol{x}, y)$.

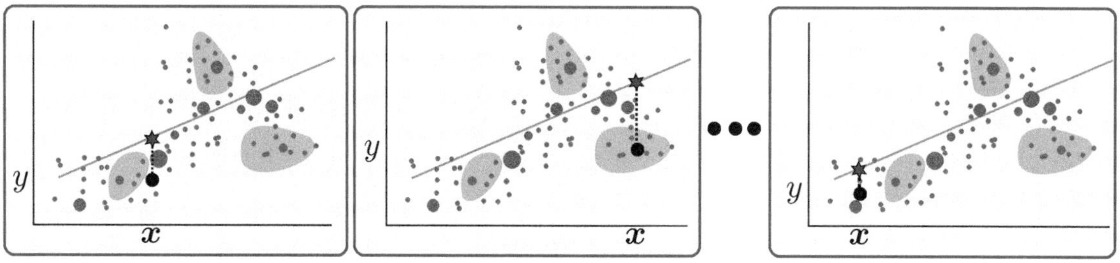

Figure 2.1: The generalization risk for a fixed training set is the weighted-average loss over all possible pairs $(\boldsymbol{x}, y)$.

For a *random* training set $\mathcal{T}$, the generalization risk is thus a random variable that depends on $\mathcal{T}$ (and $\mathcal{G}$). If we average the generalization risk over all possible instances of $\mathcal{T}$, we obtain the *expected generalization risk*:

EXPECTED GENERALIZATION RISK

$$\mathbb{E}\,\ell(g_{\mathcal{T}}^{\mathcal{G}}) = \mathbb{E}\,\mathrm{Loss}(Y, g_{\mathcal{T}}^{\mathcal{G}}(\boldsymbol{X})), \tag{2.6}$$

where $(\boldsymbol{X}, Y)$ in the expectation above is independent of $\mathcal{T}$. In the discrete case, we have $\mathbb{E}\ell(g_{\mathcal{T}}^{\mathcal{G}}) = \sum_{\boldsymbol{x},y,\boldsymbol{x}_1,y_1,\dots,\boldsymbol{x}_n,y_n} \mathrm{Loss}(y, g_{\tau}^{\mathcal{G}}(\boldsymbol{x}))f(\boldsymbol{x},y)f(\boldsymbol{x}_1,y_1)\cdots f(\boldsymbol{x}_n,y_n)$. Figure 2.2 gives an illustration.

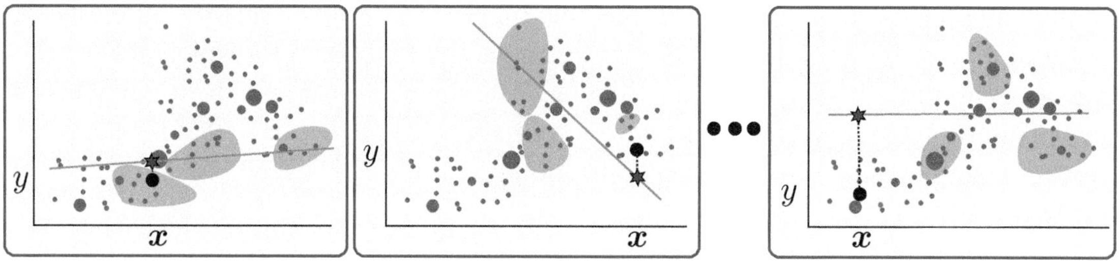

Figure 2.2: The expected generalization risk is the weighted-average loss over all possible pairs $(\boldsymbol{x}, y)$ and over all training sets.

For any outcome τ of the training data, we can estimate the generalization risk without bias by taking the sample average

$$\ell_{\mathcal{T}'}(g_{\tau}^{\mathcal{G}}) := \frac{1}{n'}\sum_{i=1}^{n'} \mathrm{Loss}(Y_i', g_{\tau}^{\mathcal{G}}(\boldsymbol{X}_i')), \tag{2.7}$$

TEST SAMPLE

TEST LOSS

where $\{(\boldsymbol{X}_1', Y_1'), \dots, (\boldsymbol{X}_{n'}', Y_{n'}')\} =: \mathcal{T}'$ is a so-called *test sample*. The test sample is completely separate from $\mathcal{T}$, but is drawn in the same way as $\mathcal{T}$; that is, via independent draws from $f(\boldsymbol{x}, y)$, for some sample size n'. We call the estimator (2.7) the *test loss*. For a random training set $\mathcal{T}$ we can define $\ell_{\mathcal{T}'}(g_{\mathcal{T}}^{\mathcal{G}})$ similarly. It is then crucial to assume that $\mathcal{T}$ is independent of $\mathcal{T}'$. Table 2.1 summarizes the main definitions and notation for supervised learning.

Table 2.1: Summary of definitions for supervised learning.

$\boldsymbol{x}$	Fixed explanatory (feature) vector.
$\boldsymbol{X}$	Random explanatory (feature) vector.
y	Fixed (real-valued) response.
Y	Random response.
$f(\boldsymbol{x}, y)$	Joint pdf of $\boldsymbol{X}$ and Y, evaluated at $(\boldsymbol{x}, y)$.
$f(y \mid \boldsymbol{x})$	Conditional pdf of Y given $\boldsymbol{X} = \boldsymbol{x}$, evaluated at y.
τ or τ_n	Fixed training data $\{(\boldsymbol{x}_i, y_i), i = 1, \ldots, n\}$.
$\mathcal{T}$ or $\mathcal{T}_n$	Random training data $\{(\boldsymbol{X}_i, Y_i), i = 1, \ldots, n\}$.
$\mathbf{X}$	Matrix of explanatory variables, with n rows $\boldsymbol{x}_i^\top, i = 1, \ldots, n$ and $\dim(\boldsymbol{x})$ feature columns; one of the features may be the constant 1.
$\boldsymbol{y}$	Vector of response variables $(y_1, \ldots, y_n)^\top$.
g	Prediction (guess) function.
$\text{Loss}(y, \widehat{y})$	Loss incurred when predicting response y with $\widehat{y}$.
$\ell(g)$	Risk for prediction function g; that is, $\mathbb{E}\,\text{Loss}(Y, g(\boldsymbol{X}))$.
g^*	Optimal prediction function; that is, $\text{argmin}_g\, \ell(g)$.
$g^{\mathcal{G}}$	Optimal prediction function in function class $\mathcal{G}$; that is, $\text{argmin}_{g \in \mathcal{G}}\, \ell(g)$.
$\ell_\tau(g)$	Training loss for prediction function g; that is, the sample average estimate of $\ell(g)$ based on a fixed training sample τ.
$\ell_{\mathcal{T}}(g)$	The same as $\ell_\tau(g)$, but now for a random training sample $\mathcal{T}$.
$g_\tau^{\mathcal{G}}$ or g_τ	The *learner*: $\text{argmin}_{g \in \mathcal{G}}\, \ell_\tau(g)$. That is, the optimal prediction function based on a fixed training set τ and function class $\mathcal{G}$. We suppress the superscript $\mathcal{G}$ if the function class is implicit.
$g_{\mathcal{T}}^{\mathcal{G}}$ or $g_{\mathcal{T}}$	The learner, where we have replaced τ with a random training set $\mathcal{T}$.

To compare the predictive performance of various learners in the function class $\mathcal{G}$, as measured by the test loss, we can use the *same* fixed training set τ and test set τ' for all learners. When there is an abundance of data, the "overall" data set is usually (randomly) divided into a training and test set, as depicted in Figure 2.3. We then use the training data to construct various learners $g_\tau^{\mathcal{G}_1}, g_\tau^{\mathcal{G}_2}, \ldots$, and use the test data to select the best (with the smallest test loss) among these learners. In this context the test set is called the *validation set*. Once the best learner has been chosen, a third "test" set can be used to assess the predictive performance of the best learner. The training, validation, and test sets can again be obtained from the overall data set via a random allocation. When the overall data set is of modest size, it is customary to perform the validation phase (model selection) on the training set only, using cross-validation. This is the topic of Section 2.5.2.

VALIDATION SET

☞ 37

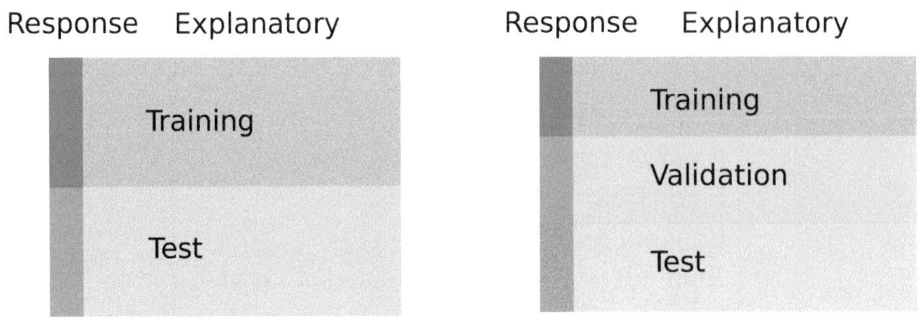

Figure 2.3: Statistical learning algorithms often require the data to be divided into training and test data. If the latter is used for model selection, a third set is needed for testing the performance of the selected model.

We next consider a concrete example that illustrates the concepts introduced so far.

■ **Example 2.1 (Polynomial Regression)** In what follows, it will appear that we have arbitrarily replaced the symbols $\boldsymbol{x}, g, \mathcal{G}$ with $u, h, \mathcal{H}$, respectively. The reason for this switch of notation will become clear at the end of the example.

POLYNOMIAL REGRESSION MODEL

The data (depicted as dots) in Figure 2.4 are $n = 100$ points $(u_i, y_i), i = 1, \ldots, n$ drawn from iid random points $(U_i, Y_i), i = 1, \ldots, n$, where the $\{U_i\}$ are uniformly distributed on the interval $(0, 1)$ and, given $U_i = u_i$, the random variable Y_i has a normal distribution with expectation $10 - 140u_i + 400u_i^2 - 250u_i^3$ and variance $\ell^* = 25$. This is an example of a *polynomial regression model*. Using a squared-error loss, the optimal prediction function $h^*(u) = \mathbb{E}[Y \mid U = u]$ is thus

$$h^*(u) = 10 - 140u + 400u^2 - 250u^3,$$

which is depicted by the dashed curve in Figure 2.4.

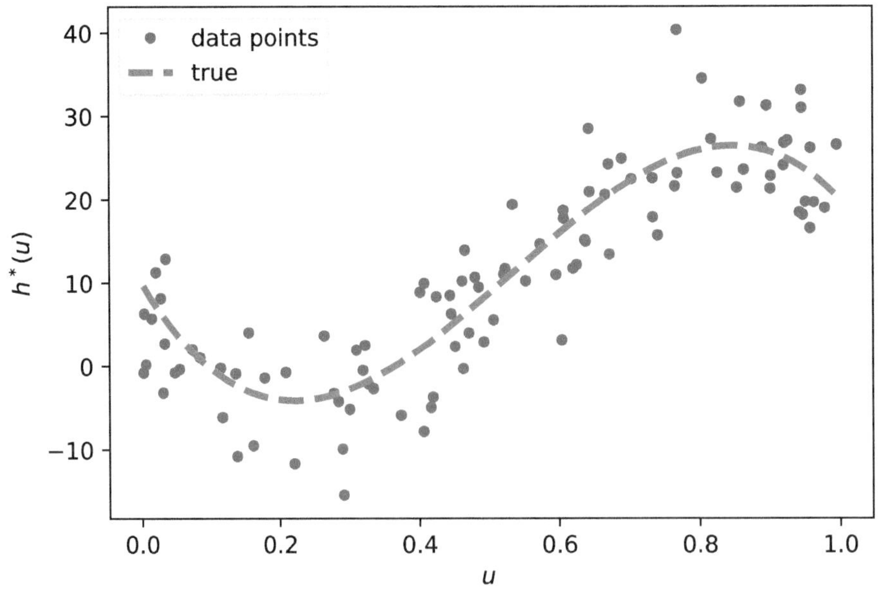

Figure 2.4: Training data and the optimal polynomial prediction function h^*.

To obtain a good estimate of $h^*(u)$ based on the training set $\tau = \{(u_i, y_i), i = 1, \ldots, n\}$, we minimize the outcome of the training loss (2.3):

$$\ell_\tau(h) = \frac{1}{n}\sum_{i=1}^{n}(y_i - h(u_i))^2, \tag{2.8}$$

over a suitable set $\mathcal{H}$ of candidate functions. Let us take the set $\mathcal{H}_p$ of polynomial functions in u of order $p-1$:

$$h(u) := \beta_1 + \beta_2 u + \beta_3 u^2 + \cdots + \beta_p u^{p-1} \tag{2.9}$$

for $p = 1, 2, \ldots$ and parameter vector $\boldsymbol{\beta} = [\beta_1, \beta_2, \ldots, \beta_p]^\top$. This function class contains the best possible $h^*(u) = \mathbb{E}[Y \mid U = u]$ for $p \geqslant 4$. Note that optimization over $\mathcal{H}_p$ is a parametric optimization problem, in that we need to find the best $\boldsymbol{\beta}$. Optimization of (2.8) over $\mathcal{H}_p$ is not straightforward, unless we notice that (2.9) is a *linear* function in $\boldsymbol{\beta}$. In particular, if we map each feature u to a feature vector $\boldsymbol{x} = [1, u, u^2, \ldots, u^{p-1}]^\top$, then the right-hand side of (2.9) can be written as the function

$$g(\boldsymbol{x}) = \boldsymbol{x}^\top\boldsymbol{\beta},$$

which is linear in $\boldsymbol{x}$ (as well as $\boldsymbol{\beta}$). The optimal $h^*(u)$ in $\mathcal{H}_p$ for $p \geqslant 4$ then corresponds to the function $g^*(\boldsymbol{x}) = \boldsymbol{x}^\top\boldsymbol{\beta}^*$ in the set $\mathcal{G}_p$ of linear functions from $\mathbb{R}^p$ to $\mathbb{R}$, where $\boldsymbol{\beta}^* = [10, -140, 400, -250, 0, \ldots, 0]^\top$. Thus, instead of working with the set $\mathcal{H}_p$ of polynomial functions we may prefer to work with the set $\mathcal{G}_p$ of linear functions. This brings us to a very important idea in statistical learning:

> Expand the feature space to obtain a *linear* prediction function.

Let us now reformulate the learning problem in terms of the new explanatory (feature) variables $\boldsymbol{x}_i = [1, u_i, u_i^2, \ldots, u_i^{p-1}]^\top$, $i = 1, \ldots, n$. It will be convenient to arrange these feature vectors into a matrix $\mathbf{X}$ with rows $\boldsymbol{x}_1^\top, \ldots, \boldsymbol{x}_n^\top$:

$$\mathbf{X} = \begin{bmatrix} 1 & u_1 & u_1^2 & \cdots & u_1^{p-1} \\ 1 & u_2 & u_2^2 & \cdots & u_2^{p-1} \\ \vdots & \vdots & \vdots & \ddots & \vdots \\ 1 & u_n & u_n^2 & \cdots & u_n^{p-1} \end{bmatrix}. \tag{2.10}$$

Collecting the responses $\{y_i\}$ into a column vector $\boldsymbol{y}$, the training loss (2.3) can now be written compactly as

$$\frac{1}{n}\|\boldsymbol{y} - \mathbf{X}\boldsymbol{\beta}\|^2. \tag{2.11}$$

To find the optimal learner (2.4) in the class $\mathcal{G}_p$ we need to find the minimizer of (2.11):

$$\widehat{\boldsymbol{\beta}} = \operatorname*{argmin}_{\boldsymbol{\beta}} \|\boldsymbol{y} - \mathbf{X}\boldsymbol{\beta}\|^2, \tag{2.12}$$

which is called the *ordinary least-squares* solution. As is illustrated in Figure 2.5, to find $\widehat{\boldsymbol{\beta}}$, we choose $\mathbf{X}\widehat{\boldsymbol{\beta}}$ to be equal to the orthogonal projection of $\boldsymbol{y}$ onto the linear space spanned by the columns of the matrix $\mathbf{X}$; that is, $\mathbf{X}\widehat{\boldsymbol{\beta}} = \mathbf{P}\boldsymbol{y}$, where $\mathbf{P}$ is the *projection matrix*.

ORDINARY LEAST-SQUARES

PROJECTION MATRIX

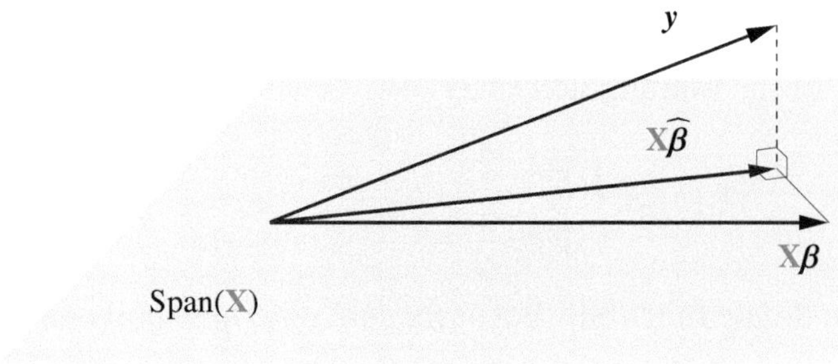

Figure 2.5: $\mathbf{X}\widehat{\boldsymbol{\beta}}$ is the orthogonal projection of $\boldsymbol{y}$ onto the linear space spanned by the columns of the matrix $\mathbf{X}$.

☞ 364

According to Theorem A.4, the projection matrix is given by

$$\mathbf{P} = \mathbf{X}\,\mathbf{X}^{+}, \tag{2.13}$$

☞ 362

PSEUDO-INVERSE

☞ 358

where the $p \times n$ matrix $\mathbf{X}^{+}$ in (2.13) is the *pseudo-inverse* of $\mathbf{X}$. If $\mathbf{X}$ happens to be of *full column rank* (so that none of the columns can be expressed as a linear combination of the other columns), then $\mathbf{X}^{+} = (\mathbf{X}^{\top}\mathbf{X})^{-1}\mathbf{X}^{\top}$.

NORMAL EQUATIONS

In any case, from $\mathbf{X}\widehat{\boldsymbol{\beta}} = \mathbf{P}\boldsymbol{y}$ and $\mathbf{P}\mathbf{X} = \mathbf{X}$, we can see that $\widehat{\boldsymbol{\beta}}$ satisfies the *normal equations*:

$$\mathbf{X}^{\top}\mathbf{X}\boldsymbol{\beta} = \mathbf{X}^{\top}\mathbf{P}\boldsymbol{y} = (\mathbf{P}\mathbf{X})^{\top}\boldsymbol{y} = \mathbf{X}^{\top}\boldsymbol{y}. \tag{2.14}$$

This is a set of linear equations, which can be solved very fast and whose solution can be written explicitly as:

$$\widehat{\boldsymbol{\beta}} = \mathbf{X}^{+}\boldsymbol{y}. \tag{2.15}$$

Figure 2.6 shows the trained learners for various values of p:

$$h_{\tau}^{\mathcal{H}_p}(u) = g_{\tau}^{\mathcal{G}_p}(\boldsymbol{x}) = \boldsymbol{x}^{\top}\widehat{\boldsymbol{\beta}}$$

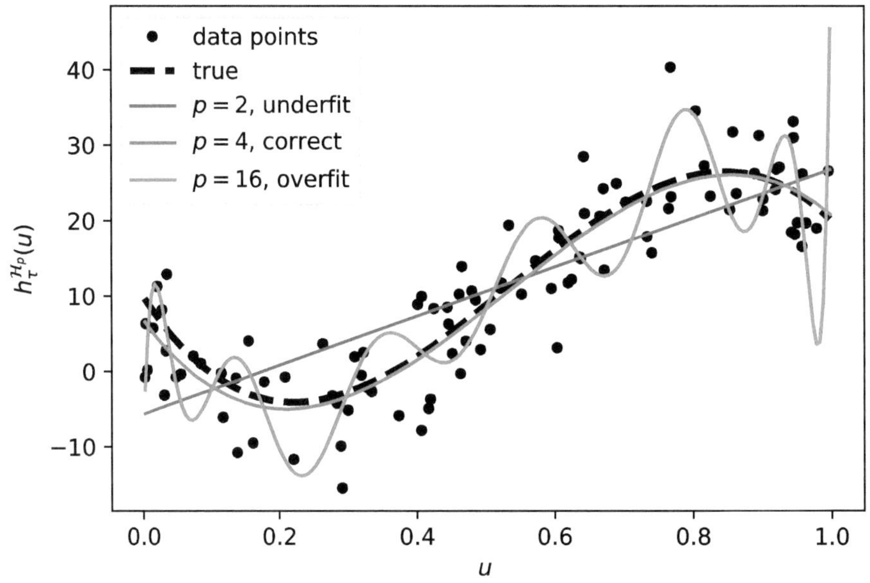

Figure 2.6: Training data with fitted curves for $p = 2, 4$, and 16. The true cubic polynomial curve for $p = 4$ is also plotted (dashed line).

We see that for $p = 16$ the fitted curve lies closer to the data points, but is further away from the dashed true polynomial curve, indicating that we overfit. The choice $p = 4$ (the true cubic polynomial) is much better than $p = 16$, or indeed $p = 2$ (straight line).

Each function class $\mathcal{G}_p$ gives a different learner $g_\tau^{\mathcal{G}_p}$, $p = 1, 2, \ldots$. To assess which is better, we should not simply take the one that gives the smallest training loss. We can always get a *zero* training loss by taking $p = n$, because for any set of n points there exists a polynomial of degree $n - 1$ that interpolates all points!

Instead, we assess the predictive performance of the learners using the test loss (2.7), computed from a test data set. If we collect all n' test feature vectors in a matrix $\mathbf{X}'$ and the corresponding test responses in a vector $\boldsymbol{y}'$, then, similar to (2.11), the test loss can be written compactly as

$$\ell_{\tau'}(g_\tau^{\mathcal{G}_p}) = \frac{1}{n'} \|\boldsymbol{y}' - \mathbf{X}'\widehat{\boldsymbol{\beta}}\|^2,$$

where $\widehat{\boldsymbol{\beta}}$ is given by (2.15), using the training data.

Figure 2.7 shows a plot of the test loss against the number of parameters in the vector $\boldsymbol{\beta}$; that is, p. The graph has a characteristic "bath-tub" shape and is at its lowest for $p = 4$, correctly identifying the polynomial order 3 for the true model. Note that the test loss, as an estimate for the generalization risk (2.7), becomes numerically unreliable after $p = 16$ (the graph goes down, where it should go up). The reader may check that the graph for the training loss exhibits a similar numerical instability for large p, and in fact fails to numerically decrease to 0 for large p, contrary to what it should do in theory. The numerical problems arise from the fact that for large p the columns of the (Vandermonde) matrix $\mathbf{X}$ are of vastly different magnitudes and so floating point errors quickly become very large.

Finally, observe that the lower bound for the test loss is here around 21, which corresponds to an estimate of the minimal (squared-error) risk $\ell^* = 25$.

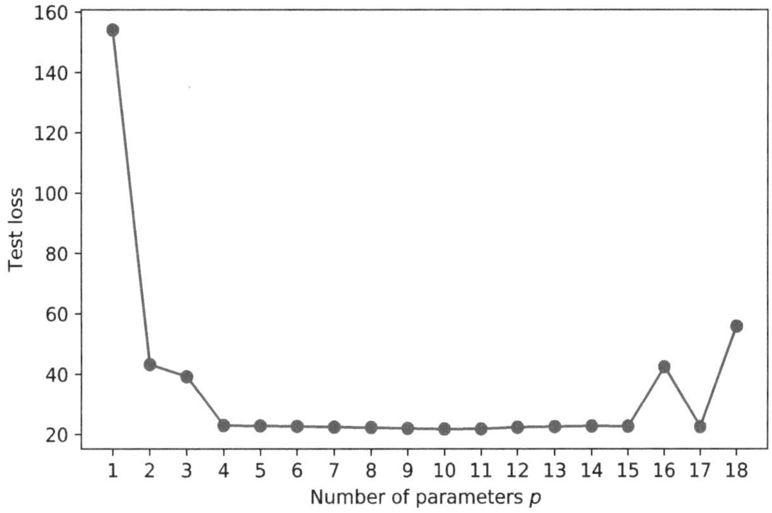

Figure 2.7: Test loss as function of the number of parameters p of the model.

This script shows how the training data were generated and plotted in Python:

polyreg1.py

```
import numpy as np
from numpy.random import rand , randn
from numpy.linalg import norm , solve
import matplotlib.pyplot as plt
def generate_data(beta , sig, n):
   u = np.random.rand(n, 1)
   y = (u ** np.arange(0, 4)) @ beta + sig * np.random.randn(n, 1)
   return u, y

np.random.seed(12)
beta = np.array([[10, -140, 400, -250]]).T
n = 100
sig = 5
u, y = generate_data(beta , sig, n)
xx = np.arange(np.min(u), np.max(u)+5e-3, 5e-3)
yy = np.polyval(np.flip(beta), xx)
plt.plot(u, y, '.', markersize=8)
plt.plot(xx, yy, '--',linewidth=3)
plt.xlabel(r'$u$')
plt.ylabel(r'$h^*(u)$')
plt.legend(['data points','true'])
plt.show()
```

The following code, which imports the code above, fits polynomial models with $p = 1, \ldots, K = 18$ parameters to the training data and plots a selection of fitted curves, as shown in Figure 2.6.

polyreg2.py

```
from polyreg1 import *

max_p = 18
p_range = np.arange(1, max_p + 1, 1)
X = np.ones((n, 1))
betahat, trainloss = {}, {}

for p in p_range:  # p is the number of parameters
    if p > 1:
        X = np.hstack((X, u**(p-1)))  # add column to matrix

    betahat[p] = solve(X.T @ X, X.T @ y)
    trainloss[p] = (norm(y - X @ betahat[p])**2/n)

p = [2, 4, 16]  # select three curves

#replot the points and true line and store in the list "plots"
plots = [plt.plot(u, y, 'k.', markersize=8)[0],
         plt.plot(xx, yy, 'k--',linewidth=3)[0]]
# add the three curves
for i in p:
    yy = np.polyval(np.flip(betahat[i]), xx)
    plots.append(plt.plot(xx, yy)[0])
```

```
plt.xlabel(r'$u$')
plt.ylabel(r'$h^{\mathcal{H}_p}_{\tau}(u)$')
plt.legend(plots,('data points', 'true','$p=2$, underfit',
                  '$p=4$, correct','$p=16$, overfit'))
plt.savefig('polyfitpy.pdf',format='pdf')
plt.show()
```

The last code snippet which imports the previous code, generates the test data and plots the graph of the test loss, as shown in Figure 2.7.

polyreg3.py

```
from polyreg2 import *

# generate test data
u_test, y_test = generate_data(beta, sig, n)

MSE = []
X_test = np.ones((n, 1))

for p in p_range:
    if p > 1:
        X_test = np.hstack((X_test, u_test**(p-1)))

    y_hat = X_test @ betahat[p]  # predictions
    MSE.append(np.sum((y_test - y_hat)**2/n))

plt.plot(p_range, MSE, 'b', p_range, MSE, 'bo')
plt.xticks(ticks=p_range)
plt.xlabel('Number of parameters $p$')
plt.ylabel('Test loss')
```

■

2.4 Tradeoffs in Statistical Learning

The art of machine learning in the supervised case is to make the generalization risk (2.5) or expected generalization risk (2.6) as small as possible, while using as few computational resources as possible. In pursuing this goal, a suitable class $\mathcal{G}$ of prediction functions has to be chosen. This choice is driven by various factors, such as

- the complexity of the class (e.g., is it rich enough to adequately approximate, or even contain, the optimal prediction function g^*?),
- the ease of training the learner via the optimization program (2.4),
- how accurately the training loss (2.3) estimates the risk (2.1) within class $\mathcal{G}$,
- the feature types (categorical, continuous, etc.).

As a result, the choice of a suitable function class $\mathcal{G}$ usually involves a tradeoff between conflicting factors. For example, a learner from a simple class $\mathcal{G}$ can be trained very quickly, but may not approximate g^* very well, whereas a learner from a rich class $\mathcal{G}$ that contains g^* may require a lot of computing resources to train.

To better understand the relation between model complexity, computational simplicity, and estimation accuracy, it is useful to decompose the generalization risk into several parts, so that the tradeoffs between these parts can be studied. We will consider two such decompositions: the approximation–estimation tradeoff and the bias–variance tradeoff.

We can decompose the generalization risk (2.5) into the following three components:

$$\ell(g_\tau^{\mathcal{G}}) = \underbrace{\ell^*}_{\text{irreducible risk}} + \underbrace{\ell(g^{\mathcal{G}}) - \ell^*}_{\text{approximation error}} + \underbrace{\ell(g_\tau^{\mathcal{G}}) - \ell(g^{\mathcal{G}})}_{\text{statistical error}}, \tag{2.16}$$

IRREDUCIBLE RISK

where $\ell^* := \ell(g^*)$ is the *irreducible risk* and $g^{\mathcal{G}} := \operatorname{argmin}_{g \in \mathcal{G}} \ell(g)$ is the best learner within class $\mathcal{G}$. No learner can predict a new response with a smaller risk than ℓ^*.

APPROXIMATION ERROR

The second component is the *approximation error*; it measures the difference between the irreducible risk and the best possible risk that can be obtained by selecting the best prediction function in the selected class of functions $\mathcal{G}$. Determining a suitable class $\mathcal{G}$ and minimizing $\ell(g)$ over this class is purely a problem of numerical and functional analysis, as the training data τ are not present. For a fixed $\mathcal{G}$ that does not contain the optimal g^*, the approximation error cannot be made arbitrarily small and may be the dominant component in the generalization risk. The only way to reduce the approximation error is by expanding the class $\mathcal{G}$ to include a larger set of possible functions.

STATISTICAL (ESTIMATION) ERROR

☞ 441

The third component is the *statistical (estimation) error*. It depends on the training set τ and, in particular, on how well the learner $g_\tau^{\mathcal{G}}$ estimates the best possible prediction function, $g^{\mathcal{G}}$, within class $\mathcal{G}$. For any sensible estimator this error should decay to zero (in probability or expectation) as the training size tends to infinity.

APPROXIMATION–ESTIMATION TRADEOFF

The *approximation–estimation tradeoff* pits two competing demands against each other. The first is that the class $\mathcal{G}$ has to be simple enough so that the statistical error is not too large. The second is that the class $\mathcal{G}$ has to be rich enough to ensure a small approximation error. Thus, there is a tradeoff between the approximation and estimation errors.

For the special case of the squared-error loss, the generalization risk is equal to $\ell(g_\tau^{\mathcal{G}}) = \mathbb{E}(Y - g_\tau^{\mathcal{G}}(\boldsymbol{X}))^2$; that is, the expected squared error[1] between the predicted value $g_\tau^{\mathcal{G}}(\boldsymbol{X})$ and the response Y. Recall that in this case the optimal prediction function is given by $g^*(\boldsymbol{x}) = \mathbb{E}[Y \mid \boldsymbol{X} = \boldsymbol{x}]$. The decomposition (2.16) can now be interpreted as follows.

1. The first component, $\ell^* = \mathbb{E}(Y - g^*(\boldsymbol{X}))^2$, is the *irreducible error*, as no prediction function will yield a smaller expected squared error.

2. The second component, the approximation error $\ell(g^{\mathcal{G}}) - \ell(g^*)$, is equal to $\mathbb{E}(g^{\mathcal{G}}(\boldsymbol{X}) - g^*(\boldsymbol{X}))^2$. We leave the proof (which is similar to that of Theorem 2.1) as an exercise; see Exercise 2. Thus, the approximation error (defined as a risk difference) can here be interpreted as the expected squared error between the optimal predicted value and the optimal predicted value within the class $\mathcal{G}$.

3. For the third component, the statistical error, $\ell(g_\tau^{\mathcal{G}}) - \ell(g^{\mathcal{G}})$ there is no direct interpretation as an expected squared error *unless* $\mathcal{G}$ is the class of *linear* functions; that is, $g(\boldsymbol{x}) = \boldsymbol{x}^\top \boldsymbol{\beta}$ for some vector $\boldsymbol{\beta}$. In this case we can write (see Exercise 3) the statistical error as $\ell(g_\tau^{\mathcal{G}}) - \ell(g^{\mathcal{G}}) = \mathbb{E}(g_\tau^{\mathcal{G}}(\boldsymbol{X}) - g^{\mathcal{G}}(\boldsymbol{X}))^2$.

[1]Colloquially called *mean squared error*.

Thus, when using a squared-error loss, the generalization risk for a linear class $\mathcal{G}$ can be decomposed as:

$$\ell(g_\tau^{\mathcal{G}}) = \mathbb{E}(g_\tau^{\mathcal{G}}(\boldsymbol{X}) - Y)^2 = \ell^* + \underbrace{\mathbb{E}(g^{\mathcal{G}}(\boldsymbol{X}) - g^*(\boldsymbol{X}))^2}_{\text{approximation error}} + \underbrace{\mathbb{E}(g_\tau^{\mathcal{G}}(\boldsymbol{X}) - g^{\mathcal{G}}(\boldsymbol{X}))^2}_{\text{statistical error}}. \tag{2.17}$$

Note that in this decomposition the statistical error is the only term that depends on the training set.

■ **Example 2.2 (Polynomial Regression (cont.))** We continue Example 2.1. Here $\mathcal{G} = \mathcal{G}_p$ is the class of linear functions of $\boldsymbol{x} = [1, u, u^2, \dots, u^{p-1}]^\top$, and $g^*(\boldsymbol{x}) = \boldsymbol{x}^\top \boldsymbol{\beta}^*$. Conditional on $\boldsymbol{X} = \boldsymbol{x}$ we have that $Y = g^*(\boldsymbol{x}) + \varepsilon(\boldsymbol{x})$, with $\varepsilon(\boldsymbol{x}) \sim \mathcal{N}(0, \ell^*)$, where $\ell^* = \mathbb{E}(Y - g^*(\boldsymbol{X}))^2 = 25$ is the irreducible error. We wish to understand how the approximation and statistical errors behave as we change the complexity parameter p.

First, we consider the approximation error. Any function $g \in \mathcal{G}_p$ can be written as

$$g(\boldsymbol{x}) = h(u) = \beta_1 + \beta_2 u + \cdots + \beta_p u^{p-1} = [1, u, \dots, u^{p-1}]\boldsymbol{\beta},$$

and so $g(\boldsymbol{X})$ is distributed as $[1, U, \dots, U^{p-1}]\boldsymbol{\beta}$, where $U \sim \mathcal{U}(0,1)$. Similarly, $g^*(\boldsymbol{X})$ is distributed as $[1, U, U^2, U^3]\boldsymbol{\beta}^*$. It follows that an expression for the approximation error is: $\int_0^1 \left([1, u, \dots, u^{p-1}]\boldsymbol{\beta} - [1, u, u^2, u^3]\boldsymbol{\beta}^*\right)^2 \mathrm{d}u$. To minimize this error, we set the gradient with respect to $\boldsymbol{\beta}$ to zero and obtain the p linear equations ☞ 399

$$\begin{aligned}
\int_0^1 \left([1, u, \dots, u^{p-1}]\boldsymbol{\beta} - [1, u, u^2, u^3]\boldsymbol{\beta}^*\right) \mathrm{d}u &= 0,\\
\int_0^1 \left([1, u, \dots, u^{p-1}]\boldsymbol{\beta} - [1, u, u^2, u^3]\boldsymbol{\beta}^*\right) u \,\mathrm{d}u &= 0,\\
&\vdots\\
\int_0^1 \left([1, u, \dots, u^{p-1}]\boldsymbol{\beta} - [1, u, u^2, u^3]\boldsymbol{\beta}^*\right) u^{p-1} \mathrm{d}u &= 0.
\end{aligned}$$

Let

$$\mathbf{H}_p = \int_0^1 [1, u, \dots, u^{p-1}]^\top [1, u, \dots, u^{p-1}] \,\mathrm{d}u$$

be the $p \times p$ *Hilbert matrix*, which has (i, j)-th entry given by $\int_0^1 u^{i+j-2}\,\mathrm{d}u = 1/(i+j-1)$. Then, the above system of linear equations can be written as $\mathbf{H}_p \boldsymbol{\beta} = \widetilde{\mathbf{H}}\boldsymbol{\beta}^*$, where $\widetilde{\mathbf{H}}$ is the $p \times 4$ upper left sub-block of $\mathbf{H}_{\widetilde{p}}$ and $\widetilde{p} = \max\{p, 4\}$. The solution, which we denote by $\boldsymbol{\beta}_p$, is: HILBERT MATRIX

$$\boldsymbol{\beta}_p = \begin{cases} \frac{65}{6}, & p = 1,\\ [-\frac{20}{3}, 35]^\top, & p = 2,\\ [-\frac{5}{2}, 10, 25]^\top, & p = 3,\\ [10, -140, 400, -250, 0, \dots, 0]^\top, & p \geqslant 4. \end{cases} \tag{2.18}$$

Hence, the approximation error $\mathbb{E}\left(g^{\mathcal{G}_p}(\boldsymbol{X}) - g^*(\boldsymbol{X})\right)^2$ is given by

$$\int_0^1 \left([1, u, \dots, u^{p-1}]\boldsymbol{\beta}_p - [1, u, u^2, u^3]\boldsymbol{\beta}^*\right)^2 \mathrm{d}u = \begin{cases} \frac{32225}{252} \approx 127.9, & p = 1,\\ \frac{1625}{63} \approx 25.8, & p = 2,\\ \frac{625}{28} \approx 22.3, & p = 3,\\ 0, & p \geqslant 4. \end{cases} \tag{2.19}$$

Notice how the approximation error becomes smaller as p increases. In this particular example the approximation error is in fact zero for $p \geqslant 4$. In general, as the class of approximating functions $\mathcal{G}$ becomes more complex, the approximation error goes down.

Next, we illustrate the typical behavior of the statistical error. Since $g_\tau(\boldsymbol{x}) = \boldsymbol{x}^\top\widehat{\boldsymbol{\beta}}$, the statistical error can be written as

$$\int_0^1 \left([1,\ldots,u^{p-1}](\widehat{\boldsymbol{\beta}} - \boldsymbol{\beta}_p)\right)^2 \mathrm{d}u = (\widehat{\boldsymbol{\beta}} - \boldsymbol{\beta}_p)^\top \mathbf{H}_p(\widehat{\boldsymbol{\beta}} - \boldsymbol{\beta}_p). \tag{2.20}$$

Figure 2.8 illustrates the decomposition (2.17) of the generalization risk for the *same* training set that was used to compute the test loss in Figure 2.7. Recall that test loss gives an estimate of the generalization risk, using independent test data. Comparing the two figures, we see that in this case the two match closely. The global minimum of the statistical error is approximately 0.28, with minimizer $p = 4$. Since the approximation error is monotonically decreasing to zero, $p = 4$ is also the global minimizer of the generalization risk.

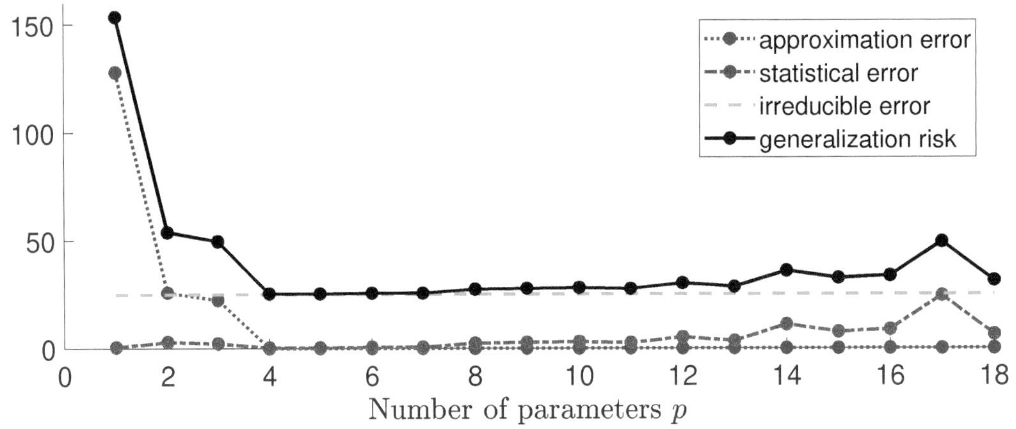

Figure 2.8: The generalization risk for a particular training set is the sum of the irreducible error, the approximation error, and the statistical error. The approximation error decreases to zero as p increases, whereas the statistical error has a tendency to increase after $p = 4$.

Note that the statistical error depends on the estimate $\widehat{\boldsymbol{\beta}}$, which in its turn depends on the training set τ. We can obtain a better understanding of the statistical error by considering its *expected* behavior; that is, averaged over many training sets. This is explored in Exercise 11. ■

Using again a squared-error loss, a second decomposition (for general $\mathcal{G}$) starts from

$$\ell(g_\tau^{\mathcal{G}}) = \ell^* + \ell(g_\tau^{\mathcal{G}}) - \ell(g^*),$$

where the statistical error and approximation error are combined. Using similar reasoning as in the proof of Theorem 2.1, we have

$$\ell(g_\tau^{\mathcal{G}}) = \mathbb{E}(g_\tau^{\mathcal{G}}(\boldsymbol{X}) - Y)^2 = \ell^* + \mathbb{E}\left(g_\tau^{\mathcal{G}}(\boldsymbol{X}) - g^*(\boldsymbol{X})\right)^2 = \ell^* + \mathbb{E}D^2(\boldsymbol{X},\tau),$$

where $D(\mathbf{x}, \tau) := g_\tau^{\mathcal{G}}(\mathbf{x}) - g^*(\mathbf{x})$. Now consider the random variable $D(\mathbf{x}, \mathcal{T})$ for a random training set $\mathcal{T}$. The expectation of its square is:

$$\begin{aligned}\mathbb{E}\left(g_{\mathcal{T}}^{\mathcal{G}}(\mathbf{x}) - g^*(\mathbf{x})\right)^2 &= \mathbb{E}D^2(\mathbf{x}, \mathcal{T}) = (\mathbb{E}D(\mathbf{x}, \mathcal{T}))^2 + \mathbb{V}\mathrm{ar}\, D(\mathbf{x}, \mathcal{T}) \\ &= \underbrace{(\mathbb{E}g_{\mathcal{T}}^{\mathcal{G}}(\mathbf{x}) - g^*(\mathbf{x}))^2}_{\text{pointwise squared bias}} + \underbrace{\mathbb{V}\mathrm{ar}\, g_{\mathcal{T}}^{\mathcal{G}}(\mathbf{x})}_{\text{pointwise variance}}.\end{aligned} \tag{2.21}$$

If we view the learner $g_{\mathcal{T}}^{\mathcal{G}}(\mathbf{x})$ as a function of a random training set, then the *pointwise squared bias* term is a measure for how close $g_{\mathcal{T}}^{\mathcal{G}}(\mathbf{x})$ is on average to the true $g^*(\mathbf{x})$, whereas the *pointwise variance* term measures the deviation of $g_{\mathcal{T}}^{\mathcal{G}}(\mathbf{x})$ from its expected value $\mathbb{E}g_{\mathcal{T}}^{\mathcal{G}}(\mathbf{x})$. The squared bias can be reduced by making the class of functions $\mathcal{G}$ more complex. However, decreasing the bias by increasing the complexity often leads to an increase in the variance term. We are thus seeking learners that provide an optimal balance between the bias and variance, as expressed via a minimal generalization risk. This is called the *bias–variance tradeoff*.

POINTWISE SQUARED BIAS

POINTWISE VARIANCE

BIAS–VARIANCE TRADEOFF

Note that the *expected* generalization risk (2.6) can be written as $\ell^* + \mathbb{E}D^2(\mathbf{X}, \mathcal{T})$, where $\mathbf{X}$ and $\mathcal{T}$ are independent. It therefore decomposes as

$$\mathbb{E}\,\ell(g_{\mathcal{T}}^{\mathcal{G}}) = \ell^* + \underbrace{\mathbb{E}\,(\mathbb{E}[g_{\mathcal{T}}^{\mathcal{G}}(\mathbf{X}) \,|\, \mathbf{X}] - g^*(\mathbf{X}))^2}_{\text{expected squared bias}} + \underbrace{\mathbb{E}[\mathbb{V}\mathrm{ar}[g_{\mathcal{T}}^{\mathcal{G}}(\mathbf{X}) \,|\, \mathbf{X}]]}_{\text{expected variance}}. \tag{2.22}$$

2.5 Estimating Risk

The most straightforward way to quantify the generalization risk (2.5) is to estimate it via the test loss (2.7). However, the generalization risk depends inherently on the training set, and so different training sets may yield significantly different estimates. Moreover, when there is a limited amount of data available, reserving a substantial proportion of the data for testing rather than training may be uneconomical. In this section we consider different methods for estimating risk measures which aim to circumvent these difficulties.

2.5.1 In-Sample Risk

We mentioned that, due to the phenomenon of overfitting, the training loss of the learner, $\ell_\tau(g_\tau)$ (for simplicity, here we omit $\mathcal{G}$ from $g_\tau^{\mathcal{G}}$), is not a good estimate of the generalization risk $\ell(g_\tau)$ of the learner. One reason for this is that we use the same data for both training the model and assessing its risk. How should we then estimate the generalization risk or expected generalization risk?

To simplify the analysis, suppose that we wish to estimate the average accuracy of the predictions of the learner g_τ at the n feature vectors $\mathbf{x}_1, \ldots, \mathbf{x}_n$ (these are part of the training set τ). In other words, we wish to estimate the *in-sample risk* of the learner g_τ:

IN-SAMPLE RISK

$$\ell_{\text{in}}(g_\tau) = \frac{1}{n}\sum_{i=1}^{n} \mathbb{E}\,\mathrm{Loss}(Y_i', g_\tau(\mathbf{x}_i)), \tag{2.23}$$

where each response Y_i' is drawn from $f(y \,|\, \mathbf{x}_i)$, independently. Even in this simplified setting, the training loss of the learner will be a poor estimate of the in-sample risk. Instead, the proper way to assess the prediction accuracy of the learner at the feature vectors $\boldsymbol{x}_1, \ldots, \boldsymbol{x}_n$, is to draw new response values $Y_i' \sim f(y \mid \boldsymbol{x}_i)$, $i = 1, \ldots, n$, that are independent from the responses $y_1, \ldots, y_n$ in the training data, and then estimate the in-sample risk of g_τ via

$$\frac{1}{n}\sum_{i=1}^{n} \text{Loss}(Y_i', g_\tau(\boldsymbol{x}_i)).$$

For a fixed training set τ, we can compare the training loss of the learner with the in-sample risk. Their difference,

$$\text{op}_\tau = \ell_{\text{in}}(g_\tau) - \ell_\tau(g_\tau),$$

is called the *optimism* (of the training loss), because it measures how much the training loss underestimates (is optimistic about) the unknown in-sample risk. Mathematically, it is simpler to work with the *expected optimism*:

EXPECTED OPTIMISM

$$\mathbb{E}[\text{op}_{\mathcal{T}} \mid \boldsymbol{X}_1 = \boldsymbol{x}_1, \ldots, \boldsymbol{X}_n = \boldsymbol{x}_n] =: \mathbb{E}_{\mathbf{X}}\, \text{op}_{\mathcal{T}},$$

where the expectation is taken over a random training set $\mathcal{T}$, conditional on $\boldsymbol{X}_i = \boldsymbol{x}_i$, $i = 1, \ldots, n$. For ease of notation, we have abbreviated the expected optimism to $\mathbb{E}_{\mathbf{X}}\, \text{op}_{\mathcal{T}}$, where $\mathbb{E}_{\mathbf{X}}$ denotes the expectation operator conditional on $\boldsymbol{X}_i = \boldsymbol{x}_i, i = 1, \ldots, n$. As in Example 2.1, the feature vectors are stored as the rows of an $n \times p$ matrix $\mathbf{X}$. It turns out that the expected optimism for various loss functions can be expressed in terms of the (conditional) covariance between the observed and predicted response.

Theorem 2.2: Expected Optimism

For the squared-error loss and 0–1 loss with 0–1 response, the expected optimism is

$$\mathbb{E}_{\mathbf{X}}\, \text{op}_{\mathcal{T}} = \frac{2}{n}\sum_{i=1}^{n} \mathbb{C}\text{ov}_{\mathbf{X}}(g_{\mathcal{T}}(\boldsymbol{x}_i), Y_i). \tag{2.24}$$

Proof: In what follows, all expectations are taken conditional on $\boldsymbol{X}_1 = \boldsymbol{x}_1, \ldots, \boldsymbol{X}_n = \boldsymbol{x}_n$. Let Y_i be the response for $\boldsymbol{x}_i$ and let $\widehat{Y}_i = g_{\mathcal{T}}(\boldsymbol{x}_i)$ be the predicted value. Note that the latter depends on $Y_1, \ldots, Y_n$. Also, let Y_i' be an independent copy of Y_i for the *same* $\boldsymbol{x}_i$, as in (2.23). In particular, Y_i' has the same distribution as Y_i and is statistically independent of all $\{Y_j\}$, including Y_i, and therefore is also independent of $\widehat{Y}_i$. We have

$$\begin{aligned}\mathbb{E}_{\mathbf{X}}\, \text{op}_{\mathcal{T}} &= \frac{1}{n}\sum_{i=1}^{n} \mathbb{E}_{\mathbf{X}}\left[(Y_i' - \widehat{Y}_i)^2 - (Y_i - \widehat{Y}_i)^2\right] = \frac{2}{n}\sum_{i=1}^{n} \mathbb{E}_{\mathbf{X}}\left[(Y_i - Y_i')\widehat{Y}_i\right] \\ &= \frac{2}{n}\sum_{i=1}^{n}\left(\mathbb{E}_{\mathbf{X}}[Y_i\widehat{Y}_i] - \mathbb{E}_{\mathbf{X}} Y_i\, \mathbb{E}_{\mathbf{X}}\widehat{Y}_i\right) = \frac{2}{n}\sum_{i=1}^{n} \mathbb{C}\text{ov}_{\mathbf{X}}(\widehat{Y}_i, Y_i).\end{aligned}$$

The proof for the 0–1 loss with 0–1 response is left as Exercise 4. □

In summary, the expected optimism indicates how much, on average, the training loss deviates from the expected in-sample risk. Since the covariance of independent random variables is zero, the expected optimism is zero if the learner $g_{\mathcal{T}}$ is statistically independent from the responses $Y_1, \ldots, Y_n$.

■ **Example 2.3 (Polynomial Regression (cont.))** We continue Example 2.2, where the components of the response vector $\boldsymbol{Y} = [Y_1, \ldots, Y_n]^\top$ are independent and normally distributed with variance $\ell^* = 25$ (the irreducible error) and expectations $\mathbb{E}_{\mathbf{X}} Y_i = g^*(\boldsymbol{x}_i) = \boldsymbol{x}_i^\top \boldsymbol{\beta}^*$, $i = 1, \ldots, n$. Using the formula (2.15) for the least-squares estimator $\widehat{\boldsymbol{\beta}}$, the expected optimism (2.24) is

$$\frac{2}{n}\sum_{i=1}^{n} \mathbb{C}\mathrm{ov}_{\mathbf{X}}\left(\boldsymbol{x}_i^\top\widehat{\boldsymbol{\beta}}, Y_i\right) = \frac{2}{n}\mathrm{tr}\left(\mathbb{C}\mathrm{ov}_{\mathbf{X}}\left(\mathbf{X}\widehat{\boldsymbol{\beta}}, \boldsymbol{Y}\right)\right) = \frac{2}{n}\mathrm{tr}\left(\mathbb{C}\mathrm{ov}_{\mathbf{X}}\left(\mathbf{X}\mathbf{X}^+\boldsymbol{Y}, \boldsymbol{Y}\right)\right)$$
$$= \frac{2\mathrm{tr}\left(\mathbf{X}\mathbf{X}^+\mathbb{C}\mathrm{ov}_{\mathbf{X}}\left(\boldsymbol{Y}, \boldsymbol{Y}\right)\right)}{n} = \frac{2\ell^*\mathrm{tr}\left(\mathbf{X}\mathbf{X}^+\right)}{n} = \frac{2\ell^* p}{n}.$$

In the last equation we used the cyclic property of the trace (Theorem A.1): $\mathrm{tr}(\mathbf{X}\mathbf{X}^+) = \mathrm{tr}(\mathbf{X}^+\mathbf{X}) = \mathrm{tr}(\mathbf{I}_p)$, assuming that $\mathrm{rank}(\mathbf{X}) = p$. Therefore, an estimate for the in-sample risk (2.23) is: ☞ 359

$$\widehat{\ell_{\mathrm{in}}}(g_\tau) = \ell_\tau(g_\tau) + 2\ell^* p/n, \tag{2.25}$$

where we have assumed that the irreducible risk ℓ^* is known. Figure 2.9 shows that this estimate is very close to the test loss from Figure 2.7. Hence, instead of computing the test loss to assess the best model complexity p, we could simply have minimized the training loss plus the correction term $2\ell^* p/n$. In practice, ℓ^* also has to be estimated somehow.

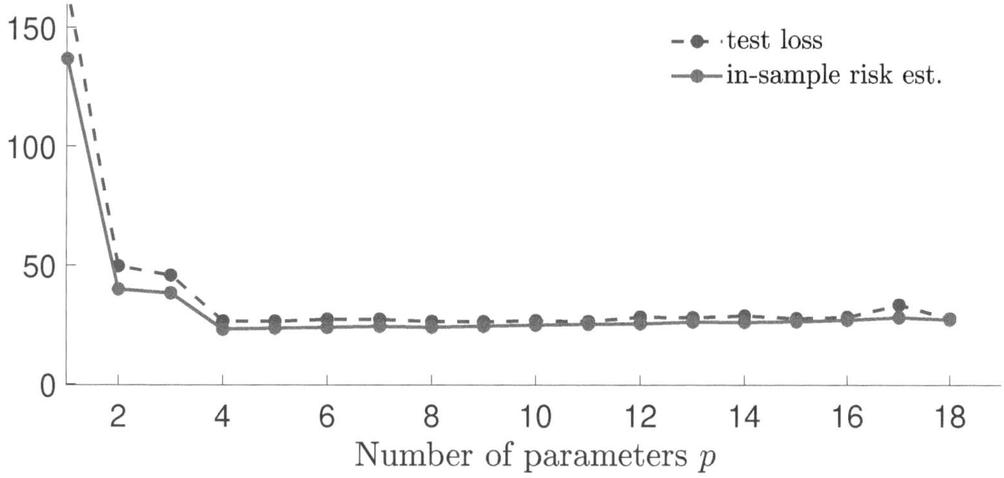

Figure 2.9: In-sample risk estimate $\widehat{\ell_{\mathrm{in}}}(g_\tau)$ as a function of the number of parameters p of the model. The test loss is superimposed as a blue dashed curve.

■

2.5.2 Cross-Validation

In general, for complex function classes $\mathcal{G}$, it is very difficult to derive simple formulas of the approximation and statistical errors, let alone for the generalization risk or expected generalization risk. As we saw, when there is an abundance of data, the easiest way to assess the generalization risk for a given training set τ is to obtain a test set τ' and evaluate the test loss (2.7). When a sufficiently large test set is not available but computational resources are cheap, one can instead gain direct knowledge of the expected generalization risk via a computationally intensive method called *cross-validation*. ☞ 24

CROSS-VALIDATION

The idea is to make multiple identical copies of the data set, and to partition each copy into different training and test sets, as illustrated in Figure 2.10. Here, there are four copies of the data set (consisting of response and explanatory variables). Each copy is divided into a test set (colored blue) and training set (colored pink). For each of these sets, we estimate the model parameters using only training data and then predict the responses for the test set. The average loss between the predicted and observed responses is then a measure for the predictive power of the model.

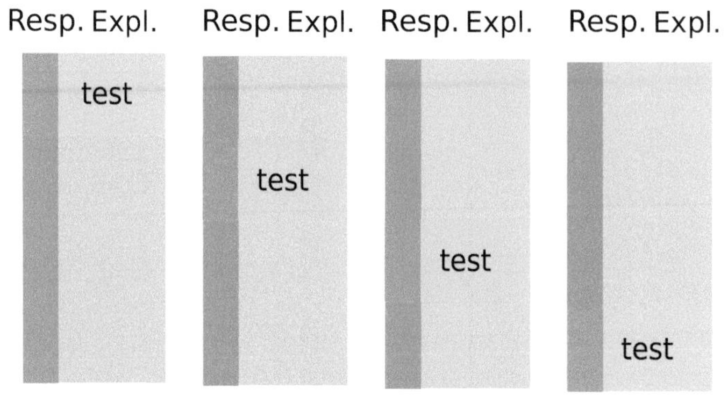

Figure 2.10: An illustration of four-fold cross-validation, representing four copies of the same data set. The data in each copy is partitioned into a training set (pink) and a test set (blue). The darker columns represent the response variable and the lighter ones the explanatory variables.

FOLDS

In particular, suppose we partition a data set $\mathcal{T}$ of size n into K *folds* $C_1, \ldots, C_K$ of sizes $n_1, \ldots, n_K$ (hence, $n_1 + \cdots + n_K = n$). Typically $n_k \approx n/K$, $k = 1, \ldots, K$.

Let ℓ_{C_k} be the test loss when using C_k as test data and all remaining data, denoted $\mathcal{T}_{-k}$, as training data. Each ℓ_{C_k} is an unbiased estimator of the generalization risk for training set $\mathcal{T}_{-k}$; that is, for $\ell(g_{\mathcal{T}_{-k}})$.

K-FOLD CROSS-VALIDATION

The *K-fold cross-validation* loss is the weighted average of these risk estimators:

$$\begin{aligned}\mathrm{CV}_K &= \sum_{k=1}^{K} \frac{n_k}{n} \ell_{C_k}(g_{\mathcal{T}_{-k}}) \\ &= \frac{1}{n} \sum_{k=1}^{K} \sum_{i \in C_k} \mathrm{Loss}(g_{\mathcal{T}_{-k}}(\boldsymbol{x}_i), y_i) \\ &= \frac{1}{n} \sum_{i=1}^{n} \mathrm{Loss}(g_{\mathcal{T}_{-\kappa(i)}}(\boldsymbol{x}_i), y_i),\end{aligned}$$

where the function $\kappa : \{1, \ldots, n\} \mapsto \{1, \ldots, K\}$ indicates to which of the K folds each of the n observations belongs. As the average is taken over varying training sets $\{\mathcal{T}_{-k}\}$, it estimates the expected generalization risk $\mathbb{E}\, \ell(g_{\mathcal{T}})$, rather than the generalization risk $\ell(g_\tau)$ for the particular training set τ.

■ **Example 2.4 (Polynomial Regression (cont.))** For the polynomial regression example, we can calculate a K-fold cross-validation loss with a nonrandom partitioning of the training set using the following code, which imports the previous code for the polynomial regression example. We omit the full plotting code.

polyregCV.py

```
from polyreg3 import *

K_vals = [5, 10, 100]  # number of folds
cv = np.zeros((len(K_vals), max_p))   # cv loss
X = np.ones((n, 1))

for p in p_range:
  if p > 1:
    X = np.hstack((X, u**(p-1)))
  j = 0
  for K in K_vals:
    loss = []
    for k in range(1, K+1):
       # integer indices of test samples
       test_ind = ((n/K)*(k-1) + np.arange(1,n/K+1)-1).astype('int')
       train_ind = np.setdiff1d(np.arange(n), test_ind)

       X_train, y_train = X[train_ind, :], y[train_ind, :]
       X_test, y_test = X[test_ind, :], y[test_ind]

       # fit model and evaluate test loss
       betahat = solve(X_train.T @ X_train, X_train.T @ y_train)
       loss.append(norm(y_test - X_test @ betahat) ** 2)

    cv[j, p-1] = sum(loss)/n
    j += 1

# basic plotting
plt.plot(p_range, cv[0, :], 'k-.')
plt.plot(p_range, cv[1, :], 'r')
plt.plot(p_range, cv[2, :], 'b--')
plt.show()
```

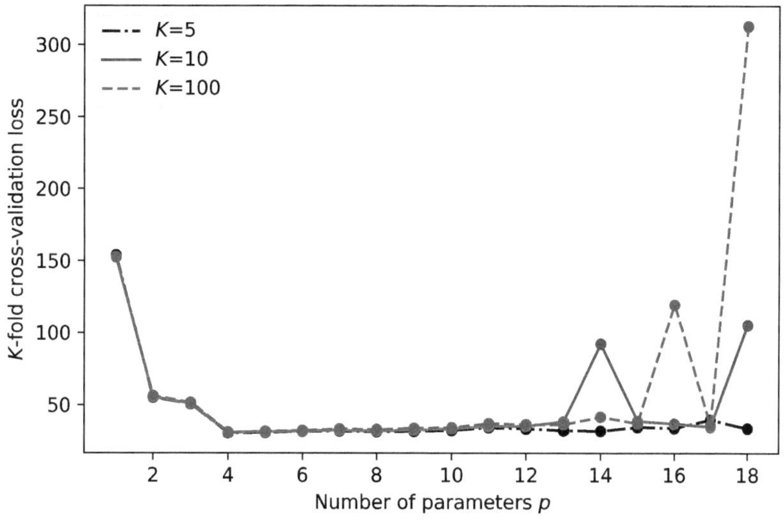

Figure 2.11: *K*-fold cross-validation for the polynomial regression example.

Figure 2.11 shows the cross-validation loss for $K \in \{5, 10, 100\}$. The case $K = 100$ corresponds to the *leave-one-out cross-validation*, which can be computed more efficiently using the formula in Theorem 5.1. ■

LEAVE-ONE-OUT CROSS-VALIDATION

☞ 174

2.6 Modeling Data

MODEL

The first step in any data analysis is to *model* the data in one form or another. For example, in an *unsupervised* learning setting with data represented by a vector $\boldsymbol{x} = [x_1, \ldots, x_p]^\top$, a very general model is to assume that $\boldsymbol{x}$ is the outcome of a random vector $\boldsymbol{X} = [X_1, \ldots, X_p]^\top$ with some unknown pdf f. The model can then be refined by assuming a specific form of f.

When given a sequence of such data vectors $\boldsymbol{x}_1, \ldots, \boldsymbol{x}_n$, one of the simplest models is to assume that the corresponding random vectors $\boldsymbol{X}_1, \ldots, \boldsymbol{X}_n$ are *independent and identically distributed (iid)*. We write

☞ 431

$$\boldsymbol{X}_1, \ldots, \boldsymbol{X}_n \overset{\text{iid}}{\sim} f \quad \text{or} \quad \boldsymbol{X}_1, \ldots, \boldsymbol{X}_n \overset{\text{iid}}{\sim} \mathsf{Dist},$$

to indicate that the random vectors form an iid sample from a sampling pdf f or sampling distribution Dist. This model formalizes the notion that the knowledge about one variable does not provide extra information about another variable. The main theoretical use of independent data models is that the joint density of the random vectors $\boldsymbol{X}_1, \ldots, \boldsymbol{X}_n$ is simply the *product* of the marginal ones; see Theorem C.1. Specifically,

☞ 431

$$f_{\boldsymbol{X}_1, \ldots, \boldsymbol{X}_n}(\boldsymbol{x}_1, \ldots, \boldsymbol{x}_n) = f(\boldsymbol{x}_1) \cdots f(\boldsymbol{x}_n).$$

In most models of this kind, our approximation or model for the sampling distribution is specified up to a small number of parameters. That is, $g(\boldsymbol{x})$ is of the form $g(\boldsymbol{x} \,|\, \boldsymbol{\beta})$ which is known up to some parameter vector $\boldsymbol{\beta}$. Examples for the one-dimensional case ($p = 1$) include the $\mathcal{N}(\mu, \sigma^2)$, $\mathsf{Bin}(n, p)$, and $\mathsf{Exp}(\lambda)$ distributions. See Tables C.1 and C.2 for other common sampling distributions.

☞ 427

Typically, the parameters are unknown and must be estimated from the data. In a non-parametric setting the whole sampling distribution would be unknown. To visualize the underlying sampling distribution from outcomes $\boldsymbol{x}_1, \ldots, \boldsymbol{x}_n$ one can use graphical representations such as histograms, density plots, and empirical cumulative distribution functions, as discussed in Chapter 1.

☞ 11

If the order in which the data were collected (or their labeling) is not informative or relevant, then the joint pdf of $\boldsymbol{X}_1, \ldots, \boldsymbol{X}_n$ satisfies the symmetry:

$$f_{\boldsymbol{X}_1, \ldots, \boldsymbol{X}_n}(\boldsymbol{x}_1, \ldots, \boldsymbol{x}_n) = f_{\boldsymbol{X}_{\pi_1}, \ldots, \boldsymbol{X}_{\pi_n}}(\boldsymbol{x}_{\pi_1}, \ldots, \boldsymbol{x}_{\pi_n}) \tag{2.26}$$

for any permutation $\pi_1, \ldots, \pi_n$ of the integers $1, \ldots, n$. We say that the infinite sequence $\boldsymbol{X}_1, \boldsymbol{X}_2, \ldots$ is *exchangeable* if this permutational invariance (2.26) holds for any finite subset of the sequence. As we shall see in Section 2.9 on Bayesian learning, it is common to assume that the random vectors $\boldsymbol{X}_1, \ldots, \boldsymbol{X}_n$ are a subset of an exchangeable sequence and thus satisfy (2.26). Note that while iid random variables are exchangeable, the converse is not necessarily true. Thus, the assumption of an exchangeable sequence of random vectors is weaker than the assumption of iid random vectors.

EXCHANGEABLE

Figure 2.12 illustrates the modeling tradeoffs. The keywords within the triangle represent various modeling paradigms. A few keywords have been highlighted, symbolizing their importance in modeling. The specific meaning of the keywords does not concern us here, but the point is there are many models to choose from, depending on what assumptions are made about the data.

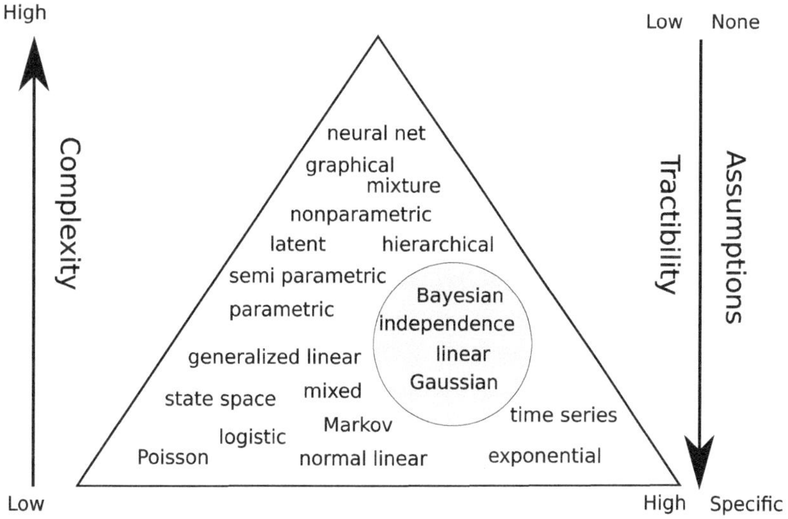

Figure 2.12: Illustration of the modeling dilemma. Complex models are more generally applicable, but may be difficult to analyze. Simple models may be highly tractable, but may not describe the data accurately. The triangular shape signifies that there are a great many specific models but not so many generic ones.

On the one hand, models that make few assumptions are more widely applicable, but at the same time may not be very mathematically tractable or provide insight into the nature of the data. On the other hand, very specific models may be easy to handle and interpret, but may not match the data very well. This tradeoff between the tractability and applicability of the model is very similar to the approximation–estimation tradeoff described in Section 2.4.

In the typical *unsupervised* setting we have a training set $\tau = \{\boldsymbol{x}_1, \ldots, \boldsymbol{x}_n\}$ that is viewed as the outcome of n iid random variables $\boldsymbol{X}_1, \ldots, \boldsymbol{X}_n$ from some unknown pdf f. The objective is then to learn or estimate f from the finite training data. To put the learning in a similar framework as for supervised learning discussed in the preceding Sections 2.3–2.5, we begin by specifying a class of probability density functions $\mathcal{G}_p := \{g(\cdot \,|\, \boldsymbol{\theta}), \boldsymbol{\theta} \in \boldsymbol{\Theta}\}$, where $\boldsymbol{\theta}$ is a parameter in some subset $\boldsymbol{\Theta}$ of $\mathbb{R}^p$. We now seek the best g in $\mathcal{G}_p$ to minimize some risk. Note that $\mathcal{G}_p$ may not necessarily contain the true f even for very large p.

⚠ We stress that our notation $g(\boldsymbol{x})$ has a different meaning in the supervised and unsupervised case. In the supervised case, g is interpreted as a prediction function for a response y; in the unsupervised setting, g is an approximation of a density f.

For each $\boldsymbol{x}$ we measure the discrepancy between the true model $f(\boldsymbol{x})$ and the hypothesized model $g(\boldsymbol{x} \,|\, \boldsymbol{\theta})$ using the loss function

$$\text{Loss}(f(\boldsymbol{x}), g(\boldsymbol{x} \,|\, \boldsymbol{\theta})) = \ln \frac{f(\boldsymbol{x})}{g(\boldsymbol{x} \,|\, \boldsymbol{\theta})} = \ln f(\boldsymbol{x}) - \ln g(\boldsymbol{x} \,|\, \boldsymbol{\theta}).$$

The expected value of this loss (that is, the risk) is thus

$$\ell(g) = \mathbb{E} \ln \frac{f(\boldsymbol{X})}{g(\boldsymbol{X} \mid \boldsymbol{\theta})} = \int f(\boldsymbol{x}) \ln \frac{f(\boldsymbol{x})}{g(\boldsymbol{x} \mid \boldsymbol{\theta})} \, \mathrm{d}\boldsymbol{x}. \tag{2.27}$$

KULLBACK–LEIBLER DIVERGENCE

The integral in (2.27) provides a fundamental way to measure the distance between two densities and is called the *Kullback–Leibler (KL) divergence*[2] between f and $g(\cdot \mid \boldsymbol{\theta})$. Note that the KL divergence is not symmetric in f and $g(\cdot \mid \boldsymbol{\theta})$. Moreover, it is always greater than or equal to 0 (see Exercise 15) and equal to 0 when $f = g(\cdot \mid \boldsymbol{\theta})$.

Using similar notation as for the supervised learning setting in Table 2.1, define $g^{\mathcal{G}_p}$ as the global minimizer of the risk in the class $\mathcal{G}_p$; that is, $g^{\mathcal{G}_p} = \operatorname{argmin}_{g \in \mathcal{G}_p} \ell(g)$. If we define

$$\begin{aligned}
\boldsymbol{\theta}^* &= \underset{\boldsymbol{\theta}}{\operatorname{argmin}}\, \mathbb{E}\, \mathrm{Loss}(f(\boldsymbol{X}), g(\boldsymbol{X} \mid \boldsymbol{\theta})) = \underset{\boldsymbol{\theta}}{\operatorname{argmin}} \int (\ln f(\boldsymbol{x}) - \ln g(\boldsymbol{x} \mid \boldsymbol{\theta})) f(\boldsymbol{x}) \, \mathrm{d}\boldsymbol{x} \\
&= \underset{\boldsymbol{\theta}}{\operatorname{argmax}} \int f(\boldsymbol{x}) \ln g(\boldsymbol{x} \mid \boldsymbol{\theta}) \, \mathrm{d}\boldsymbol{x} = \underset{\boldsymbol{\theta}}{\operatorname{argmax}}\, \mathbb{E} \ln g(\boldsymbol{X} \mid \boldsymbol{\theta}),
\end{aligned}$$

then $g^{\mathcal{G}_p} = g(\cdot \mid \boldsymbol{\theta}^*)$ and learning $g^{\mathcal{G}_p}$ is equivalent to learning (or estimating) $\boldsymbol{\theta}^*$. To learn $\boldsymbol{\theta}^*$ from a training set $\tau = \{\boldsymbol{x}_1, \ldots, \boldsymbol{x}_n\}$ we then minimize the training loss,

$$\frac{1}{n} \sum_{i=1}^{n} \mathrm{Loss}(f(\boldsymbol{x}_i), g(\boldsymbol{x}_i \mid \boldsymbol{\theta})) = -\frac{1}{n} \sum_{i=1}^{n} \ln g(\boldsymbol{x}_i \mid \boldsymbol{\theta}) + \frac{1}{n} \sum_{i=1}^{n} \ln f(\boldsymbol{x}_i),$$

giving:

$$\widehat{\boldsymbol{\theta}}_n := \underset{\boldsymbol{\theta}}{\operatorname{argmax}} \frac{1}{n} \sum_{i=1}^{n} \ln g(\boldsymbol{x}_i \mid \boldsymbol{\theta}). \tag{2.28}$$

As the logarithm is an increasing function, this is equivalent to

$$\widehat{\boldsymbol{\theta}}_n := \underset{\boldsymbol{\theta}}{\operatorname{argmax}} \prod_{i=1}^{n} g(\boldsymbol{x}_i \mid \boldsymbol{\theta}),$$

where $\prod_{i=1}^{n} g(\boldsymbol{x}_i \mid \boldsymbol{\theta})$ is the *likelihood* of the data; that is, the joint density of the $\{\boldsymbol{X}_i\}$ evaluated at the points $\{\boldsymbol{x}_i\}$. We therefore have recovered the classical *maximum likelihood estimate* of $\boldsymbol{\theta}^*$.

MAXIMUM LIKELIHOOD ESTIMATE

☞ 458

When the risk $\ell(g(\cdot \mid \boldsymbol{\theta}))$ is convex in $\boldsymbol{\theta}$ over a convex set $\boldsymbol{\Theta}$, we can find the maximum likelihood estimator by setting the gradient of the training loss to zero; that is, we solve

$$-\frac{1}{n} \sum_{i=1}^{n} \boldsymbol{S}(\boldsymbol{x}_i \mid \boldsymbol{\theta}) = \boldsymbol{0},$$

where $\boldsymbol{S}(\boldsymbol{x} \mid \boldsymbol{\theta}) := \frac{\partial \ln g(\boldsymbol{x} \mid \boldsymbol{\theta})}{\partial \boldsymbol{\theta}}$ is the gradient of $\ln g(\boldsymbol{x} \mid \boldsymbol{\theta})$ with respect to $\boldsymbol{\theta}$ and is often called the *score*.

SCORE

■ **Example 2.5 (Exponential Model)** Suppose we have the training data $\tau_n = \{x_1, \ldots, x_n\}$, which is modeled as a realization of n positive iid random variables: $X_1, \ldots, X_n \sim_{\text{iid}} f(x)$. We select the class of approximating functions $\mathcal{G}$ to be the parametric class $\{g : g(x \mid \theta) =$

[2]Sometimes called cross-entropy distance.

$\theta \exp(-x\theta), x > 0, \theta > 0\}$. In other words, we look for the best $g^{\mathcal{G}}$ within the family of exponential distributions with unknown parameter $\theta > 0$. The likelihood of the data is

$$\prod_{i=1}^{n} g(x_i \,|\, \theta) = \prod_{i=1}^{n} \theta \exp(-\theta x_i) = \exp(-\theta\, n\, \overline{x}_n + n \ln \theta)$$

and the score is $S(x \,|\, \theta) = -x + \theta^{-1}$. Thus, maximizing the likelihood with respect to θ is the same as maximizing $-\theta\, n\, \overline{x}_n + n \ln \theta$ or solving $-\sum_{i=1}^{n} S(x_i \,|\, \theta)/n = \overline{x}_n - \theta^{-1} = 0$. In other words, the solution to (2.28) is the maximum likelihood estimate $\widehat{\theta}_n = 1/\overline{x}_n$. ■

In a *supervised* setting, where the data is represented by a vector $\boldsymbol{x}$ of explanatory variables and a response y, the general model is that $(\boldsymbol{x}, y)$ is an outcome of $(\boldsymbol{X}, Y) \sim f$ for some unknown f. And for a training sequence $(\boldsymbol{x}_1, y_1), \ldots, (\boldsymbol{x}_n, y_n)$ the default model assumption is that $(\boldsymbol{X}_1, Y_1), \ldots, (\boldsymbol{X}_n, Y_n) \sim_{\text{iid}} f$. As explained in Section 2.2, the analysis primarily involves the conditional pdf $f(y \,|\, \boldsymbol{x})$ and in particular (when using the squared-error loss) the conditional expectation $g^*(\boldsymbol{x}) = \mathbb{E}[Y \,|\, \boldsymbol{X} = \boldsymbol{x}]$. The resulting representation (2.2) allows us to then write the response at $\boldsymbol{X} = \boldsymbol{x}$ as a function of the feature $\boldsymbol{x}$ plus an error term: $Y = g^*(\boldsymbol{x}) + \varepsilon(\boldsymbol{x})$.

This leads to the simplest and most important model for supervised learning, where we choose a *linear* class $\mathcal{G}$ of prediction or guess functions and assume that it is rich enough to contain the true g^*. If we further assume that, conditional on $\boldsymbol{X} = \boldsymbol{x}$, the error term ε does not depend on $\boldsymbol{x}$, that is, $\mathbb{E}\,\varepsilon = 0$ and $\mathbb{V}\text{ar}\,\varepsilon = \sigma^2$, then we obtain the following model.

Definition 2.1: Linear Model

In a *linear model* the response Y depends on a p-dimensional explanatory variable $\boldsymbol{x} = [x_1, \ldots, x_p]^\top$ via the linear relationship LINEAR MODEL

$$Y = \boldsymbol{x}^\top \boldsymbol{\beta} + \varepsilon, \qquad (2.29)$$

where $\mathbb{E}\,\varepsilon = 0$ and $\mathbb{V}\text{ar}\,\varepsilon = \sigma^2$.

Note that (2.29) is a model for a single pair $(\boldsymbol{x}, Y)$. The model for the training set $\{(\boldsymbol{x}_i, Y_i)\}$ is simply that each Y_i satisfies (2.29) (with $\boldsymbol{x} = \boldsymbol{x}_i$) and that the $\{Y_i\}$ are independent. Gathering all responses in the vector $\boldsymbol{Y} = [Y_1, \ldots, Y_n]^\top$, we can write

$$\boldsymbol{Y} = \mathbf{X}\boldsymbol{\beta} + \boldsymbol{\varepsilon}, \qquad (2.30)$$

where $\boldsymbol{\varepsilon} = [\varepsilon_1, \ldots, \varepsilon_n]^\top$ is a vector of iid copies of ε and $\mathbf{X}$ is the so-called *model matrix*, with rows $\boldsymbol{x}_1^\top, \ldots, \boldsymbol{x}_n^\top$. Linear models are fundamental building blocks of statistical learning algorithms. For this reason, a large part of Chapter 5 is devoted to linear regression models. MODEL MATRIX ☞ 167

■ **Example 2.6 (Polynomial Regression (cont.))** For our running Example 2.1, we see that the data is described by a linear model of the form (2.30), with model matrix $\mathbf{X}$ given in (2.10). ■ ☞ 26

Before we discuss a few other models in the following sections, we would like to emphasize a number of points about modeling.

- *Any* model for data is likely to be *wrong*. For example, real data (as opposed to computer-generated data) are often assumed to come from a normal distribution, which is never exactly true. However, an important advantage of using a normal distribution is that it has many nice mathematical properties, as we will see in Section 2.7.

- Most data models depend on a number of unknown parameters, which need to be estimated from the observed data.

- Any model for real-life data needs to be *checked* for suitability. An important criterion is that data simulated from the model should resemble the observed data, at least for a certain choice of model parameters.

Here are some guidelines for choosing a model. Think of the data as a spreadsheet or data frame, as in Chapter 1, where rows represent the data units and the columns the data features (variables, groups).

- First establish the *type* of the features (quantitative, qualitative, discrete, continuous, etc.).

- Assess whether the data can be assumed to be independent across rows or columns.

- Decide on the level of generality of the model. For example, should we use a simple model with a few unknown parameters or a more generic model that has a large number of parameters? Simple specific models are easier to fit to the data (low estimation error) than more general models, but the fit itself may not be accurate (high approximation error). The tradeoffs discussed in Section 2.4 play an important role here.

- Decide on using a classical (frequentist) or Bayesian model. Section 2.9 gives a short introduction to Bayesian learning. ☞ 47

2.7 Multivariate Normal Models

A standard model for numerical observations $x_1, \dots, x_n$ (forming, e.g., a column in a spreadsheet or data frame) is that they are the outcomes of iid normal random variables

$$X_1, \dots, X_n \overset{\text{iid}}{\sim} \mathcal{N}(\mu, \sigma^2).$$

It is helpful to view a normally distributed random variable as a simple transformation of a standard normal random variable. To wit, if Z has a standard normal distribution, then $X = \mu + \sigma Z$ has a $\mathcal{N}(\mu, \sigma^2)$ distribution. The generalization to n dimensions is discussed in Appendix C.7. We summarize the main points: Let $Z_1, \dots, Z_n \overset{\text{iid}}{\sim} \mathcal{N}(0, 1)$. The pdf of $\mathbf{Z} = [Z_1, \dots, Z_n]^\top$ (that is, the joint pdf of $Z_1, \dots, Z_n$) is given by ☞ 436

$$f_{\mathbf{Z}}(\mathbf{z}) = \prod_{i=1}^{n} \frac{1}{\sqrt{2\pi}} \mathrm{e}^{-\frac{1}{2} z_i^2} = (2\pi)^{-\frac{n}{2}} \mathrm{e}^{-\frac{1}{2} \mathbf{z}^\top \mathbf{z}}, \quad \mathbf{z} \in \mathbb{R}^n. \tag{2.31}$$

We write $\mathbf{Z} \sim \mathcal{N}(\mathbf{0}, \mathbf{I}_n)$ and say that $\mathbf{Z}$ has a standard normal distribution in $\mathbb{R}^n$. Let

$$\boldsymbol{X} = \boldsymbol{\mu} + \mathbf{B}\,\mathbf{Z} \tag{2.32}$$

for some $m \times n$ matrix $\mathbf{B}$ and m-dimensional vector $\boldsymbol{\mu}$. Then $\boldsymbol{X}$ has expectation vector $\boldsymbol{\mu}$ and covariance matrix $\boldsymbol{\Sigma} = \mathbf{B}\mathbf{B}^\top$; see (C.20) and (C.21). This leads to the following definition. ☞ 434

Definition 2.2: Multivariate Normal Distribution

An m-dimensional random vector $\boldsymbol{X}$ that can be written in the form (2.32) for some m-dimensional vector $\boldsymbol{\mu}$ and $m \times n$ matrix $\mathbf{B}$, with $\mathbf{Z} \sim \mathcal{N}(\mathbf{0}, \mathbf{I}_n)$, is said to have a *multivariate normal* or *multivariate Gaussian* distribution with mean vector $\boldsymbol{\mu}$ and covariance matrix $\boldsymbol{\Sigma} = \mathbf{B}\mathbf{B}^\top$. We write $\boldsymbol{X} \sim \mathcal{N}(\boldsymbol{\mu}, \boldsymbol{\Sigma})$.

MULTIVARIATE NORMAL

The m-dimensional density of a multivariate normal distribution has a very similar form to the density of the one-dimensional normal distribution and is given in the next theorem. We leave the proof as an exercise; see Exercise 5. ☞ 59

Theorem 2.3: Density of a Multivariate Random Vector

Let $\boldsymbol{X} \sim \mathcal{N}(\boldsymbol{\mu}, \boldsymbol{\Sigma})$, where the $m \times m$ covariance matrix $\boldsymbol{\Sigma}$ is invertible. Then $\boldsymbol{X}$ has pdf

$$f_{\boldsymbol{X}}(\boldsymbol{x}) = \frac{1}{\sqrt{(2\pi)^m\,|\boldsymbol{\Sigma}|}}\,\mathrm{e}^{-\frac{1}{2}\,(\boldsymbol{x}-\boldsymbol{\mu})^\top\boldsymbol{\Sigma}^{-1}(\boldsymbol{x}-\boldsymbol{\mu})}, \quad \boldsymbol{x} \in \mathbb{R}^m. \tag{2.33}$$

Figure 2.13 shows the pdfs of two bivariate (that is, two-dimensional) normal distributions. In both cases the mean vector is $\boldsymbol{\mu} = [0, 0]^\top$ and the variances (the diagonal elements of $\boldsymbol{\Sigma}$) are 1. The correlation coefficients (or, equivalently here, the covariances) are respectively $\varrho = 0$ and $\varrho = 0.8$.

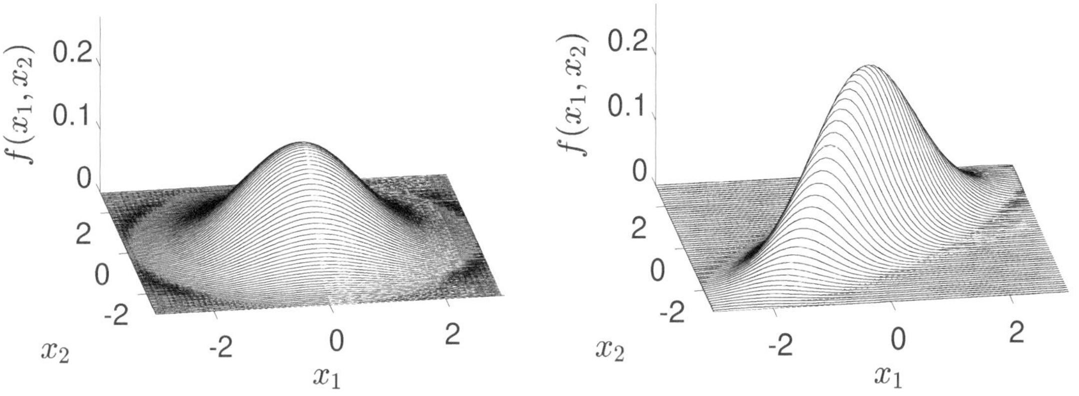

Figure 2.13: Pdfs of bivariate normal distributions with means zero, variances 1, and correlation coefficients 0 (left) and 0.8 (right).

The main reason why the multivariate normal distribution plays an important role in data science and machine learning is that it satisfies the following properties, the details and proofs of which can be found in Appendix C.7:

☞ 436

1. Affine combinations are normal.
2. Marginal distributions are normal.
3. Conditional distributions are normal.

2.8 Normal Linear Models

Normal linear models combine the simplicity of the linear model with the tractability of the Gaussian distribution. They are the principal model for traditional statistics, and include the classic linear regression and analysis of variance models.

Definition 2.3: Normal Linear Model

NORMAL LINEAR MODEL

In a *normal linear model* the response Y depends on a p-dimensional explanatory variable $\boldsymbol{x} = [x_1, \ldots, x_p]^\top$, via the linear relationship

$$Y = \boldsymbol{x}^\top \boldsymbol{\beta} + \varepsilon, \tag{2.34}$$

where $\varepsilon \sim \mathcal{N}(0, \sigma^2)$.

Thus, a normal linear model is a linear model (in the sense of Definition 2.1) with normal error terms. Similar to (2.30), the corresponding normal linear model for the whole training set $\{(\boldsymbol{x}_i, Y_i)\}$ has the form

$$\boldsymbol{Y} = \mathbf{X}\boldsymbol{\beta} + \boldsymbol{\varepsilon}, \tag{2.35}$$

where $\mathbf{X}$ is the model matrix comprised of rows $\boldsymbol{x}_1^\top, \ldots, \boldsymbol{x}_n^\top$ and $\boldsymbol{\varepsilon} \sim \mathcal{N}(\mathbf{0}, \sigma^2 \mathbf{I}_n)$. Consequently, $\boldsymbol{Y}$ can be written as $\boldsymbol{Y} = \mathbf{X}\boldsymbol{\beta} + \sigma \boldsymbol{Z}$, where $\boldsymbol{Z} \sim \mathcal{N}(\mathbf{0}, \mathbf{I}_n)$, so that $\boldsymbol{Y} \sim \mathcal{N}(\mathbf{X}\boldsymbol{\beta}, \sigma^2 \mathbf{I}_n)$. It follows from (2.33) that its joint density is given by

☞ 45

$$g(\boldsymbol{y} \,|\, \boldsymbol{\beta}, \sigma^2, \mathbf{X}) = (2\pi\sigma^2)^{-\frac{n}{2}} \, \mathrm{e}^{-\frac{1}{2\sigma^2}\|\boldsymbol{y} - \mathbf{X}\boldsymbol{\beta}\|^2}. \tag{2.36}$$

Estimation of the parameter $\boldsymbol{\beta}$ can be performed via the least-squares method, as discussed in Example 2.1. An estimate can also be obtained via the maximum likelihood method. This simply means finding the parameters σ^2 and $\boldsymbol{\beta}$ that maximize the likelihood of the outcome $\boldsymbol{y}$, given by the right-hand side of (2.36). It is clear that for every value of σ^2 the likelihood is maximal when $\|\boldsymbol{y} - \mathbf{X}\boldsymbol{\beta}\|^2$ is minimal. As a consequence, the maximum likelihood estimate for $\boldsymbol{\beta}$ is the same as the least-squares estimate (2.15). We leave it as an exercise (see Exercise 18) to show that the maximum likelihood estimate of σ^2 is equal to

☞ 63

$$\widehat{\sigma^2} = \frac{\|\boldsymbol{y} - \mathbf{X}\widehat{\boldsymbol{\beta}}\|^2}{n}, \tag{2.37}$$

where $\widehat{\boldsymbol{\beta}}$ is the maximum likelihood estimate (least squares estimate in this case) of $\boldsymbol{\beta}$.

2.9 Bayesian Learning

In Bayesian unsupervised learning, we seek to approximate the unknown joint density $f(\boldsymbol{x}_1, \ldots, \boldsymbol{x}_n)$ of the training data $\mathcal{T}_n = \{\boldsymbol{X}_1, \ldots, \boldsymbol{X}_n\}$ via a joint pdf of the form

$$\int \left(\prod_{i=1}^{n} g(\boldsymbol{x}_i \,|\, \boldsymbol{\theta}) \right) w(\boldsymbol{\theta}) \, \mathrm{d}\boldsymbol{\theta}, \tag{2.38}$$

where $g(\cdot \,|\, \boldsymbol{\theta})$ belongs to a family of parametric densities $\mathcal{G}_p := \{g(\cdot \,|\, \boldsymbol{\theta}),\ \boldsymbol{\theta} \in \boldsymbol{\Theta}\}$ (viewed as a family of pdfs conditional on a parameter $\boldsymbol{\theta}$ in some set $\boldsymbol{\Theta} \subset \mathbb{R}^p$) and $w(\boldsymbol{\theta})$ is a pdf that belongs to a (possibly different) family of densities $\mathcal{W}_p$. Note how the joint pdf (2.38) satisfies the permutational invariance (2.26) and can thus be useful as a model for training data which is part of an exchangeable sequence of random variables.

> ⚠ Following standard practice in a Bayesian context, instead of writing $f_X(x)$ and $f_{X \,|\, Y}(x \,|\, y)$ for the pdf of X and the conditional pdf of X given Y, one simply writes $f(x)$ and $f(x \,|\, y)$. If Y is a different random variable, its pdf (at y) is thus denoted by $f(y)$.

Thus, we will use the same symbol g for different (conditional) approximating probability densities and f for the different (conditional) true and unknown probability densities. Using Bayesian notation, we can write $g(\tau \,|\, \boldsymbol{\theta}) = \prod_{i=1}^{n} g(\boldsymbol{x}_i \,|\, \boldsymbol{\theta})$ and thus the approximating joint pdf (2.38) can then be written as $\int g(\tau \,|\, \boldsymbol{\theta})\, w(\boldsymbol{\theta})\, \mathrm{d}\boldsymbol{\theta}$ and the true unknown joint pdf as $f(\tau) = f(\boldsymbol{x}_1, \ldots, \boldsymbol{x}_n)$.

Once $\mathcal{G}_p$ and $\mathcal{W}_p$ are specified, selecting an approximating function $g(\boldsymbol{x})$ of the form

$$g(\boldsymbol{x}) = \int g(\boldsymbol{x} \,|\, \boldsymbol{\theta})\, w(\boldsymbol{\theta})\, \mathrm{d}\boldsymbol{\theta}$$

is equivalent to selecting a suitable w from $\mathcal{W}_p$. Similar to (2.27), we can use the Kullback–Leibler risk to measure the discrepancy between the proposed approximation (2.38) and the true $f(\tau)$:

$$\ell(g) = \mathbb{E} \ln \frac{f(\mathcal{T})}{\int g(\mathcal{T} \,|\, \boldsymbol{\theta})\, w(\boldsymbol{\theta})\, \mathrm{d}\boldsymbol{\theta}} = \int f(\tau) \ln \frac{f(\tau)}{\int g(\tau \,|\, \boldsymbol{\theta})\, w(\boldsymbol{\theta})\, \mathrm{d}\boldsymbol{\theta}} \, \mathrm{d}\tau. \tag{2.39}$$

The main difference with (2.27) is that since the training data is not necessarily iid (it may be exchangeable, for example), the expectation must be with respect to the joint density of $\mathcal{T}$, not with respect to the marginal $f(\boldsymbol{x})$ (as in the iid case). ☞ 40

Minimizing the training loss is equivalent to maximizing the likelihood of the training data τ; that is, solving the optimization problem

$$\max_{w \in \mathcal{W}_p} \int g(\tau \,|\, \boldsymbol{\theta})\, w(\boldsymbol{\theta})\, \mathrm{d}\boldsymbol{\theta},$$

where the maximization is over an appropriate class $\mathcal{W}_p$ of density functions that is believed to result in the smallest KL risk.

Suppose that we have a rough guess, denoted $w_0(\boldsymbol{\theta})$, for the best $w \in \mathcal{W}_p$ that minimizes the Kullback–Leibler risk. We can always increase the resulting likelihood $L_0 := \int g(\tau \mid \boldsymbol{\theta})\, w_0(\boldsymbol{\theta})\, \mathrm{d}\boldsymbol{\theta}$ by instead using the density $w_1(\boldsymbol{\theta}) := w_0(\boldsymbol{\theta})\, g(\tau \mid \boldsymbol{\theta})/L_0$, giving a likelihood $L_1 := \int g(\tau \mid \boldsymbol{\theta})\, w_1(\boldsymbol{\theta})\, \mathrm{d}\boldsymbol{\theta}$. To see this, write L_0 and L_1 as expectations with respect to w_0. In particular, we can write

$$L_0 = \mathbb{E}_{w_0}\, g(\tau \mid \boldsymbol{\theta}) \quad \text{and} \quad L_1 = \mathbb{E}_{w_1}\, g(\tau \mid \boldsymbol{\theta}) = \mathbb{E}_{w_0} g^2(\tau \mid \boldsymbol{\theta})/L_0.$$

It follows that

$$L_1 - L_0 = \frac{1}{L_0}\mathbb{E}_{w_0}\left[g^2(\tau \mid \boldsymbol{\theta}) - L_0^2\right] = \frac{1}{L_0}\mathbb{V}\mathrm{ar}_{w_0}[g(\tau \mid \boldsymbol{\theta})] \geqslant 0. \tag{2.40}$$

We may thus expect to obtain better predictions using w_1 instead of w_0, because w_1 has taken into account the observed data τ and increased the likelihood of the model. In fact, if we iterate this process (see Exercise 20) and create a sequence of densities $w_1, w_2, \ldots$ such that $w_t(\boldsymbol{\theta}) \propto w_{t-1}(\boldsymbol{\theta})\, g(\tau \mid \boldsymbol{\theta})$, then $w_t(\boldsymbol{\theta})$ concentrates more and more of its probability mass at the maximum likelihood estimator $\widehat{\boldsymbol{\theta}}$ (see (2.28)) and in the limit equals a (degenerate) point-mass pdf at $\widehat{\boldsymbol{\theta}}$. In other words, in the limit we recover the maximum likelihood method: $g_\tau(\boldsymbol{x}) = g(\boldsymbol{x} \mid \widehat{\boldsymbol{\theta}})$. Thus, unless the class of densities $\mathcal{W}_p$ is restricted to be non-degenerate, maximizing the likelihood as much as possible leads to a degenerate choice for $w(\boldsymbol{\theta})$.

In many situations, the maximum likelihood estimate $g(\tau \mid \widehat{\boldsymbol{\theta}})$ is either not an appropriate approximation to $f(\tau)$ (see Example 2.9), or simply fails to exist (see Exer-
☞ 161 cise 10 in Chapter 4). In such cases, given an initial non-degenerate guess $w_0(\boldsymbol{\theta}) = g(\boldsymbol{\theta})$, one can obtain a more appropriate and non-degenerate approximation to $f(\tau)$ by taking $w(\boldsymbol{\theta}) = w_1(\boldsymbol{\theta}) \propto g(\tau \mid \boldsymbol{\theta})\, g(\boldsymbol{\theta})$ in (2.38), giving the following Bayesian learner of $f(\boldsymbol{x})$:

$$g_\tau(\boldsymbol{x}) := \int g(\boldsymbol{x} \mid \boldsymbol{\theta}) \frac{g(\tau \mid \boldsymbol{\theta})\, g(\boldsymbol{\theta})}{\int g(\tau \mid \boldsymbol{\vartheta})\, g(\boldsymbol{\vartheta})\, \mathrm{d}\boldsymbol{\vartheta}}\, \mathrm{d}\boldsymbol{\theta}, \tag{2.41}$$

☞ 430 where $\int g(\tau \mid \boldsymbol{\vartheta})\, g(\boldsymbol{\vartheta})\, \mathrm{d}\boldsymbol{\vartheta} = g(\tau)$. Using Bayes' formula for probability densities,

$$g(\boldsymbol{\theta} \mid \tau) = \frac{g(\tau \mid \boldsymbol{\theta})\, g(\boldsymbol{\theta})}{g(\tau)}, \tag{2.42}$$

we can write $w_1(\boldsymbol{\theta}) = g(\boldsymbol{\theta} \mid \tau)$. With this notation, we have the following definitions.

Definition 2.4: Prior, Likelihood, and Posterior

Let τ and $\mathcal{G}_p := \{g(\cdot \mid \boldsymbol{\theta}), \boldsymbol{\theta} \in \boldsymbol{\Theta}\}$ be the training set and family of approximating functions.

PRIOR
- A pdf $g(\boldsymbol{\theta})$ that reflects our *a priori* beliefs about $\boldsymbol{\theta}$ is called the *prior* pdf.

LIKELIHOOD
- The conditional pdf $g(\tau \mid \boldsymbol{\theta})$ is called the *likelihood.*

POSTERIOR
- Inference about $\boldsymbol{\theta}$ is given by the *posterior* pdf $g(\boldsymbol{\theta} \mid \tau)$, which is proportional to the product of the prior and the likelihood:

$$g(\boldsymbol{\theta} \mid \tau) \propto g(\tau \mid \boldsymbol{\theta})\, g(\boldsymbol{\theta}).$$

■ **Remark 2.1 (Early Stopping)** Bayes iteration is an example of an "early stopping" heuristic for maximum likelihood optimization, where we exit after only one step. As observed above, if we keep iterating, we obtain the maximum likelihood estimate (MLE). In a sense the Bayes rule provides a regularization of the MLE. Regularization is discussed in more detail in Chapter 6; see also Example 2.9. The early stopping rule is also of benefit in regularization; see Exercise 20 in Chapter 6. ■

On the one hand, the initial guess $g(\boldsymbol{\theta})$ conveys the *a priori* (prior to training the Bayesian learner) information about the optimal density in $\mathcal{W}_p$ that minimizes the KL risk. Using this prior $g(\boldsymbol{\theta})$, the Bayesian approximation to $f(\boldsymbol{x})$ is the *prior predictive density*:

PRIOR PREDICTIVE DENSITY

$$g(\boldsymbol{x}) = \int g(\boldsymbol{x} \,|\, \boldsymbol{\theta})\, g(\boldsymbol{\theta})\, \mathrm{d}\boldsymbol{\theta}.$$

On the other hand, the posterior pdf conveys improved knowledge about this optimal density in $\mathcal{W}_p$ after training with τ. Using the posterior $g(\boldsymbol{\theta} \,|\, \tau)$, the Bayesian learner of $f(\boldsymbol{x})$ is the *posterior predictive density*:

POSTERIOR PREDICTIVE DENSITY

$$g_\tau(\boldsymbol{x}) = g(\boldsymbol{x} \,|\, \tau) = \int g(\boldsymbol{x} \,|\, \boldsymbol{\theta})\, g(\boldsymbol{\theta} \,|\, \tau)\, \mathrm{d}\boldsymbol{\theta},$$

where we have assumed that $g(\boldsymbol{x} \,|\, \boldsymbol{\theta}, \tau) = g(\boldsymbol{x} \,|\, \boldsymbol{\theta})$; that is, the likelihood depends on τ only through the parameter $\boldsymbol{\theta}$.

The choice of the prior is typically governed by two considerations:

1. the prior should be simple enough to facilitate the computation or simulation of the posterior pdf;
2. the prior should be general enough to model ignorance of the parameter of interest.

Priors that do not convey much knowledge of the parameter are said to be *uninformative*. The uniform or *flat* prior in Example 2.9 (to follow) is frequently used.

UNINFORMATIVE PRIOR

> For the purpose of analytical and numerical computations, we can view $\boldsymbol{\theta}$ as a random vector with prior density $g(\boldsymbol{\theta})$, which after training is updated to the posterior density $g(\boldsymbol{\theta} \,|\, \tau)$.

The above thinking allows us to write $g(\boldsymbol{x} \,|\, \tau) \propto \int g(\boldsymbol{x} \,|\, \boldsymbol{\theta})\, g(\tau \,|\, \boldsymbol{\theta})\, g(\boldsymbol{\theta})\, \mathrm{d}\boldsymbol{\theta}$, for example, thus ignoring any constants that do not depend on the argument of the densities.

■ **Example 2.7 (Normal Model)** Suppose that the training data $\mathcal{T} = \{X_1, \ldots, X_n\}$ is modeled using the likelihood $g(x \,|\, \boldsymbol{\theta})$ that is the pdf of

$$X \,|\, \boldsymbol{\theta} \sim \mathcal{N}(\mu, \sigma^2),$$

where $\boldsymbol{\theta} := [\mu, \sigma^2]^\top$. Next, we need to specify the prior distribution of $\boldsymbol{\theta}$ to complete the model. We can specify prior distributions for μ and σ^2 separately and then take their product to obtain the prior for vector $\boldsymbol{\theta}$ (assuming independence). A possible prior distribution for μ is

$$\mu \sim \mathcal{N}(\nu, \phi^2). \tag{2.43}$$

HYPERPARAMETERS

It is typical to refer to any parameters of the prior density as *hyperparameters* of the Bayesian model. Instead of giving directly a prior for σ^2 (or σ), it turns out to be convenient to give the following prior distribution to $1/\sigma^2$:

$$\frac{1}{\sigma^2} \sim \mathsf{Gamma}(\alpha, \beta). \tag{2.44}$$

INVERSE GAMMA

☞ 63

The smaller α and β are, the less informative is the prior. Under this prior, σ^2 is said to have an *inverse gamma*[3] distribution. If $1/Z \sim \mathsf{Gamma}(\alpha, \beta)$, then the pdf of Z is proportional to $\exp(-\beta/z)/z^{\alpha+1}$ (Exercise 19). The Bayesian posterior is then given by:

$$\begin{aligned} g(\mu, \sigma^2 \mid \tau) &\propto g(\mu) \times g(\sigma^2) \times g(\tau \mid \mu, \sigma^2) \\ &\propto \exp\left\{-\frac{(\mu - \nu)^2}{2\phi^2}\right\} \times \frac{\exp\left\{-\beta/\sigma^2\right\}}{(\sigma^2)^{\alpha+1}} \times \frac{\exp\left\{-\sum_i (x_i - \mu)^2/(2\sigma^2)\right\}}{(\sigma^2)^{n/2}} \\ &\propto (\sigma^2)^{-n/2-\alpha-1} \exp\left\{-\frac{(\mu - \nu)^2}{2\phi^2} - \frac{\beta}{\sigma^2} - \frac{(\mu - \overline{x}_n)^2 + S_n^2}{2\sigma^2/n}\right\}, \end{aligned}$$

where $S_n^2 := \frac{1}{n}\sum_i x_i^2 - \overline{x}_n^2 = \frac{1}{n}\sum_i (x_i - \overline{x}_n)^2$ is the (scaled) sample variance. All inference about (μ, σ^2) is then represented by the posterior pdf. To facilitate computations it is helpful to find out if the posterior belongs to a recognizable family of distributions. For example, the conditional pdf of μ given σ^2 and τ is

$$g(\mu \mid \sigma^2, \tau) \propto \exp\left\{-\frac{(\mu - \nu)^2}{2\phi^2} - \frac{(\mu - \overline{x}_n)^2}{2\sigma^2/n}\right\},$$

which after simplification can be recognized as the pdf of

$$(\mu \mid \sigma^2, \tau) \sim \mathcal{N}\left(\gamma_n \overline{x}_n + (1 - \gamma_n)\nu,\ \gamma_n \sigma^2/n\right), \tag{2.45}$$

where we have defined the weight parameter: $\gamma_n := \frac{n}{\sigma^2}\Big/\left(\frac{1}{\phi^2} + \frac{n}{\sigma^2}\right)$. We can then see that the posterior mean $\mathbb{E}[\mu \mid \sigma^2, \tau] = \gamma_n \overline{x}_n + (1 - \gamma_n)\nu$ is a weighted linear combination of the prior mean ν and the sample average $\overline{x}_n$. Further, as $n \to \infty$, the weight $\gamma_n \to 1$ and thus the posterior mean approaches the maximum likelihood estimate $\overline{x}_n$. ■

It is sometimes possible to use a prior $g(\boldsymbol{\theta})$ that is not a *bona fide* probability density, in the sense that $\int g(\boldsymbol{\theta})\, \mathrm{d}\boldsymbol{\theta} = \infty$, as long as the resulting posterior $g(\boldsymbol{\theta} \mid \tau) \propto g(\tau \mid \boldsymbol{\theta}) g(\boldsymbol{\theta})$ is a proper pdf. Such a prior is called an *improper prior*.

IMPROPER PRIOR

■ **Example 2.8 (Normal Model (cont.))** An example of an improper prior is obtained from (2.43) when we let $\phi \to \infty$ (the larger ϕ is, the more uninformative is the prior). Then, $g(\mu) \propto 1$ is a flat prior, but $\int g(\mu)\, \mathrm{d}\mu = \infty$, making it an improper prior. Nevertheless, the posterior is a proper density, and in particular the conditional posterior of $(\mu \mid \sigma^2, \tau)$ simplifies to

$$(\mu \mid \sigma^2, \tau) \sim \mathcal{N}\left(\overline{x}_n, \sigma^2/n\right),$$

[3] Reciprocal gamma distribution would have been a better name.

because the weight parameter γ_n goes to 1 as $\phi \to \infty$. The improper prior $g(\mu) \propto 1$ also allows us to simplify the posterior marginal for σ^2:

$$g(\sigma^2 \,|\, \tau) = \int g(\mu, \sigma^2 \,|\, \tau)\, \mathrm{d}\mu \propto (\sigma^2)^{-(n-1)/2-\alpha-1} \exp\left\{-\frac{\beta + nS_n^2/2}{\sigma^2}\right\},$$

which we recognize as the density corresponding to

$$\frac{1}{\sigma^2} \,\Big|\, \tau \sim \mathsf{Gamma}\left(\alpha + \frac{n-1}{2},\, \beta + \frac{n}{2}S_n^2\right).$$

In addition to $g(\mu) \propto 1$, we can also use an improper prior for σ^2. If we take the limit $\alpha \to 0$ and $\beta \to 0$ in (2.44), then we also obtain the improper prior $g(\sigma^2) \propto 1/\sigma^2$ (or equivalently $g(1/\sigma^2) \propto 1/\sigma^2$). In this case, the posterior marginal density for σ^2 implies that:

$$\frac{nS_n^2}{\sigma^2} \,\Big|\, \tau \sim \chi^2_{n-1}$$

and the posterior marginal density for μ implies that:

$$\frac{\mu - \overline{x}_n}{S_n/\sqrt{n-1}} \,\Big|\, \tau \sim \mathsf{t}_{n-1}. \tag{2.46}$$

In general, deriving a simple formula for the posterior density of $\boldsymbol{\theta}$ is either impossible or too tedious. Instead, the Monte Carlo methods in Chapter 3 can be used to simulate (approximately) from the posterior for the purposes of inference and prediction. ■

One way in which a distributional result such as (2.46) can be useful is in the construction of a 95% *credible interval* $\mathcal{I}$ for the parameter μ; that is, an interval $\mathcal{I}$ such that the probability $\mathbb{P}[\mu \in \mathcal{I} \,|\, \tau]$ is equal to 0.95. For example, the symmetric 95% credible interval is

CREDIBLE INTERVAL

$$\mathcal{I} = \left[\overline{x}_n - \frac{S_n}{\sqrt{n-1}}\gamma,\; \overline{x}_n + \frac{S_n}{\sqrt{n-1}}\gamma\right],$$

where γ is the 0.975-quantile of the t_{n-1} distribution. Note that the credible interval is not a random object and that the parameter μ is interpreted as a random variable with a distribution. This is unlike the case of classical confidence intervals, where the parameter is nonrandom, but the interval is (the outcome of) a random object.

☞ 459

As a generalization of the 95% Bayesian credible interval we can define a $1-\alpha$ *credible region*, which is any set $\mathcal{R}$ satisfying

CREDIBLE REGION

$$\mathbb{P}[\boldsymbol{\theta} \in \mathcal{R} \,|\, \tau] = \int_{\boldsymbol{\theta} \in \mathcal{R}} g(\boldsymbol{\theta} \,|\, \tau)\, \mathrm{d}\boldsymbol{\theta} \geqslant 1 - \alpha. \tag{2.47}$$

■ **Example 2.9 (Bayesian Regularization of Maximum Likelihood)** Consider modeling the number of deaths during birth in a maternity ward. Suppose that the hospital data consists of $\tau = \{x_1, \ldots, x_n\}$, with $x_i = 1$ if the i-th baby has died during birth and $x_i = 0$ otherwise, for $i = 1, \ldots, n$. A possible Bayesian model for the data is $\theta \sim \mathcal{U}(0, 1)$ (uniform prior) with $(X_1, \ldots, X_n \mid \theta) \stackrel{\text{iid}}{\sim} \mathsf{Ber}(\theta)$. The likelihood is therefore

$$g(\tau \mid \theta) = \prod_{i=1}^{n} \theta^{x_i}(1-\theta)^{1-x_i} = \theta^s (1-\theta)^{n-s},$$

where $s = x_1 + \cdots + x_n$ is the total number of deaths. Since $g(\theta) = 1$, the posterior pdf is

$$g(\theta \mid \tau) \propto \theta^s (1-\theta)^{n-s}, \quad \theta \in [0, 1],$$

which is the pdf of the $\mathsf{Beta}(s+1, n-s+1)$ distribution. The normalization constant is $(n+1)\binom{n}{s}$. The posterior pdf is shown in Figure 2.14 for $(s, n) = (0, 100)$. It is not difficult

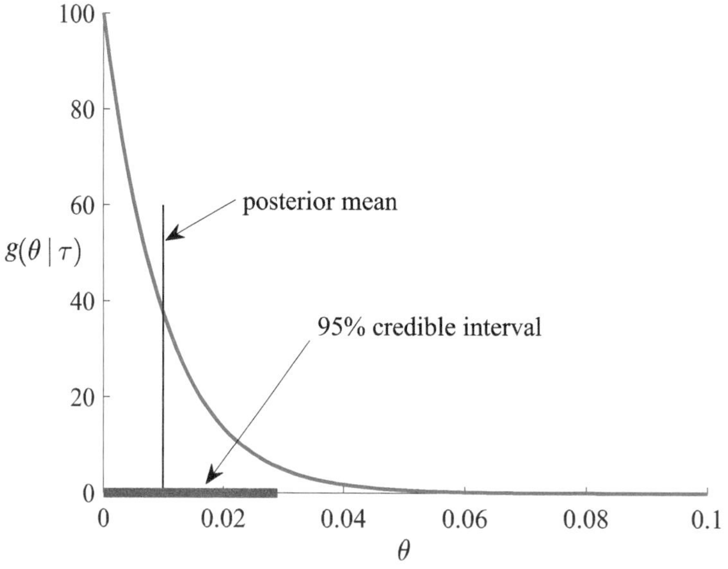

Figure 2.14: Posterior pdf for θ, with $n = 100$ and $s = 0$.

MAXIMUM A POSTERIORI

to see that the *maximum a posteriori* (MAP) estimate of θ (the mode or maximizer of the posterior density) is

$$\underset{\theta}{\operatorname{argmax}}\, g(\theta \mid \tau) = \frac{s}{n},$$

which agrees with the maximum likelihood estimate. Figure 2.14 also shows that the left one-sided 95% credible interval for θ is $[0, 0.0292]$, where 0.0292 is the 0.95 quantile (rounded) of the $\mathsf{Beta}(1, 101)$ distribution.

Observe that when $(s, n) = (0, 100)$ the maximum likelihood estimate $\widehat{\theta} = 0$ infers that deaths at birth are not possible. We know that this inference is wrong — the probability of death can never be zero, it is simply (and fortunately) too small to be inferred accurately from a sample size of $n = 100$. In contrast to the maximum likelihood estimate, the posterior mean $\mathbb{E}[\theta \mid \tau] = (s+1)/(n+2)$ is not zero for $(s, n) = (0, 100)$ and provides the more reasonable point estimate of 0.0098 for the probability of death.

In addition, while computing a Bayesian credible interval poses no conceptual difficulties, it is not simple to derive a confidence interval for the maximum likelihood estimate of $\widehat{\theta}$, because the likelihood as a function of θ is not differentiable at $\theta = 0$. As a result of this lack of smoothness, the usual confidence intervals based on the normal approximation cannot be used. ■

We now return to the unsupervised learning setting of Section 2.6, but consider this from a Bayesian perspective. Recall from (2.39) that the Kullback–Leibler risk for an approximating function g is

$$\ell(g) = \int f(\tau'_n)[\ln f(\tau'_n) - \ln g(\tau'_n)]\,\mathrm{d}\tau'_n,$$

where τ'_n denotes the test data. Since $\int f(\tau'_n) \ln f(\tau'_n)\,\mathrm{d}\tau'_n$ plays no role in minimizing the risk, we consider instead the *cross-entropy risk*, defined as ☞ 122

$$\ell(g) = -\int f(\tau'_n) \ln g(\tau'_n)\,\mathrm{d}\tau'_n.$$

Note that the smallest possible cross-entropy risk is $\ell^*_n = -\int f(\tau'_n) \ln f(\tau'_n)\,\mathrm{d}\tau'_n$. The expected generalization risk of the Bayesian learner can then be decomposed as

$$\mathbb{E}\,\ell(g_{\mathcal{T}_n}) = \ell^*_n + \underbrace{\int f(\tau'_n) \ln \frac{f(\tau'_n)}{\mathbb{E}\,g(\tau'_n \,|\, \mathcal{T}_n)}\,\mathrm{d}\tau'_n}_{\text{"bias" component}} + \underbrace{\mathbb{E}\int f(\tau'_n) \ln \frac{\mathbb{E}\,g(\tau'_n \,|\, \mathcal{T}_n)}{g(\tau'_n \,|\, \mathcal{T}_n)}\,\mathrm{d}\tau'_n}_{\text{"variance" component}},$$

where $g_{\mathcal{T}_n}(\tau'_n) = g(\tau'_n \,|\, \mathcal{T}_n) = \int g(\tau'_n \,|\, \boldsymbol{\theta})\, g(\boldsymbol{\theta} \,|\, \mathcal{T}_n)\,\mathrm{d}\boldsymbol{\theta}$ is the posterior predictive density after observing $\mathcal{T}_n$.

Assuming that the sets $\mathcal{T}_n$ and $\mathcal{T}'_n$ are comprised of $2n$ iid random variables with density f, we can show (Exercise 23) that the expected generalization risk simplifies to

$$\mathbb{E}\,\ell(g_{\mathcal{T}_n}) = \mathbb{E} \ln g(\mathcal{T}_n) - \mathbb{E} \ln g(\mathcal{T}_{2n}), \tag{2.48}$$

where $g(\tau_n)$ and $g(\tau_{2n})$ are the prior predictive densities of τ_n and τ_{2n}, respectively.

Let $\overline{\boldsymbol{\theta}}_n = \operatorname{argmax}_{\boldsymbol{\theta}}\, g(\boldsymbol{\theta} \,|\, \mathcal{T}_n)$ be the MAP estimator of $\boldsymbol{\theta}^* := \operatorname{argmax}_{\boldsymbol{\theta}} \mathbb{E} \ln g(\boldsymbol{X} \,|\, \boldsymbol{\theta})$. Assuming that $\overline{\boldsymbol{\theta}}_n$ converges to $\boldsymbol{\theta}^*$ (with probability one) and $\frac{1}{n}\mathbb{E} \ln g(\mathcal{T}_n \,|\, \overline{\boldsymbol{\theta}}_n) = \mathbb{E} \ln g(\boldsymbol{X} \,|\, \boldsymbol{\theta}^*) + \mathcal{O}(1/n)$, we can use the following large-sample approximation of the expected generalization risk.

Theorem 2.4: Approximating the Bayesian Cross-Entropy Risk

For $n \to \infty$, the expected cross-entropy generalization risk satisfies:

$$\mathbb{E}\ell(g_{\mathcal{T}_n}) \simeq -\mathbb{E} \ln g(\mathcal{T}_n) - \frac{p}{2} \ln n, \tag{2.49}$$

where (with p the dimension of the parameter vector $\boldsymbol{\theta}$ and $\overline{\boldsymbol{\theta}}_n$ the MAP estimator):

$$\mathbb{E} \ln g(\mathcal{T}_n) \simeq \mathbb{E} \ln g(\mathcal{T}_n \,|\, \overline{\boldsymbol{\theta}}_n) - \frac{p}{2} \ln n. \tag{2.50}$$

☞ 452

Proof: To show (2.50), we apply Theorem C.21 to $\ln \int \mathrm{e}^{-nr_n(\boldsymbol{\theta})} g(\boldsymbol{\theta})\, \mathrm{d}\boldsymbol{\theta}$, where

$$r_n(\boldsymbol{\theta}) := -\frac{1}{n} \ln g(\mathcal{T}_n \,|\, \boldsymbol{\theta}) = -\frac{1}{n} \sum_{i=1}^{n} \ln g(\boldsymbol{X}_i \,|\, \boldsymbol{\theta}) \xrightarrow{\text{a.s.}} -\mathbb{E} \ln g(\boldsymbol{X} \,|\, \boldsymbol{\theta}) =: r(\boldsymbol{\theta}) < \infty.$$

This gives (with probability one)

$$\ln \int g(\mathcal{T}_n \,|\, \boldsymbol{\theta})\, g(\boldsymbol{\theta})\, \mathrm{d}\boldsymbol{\theta} \simeq -nr(\boldsymbol{\theta}^*) - \frac{p}{2} \ln(n).$$

Taking expectations on both sides and using $nr(\boldsymbol{\theta}^*) = n\mathbb{E}[r_n(\overline{\boldsymbol{\theta}}_n)] + \mathcal{O}(1)$, we deduce (2.50). To demonstrate (2.49), we derive the asymptotic approximation of $\mathbb{E} \ln g(\mathcal{T}_{2n})$ by repeating the argument for (2.50), but replacing n with $2n$, where necessary. Thus, we obtain:

$$\mathbb{E} \ln g(\mathcal{T}_{2n}) \simeq -2nr(\boldsymbol{\theta}^*) - \frac{p}{2} \ln(2n).$$

Then, (2.49) follows from the identity (2.48). □

The results of Theorem 2.4 have two major implications for model selection and assessment. First, (2.49) suggests that $-\ln g(\mathcal{T}_n)$ can be used as a crude (leading-order) asymptotic approximation to the expected generalization risk for large n and fixed p. In this context, the prior predictive density $g(\mathcal{T}_n)$ is usually called the *model evidence* or *marginal likelihood* for the class $\mathcal{G}_p$. Since the integral $\int g(\mathcal{T}_n \,|\, \boldsymbol{\theta})\, g(\boldsymbol{\theta})\, \mathrm{d}\boldsymbol{\theta}$ is rarely available in closed form, the exact computation of the model evidence is typically not feasible and may require Monte Carlo estimation methods.

MODEL EVIDENCE

☞ 78

Second, when the model evidence is difficult to compute via Monte Carlo methods or otherwise, (2.50) suggests that we can use the following large-sample approximation:

$$-2\mathbb{E} \ln g(\mathcal{T}_n) \simeq -2 \ln g(\mathcal{T}_n \,|\, \overline{\boldsymbol{\theta}}_n) + p \ln(n). \tag{2.51}$$

The asymptotic approximation on the right-hand side of (2.51) is called the *Bayesian information criterion* (BIC). We prefer the class $\mathcal{G}_p$ with the smallest BIC. The BIC is typically used when the model evidence is difficult to compute and n is sufficiently larger than p. For a fixed p, and as n becomes larger and larger, the BIC becomes a more and more accurate estimator of $-2\mathbb{E} \ln g(\mathcal{T}_n)$. Note that the BIC approximation is valid even when the true density $f \notin \mathcal{G}_p$. The BIC provides an alternative to the *Akaike information criterion* (AIC) for model selection. However, while the BIC approximation does not assume that the true model f belongs to the parametric class under consideration, the AIC assumes that $f \in \mathcal{G}_p$. Thus, the AIC is merely a *heuristic* approximation based on the asymptotic approximations in Theorem 4.1.

BAYESIAN INFORMATION CRITERION

☞ 126

Although the above Bayesian theory has been presented in an unsupervised learning setting, it can be readily extended to the supervised case. We only need to relabel the training set $\mathcal{T}_n$. In particular, when (as is typical for regression models) the training responses $Y_1, \ldots, Y_n$ are considered as random variables but the corresponding feature vectors $\boldsymbol{x}_1, \ldots, \boldsymbol{x}_n$ are viewed as being fixed, then $\mathcal{T}_n$ is the collection of random responses $\{Y_1, \ldots, Y_n\}$. Alternatively, we can simply identify $\mathcal{T}_n$ with the response vector $\boldsymbol{Y} = [Y_1, \ldots, Y_n]^\top$. We will adopt this notation in the next example.

■ **Example 2.10 (Polynomial Regression (cont.))** Consider Example 2.2 once again, but now in a Bayesian framework, where the prior knowledge on $(\sigma^2, \boldsymbol{\beta})$ is specified by $g(\sigma^2) = 1/\sigma^2$ and $\boldsymbol{\beta} \mid \sigma^2 \sim \mathcal{N}(\mathbf{0}, \sigma^2\mathbf{D})$, and $\mathbf{D}$ is a (matrix) hyperparameter. Let $\boldsymbol{\Sigma} := (\mathbf{X}^\top\mathbf{X} + \mathbf{D}^{-1})^{-1}$. Then the posterior can be written as:

$$g(\boldsymbol{\beta}, \sigma^2 \mid \mathbf{y}) = \frac{\exp\left(-\frac{\|\mathbf{y}-\mathbf{X}\boldsymbol{\beta}\|^2}{2\sigma^2}\right)}{(2\pi\sigma^2)^{n/2}} \times \frac{\exp\left(-\frac{\boldsymbol{\beta}^\top\mathbf{D}^{-1}\boldsymbol{\beta}}{2\sigma^2}\right)}{(2\pi\sigma^2)^{p/2}\,|\mathbf{D}|^{1/2}} \times \frac{1}{\sigma^2} \Big/ g(\mathbf{y})$$

$$= \frac{(\sigma^2)^{-(n+p)/2-1}}{(2\pi)^{(n+p)/2}\,|\mathbf{D}|^{1/2}} \exp\left(-\frac{\|\Sigma^{-1/2}(\boldsymbol{\beta} - \overline{\boldsymbol{\beta}})\|^2}{2\sigma^2} - \frac{(n+p+2)\,\overline{\sigma}^2}{2\sigma^2}\right) \Big/ g(\mathbf{y}),$$

where $\overline{\boldsymbol{\beta}} := \boldsymbol{\Sigma}\mathbf{X}^\top\mathbf{y}$ and $\overline{\sigma}^2 := \mathbf{y}^\top(\mathbf{I} - \mathbf{X}\boldsymbol{\Sigma}\mathbf{X}^\top)\mathbf{y}/(n+p+2)$ are the MAP estimates of $\boldsymbol{\beta}$ and σ^2, and $g(\mathbf{y})$ is the model evidence for $\mathcal{G}_p$:

$$g(\mathbf{y}) = \iint g(\boldsymbol{\beta}, \sigma^2, \mathbf{y})\, \mathrm{d}\boldsymbol{\beta}\, \mathrm{d}\sigma^2$$

$$= \frac{|\boldsymbol{\Sigma}|^{1/2}}{(2\pi)^{n/2}|\mathbf{D}|^{1/2}} \int_0^\infty \frac{\exp\left(-\frac{(n+p+2)\,\overline{\sigma}^2}{2\sigma^2}\right)}{(\sigma^2)^{n/2+1}}\, \mathrm{d}\sigma^2$$

$$= \frac{|\boldsymbol{\Sigma}|^{1/2}\Gamma(n/2)}{|\mathbf{D}|^{1/2}(\pi(n+p+2)\,\overline{\sigma}^2)^{n/2}}.$$

Therefore, based on (2.49), we have

$$2\mathbb{E}\ell(g_{\mathcal{T}_n}) \simeq -2\ln g(\mathbf{y}) = n\ln\left[\pi(n+p+2)\,\overline{\sigma}^2\right] - 2\ln\Gamma(n/2) + \ln|\mathbf{D}| - \ln|\boldsymbol{\Sigma}|.$$

On the other hand, the minus of the log-likelihood of $\boldsymbol{Y}$ can be written as

$$-\ln g(\mathbf{y} \mid \boldsymbol{\beta}, \sigma^2) = \frac{\|\mathbf{y} - \mathbf{X}\boldsymbol{\beta}\|^2}{2\sigma^2} + \frac{n}{2}\ln(2\pi\sigma^2)$$

$$= \frac{\|\boldsymbol{\Sigma}^{-1/2}(\boldsymbol{\beta} - \overline{\boldsymbol{\beta}})\|^2}{2\sigma^2} + \frac{(n+p+2)\,\overline{\sigma}^2}{2\sigma^2} + \frac{n}{2}\ln(2\pi\sigma^2).$$

Therefore, the BIC approximation (2.51) is

$$-2\ln g(\mathbf{y} \mid \overline{\boldsymbol{\beta}}, \overline{\sigma}^2) + (p+1)\ln(n) = n[\ln(2\pi\,\overline{\sigma}^2) + 1] + (p+1)\ln(n) + (p+2), \tag{2.52}$$

where the extra $\ln(n)$ term in $(p+1)\ln(n)$ is due to the inclusion of σ^2 in $\boldsymbol{\theta} = (\sigma^2, \boldsymbol{\beta})$. Figure 2.15 shows the model evidence and its BIC approximation, where we used a hyperparameter $\mathbf{D} = 10^4 \times \mathbf{I}_p$ for the prior density of $\boldsymbol{\beta}$. We can see that both approximations exhibit a pronounced minimum at $p = 4$, thus identifying the true polynomial regression model. Compare the overall qualitative shape of the cross-entropy risk estimate with the shape of the square-error risk estimate in Figure 2.11.

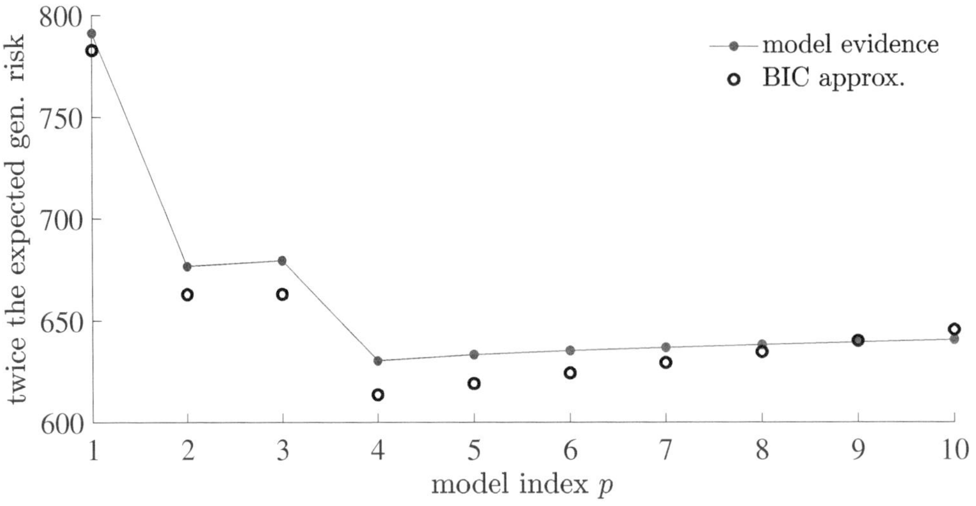

Figure 2.15: The BIC and marginal likelihood used for model selection.

■

It is possible to give the model complexity parameter p a Bayesian treatment, in which we define a prior density on the set of all models under consideration. For example, let $g(p),\ p = 1, \ldots, m$ be a prior density on m candidate models. Treating the model complexity index p as an additional parameter to $\boldsymbol{\theta} \in \mathbb{R}^p$, and applying Bayes' formula, the posterior for $(\boldsymbol{\theta}, p)$ can be written as:

$$\begin{aligned} g(\boldsymbol{\theta}, p \mid \tau) &= g(\boldsymbol{\theta} \mid p, \tau) \times g(p \mid \tau) \\ &= \underbrace{\frac{g(\tau \mid \boldsymbol{\theta}, p)\, g(\boldsymbol{\theta} \mid p)}{g(\tau \mid p)}}_{\text{posterior of } \boldsymbol{\theta} \text{ given model } p} \times \underbrace{\frac{g(\tau \mid p)\, g(p)}{g(\tau)}}_{\text{posterior of model } p}. \end{aligned}$$

The model evidence for a fixed p is now interpreted as the prior predictive density of τ, conditional on the model p:

$$g(\tau \mid p) = \int g(\tau \mid \boldsymbol{\theta}, p)\, g(\boldsymbol{\theta} \mid p)\, \mathrm{d}\boldsymbol{\theta},$$

and the quantity $g(\tau) = \sum_{p=1}^{m} g(\tau \mid p)\, g(p)$ is interpreted as the marginal likelihood of all the m candidate models. Finally, a simple method for model selection is to pick the index $\widehat{p}$ with the largest posterior probability:

$$\widehat{p} = \operatorname*{argmax}_{p}\, g(p \mid \tau) = \operatorname*{argmax}_{p}\, g(\tau \mid p)\, g(p).$$

■ **Example 2.11 (Polynomial Regression (cont.))** Let us revisit Example 2.10 by giving the parameter $p = 1, \ldots, m$, with $m = 10$, a Bayesian treatment. Recall that we used the notation $\tau = \mathbf{y}$ in that example. We assume that the prior $g(p) = 1/m$ is flat and uninformative so that the posterior is given by

$$g(p \mid \mathbf{y}) \propto g(\mathbf{y} \mid p) = \frac{|\boldsymbol{\Sigma}|^{1/2}\, \Gamma(n/2)}{|\mathbf{D}|^{1/2} (\pi (n + p + 2)\, \overline{\sigma^2})^{n/2}},$$

where all quantities in $g(\mathbf{y} \mid p)$ are computed using the first p columns of $\mathbf{X}$. Figure 2.16 shows the resulting posterior density $g(p \mid \mathbf{y})$. The figure also shows the posterior density $\widehat{g}(\mathbf{y} \mid p) / \sum_{p=1}^{10} \widehat{g}(\mathbf{y} \mid p)$, where

$$\widehat{g}(\mathbf{y} \mid p) := \exp\left(-\frac{n[\ln(2\pi\overline{\sigma}^2) + 1] + (p+1)\ln(n) + (p+2)}{2}\right)$$

is derived from the BIC approximation (2.52). In both cases, there is a clear maximum at $p = 4$, suggesting that a third-degree polynomial is the most appropriate model for the data.

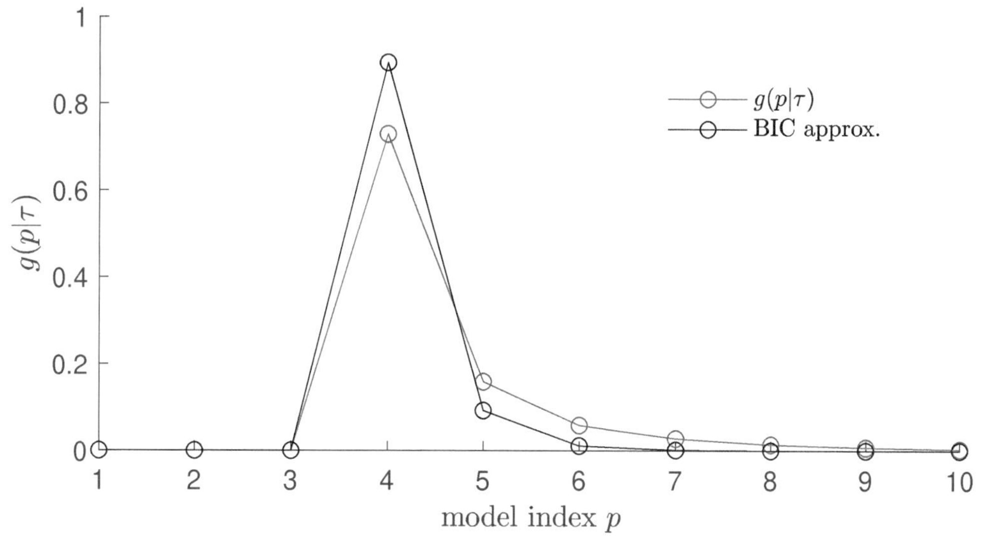

Figure 2.16: Posterior probabilities for each polynomial model of degree $p - 1$.

■

Suppose that we wish to compare two models, say model $p = 1$ and model $p = 2$. Instead of computing the posterior $g(p \mid \tau)$ explicitly, we can compare the posterior odds ratio:

$$\frac{g(p=1 \mid \tau)}{g(p=2 \mid \tau)} = \frac{g(p=1)}{g(p=2)} \times \underbrace{\frac{g(\tau \mid p=1)}{g(\tau \mid p=2)}}_{\text{Bayes factor } B_{1\mid 2}}.$$

This gives rise to the *Bayes factor* $B_{i \mid j}$, whose value signifies the strength of the evidence in favor of model i over model j. In particular $B_{i \mid j} > 1$ means that the evidence in favor for model i is larger. BAYES FACTOR

■ **Example 2.12 (Savage–Dickey Ratio)** Suppose that we have two models. Model $p = 2$ has a likelihood $g(\tau \mid \mu, \nu, p = 2)$, depending on two parameters. Model $p = 1$ has the same functional form for the likelihood but now ν is fixed to some (known) ν_0; that is, $g(\tau \mid \mu, p = 1) = g(\tau \mid \mu, \nu = \nu_0, p = 2)$. We also assume that the prior information on μ

for model 1 is the same as that for model 2, conditioned on $\nu = \nu_0$. That is, we assume $g(\mu \mid p = 1) = g(\mu \mid \nu = \nu_0, p = 2)$. As model 2 contains model 1 as a special case, the latter is said to be *nested* inside model 2. We can formally write (see also Exercise 26):

$$\begin{aligned} g(\tau \mid p = 1) &= \int g(\tau \mid \mu, p = 1)\, g(\mu \mid p = 1)\, \mathrm{d}\mu \\ &= \int g(\tau \mid \mu, \nu = \nu_0, p = 2)\, g(\mu \mid \nu = \nu_0, p = 2)\, \mathrm{d}\mu \\ &= g(\tau \mid \nu = \nu_0, p = 2) = \frac{g(\tau, \nu = \nu_0 \mid p = 2)}{g(\nu = \nu_0 \mid p = 2)}. \end{aligned}$$

Hence, the Bayes factor simplifies to

$$B_{1\mid 2} = \frac{g(\tau \mid p = 1)}{g(\tau \mid p = 2)} = \frac{g(\tau, \nu = \nu_0 \mid p = 2)}{g(\nu = \nu_0 \mid p = 2)} \Big/ g(\tau \mid p = 2) = \frac{g(\nu = \nu_0 \mid \tau, p = 2)}{g(\nu = \nu_0 \mid p = 2)}.$$

In other words, $B_{1\mid 2}$ is the ratio of the posterior density to the prior density of ν, evaluated at $\nu = \nu_0$ and both under the unrestricted model $p = 2$. This ratio of posterior to prior densities is called the *Savage–Dickey density ratio*. ■

SAVAGE–DICKEY DENSITY RATIO

Whether to use a classical (frequentist) or Bayesian model is largely a question of convenience. Classical inference is useful because it comes with a huge repository of ready-to-use results, and requires no (subjective) prior information on the parameters. Bayesian models are useful because the whole theory is based on the elegant Bayes' formula, and uncertainty in the inference (e.g., confidence intervals) can be quantified much more naturally (e.g., credible intervals). A usual practice is to "Bayesify" a classical model, simply by adding some prior information on the parameters.

Further Reading

A popular textbook on statistical learning is [55]. Accessible treatments of mathematical statistics can be found, for example, in [69], [74], and [124]. More advanced treatments are given in [10], [25], and [78]. A good overview of modern-day statistical inference is given in [36]. Classical references on pattern classification and machine learning are [12] and [35]. For advanced learning theory including information theory and Rademacher complexity, we refer to [28] and [109]. An applied reference for Bayesian inference is [46]. For a survey of numerical techniques relevant to computational statistics, see [90].

Exercises

1. Suppose that the loss function is the piecewise linear function

$$\mathrm{Loss}(y, \widehat{y}) = \alpha\, (\widehat{y} - y)_+ + \beta\, (y - \widehat{y})_+, \quad \alpha, \beta > 0,$$

where c_+ is equal to c if $c > 0$, and zero otherwise. Show that the minimizer of the risk $\ell(g) = \mathbb{E}\,\mathrm{Loss}(Y, g(\boldsymbol{X}))$ satisfies

$$\mathbb{P}[Y < g^*(\boldsymbol{x}) \mid \boldsymbol{X} = \boldsymbol{x}] = \frac{\beta}{\alpha + \beta}.$$

In other words, $g^*(\boldsymbol{x})$ is the $\beta/(\alpha + \beta)$ quantile of Y, conditional on $\boldsymbol{X} = \boldsymbol{x}$.

2. Show that, for the squared-error loss, the approximation error $\ell(g^{\mathcal{G}}) - \ell(g^*)$ in (2.16), is equal to $\mathbb{E}(g^{\mathcal{G}}(\boldsymbol{X}) - g^*(\boldsymbol{X}))^2$. [Hint: expand $\ell(g^{\mathcal{G}}) = \mathbb{E}(Y - g^*(\boldsymbol{X}) + g^*(\boldsymbol{X}) - g^{\mathcal{G}}(\boldsymbol{X}))^2$.]

3. Suppose $\mathcal{G}$ is the class of *linear* functions. A linear function evaluated at a feature $\boldsymbol{x}$ can be described as $g(\boldsymbol{x}) = \boldsymbol{\beta}^\top \boldsymbol{x}$ for some parameter vector $\boldsymbol{\beta}$ of appropriate dimension. Denote $g^{\mathcal{G}}(\boldsymbol{x}) = \boldsymbol{x}^\top \boldsymbol{\beta}^{\mathcal{G}}$ and $g_\tau^{\mathcal{G}}(\boldsymbol{x}) = \boldsymbol{x}^\top \widehat{\boldsymbol{\beta}}$. Show that

$$\mathbb{E}\left(g_\tau^{\mathcal{G}}(\boldsymbol{X}) - g^*(\boldsymbol{X})\right)^2 = \mathbb{E}\left(\boldsymbol{X}^\top \widehat{\boldsymbol{\beta}} - \boldsymbol{X}^\top \boldsymbol{\beta}^{\mathcal{G}}\right)^2 + \mathbb{E}\left(\boldsymbol{X}^\top \boldsymbol{\beta}^{\mathcal{G}} - g^*(\boldsymbol{X})\right)^2.$$

Hence, deduce that the statistical error in (2.16) is $\ell(g_\tau^{\mathcal{G}}) - \ell(g^{\mathcal{G}}) = \mathbb{E}\,(g_\tau^{\mathcal{G}}(\boldsymbol{X}) - g^{\mathcal{G}}(\boldsymbol{X}))^2$.

4. Show that formula (2.24) holds for the 0–1 loss with 0–1 response.

5. Let $\boldsymbol{X}$ be an n-dimensional normal random vector with mean vector $\boldsymbol{\mu}$ and covariance matrix $\boldsymbol{\Sigma}$, where the determinant of $\boldsymbol{\Sigma}$ is non-zero. Show that $\boldsymbol{X}$ has joint probability density

$$f_{\boldsymbol{X}}(\boldsymbol{x}) = \frac{1}{\sqrt{(2\pi)^n\,|\boldsymbol{\Sigma}|}}\,\mathrm{e}^{-\frac{1}{2}(\boldsymbol{x}-\boldsymbol{\mu})^\top \boldsymbol{\Sigma}^{-1}(\boldsymbol{x}-\boldsymbol{\mu})}, \quad \boldsymbol{x} \in \mathbb{R}^n.$$

6. Let $\widehat{\boldsymbol{\beta}} = \mathbf{A}^+ \boldsymbol{y}$. Using the defining properties of the pseudo-inverse, show that for any $\boldsymbol{\beta} \in \mathbb{R}^p$, ☞ 362

$$\|\mathbf{A}\widehat{\boldsymbol{\beta}} - \boldsymbol{y}\| \leqslant \|\mathbf{A}\boldsymbol{\beta} - \boldsymbol{y}\|.$$

7. Suppose that in the polynomial regression Example 2.1 we select the linear class of functions $\mathcal{G}_p$ with $p \geqslant 4$. Then, $g^* \in \mathcal{G}_p$ and the approximation error is zero, because $g^{\mathcal{G}_p}(\boldsymbol{x}) = g^*(\boldsymbol{x}) = \boldsymbol{x}^\top \boldsymbol{\beta}$, where $\boldsymbol{\beta} = [10, -140, 400, -250, 0, \ldots, 0]^\top \in \mathbb{R}^p$. Use the tower property to show that the learner $g_\tau(\boldsymbol{x}) = \boldsymbol{x}^\top \widehat{\boldsymbol{\beta}}$ with $\widehat{\boldsymbol{\beta}} = \mathbf{X}^+ \boldsymbol{y}$, assuming $\mathrm{rank}(\mathbf{X}) \geqslant 4$, is *unbiased*: ☞ 433

UNBIASED

$$\mathbb{E}\, g_{\mathcal{T}}(\boldsymbol{x}) = g^*(\boldsymbol{x}).$$

8. (Exercise 7 continued.) Observe that the learner $g_{\mathcal{T}}$ can be written as a linear combination of the response variable: $g_{\mathcal{T}}(\boldsymbol{x}) = \boldsymbol{x}^\top \mathbf{X}^+ \boldsymbol{Y}$. Prove that for any learner of the form $\boldsymbol{x}^\top \mathbf{A}\boldsymbol{y}$, where $\mathbf{A} \in \mathbb{R}^{p \times n}$ is some matrix and that satisfies $\mathbb{E}_{\mathbf{X}}[\boldsymbol{x}^\top \mathbf{A} \boldsymbol{Y}] = g^*(\boldsymbol{x})$, we have

$$\mathbb{V}\mathrm{ar}_{\mathbf{X}}[\boldsymbol{x}^\top \mathbf{X}^+ \boldsymbol{Y}] \leqslant \mathbb{V}\mathrm{ar}_{\mathbf{X}}[\boldsymbol{x}^\top \mathbf{A} \boldsymbol{Y}],$$

where the equality is achieved for $\mathbf{A} = \mathbf{X}^+$. This is called the *Gauss–Markov inequality*. Hence, using the Gauss–Markov inequality deduce that for the unconditional variance:

GAUSS–MARKOV INEQUALITY

$$\mathbb{V}\mathrm{ar}\, g_{\mathcal{T}}(\boldsymbol{x}) \leqslant \mathbb{V}\mathrm{ar}[\boldsymbol{x}^\top \mathbf{A} \boldsymbol{Y}].$$

Deduce that $\mathbf{A} = \mathbf{X}^+$ also minimizes the expected generalization risk.

9. Consider again the polynomial regression Example 2.1. Use the fact that $\mathbb{E}_{\mathbf{X}} \widehat{\boldsymbol{\beta}} = \mathbf{X}^+ \boldsymbol{h}^*(\boldsymbol{u})$, where $\boldsymbol{h}^*(\boldsymbol{u}) = \mathbb{E}[\boldsymbol{Y} \,|\, \boldsymbol{U} = \boldsymbol{u}] = [h^*(u_1), \ldots, h^*(u_n)]^\top$, to show that the expected in-sample risk is:

$$\mathbb{E}_{\mathbf{X}}\, \ell_{\mathrm{in}}(g_{\mathcal{T}}) = \ell^* + \frac{\|\boldsymbol{h}^*(\boldsymbol{u})\|^2 - \|\mathbf{X}\mathbf{X}^+ \boldsymbol{h}^*(\boldsymbol{u})\|^2}{n} + \frac{\ell^* p}{n}.$$

Also, use Theorem C.2 to show that the expected statistical error is: ☞ 432

$$\mathbb{E}_{\mathbf{X}}\,(\widehat{\boldsymbol{\beta}} - \boldsymbol{\beta})^\top \mathbf{H}_p (\widehat{\boldsymbol{\beta}} - \boldsymbol{\beta}) = \ell^* \mathrm{tr}(\mathbf{X}^+ (\mathbf{X}^+)^\top \mathbf{H}_p) + (\mathbf{X}^+ \boldsymbol{h}^*(\boldsymbol{u}) - \boldsymbol{\beta})^\top \mathbf{H}_p (\mathbf{X}^+ \boldsymbol{h}^*(\boldsymbol{u}) - \boldsymbol{\beta}).$$

10. Consider the setting of the polynomial regression in Example 2.2. Use Theorem C.19 to prove that

☞ 451

$$\sqrt{n}\,(\widehat{\boldsymbol{\beta}}_n - \boldsymbol{\beta}_p) \xrightarrow{\text{d}} \mathcal{N}\left(\mathbf{0},\ \ell^* \mathbf{H}_p^{-1} + \mathbf{H}_p^{-1}\mathbf{M}_p\mathbf{H}_p^{-1}\right), \tag{2.53}$$

where $\mathbf{M}_p := \mathbb{E}[\boldsymbol{X}\boldsymbol{X}^\top(g^*(\boldsymbol{X}) - g^{\mathcal{G}_p}(\boldsymbol{X}))^2]$ is the matrix with (i, j)-th entry:

$$\int_0^1 u^{i+j-2}(h^{\mathcal{H}_p}(u) - h^*(u))^2 \,\mathrm{d}u,$$

INVERSE HILBERT MATRIX

and $\mathbf{H}_p^{-1}$ is the $p \times p$ *inverse Hilbert matrix* with (i, j)-th entry:

$$(-1)^{i+j}(i + j - 1)\binom{p+i-1}{p-j}\binom{p+j-1}{p-i}\binom{i+j-2}{i-1}^2.$$

Observe that $\mathbf{M}_p = \mathbf{0}$ for $p \geqslant 4$, so that the matrix $\mathbf{M}_p$ term is due to choosing a restrictive class $\mathcal{G}_p$ that does not contain the true prediction function.

11. In Example 2.2 we saw that the statistical error can be expressed (see (2.20)) as

$$\int_0^1 \left([1, \ldots, u^{p-1}](\widehat{\boldsymbol{\beta}} - \boldsymbol{\beta}_p)\right)^2 \mathrm{d}u = (\widehat{\boldsymbol{\beta}} - \boldsymbol{\beta}_p)^\top \mathbf{H}_p(\widehat{\boldsymbol{\beta}} - \boldsymbol{\beta}_p).$$

By Exercise 10 the random vector $\boldsymbol{Z}_n := \sqrt{n}(\widehat{\boldsymbol{\beta}}_n - \boldsymbol{\beta}_p)$ has asymptotically a multivariate normal distribution with mean vector $\mathbf{0}$ and covariance matrix $\mathbf{V} := \ell^*\mathbf{H}_p^{-1} + \mathbf{H}_p^{-1}\mathbf{M}_p\mathbf{H}_p^{-1}$. Use Theorem C.2 to show that the *expected* statistical error is asymptotically

☞ 432

$$\mathbb{E}\,(\widehat{\boldsymbol{\beta}} - \boldsymbol{\beta}_p)^\top \mathbf{H}_p(\widehat{\boldsymbol{\beta}} - \boldsymbol{\beta}_p) \simeq \frac{\ell^* p}{n} + \frac{\operatorname{tr}(\mathbf{M}_p\mathbf{H}_p^{-1})}{n}, \quad n \to \infty. \tag{2.54}$$

Plot this large-sample approximation of the expected statistical error and compare it with the outcome of the statistical error.

We note a subtle technical detail: In general, convergence in distribution does not imply convergence in L_p-norm (see Example C.6), and so here we have implicitly assumed that $\|\boldsymbol{Z}_n\| \xrightarrow{\text{d}} \text{Dist.} \Rightarrow \|\boldsymbol{Z}_n\| \xrightarrow{L_2} \text{constant} := \lim_{n\uparrow\infty} \mathbb{E}\|\boldsymbol{Z}_n\|$.

☞ 444

12. Consider again Example 2.2. The result in (2.53) suggests that $\mathbb{E}\widehat{\boldsymbol{\beta}} \to \boldsymbol{\beta}_p$ as $n \to \infty$, where $\boldsymbol{\beta}_p$ is the solution in the class $\mathcal{G}_p$ given in (2.18). Thus, the large-sample approximation of the pointwise bias of the learner $g_{\mathcal{T}}^{\mathcal{G}_p}(\boldsymbol{x}) = \boldsymbol{x}^\top\widehat{\boldsymbol{\beta}}$ at $\boldsymbol{x} = [1, \ldots, u^{p-1}]^\top$ is

$$\mathbb{E}\, g_{\mathcal{T}}^{\mathcal{G}_p}(\boldsymbol{x}) - g^*(\boldsymbol{x}) \simeq [1, \ldots, u^{p-1}]\boldsymbol{\beta}_p - [1, u, u^2, u^3]\boldsymbol{\beta}^*, \quad n \to \infty.$$

Use Python to reproduce Figure 2.17, which shows the (large-sample) pointwise squared bias of the learner for $p \in \{1, 2, 3\}$. Note how the bias is larger near the endpoints $u = 0$ and $u = 1$. Explain why the areas under the curves correspond to the approximation errors.

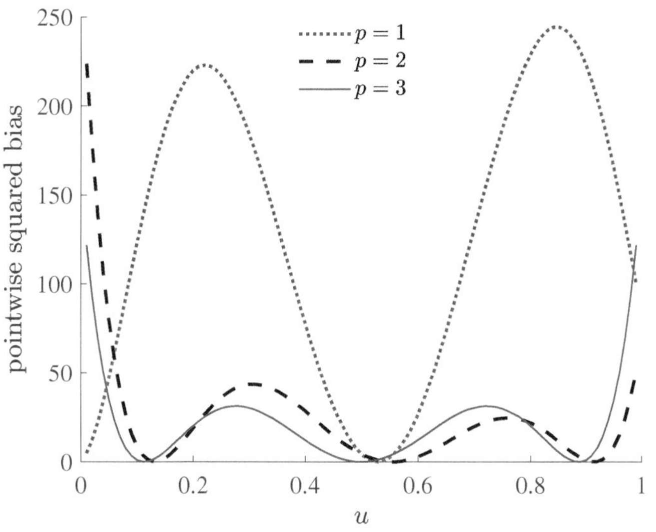

Figure 2.17: The large-sample pointwise squared bias of the learner for $p = 1, 2, 3$. The bias is zero for $p \geqslant 4$.

13. For our running Example 2.2 we can use (2.53) to derive a large-sample approximation of the pointwise variance of the learner $g_{\mathcal{T}}(\boldsymbol{x}) = \boldsymbol{x}^\top \widehat{\boldsymbol{\beta}}_n$. In particular, show that for large n

$$\mathbb{V}\mathrm{ar}\, g_{\mathcal{T}}(\boldsymbol{x}) \simeq \frac{\ell^* \, \boldsymbol{x}^\top \mathbf{H}_p^{-1} \boldsymbol{x}}{n} + \frac{\boldsymbol{x}^\top \mathbf{H}_p^{-1} \mathbf{M}_p \mathbf{H}_p^{-1} \boldsymbol{x}}{n}, \quad n \to \infty. \tag{2.55}$$

Figure 2.18 shows this (large-sample) variance of the learner for different values of the predictor u and model index p. Observe that the variance ultimately increases in p and that it is smaller at $u = 1/2$ than closer to the endpoints $u = 0$ or $u = 1$. Since the bias is also

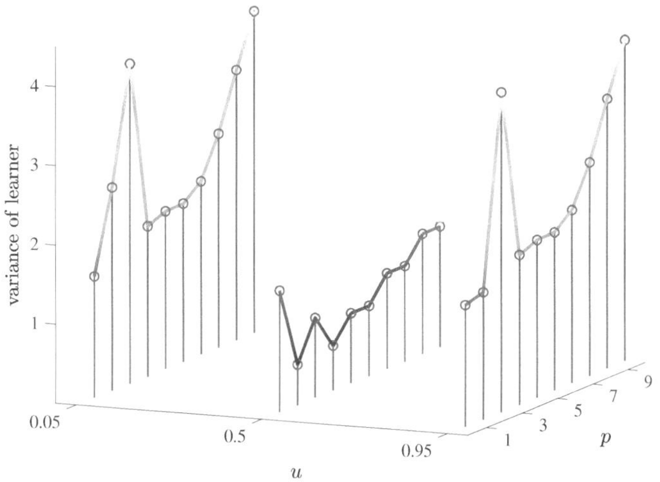

Figure 2.18: The pointwise variance of the learner for various pairs of p and u.

larger near the endpoints, we deduce that the pointwise mean squared error (2.21) is larger near the endpoints of the interval $[0, 1]$ than near its middle. In other words, the error is much smaller in the center of the data cloud than near its periphery.

☞ 405
JENSEN'S INEQUALITY

14. Let $h : \boldsymbol{x} \mapsto \mathbb{R}$ be a convex function and let $\boldsymbol{X}$ be a random variable. Use the subgradient definition of convexity to prove *Jensen's inequality*:

$$\mathbb{E}\, h(\boldsymbol{X}) \geqslant h(\mathbb{E}\boldsymbol{X}). \tag{2.56}$$

15. Using Jensen's inequality, show that the Kullback–Leibler divergence between probability densities f and g is always positive; that is,

$$\mathbb{E} \ln \frac{f(\boldsymbol{X})}{g(\boldsymbol{X})} \geqslant 0,$$

where $\boldsymbol{X} \sim f$.

VAPNIK–CHERNOVENKIS BOUND

16. The purpose of this exercise is to prove the following *Vapnik–Chernovenkis bound*: for any *finite* class $\mathcal{G}$ (containing only a finite number $|\mathcal{G}|$ of possible functions) and a general *bounded* loss function, $l \leqslant \text{Loss} \leqslant u$, the expected statistical error is bounded from above according to:

$$\mathbb{E}\, \ell(g^{\mathcal{G}}_{\mathcal{T}_n}) - \ell(g^{\mathcal{G}}) \leqslant \frac{(u-l)\sqrt{2\ln(2|\mathcal{G}|)}}{\sqrt{n}}. \tag{2.57}$$

Note how this bound conveniently does not depend on the distribution of the training set $\mathcal{T}_n$ (which is typically unknown), but only on the complexity (i.e., cardinality) of the class $\mathcal{G}$. We can break up the proof of (2.57) into the following four parts:

(a) For a general function class $\mathcal{G}$, training set $\mathcal{T}$, risk function ℓ, and training loss $\ell_{\mathcal{T}}$, we have, by definition, $\ell(g^{\mathcal{G}}) \leqslant \ell(g)$ and $\ell_{\mathcal{T}}(g^{\mathcal{G}}_{\mathcal{T}}) \leqslant \ell_{\mathcal{T}}(g)$ for all $g \in \mathcal{G}$. Show that

$$\ell(g^{\mathcal{G}}_{\mathcal{T}}) - \ell(g^{\mathcal{G}}) \leqslant \sup_{g\in\mathcal{G}} |\ell_{\mathcal{T}}(g) - \ell(g)| + \ell_{\mathcal{T}}(g^{\mathcal{G}}) - \ell(g^{\mathcal{G}}),$$

where we used the notation sup (supremum) for the least upper bound. Since $\mathbb{E}\ell_{\mathcal{T}}(g) = \mathbb{E}\ell(g)$, we obtain, after taking expectations on both sides of the inequality above:

$$\mathbb{E}\, \ell(g^{\mathcal{G}}_{\mathcal{T}}) - \ell(g^{\mathcal{G}}) \leqslant \mathbb{E} \sup_{g\in\mathcal{G}} |\ell_{\mathcal{T}}(g) - \ell(g)|.$$

HOEFFDING'S INEQUALITY

(b) If X is a zero-mean random variable taking values in the interval $[l, u]$, then the following *Hoeffding's inequality* states that the moment generating function satisfies

$$\mathbb{E}\, \mathrm{e}^{tX} \leqslant \exp\left(\frac{t^2(u-l)^2}{8}\right), \quad t \in \mathbb{R}. \tag{2.58}$$

Prove this result by using the fact that the line segment joining points $(l, \exp(tl))$ and $(u, \exp(tu))$ bounds the convex function $x \mapsto \exp(tx)$ for $x \in [l, u]$; that is:

$$\mathrm{e}^{tx} \leqslant \mathrm{e}^{tl}\frac{u-x}{u-l} + \mathrm{e}^{tu}\frac{x-l}{u-l}, \quad x \in [l, u].$$

☞ 429

(c) Let $Z_1, \ldots, Z_n$ be (possibly dependent and non-identically distributed) zero-mean random variables with moment generating functions that satisfy $\mathbb{E}\exp(tZ_k) \leqslant \exp(t^2\eta^2/2)$ for all k and some parameter η. Use Jensen's inequality (2.56) to prove that for any

$t > 0$,

$$\mathbb{E} \max_k Z_k = \frac{1}{t} \mathbb{E} \ln \max_k \mathrm{e}^{tZ_k} \leqslant \frac{1}{t} \ln n + \frac{t\eta^2}{2}.$$

From this derive that

$$\mathbb{E} \max_k Z_k \leqslant \eta \sqrt{2 \ln n}.$$

Finally, show that this last inequality implies that

$$\mathbb{E} \max_k |Z_k| \leqslant \eta \sqrt{2 \ln(2n)}. \tag{2.59}$$

(d) Returning to the objective of this exercise, denote the elements of $\mathcal{G}$ by $g_1, \ldots, g_{|\mathcal{G}|}$, and let $Z_k = \ell_{\mathcal{T}_n}(g_k) - \ell(g_k)$. By part (a) it is sufficient to bound $\mathbb{E} \max_k |Z_k|$. Show that the $\{Z_k\}$ satisfy the conditions of (c) with $\eta = (u - l)/\sqrt{n}$. For this you will need to apply part (b) to the random variable $\text{Loss}(g(\boldsymbol{X}), Y) - \ell(g)$, where $(\boldsymbol{X}, Y)$ is a generic data point. Now complete the proof of (2.57).

17. Consider the problem in Exercise 16a above. Show that

$$|\ell_{\mathcal{T}}(g_{\mathcal{T}}^{\mathcal{G}}) - \ell(g^{\mathcal{G}})| \leqslant 2 \sup_{g \in \mathcal{G}} |\ell_{\mathcal{T}}(g) - \ell(g)| + \ell_{\mathcal{T}}(g^{\mathcal{G}}) - \ell(g^{\mathcal{G}}).$$

From this, conclude:

$$\mathbb{E}\, |\ell_{\mathcal{T}}(g_{\mathcal{T}}^{\mathcal{G}}) - \ell(g^{\mathcal{G}})| \leqslant 2\mathbb{E} \sup_{g \in \mathcal{G}} |\ell_{\mathcal{T}}(g) - \ell(g)|.$$

The last bound allows us to assess how close the training loss $\ell_{\mathcal{T}}(g_{\mathcal{T}}^{\mathcal{G}})$ is to the optimal risk $\ell(g^{\mathcal{G}})$ within class $\mathcal{G}$.

18. Show that for the normal linear model $\boldsymbol{Y} \sim \mathcal{N}(\mathbf{X}\boldsymbol{\beta}, \sigma^2 \mathbf{I}_n)$, the maximum likelihood estimator of σ^2 is identical to the method of moments estimator (2.37).

19. Let $X \sim \mathsf{Gamma}(\alpha, \lambda)$. Show that the pdf of $Z = 1/X$ is equal to

$$\frac{\lambda^{\alpha} (z)^{-\alpha-1} \mathrm{e}^{-\lambda (z)^{-1}}}{\Gamma(\alpha)}, \quad z > 0.$$

20. Consider the sequence $w_0, w_1, \ldots$, where $w_0 = g(\boldsymbol{\theta})$ is a non-degenerate initial guess and $w_t(\boldsymbol{\theta}) \propto w_{t-1}(\boldsymbol{\theta}) g(\tau \,|\, \boldsymbol{\theta})$, $t > 1$. We assume that $g(\tau \,|\, \boldsymbol{\theta})$ is not the constant function (with respect to $\boldsymbol{\theta}$) and that the maximum likelihood value

$$g(\tau \,|\, \widehat{\boldsymbol{\theta}}) = \max_{\boldsymbol{\theta}} g(\tau \,|\, \boldsymbol{\theta}) < \infty$$

exists (is bounded). Let

$$l_t := \int g(\tau \,|\, \boldsymbol{\theta}) w_t(\boldsymbol{\theta}) \, \mathrm{d}\boldsymbol{\theta}.$$

Show that $\{l_t\}$ is a strictly increasing and bounded sequence. Hence, conclude that its limit is $g(\tau \,|\, \widehat{\boldsymbol{\theta}})$.

21. Consider the Bayesian model for $\tau = \{x_1, \ldots, x_n\}$ with likelihood $g(\tau \mid \mu)$ such that $(X_1, \ldots, X_n \mid \mu) \sim_{\text{iid}} \mathcal{N}(\mu, 1)$ and prior pdf $g(\mu)$ such that $\mu \sim \mathcal{N}(\nu, 1)$ for some hyperparameter ν. Define a sequence of densities $w_t(\mu), t \geqslant 2$ via $w_t(\mu) \propto w_{t-1}(\mu)\, g(\tau \mid \mu)$, starting with $w_1(\mu) = g(\mu)$. Let a_t and b_t denote the mean and precision[4] of μ under the posterior $g_t(\mu \mid \tau) \propto g(\tau \mid \mu) w_t(\mu)$. Show that $g_t(\mu \mid \tau)$ is a normal density with precision $b_t = b_{t-1} + n$, $b_0 = 1$ and mean $a_t = (1 - \gamma_t) a_{t-1} + \gamma_t \overline{x}_n$, $a_0 = \nu$, where $\gamma_t := n/(b_{t-1} + n)$. Hence, deduce that $g_t(\mu \mid \tau)$ converges to a degenerate density with a point-mass at $\overline{x}_n$.

22. Consider again Example 2.8, where we have a normal model with improper prior $g(\boldsymbol{\theta}) = g(\mu, \sigma^2) \propto 1/\sigma^2$. Show that the prior predictive pdf is an improper density $g(x) \propto 1$, but that the posterior predictive density is

$$g(x \mid \tau) \propto \left(1 + \frac{(x - \overline{x}_n)^2}{(n+1) S_n^2}\right)^{-n/2}.$$

Deduce that $\frac{X - \overline{x}_n}{S_n \sqrt{(n+1)/(n-1)}} \sim \mathsf{t}_{n-1}$.

23. Assuming that $\boldsymbol{X}_1, \ldots, \boldsymbol{X}_n \overset{\text{iid}}{\sim} f$, show that (2.48) holds and that $\ell_n^* = -n \, \mathbb{E} \ln f(\boldsymbol{X})$.

24. Suppose that $\tau = \{x_1, \ldots, x_n\}$ are observations of iid continuous and strictly positive random variables, and that there are two possible models for their pdf. The first model $p = 1$ is

$$g(x \mid \theta, p = 1) = \theta \exp(-\theta x)$$

and the second $p = 2$ is

$$g(x \mid \theta, p = 2) = \left(\frac{2\theta}{\pi}\right)^{1/2} \exp\left(-\frac{\theta x^2}{2}\right).$$

For both models, assume that the prior for θ is a gamma density

$$g(\theta) = \frac{b^t}{\Gamma(t)} \theta^{t-1} \exp(-b\theta),$$

with the same hyperparameters b and t. Find a formula for the Bayes factor, $g(\tau \mid p = 1)/g(\tau \mid p = 2)$, for comparing these models.

25. Suppose that we have a total of m possible models with prior probabilities $g(p), p = 1, \ldots, m$. Show that the posterior probability of model $g(p \mid \tau)$ can be expressed in terms of all the $p(p-1)$ Bayes factors:

$$g(p = i \mid \tau) = \left(1 + \sum_{j \neq i} \frac{g(p = j)}{g(p = i)} B_{j \mid i}\right)^{-1}.$$

[4]The precision is the reciprocal of the variance.

26. Given the data $\tau = \{x_1, \ldots, x_n\}$, suppose that we use the likelihood $(X \mid \boldsymbol{\theta}) \sim \mathcal{N}(\mu, \sigma^2)$ with parameter $\boldsymbol{\theta} = (\mu, \sigma^2)^\top$ and wish to compare the following two nested models.

(a) Model $p = 1$, where $\sigma^2 = \sigma_0^2$ is known and this is incorporated via the prior

$$g(\boldsymbol{\theta} \mid p = 1) = g(\mu \mid \sigma^2, p = 1)\, g(\sigma^2 \mid p = 1) = \frac{1}{\sqrt{2\pi}\sigma} \mathrm{e}^{-\frac{(\mu - x_0)^2}{2\sigma^2}} \times \delta(\sigma^2 - \sigma_0^2).$$

(b) Model $p = 2$, where both mean and variance are unknown with prior

$$g(\boldsymbol{\theta} \mid p = 2) = g(\mu \mid \sigma^2)\, g(\sigma^2) = \frac{1}{\sqrt{2\pi}\sigma} \mathrm{e}^{-\frac{(\mu - x_0)^2}{2\sigma^2}} \times \frac{b^t (\sigma^2)^{-t-1} \mathrm{e}^{-b/\sigma^2}}{\Gamma(t)}.$$

Show that the prior $g(\boldsymbol{\theta} \mid p = 1)$ can be viewed as the limit of the prior $g(\boldsymbol{\theta} \mid p = 2)$ when $t \to \infty$ and $b = t\sigma_0^2$. Hence, conclude that

$$g(\tau \mid p = 1) = \lim_{\substack{t \to \infty \\ b = t\sigma_0^2}} g(\tau \mid p = 2)$$

and use this result to calculate $B_{1 \mid 2}$. Check that the formula for $B_{1 \mid 2}$ agrees with the Savage–Dickey density ratio:

$$\frac{g(\tau \mid p = 1)}{g(\tau \mid p = 2)} = \frac{g(\sigma^2 = \sigma_0^2 \mid \tau)}{g(\sigma^2 = \sigma_0^2)},$$

where $g(\sigma^2 \mid \tau)$ and $g(\sigma^2)$ are the posterior and prior, respectively, under model $p = 2$.

CHAPTER 3

MONTE CARLO METHODS

Many algorithms in machine learning and data science make use of Monte Carlo techniques. This chapter gives an introduction to the three main uses of Monte Carlo simulation: to (1) simulate random objects and processes in order to observe their behavior, (2) estimate numerical quantities by repeated sampling, and (3) solve complicated optimization problems through randomized algorithms.

3.1 Introduction

Briefly put, *Monte Carlo simulation* is the generation of random data by means of a computer. These data could arise from simple models, such as those described in Chapter 2, or from very complicated models describing real-life systems, such as the positions of vehicles on a complex road network, or the evolution of security prices in the stock market. In many cases, Monte Carlo simulation simply involves random sampling from certain probability distributions. The idea is to repeat the random experiment that is described by the model many times to obtain a large quantity of data that can be used to answer questions about the model. The three main uses of Monte Carlo simulation are:

MONTE CARLO SIMULATION

Sampling. Here the objective is to gather information about a random object by observing many realizations of it. For instance, this could be a random process that mimics the behavior of some real-life system such as a production line or telecommunications network. Another usage is found in Bayesian statistics, where Markov chains are often used to sample from a posterior distribution. ☞ 48

Estimation. In this case the emphasis is on estimating certain numerical quantities related to a simulation model. An example is the evaluation of multidimensional integrals via Monte Carlo techniques. This is achieved by writing the integral as the expectation of a random variable, which is then approximated by the sample mean. Appealing to the Law of Large Numbers guarantees that this approximation will eventually converge when the sample size becomes large. ☞ 448

Optimization. Monte Carlo simulation is a powerful tool for the optimization of complicated objective functions. In many applications these functions are deterministic and randomness is introduced artificially in order to more efficiently search the domain of the objective function. Monte Carlo techniques are also used to optimize noisy functions, where the function itself is random; for example, when the objective function is the output of a Monte Carlo simulation.

The Monte Carlo method dramatically changed the way in which statistics is used in today's analysis of data. The ever-increasing complexity of data requires radically different statistical models and analysis techniques from those that were used 20 to 100 years ago. By using Monte Carlo techniques, the data analyst is no longer restricted to using basic (and often inappropriate) models to describe data. Now, any probabilistic model that can be simulated on a computer can serve as the basis for statistical analysis. This Monte Carlo revolution has had an impact on both Bayesian and frequentist statistics. In particular, in frequentist statistics, Monte Carlo methods are often referred to as resampling techniques. An important example is the well-known bootstrap method [37], where statistical quantities such as confidence intervals and P-values for statistical tests can simply be determined by simulation without the need of a sophisticated analysis of the underlying probability distributions; see, for example, [69] for basic applications. The impact on Bayesian statistics has been even more profound, through the use of Markov chain Monte Carlo (MCMC) techniques [87, 48]. MCMC samplers construct a Markov process which converges in distribution to a desired (often high-dimensional) density. This convergence in distribution justifies using a finite run of the Markov process as an approximate random realization from the target density. The MCMC approach has rapidly gained popularity as a versatile heuristic approximation, partly due to its simple computer implementation and inbuilt mechanism to tradeoff between computational cost and accuracy; namely, the longer one runs the Markov process, the better the approximation. Nowadays, MCMC methods are indispensable for analyzing posterior distributions for inference and model selection; see also [50, 99].

The following three sections elaborate on these three uses of Monte Carlo simulation in turn.

3.2 Monte Carlo Sampling

In this section we describe a variety of Monte Carlo sampling methods, from the building block of simulating uniform random numbers to MCMC samplers.

3.2.1 Generating Random Numbers

RANDOM NUMBER GENERATOR

At the heart of any Monte Carlo method is a *random number generator*: a procedure that produces a stream of uniform random numbers on the interval (0,1). Since such numbers are usually produced via deterministic algorithms, they are not truly random. However, for most applications all that is required is that such pseudo-random numbers are statistically indistinguishable from genuine random numbers $U_1, U_2, \ldots$ that are uniformly distributed on the interval (0,1) and are independent of each other; we write $U_1, U_2, \ldots \sim_{\text{iid}} \mathcal{U}(0, 1)$. For example, in Python the `rand` method of the `numpy.random` module is widely used for this purpose.

Most random number generators at present are based on linear recurrence relations. One of the most important random number generators is the *multiple-recursive generator* (MRG) of *order* k, which generates a sequence of integers $X_k, X_{k+1}, \ldots$ via the linear recurrence

MULTIPLE-RECURSIVE GENERATOR

$$X_t = (a_1 X_{t-1} + \cdots + a_k X_{t-k}) \bmod m, \quad t = k, k+1, \ldots \tag{3.1}$$

for some *modulus* m and *multipliers* $\{a_i, i = 1, \ldots, k\}$. Here "mod" refers to the modulo operation: $n \bmod m$ is the remainder when n is divided by m. The recurrence is initialized by specifying k "seeds", $X_0, \ldots, X_{k-1}$. To yield fast algorithms, all but a few of the multipliers should be 0. When m is a large integer, one can obtain a stream of pseudo-random numbers $U_k, U_{k+1}, \ldots$ between 0 and 1 from the sequence $X_k, X_{k+1}, \ldots$, simply by setting $U_t = X_t/m$. It is also possible to set a small modulus, in particular $m = 2$. The output function for such *modulo 2 generators* is then typically of the form

MODULUS

MULTIPLIERS

MODULO 2 GENERATORS

$$U_t = \sum_{i=1}^{w} X_{tw+i-1} 2^{-i}$$

for some $w \leqslant k$, e.g., $w = 32$ or 64. Examples of modulo 2 generators are the *feedback shift register* generators, the most popular of which are the *Mersenne twisters*; see, for example, [79] and [83]. MRGs with excellent statistical properties can be implemented efficiently by combining several simpler MRGs and carefully choosing their respective moduli and multipliers. One of the most successful is L'Ecuyer's `MRG32k3a` generator; see [77]. From now on, we assume that the reader has a sound random number generator available.

FEEDBACK SHIFT REGISTER

MERSENNE TWISTERS

3.2.2 Simulating Random Variables

Simulating a random variable X from an arbitrary (that is, not necessarily uniform) distribution invariably involves the following two steps:

1. Simulate uniform random numbers $U_1, \ldots, U_k$ on $(0, 1)$ for some $k = 1, 2, \ldots$.
2. Return $X = g(U_1, \ldots, U_k)$, where g is some real-valued function.

The construction of suitable functions g is as much of an art as a science. Many simulation methods may be found, for example, in [71] and the accompanying website `www.montecarlohandbook.org`. Two of the most useful general procedures for generating random variables are the *inverse-transform* method and the *acceptance–rejection* method. Before we discuss these, we show one possible way to simulate standard normal random variables. In Python we can generate standard normal random variables via the **`randn`** method of the `numpy.random` module.

■ **Example 3.1 (Simulating Standard Normal Random Variables)** If X and Y are independent standard normally distributed random variables (that is, $X, Y \sim_{\text{iid}} \mathcal{N}(0, 1)$), then their joint pdf is

$$f(x, y) = \frac{1}{2\pi} e^{-\frac{1}{2}(x^2+y^2)}, \quad (x, y) \in \mathbb{R}^2,$$

which is a radially symmetric function. In Example C.2 we see that, in polar coordinates, the angle Θ that the random vector $[X, Y]^\top$ makes with the positive x-axis is $\mathcal{U}(0, 2\pi)$ ☞ 435 distributed (as would be expected from the radial symmetry) and the radius R has pdf $f_R(r) = r\,\mathrm{e}^{-r^2/2}, r > 0$. Moreover, R and Θ are independent. We will see shortly, in Example 3.4, that R has the same distribution as $\sqrt{-2 \ln U}$ with $U \sim \mathcal{U}(0,1)$. So, to simulate $X, Y \sim_{\text{iid}} \mathcal{N}(0,1)$, the idea is to first simulate R and Θ independently and then return $X = R\cos(\Theta)$ and $Y = R\sin(\Theta)$ as a pair of independent standard normal random variables. This leads to the Box–Muller approach for generating standard normal random variables.

☞ 72

Algorithm 3.2.1: Normal Random Variable Simulation: Box–Muller Approach

output: Independent standard normal random variables X and Y.
1 Simulate two independent random variables, U_1 and U_2, from $\mathcal{U}(0,1)$.
2 $X \leftarrow (-2 \ln U_1)^{1/2} \cos(2\pi U_2)$
3 $Y \leftarrow (-2 \ln U_1)^{1/2} \sin(2\pi U_2)$
4 **return** X, Y

■

Once a standard normal number generator is available, simulation from any n-dimensional normal distribution $\mathcal{N}(\boldsymbol{\mu}, \boldsymbol{\Sigma})$ is relatively straightforward. The first step is to find an $n \times n$ matrix $\mathbf{B}$ that decomposes $\boldsymbol{\Sigma}$ into the matrix product $\mathbf{B}\mathbf{B}^\top$. In fact there exist many such decompositions. One of the more important ones is the *Cholesky decomposition*, which is a special case of the LU decomposition; see Section A.6.1 for more information on such decompositions. In Python, the function `cholesky` of `numpy.linalg` can be used to produce such a matrix $\mathbf{B}$.

CHOLESKY DECOMPOSITION

☞ 370

Once the Cholesky factorization is determined, it is easy to simulate $\mathbf{X} \sim \mathcal{N}(\boldsymbol{\mu}, \boldsymbol{\Sigma})$ as, by definition, it is the affine transformation $\boldsymbol{\mu} + \mathbf{B}\mathbf{Z}$ of an n-dimensional standard normal random vector.

Algorithm 3.2.2: Normal Random Vector Simulation

input: $\boldsymbol{\mu}, \boldsymbol{\Sigma}$
output: $\mathbf{X} \sim \mathcal{N}(\boldsymbol{\mu}, \boldsymbol{\Sigma})$
1 Determine the Cholesky factorization $\boldsymbol{\Sigma} = \mathbf{B}\mathbf{B}^\top$.
2 Simulate $\mathbf{Z} = [Z_1, \ldots, Z_n]^\top$ by drawing $Z_1, \ldots, Z_n \sim_{\text{iid}} \mathcal{N}(0,1)$.
3 $\mathbf{X} \leftarrow \boldsymbol{\mu} + \mathbf{B}\mathbf{Z}$
4 **return** $\mathbf{X}$

■ **Example 3.2 (Simulating from a Bivariate Normal Distribution)** The Python code below draws $N = 1000$ iid samples from the two bivariate ($n = 2$) normal pdfs in Figure 2.13. The resulting point clouds are given in Figure 3.1.

☞ 45

bvnormal.py

```
import numpy as np
from numpy.random import randn
import matplotlib.pyplot as plt

N = 1000
r = 0.0   #change to 0.8 for other plot
Sigma = np.array([[1, r], [r, 1]])
```

```
B = np.linalg.cholesky(Sigma)
x = B @ randn(2,N)
plt.scatter([x[0,:]],[x[1,:]], alpha =0.4, s = 4)
```

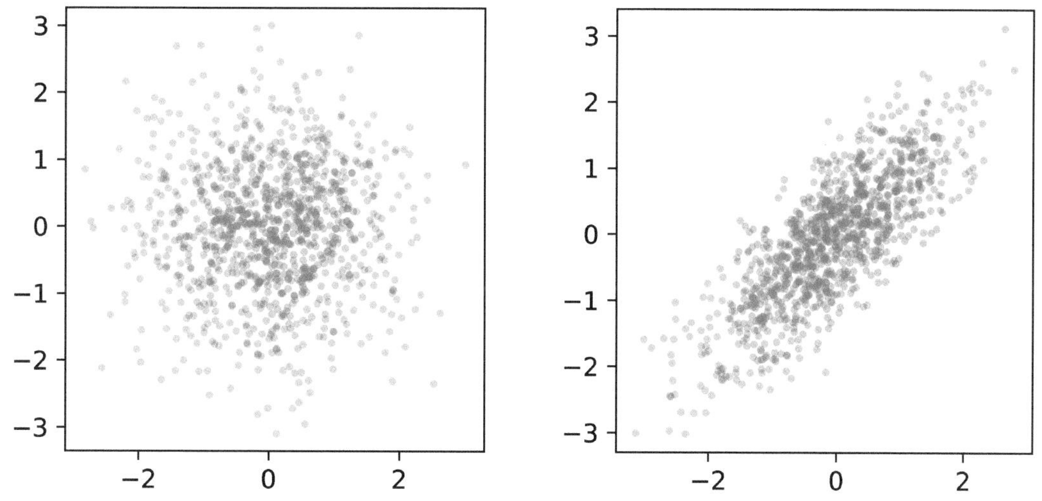

Figure 3.1: 1000 realizations of bivariate normal distributions with means zero, variances 1, and correlation coefficients 0 (left) and 0.8 (right).

■

In some cases, the covariance matrix $\boldsymbol{\Sigma}$ has special structure which can be exploited to create even faster generation algorithms, as illustrated in the following example.

■ **Example 3.3 (Simulating Normal Vectors in $O(n^2)$ Time)** Suppose that the random vector $\boldsymbol{X} = [X_1, \ldots, X_n]^\top$ represents the values at times $t_0 + k\delta$, $k = 0, \ldots, n-1$ of a zero-mean *Gaussian process* $(X(t), t \geqslant 0)$ that is *weakly stationary*, meaning that $\mathbb{C}\text{ov}(X(s), X(t))$ depends only on $t-s$. Then clearly the covariance matrix of $\boldsymbol{X}$, say $\mathbf{A}_n$, is a symmetric Toeplitz matrix. Suppose for simplicity that $\mathbb{V}\text{ar}\, X(t) = 1$. Then the covariance matrix is in fact a correlation matrix, and will have the following structure:

☞ 239

☞ 381

$$\mathbf{A}_n := \begin{bmatrix} 1 & a_1 & \ldots & a_{n-2} & a_{n-1} \\ a_1 & 1 & \ddots & & a_{n-2} \\ \vdots & \ddots & \ddots & \ddots & \vdots \\ a_{n-2} & & \ddots & \ddots & a_1 \\ a_{n-1} & a_{n-2} & \cdots & a_1 & 1 \end{bmatrix}.$$

Using the Levinson–Durbin algorithm we can compute a lower diagonal matrix $\mathbf{L}_n$ and a diagonal matrix $\mathbf{D}_n$ in $O(n^2)$ time such that $\mathbf{L}_n \mathbf{A}_n \mathbf{L}_n^\top = \mathbf{D}_n$; see Theorem A.14. If we simulate $\boldsymbol{Z}_n \sim \mathcal{N}(\mathbf{0}, \mathbf{I}_n)$, then the solution $\boldsymbol{X}$ of the linear system:

☞ 385

$$\mathbf{L}_n \boldsymbol{X} = \mathbf{D}_n^{1/2} \boldsymbol{Z}_n$$

has the desired distribution $\mathcal{N}(\mathbf{0}, \mathbf{A}_n)$. The linear system is solved in $O(n^2)$ time via forward substitution. ■

3.2.2.1 Inverse-Transform Method

Let X be a random variable with cumulative distribution function (cdf) F. Let F^{-1} denote the inverse[1] of F and $U \sim \mathcal{U}(0,1)$. Then,

$$\mathbb{P}[F^{-1}(U) \leqslant x] = \mathbb{P}[U \leqslant F(x)] = F(x). \tag{3.2}$$

This leads to the following method to simulate a random variable X with cdf F:

Algorithm 3.2.3: Inverse-Transform Method

input: Cumulative distribution function F.
output: Random variable X distributed according to F.
1 Generate U from $\mathcal{U}(0,1)$.
2 $X \leftarrow F^{-1}(U)$
3 **return** X

The inverse-transform method works both for continuous and discrete distributions. After importing `numpy` as `np`, simulating numbers $0,\ldots,k-1$ according to probabilities $p_0,\ldots,p_{k-1}$ can be done via `np.min(np.where(np.cumsum(p) > np.random.rand()))`, where `p` is the vector of the probabilities.

■ **Example 3.4 (Example 3.1 (cont.))** One remaining issue in Example 3.1 was how to simulate the radius R when we only know its density $f_R(r) = r\,\mathrm{e}^{-r^2/2}, r > 0$. We can use the inverse-transform method for this, but first we need to determine its cdf. The cdf of R is, by integration of the pdf,

$$F_R(r) = 1 - \mathrm{e}^{-\frac{1}{2}r^2}, \quad r > 0,$$

and its inverse is found by solving $u = F_R(r)$ in terms of r, giving

$$F_R^{-1}(u) = \sqrt{-2\ln(1-u)}, \quad u \in (0,1).$$

Thus R has the same distribution as $\sqrt{-2\ln(1-U)}$, with $U \sim \mathcal{U}(0,1)$. Since $1-U$ also has a $\mathcal{U}(0,1)$ distribution, R has also the same distribution as $\sqrt{-2\ln U}$. ■

3.2.2.2 Acceptance–Rejection Method

The acceptance–rejection method is used to sample from a "difficult" probability density function (pdf) $f(x)$ by generating instead from an "easy" pdf $g(x)$ satisfying $f(x) \leqslant C\,g(x)$ for some constant $C \geqslant 1$ (for example, via the inverse-transform method), and then accepting or rejecting the drawn sample with a certain probability. Algorithm 3.2.4 gives the pseudo-code.

The idea of the algorithm is to generate uniformly a point (X,Y) under the graph of the function Cg, by first drawing $X \sim g$ and then $Y \sim \mathcal{U}(0, Cg(X))$. If this point lies under the graph of f, then we accept X as a sample from f; otherwise, we try again. The efficiency of the acceptance–rejection method is usually expressed in terms of the probability of acceptance, which is $1/C$.

[1]Every cdf has a unique inverse function defined by $F^{-1}(u) = \inf\{x : F(x) \geqslant u\}$. If, for each u, the equation $F(x) = u$ has a unique solution x, this definition coincides with the usual interpretation of the inverse function.

Algorithm 3.2.4: Acceptance–Rejection Method

input: Pdf g and constant C such that $Cg(x) \geqslant f(x)$ for all x.
output: Random variable X distributed according to pdf f.

```
1 found ← false
2 while not found do
3     Generate X from g.
4     Generate U from U(0, 1) independently of X.
5     Y ← UCg(X)
6     if Y ⩽ f(X) then found ← true
7 return X
```

■ **Example 3.5 (Simulating Gamma Random Variables)** Simulating random variables from a Gamma(α, λ) distribution is generally done via the acceptance–rejection method. Consider, for example, the Gamma distribution with $\alpha = 1.3$ and $\lambda = 5.6$. Its pdf, ☞ 427

$$f(x) = \frac{\lambda^\alpha x^{\alpha-1} \mathrm{e}^{-\lambda x}}{\Gamma(\alpha)}, \quad x \geqslant 0,$$

where Γ is the gamma function $\Gamma(\alpha) := \int_0^\infty \mathrm{e}^{-x} x^{\alpha-1} \, \mathrm{d}x$, $\alpha > 0$, is depicted by the blue solid curve in Figure 3.2.

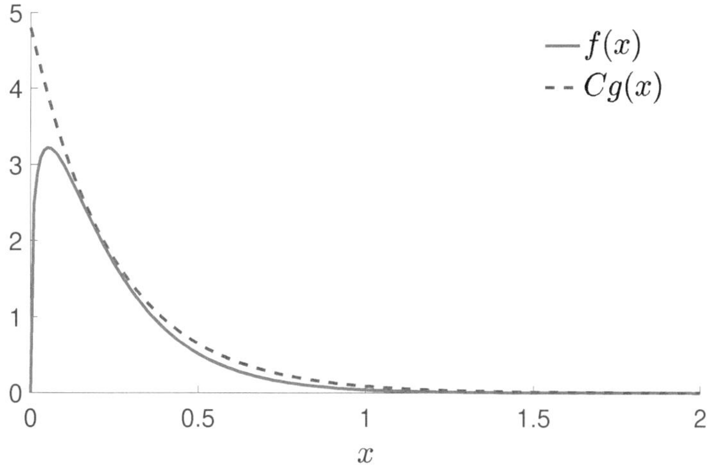

Figure 3.2: The pdf g of the Exp(4) distribution multiplied by $C = 1.2$ dominates the pdf f of the Gamma(1.3, 5.6) distribution.

This pdf happens to lie completely under the graph of $Cg(x)$, where $C = 1.2$ and $g(x) = 4\exp(-4x), x \geqslant 0$ is the pdf of the exponential distribution Exp(4). Hence, we can simulate from this particular Gamma distribution by accepting or rejecting a sample from the Exp(4) distribution according to Step 6 of Algorithm 3.2.4. Simulating from the Exp(4) distribution can be done via the inverse-transform method: simulate $U \sim \mathcal{U}(0, 1)$ and return $X = -\ln(U)/4$. The following Python code implements Algorithm 3.2.4 for this example. ☞ 427 accrejgamma.py

```python
from math import exp, gamma, log
from numpy.random import rand

alpha = 1.3
lam = 5.6
f = lambda x: lam**alpha * x**(alpha-1) * exp(-lam*x)/gamma(alpha)
g = lambda x: 4*exp(-4*x)
C = 1.2

found = False
while not found:
    x = - log(rand())/4
    if C*g(x)*rand() <= f(x):
        found = True

print(x)
```

■

3.2.3 Simulating Random Vectors and Processes

Techniques for generating random vectors and processes are as diverse as the class of random processes themselves; see, for example, [71]. We highlight a few general scenarios.

When $X_1, \ldots, X_n$ are *independent* random variables with pdfs f_i, $i = 1, \ldots, n$, so that their joint pdf is $f(\boldsymbol{x}) = f_1(x_1) \cdots f_n(x_n)$, the random vector $\boldsymbol{X} = [X_1, \ldots, X_n]^\top$ can be simply simulated by drawing each component $X_i \sim f_i$ individually — for example, via the inverse-transform method or acceptance–rejection.

☞ 431

For *dependent* components $X_1, \ldots, X_n$, we can, as a consequence of the *product rule* of probability, represent the joint pdf $f(\boldsymbol{x})$ as

☞ 433

$$f(\boldsymbol{x}) = f(x_1, \ldots, x_n) = f_1(x_1)\, f_2(x_2 \,|\, x_1) \cdots f_n(x_n \,|\, x_1, \ldots, x_{n-1}), \tag{3.3}$$

where $f_1(x_1)$ is the marginal pdf of X_1 and $f_k(x_k \,|\, x_1, \ldots, x_{k-1})$ is the conditional pdf of X_k given $X_1 = x_1, X_2 = x_2, \ldots, X_{k-1} = x_{k-1}$. Provided the conditional pdfs are known, one can generate $\boldsymbol{X}$ by first generating X_1, then, given $X_1 = x_1$, generate X_2 from $f_2(x_2 \,|\, x_1)$, and so on, until generating X_n from $f_n(x_n \,|\, x_1, \ldots, x_{n-1})$.

The latter method is particularly applicable for generating Markov chains. Recall from Section C.10 that a *Markov chain* is a stochastic process $\{X_t, t = 0, 1, 2, \ldots\}$ that satisfies the *Markov property*; meaning that for all t and s the conditional distribution of X_{t+s} given $X_u, u \leqslant t$, is the same as that of X_{t+s} given only X_t. As a result, each conditional density $f_t(x_t \,|\, x_1, \ldots, x_{t-1})$ can be written as a one-step *transition density* $q_t(x_t \,|\, x_{t-1})$; that is, the probability density to go from state x_{t-1} to state x_t in one step. In many cases of interest the chain is *time-homogeneous*, meaning that the transition density q_t does not depend on t. Such Markov chains can be generated *sequentially*, as given in Algorithm 3.2.5.

☞ 453

MARKOV CHAIN

Algorithm 3.2.5: Simulate a Markov Chain

---

input: Number of steps N, initial pdf f_0, transition density q.

Draw X_0 from the initial pdf f_0.

for $t = 1$ **to** N **do**

  Draw X_t from the distribution corresponding to the density $q(\cdot \,|\, X_{t-1})$

return $X_0, \ldots, X_N$

■ **Example 3.6 (Markov Chain Simulation)** For time-homogeneous Markov chains with a discrete state space, we can visualize the one-step transitions by means of a *transition graph*, where arrows indicate possible transitions between states and the labels describe the corresponding probabilities. Figure 3.3 shows (on the left) the transition graph of the Markov chain $\{X_t, t = 0, 1, 2, \ldots\}$ with state space $\{1, 2, 3, 4\}$ and one-step transition matrix

TRANSITION GRAPH

$$\mathbf{P} = \begin{bmatrix} 0 & 0.2 & 0.5 & 0.3 \\ 0.5 & 0 & 0.5 & 0 \\ 0.3 & 0.7 & 0 & 0 \\ 0.1 & 0 & 0 & 0.9 \end{bmatrix}.$$

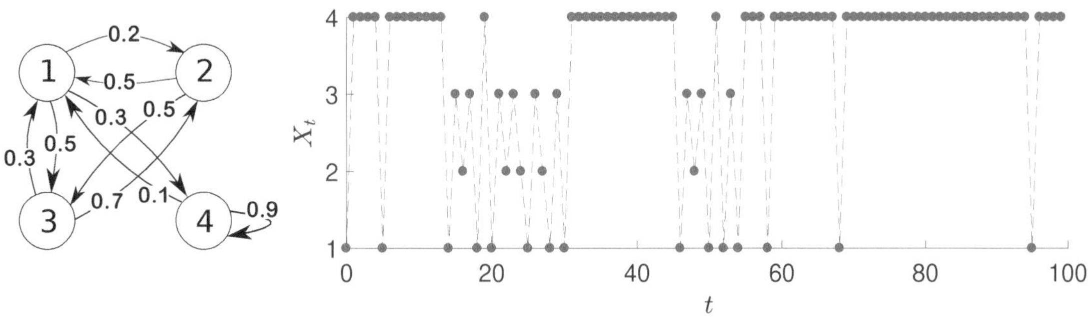

Figure 3.3: The transition graph (left) and a typical path (right) of the Markov chain.

In the same figure (on the right) a typical outcome (path) of the Markov chain is shown. The path was simulated using the Python program below. In this implementation the Markov chain always starts in state 1. We will revisit Markov chains, and in particular Markov chains with continuous state spaces, in Section 3.2.5.

☞ 78

MCsim.py

```
import numpy as np
import matplotlib.pyplot as plt

n = 101
P = np.array([[0, 0.2, 0.5, 0.3],
              [0.5, 0, 0.5, 0],
              [0.3, 0.7, 0, 0],
              [0.1, 0, 0, 0.9]])
x = np.array(np.ones(n, dtype=int))
x[0] = 0
for t in range(0,n-1):
```

```
    x[t+1] = np.min(np.where(np.cumsum(P[x[t],:]) >
                np.random.rand()))
x = x + 1  #add 1 to all elements of the vector x
plt.plot(np.array(range(0,n)),x, 'o')
plt.plot(np.array(range(0,n)),x, '--')
plt.show()
```

■

3.2.4 Resampling

RESAMPLING

The idea behind *resampling* is very simple: an iid sample $\tau := \{x_1, \ldots, x_n\}$ from some unknown cdf F represents our best knowledge of F if we make no further *a priori* assumptions about it. If it is not possible to simulate more samples from F, the best way to "repeat" the experiment is to *resample* from the original data by drawing from the empirical cdf F_n; see (1.2). That is, we draw each x_i with equal probability and repeat this N times, according to Algorithm 3.2.6 below. As we draw here "with replacement", multiple instances of the original data points may occur in the resampled data.

☞ 11

Algorithm 3.2.6: Sampling from an Empirical Cdf.

input: Original iid sample $x_1, \ldots, x_n$ and sample size N.
output: Iid sample $X_1^*, \ldots, X_N^*$ from the empirical cdf.
1 **for** $t = 1$ **to** N **do**
2   Draw $U \sim \mathcal{U}(0, 1)$
3   Set $I \leftarrow \lceil nU \rceil$
4   Set $X_t^* \leftarrow x_I$
5 **return** $X_1^*, \ldots, X_N^*$

In Step 3, $\lceil nU \rceil$ returns the *ceiling* of nU; that is, it is the smallest integer larger than or equal to nU. Consequently, I is drawn uniformly at random from the set of indices $\{1, \ldots, n\}$.

By sampling from the empirical cdf we can thus (approximately) repeat the experiment that gave us the original data as many times as we like. This is useful if we want to assess the properties of certain statistics obtained from the data. For example, suppose that the original data τ gave the statistic $t(\tau)$. By resampling we can gain information about the *distribution* of the corresponding random variable $t(\mathcal{T})$.

■ **Example 3.7 (Quotient of Uniforms)** Let $U_1, \ldots, U_n, V_1, \ldots, V_n$ be iid $\mathcal{U}(0, 1)$ random variables and define $X_i = U_i/V_i$, $i = 1, \ldots, n$. Suppose we wish to investigate the distribution of the sample median $\widetilde{X}$ and sample mean $\overline{X}$ of the (random) data $\mathcal{T} := \{X_1, \ldots, X_n\}$. Since we know the model for $\mathcal{T}$ exactly, we can generate a large number, N say, of independent copies of it, and for each of these copies evaluate the sample medians $\widetilde{X}_1, \ldots, \widetilde{X}_N$ and sample means $\overline{X}_1, \ldots, \overline{X}_N$. For $n = 100$ and $N = 1000$ the empirical cdfs might look like the left and right curves in Figure 3.4, respectively. Contrary to what you might have expected, the distributions of the sample median and sample mean do not match at all. The sample median is quite concentrated around 1, whereas the distribution of the sample mean is much more spread out.

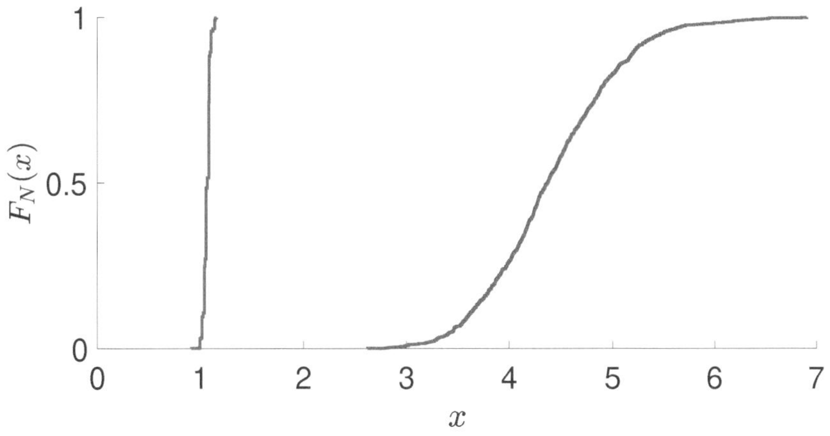

Figure 3.4: Empirical cdfs of the medians of the resampled data (left curve) and sample means (right curve) of the resampled data.

Instead of sampling completely new data, we could also *reuse* the original data by resampling them via Algorithm 3.2.6. This gives independent copies $\widetilde{X}_1^*, \ldots, \widetilde{X}_N^*$ and $\overline{X}_1^*, \ldots, \overline{X}_N^*$, for which we can again plot the empirical cdf. The results will be similar to the previous case. In fact, in Figure 3.4 the cdf of the *resampled* sample medians and sample means are plotted. The corresponding Python code is given below. The essential point of this example is that resampling of data can greatly add to the understanding of the probabilistic properties of certain measurements on the data, *even if the underlying model is not known*. See Exercise 12 for a further investigation of this example. ☞ 116

quotunif.py

```
import numpy as np
from numpy.random import rand, choice
import matplotlib.pyplot as plt
from statsmodels.distributions.empirical_distribution import ECDF

n = 100
N = 1000
x = rand(n)/rand(n)  # data
med = np.zeros(N)
ave = np.zeros(N)
for i in range(0,N):
    s = choice(x, n, replace=True) # resampled data
    med[i] = np.median(s)
    ave[i] = np.mean(s)

med_cdf = ECDF(med)
ave_cdf = ECDF(ave)
plt.plot(med_cdf.x, med_cdf.y)
plt.plot(ave_cdf.x, ave_cdf.y)
plt.show()
```

■

3.2.5 Markov Chain Monte Carlo

MARKOV CHAIN MONTE CARLO

TARGET

☞ 455

BURN-IN PERIOD

Markov chain Monte Carlo (MCMC) is a Monte Carlo sampling technique for (approximately) generating samples from an arbitrary distribution — often referred to as the *target* distribution. The basic idea is to run a Markov chain long enough such that its limiting distribution is close to the target distribution. Often such a Markov chain is constructed to be reversible, so that the detailed balance equations (C.43) can be used. Depending on the starting position of the Markov chain, the initial random variables in the Markov chain may have a distribution that is significantly different from the target (limiting) distribution. The random variables that are generated during this *burn-in period* are often discarded. The remaining random variables form an *approximate* and *dependent* sample from the target distribution.

In the next two sections we discuss two popular MCMC samplers: the Metropolis–Hastings sampler and the Gibbs sampler.

3.2.5.1 Metropolis–Hastings Sampler

☞ 72

PROPOSAL

ACCEPTANCE PROBABILITY

The Metropolis–Hastings sampler [87] is similar to the acceptance–rejection method in that it simulates a trial state, which is then accepted or rejected according to some random mechanism. Specifically, suppose we wish to sample from a target pdf $f(\boldsymbol{x})$, where $\boldsymbol{x}$ takes values in some d-dimensional set. The aim is to construct a Markov chain $\{\boldsymbol{X}_t, t = 0, 1, \ldots\}$ in such a way that its limiting pdf is f. Suppose the Markov chain is in state $\boldsymbol{x}$ at time t. A transition of the Markov chain from state $\boldsymbol{x}$ is carried out in two phases. First a *proposal* state $\boldsymbol{Y}$ is drawn from a transition density $q(\cdot \,|\, \boldsymbol{x})$. This state is accepted as the new state, with *acceptance probability*

$$\alpha(\boldsymbol{x}, \boldsymbol{y}) = \min\left\{\frac{f(\boldsymbol{y})\, q(\boldsymbol{x} \,|\, \boldsymbol{y})}{f(\boldsymbol{x})\, q(\boldsymbol{y} \,|\, \boldsymbol{x})}, 1\right\}, \tag{3.4}$$

or rejected otherwise. In the latter case the chain remains in state $\boldsymbol{x}$. The algorithm just described can be summarized as follows.

Algorithm 3.2.7: Metropolis–Hastings Sampler

input: Initial state $\boldsymbol{X}_0$, sample size N, target pdf $f(\boldsymbol{x})$, proposal function $q(\boldsymbol{y} \,|\, \boldsymbol{x})$.
output: $\boldsymbol{X}_1, \ldots, \boldsymbol{X}_N$ (dependent), approximately distributed according to $f(\boldsymbol{x})$.

for $t = 0$ **to** $N - 1$ **do**
    Draw $\boldsymbol{Y} \sim q(\boldsymbol{y} \,|\, \boldsymbol{X}_t)$ `// draw a proposal`
    $\alpha \leftarrow \alpha(\boldsymbol{X}_t, \boldsymbol{Y})$ `// acceptance probability as in (3.4)`
    Draw $U \sim \mathcal{U}(0, 1)$
    if $U \leqslant \alpha$ **then** $\boldsymbol{X}_{t+1} \leftarrow \boldsymbol{Y}$
    else $\boldsymbol{X}_{t+1} \leftarrow \boldsymbol{X}_t$
return $\boldsymbol{X}_1, \ldots, \boldsymbol{X}_N$

The fact that the limiting distribution of the Metropolis–Hastings Markov chain is equal to the target distribution (under general conditions) is a consequence of the following result.

Theorem 3.1: Local Balance for the Metropolis–Hastings Sampler

The transition density of the Metropolis–Hastings Markov chain satisfies the detailed balance equations. ☞ 455

Proof: We prove the theorem for the discrete case only. Because a transition of the Metropolis–Hastings Markov chain consists of two steps, the one-step transition probability to go from $\boldsymbol{x}$ to $\boldsymbol{y}$ is not $q(\boldsymbol{y} \,|\, \boldsymbol{x})$ but

$$\widetilde{q}(\boldsymbol{y} \,|\, \boldsymbol{x}) = \begin{cases} q(\boldsymbol{y} \,|\, \boldsymbol{x})\, \alpha(\boldsymbol{x}, \boldsymbol{y}), & \text{if } \boldsymbol{y} \neq \boldsymbol{x}, \\ 1 - \sum_{\boldsymbol{z} \neq \boldsymbol{x}} q(\boldsymbol{z} \,|\, \boldsymbol{x})\, \alpha(\boldsymbol{x}, \boldsymbol{z}), & \text{if } \boldsymbol{y} = \boldsymbol{x}. \end{cases} \tag{3.5}$$

We thus need to show that

$$f(\boldsymbol{x})\, \widetilde{q}(\boldsymbol{y} \,|\, \boldsymbol{x}) = f(\boldsymbol{y})\, \widetilde{q}(\boldsymbol{x} \,|\, \boldsymbol{y}) \quad \text{for all } \boldsymbol{x}, \boldsymbol{y}. \tag{3.6}$$

With the acceptance probability as in (3.4), we need to check (3.6) for three cases:

(a) $\boldsymbol{x} = \boldsymbol{y}$,

(b) $\boldsymbol{x} \neq \boldsymbol{y}$ and $f(\boldsymbol{y})q(\boldsymbol{x} \,|\, \boldsymbol{y}) \leqslant f(\boldsymbol{x})q(\boldsymbol{y} \,|\, \boldsymbol{x})$, and

(c) $\boldsymbol{x} \neq \boldsymbol{y}$ and $f(\boldsymbol{y})q(\boldsymbol{x} \,|\, \boldsymbol{y}) > f(\boldsymbol{x})q(\boldsymbol{y} \,|\, \boldsymbol{x})$.

Case (a) holds trivially. For case (b), $\alpha(\boldsymbol{x}, \boldsymbol{y}) = f(\boldsymbol{y})q(\boldsymbol{x} \,|\, \boldsymbol{y})/(f(\boldsymbol{x})q(\boldsymbol{y} \,|\, \boldsymbol{x}))$ and $\alpha(\boldsymbol{y}, \boldsymbol{x}) = 1$. Consequently,

$$\widetilde{q}(\boldsymbol{y} \,|\, \boldsymbol{x}) = f(\boldsymbol{y})q(\boldsymbol{x} \,|\, \boldsymbol{y})/f(\boldsymbol{x}) \quad \text{and} \quad \widetilde{q}(\boldsymbol{x} \,|\, \boldsymbol{y}) = q(\boldsymbol{x} \,|\, \boldsymbol{y}),$$

so that (3.6) holds. Similarly, for case (c) we have $\alpha(\boldsymbol{x}, \boldsymbol{y}) = 1$ and $\alpha(\boldsymbol{y}, \boldsymbol{x}) = f(\boldsymbol{x})q(\boldsymbol{y} \,|\, \boldsymbol{x})/ (f(\boldsymbol{y})q(\boldsymbol{x} \,|\, \boldsymbol{y}))$. It follows that,

$$\widetilde{q}(\boldsymbol{y} \,|\, \boldsymbol{x}) = q(\boldsymbol{y} \,|\, \boldsymbol{x}) \quad \text{and} \quad \widetilde{q}(\boldsymbol{x} \,|\, \boldsymbol{y}) = f(\boldsymbol{x})q(\boldsymbol{y} \,|\, \boldsymbol{x})/f(\boldsymbol{y}),$$

so that (3.6) holds again. □

Thus if the Metropolis–Hastings Markov chain is ergodic, then its limiting pdf is $f(\boldsymbol{x})$. ☞ 454 A fortunate property of the algorithm, which is important in many applications, is that in order to evaluate the acceptance probability $\alpha(\boldsymbol{x}, \boldsymbol{y})$ in (3.4), one only needs to know the target pdf $f(\boldsymbol{x})$ *up to a constant*; that is $f(\boldsymbol{x}) = c\,\overline{f}(\boldsymbol{x})$ for some known function $\overline{f}(\boldsymbol{x})$ but unknown constant c.

The efficiency of the algorithm depends of course on the choice of the proposal transition density $q(\boldsymbol{y} \,|\, \boldsymbol{x})$. Ideally, we would like $q(\boldsymbol{y} \,|\, \boldsymbol{x})$ to be "close" to the target $f(\boldsymbol{y})$, irrespective of $\boldsymbol{x}$. We discuss two common approaches.

1. Choose the proposal transition density $q(\boldsymbol{y} \,|\, \boldsymbol{x})$ independent of $\boldsymbol{x}$; that is, $q(\boldsymbol{y} \,|\, \boldsymbol{x}) = g(\boldsymbol{y})$ for some pdf $g(\boldsymbol{y})$. An MCMC sampler of this type is called an *independence sampler*. The acceptance probability is thus

INDEPENDENCE SAMPLER

$$\alpha(\boldsymbol{x}, \boldsymbol{y}) = \min\left\{\frac{f(\boldsymbol{y})\, g(\boldsymbol{x})}{f(\boldsymbol{x})\, g(\boldsymbol{y})},\ 1\right\}.$$

2. If the proposal transition density is symmetric (that is, $q(\boldsymbol{y} \mid \boldsymbol{x}) = q(\boldsymbol{x} \mid \boldsymbol{y})$), then the acceptance probability has the simple form

$$\alpha(\boldsymbol{x}, \boldsymbol{y}) = \min\left\{\frac{f(\boldsymbol{y})}{f(\boldsymbol{x})}, 1\right\}, \tag{3.7}$$

RANDOM WALK SAMPLER

and the MCMC algorithm is called a *random walk sampler*. A typical example is when, for a given current state $\boldsymbol{x}$, the proposal state $\boldsymbol{Y}$ is of the form $\boldsymbol{Y} = \boldsymbol{x} + \boldsymbol{Z}$, where $\boldsymbol{Z}$ is generated from some spherically symmetric distribution, such as $\mathcal{N}(\mathbf{0}, \mathbf{I})$.

We now give an example illustrating the second approach.

■ **Example 3.8 (Random Walk Sampler)** Consider the two-dimensional pdf

$$f(x_1, x_2) = c\, \mathrm{e}^{-\frac{1}{4}\sqrt{x_1^2+x_2^2}}\left(\sin\left(2\sqrt{x_1^2+x_2^2}\right)+1\right), \quad -2\pi < x_1 < 2\pi,\ -2\pi < x_2 < 2\pi, \tag{3.8}$$

where c is an unknown normalization constant. The graph of this pdf (unnormalized) is depicted in the left panel of Figure 3.5.

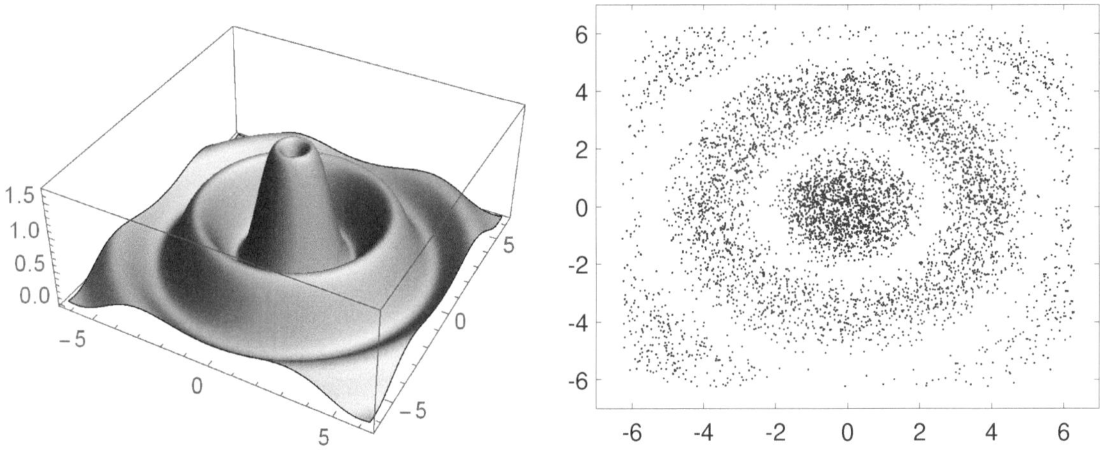

Figure 3.5: Left panel: the two-dimensional target pdf. Right panel: points from the random walk sampler are approximately distributed according to the target pdf.

The following Python program implements a random walk sampler to (approximately) draw $N = 10^4$ dependent samples from the pdf f. At each step, given a current state $\boldsymbol{x}$, a proposal $\boldsymbol{Y}$ is drawn from the $\mathcal{N}(\boldsymbol{x}, \mathbf{I})$ distribution. That is, $\boldsymbol{Y} = \boldsymbol{x} + \boldsymbol{Z}$, with $\boldsymbol{Z}$ bivariate standard normal. We see in the right panel of Figure 3.5 that the sampler works correctly. The starting point for the Markov chain is chosen as $(0, 0)$. Note that the normalization constant c is never required to be specified in the program.

rwsamp.py

```
import numpy as np
import matplotlib.pyplot as plt
from numpy import pi, exp, sqrt, sin
from numpy.random import rand, randn
```

```
N = 10000
a = lambda x: -2*pi < x
b = lambda x: x < 2*pi
f = lambda x1, x2: exp(-sqrt(x1**2+x2**2)/4)*(
        sin(2*sqrt(x1**2+x2**2))+1)*a(x1)*b(x1)*a(x2)*b(x2)

xx = np.zeros((N,2))
x = np.zeros((1,2))
for i in range(1,N):
    y = x + randn(1,2)
    alpha = np.amin((f(y[0][0],y[0][1])/f(x[0][0],x[0][1]),1))
    r = rand() < alpha
    x = r*y + (1-r)*x
    xx[i,:] = x

plt.scatter(xx[:,0], xx[:,1], alpha =0.4,s =2)
plt.axis('equal')
plt.show()
```

■

3.2.5.2 Gibbs Sampler

The *Gibbs sampler* [48] uses a somewhat different methodology from the Metropolis–Hastings algorithm and is particularly useful for generating n-dimensional random vectors. The key idea of the Gibbs sampler is to update the components of the random vector one at a time, by sampling them from *conditional* pdfs. Thus, Gibbs sampling can be advantageous if it is easier to sample from the conditional distributions than from the joint distribution.

GIBBS SAMPLER

Specifically, suppose that we wish to sample a random vector $\boldsymbol{X} = [X_1, \ldots, X_n]^\top$ according to a target pdf $f(\boldsymbol{x})$. Let $f(x_i \,|\, x_1, \ldots, x_{i-1}, x_{i+1}, \ldots, x_n)$ represent the conditional pdf[2] of the i-th component, X_i, given the other components $x_1, \ldots, x_{i-1}, x_{i+1}, \ldots, x_n$. The Gibbs sampling algorithm is as follows.

Algorithm 3.2.8: Gibbs Sampler

---

input: Initial point $\boldsymbol{X}_0$, sample size N, and target pdf f.
output: $\boldsymbol{X}_1, \ldots, \boldsymbol{X}_N$ approximately distributed according to f.

```
1 for t = 0 to N − 1 do
2     Draw Y_1 from the conditional pdf f(y_1 | X_{t,2}, ..., X_{t,n}).
3     for i = 2 to n do
4         Draw Y_i from the conditional pdf f(y_i | Y_1, ..., Y_{i−1}, X_{t,i+1}, ..., X_{t,n}).
5     X_{t+1} ← Y
6 return X_1, ..., X_N
```

There exist many variants of the Gibbs sampler, depending on the steps required to update $\boldsymbol{X}_t$ to $\boldsymbol{X}_{t+1}$ — called the *cycle* of the Gibbs algorithm. In the algorithm above, the

CYCLE

[2] In this section we employ a Bayesian notation style, using the same letter f for different (conditional) densities.

cycle consists of Steps 2–5, in which the components are updated in a fixed order $1 \to 2 \to \cdots \to n$. For this reason Algorithm 3.2.8 is also called the *systematic Gibbs sampler*.

SYSTEMATIC GIBBS SAMPLER

RANDOM-ORDER GIBBS SAMPLER

☞ 115

In the *random-order Gibbs sampler*, the order in which the components are updated in each cycle is a random permutation of $\{1, \ldots, n\}$ (see Exercise 9). Other modifications are to update the components in blocks (i.e., several at the same time), or to update only a random selection of components. The variant where in each cycle only a single random component is updated is called the *random Gibbs sampler*. In the *reversible Gibbs sampler* a cycle consists of the coordinate-wise updating $1 \to 2 \to \cdots \to n-1 \to n \to n-1 \to \cdots \to 2 \to 1$. In all cases, except for the systematic Gibbs sampler, the resulting Markov chain $\{\boldsymbol{X}_t, t = 1, 2, \ldots\}$ is *reversible* and hence its limiting distribution is precisely $f(\boldsymbol{x})$.

RANDOM GIBBS SAMPLER

REVERSIBLE GIBBS SAMPLER

☞ 454

Unfortunately, the systematic Gibbs Markov chain is not reversible and so the detailed balance equations are not satisfied. However, a similar result holds, due to Hammersley and Clifford, under the so-called *positivity condition*: if at a point $\boldsymbol{x} = (x_1, \ldots, x_n)$ all marginal densities $f(x_i) > 0, i = 1, \ldots, n$, then the joint density $f(\boldsymbol{x}) > 0$.

Theorem 3.2: Hammersley–Clifford Balance for the Gibbs Sampler

Let $q_{1\to n}(\boldsymbol{y} \,|\, \boldsymbol{x})$ denote the transition density of the systematic Gibbs sampler, and let $q_{n\to 1}(\boldsymbol{x} \,|\, \boldsymbol{y})$ be the transition density of the reverse move, in the order $n \to n-1 \to \cdots \to 1$. Then, if the positivity condition holds,

$$f(\boldsymbol{x})\, q_{1\to n}(\boldsymbol{y} \,|\, \boldsymbol{x}) = f(\boldsymbol{y})\, q_{n\to 1}(\boldsymbol{x} \,|\, \boldsymbol{y}). \tag{3.9}$$

Proof: For the forward move we have:

$$q_{1\to n}(\boldsymbol{y} \,|\, \boldsymbol{x}) = f(y_1 \,|\, x_2, \ldots, x_n) f(y_2 \,|\, y_1, x_3, \ldots, x_n) \cdots f(y_n \,|\, y_1, \ldots, y_{n-1}),$$

and for the reverse move:

$$q_{n\to 1}(\boldsymbol{x} \,|\, \boldsymbol{y}) = f(x_n \,|\, y_1, \ldots, y_{n-1}) f(x_{n-1} \,|\, y_1, \ldots, y_{n-2}, x_n) \cdots f(x_1 \,|\, x_2, \ldots, x_n).$$

Consequently,

$$\begin{aligned}
\frac{q_{1\to n}(\boldsymbol{y} \,|\, \boldsymbol{x})}{q_{n\to 1}(\boldsymbol{x} \,|\, \boldsymbol{y})} &= \prod_{i=1}^{n} \frac{f(y_i \,|\, y_1, \ldots, y_{i-1}, x_{i+1}, \ldots, x_n)}{f(x_i \,|\, y_1, \ldots, y_{i-1}, x_{i+1}, \ldots, x_n)} \\
&= \prod_{i=1}^{n} \frac{f(y_1, \ldots, y_i, x_{i+1}, \ldots, x_n)}{f(y_1, \ldots, y_{i-1}, x_i, \ldots, x_n)} \\
&= \frac{f(\boldsymbol{y}) \prod_{i=1}^{n-1} f(y_1, \ldots, y_i, x_{i+1}, \ldots, x_n)}{f(\boldsymbol{x}) \prod_{j=2}^{n} f(y_1, \ldots, y_{j-1}, x_j, \ldots, x_n)} \\
&= \frac{f(\boldsymbol{y}) \prod_{i=1}^{n-1} f(y_1, \ldots, y_i, x_{i+1}, \ldots, x_n)}{f(\boldsymbol{x}) \prod_{j=1}^{n-1} f(y_1, \ldots, y_j, x_{j+1}, \ldots, x_n)} = \frac{f(\boldsymbol{y})}{f(\boldsymbol{x})}.
\end{aligned}$$

The result follows by rearranging the last identity. The positivity condition ensures that we do not divide by 0 along the line. □

Intuitively, the long-run proportion of transitions $\boldsymbol{x} \to \boldsymbol{y}$ for the "forward move" chain is equal to the long-run proportion of transitions $\boldsymbol{y} \to \boldsymbol{x}$ for the "reverse move" chain.

To verify that the Markov chain $X_0, X_1, \ldots$ for the systematic Gibbs sampler indeed has limiting pdf $f(x)$, we need to check that the global balance equations (C.42) hold. By integrating (in the continuous case) both sides in (3.9) with respect to x, we see that indeed ☞ 454

$$\int f(\boldsymbol{x})\, q_{1\to n}(\boldsymbol{y} \mid \boldsymbol{x})\, \mathrm{d}\boldsymbol{x} = f(\boldsymbol{y}).$$

■ **Example 3.9 (Gibbs Sampler for the Bayesian Normal Model)** Gibbs samplers are often applied in Bayesian statistics, to sample from the posterior pdf. Consider for instance the Bayesian normal model ☞ 50

$$f(\mu, \sigma^2) = 1/\sigma^2$$
$$(\boldsymbol{x} \mid \mu, \sigma^2) \sim \mathcal{N}(\mu\mathbf{1}, \sigma^2\mathbf{I}).$$

Here the prior for (μ, σ^2) is *improper*. That is, it is not a pdf in itself, but by obstinately applying Bayes' formula it does yield a proper posterior pdf. In some sense this prior conveys the least amount of information about μ and σ^2. Following the same procedure as in Example 2.8, we find the posterior pdf: IMPROPER PRIOR

$$f(\mu, \sigma^2 \mid \boldsymbol{x}) \propto \left(\sigma^2\right)^{-n/2-1} \exp\left\{-\frac{1}{2}\frac{\sum_i (x_i - \mu)^2}{\sigma^2}\right\}. \tag{3.10}$$

Note that μ and σ^2 here are the "variables" and $\boldsymbol{x}$ is a fixed data vector. To simulate samples μ and σ^2 from (3.10) using the Gibbs sampler, we need the distributions of both $(\mu \mid \sigma^2, \boldsymbol{x})$ and $(\sigma^2 \mid \mu, \boldsymbol{x})$. To find $f(\mu \mid \sigma^2, \boldsymbol{x})$, view the right-hand side of (3.10) as a function of μ only, regarding σ^2 as a constant. This gives

$$f(\mu \mid \sigma^2, \boldsymbol{x}) \propto \exp\left\{-\frac{n\mu^2 - 2\mu \sum_i x_i}{2\sigma^2}\right\} = \exp\left\{-\frac{\mu^2 - 2\mu\overline{x}}{2(\sigma^2/n)}\right\}$$
$$\propto \exp\left\{-\frac{1}{2}\frac{(\mu - \overline{x})^2}{\sigma^2/n}\right\}. \tag{3.11}$$

This shows that $(\mu \mid \sigma^2, \boldsymbol{x})$ has a normal distribution with mean $\overline{x}$ and variance σ^2/n.

Similarly, to find $f(\sigma^2 \mid \mu, \boldsymbol{x})$, view the right-hand side of (3.10) as a function of σ^2, regarding μ as a constant. This gives

$$f(\sigma^2 \mid \mu, \boldsymbol{x}) \propto (\sigma^2)^{-n/2-1} \exp\left\{-\frac{1}{2}\sum_{i=1}^{n}(x_i - \mu)^2/\sigma^2\right\}, \tag{3.12}$$

showing that $(\sigma^2 \mid \mu, \boldsymbol{x})$ has an inverse-gamma distribution with parameters $n/2$ and $\sum_{i=1}^{n}(x_i - \mu)^2/2$. The Gibbs sampler thus involves the repeated simulation of ☞ 427

$$(\mu \mid \sigma^2, \boldsymbol{x}) \sim \mathcal{N}\left(\overline{x},\, \sigma^2/n\right) \quad \text{and} \quad (\sigma^2 \mid \mu, \boldsymbol{x}) \sim \mathsf{InvGamma}\left(n/2,\, \sum_{i=1}^{n}(x_i - \mu)^2/2\right).$$

Simulating $X \sim \mathsf{InvGamma}(\alpha, \lambda)$ is achieved by first generating $Z \sim \mathsf{Gamma}(\alpha, \lambda)$ and then returning $X = 1/Z$.

> In our parameterization of the $\mathsf{Gamma}(\alpha, \lambda)$ distribution, λ is the *rate* parameter. Many software packages instead use the *scale* parameter $c = 1/\lambda$. Be aware of this when simulating Gamma random variables.

The Python script below defines a small data set of size $n = 10$ (which was randomly simulated from a standard normal distribution), and implements the systematic Gibbs sampler to simulate from the posterior distribution, using $N = 10^5$ samples.

gibbsamp.py

```
import numpy as np
import matplotlib.pyplot as plt

x = np.array([[-0.9472, 0.5401, -0.2166, 1.1890, 1.3170,
               -0.4056, -0.4449, 1.3284, 0.8338, 0.6044]])
n=x.size
sample_mean = np.mean(x)
sample_var = np.var(x)
sig2 = np.var(x)
mu=sample_mean

N=10**5
gibbs_sample = np.array(np.zeros((N, 2)))
for k in range(N):
    mu=sample_mean + np.sqrt(sig2/n)*np.random.randn()
    V=np..sum((x-mu)**2)/2
    sig2 = 1/np.random.gamma(n/2, 1/V)
    gibbs_sample[k,:]= np.array([mu, sig2])
plt.scatter(gibbs_sample[:,0], gibbs_sample[:,1],alpha =0.1,s =1)
plt.plot(np.mean(x), np.var(x),'wo')
plt.show()
```

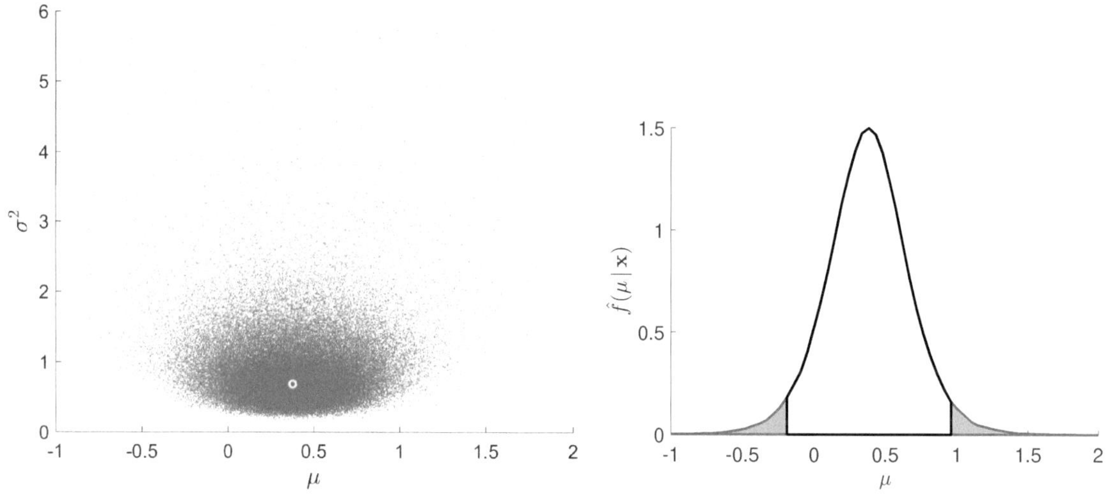

Figure 3.6: Left: approximate draws from the posterior pdf $f(\mu, \sigma^2 \mid \boldsymbol{x})$ obtained via the Gibbs sampler. Right: estimate of the posterior pdf $f(\mu \mid \boldsymbol{x})$.

The left panel of Figure 3.6 shows the (μ, σ^2) points generated by the Gibbs sampler. Also shown, via the white circle, is the point $(\overline{x}, s^2)$, where $\overline{x} = 0.3798$ is the sample mean and $s^2 = 0.6810$ the sample variance. This posterior point cloud visualizes the considerable uncertainty in the estimates. By projecting the (μ, σ^2) points onto the μ-axis — that is, by ignoring the σ^2 values — one obtains (approximate) samples from the posterior pdf of μ; that is, $f(\mu \,|\, \boldsymbol{x})$. The right panel of Figure 3.6 shows a kernel density estimate (see Section 4.4) of this pdf. The corresponding 0.025 and 0.975 sample quantiles were found to be -0.2054 and 0.9662, respectively, giving the 95% credible interval $(-0.2054, 0.9662)$ for μ, which contains the true expectation 0. Similarly, an estimated 95% credible interval for σ^2 is $(0.3218, 2.2485)$, which contains the true variance 1.

☞ 134

3.3 Monte Carlo Estimation

In this section we describe how Monte Carlo simulation can be used to estimate complicated integrals, probabilities, and expectations. A number of variance reduction techniques are introduced as well, including the recent cross-entropy method.

3.3.1 Crude Monte Carlo

The most common setting for Monte Carlo estimation is the following: Suppose we wish to compute the expectation $\mu = \mathbb{E}Y$ of some (say continuous) random variable Y with pdf f, but the integral $\mathbb{E}Y = \int y f(y)\, \mathrm{d}y$ is difficult to evaluate. For example, if Y is a complicated function of other random variables, it would be difficult to obtain an exact expression for $f(y)$. The idea of *crude Monte Carlo* — sometimes abbreviated as CMC — is to approximate μ by simulating many independent copies $Y_1, \ldots, Y_N$ of Y and then take their *sample mean* $\overline{Y}$ as an estimator of μ. All that is needed is an algorithm to simulate such copies.

CRUDE MONTE CARLO

By the Law of Large Numbers, $\overline{Y}$ converges to μ as $N \to \infty$, provided the expectation of Y exists. Moreover, by the Central Limit Theorem, $\overline{Y}$ approximately has a $\mathcal{N}(\mu, \sigma^2/N)$ distribution for large N, provided that the variance $\sigma^2 = \mathbb{V}\mathrm{ar}Y < \infty$. This enables the construction of an approximate $(1 - \alpha)$ *confidence interval* for μ:

☞ 448
☞ 449

CONFIDENCE INTERVAL

$$\left(\overline{Y} - z_{1-\alpha/2} \frac{S}{\sqrt{N}}, \quad \overline{Y} + z_{1-\alpha/2} \frac{S}{\sqrt{N}} \right), \tag{3.13}$$

where S is the sample standard deviation of the $\{Y_i\}$ and z_γ denotes the γ-quantile of the $\mathcal{N}(0, 1)$ distribution; see also Section C.13. Instead of specifying the confidence interval, one often reports only the sample mean and the *estimated standard error*: $S/\sqrt{N}$, or the *estimated relative error*: $S/(\overline{Y}\sqrt{N})$. The basic estimation procedure for independent data is summarized in Algorithm 3.3.1 below.

☞ 459

ESTIMATED STANDARD ERROR

ESTIMATED RELATIVE ERROR

It is often the case that the output Y is a function of some underlying random vector or stochastic process; that is, $Y = H(\boldsymbol{X})$, where H is a real-valued function and $\boldsymbol{X}$ is a random vector or process. The beauty of Monte Carlo for estimation is that (3.13) holds regardless of the dimension of $\boldsymbol{X}$.

Algorithm 3.3.1: Crude Monte Carlo for Independent Data

input: Simulation algorithm for $Y \sim f$, sample size N, confidence level $1 - \alpha$.
output: Point estimate and approximate $(1 - \alpha)$ confidence interval for $\mu = \mathbb{E}Y$.

1. Simulate $Y_1, \ldots, Y_N \overset{\text{iid}}{\sim} f$.
2. $\overline{Y} \leftarrow \frac{1}{N} \sum_{i=1}^{N} Y_i$
3. $S^2 \leftarrow \frac{1}{N-1} \sum_{i=1}^{N} (Y_i - \overline{Y})^2$
4. **return** $\overline{Y}$ and the interval (3.13).

MONTE CARLO INTEGRATION

■ **Example 3.10 (Monte Carlo Integration)** In *Monte Carlo integration*, simulation is used to evaluate complicated integrals. Consider, for example, the integral

$$\mu = \int_{-\infty}^{\infty} \int_{-\infty}^{\infty} \int_{-\infty}^{\infty} \sqrt{|x_1 + x_2 + x_3|}\, \mathrm{e}^{-(x_1^2+x_2^2+x_3^2)/2}\, \mathrm{d}x_1\, \mathrm{d}x_2\, \mathrm{d}x_3.$$

Defining $Y = |X_1 + X_2 + X_3|^{1/2}(2\pi)^{3/2}$, with $X_1, X_2, X_3 \overset{\text{iid}}{\sim} \mathcal{N}(0, 1)$, we can write $\mu = \mathbb{E}Y$. Using the following Python program, with a sample size of $N = 10^6$, we obtained an estimate $\overline{Y} = 17.031$ with an approximate 95% confidence interval (17.017, 17.046).

mcint.py

```
import numpy as np
from numpy import pi

c = (2*pi)**(3/2)
H = lambda x: c*np.sqrt(np.abs(np.sum(x,axis=1)))
N = 10**6
z = 1.96
x = np.random.randn(N,3)
y = H(x)
mY = np.mean(y)
sY = np.std(y)
RE = sY/mY/np.sqrt(N)
print('Estimate = {:3.3f}, CI = ({:3.3f},{:3.3f})'.format(
        mY, mY*(1-z*RE), mY*(1+z*RE)))
```

```
Estimate = 17.031, CI = (17.017,17.046)
```

■

☞ 26
☞ 29
☞ 23

■ **Example 3.11 (Example 2.1 (cont.))** We return to the bias–variance tradeoff in Example 2.1. Figure 2.7 gives estimates of the (squared-error) generalization risk (2.5) as a function of the number of parameters in the model. But how accurate are these estimates? Because we know in this case the exact model for the data, we can use Monte Carlo simulation to estimate the generalization risk (for a fixed training set) and the expected generalization risk (averaged over all training sets) precisely. All we need to do is repeat the data generation, fitting, and validation steps many times and then take averages of the results. The following Python code repeats 100 times:

1. Simulate the training set of size $n = 100$.
2. Fit models up to size $k = 8$.

3. Estimate the test loss using a test set with the same sample size $n = 100$.

Figure 3.7 shows that there is some variation in the test losses, due to the randomness in both the training and test sets. To obtain an accurate estimate of the expected generalization risk (2.6), take the average of the test losses. We see that for $k \leqslant 8$ the estimate in Figure 2.7 is close to the true expected generalization risk.

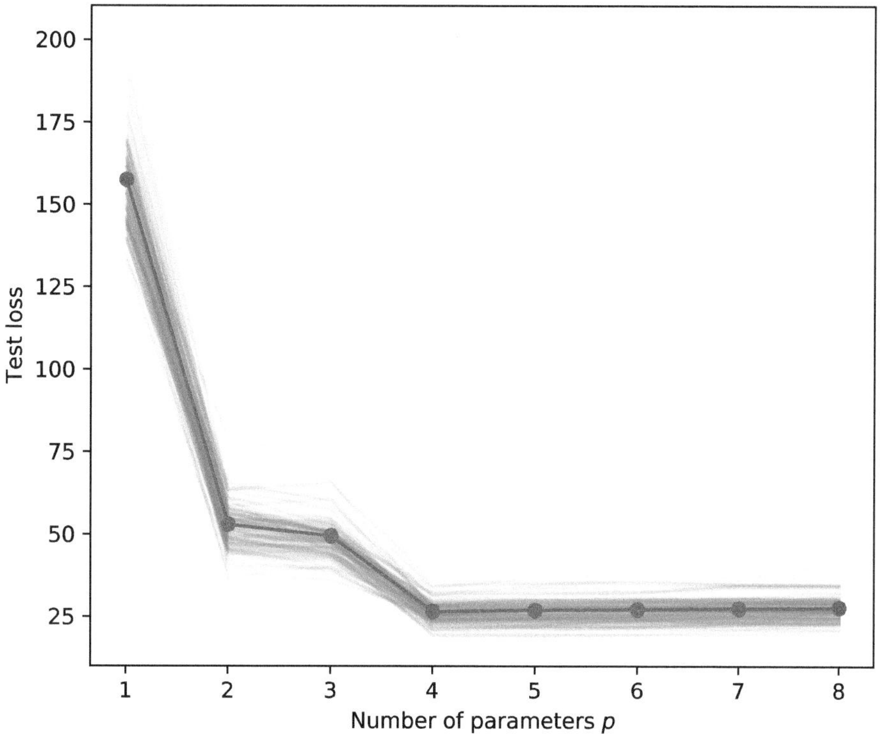

Figure 3.7: Independent estimates of the test loss show some variability.

CMCtestloss.py

```
import numpy as np, matplotlib.pyplot as plt
from numpy.random import rand, randn
from numpy.linalg import solve

def generate_data(beta, sig, n):
    u = rand(n, 1)
    y = (u ** np.arange(0, 4)) @ beta + sig * randn(n, 1)
    return u, y

beta = np.array([[10, -140, 400, -250]]).T
n = 100
sig = 5
betahat = {}
plt.figure(figsize=[6,5])
totMSE = np.zeros(8)
max_p = 8
p_range = np.arange(1, max_p + 1, 1)

for N in range(0,100):
```

```
    u, y = generate_data(beta, sig, n)  #training data
    X = np.ones((n, 1))
    for p in p_range:
       if p > 1:
         X = np.hstack((X, u**(p-1)))
       betahat[p] = solve(X.T @ X, X.T @ y)

    u_test, y_test = generate_data(beta, sig, n)  #test data
    MSE = []
    X_test = np.ones((n, 1))
    for p in p_range:
       if p > 1:
           X_test = np.hstack((X_test, u_test**(p-1)))
       y_hat = X_test @ betahat[p] # predictions
       MSE.append(np.sum((y_test - y_hat)**2/n))

    totMSE = totMSE + np.array(MSE)
    plt.plot(p_range, MSE,'C0',alpha=0.1)

plt.plot(p_range,totMSE/N,'r-o')
plt.xticks(ticks=p_range)
plt.xlabel('Number of parameters $p$')
plt.ylabel('Test loss')
plt.tight_layout()
plt.savefig('MSErepeatpy.pdf',format='pdf')
plt.show()
```

■

3.3.2 Bootstrap Method

☞ 76

The bootstrap method [37] combines CMC estimation with the resampling procedure of Section 3.2.4. The idea is as follows: Suppose we wish to estimate a number μ via some estimator $Y = H(\mathcal{T})$, where $\mathcal{T} := \{X_1, \ldots, X_n\}$ is an iid sample from some unknown cdf F. It is assumed that Y does not depend on the order of the $\{X_i\}$. To assess the quality (for example, accuracy) of the estimator Y, one could draw independent replications $\mathcal{T}_1, \ldots, \mathcal{T}_N$ of $\mathcal{T}$ and find sample estimates for quantities such as the variance $\mathbb{V}\mathrm{ar}Y$, the bias $\mathbb{E}Y - \mu$, and the mean squared error $\mathbb{E}(Y - \mu)^2$. However, it may be too time-consuming or simply not feasible to obtain such replications. An alternative is to *resample* the original data. To reiterate, given an outcome $\tau = \{x_1, \ldots, x_n\}$ of $\mathcal{T}$, we simulate an iid sample $\mathcal{T}^* := \{X_1^*, \ldots, X_n^*\}$ from the empirical cdf F_n, via Algorithm 3.2.6 (hence the resampling size is $N = n$ here).

☞ 76

The rationale is that the empirical cdf F_n is close to the actual cdf F and gets closer as n gets larger. Hence, any quantities depending on F, such as $\mathbb{E}_F g(Y)$, where g is a function, can be approximated by $\mathbb{E}_{F_n} g(Y)$. The latter is usually still difficult to evaluate, but it can be simply estimated via CMC as

$$\frac{1}{K}\sum_{i=1}^{K} g(Y_i^*),$$

where $Y_1^*, \ldots, Y_K^*$ are independent random variables, each distributed as $Y^* = H(\mathcal{T}^*)$. This seemingly self-referent procedure is called *bootstrapping* — alluding to Baron von Mün-

BOOTSTRAPPING

chausen, who pulled himself out of a swamp by his own bootstraps. As an example, the bootstrap estimate of the expectation of Y is

$$\widehat{\mathbb{E}Y} = \overline{Y}^* = \frac{1}{K}\sum_{i=1}^{K} Y_i^*,$$

which is simply the sample mean of $\{Y_i^*\}$. Similarly, the bootstrap estimate for $\mathbb{V}\mathrm{ar}Y$ is the sample variance

$$\widehat{\mathbb{V}\mathrm{ar}Y} = \frac{1}{K-1}\sum_{i=1}^{K}(Y_i^* - \overline{Y}^*)^2. \tag{3.14}$$

Bootstrap estimators for the bias and MSE are $\overline{Y}^* - Y$ and $\frac{1}{K}\sum_{i=1}^{K}(Y_i^* - Y)^2$, respectively. Note that for these estimators the unknown quantity μ is replaced with its original estimator Y. Confidence intervals can be constructed in the same fashion. We mention two variants: the *normal method* and the *percentile method*. In the normal method, a $1-\alpha$ confidence interval for μ is given by

NORMAL METHOD

PERCENTILE METHOD

$$(Y \pm z_{1-\alpha/2}S^*),$$

where S^* is the bootstrap estimate of the standard deviation of Y; that is, the square root of (3.14). In the percentile method, the upper and lower bounds of the $1-\alpha$ confidence interval for μ are given by the $1-\alpha/2$ and $\alpha/2$ quantiles of Y, which in turn are estimated via the corresponding sample quantiles of the bootstrap sample $\{Y_i^*\}$.

The following example illustrates the usefulness of the bootstrap method for *ratio estimation* and also introduces the *renewal reward process* model for data.

■ **Example 3.12 (Bootstrapping the Ratio Estimator)** A common scenario in stochastic simulation is that the output of the simulation consists of independent pairs of data $(C_1, R_1), (C_2, R_2), \ldots$, where each C is interpreted as the length of a period of time — a so-called *cycle* — and R is the *reward* obtained during that cycle. Such a collection of random variables $\{(C_i, R_i)\}$ is called a *renewal reward process*. Typically, the reward R_i depends on the cycle length C_i. Let A_t be the *average reward* earned by time t; that is, $A_t = \sum_{i=1}^{N_t} R_i/t$, where $N_t = \max\{n : C_1 + \cdots + C_n \leqslant t\}$ counts the number of complete cycles at time t. It can be shown, see Exercise 20, that if the expectations of the cycle length and reward are finite, then A_t converges to the constant $\mathbb{E}R/\mathbb{E}C$. This ratio can thus be interpreted as the *long-run average reward*.

RENEWAL REWARD PROCESS

☞ 118

LONG-RUN AVERAGE REWARD

Estimation of the ratio $\mathbb{E}R/\mathbb{E}C$ from data $(C_1, R_1), \ldots, (C_n, R_n)$ is easy: take the *ratio estimator*

RATIO ESTIMATOR

$$A = \frac{\overline{R}}{\overline{C}}.$$

However, this estimator A is not unbiased and it is not obvious how to derive confidence intervals. Fortunately, the bootstrap method can come to the rescue: simply resample the pairs $\{(C_i, R_i)\}$, obtain ratio estimators $A_1^*, \ldots, A_K^*$, and from these compute quantities of interest such as confidence intervals.

As a concrete example, let us return to the Markov chain in Example 3.6. Recall that the chain starts at state 1 at time 0. After a certain amount of time T_1, the process returns to state 1. The time steps $0, \ldots, T_1 - 1$ form a natural "cycle" for this process, as from time T_1 onwards the process behaves probabilistically *exactly the same* as when it started,

☞ 75

independently of $X_0, \ldots, X_{T_1-1}$. Thus, if we define $T_0 = 0$, and let T_i be the i-th time that the chain returns to state 1, then we can break up the time interval into independent cycles of lengths $C_i = T_i - T_{i-1}$, $i = 1, 2, \ldots$. Now suppose that during the i-th cycle a reward

$$R_i = \sum_{t=T_{i-1}}^{T_i-1} \varrho^{t-T_{i-1}}\, r(X_t)$$

is received, where $r(i)$ is some fixed reward for visiting state $i \in \{1, 2, 3, 4\}$ and $\varrho \in (0, 1)$ is a discounting factor. Clearly, $\{(C_i, R_i)\}$ is a renewal reward process. Figure 3.8 shows the outcomes of 1000 pairs (C, R), using $r(1) = 4, r(2) = 3, r(3) = 10, r(4) = 1$, and $\varrho = 0.9$.

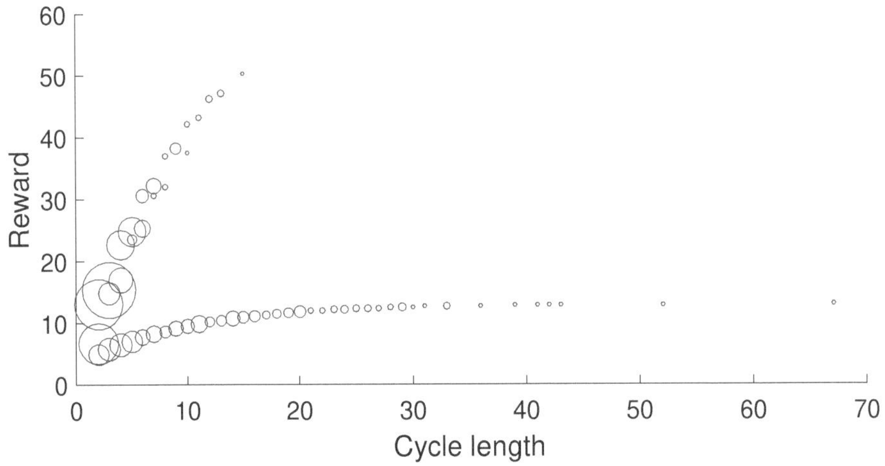

Figure 3.8: Each circle represents a (cycle length, reward) pair. The varying circle sizes indicate the number of occurrences for a given pair. For example, (2,15.43) is the most likely pair here, occurring 186 out of a 1000 times. It corresponds to the cycle path $1 \to 3 \to 2 \to 1$.

The long-run average reward is estimated as 2.50 for our data. But how accurate is this estimate? Figure 3.9 shows a density plot of the bootstrapped ratio estimates, where we independently resampled the data pairs 1000 times.

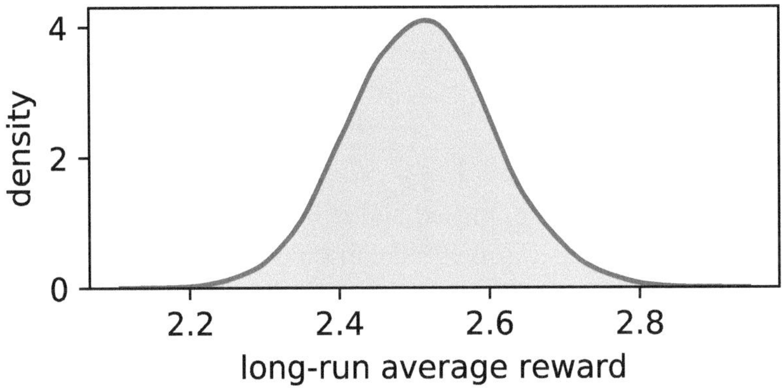

Figure 3.9: Density plot of the bootstrapped ratio estimates for the Markov chain renewal reward process.

Figure 3.9 indicates that the true long-run average reward lies between 2.2 and 2.8 with high confidence. More precisely, the 99% bootstrap confidence interval (percentile method) is here (2.27, 2.77). The following Python script spells out the procedure.

ratioest.py

```
import numpy as np, matplotlib.pyplot as plt, seaborn as sns
from numba import jit

np.random.seed(123)
n = 1000
P = np.array([[0, 0.2, 0.5, 0.3],
              [0.5 ,0, 0.5, 0],
              [0.3, 0.7, 0, 0],
              [0.1, 0, 0, 0.9]])
r = np.array([4,3,10,1])
Corg = np.array(np.zeros((n,1)))
Rorg = np.array(np.zeros((n,1)))
rho=0.9

@jit()   #for speed-up; see Appendix
def generate_cyclereward(n):
   for i in range(n):
       t=1
       xreg = 1   #regenerative state  (out of 1,2,3,4)
       reward = r[0]
       x= np.amin(np.argwhere(np.cumsum(P[xreg-1,:]) > np.random.
          rand())) + 1
       while x != xreg:
           t += 1
           reward += rho**(t-1)*r[x-1]
           x = np.amin(np.where(np.cumsum(P[x-1,:]) > np.random.rand
              ())) + 1
       Corg[i] = t
       Rorg[i] = reward
   return Corg, Rorg

Corg, Rorg = generate_cyclereward(n)

Aorg = np.mean(Rorg)/np.mean(Corg)
K = 5000
A = np.array(np.zeros((K,1)))
C = np.array(np.zeros((n,1)))
R = np.array(np.zeros((n,1)))
for i in range(K):
    ind = np.ceil(n*np.random.rand(1,n)).astype(int)[0]-1
    C = Corg[ind]
    R = Rorg[ind]
    A[i] = np.mean(R)/np.mean(C)

plt.xlabel('long-run average reward')
plt.ylabel('density')
sns.kdeplot(A.flatten(),shade=True)
plt.show()
```

■

3.3.3 Variance Reduction

The estimation of performance measures in Monte Carlo simulation can be made more efficient by utilizing known information about the simulation model. Variance reduction techniques include antithetic variables, control variables, importance sampling, conditional Monte Carlo, and stratified sampling; see, for example, [71, Chapter 9]. We shall only deal with control variables and importance sampling here.

CONTROL VARIABLE

Suppose Y is the output of a simulation experiment. A random variable $\widetilde{Y}$, obtained from the same simulation run, is called a *control variable* for Y if Y and $\widetilde{Y}$ are correlated (negatively or positively) and the expectation of $\widetilde{Y}$ is known. The use of control variables for variance reduction is based on the following theorem. We leave its proof to Exercise 21.

☞ 118

Theorem 3.3: Control Variable Estimation

Let $Y_1, \ldots, Y_N$ be the output of N independent simulation runs and let $\widetilde{Y}_1, \ldots, \widetilde{Y}_N$ be the corresponding control variables, with $\mathbb{E}\widetilde{Y}_k = \widetilde{\mu}$ known. Let $\varrho_{Y,\widetilde{Y}}$ be the correlation coefficient between each Y_k and $\widetilde{Y}_k$. For each $\alpha \in \mathbb{R}$ the estimator

$$\widehat{\mu}^{(c)} = \frac{1}{N} \sum_{k=1}^{N} \left[Y_k - \alpha \left(\widetilde{Y}_k - \widetilde{\mu} \right) \right] \tag{3.15}$$

is an unbiased estimator for $\mu = \mathbb{E}Y$. The minimal variance of $\widehat{\mu}^{(c)}$ is

$$\mathbb{V}\mathrm{ar}\, \widehat{\mu}^{(c)} = \frac{1}{N} (1 - \varrho^2_{Y,\widetilde{Y}})\, \mathbb{V}\mathrm{ar}\, Y, \tag{3.16}$$

which is obtained for $\alpha = \varrho_{Y,\widetilde{Y}} \sqrt{\mathbb{V}\mathrm{ar}Y / \mathbb{V}\mathrm{ar}\widetilde{Y}}$.

From (3.16) we see that, by using the optimal α in (3.15), the variance of the control variate estimator is a factor $1 - \varrho^2_{Y,\widetilde{Y}}$ smaller than the variance of the crude Monte Carlo estimator. Thus, if $\widetilde{Y}$ is highly correlated with Y, a significant variance reduction can be achieved. The optimal α is usually unknown, but it can be easily estimated from the sample covariance matrix of $\{(Y_k, \widetilde{Y}_k)\}$.

☞ 458

In the next example, we estimate the multiple integral in Example 3.10 using control variables.

☞ 86

■ **Example 3.13 (Monte Carlo Integration (cont.))** The random variable $Y = |X_1 + X_2 + X_3|^{1/2}(2\pi)^{3/2}$ is positively correlated with the random variable $\widetilde{Y} = X_1^2 + X_2^2 + X_3^2$, for the same choice of $X_1, X_2, X_3 \overset{\text{iid}}{\sim} \mathcal{N}(0, 1)$. As $\mathbb{E}\widetilde{Y} = \mathbb{V}\mathrm{ar}(X_1 + X_2 + X_3) = 3$, we can use it as a control variable to estimate the expectation of Y. The following Python program is based on Theorem 3.3. It imports the crude Monte Carlo sampling code from Example 3.10.

mcintCV.py
```
from mcint import *

Yc = np.sum(x**2, axis=1) # control variable data
yc = 3  # true expectation of control variable
C = np.cov(y,Yc) # sample covariance matrix
cor = C[0][1]/np.sqrt(C[0][0]*C[1][1])
alpha = C[0][1]/C[1][1]

est = np.mean(y-alpha*(Yc-yc))
RECV = np.sqrt((1-cor**2)*C[0][0]/N)/est  #relative error

print('Estimate = {:3.3f}, CI = ({:3.3f},{:3.3f}), Corr = {:3.3f}'.
      format(est, est*(1-z*RECV), est*(1+z*RECV),cor))
```

```
Estimate = 17.045, CI = (17.032,17.057), Corr = 0.480
```

A typical estimate of the correlation coefficient $\varrho_{Y,\widetilde{Y}}$ is 0.48, which gives a reduction of the variance with a factor $1-0.48^2 \approx 0.77$ — a simulation speed-up of 23% compared with crude Monte Carlo. Although the gain is small in this case, due to the modest correlation between Y and $\widetilde{Y}$, little extra work was required to achieve this variance reduction. ■

One of the most important variance reduction techniques is *importance sampling*. This technique is especially useful for the estimation of very small probabilities. The standard setting is the estimation of a quantity

IMPORTANCE SAMPLING

$$\mu = \mathbb{E}_f H(\boldsymbol{X}) = \int H(\boldsymbol{x})\, f(\boldsymbol{x})\, \mathrm{d}\boldsymbol{x}, \tag{3.17}$$

where H is a real-valued function and f the probability density of a random vector $\boldsymbol{X}$, called the *nominal pdf*. The subscript f is added to the expectation operator to indicate that it is taken with respect to the density f.

NOMINAL PDF

Let g be another probability density such that $g(\boldsymbol{x}) = 0$ implies that $H(\boldsymbol{x})\, f(\boldsymbol{x}) = 0$. Using the density g we can represent μ as

$$\mu = \int H(\boldsymbol{x}) \frac{f(\boldsymbol{x})}{g(\boldsymbol{x})}\, g(\boldsymbol{x})\, \mathrm{d}\boldsymbol{x} = \mathbb{E}_g \left[H(\boldsymbol{X}) \frac{f(\boldsymbol{X})}{g(\boldsymbol{X})} \right]. \tag{3.18}$$

Consequently, if $\boldsymbol{X}_1, \ldots, \boldsymbol{X}_N \sim_{\text{iid}} g$, then

$$\widehat{\mu} = \frac{1}{N} \sum_{k=1}^{N} H(\boldsymbol{X}_k) \frac{f(\boldsymbol{X}_k)}{g(\boldsymbol{X}_k)} \tag{3.19}$$

is an unbiased estimator of μ. This estimator is called the *importance sampling estimator* and g is called the importance sampling density. The ratio of densities, $f(\boldsymbol{x})/g(\boldsymbol{x})$, is called the *likelihood ratio*. The importance sampling pseudo-code is given in Algorithm 3.3.2.

IMPORTANCE SAMPLING ESTIMATOR

LIKELIHOOD RATIO

Algorithm 3.3.2: Importance Sampling Estimation

input: Function H, importance sampling density g such that $g(\boldsymbol{x}) = 0$ for all $\boldsymbol{x}$ for which $H(\boldsymbol{x})f(\boldsymbol{x}) = 0$, sample size N, confidence level $1-\alpha$.

output: Point estimate and approximate $(1-\alpha)$ confidence interval for $\mu = \mathbb{E}H(\boldsymbol{X})$, where $\boldsymbol{X} \sim f$.

1 Simulate $\boldsymbol{X}_1, \ldots, \boldsymbol{X}_N \overset{\text{iid}}{\sim} g$ and let $Y_i = H(\boldsymbol{X}_i)f(\boldsymbol{X}_i)/g(\boldsymbol{X}_i)$, $i = 1, \ldots, N$.

2 Estimate μ via $\widehat{\mu} = \overline{Y}$ and determine an approximate $(1-\alpha)$ confidence interval as

$$\mathcal{I} := \left(\widehat{\mu} - z_{1-\alpha/2}\frac{S}{\sqrt{N}},\ \widehat{\mu} + z_{1-\alpha/2}\frac{S}{\sqrt{N}}\right),$$

where z_γ denotes the γ-quantile of the $\mathcal{N}(0,1)$ distribution and S is the sample standard deviation of $Y_1, \ldots, Y_N$.

3 **return** $\widehat{\mu}$ and the interval $\mathcal{I}$.

■ **Example 3.14 (Importance Sampling)** Let us examine the workings of importance sampling by estimating the area, μ say, under the graph of the function

$$M(x_1, x_2) = \mathrm{e}^{-\frac{1}{4}\sqrt{x_1^2+x_2^2}}\left(\sin\left(2\sqrt{x_1^2+x_2^2}\right)+1\right), \quad (x_1, x_2) \in \mathbb{R}^2. \tag{3.20}$$

☞ 80

We saw a similar function in Example 3.8 (but note the different domain). A natural approach to estimate the area is to truncate the domain to the square $[-b, b]^2$, for large enough b, and to estimate the integral

$$\mu_b = \int_{-b}^{b}\int_{-b}^{b} \underbrace{(2b)^2 M(\boldsymbol{x})}_{H(\boldsymbol{x})} f(\boldsymbol{x})\,\mathrm{d}\boldsymbol{x} = \mathbb{E}_f H(\boldsymbol{X})$$

via crude Monte Carlo, where $f(\boldsymbol{x}) = 1/(2b)^2, \boldsymbol{x} \in [-b, b]^2$, is the pdf of the uniform distribution on $[-b, b]^2$. Here is the Python code which does just that.

impsamp1.py

```
import numpy as np
from numpy import exp, sqrt, sin, pi, log, cos
from numpy.random import rand

b = 1000
H = lambda x1, x2: (2*b)**2 * exp(-sqrt(x1**2+x2**2)/4)*(sin(2*sqrt(
        x1**2+x2**2))+1)*(x1**2 + x2**2 < b**2)
f = 1/((2*b)**2)
N = 10**6
X1 = -b + 2*b*rand(N,1)
X2 = -b + 2*b*rand(N,1)
Z = H(X1,X2)
estCMC = np.mean(Z).item()  # to obtain scalar
RECMC = np.std(Z)/estCMC/sqrt(N).item()
print('CI = ({:3.3f},{:3.3f}), RE = {: 3.3f}'.format(estCMC*(1-1.96*
    RECMC), estCMC*(1+1.96*RECMC),RECMC))
```

```
CI = (82.663,135.036), RE =  0.123
```

For a truncation level of $b = 1000$ and a sample size of $N = 10^6$, a typical estimate is 108.8, with an estimated relative error of 0.123. We have two sources of error here. The first is the error in approximating μ by μ_b. However, as the function H decays exponentially fast, $b = 1000$ is more than enough to ensure this error is negligible. The second type of error is the statistical error, due to the estimation process itself. This can be quantified by the estimated relative error, and can be reduced by increasing the sample size.

Let us now consider an importance sampling approach in which the importance sampling pdf g is radially symmetric and decays exponentially in the radius, similar to the function H. In particular, we simulate (X_1, X_2) in a way akin to Example 3.1, by first generating a radius $R \sim \mathsf{Exp}(\lambda)$ and an angle $\Theta \sim \mathcal{U}(0, 2\pi)$, and then returning $X_1 = R\cos(\Theta)$ and $X_2 = R\sin(\Theta)$. By the Transformation Rule (Theorem C.4) we then have ☞ 69 ☞ 435

$$g(\boldsymbol{x}) = f_{R,\Theta}(r,\theta)\frac{1}{r} = \lambda\,\mathrm{e}^{-\lambda r}\frac{1}{2\pi}\frac{1}{r} = \frac{\lambda \mathrm{e}^{-\lambda\sqrt{x_1^2+x_2^2}}}{2\pi\sqrt{x_1^2+x_2^2}}, \quad \boldsymbol{x} \in \mathbb{R}^2 \setminus \{\boldsymbol{0}\}.$$

The following code, which imports the one given above, implements the importance sampling steps, using the parameter $\lambda = 0.1$.

impsamp2.py

```
from impsamp1 import *

lam = 0.1;
g = lambda x1, x2: lam*exp(-sqrt(x1**2 + x2**2)*lam)/sqrt(x1**2 + x2
    **2)/(2*pi);
U = rand(N,1); V = rand(N,1)
R = -log(U)/lam
X1 = R*cos(2*pi*V)
X2 = R*sin(2*pi*V)
Z = H(X1,X2)*f/g(X1,X2)
estIS = np.mean(Z).item()  # obtain scalar
REIS = np.std(Z)/estIS/sqrt(N).item()
print('CI = ({:3.3f},{:3.3f}), RE = {: 3.3f}'.format(estIS*(1-1.96*
    REIS), estIS*(1+1.96*REIS),REIS))
```

```
CI = (100.723,101.077), RE =  0.001
```

A typical estimate is 100.90 with an estimated relative error of $1 \cdot 10^{-4}$, which gives a substantial variance reduction. In terms of approximate 95% confidence intervals, we have (82.7,135.0) in the CMC case versus (100.7,101.1) in the importance sampling case. Of course, we could have reduced the truncation level b to improve the performance of CMC, but then the approximation error might become more significant. For the importance sampling case, the relative error is hardly affected by the threshold level, but does depend on the choice of λ. We chose λ such that the decay rate is slower than the decay rate of the function H, which is 0.25. ■

As illustrated in the above example, a main difficulty in importance sampling is how to choose the importance sampling distribution. A poor choice of g may seriously affect the accuracy of both the estimate and the confidence interval. The theoretically optimal choice

g^* for the importance sampling density minimizes the variance of $\widehat{\mu}$ and is therefore the solution to the functional minimization program

$$\min_g \mathbb{V}\mathrm{ar}_g \left(H(\boldsymbol{X}) \frac{f(\boldsymbol{X})}{g(\boldsymbol{X})} \right). \tag{3.21}$$

☞ 118

OPTIMAL IMPORTANCE SAMPLING PDF

It is not difficult to show, see also Exercise 22, that if either $H(\boldsymbol{x}) \geqslant 0$ or $H(\boldsymbol{x}) \leqslant 0$ for all $\boldsymbol{x}$, then the *optimal importance sampling pdf* is

$$g^*(\boldsymbol{x}) = \frac{H(\boldsymbol{x})\, f(\boldsymbol{x})}{\mu}. \tag{3.22}$$

Namely, in this case $\mathbb{V}\mathrm{ar}_{g^*}\widehat{\mu} = \mathbb{V}\mathrm{ar}_{g^*}(H(\boldsymbol{X})f(\boldsymbol{X})/g(\boldsymbol{X})) = \mathbb{V}\mathrm{ar}_{g^*}\mu = 0$, so that the estimator $\widehat{\mu}$ is *constant* under g^*. An obvious difficulty is that the evaluation of the optimal importance sampling density g^* is usually not possible, since $g^*(\boldsymbol{x})$ in (3.22) depends on the unknown quantity μ. Nevertheless, one can typically choose a good importance sampling density g "close" to the minimum variance density g^*.

One of the main considerations for choosing a good importance sampling pdf is that the estimator (3.19) should have finite variance. This is equivalent to the requirement that

$$\mathbb{E}_g \left[H^2(\boldsymbol{X}) \frac{f^2(\boldsymbol{X})}{g^2(\boldsymbol{X})} \right] = \mathbb{E}_f \left[H^2(\boldsymbol{X}) \frac{f(\boldsymbol{X})}{g(\boldsymbol{X})} \right] < \infty. \tag{3.23}$$

This suggests that g should not have lighter tails than f and that, preferably, the likelihood ratio, f/g, should be bounded.

3.4 Monte Carlo for Optimization

In this section we describe several Monte Carlo methods for optimization. Such randomized algorithms can be useful for solving optimization problems with many local optima and complicated constraints, possibly involving a mix of continuous and discrete variables. Randomized algorithms are also used to solve *noisy* optimization problems, in which the objective function is unknown and has to be obtained via Monte Carlo simulation.

3.4.1 Simulated Annealing

SIMULATED ANNEALING

Simulated annealing is a Monte Carlo technique for minimization that emulates the physical state of atoms in a metal when the metal is heated up and then slowly cooled down. When the cooling is performed very slowly, the atoms settle down to a minimum-energy state. Denoting the state as $\boldsymbol{x}$ and the energy of a state as $S(\boldsymbol{x})$, the probability distribution of the (random) states is described by the *Boltzmann pdf*

$$f(\boldsymbol{x}) \propto \mathrm{e}^{-\frac{S(\boldsymbol{x})}{kT}}, \quad \boldsymbol{x} \in \mathcal{X},$$

where k is Boltzmann's constant and T is the temperature.

Going beyond the physical interpretation, suppose that $S(\boldsymbol{x})$ is an arbitrary function to be minimized, with $\boldsymbol{x}$ taking values in some discrete or continuous set $\mathcal{X}$. The *Gibbs pdf* corresponding to $S(\boldsymbol{x})$ is defined as

GIBBS PDF

$$f_T(\boldsymbol{x}) = \frac{\mathrm{e}^{-\frac{S(\boldsymbol{x})}{T}}}{z_T}, \quad \boldsymbol{x} \in \mathcal{X},$$

provided that the normalization constant $z_T := \sum_{\boldsymbol{x}} \exp(-S(\boldsymbol{x})/T)$ is finite. Note that this is simply the Boltzmann pdf with the Boltzmann constant k removed. As $T \to 0$, the pdf becomes more and more peaked around the set of global minimizers of S.

The idea of simulated annealing is to create a sequence of points $\boldsymbol{X}_1, \boldsymbol{X}_2, \ldots$ that are approximately distributed according to pdfs $f_{T_1}(\boldsymbol{x}), f_{T_2}(\boldsymbol{x}), \ldots$, where $T_1, T_2, \ldots$ is a sequence of "temperatures" that decreases (is "cooled") to 0 — known as the *annealing schedule*. If each $\boldsymbol{X}_t$ were sampled *exactly* from f_{T_t}, then $\boldsymbol{X}_t$ would converge to a global minimum of $S(\boldsymbol{x})$ as $T_t \to 0$. However, in practice sampling is *approximate* and convergence to a global minimum is not assured. A generic simulated annealing algorithm is as follows.

ANNEALING SCHEDULE

Algorithm 3.4.1: Simulated Annealing

input: Annealing schedule $T_0, T_1, \ldots,$, function S, initial value $\boldsymbol{x}_0$.
output: Approximations to the global minimizer $\boldsymbol{x}^*$ and minimum value $S(\boldsymbol{x}^*)$.

```
Set X_0 ← x_0 and t ← 1.
while not stopping do
    Approximately simulate X_t from f_{T_t}(x).
    t ← t + 1
return X_t, S(X_t)
```

A popular annealing schedule is *geometric cooling*, where $T_t = \beta T_{t-1}$, $t = 1, 2, \ldots$, for a given initial temperature T_0 and a *cooling factor* $\beta \in (0, 1)$. Appropriate values for T_0 and β are problem-dependent and this has traditionally required tuning on the part of the user. A possible stopping criterion is to stop after a fixed number of iterations, or when the temperature is "small enough".

GEOMETRIC COOLING

COOLING FACTOR

Approximate sampling from a Gibbs distribution is most often carried out via Markov chain Monte Carlo. For each iteration t, the Markov chain should theoretically run for a large number of steps to accurately sample from the Gibbs pdf f_{T_t}. However, in practice, one often only runs a *single* step of the Markov chain, before updating the temperature, as in Algorithm 3.4.2 below.

To sample from a Gibbs distribution f_T, this algorithm uses a random walk Metropolis–Hastings sampler. From (3.7), the acceptance probability of a proposal $\boldsymbol{y}$ is thus ☞ 80

$$\alpha(\boldsymbol{x}, \boldsymbol{y}) = \min\left\{\frac{\mathrm{e}^{-\frac{1}{T}S(\boldsymbol{y})}}{\mathrm{e}^{-\frac{1}{T}S(\boldsymbol{x})}},\ 1\right\} = \min\left\{\mathrm{e}^{-\frac{1}{T}(S(\boldsymbol{y})-S(\boldsymbol{x}))},\ 1\right\}.$$

Hence, if $S(\boldsymbol{y}) < S(\boldsymbol{x})$, then the proposal is aways accepted. Otherwise, the proposal is accepted with probability $\exp(-\frac{1}{T}(S(\boldsymbol{y}) - S(\boldsymbol{x})))$.

Algorithm 3.4.2: Simulated Annealing with a Random Walk Sampler

input: Objective function S, starting state $\boldsymbol{X}_0$, initial temperature T_0, number of iterations N, symmetric proposal density $q(\boldsymbol{y} \mid \boldsymbol{x})$, constant β.
output: Approximate minimizer and minimum value of S.

for $t = 0$ **to** $N - 1$ **do**
  Simulate a new state $\boldsymbol{Y}$ from the symmetric proposal $q(\boldsymbol{y} \mid \boldsymbol{X}_t)$.
  if $S(\boldsymbol{Y}) < S(\boldsymbol{X}_t)$ **then**
    $\boldsymbol{X}_{t+1} \leftarrow \boldsymbol{Y}$
  else
    Draw $U \sim \mathcal{U}(0, 1)$.
    if $U \leqslant \mathrm{e}^{-(S(\boldsymbol{Y})-S(\boldsymbol{X}_t))/T_t}$ **then**
      $\boldsymbol{X}_{t+1} \leftarrow \boldsymbol{Y}$
    else
      $\boldsymbol{X}_{t+1} \leftarrow \boldsymbol{X}_t$
  $T_{t+1} \leftarrow \beta T_t$
return $\boldsymbol{X}_N$ and $S(\boldsymbol{X}_N)$

■ **Example 3.15 (Simulated Annealing for Minimization)** Let us minimize the "wiggly" function depicted in the bottom panel of Figure 3.10 and given by:

$$S(x) = \begin{cases} -\mathrm{e}^{-x^2/100} \sin(13x - x^4)^5 \sin(1 - 3x^2)^2, & \text{if } -2 \leqslant x \leqslant 2, \\ \infty, & \text{otherwise.} \end{cases}$$

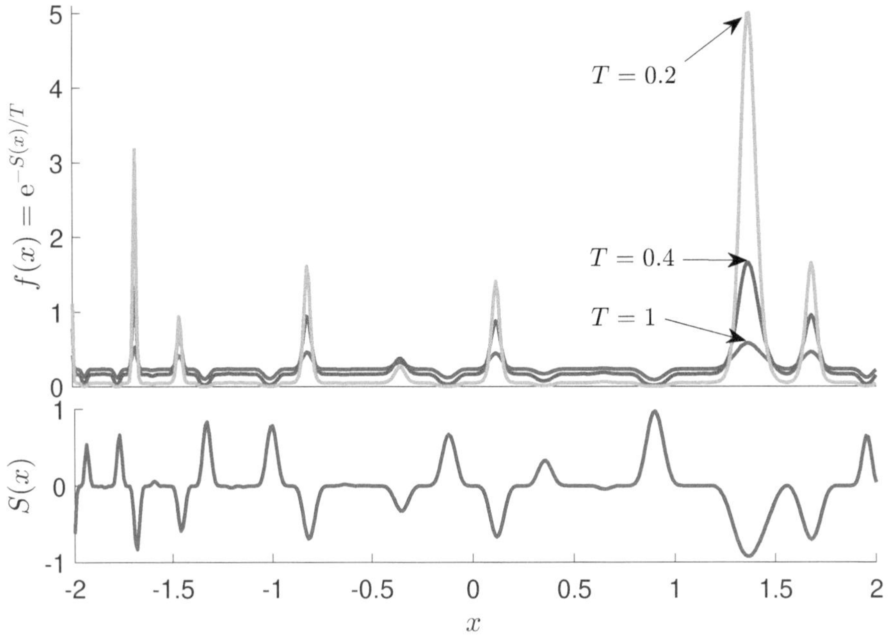

Figure 3.10: Lower panel: the "wiggly" function $S(x)$. Upper panel: three (normalized) Gibbs pdfs for temperatures $T = 1, 0.4, 0.2$. As the temperature decreases, the Gibbs pdf converges to the pdf that has all its mass concentrated at the minimizer of S.

The function has many local minima and maxima, with a global minimum around 1.4. The figure also illustrates the relationship between S and the (unnormalized) Gibbs pdf f_T.

The following Python code implements a slight variant of Algorithm 3.4.2 where, instead of stopping after a fixed number of iterations, the algorithm stops when the temperature is lower than some threshold (here 10^{-3}).

> Instead of stopping after a fixed number N of iterations or when the temperature is low enough, it is useful to stop when consecutive function values are closer than some distance ε to each other, or when the best found function value has not changed over a fixed number d of iterations.

For a "current" state x, the proposal state Y is here drawn from the $\mathcal{N}(x, 0.5^2)$ distribution. We use geometric cooling with decay parameter $\beta = 0.999$ and initial temperature $T_0 = 1$. We set the initial state to $x_0 = 0$. Figure 3.11 depicts a realization of the sequence of states x_t for $t = 0, 1, \ldots$. After initially fluctuating wildly, the sequence settles down to a value around 1.37, with $S(1.37) = -0.92$, corresponding to the global optimizer and minimum, respectively.

simann.py

```
import numpy as np
import matplotlib.pyplot as plt

def wiggly(x):
    y = -np.exp(x**2/100)*np.sin(13*x-x**4)**5*np.sin(1-3*x**2)**2
    ind = np.vstack((np.argwhere(x<-2),np.argwhere(x>2)))
    y[ind]=float('inf')
    return y

S = wiggly
beta = 0.999
sig = 0.5
T=1
x= np.array([0])
xx=[]
Sx=S(x)
while T>10**(-3):
    T=beta*T
    y = x+sig*np.random.randn()
    Sy = S(y)
    alpha = np.amin((np.exp(-(Sy-Sx)/T),1))
    if np.random.uniform()<alpha:
        x=y
        Sx=Sy
    xx=np.hstack((xx,x))

print('minimizer = {:3.3f}, minimum ={:3.3f}'.format(x[0],Sx[0]))
plt.plot(xx)
plt.show()
```

```
minimizer = 1.365, minimum = -0.958
```

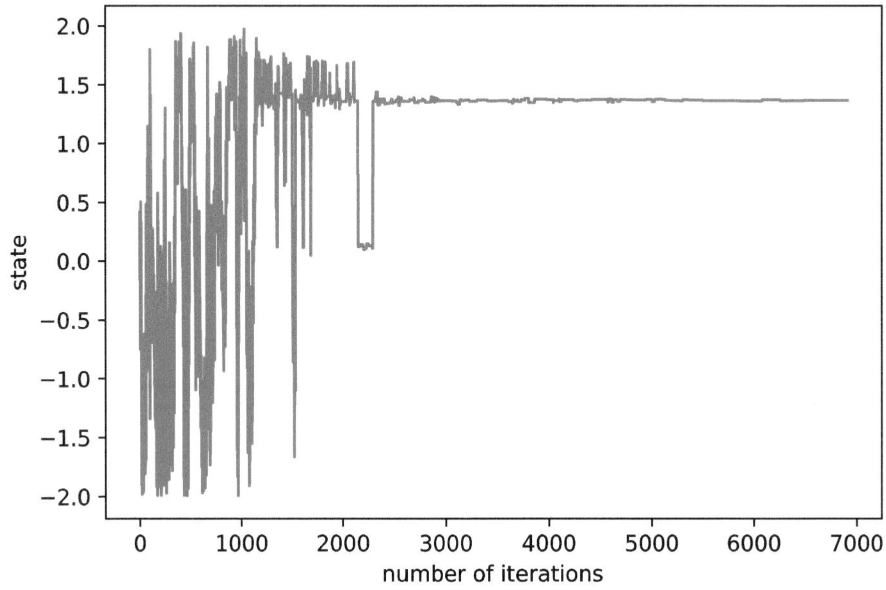

Figure 3.11: Typical states generated by the simulated annealing algorithm.

■

3.4.2 Cross-Entropy Method

CROSS-ENTROPY

The *cross-entropy* (CE) method [103] is a simple Monte Carlo algorithm that can be used for both optimization and estimation.

The basic idea of the CE method for minimizing a function S on a set $\mathcal{X}$ is to define a parametric family of probability densities $\{f(\cdot \mid \boldsymbol{v}), \boldsymbol{v} \in \mathcal{V}\}$ on $\mathcal{X}$ and to iteratively update the parameter $\boldsymbol{v}$ so that $f(\cdot \mid \boldsymbol{v})$ places more mass on states $\boldsymbol{x}$ that have smaller S values than on the previous iteration. In particular, the CE algorithm has two basic phases:

- *Sampling*: Samples $\boldsymbol{X}_1, \ldots, \boldsymbol{X}_N$ are drawn independently according to $f(\cdot \mid \boldsymbol{v})$. The objective function S is evaluated at these points.

- *Updating*: A new parameter $\boldsymbol{v}'$ is selected on the basis of those $\boldsymbol{X}_i$ for which $S(\boldsymbol{X}_i) \leqslant \gamma$ for some level γ. These $\{\boldsymbol{X}_i\}$ form the *elite sample* set, $\mathcal{E}$.

ELITE SAMPLE

RARITY PARAMETER

SMOOTHING PARAMETER

At each iteration the level parameter γ is chosen as the worst of the $N^{\text{elite}} := \lceil \varrho N \rceil$ best performing samples, where $\varrho \in (0, 1)$ is the *rarity parameter* — typically, $\varrho = 0.1$ or $\varrho = 0.01$. The parameter $\boldsymbol{v}$ is updated as a smoothed average $\alpha \boldsymbol{v}' + (1-\alpha)\boldsymbol{v}$, where $\alpha \in (0, 1)$ is the *smoothing parameter* and

$$\boldsymbol{v}' := \underset{\boldsymbol{v} \in \mathcal{V}}{\operatorname{argmax}} \sum_{\boldsymbol{X} \in \mathcal{E}} \ln f(\boldsymbol{X} \mid \boldsymbol{v}). \tag{3.24}$$

The updating rule (3.24) is the result of minimizing the Kullback–Leibler divergence between the conditional density of $\boldsymbol{X} \sim f(\boldsymbol{x} \mid \boldsymbol{v})$ given $S(\boldsymbol{X}) \leqslant \gamma$, and $f(\boldsymbol{x}; \boldsymbol{v})$; see [103]. Note that (3.24) yields the *maximum likelihood estimator* (MLE) of $\boldsymbol{v}$ based on the elite samples. Hence, for many specific families of distributions, explicit solutions can be found. An important example is where $\boldsymbol{X} \sim \mathcal{N}(\boldsymbol{\mu}, \operatorname{diag}(\boldsymbol{\sigma}^2))$; that is, $\boldsymbol{X}$ has independent Gaussian

☞ 458

components. In this case, the mean vector $\boldsymbol{\mu}$ and the vector of variances $\boldsymbol{\sigma}^2$ are simply updated via the sample mean and sample variance of the elite samples. This is known as *normal updating*. A generic CE procedure for minimization is given in Algorithm 3.4.3.

NORMAL UPDATING

Algorithm 3.4.3: Cross-Entropy Method for Minimization

input: Function S, initial sampling parameter $\boldsymbol{v}_0$, sample size N, rarity parameter ϱ, smoothing parameter α.

output: Approximate minimum of S and optimal sampling parameter $\boldsymbol{v}$.

1 Initialize $\boldsymbol{v}_0$, set $N^{\text{elite}} \leftarrow \lceil \varrho N \rceil$ and $t \leftarrow 0$.
2 **while** a stopping criterion is not met **do**
3 $\quad t \leftarrow t + 1$
4 $\quad$ Simulate an iid sample $\boldsymbol{X}_1, \ldots, \boldsymbol{X}_N$ from the density $f(\cdot \,|\, \boldsymbol{v}_{t-1})$.
5 $\quad$ Evaluate the performances $S(\boldsymbol{X}_1), \ldots, S(\boldsymbol{X}_N)$ and sort them from smallest to largest: $S_{(1)}, \ldots, S_{(N)}$.
6 $\quad$ Let γ_t be the sample ϱ-quantile of the performances:
$$\gamma_t \leftarrow S_{(N^{\text{elite}})}. \tag{3.25}$$
7 $\quad$ Determine the set of elite samples $\mathcal{E}_t = \{\boldsymbol{X}_i : S(\boldsymbol{X}_i) \leqslant \gamma_t\}$.
8 $\quad$ Let $\boldsymbol{v}_t'$ be the MLE of the elite samples:
$$\boldsymbol{v}_t' \leftarrow \underset{\boldsymbol{v}}{\operatorname{argmax}} \sum_{\boldsymbol{X} \in \mathcal{E}_t} \ln f(\boldsymbol{X} \,|\, \boldsymbol{v}). \tag{3.26}$$
9 $\quad$ Update the sampling parameter as
$$\boldsymbol{v}_t \leftarrow \alpha \boldsymbol{v}_t' + (1 - \alpha)\boldsymbol{v}_{t-1}. \tag{3.27}$$
10 **return** γ_t, $\boldsymbol{v}_t$

The CE algorithm produces a sequence of pairs $(\gamma_1, \boldsymbol{v}_1), (\gamma_2, \boldsymbol{v}_2), \ldots$, such that γ_t converges (approximately) to the minimal function value, and $f(\cdot \,|\, \boldsymbol{v}_t)$ to a degenerate pdf that (approximately) concentrates all its mass at a minimizer of S, as $t \to \infty$. A possible stopping condition is to stop when the sampling distribution $f(\cdot \,|\, \boldsymbol{v}_t)$ is sufficiently close to a degenerate distribution. For normal updating this means that the standard deviation is sufficiently small.

> The output of the CE algorithm could also include the overall best function value and corresponding solution.

In the following example, we minimize the same function as in Example 3.15, but instead use the CE algorithm. ☞ 97

■ **Example 3.16 (Cross-Entropy Method for Minimization)** In this case we take the family of normal distributions $\{\mathcal{N}(\mu, \sigma^2)\}$ for the sampling step (Step 4 of Algorithm 3.4.3), starting with $\mu = 0$ and $\sigma = 3$. The choice of the initial parameter is quite arbitrary, as long as σ is large enough to sample a wide range of points. We take $N = 100$ samples at each iteration, set $\varrho = 0.1$, and keep the $N^{\text{elite}} = 10 = \lceil N\varrho \rceil$ smallest ones as the elite samples. The parameters μ and σ are then updated via the sample mean and sample standard deviation of the elite samples. In this case we do not use any smoothing ($\alpha = 1$). In the following Python code the 100×2 matrix Sx stores the x-values in the first column and the function values in the second column. The rows of this matrix are sorted in ascending order according to the function values, giving the matrix sortSx. The first $N^{\text{elite}} = 10$ rows of this sorted matrix correspond to the elite samples and their function values. The updating of μ and σ is done in Lines 14 and 15. Figure 3.12 shows how the pdfs of the $\mathcal{N}(\mu_t, \sigma_t^2)$ sampling distributions degenerate to the point mass at the global minimizer 1.366.

CEmethod.py

```python
from simann import wiggly
import numpy as np
np.set_printoptions(precision=3)
mu, sigma = 0, 3
N, Nel = 100, 10
eps = 10**-5
S = wiggly
while sigma > eps:
    X = np.random.randn(N,1)*sigma + np.array(np.ones((N,1)))*mu
    Sx = np.hstack((X, S(X)))
    sortSx = Sx[Sx[:,1].argsort(),]
    Elite = sortSx[0:Nel,:-1]
    mu = np.mean(Elite, axis=0)
    sigma = np.std(Elite, axis=0)
    print('S(mu)= {}, mu: {}, sigma: {}\n'.format(S(mu), mu, sigma))
```

```
S(mu)= [0.071], mu: [0.414], sigma: [0.922]
S(mu)= [0.063], mu: [0.81], sigma: [0.831]
S(mu)= [-0.033], mu: [1.212], sigma: [0.69]
S(mu)= [-0.588], mu: [1.447], sigma: [0.117]
S(mu)= [-0.958], mu: [1.366], sigma: [0.007]
S(mu)= [-0.958], mu: [1.366], sigma: [0.]
S(mu)= [-0.958], mu: [1.366], sigma: [3.535e-05]
S(mu)= [-0.958], mu: [1.366], sigma: [2.023e-06]
```

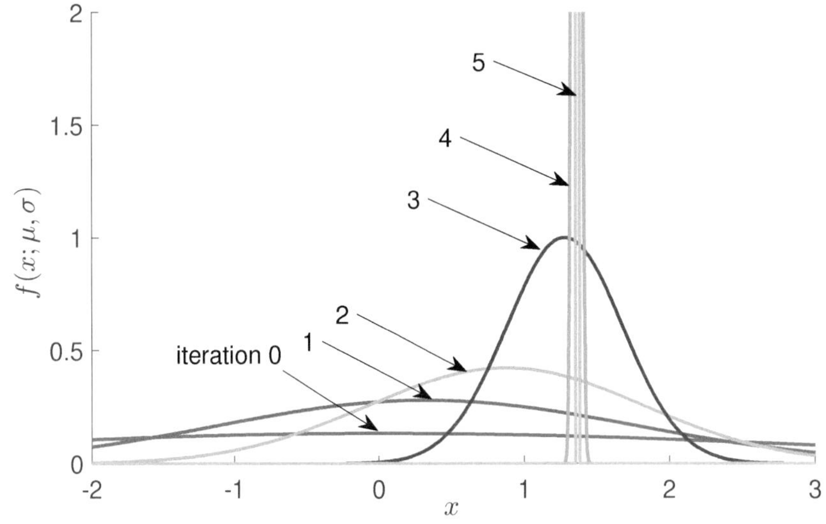

Figure 3.12: The normal pdfs of the first six sampling distributions, truncated to the interval $[-2, 3]$. The initial sampling distribution is $\mathcal{N}(0, 3^2)$.

3.4.3 Splitting for Optimization

Minimizing a function $S(\boldsymbol{x})$, $\boldsymbol{x} \in \mathcal{X}$ is closely related to drawing a random sample from a *level set* of the form $\{\boldsymbol{x} \in \mathcal{X} : S(\boldsymbol{x}) \leqslant \gamma\}$. Suppose S has minimum value γ^* attained at $\boldsymbol{x}^*$. As long as $\gamma \geqslant \gamma^*$, this level set contains the minimizer. Moreover, if γ is close to γ^*, the volume of this level set will be small. So, a randomly selected point from this set is expected to be close to $\boldsymbol{x}^*$. Thus, by gradually decreasing the level parameter γ, the level sets will gradually shrink towards the set $\{\boldsymbol{x}^*\}$. Indeed, the CE method was developed with exactly this connection in mind; see, e.g., [102]. Note that the CE method employs a *parametric* sampling distribution to obtain samples from the level sets (the elite samples). In [34] a *non-parametric* sampling mechanism is introduced that uses an evolving collection of particles. The resulting optimization algorithm, called *splitting for continuous optimization* (SCO), provides a fast and accurate way to optimize complicated continuous functions. The details of SCO are given in Algorithm 3.4.4.

LEVEL SET

SPLITTING FOR CONTINUOUS OPTIMIZATION

Algorithm 3.4.4: Splitting for Continuous Optimization (SCO)

---

input: Objective function S, sample size N, rarity parameter ϱ, scale factor w, bounded region $\mathcal{B} \subset \mathcal{X}$ that is known to contain a global minimizer, and maximum number of attempts `MaxTry`.

output: Final iteration number t and sequence $(\boldsymbol{X}_{\text{best},1}, b_1), \ldots, (\boldsymbol{X}_{\text{best},t}, b_t)$ of best solutions and function values at each iteration.

Simulate $\mathcal{Y}_0 = \{\boldsymbol{Y}_1, \ldots, \boldsymbol{Y}_N\}$ uniformly on $\mathcal{B}$. Set $t \leftarrow 0$ and $N^{\text{elite}} \leftarrow \lceil N\varrho \rceil$.

while stopping condition is not satisfied **do**

  Determine the N^{elite} smallest values, $S_{(1)} \leqslant \cdots \leqslant S_{(N^{\text{elite}})}$, of $\{S(\boldsymbol{X}), \boldsymbol{X} \in \mathcal{Y}_t\}$, and store the corresponding vectors, $\boldsymbol{X}_{(1)}, \ldots, \boldsymbol{X}_{(N^{\text{elite}})}$, in $\mathcal{X}_{t+1}$. Set $b_{t+1} \leftarrow S_{(1)}$ and $\boldsymbol{X}_{\text{best},t+1} \leftarrow \boldsymbol{X}_{(1)}$.

  Draw $B_i \sim \mathsf{Bernoulli}(\frac{1}{2})$, $i = 1, \ldots, N^{\text{elite}}$, with $\sum_{i=1}^{N^{\text{elite}}} B_i = N \bmod N^{\text{elite}}$.

  for $i = 1$ **to** N^{elite} **do**

    $R_i \leftarrow \left\lfloor \frac{N}{N^{\text{elite}}} \right\rfloor + B_i$ `// random splitting factor`

    $\boldsymbol{Y} \leftarrow \boldsymbol{X}_{(i)}$; $\boldsymbol{Y}' \leftarrow \boldsymbol{Y}$

    for $j = 1$ **to** R_i **do**

      Draw $I \in \{1, \ldots, N^{\text{elite}}\} \setminus \{i\}$ uniformly and let $\boldsymbol{\sigma}_i \leftarrow w|\boldsymbol{X}^{(i)} - \boldsymbol{X}^{(I)}|$.

      Simulate a uniform permutation $\boldsymbol{\pi} = (\pi_1, \ldots, \pi_n)$ of $(1, \ldots, n)$.

      for $k = 1$ **to** n **do**

        for `Try` $= 1$ **to** `MaxTry` **do**

          $\boldsymbol{Y}'(\pi_k) \leftarrow \boldsymbol{Y}(\pi_k) + \boldsymbol{\sigma}_i(\pi_k)Z, \quad Z \sim \mathcal{N}(0, 1)$

          if $S(\boldsymbol{Y}') < S(\boldsymbol{Y})$ **then** $\boldsymbol{Y} \leftarrow \boldsymbol{Y}'$ and **break**.

      Add $\boldsymbol{Y}$ to $\mathcal{Y}_{t+1}$

  $t \leftarrow t + 1$

return $\{(\boldsymbol{X}_{\text{best},k}, b_k), k = 1, \ldots, t\}$

At iteration $t = 0$, the algorithm starts with a population of particles $\mathcal{Y}_0 = \{\boldsymbol{Y}_1, \ldots, \boldsymbol{Y}_N\}$ that are uniformly generated on some bounded region $\mathcal{B}$, which is large enough to contain a global minimizer. The function values of all particles in $\mathcal{Y}_0$ are sorted, and the best

$N^{\text{elite}} = \lceil N\varrho \rceil$ form the elite particle set $\mathcal{X}_1$, exactly as in the CE method. Next, the elite particles are "split" into $\lfloor N/N^{\text{elite}} \rfloor$ children particles, adding one extra child to some of the elite particles to ensure that the total number of children is again N. The purpose of Line 4 is to randomize which elite particles receive an extra child. Lines 8–15 describe how the children of the i-th elite particle are generated. First, in Line 9, we select one of the *other* elite particles uniformly at random. The same line defines an n-dimensional vector $\boldsymbol{\sigma}_i$ whose components are the absolute differences between the vectors $\boldsymbol{X}_{(i)}$ and $\boldsymbol{X}_{(I)}$, multiplied by a constant w. That is,

$$\boldsymbol{\sigma}_i = w\,|\boldsymbol{X}_{(i)} - \boldsymbol{X}_{(I)}| := w \begin{bmatrix} |X_{(i),1} - X_{(I),1}| \\ |X_{(i),2} - X_{(I),2}| \\ \vdots \\ |X_{(i),n} - X_{(I),n}| \end{bmatrix}.$$

☞ 115 Next, a uniform random permutation π of $(1, \dots, n)$ is simulated (see Exercise 9). Lines 11–14 describe how, starting from a candidate child point $\boldsymbol{Y}$, each coordinate of $\boldsymbol{Y}$ is resampled, in the order determined by π, by adding a standard normal random variable to that component, multiplied by the corresponding component of $\boldsymbol{\sigma}_i$ (Line 13). If the resulting $\boldsymbol{Y}'$ has a function value that is less than that of $\boldsymbol{Y}$, then the new candidate is accepted. Otherwise, the *same* coordinate is tried again. If no improvement is found in `MaxTry` attempts, the original component is retained. This process is performed for all elite samples, to produce the first-generation population $\mathcal{Y}_1$. The procedure is then repeated for iterations $t = 1, 2, \dots$, until some stopping criterion is met, e.g., when the best found function value does not change for a number of consecutive iterations, or when the total number of function evaluations exceeds some threshold. The best found function value and corresponding argument (particle) are returned at the conclusion of the algorithm.

The input variable `MaxTry` governs how much computational time is dedicated to updating a component. In most cases we have encountered, the choices $w = 0.5$ and `MaxTry` $= 5$ work well. Empirically, relatively high value for ϱ work well, such as $\varrho = 0.4, 0.8$, or even $\varrho = 1$. The latter case means that at each stage t *all* samples from $\mathcal{Y}_{t-1}$ carry over to the elite set $\mathcal{X}_t$.

■ **Example 3.17 (Test Problem 112)** Hock and Schittkowski [58] provide a rich source of test problems for multiextremal optimization. A challenging one is Problem 112, where the goal is to find $\boldsymbol{x}$ so as to minimize the function

$$S(\boldsymbol{x}) = \sum_{j=1}^{10} x_j \left(c_j + \ln \frac{x_j}{x_1 + \cdots + x_{10}} \right),$$

subject to the following set of constraints:

$$\begin{aligned} x_1 + 2x_2 + 2x_3 + x_6 + x_{10} - 2 &= 0, \\ x_4 + 2x_5 + x_6 + x_7 - 1 &= 0, \\ x_3 + x_7 + x_8 + 2x_9 + x_{10} - 1 &= 0, \\ x_j &\geqslant 0.000001, \quad j = 1, \dots, 10, \end{aligned}$$

where the constants $\{c_i\}$ are given in Table 3.1.

Table 3.1: Constants for Test Problem 112.

$c_1 = -6.089$	$c_2 = -17.164$	$c_3 = -34.054$	$c_4 = -5.914$	$c_5 = -24.721$
$c_6 = -14.986$	$c_7 = -24.100$	$c_8 = -10.708$	$c_9 = -26.662$	$c_{10} = -22.179$

The best known minimal value in [58] was −47.707579. In [89] a better solution was found, −47.760765, using a genetic algorithm. The corresponding solution vector was completely different from the one in [58]. A further improvement,−47.76109081, was found in [70], using the CE method, giving a similar solution vector to that in [89]:

0.04067247	0.14765159	0.78323637	0.00141368	0.48526222
0.00069291	0.02736897	0.01794290	0.03729653	0.09685870

To obtain a solution with SCO, we first converted this 10-dimensional problem into a 7-dimensional one by defining the objective function

$$S_7(\boldsymbol{y}) = S(\boldsymbol{x}),$$

where $x_2 = y_1, x_3 = y_2, x_5 = y_3, x_6 = y_4, x_7 = y_5, x_9 = y_6, x_{10} = y_7$, and

$$\begin{aligned} x_1 &= 2 - (2y_1 + 2y_2 + y_4 + x_7), \\ x_4 &= 1 - (2y_3 + y_4 + y_5), \\ x_8 &= 1 - (y_2 + y_5 + 2y_6 + y_7), \end{aligned}$$

subject to $x_1, \ldots, x_{10} \geqslant 0.000001$, where the $\{x_i\}$ are taken as functions of the $\{y_i\}$. We then adopted a penalty approach (see Section B.4) by adding a penalty function to the original objective function: ☞ 417

$$\widetilde{S}_7(\boldsymbol{y}) = S(\boldsymbol{x}) + 1000 \sum_{i=1}^{10} \max\{-(x_i - 0.000001), 0\},$$

where, again, the $\{x_i\}$ are defined in terms of the $\{y_i\}$ as above.

Optimizing this last function with SCO, we found, in less time than the other algorithms, a slightly smaller function value: −47.761090859365858, with solution vector

0.040668102417464	0.147730393049955	0.783153291185250	0.001414221643059
0.485246633088859	0.000693172682617	0.027399339496606	0.017947274343948
0.037314369272343	0.096871356429511		

in line with the earlier solutions. ■

3.4.4 Noisy Optimization

In *noisy optimization*, the objective function is unknown, but estimates of function values are available, e.g., via simulation. For example, to find an optimal prediction function g in supervised learning, the exact risk $\ell(g) = \mathbb{E}\,\text{Loss}(Y, g(\boldsymbol{x}))$ is usually unknown and only estimates of the risk are available. Optimizing the risk is thus typically a noisy optimization problem. Noisy optimization features prominently in simulation studies where

NOISY OPTIMIZATION

☞ 20

the behavior of some system (e.g., vehicles on a road network) is simulated under certain parameters (e.g., the lengths of the traffic light intervals) and the aim is to choose those parameters optimally (e.g., to maximize the traffic throughput). For each parameter setting the exact value for the objective function is unknown but estimates can be obtained via the simulation.

In general, suppose the goal is to minimize a function S, where S is unknown, but an estimate of $S(\boldsymbol{x})$ can be obtained for any choice of $\boldsymbol{x} \in \mathcal{X}$. Because the gradient ∇S is unknown, one cannot directly apply classical optimization methods. The *stochastic approximation* method mimics the classical gradient descent method by replacing a deterministic gradient with an estimate $\widehat{\nabla S}(\boldsymbol{x})$.

STOCHASTIC APPROXIMATION

A simple estimator for the i-th component of $\nabla S(\boldsymbol{x})$ (that is, $\partial S(\boldsymbol{u})/\partial x_i$), is the *central difference estimator*

CENTRAL DIFFERENCE ESTIMATOR

$$\frac{\widehat{S}(\boldsymbol{x} + \boldsymbol{e}_i\,\delta/2) - \widehat{S}(\boldsymbol{x} - \boldsymbol{e}_i\,\delta/2)}{\delta}, \tag{3.28}$$

where $\boldsymbol{e}_i$ denotes the i-th unit vector, and $\widehat{S}(\boldsymbol{x}+\boldsymbol{e}_i\,\delta/2)$ and $\widehat{S}(\boldsymbol{x}-\boldsymbol{e}_i\,\delta/2)$ can be any estimators of $S(\boldsymbol{x} + \boldsymbol{e}_i\,\delta/2)$ and $S(\boldsymbol{x} - \boldsymbol{e}_i\,\delta/2)$, respectively. The difference parameter $\delta > 0$ should be small enough to reduce the bias of the estimator, but large enough to keep the variance of the estimator small.

> To reduce the variance in the estimator (3.28) it is important to have $\widehat{S}(\boldsymbol{x} + \boldsymbol{e}_i\,\delta/2)$ and $\widehat{S}(\boldsymbol{x} - \boldsymbol{e}_i\,\delta/2)$ positively correlated. This can for example be achieved by using *common random numbers* in the simulation.

COMMON RANDOM NUMBERS

☞ 414

In direct analogy to gradient descent methods, the stochastic approximation method produces a sequence of iterates, starting with some $\boldsymbol{x}_1 \in \mathcal{X}$, via

$$\boldsymbol{x}_{t+1} = \boldsymbol{x}_t - \beta_t\,\widehat{\nabla S}(\boldsymbol{x}_t), \tag{3.29}$$

where $\beta_1, \beta_2, \ldots$ is a sequence of strictly positive step sizes. A generic stochastic approximation algorithm for minimizing a function S is thus as follows.

Algorithm 3.4.5: Stochastic Approximation

input: A mechanism to estimate any gradient $\nabla S(\boldsymbol{x})$ and step sizes $\beta_1, \beta_2, \ldots$.
output: Approximate optimizer of S.

Initialize $\boldsymbol{x}_1 \in \mathcal{X}$. Set $t \leftarrow 1$.
while a stopping criterion is not met **do**
  Obtain an estimated gradient $\widehat{\nabla S}(\boldsymbol{x}_t)$ of S at $\boldsymbol{x}_t$.
  Determine a step size β_t.
  Set $\boldsymbol{x}_{t+1} \leftarrow \boldsymbol{x}_t - \beta_t\,\widehat{\nabla S}(\boldsymbol{x}_t)$.
  $t \leftarrow t + 1$
return $\boldsymbol{x}_t$

When $\widehat{\nabla S}(\boldsymbol{x}_t)$ is an *unbiased* estimator of $\nabla S(\boldsymbol{x}_t)$ in (3.29) the stochastic approximation Algorithm 3.4.5 is referred to as the *Robbins–Monro* algorithm. When finite differences are used to estimate $\widehat{\nabla S}(\boldsymbol{x}_t)$, as in (3.28), the resulting algorithm is known as the

ROBBINS–MONRO

Kiefer–Wolfowitz algorithm. In Section 9.4.1 we will see how stochastic gradient descent is employed in deep learning to minimize the training loss, based on a "minibatch" of training data.

KIEFER–WOLFOWITZ
☞ 336

It can be shown [72] that, under certain regularity conditions on S, the sequence $\boldsymbol{x}_1, \boldsymbol{x}_2, \ldots$ converges to the true minimizer $\boldsymbol{x}^*$ when the step sizes decrease slowly enough to 0; in particular, when

$$\sum_{t=1}^{\infty} \beta_t = \infty \quad \text{and} \quad \sum_{t=1}^{\infty} \beta_t^2 < \infty. \tag{3.30}$$

> In practice, one rarely uses step sizes that satisfy (3.30), as the convergence of the sequence will be too slow to be of practical use.

An alternative approach to stochastic approximation is the *stochastic counterpart* method, also called *sample average approximation*. It can be applied in situations where the noisy objective function is of the form

STOCHASTIC COUNTERPART

$$S(\boldsymbol{x}) = \mathbb{E}\widetilde{S}(\boldsymbol{x}, \boldsymbol{\xi}), \quad \boldsymbol{x} \in \mathcal{X}, \tag{3.31}$$

where $\boldsymbol{\xi}$ is a random vector that can be simulated and $\widetilde{S}(\boldsymbol{x}, \boldsymbol{\xi})$ can be evaluated exactly. The idea is to replace the optimization of (3.31) with that of the sample average

$$\widehat{S}(\boldsymbol{x}) = \frac{1}{N} \sum_{i=1}^{N} \widetilde{S}(\boldsymbol{x}, \boldsymbol{\xi}_i), \quad \boldsymbol{x} \in \mathcal{X}, \tag{3.32}$$

where $\boldsymbol{\xi}_1, \ldots, \boldsymbol{\xi}_N$ are iid copies of $\boldsymbol{\xi}$. Note that $\widehat{S}$ is a deterministic function of $\boldsymbol{x}$ and so can be optimized using any optimization algorithm. A solution to this sample average version is taken to be an estimator of a solution $\boldsymbol{x}^*$ to the original problem (3.31).

■ **Example 3.18 (Determining Good Importance Sampling Parameters)** The selection of good importance sampling parameters can be viewed as a stochastic optimization problem. Consider, for instance, the importance sampling estimator in Example 3.14. Recall that the nominal distribution is the uniform distribution on the square $[-b, b]^2$, with pdf

☞ 94

$$f_b(\boldsymbol{x}) = \frac{1}{(2b)^2}, \quad \boldsymbol{x} \in [-b, b]^2,$$

where b is large enough to ensure that μ_b is close to μ; in that example, we chose $b = 1000$. The importance sampling pdf is

$$g_\lambda(\boldsymbol{x}) = f_{R,\Theta}(r, \theta)\frac{1}{r} = \lambda \mathrm{e}^{-\lambda r}\frac{1}{2\pi}\frac{1}{r} = \frac{\lambda \mathrm{e}^{-\lambda\sqrt{x_1^2+x_2^2}}}{2\pi\sqrt{x_1^2+x_2^2}}, \quad \boldsymbol{x} = (x_1, x_2) \in \mathbb{R}^2 \setminus \{\boldsymbol{0}\},$$

which depends on a free parameter λ. In the example we chose $\lambda = 0.1$. Is this the best choice? Maybe $\lambda = 0.05$ or 0.2 would have resulted in a more accurate estimate. The important thing to realize is that the "effectiveness" of λ can be measured in terms of the variance of the estimator $\widehat{\mu}$ in (3.19), which is given by

☞ 93

$$\frac{1}{N}\mathbb{V}\mathrm{ar}_{g_\lambda}\left(H(\boldsymbol{X})\frac{f(\boldsymbol{X})}{g_\lambda(\boldsymbol{X})}\right) = \frac{1}{N}\mathbb{E}_{g_\lambda}\left[H^2(\boldsymbol{X})\frac{f^2(\boldsymbol{X})}{g_\lambda^2(\boldsymbol{X})}\right] - \frac{\mu^2}{N} = \frac{1}{N}\mathbb{E}_f\left[H^2(\boldsymbol{X})\frac{f(\boldsymbol{X})}{g_\lambda(\boldsymbol{X})}\right] - \frac{\mu^2}{N}.$$

Hence, the optimal parameter λ^* minimizes the function $S(\lambda) = \mathbb{E}_f[H^2(\boldsymbol{X})f(\boldsymbol{X})/g_\lambda(\boldsymbol{X})]$, which is unknown, but can be estimated from simulation. To solve this stochastic minimization problem, we first use stochastic approximation. Thus, at each step of the algorithm, the gradient of $S(\lambda)$ is estimated from realizations of $\widehat{S}(\lambda) = H^2(\boldsymbol{X})f(\boldsymbol{X})/g_\lambda(\boldsymbol{X})$, where $\boldsymbol{X} \sim f_b$. As in the original problem (that is, the estimation of μ), the parameter b should be large enough to avoid any bias in the estimator of λ^*, but also small enough to ensure a small variance. The following Python code implements a particular instance of Algorithm 3.4.5. For sampling from f_b here, we used $b = 100$ instead of $b = 1000$, as this will improve the crude Monte Carlo estimation of λ^*, without noticeably affecting the bias. The gradient of $S(\lambda)$ is estimated in Lines 11–17, using the central difference estimator (3.28). Notice how for the $S(\lambda-\delta/2)$ and $S(\lambda+\delta/2)$ the *same* random vector $\boldsymbol{X} = [X_1, X_2]^\top$ is used. This significantly reduces the variance of the gradient estimator; see also Exercise 23. The step size β_t should be such that $\beta_t\widehat{\nabla S}(\boldsymbol{x}_t) \approx \lambda_t$. Given the large gradient here, we choose $\beta_0 = 10^{-7}$ and decrease it each step by a factor of 0.99. Figure 3.13 shows how the sequence $\lambda_0, \lambda_1, \ldots$ decreases towards approximately 0.125, which we take as an estimator for the optimal importance sampling parameter λ^*.

☞ 118

stochapprox.py

```
import numpy as np
from numpy import pi
import matplotlib.pyplot as plt

b=100    # choose b large enough, but not too large
delta = 0.01
H = lambda x1, x2: (2*b)**2*np.exp(-np.sqrt(x1**2 + x2**2)/4)*(np.
    sin(2*np.sqrt(x1**2+x2**2)+1))*(x1**2+x2**2<b**2)
f = 1/(2*b)**2
g = lambda x1, x2, lam: lam*np.exp(-np.sqrt(x1**2+x2**2)*lam)/np.
    sqrt(x1**2+x2**2)/(2*pi)
beta = 10**-7   #step size very small, as the gradient is large
lam=0.25
lams = np.array([lam])
N=10**4
for i in range(200):
    x1 = -b + 2*b*np.random.rand(N,1)
    x2 = -b + 2*b*np.random.rand(N,1)
    lamL = lam - delta/2
    lamR = lam + delta/2
    estL = np.mean(H(x1,x2)**2*f/g(x1, x2, lamL))
    estR = np.mean(H(x1,x2)**2*f/g(x1, x2, lamR))  #use SAME x1,x2
    gr = (estR-estL)/delta  #gradient
    lam = lam - gr*beta  #gradient descend
    lams = np.hstack((lams, lam))
    beta = beta*0.99

lamsize=range(0, (lams.size))
plt.plot(lamsize, lams)
plt.show()
```

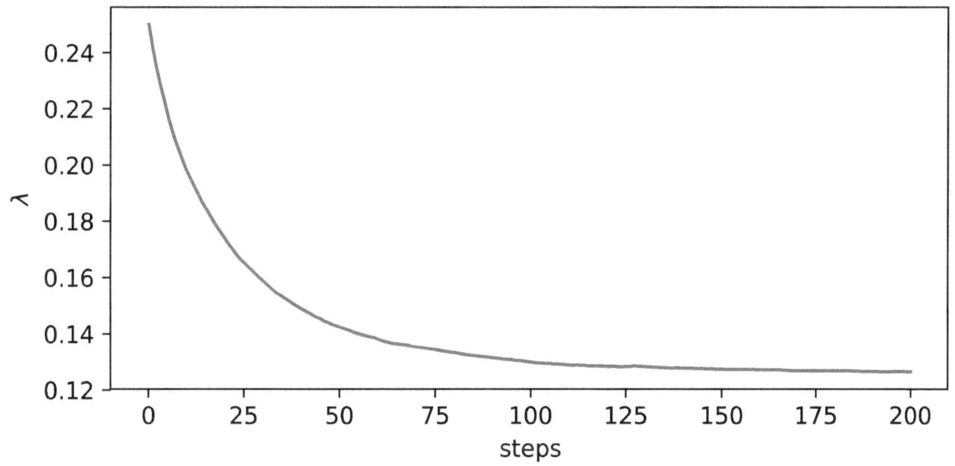

Figure 3.13: The stochastic optimization algorithm produces a sequence $\lambda_t, t = 0, 1, 2, \ldots$ that tends to an approximate estimate of the optimal importance sampling parameter $\lambda^* \approx 0.125$.

Next, we estimate λ^* using a stochastic counterpart approach. As the objective function $S(\lambda)$ is of the form (3.31) (with λ taking the role of $\boldsymbol{x}$ and $\boldsymbol{X}$ the role of $\boldsymbol{\xi}$), we obtain the sample average

$$\widehat{S}(\lambda) = \frac{1}{N} \sum_{i=1}^{N} H^2(\boldsymbol{X}_i) \frac{f(\boldsymbol{X}_i)}{g_\lambda(\boldsymbol{X}_i)}, \tag{3.33}$$

where $\boldsymbol{X}_1, \ldots, \boldsymbol{X}_N \sim_{\text{iid}} f_b$. Once the $\boldsymbol{X}_1, \ldots, \boldsymbol{X}_N \sim_{\text{iid}} f_b$ have been simulated, $\widehat{S}(\lambda)$ is a deterministic function of λ, which can be optimized by any means. We take the most basic approach and simply evaluate the function for $\lambda = 0.01, 0.02, \ldots, 0.3$ and select the minimizing λ on this grid. The code is given below and Figure 3.14 shows $\widehat{S}(\lambda)$ as a function of λ. The minimum value found was $0.60 \cdot 10^4$ for minimizer $\widehat{\lambda}^* = 0.12$, which is in accordance with the value obtained via stochastic approximation. The sensitivity of this estimate can be assessed from the graph: for a wide range of values (say from 0.04 to 0.15) $\widehat{S}$ stays rather flat. So any of these values could be used in an importance sampling procedure to estimate μ. However, very small values (less than 0.02) and large values (greater than 0.25) should be avoided. Our original choice of $\lambda = 0.1$ was therefore justified and we could not have done much better.

stochcounterpart.py

```
from stochapprox import *

lams = np.linspace(0.01, 0.31, 1000)
res=[]
res = np.array(res)
for i in range(lams.size):
    lam = lams[i]
    np.random.seed(1)
    g = lambda x1, x2: lam*np.exp(-np.sqrt(x1**2+x2**2)*lam)/np.sqrt
        (x1**2+x2**2)/(2*pi)
```

```
    X=-b+2*b*np.random.rand(N,1)
    Y=-b+2*b*np.random.rand(N,1)
    Z=H(X,Y)**2*f/g(X,Y)
    estCMC = np.mean(Z)
    res = np.hstack((res, estCMC))

plt.plot(lams, res)
plt.xlabel(r'$\lambda$')
plt.ylabel(r'$\hat{S}(\lambda)$')
plt.ticklabel_format(style='sci', axis='y', scilimits=(0,0))
plt.show()
```

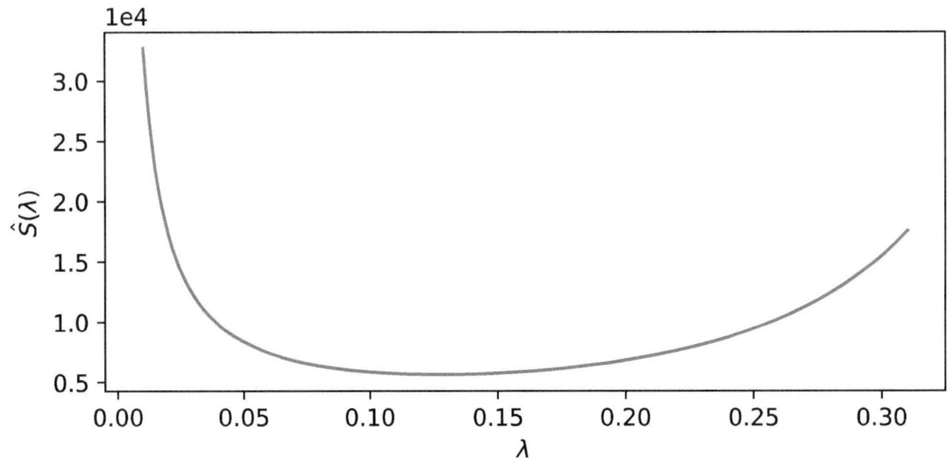

Figure 3.14: The stochastic counterpart method replaces the unknown $S(\lambda)$ (that is, the scaled variance of the importance sampling estimator) with its sample average, $\widehat{S}(\lambda)$. The minimum value of $\widehat{S}$ is attained around $\lambda = 0.12$.

■

☞ 101

A third method for stochastic optimization is the cross-entropy method. In particular, Algorithm 3.4.3 can easily be modified to minimize *noisy* functions $S(\boldsymbol{x}) = \mathbb{E}\widetilde{S}(\boldsymbol{x}, \boldsymbol{\xi})$, as defined in (3.31). The only change required in the algorithm is that every function value $S(\boldsymbol{x})$ be replaced by its estimate $\widehat{S}(\boldsymbol{x})$. Depending on the level of noise in the function, the sample size N might have to be increased considerably.

■ **Example 3.19 (Cross-Entropy Method for Noisy Optimization)** To explore the use of the CE method for noisy optimization, take the following noisy discrete optimization problem. Suppose there is a "black box" that contains an unknown binary sequence of n bits. If one feeds the black box any input vector, it will first scramble the input by independently flipping the bits (changing 0 to 1 and 1 to 0) with a probability θ and then return the number of bits that do not match the true (unknown) binary sequence. This is illustrated in Figure 3.15 for $n = 10$.

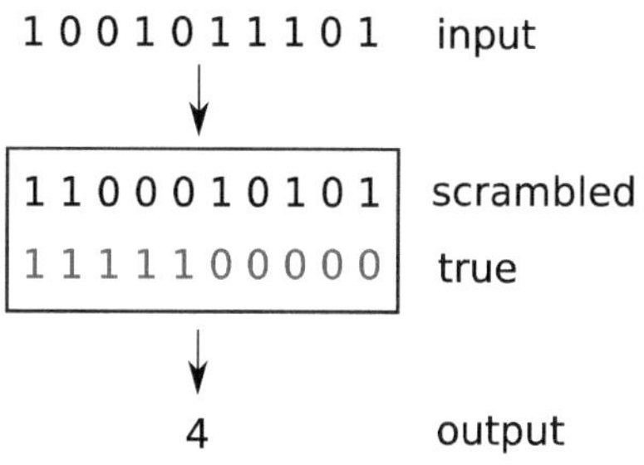

Figure 3.15: A noisy optimization function as a black box. The input to the black box is a binary vector. Inside the black box the digits of the input vector are scrambled by flipping bits with probability θ. The output is the number of bits of the scrambled vector that do not match the true (unknown) binary vector.

Denoting by $S(\boldsymbol{x})$ the true number of matching digits for a binary input vector $\boldsymbol{x}$, the black box thus returns a noisy estimate $\widehat{S}(\boldsymbol{x})$. The objective is to estimate the binary sequence inside the black box, by feeding it with many input vectors and observing their output. Or, to put it in a different way, to minimize $S(\boldsymbol{x})$ using $\widehat{S}(\boldsymbol{x})$ as a proxy. Since there are 2^n possible input vectors, it is infeasible to try all possible vectors $\boldsymbol{x}$ even for moderate n.

The following Python program implements the noisy function $\widehat{S}(\boldsymbol{x})$ for $n = 100$. Each input bit is flipped with a rather high probability $\theta = 0.4$, so that the output is a poor indicator of how many bits actually match the true vector. This true vector has 1s at positions $1, \ldots, 50$ and 0s at $51, \ldots, 100$.

Snoisy.py

```
import numpy as np

def Snoisy(X):   #takes a matrix
    n = X.shape[1]
    N = X.shape[0]
    # true binary vector
    xorg = np.hstack((np.ones((1,n//2)), np.zeros((1,n//2))))
    theta = 0.4 # probability to flip the input
    # storing the number of bits unequal to the true vector
    s = np.zeros(N)
    for i in range(0,N):
       # determine which bits to flip
       flip = (np.random.uniform(size=(n)) < theta).astype(int)
       ind = flip>0
       X[i][ind] = 1-X[i][ind]
       s[i] = (X[i] != xorg).sum()
    return s
```

The CE code below to optimize $S(\boldsymbol{x})$ is quite similar to the continuous optimization code in Example 3.16. However, instead of sampling iid random variables $X_1, \ldots, X_N$ from a normal distribution, we now sample iid binary vectors $\boldsymbol{X}_1, \ldots, \boldsymbol{X}_N$ from a $\mathsf{Ber}(\boldsymbol{p})$ distribution. More precisely, given a row vector of probabilities $\boldsymbol{p} = [p_1, \ldots, p_n]$, we independently simulate the components $X_1, \ldots, X_n$ of each binary vector $\boldsymbol{X}$ according to $X_i \sim \mathsf{Ber}(p_i)$, $i = 1, \ldots, n$. After each iteration, the vector $\boldsymbol{p}$ is updated as the (vector) mean of the elite ☞ 101 samples. The sample size is $N = 1000$ and the number of elite samples is 100. The components of the initial sampling vector $\boldsymbol{p}$ are all equal to $1/2$; that is, the $\boldsymbol{X}$ are initially uniformly sampled from the set of all binary vectors of length $n = 100$. At each subsequent iteration the parameter vector is updated via the mean of the elite samples and evolves towards a degenerate vector $\boldsymbol{p}^*$ with only 1s and 0s. Sampling from such a $\mathsf{Ber}(\boldsymbol{p}^*)$ distribution gives an outcome $\boldsymbol{x}^* = \boldsymbol{p}^*$, which can be taken as an estimate for the minimizer of S; that is, the true binary vector hidden in the black box. The algorithm stops when $\boldsymbol{p}$ has degenerated sufficiently.

Figure 3.16 shows the evolution of the vector of probabilities $\boldsymbol{p}$. This figure may be seen as the discrete analogue of Figure 3.12. We see that, despite the high noise, the CE method is able to find the true state of the black box, and hence the minimum value of S.

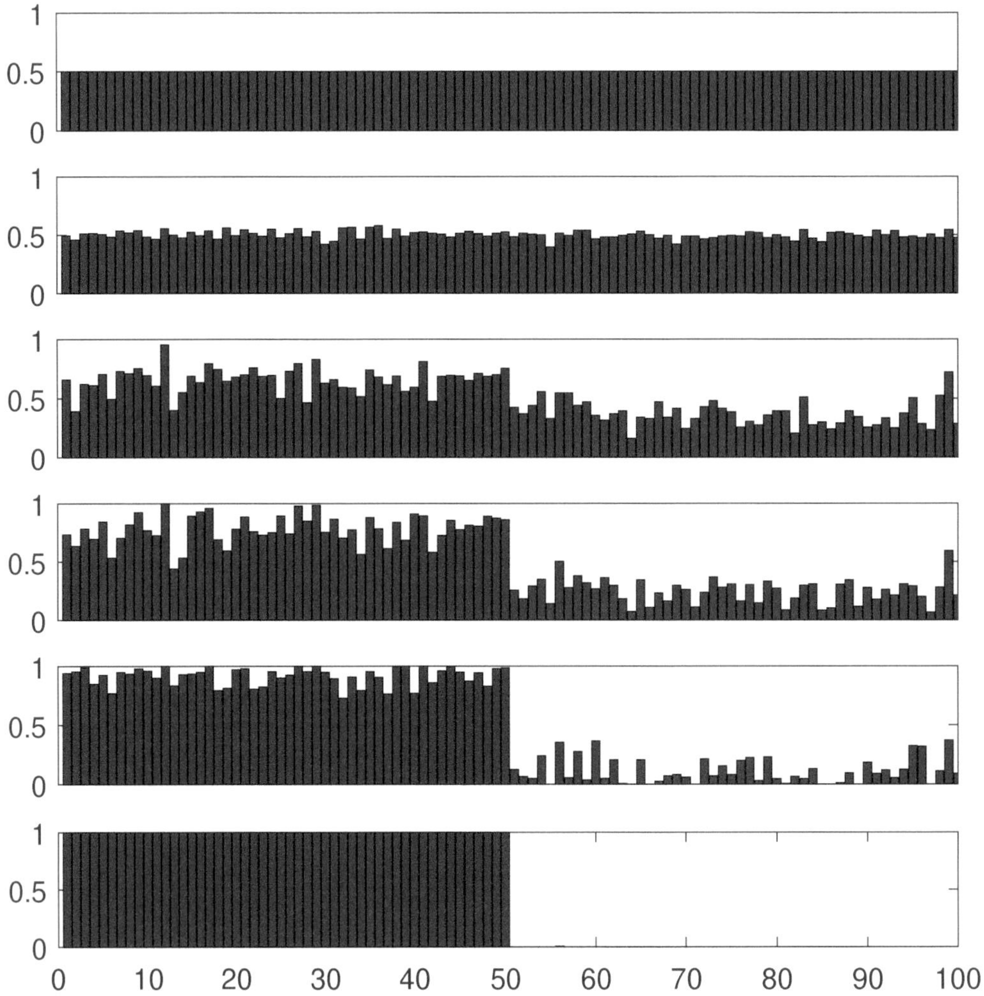

Figure 3.16: Evolution of the vector of probabilities $\boldsymbol{p} = [p_1, \ldots, p_n]$ towards the degenerate solution.

CEnoisy.py

```
from Snoisy import Snoisy
import numpy as np
n = 100
rho = 0.1
N = 1000; Nel = int(N*rho); eps = 0.01
p = 0.5*np.ones(n)
i = 0
pstart = p
ps = np.zeros((1000,n))
ps[0] = pstart
pdist = np.zeros((1,1000))
while np.max(np.minimum(p,1-p)) > eps:
    i += 1
    X = (np.random.uniform(size=(N,n)) < p).astype(int)
    X_tmp = np.array(X, copy=True)
    SX = Snoisy(X_tmp)
    ids = np.argsort(SX,axis=0)
    Elite = X[ids[0:Nel],:]
    p = np.mean(Elite,axis=0)
    ps[i] = p
print(p)
```

■

Further Reading

The article [68] explores why the Monte Carlo method is so important in today's quantitative investigations. The *Handbook of Monte Carlo Methods* [71] provides a comprehensive overview of Monte Carlo simulation that explores the latest topics, techniques, and real-world applications. Popular books on simulation and the Monte Carlo method include [42], [75], and [104]. A classic reference on random variable generation is [32]. Easy introductions to stochastic simulation are given in [49], [98], and [100]. More advanced theory can be found in [5]. Markov chain Monte Carlo is detailed in [50] and [99]. The research monograph on the cross-entropy method is [103] and a tutorial is provided in [30]. A range of optimization applications of the CE method is given in [16]. Theoretical results on adaptive tuning schemes for simulated annealing may be found, for example, in [111]. There are several established ways for gradient estimation. These include the finite difference method, infinitesimal perturbation analysis, the score function method, and the method of weak derivatives; see, for example, [51, Chapter 7].

Exercises

1. We can modify the Box–Muller method in Example 3.1 to draw X and Y uniformly on the unit disc, $\{(x, y) \in \mathbb{R}^2 : x^2 + y^2 \leqslant 1\}$, in the following way: Independently draw a radius R and an angle $\Theta \sim \mathcal{U}(0, 2\pi)$, and return $X = R\cos(\Theta)$, $Y = R\sin(\Theta)$. The question is how to draw R. ☞ 69

    (a) Show that the cdf of R is given by $F_R(r) = r^2$ for $0 \leqslant r \leqslant 1$ (with $F_R(r) = 0$ and

$F_R(r) = 1$ for $r < 0$ and $r > 1$, respectively).

(b) Explain how to simulate R using the inverse-transform method.

(c) Simulate 100 independent draws of $[X, Y]^\top$ according to the method described above.

2. A simple acceptance–rejection method to simulate a vector $\boldsymbol{X}$ in the unit d-ball $\{\boldsymbol{x} \in \mathbb{R}^d : \|\boldsymbol{x}\| \leqslant 1\}$ is to first generate $\boldsymbol{X}$ uniformly in the hyper cube $[-1, 1]^d$ and then to accept the point only if $\|\boldsymbol{X}\| \leqslant 1$. Determine an analytic expression for the probability of acceptance as a function of d and plot this for $d = 1, \ldots, 50$.

3. Let the random variable X have pdf

$$f(x) = \begin{cases} \frac{1}{2}\,x\,, & 0 \leqslant x < 1, \\ \frac{1}{2}\,, & 1 \leqslant x \leqslant \frac{5}{2}\,. \end{cases}$$

Simulate a random variable from $f(x)$, using

(a) the inverse-transform method;

(b) the acceptance–rejection method, using the proposal density

$$g(x) = \frac{8}{25}\,x\,, \quad 0 \leqslant x \leqslant \frac{5}{2}\,.$$

4. Construct simulation algorithms for the following distributions:

(a) The $\mathsf{Weib}(\alpha, \lambda)$ distribution, with cdf $F(x) = 1 - \mathrm{e}^{-(\lambda x)^\alpha}$, $x \geqslant 0$, where $\lambda > 0$ and $\alpha > 0$.

(b) The $\mathsf{Pareto}(\alpha, \lambda)$ distribution, with pdf $f(x) = \alpha\lambda(1 + \lambda x)^{-(\alpha+1)}$, $x \geqslant 0$, where $\lambda > 0$ and $\alpha > 0$.

5. We wish to sample from the pdf

$$f(x) = x\,\mathrm{e}^{-x}, \quad x \geqslant 0,$$

using acceptance–rejection with the proposal pdf $g(x) = \mathrm{e}^{-x/2}/2$, $x \geqslant 0$.

(a) Find the smallest C for which $Cg(x) \geqslant f(x)$ for all x.

(b) What is the efficiency of this acceptance–rejection method?

6. Let $[X, Y]^\top$ be uniformly distributed on the triangle with corners $(0, 0), (1, 2)$, and $(-1, 1)$. Give the distribution of $[U, V]^\top$ defined by the linear transformation

$$\begin{bmatrix} U \\ V \end{bmatrix} = \begin{bmatrix} 1 & 2 \\ 3 & 4 \end{bmatrix} \begin{bmatrix} X \\ Y \end{bmatrix}.$$

7. Explain how to generate a random variable from the *extreme value distribution*, which has cdf

$$F(x) = 1 - \mathrm{e}^{-\exp(\frac{x-\mu}{\sigma})}\,, \quad -\infty < x < \infty, \quad (\sigma > 0),$$

via the inverse-transform method.

8. Write a program that generates and displays 100 random vectors that are uniformly distributed within the ellipse

$$5\,x^2 + 21\,x\,y + 25\,y^2 = 9.$$

[Hint: Consider generating uniformly distributed samples within the circle of radius 3 and use the fact that linear transformations preserve uniformity to transform the circle to the given ellipse.]

9. Suppose that $X_i \sim \mathsf{Exp}(\lambda_i)$, independently, for all $i = 1, \ldots, n$. Let $\mathbf{\Pi} = [\Pi_1, \ldots, \Pi_n]^\top$ be the random permutation induced by the ordering $X_{\Pi_1} < X_{\Pi_2} < \cdots < X_{\Pi_n}$, and define $Z_1 := X_{\Pi_1}$ and $Z_j := X_{\Pi_j} - X_{\Pi_{j-1}}$ for $j = 2, \ldots, n$.

   (a) Determine an $n \times n$ matrix $\mathbf{A}$ such that $\mathbf{Z} = \mathbf{A}\boldsymbol{X}$ and show that $\det(\mathbf{A}) = 1$.

   (b) Denote the joint pdf of $\boldsymbol{X}$ and $\mathbf{\Pi}$ as

$$f_{\boldsymbol{X},\mathbf{\Pi}}(\boldsymbol{x}, \boldsymbol{\pi}) = \prod_{i=1}^{n} \lambda_{\pi_i} \exp\left(-\lambda_{\pi_i} x_{\pi_i}\right) \times \mathbb{1}\{x_{\pi_1} < \cdots < x_{\pi_n}\}, \quad \boldsymbol{x} \geqslant \mathbf{0},\ \boldsymbol{\pi} \in \mathcal{P}_n,$$

   where $\mathcal{P}_n$ is the set of all $n!$ permutations of $\{1, \ldots, n\}$. Use the multivariate transformation formula (C.22) to show that ☞ 434

$$f_{\boldsymbol{Z},\mathbf{\Pi}}(\boldsymbol{z}, \boldsymbol{\pi}) = \exp\left(-\sum_{i=1}^{n} z_i \sum_{k \geqslant i} \lambda_{\pi_k}\right) \prod_{i=1}^{n} \lambda_i, \qquad \boldsymbol{z} \geqslant \mathbf{0},\ \boldsymbol{\pi} \in \mathcal{P}_n.$$

   Hence, conclude that the probability mass function of the random permutation $\mathbf{\Pi}$ is:

$$\mathbb{P}[\mathbf{\Pi} = \boldsymbol{\pi}] = \prod_{i=1}^{n} \frac{\lambda_{\pi_i}}{\sum_{k \geqslant i} \lambda_{\pi_k}}, \qquad \boldsymbol{\pi} \in \mathcal{P}_n.$$

   (c) Write pseudo-code to simulate a *uniform* random permutation $\mathbf{\Pi} \in \mathcal{P}_n$; that is, such that $\mathbb{P}[\mathbf{\Pi} = \boldsymbol{\pi}] = \frac{1}{n!}$, and explain how this uniform random permutation can be used to reshuffle a training set τ_n.

10. Consider the Markov chain with transition graph given in Figure 3.17, starting in state 1.

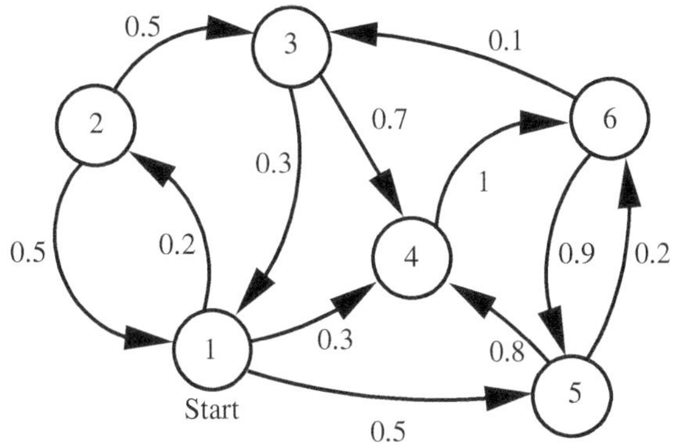

Figure 3.17: The transition graph for the Markov chain $\{X_t, t = 0, 1, 2, \ldots\}$.

(a) Construct a computer program to simulate the Markov chain, and show a realization for $N = 100$ steps.

☞ 454

(b) Compute the limiting probabilities that the Markov chain is in state 1,2,...,6, by solving the global balance equations (C.42).

(c) Verify that the exact limiting probabilities correspond to the average fraction of times that the Markov process visits states 1,2,...,6, for a large number of steps N.

☞ 455

11. As a generalization of Example C.9, consider a random walk on an arbitrary undirected connected graph with a finite vertex set $\mathcal{V}$. For any vertex $v \in \mathcal{V}$, let $d(v)$ be the number of neighbors of v — called the *degree* of v. The random walk can jump to each one of the neighbors with probability $1/d(v)$ and can be described by a Markov chain. Show that, if the chain is *aperiodic*, the limiting probability that the chain is in state v is equal to $d(v)/\sum_{v' \in \mathcal{V}} d(v')$.

☞ 76

12. Let $U, V \sim_{\text{iid}} \mathcal{U}(0, 1)$. The reason why in Example 3.7 the sample mean and sample median behave very differently is that $\mathbb{E}[U/V] = \infty$, while the median of U/V is finite. Show this, and compute the median. [Hint: start by determining the cdf of $Z = U/V$ by writing it as an expectation of an indicator function.]

13. Consider the problem of generating samples from $Y \sim \mathsf{Gamma}(2, 10)$.

(a) Direct simulation: Let $U_1, U_2 \sim_{\text{iid}} \mathcal{U}(0, 1)$. Show that $-\ln(U_1)/10 - \ln(U_2)/10 \sim \mathsf{Gamma}(2, 10)$. [Hint: derive the distribution of $-\ln(U_1)/10$ and use Example C.1.]

☞ 429

(b) Simulation via MCMC: Implement an independence sampler to simulate from the $\mathsf{Gamma}(2, 10)$ target pdf

$$f(x) = 100\, x\, \mathrm{e}^{-10x}, \quad x \geqslant 0,$$

using proposal transition density $q(y \mid x) = g(y)$, where $g(y)$ is the pdf of an $\mathsf{Exp}(5)$ random variable. Generate $N = 500$ samples, and compare the true cdf with the empirical cdf of the data.

14. Let $\boldsymbol{X} = [X, Y]^\top$ be a random column vector with a bivariate normal distribution with expectation vector $\boldsymbol{\mu} = [1, 2]^\top$ and covariance matrix

$$\boldsymbol{\Sigma} = \begin{bmatrix} 1 & a \\ a & 4 \end{bmatrix}.$$

(a) What are the conditional distributions of $(Y \mid X = x)$ and $(X \mid Y = y)$? [Hint: use Theorem C.8.]

☞ 438

(b) Implement a Gibbs sampler to draw 10^3 samples from the bivariate distribution $\mathcal{N}(\boldsymbol{\mu}, \Sigma)$ for $a = 0$, 1, and 1.75, and plot the resulting samples.

15. Here the objective is to sample from the 2-dimensional pdf

$$f(x, y) = c\, \mathrm{e}^{-(xy+x+y)}, \quad x \geqslant 0, \quad y \geqslant 0,$$

for some normalization constant c, using a Gibbs sampler. Let $(X, Y) \sim f$.

(a) Find the conditional pdf of X given $Y = y$, and the conditional pdf of Y given $X = x$.

(b) Write working Python code that implements the Gibbs sampler and outputs 1000 points that are approximately distributed according to f.

(c) Describe how the normalization constant c could be estimated via Monte Carlo simulation, using random variables $X_1, \ldots, X_N, Y_1, \ldots, Y_N \overset{\text{iid}}{\sim} \mathsf{Exp}(1)$.

16. We wish to estimate $\mu = \int_{-2}^{2} \mathrm{e}^{-x^2/2}\,\mathrm{d}x = \int H(x)f(x)\,\mathrm{d}x$ via Monte Carlo simulation using two different approaches: (1) defining $H(x) = 4\,\mathrm{e}^{-x^2/2}$ and f the pdf of the $\mathcal{U}[-2, 2]$ distribution and (2) defining $H(x) = \sqrt{2\pi}\,\mathbb{1}\{-2 \leqslant x \leqslant 2\}$ and f the pdf of the $\mathcal{N}(0, 1)$ distribution.

(a) For both cases estimate μ via the estimator $\widehat{\mu}$

$$\widehat{\mu} = N^{-1} \sum_{i=1}^{N} H(\boldsymbol{X}_i). \tag{3.34}$$

Use a sample size of $N = 1000$.

(b) For both cases estimate the relative error κ of $\widehat{\mu}$ using $N = 100$.

(c) Give a 95% confidence interval for μ for both cases using $N = 100$.

(d) From part (b), assess how large N should be such that the relative width of the confidence interval is less than 0.01, and carry out the simulation with this N. Compare the result with the true value of μ.

17. Consider estimation of the tail probability $\mu = \mathbb{P}[X \geqslant \gamma]$ of some random variable X, where γ is large. The crude Monte Carlo estimator of μ is

$$\widehat{\mu} = \frac{1}{N} \sum_{i=1}^{N} Z_i, \tag{3.35}$$

where $X_1, \ldots, X_N$ are iid copies of X and $Z_i = \mathbb{1}\{X_i \geqslant \gamma\}$, $i = 1, \ldots, N$.

(a) Show that $\widehat{\mu}$ is unbiased; that is, $\mathbb{E}\,\widehat{\mu} = \mu$.

(b) Express the relative error of $\widehat{\mu}$, i.e.,

$$\mathrm{RE} = \frac{\sqrt{\mathbb{V}\mathrm{ar}\,\widehat{\mu}}}{\mathbb{E}\,\widehat{\mu}},$$

in terms of N and μ.

(c) Explain how to estimate the relative error of $\widehat{\mu}$ from outcomes $x_1, \ldots, x_N$ of $X_1, \ldots, X_N$, and how to construct a 95% confidence interval for μ.

(d) An unbiased estimator Z of μ is said to be *logarithmically efficient* if

$$\lim_{\gamma \to \infty} \frac{\ln \mathbb{E} Z^2}{\ln \mu^2} = 1. \tag{3.36}$$

Show that the CMC estimator (3.35) with $N = 1$ is not logarithmically efficient.

18. One of the test cases in [70] involves the minimization of the *Hougen* function. Implement a cross-entropy and a simulated annealing algorithm to carry out this optimization task.

19. In the *binary knapsack problem*, the goal is to solve the optimization problem:

$$\max_{x \in \{0,1\}^n} \boldsymbol{p}^\top \boldsymbol{x},$$

subject to the constraints

$$\mathbf{A}\boldsymbol{x} \leqslant \boldsymbol{c},$$

where $\boldsymbol{p}$ and $\boldsymbol{w}$ are $n \times 1$ vectors of non-negative numbers, $\mathbf{A} = (a_{ij})$ is an $m \times n$ matrix, and $\boldsymbol{c}$ is an $m \times 1$ vector. The interpretation is that $x_j = 1$ or 0 depending on whether item j with value p_j is packed into the knapsack or not , $j = 1, \ldots, n$; The variable a_{ij} represents the i-th attribute (e.g., volume, weight) of the j-th item. Associated with each attribute is a maximal capacity, e.g., c_1 could be the maximum volume of the knapsack, c_2 the maximum weight, etc.

Write a CE program to solve the `Sento1.dat` knapsack problem at `http://people.brunel.ac.uk/~mastjjb/jeb/orlib/files/mknap2.txt`, as described in [16].

20. Let $(C_1, R_1), (C_2, R_2), \ldots$ be a renewal reward process, with $\mathbb{E}R_1 < \infty$ and $\mathbb{E}C_1 < \infty$. Let $A_t = \sum_{i=1}^{N_t} R_i/t$ be the average reward at time $t = 1, 2, \ldots$, where $N_t = \max\{n : T_n \leqslant t\}$ and we have defined $T_n = \sum_{i=1}^{n} C_i$ as the time of the n-th renewal.

    (a) Show that $T_n/n \xrightarrow{\text{a.s.}} \mathbb{E}C_1$ as $n \to \infty$.

    (b) Show that $N_t \xrightarrow{\text{a.s.}} \infty$ as $t \to \infty$.

    (c) Show that $N_t/t \xrightarrow{\text{a.s.}} 1/\mathbb{E}C_1$ as $t \to \infty$. [Hint: Use the fact that $T_{N_t} \leqslant t \leqslant T_{N_t+1}$ for all $t = 1, 2, \ldots$.]

    (d) Show that

$$A_t \xrightarrow{\text{a.s.}} \frac{\mathbb{E}R_1}{\mathbb{E}C_1} \quad \text{as} \quad t \to \infty.$$

☞ 92 21. Prove Theorem 3.3.

☞ 96 22. Prove that if $H(\boldsymbol{x}) \geqslant 0$ the importance sampling pdf g^* in (3.22) gives the zero-variance importance sampling estimator $\widehat{\mu} = \mu$.

23. Let X and Y be random variables (not necessarily independent) and suppose we wish to estimate the expected difference $\mu = \mathbb{E}[X - Y] = \mathbb{E}X - \mathbb{E}Y$.

    (a) Show that if X and Y are *positively correlated*, the variance of $X - Y$ is smaller than if X and Y are *independent*.

    (b) Suppose now that X and Y have cdfs F and G, respectively, and are simulated via the inverse-transform method: $X = F^{-1}(U)$, $Y = G^{-1}(V)$, with $U, V \sim \mathcal{U}(0, 1)$, not necessarily independent. Intuitively, one might expect that if U and V are positively correlated, the variance of $X-Y$ would be smaller than if U and V are independent. Show that this is not always the case by providing a counter-example.

(c) Continuing (b), assume now that F and G are continuous. Show that the variance of $X - Y$ by taking *common random numbers* $U = V$ is no larger than when U and V are independent. [Hint: Use the following lemma of Hoeffding [41]: If (X, Y) have joint cdf H with marginal cdfs of X and Y being F and G, respectively, then

$$\mathbb{C}\text{ov}(X, Y) = \int_{-\infty}^{\infty} \int_{-\infty}^{\infty} (H(x, y) - F(x)\, G(y))\, \mathrm{d}x\, \mathrm{d}y,$$

provided $\mathbb{C}\text{ov}(X, Y)$ exists.]

CHAPTER 4

UNSUPERVISED LEARNING

When there is no distinction between response and explanatory variables, unsupervised methods are required to learn the structure of the data. In this chapter we look at various unsupervised learning techniques, such as density estimation, clustering, and principal component analysis. Important tools in unsupervised learning include the cross-entropy training loss, mixture models, the Expectation–Maximization algorithm, and the Singular Value Decomposition.

4.1 Introduction

In contrast to supervised learning, where an "output" (response) variable y is explained by an "input" (explanatory) vector $\boldsymbol{x}$, in unsupervised learning there is no response variable and the overall goal is to extract useful information and patterns from the data, e.g., in the form $\tau = \{\boldsymbol{x}_1, \ldots, \boldsymbol{x}_n\}$ or as a matrix $\mathbf{X}^\top = [\boldsymbol{x}_1, \ldots, \boldsymbol{x}_n]$. In essence, the objective of unsupervised learning is to learn about the underlying probability distribution of the data.

We start in Section 4.2 by setting up a framework for unsupervised learning that is similar to the framework used for supervised learning in Section 2.3. That is, we formulate ☞ 23 unsupervised learning in terms of risk and loss minimization; but now involving the cross-entropy risk, rather than the squared-error risk. In a natural way this leads to fundamental learning concepts such as likelihood, Fisher information, and the Akaike information criterion. Section 4.3 introduces the Expectation–Maximization (EM) algorithm as a useful method for maximizing likelihood functions when their solution cannot be found easily in closed form.

If the data forms an iid sample from some unknown distribution, the "empirical distribution" of the data provides valuable information about the unknown distribution. In Section 4.4 we formalize the concept of the empirical distribution (a generalization of the empirical cdf) and explain how we can produce an estimate of the underlying probability ☞ 11 density function of the data using kernel density estimators.

Most unsupervised learning techniques focus on identifying certain traits of the underlying distribution, such as its local maximizers. A related idea is to partition the data into clusters of points that are in some sense "similar" to each other. In Section 4.5 we formulate the clustering problem in terms of a mixture model. In particular, the data are assumed ☞ 135 to come from a mixture of (usually Gaussian) distributions, and the objective is to recover the parameters of the mixture distributions from the data. The principal tool for parameter estimation in mixture models is the EM algorithm.

Section 4.6 discusses a more heuristic approach to clustering, where the data are grouped according to certain "cluster centers", whose positions are found by solving an optimization problem. Section 4.7 describes how clusters can be constructed in a hierarchical manner.

Finally, in Section 4.8 we discuss the unsupervised learning technique called Principal Component Analysis (PCA), which is an important tool for reducing the dimensionality of the data.

We will revisit various unsupervised learning techniques in subsequent chapters on *supervised* learning. For example, cross-entropy training loss minimization will be important in logistic regression (Section 5.7) and classification (Chapter 7), and PCA can be used for variable selection and dimensionality reduction, to make models easier to train and increase their predictive power; see e.g., Sections 6.8 and 7.4.

☞ 204
☞ 253

4.2 Risk and Loss in Unsupervised Learning

In unsupervised learning, the training data $\mathcal{T} := \{\boldsymbol{X}_1, \ldots, \boldsymbol{X}_n\}$ only consists of (what are usually assumed to be) independent copies of a feature vector $\boldsymbol{X}$; there is no response data. Suppose our objective is to learn the unknown pdf f of $\boldsymbol{X}$ based on an outcome $\tau = \{\boldsymbol{x}_1, \ldots, \boldsymbol{x}_n\}$ of the training data $\mathcal{T}$. Conveniently, we can follow the same line of reasoning as for *supervised* learning, discussed in Sections 2.3–2.5. Table 4.1 gives a summary of definitions for the case of unsupervised learning. Compare this with Table 2.1 for the supervised case.

☞ 23
☞ 25

Similar to supervised learning, we wish to find a function g, which is now a probability density (continuous or discrete), that best approximates the pdf f in terms of minimizing a risk

$$\ell(g) := \mathbb{E}\,\mathrm{Loss}(f(\boldsymbol{X}), g(\boldsymbol{X})), \tag{4.1}$$

where Loss is a loss function. In (2.27), we already encountered the Kullback–Leibler risk

$$\ell(g) := \mathbb{E} \ln \frac{f(\boldsymbol{X})}{g(\boldsymbol{X})} = \mathbb{E} \ln f(\boldsymbol{X}) - \mathbb{E} \ln g(\boldsymbol{X}). \tag{4.2}$$

If $\mathcal{G}$ is a class of functions that contains f, then minimizing the Kullback–Leibler risk over $\mathcal{G}$ will yield the (correct) minimizer f. Of course, the problem is that minimization of (4.2) depends on f, which is generally not known. However, since the term $\mathbb{E} \ln f(\boldsymbol{X})$ does not depend on g, it plays no role in the minimization of the Kullback–Leibler risk. By removing this term, we obtain the *cross-entropy risk* (for discrete $\boldsymbol{X}$ replace the integral with a sum):

CROSS-ENTROPY RISK

$$\ell(g) := -\mathbb{E} \ln g(\boldsymbol{X}) = -\int f(\boldsymbol{x}) \ln g(\boldsymbol{x})\, \mathrm{d}\boldsymbol{x}. \tag{4.3}$$

Thus, minimizing the cross-entropy risk (4.3) over all $g \in \mathcal{G}$, again gives the minimizer f, provided that $f \in \mathcal{G}$. Unfortunately, solving (4.3) is also infeasible in general, as it still

Table 4.1: Summary of definitions for unsupervised learning.

$\boldsymbol{x}$	Fixed feature vector.
$\boldsymbol{X}$	Random feature vector.
$f(\boldsymbol{x})$	Pdf of $\boldsymbol{X}$ evaluated at the point $\boldsymbol{x}$.
τ or τ_n	Fixed training data $\{\boldsymbol{x}_i, i = 1, \ldots, n\}$.
$\mathcal{T}$ or $\mathcal{T}_n$	Random training data $\{\boldsymbol{X}_i, i = 1, \ldots, n\}$.
g	Approximation of the pdf f.
$\text{Loss}(f(\boldsymbol{x}), g(\boldsymbol{x}))$	Loss incurred when approximating $f(\boldsymbol{x})$ with $g(\boldsymbol{x})$.
$\ell(g)$	Risk for approximation function g; that is, $\mathbb{E}\,\text{Loss}(f(\boldsymbol{X}), g(\boldsymbol{X}))$.
$g^{\mathcal{G}}$	Optimal approximation function in function class $\mathcal{G}$; that is, $\text{argmin}_{g\in\mathcal{G}}\, \ell(g)$.
$\ell_\tau(g)$	Training loss for approximation function (guess) g; that is, the sample average estimate of $\ell(g)$ based on a fixed training sample τ.
$\ell_{\mathcal{T}}(g)$	The same as $\ell_\tau(g)$, but now for a random training sample $\mathcal{T}$.
$g_\tau^{\mathcal{G}}$ or g_τ	The *learner*: $\text{argmin}_{g\in\mathcal{G}}\, \ell_\tau(g)$. That is, the optimal approximation function based on a fixed training set τ and function class $\mathcal{G}$. We suppress the superscript $\mathcal{G}$ if the function class is implicit.
$g_{\mathcal{T}}^{\mathcal{G}}$ or $g_{\mathcal{T}}$	The learner for a random training set $\mathcal{T}$.

depends on f. Instead, we seek to minimize the *cross-entropy training loss*: CROSS-ENTROPY TRAINING LOSS

$$\ell_\tau(g) := \frac{1}{n}\sum_{i=1}^{n} \text{Loss}(f(\boldsymbol{x}_i), g(\boldsymbol{x}_i)) = -\frac{1}{n}\sum_{i=1}^{n} \ln g(\boldsymbol{x}_i) \tag{4.4}$$

over the class of functions $\mathcal{G}$, where $\tau = \{\boldsymbol{x}_1, \ldots, \boldsymbol{x}_n\}$ is an iid sample from f. This optimization is doable without knowing f and is equivalent to solving the maximization problem

$$\max_{g\in\mathcal{G}} \sum_{i=1}^{n} \ln g(\boldsymbol{x}_i). \tag{4.5}$$

A key step in setting up the learning procedure is to select a suitable function class $\mathcal{G}$ over which to optimize. The standard approach is to parameterize g with a parameter $\boldsymbol{\theta}$ and let $\mathcal{G}$ be the class of functions $\{g(\cdot\,|\,\boldsymbol{\theta}), \boldsymbol{\theta} \in \Theta\}$ for some p-dimensional parameter set Θ. For the remainder of Section 4.2, we will be using this function class, as well as the cross-entropy risk.

The function $\boldsymbol{\theta} \mapsto g(\boldsymbol{x}\,|\,\boldsymbol{\theta})$ is called the *likelihood function*. It gives the likelihood of the observed feature vector $\boldsymbol{x}$ under $g(\cdot\,|\,\boldsymbol{\theta})$, as a function of the parameter $\boldsymbol{\theta}$. The natural logarithm of the likelihood function is called the *log-likelihood* function and its gradient with respect to $\boldsymbol{\theta}$ is called the *score function*, denoted $\mathbf{S}(\boldsymbol{x}\,|\,\boldsymbol{\theta})$; that is, LIKELIHOOD FUNCTION SCORE FUNCTION

$$\mathbf{S}(\boldsymbol{x}\,|\,\boldsymbol{\theta}) := \frac{\partial \ln g(\boldsymbol{x}\,|\,\boldsymbol{\theta})}{\partial \boldsymbol{\theta}} = \frac{\frac{\partial g(\boldsymbol{x}\,|\,\boldsymbol{\theta})}{\partial \boldsymbol{\theta}}}{g(\boldsymbol{x}\,|\,\boldsymbol{\theta})}. \tag{4.6}$$

The random score $S(X \mid \theta)$, with $X \sim g(\cdot \mid \theta)$, is of particular interest. In many cases, its expectation is *equal to the zero vector*; namely,

$$\begin{aligned}\mathbb{E}_{\theta} S(X \mid \theta) &= \int \frac{\frac{\partial g(x \mid \theta)}{\partial \theta}}{g(x \mid \theta)} g(x \mid \theta) \, \mathrm{d}x \\ &= \int \frac{\partial g(x \mid \theta)}{\partial \theta} \, \mathrm{d}x = \frac{\partial \int g(x \mid \theta) \, \mathrm{d}x}{\partial \theta} = \frac{\partial 1}{\partial \theta} = \mathbf{0},\end{aligned} \tag{4.7}$$

provided that the interchange of differentiation and integration is justified. This is true for a large number of distributions, including the normal, exponential, and binomial distributions. Notable exceptions are distributions whose support depends on the distributional parameter; for example the $\mathcal{U}(0, \theta)$ distribution.

> It is important to see whether expectations are taken with respect to $X \sim g(\cdot \mid \theta)$ or $X \sim f$. We use the expectation symbols $\mathbb{E}_\theta$ and $\mathbb{E}$ to distinguish the two cases.

From now on we simply assume that the interchange of differentiation and integration is permitted; see, e.g., [76] for sufficient conditions. The covariance matrix of the random score $S(X \mid \theta)$ is called the *Fisher information matrix*, which we denote by $\mathbf{F}$ or $\mathbf{F}(\theta)$ to show its dependence on θ. Since the expected score is $\mathbf{0}$, we have

FISHER INFORMATION MATRIX

$$\mathbf{F}(\theta) = \mathbb{E}_{\theta}[S(X \mid \theta) \, S(X \mid \theta)^{\top}]. \tag{4.8}$$

☞ 400

A related matrix is the expected Hessian matrix of $-\ln g(X \mid \theta)$:

$$\mathbf{H}(\theta) := \mathbb{E}\left[-\frac{\partial S(X \mid \theta)}{\partial \theta}\right] = -\mathbb{E}\begin{bmatrix} \frac{\partial^2 \ln g(X \mid \theta)}{\partial^2 \theta_1} & \frac{\partial^2 \ln g(X \mid \theta)}{\partial \theta_1 \partial \theta_2} & \cdots & \frac{\partial^2 \ln g(X \mid \theta)}{\partial \theta_1 \partial \theta_p} \\ \frac{\partial^2 \ln g(X \mid \theta)}{\partial \theta_2 \partial \theta_1} & \frac{\partial^2 \ln g(X \mid \theta)}{\partial^2 \theta_2} & \cdots & \frac{\partial^2 \ln g(X \mid \theta)}{\partial \theta_2 \partial \theta_p} \\ \vdots & \vdots & \ddots & \vdots \\ \frac{\partial^2 \ln g(X \mid \theta)}{\partial \theta_p \partial \theta_1} & \frac{\partial^2 \ln g(X \mid \theta)}{\partial \theta_p \partial \theta_2} & \cdots & \frac{\partial^2 \ln g(X \mid \theta)}{\partial^2 \theta_p} \end{bmatrix}. \tag{4.9}$$

Note that the expectation here is with respect to $X \sim f$. It turns out that if $f = g(\cdot \mid \theta)$, the two matrices are the *same*; that is,

$$\mathbf{F}(\theta) = \mathbf{H}(\theta), \tag{4.10}$$

provided that we may swap the order of differentiation and integration (expectation). This result is called the *information matrix equality*. We leave the proof as Exercise 1.

INFORMATION MATRIX EQUALITY

The matrices $\mathbf{F}(\theta)$ and $\mathbf{H}(\theta)$ play important roles in approximating the cross-entropy risk for large n. To set the scene, let $g^{\mathcal{G}} = g(\cdot \mid \theta^*)$ be the minimizer of the cross-entropy risk

$$r(\theta) := -\mathbb{E} \ln g(X \mid \theta).$$

We assume that r, as a function of θ, is well-behaved; in particular, that in the neighborhood of θ^* it is strictly convex and twice continuously differentiable (this holds true, for example, if g is a Gaussian density). It follows that θ^* is a root of $\mathbb{E}\, S(X \mid \theta)$, because

$$\mathbf{0} = \frac{\partial r(\theta^*)}{\partial \theta} = -\frac{\partial \mathbb{E} \ln g(X \mid \theta^*)}{\partial \theta} = -\mathbb{E}\frac{\partial \ln g(X \mid \theta^*)}{\partial \theta} = -\mathbb{E}\, S(X \mid \theta^*),$$

again provided that the order of differentiation and integration (expectation) can be swapped. In the same way, $\mathbf{H}(\boldsymbol{\theta})$ is then the Hessian matrix of r. Let $g(\cdot\,|\,\widehat{\boldsymbol{\theta}}_n)$ be the minimizer of the training loss

$$r_{\mathcal{T}_n}(\boldsymbol{\theta}) := -\frac{1}{n}\sum_{i=1}^{n} \ln g(\boldsymbol{X}_i\,|\,\boldsymbol{\theta}),$$

where $\mathcal{T}_n = \{\boldsymbol{X}_1, \ldots, \boldsymbol{X}_n\}$ is a random training set. Let r^* be the smallest possible cross-entropy risk, taken over all functions; clearly, $r^* = -\mathbb{E}\ln f(\boldsymbol{X})$, where $\boldsymbol{X} \sim f$. Similar to the supervised learning case, we can decompose the generalization risk, $\ell(g(\cdot\,|\,\widehat{\boldsymbol{\theta}}_n)) = r(\widehat{\boldsymbol{\theta}}_n)$, into

$$r(\widehat{\boldsymbol{\theta}}_n) = r^* + \underbrace{r(\boldsymbol{\theta}^*) - r^*}_{\text{approx. error}} + \underbrace{r(\widehat{\boldsymbol{\theta}}_n) - r(\boldsymbol{\theta}^*)}_{\text{statistical error}} = r(\boldsymbol{\theta}^*) - \mathbb{E}\ln\frac{g(\boldsymbol{X}\,|\,\boldsymbol{\theta}^*)}{g(\boldsymbol{X}\,|\,\widehat{\boldsymbol{\theta}}_n)}.$$

The following theorem specifies the asymptotic behavior of the components of the generalization risk. In the proof we assume that $\widehat{\boldsymbol{\theta}}_n \xrightarrow{\mathbb{P}} \boldsymbol{\theta}^*$ as $n \to \infty$. ☞ 441

Theorem 4.1: Approximating the Cross-Entropy Risk

It holds asymptotically ($n \to \infty$) that

$$\mathbb{E}r(\widehat{\boldsymbol{\theta}}_n) - r(\boldsymbol{\theta}^*) \simeq \operatorname{tr}\left(\mathbf{F}(\boldsymbol{\theta}^*)\,\mathbf{H}^{-1}(\boldsymbol{\theta}^*)\right)/(2n), \qquad (4.11)$$

where

$$r(\boldsymbol{\theta}^*) \simeq \mathbb{E}r_{\mathcal{T}_n}(\widehat{\boldsymbol{\theta}}_n) + \operatorname{tr}\left(\mathbf{F}(\boldsymbol{\theta}^*)\,\mathbf{H}^{-1}(\boldsymbol{\theta}^*)\right)/(2n). \qquad (4.12)$$

Proof: A Taylor expansion of $r(\widehat{\boldsymbol{\theta}}_n)$ around $\boldsymbol{\theta}^*$ gives the statistical error ☞ 402

$$r(\widehat{\boldsymbol{\theta}}_n) - r(\boldsymbol{\theta}^*) = (\widehat{\boldsymbol{\theta}}_n - \boldsymbol{\theta}^*)^\top \underbrace{\frac{\partial r(\boldsymbol{\theta}^*)}{\partial \boldsymbol{\theta}}}_{=\mathbf{0}} + \frac{1}{2}(\widehat{\boldsymbol{\theta}}_n - \boldsymbol{\theta}^*)^\top \mathbf{H}(\overline{\boldsymbol{\theta}}_n)(\widehat{\boldsymbol{\theta}}_n - \boldsymbol{\theta}^*), \qquad (4.13)$$

where $\overline{\boldsymbol{\theta}}_n$ lies on the line segment between $\boldsymbol{\theta}^*$ and $\widehat{\boldsymbol{\theta}}_n$. For large n we may replace $\mathbf{H}(\overline{\boldsymbol{\theta}}_n)$ with $\mathbf{H}(\boldsymbol{\theta}^*)$ as, by assumption, $\widehat{\boldsymbol{\theta}}_n$ converges to $\boldsymbol{\theta}^*$. The matrix $\mathbf{H}(\boldsymbol{\theta}^*)$ is positive definite because $r(\boldsymbol{\theta})$ is strictly convex at $\boldsymbol{\theta}^*$ by assumption, and therefore invertible. It is important to realize that $\widehat{\boldsymbol{\theta}}_n$ is in fact an M-estimator of $\boldsymbol{\theta}^*$. In particular, in the notation of Theorem C.19, we have $\boldsymbol{\psi} = \mathbf{S}$, $\mathbf{A} = \mathbf{H}(\boldsymbol{\theta}^*)$, and $\mathbf{B} = \mathbf{F}(\boldsymbol{\theta}^*)$. Consequently, by that same theorem, ☞ 451

$$\sqrt{n}\,(\widehat{\boldsymbol{\theta}}_n - \boldsymbol{\theta}^*) \xrightarrow{\text{d}} \mathcal{N}\left(\mathbf{0}, \mathbf{H}^{-1}(\boldsymbol{\theta}^*)\,\mathbf{F}(\boldsymbol{\theta}^*)\,\mathbf{H}^{-\top}(\boldsymbol{\theta}^*)\right). \qquad (4.14)$$

Combining (4.13) with (4.14), it follows from Theorem C.2 that asymptotically the expected estimation error is given by (4.11). ☞ 432

Next, we consider a Taylor expansion of $r_{\mathcal{T}_n}(\boldsymbol{\theta}^*)$ around $\widehat{\boldsymbol{\theta}}_n$:

$$r_{\mathcal{T}_n}(\boldsymbol{\theta}^*) = r_{\mathcal{T}_n}(\widehat{\boldsymbol{\theta}}_n) + (\boldsymbol{\theta}^* - \widehat{\boldsymbol{\theta}}_n)^\top \underbrace{\frac{\partial r_{\mathcal{T}_n}(\widehat{\boldsymbol{\theta}}_n)}{\partial \boldsymbol{\theta}}}_{=\mathbf{0}} + \frac{1}{2}(\boldsymbol{\theta}^* - \widehat{\boldsymbol{\theta}}_n)^\top \mathbf{H}_{\mathcal{T}_n}(\overline{\boldsymbol{\theta}}_n)(\boldsymbol{\theta}^* - \widehat{\boldsymbol{\theta}}_n), \qquad (4.15)$$

where $\mathbf{H}_{\mathcal{T}_n}(\overline{\boldsymbol{\theta}}_n) := -\frac{1}{n}\sum_{i=1}^n \frac{\partial \boldsymbol{S}(\boldsymbol{X}_i \,|\, \overline{\boldsymbol{\theta}}_n)}{\partial \boldsymbol{\theta}}$ is the Hessian of $r_{\mathcal{T}_n}(\boldsymbol{\theta})$ at some $\overline{\boldsymbol{\theta}}_n$ between $\widehat{\boldsymbol{\theta}}_n$ and $\boldsymbol{\theta}^*$. Taking expectations on both sides of (4.15), we obtain

$$r(\boldsymbol{\theta}^*) = \mathbb{E} r_{\mathcal{T}_n}(\widehat{\boldsymbol{\theta}}_n) + \frac{1}{2}\mathbb{E}\,(\boldsymbol{\theta}^* - \widehat{\boldsymbol{\theta}}_n)^\top \mathbf{H}_{\mathcal{T}_n}(\overline{\boldsymbol{\theta}}_n)(\boldsymbol{\theta}^* - \widehat{\boldsymbol{\theta}}_n).$$

Replacing $\mathbf{H}_{\mathcal{T}_n}(\overline{\boldsymbol{\theta}}_n)$ with $\mathbf{H}(\boldsymbol{\theta}^*)$ for large n and using (4.14), we have

$$n\,\mathbb{E}\,(\boldsymbol{\theta}^* - \widehat{\boldsymbol{\theta}}_n)^\top \mathbf{H}_{\mathcal{T}_n}(\overline{\boldsymbol{\theta}}_n)(\boldsymbol{\theta}^* - \widehat{\boldsymbol{\theta}}_n) \longrightarrow \operatorname{tr}\left(\mathbf{F}(\boldsymbol{\theta}^*)\,\mathbf{H}^{-1}(\boldsymbol{\theta}^*)\right), \quad n \to \infty.$$

Therefore, asymptotically as $n \to \infty$, we have (4.12). □

Theorem 4.1 has a number of interesting consequences:

☞ 35

1. Similar to Section 2.5.1, the training loss $\ell_{\mathcal{T}_n}(g_{\mathcal{T}_n}) = r_{\mathcal{T}_n}(\widehat{\boldsymbol{\theta}}_n)$ tends to underestimate the risk $\ell(g^{\mathcal{G}}) = r(\boldsymbol{\theta}^*)$, because the training set $\mathcal{T}_n$ is used to both train $g \in \mathcal{G}$ (that is, estimate $\boldsymbol{\theta}^*$) and to estimate the risk. The relation (4.12) tells us that on average the training loss underestimates the true risk by $\operatorname{tr}(\mathbf{F}(\boldsymbol{\theta}^*)\,\mathbf{H}^{-1}(\boldsymbol{\theta}^*))/(2n)$.

2. Adding equations (4.11) and (4.12), yields the following asymptotic approximation to the expected generalization risk:

$$\mathbb{E}\, r(\widehat{\boldsymbol{\theta}}_n) \simeq \mathbb{E}\, r_{\mathcal{T}_n}(\widehat{\boldsymbol{\theta}}_n) + \frac{1}{n}\operatorname{tr}\left(\mathbf{F}(\boldsymbol{\theta}^*)\,\mathbf{H}^{-1}(\boldsymbol{\theta}^*)\right) \tag{4.16}$$

The first term on the right-hand side of (4.16) can be estimated (without bias) via the training loss $r_{\mathcal{T}_n}(\widehat{\boldsymbol{\theta}}_n)$. As for the second term, we have already mentioned that when the true model $f \in \mathcal{G}$, then $\mathbf{F}(\boldsymbol{\theta}^*) = \mathbf{H}(\boldsymbol{\theta}^*)$. Therefore, when $\mathcal{G}$ is deemed to be a sufficiently rich class of models parameterized by a p-dimensional vector $\boldsymbol{\theta}$, we may approximate the second term as $\operatorname{tr}(\mathbf{F}(\boldsymbol{\theta}^*)\mathbf{H}^{-1}(\boldsymbol{\theta}^*))/n \approx \operatorname{tr}(\mathbf{I}_p)/n = p/n$. This suggests the following heuristic approximation to the (expected) generalization risk:

$$\mathbb{E}\, r(\widehat{\boldsymbol{\theta}}_n) \approx r_{\mathcal{T}_n}(\widehat{\boldsymbol{\theta}}_n) + \frac{p}{n}. \tag{4.17}$$

3. Multiplying both sides of (4.16) by $2n$ and substituting $\operatorname{tr}\left(\mathbf{F}(\boldsymbol{\theta}^*)\mathbf{H}^{-1}(\boldsymbol{\theta}^*)\right) \approx p$, we obtain the approximation:

$$2n\, r(\widehat{\boldsymbol{\theta}}_n) \approx -2\sum_{i=1}^n \ln g(\boldsymbol{X}_i \,|\, \widehat{\boldsymbol{\theta}}_n) + 2p. \tag{4.18}$$

AKAIKE INFORMATION CRITERION

The right-hand side of (4.18) is called the *Akaike information criterion* (AIC). Just like (4.17), the AIC approximation can be used to compare the difference in generalization risk of two or more learners. We prefer the learner with the smallest (estimated) generalization risk.

Suppose that, for a training set $\mathcal{T}$, the training loss $r_{\mathcal{T}}(\boldsymbol{\theta})$ has a unique minimum point $\widehat{\boldsymbol{\theta}}$ which lies in the interior of Θ. If $r_{\mathcal{T}}(\boldsymbol{\theta})$ is a differentiable function with respect to $\boldsymbol{\theta}$, then we can find the optimal parameter $\widehat{\boldsymbol{\theta}}$ by solving

$$\frac{\partial r_{\mathcal{T}}(\boldsymbol{\theta})}{\partial \boldsymbol{\theta}} = \underbrace{\frac{1}{n}\sum_{i=1}^n \boldsymbol{S}(\boldsymbol{X}_i \,|\, \boldsymbol{\theta})}_{\boldsymbol{S}_{\mathcal{T}}(\boldsymbol{\theta})} = \mathbf{0}.$$

In other words, the maximum likelihood estimate $\widehat{\boldsymbol{\theta}}$ for $\boldsymbol{\theta}$ is obtained by solving the root of the average score function, that is, by solving

$$\boldsymbol{S}_{\mathcal{T}}(\boldsymbol{\theta}) = \mathbf{0}. \tag{4.19}$$

It is often not possible to find $\widehat{\boldsymbol{\theta}}$ in an explicit form. In that case one needs to solve the equation (4.19) numerically. There exist many standard techniques for root-finding, e.g., via *Newton's method* (see Section B.3.1), whereby, starting from an initial guess $\boldsymbol{\theta}_0$, subsequent iterates are obtained via the iterative scheme

NEWTON'S METHOD ☞ 411

$$\boldsymbol{\theta}_{t+1} = \boldsymbol{\theta}_t + \mathbf{H}_{\mathcal{T}}^{-1}(\boldsymbol{\theta}_t)\, \boldsymbol{S}_{\mathcal{T}}(\boldsymbol{\theta}_t),$$

where

$$\mathbf{H}_{\mathcal{T}}(\boldsymbol{\theta}) := \frac{-\,\partial \boldsymbol{S}_{\mathcal{T}}(\boldsymbol{\theta})}{\partial \boldsymbol{\theta}} = \frac{1}{n}\sum_{i=1}^{n} -\frac{\partial \boldsymbol{S}(\boldsymbol{X}_i \,|\, \boldsymbol{\theta})}{\partial \boldsymbol{\theta}}$$

is the average Hessian matrix of $\{-\ln g(\boldsymbol{X}_i \,|\, \boldsymbol{\theta})\}_{i=1}^n$. Under $f = g(\cdot \,|\, \boldsymbol{\theta})$, the expectation of $\mathbf{H}_{\mathcal{T}}(\boldsymbol{\theta})$ is equal to the information matrix $\mathbf{F}(\boldsymbol{\theta})$, which does not depend on the data. This suggests an alternative iterative scheme, called *Fisher's scoring method*:

FISHER'S SCORING METHOD

$$\boldsymbol{\theta}_{t+1} = \boldsymbol{\theta}_t + \mathbf{F}^{-1}(\boldsymbol{\theta}_t)\, \boldsymbol{S}_{\mathcal{T}}(\boldsymbol{\theta}_t), \tag{4.20}$$

which is not only easier to implement (if the information matrix can be readily evaluated), but also is more numerically stable.

■ **Example 4.1 (Maximum Likelihood for the Gamma Distribution)** We wish to approximate the density of the $\mathsf{Gamma}(\alpha^*, \lambda^*)$ distribution for some true but unknown parameters α^* and λ^*, on the basis of a training set $\tau = \{x_1, \ldots, x_n\}$ of iid samples from this distribution. Choosing our approximating function $g(\cdot \,|\, \alpha, \lambda)$ in the same class of gamma densities,

$$g(x \,|\, \alpha, \lambda) = \frac{\lambda^\alpha x^{\alpha-1} \mathrm{e}^{-\lambda x}}{\Gamma(\alpha)}, \quad x \geqslant 0, \tag{4.21}$$

with $\alpha > 0$ and $\lambda > 0$, we seek to solve (4.19). Taking the logarithm in (4.21), the log-likelihood function is given by

$$l(x \,|\, \alpha, \lambda) := \alpha \ln \lambda - \ln \Gamma(\alpha) + (\alpha - 1) \ln x - \lambda x.$$

It follows that

$$\boldsymbol{S}(\alpha, \lambda) = \begin{bmatrix} \frac{\partial}{\partial \alpha} l(x \,|\, \alpha, \lambda) \\ \frac{\partial}{\partial \lambda} l(x \,|\, \alpha, \lambda) \end{bmatrix} = \begin{bmatrix} \ln \lambda - \psi(\alpha) + \ln x \\ \frac{\alpha}{\lambda} - x \end{bmatrix},$$

where ψ is the derivative of $\ln \Gamma$: the so-called *digamma function*. Hence,

DIGAMMA FUNCTION

$$\mathbf{H}(\alpha, \lambda) = -\mathbb{E} \begin{bmatrix} \frac{\partial^2}{\partial \alpha^2} l(X \,|\, \alpha, \lambda) & \frac{\partial^2}{\partial \alpha \partial \lambda} l(X \,|\, \alpha, \lambda) \\ \frac{\partial^2}{\partial \alpha \partial \lambda} l(X \,|\, \alpha, \lambda) & \frac{\partial^2}{\partial \lambda^2} l(X \,|\, \alpha, \lambda) \end{bmatrix} = -\mathbb{E} \begin{bmatrix} -\psi'(\alpha) & \frac{1}{\lambda} \\ \frac{1}{\lambda} & -\frac{\alpha}{\lambda^2} \end{bmatrix} = \begin{bmatrix} \psi'(\alpha) & -\frac{1}{\lambda} \\ -\frac{1}{\lambda} & \frac{\alpha}{\lambda^2} \end{bmatrix}.$$

Fisher's scoring method (4.20) can now be used to solve (4.19), with

$$\boldsymbol{S}_\tau(\alpha, \lambda) = \begin{bmatrix} \ln \lambda - \psi(\alpha) + n^{-1} \sum_{i=1}^n \ln x_i \\ \frac{\alpha}{\lambda} - n^{-1} \sum_{i=1}^n x_i \end{bmatrix}$$

and $\mathbf{F}(\alpha, \lambda) = \mathbf{H}(\alpha, \lambda)$. ■

4.3 Expectation–Maximization (EM) Algorithm

The *Expectation–Maximization* algorithm (EM) is a general algorithm for maximization of complicated (log-)likelihood functions, through the introduction of auxiliary variables.

> To simplify the notation in this section, we use a Bayesian notation system, where the same symbol is used for different (conditional) probability densities.

As in the previous section, given independent observations $\tau = \{\boldsymbol{x}_1, \ldots, \boldsymbol{x}_n\}$ from some unknown pdf f, the objective is to find the best approximation to f in a function class $\mathcal{G} = \{g(\cdot \mid \boldsymbol{\theta}), \boldsymbol{\theta} \in \Theta\}$ by solving the maximum likelihood problem:

$$\boldsymbol{\theta}^* = \underset{\boldsymbol{\theta} \in \Theta}{\operatorname{argmax}}\, g(\tau \mid \boldsymbol{\theta}), \tag{4.22}$$

where $g(\tau \mid \boldsymbol{\theta}) := g(\boldsymbol{x}_1 \mid \boldsymbol{\theta}) \cdots g(\boldsymbol{x}_n \mid \boldsymbol{\theta})$. The key element of the EM algorithm is the augmentation of the data τ with a suitable vector of *latent variables*, $\boldsymbol{z}$, such that

LATENT VARIABLES

$$g(\tau \mid \boldsymbol{\theta}) = \int g(\tau, \boldsymbol{z} \mid \boldsymbol{\theta})\, \mathrm{d}\boldsymbol{z}.$$

COMPLETE-DATA LIKELIHOOD

The function $\boldsymbol{\theta} \mapsto g(\tau, \boldsymbol{z} \mid \boldsymbol{\theta})$ is usually referred to as the *complete-data likelihood* function. The choice of the latent variables is guided by the desire to make the maximization of $g(\tau, \boldsymbol{z} \mid \boldsymbol{\theta})$ much easier than that of $g(\tau \mid \boldsymbol{\theta})$.

Suppose p denotes an arbitrary density of the latent variables $\boldsymbol{z}$. Then, we can write:

$$\begin{aligned}
\ln g(\tau \mid \boldsymbol{\theta}) &= \int p(\boldsymbol{z}) \ln g(\tau \mid \boldsymbol{\theta})\, \mathrm{d}\boldsymbol{z} \\
&= \int p(\boldsymbol{z}) \ln \left(\frac{g(\tau, \boldsymbol{z} \mid \boldsymbol{\theta})/p(\boldsymbol{z})}{g(\boldsymbol{z} \mid \tau, \boldsymbol{\theta})/p(\boldsymbol{z})} \right) \mathrm{d}\boldsymbol{z} \\
&= \int p(\boldsymbol{z}) \ln \left(\frac{g(\tau, \boldsymbol{z} \mid \boldsymbol{\theta})}{p(\boldsymbol{z})} \right) \mathrm{d}\boldsymbol{z} - \int p(\boldsymbol{z}) \ln \left(\frac{g(\boldsymbol{z} \mid \tau, \boldsymbol{\theta})}{p(\boldsymbol{z})} \right) \mathrm{d}\boldsymbol{z} \\
&= \int p(\boldsymbol{z}) \ln \left(\frac{g(\tau, \boldsymbol{z} \mid \boldsymbol{\theta})}{p(\boldsymbol{z})} \right) \mathrm{d}\boldsymbol{z} + \mathcal{D}(p, g(\cdot \mid \tau, \boldsymbol{\theta})),
\end{aligned} \tag{4.23}$$

☞ 42

where $\mathcal{D}(p, g(\cdot \mid \tau, \boldsymbol{\theta}))$ is the Kullback–Leibler divergence from the density p to $g(\cdot \mid \tau, \boldsymbol{\theta})$. Since $\mathcal{D} \geqslant 0$, it follows that

$$\ln g(\tau \mid \boldsymbol{\theta}) \geqslant \int p(\boldsymbol{z}) \ln \left(\frac{g(\tau, \boldsymbol{z} \mid \boldsymbol{\theta})}{p(\boldsymbol{z})} \right) \mathrm{d}\boldsymbol{z} =: \mathcal{L}(p, \boldsymbol{\theta})$$

for all $\boldsymbol{\theta}$ and any density p of the latent variables. In other words, $\mathcal{L}(p, \boldsymbol{\theta})$ is a lower bound on the log-likelihood that involves the complete-data likelihood. The EM algorithm then aims to increase this lower bound as much as possible by starting with an initial guess $\boldsymbol{\theta}^{(0)}$ and then, for $t = 1, 2, \ldots$, solving the following two steps:

1. $p^{(t)} = \operatorname{argmax}_p \mathcal{L}(p, \boldsymbol{\theta}^{(t-1)})$,
2. $\boldsymbol{\theta}^{(t)} = \operatorname{argmax}_{\boldsymbol{\theta} \in \Theta} \mathcal{L}(p^{(t)}, \boldsymbol{\theta})$.

The first optimization problem can be solved explicitly. Namely, by (4.23), we have that

$$p^{(t)} = \underset{p}{\operatorname{argmin}}\, \mathcal{D}(p, g(\cdot \mid \tau, \boldsymbol{\theta}^{(t-1)})) = g(\cdot \mid \tau, \boldsymbol{\theta}^{(t-1)}).$$

That is, the optimal density is the conditional density of the latent variables given the data τ and the parameter $\boldsymbol{\theta}^{(t-1)}$. The second optimization problem can be simplified by writing $\mathcal{L}(p^{(t)}, \boldsymbol{\theta}) = Q^{(t)}(\boldsymbol{\theta}) - \mathbb{E}_{p^{(t)}} \ln p^{(t)}(\mathbf{Z})$, where

$$Q^{(t)}(\boldsymbol{\theta}) := \mathbb{E}_{p^{(t)}} \ln g(\tau, \mathbf{Z} \mid \boldsymbol{\theta})$$

is the expected complete-data log-likelihood under $\mathbf{Z} \sim p^{(t)}$. Consequently, the maximization of $\mathcal{L}(p^{(t)}, \boldsymbol{\theta})$ with respect to $\boldsymbol{\theta}$ is equivalent to finding

$$\boldsymbol{\theta}^{(t)} = \underset{\boldsymbol{\theta} \in \Theta}{\operatorname{argmax}}\, Q^{(t)}(\boldsymbol{\theta}).$$

This leads to the following generic EM algorithm.

Algorithm 4.3.1: Generic EM Algorithm

---

input: Data τ, initial guess $\boldsymbol{\theta}^{(0)}$.
output: Approximation of the maximum likelihood estimate.

$t \leftarrow 1$
while a stopping criterion is not met **do**
  Expectation Step: Find $p^{(t)}(\boldsymbol{z}) := g(\boldsymbol{z} \mid \tau, \boldsymbol{\theta}^{(t-1)})$ and compute the expectation

$$Q^{(t)}(\boldsymbol{\theta}) := \mathbb{E}_{p^{(t)}} \ln g(\tau, \mathbf{Z} \mid \boldsymbol{\theta}). \tag{4.24}$$

  Maximization Step: Let $\boldsymbol{\theta}^{(t)} \leftarrow \operatorname{argmax}_{\boldsymbol{\theta} \in \Theta} Q^{(t)}(\boldsymbol{\theta})$.
  $t \leftarrow t + 1$
return $\boldsymbol{\theta}^{(t)}$

A possible stopping criterion is to stop when

$$\left| \frac{\ln g(\tau \mid \boldsymbol{\theta}^{(t)}) - \ln g(\tau \mid \boldsymbol{\theta}^{(t-1)})}{\ln g(\tau \mid \boldsymbol{\theta}^{(t)})} \right| \leqslant \varepsilon$$

for some small tolerance $\varepsilon > 0$.

■ **Remark 4.1 (Properties of the EM Algorithm)** The identity (4.23) can be used to show that the likelihood $g(\tau \mid \boldsymbol{\theta}^{(t)})$ does not decrease with every iteration of the algorithm. This property is one of the strengths of the algorithm. For example, it can be used to debug computer implementations of the EM algorithm: if the likelihood is observed to decrease at any iteration, then one has detected a bug in the program.

The convergence of the sequence $\{\boldsymbol{\theta}^{(t)}\}$ to a global maximum (if it exists) is highly dependent on the initial value $\boldsymbol{\theta}^{(0)}$ and, in many cases, an appropriate choice of $\boldsymbol{\theta}^{(0)}$ may not be clear. Typically, practitioners run the algorithm from different random starting points over Θ, to ascertain empirically that a suitable optimum is achieved. ■

■ **Example 4.2 (Censored Data)** Suppose the lifetime (in years) of a certain type of machine is modeled via a $\mathcal{N}(\mu, \sigma^2)$ distribution. To estimate μ and σ^2, the lifetimes of n (independent) machines are recorded up to c years. Denote these *censored* lifetimes by $x_1, \ldots, x_n$. The $\{x_i\}$ are thus realizations of iid random variables $\{X_i\}$, distributed as $\min\{Y, c\}$, where $Y \sim \mathcal{N}(\mu, \sigma^2)$.

☞ 430 By the law of total probability (see (C.9)), the marginal pdf of each X can be written as:

$$g(x \mid \mu, \sigma^2) = \underbrace{\Phi((c-\mu)/\sigma)}_{\mathbb{P}[Y<c]} \frac{\varphi_{\sigma^2}(x-\mu)}{\Phi((c-\mu)/\sigma)} \mathbb{1}\{x < c\} + \underbrace{\overline{\Phi}((c-\mu)/\sigma)}_{\mathbb{P}[Y \geqslant c]} \mathbb{1}\{x = c\},$$

where $\varphi_{\sigma^2}(\cdot)$ is the pdf of the $\mathcal{N}(0, \sigma^2)$ distribution, Φ is the cdf of the standard normal distribution, and $\overline{\Phi} := 1 - \Phi$. It follows that the likelihood of the data $\tau = \{x_1, \ldots, x_n\}$ as a function of the parameter $\boldsymbol{\theta} := [\mu, \sigma^2]^\top$ is:

$$g(\tau \mid \boldsymbol{\theta}) = \prod_{i:x_i<c} \frac{\exp\left(-\frac{(x_i-\mu)^2}{2\sigma^2}\right)}{\sqrt{2\pi\sigma^2}} \times \prod_{i:x_i=c} \overline{\Phi}((c-\mu)/\sigma).$$

Let n_c be the total number of x_i such that $x_i = c$. Using n_c latent variables $\boldsymbol{z} = [z_1, \ldots, z_{n_c}]^\top$, we can write the joint pdf:

$$g(\tau, \boldsymbol{z} \mid \boldsymbol{\theta}) = \frac{1}{(2\pi\sigma^2)^{n/2}} \exp\left(-\frac{\sum_{i:x_i<c}(x_i-\mu)^2}{2\sigma^2} - \frac{\sum_{i=1}^{n_c}(z_i-\mu)^2}{2\sigma^2}\right) \mathbb{1}\left\{\min_i z_i \geqslant c\right\},$$

so that $\int g(\tau, \boldsymbol{z} \mid \boldsymbol{\theta})\, \mathrm{d}\boldsymbol{z} = g(\tau \mid \boldsymbol{\theta})$. We can thus apply the EM algorithm to maximize the likelihood, as follows.

For the E(xpectation)-step, we have for a fixed $\boldsymbol{\theta}$:

$$g(\boldsymbol{z} \mid \tau, \boldsymbol{\theta}) = \prod_{i=1}^{n_c} g(z_i \mid \tau, \boldsymbol{\theta}),$$

where $g(z \mid \tau, \boldsymbol{\theta}) = \mathbb{1}\{z \geqslant c\}\, \varphi_{\sigma^2}(z-\mu)/\overline{\Phi}((c-\mu)/\sigma)$ is simply the pdf of the $\mathcal{N}(\mu, \sigma^2)$ distribution, truncated to $[c, \infty)$.

For the M(aximization)-step, we compute the expectation of the complete log-likelihood with respect to a fixed $g(\boldsymbol{z} \mid \tau, \boldsymbol{\theta})$ and use the fact that $Z_1, \ldots, Z_{n_c}$ are iid:

$$\mathbb{E} \ln g(\tau, \boldsymbol{Z} \mid \boldsymbol{\theta}) = -\frac{\sum_{i:x_i<c}(x_i-\mu)^2}{2\sigma^2} - \frac{n_c \mathbb{E}(Z-\mu)^2}{2\sigma^2} - \frac{n}{2}\ln\sigma^2 - \frac{n}{2}\ln(2\pi),$$

where Z has a $\mathcal{N}(\mu, \sigma^2)$ distribution, truncated to $[c, \infty)$. To maximize the last expression with respect to μ we set the derivative with respect to μ to zero, and obtain:

$$\mu = \frac{n_c \mathbb{E}Z + \sum_{i:x_i<c} x_i}{n}.$$

Similarly, setting the derivative with respect to σ^2 to zero gives:

$$\sigma^2 = \frac{n_c \mathbb{E}(Z-\mu)^2 + \sum_{i:x_i<c}(x_i-\mu)^2}{n}.$$

In summary, the EM iterates for $t = 1, 2, \ldots$ are as follows.

E-step. Given the current estimate $\boldsymbol{\theta}_t := [\mu_t, \sigma_t^2]^\top$, compute the expectations $\nu_t := \mathbb{E}Z$ and $\zeta_t^2 := \mathbb{E}(Z - \mu_t)^2$, where $Z \sim \mathcal{N}(\mu_t, \sigma_t^2)$, conditional on $Z \geqslant c$; that is,

$$\nu_t := \mu_t + \sigma_t^2 \frac{\varphi_{\sigma_t^2}(c - \mu_t)}{\overline{\Phi}((c - \mu_t)/\sigma_t)}$$

$$\zeta_t^2 := \sigma_t^2 \left(1 + (c - \mu_t) \frac{\varphi_{\sigma_t^2}(c - \mu_t)}{\overline{\Phi}((c - \mu_t)/\sigma_t)}\right).$$

M-step. Update the estimate to $\boldsymbol{\theta}_{t+1} := [\mu_{t+1}, \sigma_{t+1}^2]^\top$ via the formulas:

$$\mu_{t+1} = \frac{n_c \nu_t + \sum_{i:x_i<c} x_i}{n}$$

$$\sigma_{t+1}^2 = \frac{n_c \zeta_t^2 + \sum_{i:x_i<c}(x_i - \mu_{t+1})^2}{n}.$$

■

4.4 Empirical Distribution and Density Estimation

In Section 1.5.2.3 we saw how the empirical cdf $\widehat{F}_n$, obtained from an iid training set $\tau = \{x_1, \ldots, x_n\}$ from an unknown distribution on $\mathbb{R}$, gives an estimate of the unknown cdf F of this sampling distribution. The function $\widehat{F}_n$ is a genuine cdf, as it is right-continuous, increasing, and lies between 0 and 1. The corresponding discrete probability distribution is called the *empirical distribution* of the data. A random variable X distributed according to this empirical distribution takes the values $x_1, \ldots, x_n$ with equal probability $1/n$. The concept of empirical distribution naturally generalizes to higher dimensions: a random vector $\boldsymbol{X}$ that is distributed according to the empirical distribution of $\boldsymbol{x}_1, \ldots, \boldsymbol{x}_n$ has discrete pdf $\mathbb{P}[\boldsymbol{X} = \boldsymbol{x}_i] = 1/n, i = 1, \ldots, n$. Sampling from such a distribution — in other words *resampling* the original data — was discussed in Section 3.2.4. The preeminent usage of such sampling is the bootstrap method, discussed in Section 3.3.2.

☞ 11

EMPIRICAL DISTRIBUTION

☞ 76

☞ 88

In a way, the empirical distribution is the natural answer to the unsupervised learning question: what is the underlying probability distribution of the data? However, the empirical distribution is, by definition, a discrete distribution, whereas the true sampling distribution might be continuous. For continuous data it makes sense to also consider estimation of the pdf of the data. A common approach is to estimate the density via a *kernel density estimate* (KDE), the most prevalent learner to carry this out is given next.

Definition 4.1: Gaussian KDE

Let $\boldsymbol{x}_1, \ldots, \boldsymbol{x}_n \in \mathbb{R}^d$ be the outcomes of an iid sample from a continuous pdf f. A *Gaussian kernel density estimate* of f is a mixture of normal pdfs, of the form

GAUSSIAN KERNEL DENSITY ESTIMATE

$$g_{\tau_n}(\boldsymbol{x} \,|\, \sigma) = \frac{1}{n} \sum_{i=1}^{n} \frac{1}{(2\pi)^{d/2}\sigma^d} \, \mathrm{e}^{-\frac{\|\boldsymbol{x}-\boldsymbol{x}_i\|^2}{2\sigma^2}}, \quad \boldsymbol{x} \in \mathbb{R}^d, \tag{4.25}$$

where $\sigma > 0$ is called the *bandwidth*.

We see that g_{τ_n} in (4.25) is the average of a collection of n normal pdfs, where each normal distribution is centered at the data point $\boldsymbol{x}_i$ and has covariance matrix $\sigma^2\mathbf{I}_d$. A major question is how to choose the bandwidth σ so as to best approximate the unknown pdf f. Choosing very small σ will result in a "spiky" estimate, whereas a large σ will produce an over-smoothed estimate that may not identify important peaks that are present in the unknown pdf. Figure 4.1 illustrates this phenomenon. In this case the data are comprised of 20 points uniformly drawn from the unit square. The true pdf is thus 1 on $[0, 1]^2$ and 0 elsewhere.

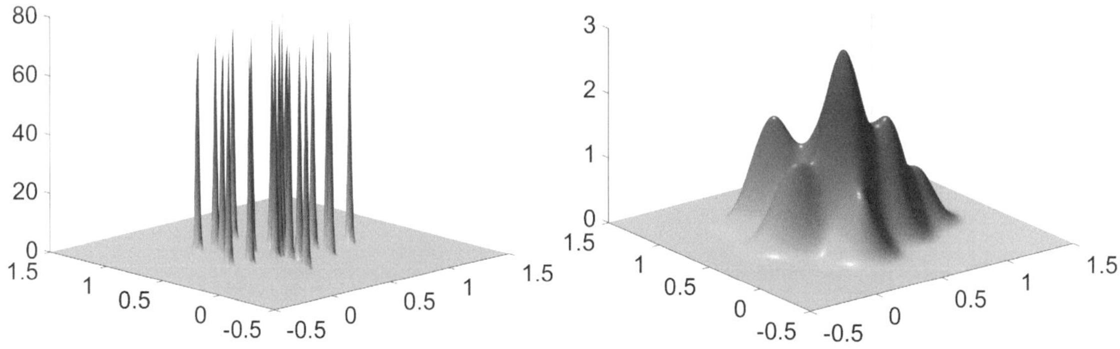

Figure 4.1: Two two-dimensional Gaussian KDEs, with $\sigma = 0.01$ (left) and $\sigma = 0.1$ (right).

Let us write the Gaussian KDE in (4.25) as

$$g_{\tau_n}(\boldsymbol{x}\,|\,\sigma) = \frac{1}{n}\sum_{i=1}^{n}\frac{1}{\sigma^d}\phi\left(\frac{\boldsymbol{x}-\boldsymbol{x}_i}{\sigma}\right), \tag{4.26}$$

where

$$\phi(\boldsymbol{z}) = \frac{1}{(2\pi)^{d/2}}\,\mathrm{e}^{-\frac{\|\boldsymbol{z}\|^2}{2}}, \quad \boldsymbol{z}\in\mathbb{R}^d \tag{4.27}$$

is the pdf of the d-dimensional standard normal distribution. By choosing a different probability density ϕ in (4.26), satisfying $\phi(\boldsymbol{x}) = \phi(-\boldsymbol{x})$ for all $\boldsymbol{x}$, we can obtain a wide variety of kernel density estimates. A simple pdf ϕ is, for example, the uniform pdf on $[-1, 1]^d$:

$$\phi(\boldsymbol{z}) = \begin{cases} 2^{-d}, & \text{if } \boldsymbol{z}\in[-1,1]^d, \\ 0, & \text{otherwise.}\end{cases}$$

Figure 4.2 shows the graph of the corresponding KDE, using the same data as in Figure 4.1 and with bandwidth $\sigma = 0.1$. We observe qualitatively similar behavior for the Gaussian and uniform KDEs. As a rule, the choice of the function ϕ is less important than the choice of the bandwidth in determining the quality of the estimate.

The important issue of bandwidth selection has been extensively studied for one-dimensional data. To explain the ideas, we use our usual setup and let $\tau = \{x_1, \ldots, x_n\}$ be the observed (one-dimensional) data from the unknown pdf f. First, we define the loss function as

$$\mathrm{Loss}(f(x), g(x)) = \frac{(f(x)-g(x))^2}{f(x)}. \tag{4.28}$$

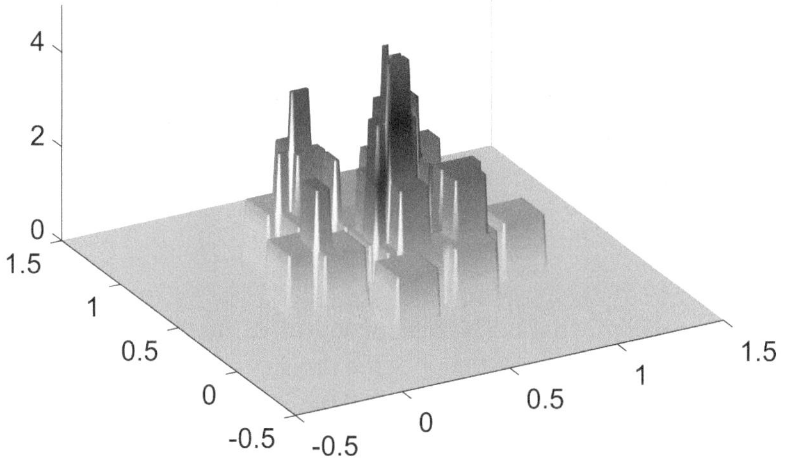

Figure 4.2: A two-dimensional uniform KDE, with bandwidth $\sigma = 0.1$.

The risk to minimize is thus $\ell(g) := \mathbb{E}_f \text{Loss}(f(X), g(X)) = \int (f(x) - g(x))^2 \,\mathrm{d}x$. We bypass the selection of a class of approximation functions by choosing the learner to be specified by (4.25) for a fixed σ. The objective is now to find a σ that minimizes the generalization risk $\ell(g_\tau(\cdot \,|\, \sigma))$ or the expected generalization risk $\mathbb{E}\ell(g_{\mathcal{T}}(\cdot \,|\, \sigma))$. The generalization risk is in this case

$$\int (f(x) - g_\tau(x \,|\, \sigma))^2 \,\mathrm{d}x = \int f^2(x) \,\mathrm{d}x - 2 \int f(x) g_\tau(x \,|\, \sigma) \,\mathrm{d}x + \int g_\tau^2(x \,|\, \sigma) \,\mathrm{d}x.$$

Minimizing this expression with respect to σ is equivalent to minimizing the last two terms, which can be written as

$$-2\,\mathbb{E}_f g_\tau(X \,|\, \sigma) + \int \left(\frac{1}{n} \sum_{i=1}^{n} \frac{1}{\sigma} \phi\left(\frac{x - x_i}{\sigma} \right) \right)^2 \mathrm{d}x.$$

This expression in turn can be estimated by using a test sample $\{x'_1 \ldots, x'_{n'}\}$ from f, yielding the following minimization problem:

$$\min_\sigma -\frac{2}{n'} \sum_{i=1}^{n'} g_\tau(x'_i \,|\, \sigma) + \frac{1}{n^2} \sum_{i=1}^{n} \sum_{j=1}^{n} \int \frac{1}{\sigma^2} \phi\left(\frac{x - x_i}{\sigma} \right) \phi\left(\frac{x - x_j}{\sigma} \right) \mathrm{d}x,$$

where $\int \frac{1}{\sigma^2} \phi\left(\frac{x-x_i}{\sigma}\right) \phi\left(\frac{x-x_j}{\sigma}\right) \mathrm{d}x = \frac{1}{\sqrt{2}\sigma} \phi\left(\frac{x_i - x_j}{\sqrt{2}\sigma}\right)$ in the case of the Gaussian kernel (4.27) with $d = 1$. To estimate σ in this way clearly requires a test sample, or at least an application of *cross-validation*. Another approach is to minimize the *expected* generalization risk, (that is, averaged over all training sets): ☞ 37

$$\mathbb{E} \int (f(x) - g_{\mathcal{T}}(x \,|\, \sigma))^2 \,\mathrm{d}x.$$

This is called the *mean integrated squared error* (MISE). It can be decomposed into an integrated squared bias and integrated variance component:

MEAN INTEGRATED SQUARED ERROR

$$\int (f(x) - \mathbb{E} g_{\mathcal{T}}(x \,|\, \sigma))^2 \,\mathrm{d}x + \int \mathbb{V}\mathrm{ar}(g_{\mathcal{T}}(x \,|\, \sigma)) \,\mathrm{d}x.$$

A typical analysis now proceeds by investigating how the MISE behaves for large n, under various assumptions on f. For example, it is shown in [114] that, for $\sigma \to 0$ and $n\sigma \to \infty$, the asymptotic approximation to the MISE of the Gaussian kernel density estimator (4.25) (for $d = 1$) is given by

$$\frac{1}{4}\sigma^4 \|f''\|^2 + \frac{1}{2n\sqrt{\pi\sigma^2}}, \tag{4.29}$$

where $\|f''\|^2 := \int (f''(x))^2 \, dx$. The asymptotically optimal value of σ is the minimizer

$$\sigma^* := \left(\frac{1}{2n\sqrt{\pi}\,\|f''\|^2}\right)^{1/5}. \tag{4.30}$$

GAUSSIAN RULE OF THUMB

To compute the optimal σ^* in (4.30), one needs to estimate the functional $\|f''\|^2$. The *Gaussian rule of thumb* is to assume that f is the density of the $\mathcal{N}(\overline{x}, s^2)$ distribution, where $\overline{x}$ and s^2 are the sample mean and variance of the data, respectively [113]. In this case $\|f''\|^2 = s^{-5}\pi^{-1/2}3/8$ and the Gaussian rule of thumb becomes:

$$\sigma_{\text{rot}} = \left(\frac{4\,s^5}{3\,n}\right)^{1/5} \approx 1.06\, s\, n^{-1/5}.$$

THETA KDE

We recommend, however, the fast and reliable *theta KDE* of [14], which chooses the bandwidth in an optimal way via a fixed-point procedure. Figures 4.1 and 4.2 illustrate a common problem with traditional KDEs: for distributions on a bounded domain, such as the uniform distribution on $[0, 1]^2$, the KDE assigns positive probability mass *outside* this domain. An additional advantage of the theta KDE is that it largely avoids this boundary effect. We illustrate the theta KDE with the following example.

■ **Example 4.3 (Comparison of Gaussian and theta KDEs)** The following Python program draws an iid sample from the Exp(1) distribution and constructs a Gaussian kernel density estimate. We see in Figure 4.3 that with an appropriate choice of the bandwidth a good fit to the true pdf can be achieved, except at the boundary $x = 0$. The theta KDE does not exhibit this boundary effect. Moreover, it chooses the bandwidth automatically, to achieve a superior fit. The theta KDE source code is available as `kde.py` on the book's GitHub site.

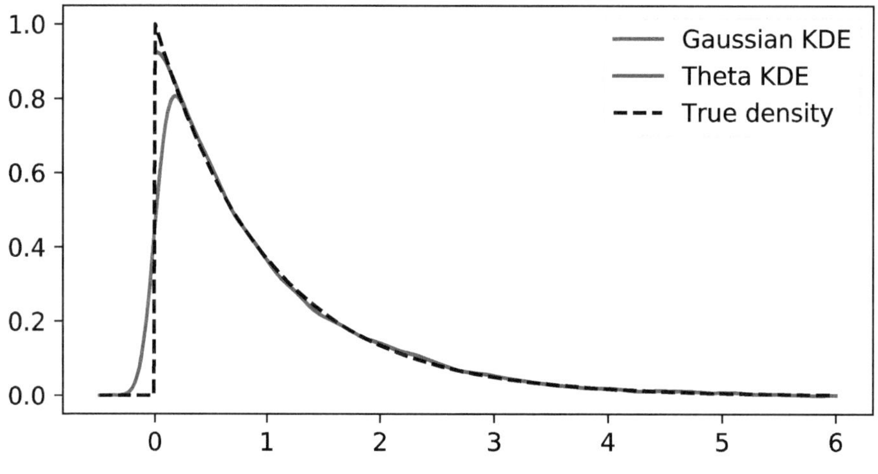

Figure 4.3: Kernel density estimates for Exp(1)-distributed data.

gausthetakde.py

```
import matplotlib.pyplot as plt
import numpy as np
from kde import *

sig = 0.1; sig2 = sig**2; c = 1/np.sqrt(2*np.pi)/sig #Constants
phi = lambda x,x0: np.exp(-(x-x0)**2/(2*sig2)) #Unscaled Kernel
f = lambda x: np.exp(-x)*(x >= 0) # True PDF
n = 10**4 # Sample Size
x = -np.log(np.random.uniform(size=n))# Generate Data via IT method
xx = np.arange(-0.5,6,0.01, dtype = "d")# Plot Range
phis = np.zeros(len(xx))
for i in range(0,n):
    phis = phis + phi(xx,x[i])
phis = c*phis/n
plt.plot(xx,phis,'r')# Plot Gaussian KDE
[bandwidth,density,xmesh,cdf] = kde(x,2**12,0,max(x))
idx = (xmesh <= 6)
plt.plot(xmesh[idx],density[idx])# Plot Theta KDE
plt.plot(xx,f(xx))# Plot True PDF
```

■

4.5 Clustering via Mixture Models

Clustering is concerned with the grouping of unlabeled feature vectors into clusters, such that samples within a cluster are more similar to each other than samples belonging to different clusters. Usually, it is assumed that the number of clusters is known in advance, but otherwise no prior information is given about the data. Applications of clustering can be found in the areas of communication, data compression and storage, database searching, pattern matching, and object recognition.

A common approach to clustering analysis is to assume that the data comes from a mixture of (usually Gaussian) distributions, and thus the objective is to estimate the parameters of the mixture model by maximizing the likelihood function for the data. Direct optimization of the likelihood function in this case is not a simple task, due to necessary constraints on the parameters (more about this later) and the complicated nature of the likelihood function, which in general has a great number of local maxima and saddle-points. A popular method to estimate the parameters of the mixture model is the EM algorithm, which was discussed in a more general setting in Section 4.3. In this section we explain the basics of mixture modeling and explain the workings of the EM method in this context. In addition, we show how direct optimization methods can be used to maximize the likelihood.

☞ 128

4.5.1 Mixture Models

Let $\mathcal{T} := \{X_1, \dots, X_n\}$ be iid random vectors taking values in some set $\mathcal{X} \subseteq \mathbb{R}^d$, each X_i being distributed according to the *mixture density*

MIXTURE DENSITY

$$g(x \mid \theta) = w_1 \phi_1(x) + \cdots + w_K \phi_K(x), \quad x \in \mathcal{X}, \tag{4.31}$$

where $\phi_1, \ldots, \phi_K$ are probability densities (discrete or continuous) on $\mathcal{X}$, and the positive *weights* $w_1, \ldots, w_K$ sum up to 1. This mixture pdf can be interpreted in the following way. Let Z be a discrete random variable taking values $1, 2, \ldots, K$ with probabilities $w_1, \ldots, w_K$, and let $\boldsymbol{X}$ be a random vector whose conditional pdf, given $Z = z$, is ϕ_z. By the product rule (C.17), the joint pdf of Z and $\boldsymbol{X}$ is given by

WEIGHTS

☞ 433

$$\phi_{Z,\boldsymbol{X}}(z, \boldsymbol{x}) = \phi_Z(z)\, \phi_{\boldsymbol{X}\,|\,Z}(\boldsymbol{x}\,|\,z) = w_z\, \phi_z(\boldsymbol{x})$$

and the marginal pdf of $\boldsymbol{X}$ is found by summing the joint pdf over the values of z, which gives (4.31). A random vector $\boldsymbol{X} \sim g$ can thus be simulated in two steps:

1. First, draw Z according to the probabilities $\mathbb{P}[Z = z] = w_z$, $z = 1, \ldots, K$.

2. Then draw $\boldsymbol{X}$ according to the pdf ϕ_Z.

As $\mathcal{T}$ only contain the $\{\boldsymbol{X}_i\}$ variables, the $\{Z_i\}$ are viewed as *latent* variables. We can interpret Z_i as the hidden label of the cluster to which $\boldsymbol{X}_i$ belongs.

Typically, each ϕ_k in (4.31) is assumed to be known up to some parameter vector $\boldsymbol{\eta}_k$. It is customary[1] in clustering analysis to work with *Gaussian* mixtures; that is, each density ϕ_k is Gaussian with some unknown expectation vector $\boldsymbol{\mu}_k$ and covariance matrix $\boldsymbol{\Sigma}_k$. We gather all unknown parameters, including the weights $\{w_k\}$, into a parameter vector $\boldsymbol{\theta}$. As usual, $\tau = \{\boldsymbol{x}_1, \ldots, \boldsymbol{x}_n\}$ denotes the outcome of $\mathcal{T}$. As the components of $\mathcal{T}$ are iid, their (joint) pdf is given by

$$g(\tau\,|\,\boldsymbol{\theta}) := \prod_{i=1}^{n} g(\boldsymbol{x}_i\,|\,\boldsymbol{\theta}) = \prod_{i=1}^{n} \sum_{k=1}^{K} w_k\, \phi_k(\boldsymbol{x}_i\,|\,\boldsymbol{\mu}_k, \boldsymbol{\Sigma}_k). \tag{4.32}$$

Following the same reasoning as for (4.5), we can estimate $\boldsymbol{\theta}$ from an outcome τ by maximizing the log-likelihood function

$$l(\boldsymbol{\theta}\,|\,\tau) := \sum_{i=1}^{n} \ln g(\boldsymbol{x}_i\,|\,\boldsymbol{\theta}) = \sum_{i=1}^{n} \ln\left(\sum_{k=1}^{K} w_k\, \phi_k(\boldsymbol{x}_i\,|\,\boldsymbol{\mu}_k, \boldsymbol{\Sigma}_k)\right). \tag{4.33}$$

However, finding the maximizer of $l(\boldsymbol{\theta}\,|\,\tau)$ is not easy in general, since the function is typically multiextremal.

■ **Example 4.4 (Clustering via Mixture Models)** The data depicted in Figure 4.4 consists of 300 data points that were independently generated from three bivariate normal distributions, whose parameters are given in that same figure. For each of these three distributions, exactly 100 points were generated. Ideally, we would like to cluster the data into three clusters that correspond to the three cases.

To cluster the data into three groups, a possible model for the data is to assume that the points are iid draws from an (unknown) mixture of three 2-dimensional Gaussian distributions. This is a sensible approach, although in reality the data were not simulated in this way. It is instructive to understand the difference between the two models. In the mixture model, each cluster label Z takes the value $\{1, 2, 3\}$ with equal probability, and hence, drawing the labels independently, the total number of points in each cluster would

[1] Other common mixture distributions include Student t and Beta distributions.

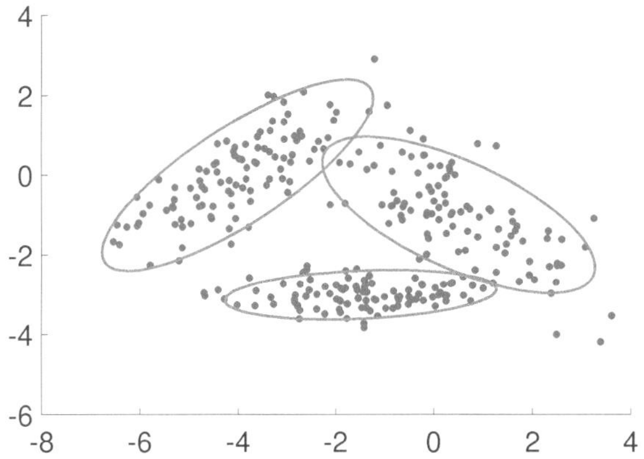

cluster	mean vector	covariance matrix
1	$\begin{bmatrix} -4 \\ 0 \end{bmatrix}$	$\begin{bmatrix} 2 & 1.4 \\ 1.4 & 1.5 \end{bmatrix}$
2	$\begin{bmatrix} 0.5 \\ -1 \end{bmatrix}$	$\begin{bmatrix} 2 & -0.95 \\ -0.95 & 1 \end{bmatrix}$
3	$\begin{bmatrix} -1.5 \\ -3 \end{bmatrix}$	$\begin{bmatrix} 2 & 0.1 \\ 0.1 & 0.1 \end{bmatrix}$

Figure 4.4: Cluster the 300 data points (left) into three clusters, without making any assumptions about the probability distribution of the data. In fact, the data were generated from three bivariate normal distributions, whose parameters are listed on the right.

be $\mathsf{Bin}(300, 1/3)$ distributed. However, in the actual simulation, the number of points in each cluster is exactly 100. Nevertheless, the mixture model would be an accurate (although not exact) model for these data. Figure 4.5 displays the "target" Gaussian mixture density for the data in Figure 4.4; that is, the mixture with equal weights and with the exact parameters as specified in Figure 4.4.

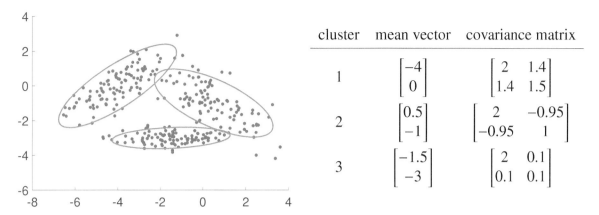

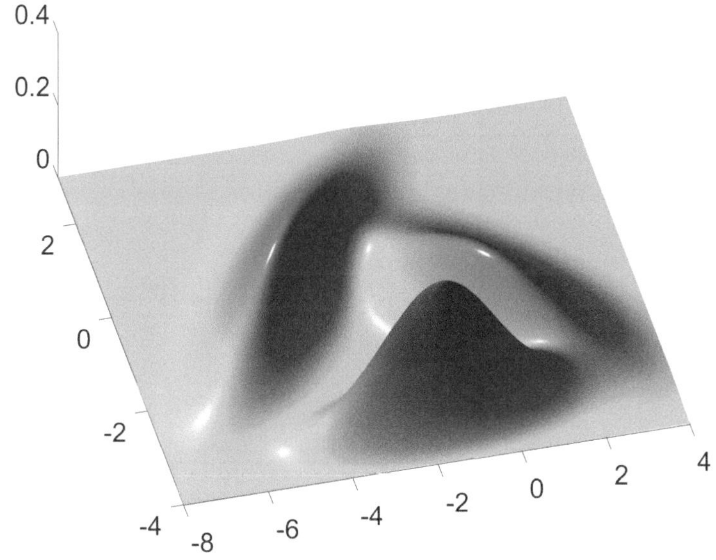

Figure 4.5: The target mixture density for the data in Figure 4.4.

In the next section we will carry out the clustering by using the EM algorithm. ■

4.5.2 EM Algorithm for Mixture Models

As we saw in Section 4.3, instead of maximizing the log-likelihood function (4.33) directly from the data $\tau = \{\boldsymbol{x}_1, \ldots, \boldsymbol{x}_n\}$, the EM algorithm first *augments* the data with the vector of latent variables — in this case the hidden cluster labels $\boldsymbol{z} = \{z_1, \ldots, z_n\}$. The idea is that τ is

DATA AUGMENTATION

only the *observed* part of the complete random data $(\mathcal{T}, \mathbf{Z})$, which were generated via the two-step procedure described above. That is, for each data point $\mathbf{X}$, first draw the cluster label $Z \in \{1, \dots, K\}$ according to probabilities $\{w_1, \dots, w_K\}$ and then, given $Z = z$, draw $\mathbf{X}$ from ϕ_z. The joint pdf of $\mathcal{T}$ and $\mathbf{Z}$ is

$$g(\tau, \mathbf{z} \,|\, \boldsymbol{\theta}) = \prod_{i=1}^{n} w_{z_i}\, \phi_{z_i}(\mathbf{x}_i),$$

which is of a much simpler form than (4.32). It follows that the *complete-data log-likelihood* function

COMPLETE-DATA LOG-LIKELIHOOD

$$\widetilde{l}(\boldsymbol{\theta} \,|\, \tau, \mathbf{z}) = \sum_{i=1}^{n} \ln[w_{z_i}\, \phi_{z_i}(\mathbf{x}_i)] \tag{4.34}$$

is often easier to maximize than the original log-likelihood (4.33), for any given $(\tau, \mathbf{z})$. But, of course the latent variables $\mathbf{z}$ are not observed and so $\widetilde{l}(\boldsymbol{\theta} \,|\, \tau, \mathbf{z})$ cannot be evaluated. In the E-step of the EM algorithm, the complete-data log-likelihood is replaced with the expectation $\mathbb{E}_p \widetilde{l}(\boldsymbol{\theta} \,|\, \tau, \mathbf{Z})$, where the subscript p in the expectation indicates that $\mathbf{Z}$ is distributed according to the conditional pdf of $\mathbf{Z}$ given $\mathcal{T} = \tau$; that is, with pdf

$$p(\mathbf{z}) = g(\mathbf{z} \,|\, \tau, \boldsymbol{\theta}) \propto g(\tau, \mathbf{z} \,|\, \boldsymbol{\theta}). \tag{4.35}$$

Note that $p(\mathbf{z})$ is of the form $p_1(z_1) \cdots p_n(z_n)$ so that, given $\mathcal{T} = \tau$, the components of $\mathbf{Z}$ are independent of each other. The EM algorithm for mixture models can now be formulated as follows.

Algorithm 4.5.1: EM Algorithm for Mixture Models

---

input: Data τ, initial guess $\boldsymbol{\theta}^{(0)}$.
output: Approximation of the maximum likelihood estimate.

1 $t \leftarrow 1$
2 **while** a stopping criterion is not met **do**
3   **Expectation Step**: Find $p^{(t)}(\mathbf{z}) := g(\mathbf{z} \,|\, \tau, \boldsymbol{\theta}^{(t-1)})$ and $Q^{(t)}(\boldsymbol{\theta}) := \mathbb{E}_{p^{(t)}} \widetilde{l}(\boldsymbol{\theta} \,|\, \tau, \mathbf{Z})$.
4   **Maximization Step**: Let $\boldsymbol{\theta}^{(t)} \leftarrow \operatorname{argmax}_{\boldsymbol{\theta}} Q^{(t)}(\boldsymbol{\theta})$.
5     $t \leftarrow t + 1$
6 **return** $\boldsymbol{\theta}^{(t)}$

A possible termination condition is to stop when $\left| l(\boldsymbol{\theta}^{(t)} \,|\, \tau) - l(\boldsymbol{\theta}^{(t-1)} \,|\, \tau) \right| / \left| l(\boldsymbol{\theta}^{(t)} \,|\, \tau) \right| < \varepsilon$ for some small tolerance $\varepsilon > 0$. As was mentioned in Section 4.3, the sequence of log-likelihood values *does not decrease* with each iteration. Under certain continuity conditions, the sequence $\{\boldsymbol{\theta}^{(t)}\}$ is guaranteed to converge to a local maximizer of the log-likelihood l. Convergence to a global maximizer (if it exists) depends on the appropriate choice for the starting value. Typically, the algorithm is run from different random starting points.

For the case of Gaussian mixtures, each $\phi_k = \phi(\cdot \,|\, \boldsymbol{\mu}_k, \boldsymbol{\Sigma}_k), k = 1, \dots, K$ is the density of a d-dimensional Gaussian distribution. Let $\boldsymbol{\theta}^{(t-1)}$ be the current guess for the optimal parameter vector, consisting of the weights $\{w_k^{(t-1)}\}$, mean vectors $\{\boldsymbol{\mu}_k^{(t-1)}\}$, and covariance matrices $\{\boldsymbol{\Sigma}_k^{(t-1)}\}$. We first determine $p^{(t)}$ — the pdf of $\mathbf{Z}$ conditional on $\mathcal{T} = \tau$ — for the given guess $\boldsymbol{\theta}^{(t-1)}$. As mentioned before, the components of $\mathbf{Z}$ given $\mathcal{T} = \tau$ are independent, so it suffices to specify the discrete pdf, $p_i^{(t)}$ say, of each Z_i given the observed point $\boldsymbol{X}_i = \boldsymbol{x}_i$. The latter can be found from Bayes' formula:

$$p_i^{(t)}(k) \propto w_k^{(t-1)} \, \phi_k(\boldsymbol{x}_i \,|\, \boldsymbol{\mu}_k^{(t-1)}, \boldsymbol{\Sigma}_k^{(t-1)}), \quad k = 1, \ldots, K. \tag{4.36}$$

Next, in view of (4.34), the function $Q^{(t)}(\boldsymbol{\theta})$ can be written as

$$Q^{(t)}(\boldsymbol{\theta}) = \mathbb{E}_{\boldsymbol{p}^{(t)}} \sum_{i=1}^{n} \left(\ln w_{Z_i} + \ln \phi_{Z_i}(\boldsymbol{x}_i \,|\, \boldsymbol{\mu}_{Z_i}, \boldsymbol{\Sigma}_{Z_i})\right) = \sum_{i=1}^{n} \mathbb{E}_{p_i^{(t)}} \left[\ln w_{Z_i} + \ln \phi_{Z_i}(\boldsymbol{x}_i \,|\, \boldsymbol{\mu}_{Z_i}, \boldsymbol{\Sigma}_{Z_i})\right],$$

where the $\{Z_i\}$ are independent and Z_i is distributed according to $p_i^{(t)}$ in (4.36). This completes the *E-step*. In the *M-step* we maximize $Q^{(t)}$ with respect to the parameter $\boldsymbol{\theta}$; that is, with respect to the $\{w_k\}$, $\{\boldsymbol{\mu}_k\}$, and $\{\boldsymbol{\Sigma}_k\}$. In particular, we maximize

$$\sum_{i=1}^{n} \sum_{k=1}^{K} p_i^{(t)}(k) \left[\ln w_k + \ln \phi_k(\boldsymbol{x}_i \,|\, \boldsymbol{\mu}_k, \boldsymbol{\Sigma}_k)\right],$$

under the condition $\sum_k w_k = 1$. Using Lagrange multipliers and the fact that $\sum_{k=1}^{K} p_i^{(t)}(k) = 1$ gives the solution for the $\{w_k\}$:

$$w_k = \frac{1}{n} \sum_{i=1}^{n} p_i^{(t)}(k), \quad k = 1, \ldots, K. \tag{4.37}$$

The solutions for $\boldsymbol{\mu}_k$ and $\boldsymbol{\Sigma}_k$ now follow from maximizing $\sum_{i=1}^{n} p_i^{(t)}(k) \ln \phi_k(\boldsymbol{x}_i \,|\, \boldsymbol{\mu}_k, \boldsymbol{\Sigma}_k)$, leading to

$$\boldsymbol{\mu}_k = \frac{\sum_{i=1}^{n} p_i^{(t)}(k) \, \boldsymbol{x}_i}{\sum_{i=1}^{n} p_i^{(t)}(k)}, \quad k = 1, \ldots, K \tag{4.38}$$

and

$$\boldsymbol{\Sigma}_k = \frac{\sum_{i=1}^{n} p_i^{(t)}(k) \, (\boldsymbol{x}_i - \boldsymbol{\mu}_k)(\boldsymbol{x}_i - \boldsymbol{\mu}_k)^\top}{\sum_{i=1}^{n} p_i^{(t)}(k)}, \quad k = 1, \ldots, K, \tag{4.39}$$

which are very similar to the well-known formulas for the MLEs of the parameters of a Gaussian distribution. After assigning the solution parameters to $\boldsymbol{\theta}^{(t)}$ and increasing the iteration counter t by 1, the steps (4.36), (4.37), (4.38), and (4.39) are repeated until convergence is reached. Convergence of the EM algorithm is very sensitive to the choice of initial parameters. It is therefore recommended to try various different starting conditions. For a further discussion of the theoretical and practical aspects of the EM algorithm we refer to [85].

■ **Example 4.5 (Clustering via EM)** We return to the data in Example 4.4, depicted in Figure 4.4, and adopt the model that the data is coming from a mixture of three bivariate Gaussian distributions.

The Python code below implements the EM procedure described in Algorithm 4.5.1. The initial mean vectors $\{\boldsymbol{\mu}_k\}$ of the bivariate Gaussian distributions are chosen (from visual inspection) to lie roughly in the middle of each cluster, in this case $[-2, -3]^\top$, $[-4, 1]^\top$, and $[0, -1]^\top$. The corresponding covariance matrices are initially chosen as identity matrices, which is appropriate given the observed spread of the data in Figure 4.4. Finally, the initial weights are $1/3, 1/3, 1/3$. For simplicity, the algorithm stops after 100 iterations, which in this case is more than enough to guarantee convergence. The code and data are available from the book's website in the GitHub folder Chapter4.

EMclust.py

```
import numpy as np
from scipy.stats import multivariate_normal

Xmat = np.genfromtxt('clusterdata.csv', delimiter=',')
K = 3
n, D = Xmat.shape

W = np.array([[1/3,1/3,1/3]])
M  = np.array([[-2.0,-4,0],[-3,1,-1]], dtype=np.float32)
# Note that if above *all* entries were written as integers, M would
# be defined to be of integer type, which will give the wrong answer

C = np.zeros((3,2,2))

C[:,0,0] = 1
C[:,1,1] = 1

p = np.zeros((3,300))

for i in range(0,100):

#E-step
    for k in range(0,K):
        mvn = multivariate_normal( M[:,k].T, C[k,:,:] )
        p[k,:] = W[0,k]*mvn.pdf(Xmat)

# M-Step
    p = (p/sum(p,0))    #normalize
    W = np.mean(p,1).reshape(1,3)

    for k in range(0,K):
        M[:,k] = (Xmat.T @ p[k,:].T)/sum(p[k,:])
        xm = Xmat.T - M[:,k].reshape(2,1)
        C[k,:,:] = xm @ (xm*p[k,:]).T/sum(p[k,:])
```

The estimated parameters of the mixture distribution are given on the right-hand side of Figure 4.6. After relabeling of the clusters, we can observe a close match with the parameters in Figure 4.4.

The ellipses on the left-hand side of Figure 4.6 show a close match between the 95% probability ellipses[2] of the original Gaussian distributions (in gray) and the estimated ones. A natural way to cluster each point $\boldsymbol{x}_i$ is to assign it to the cluster k for which the conditional probability $p_i(k)$ is maximal (with ties resolved arbitrarily). This gives the clustering of the points into red, green, and blue clusters in the figure.

[2]For each mixture component, the contour of the corresponding bivariate normal pdf is shown that encloses 95% of the probability mass.

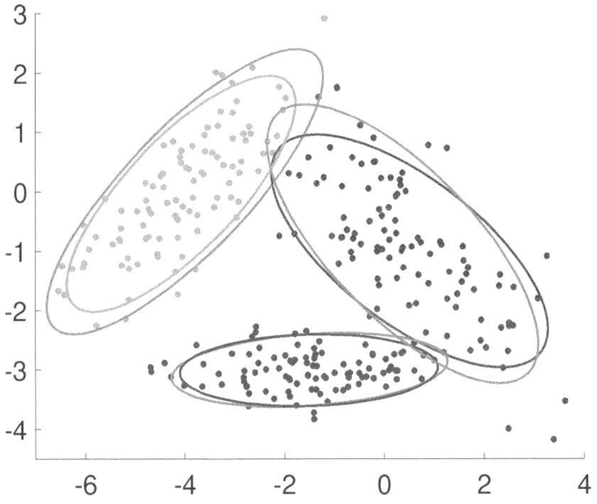

weight	mean vector	covariance matrix
0.33	$\begin{bmatrix} -1.51 \\ -3.01 \end{bmatrix}$	$\begin{bmatrix} 1.75 & 0.03 \\ 0.03 & 0.095 \end{bmatrix}$
0.32	$\begin{bmatrix} -4.08 \\ -0.033 \end{bmatrix}$	$\begin{bmatrix} 1.37 & 0.92 \\ 0.92 & 1.03 \end{bmatrix}$
0.35	$\begin{bmatrix} 0.36 \\ -0.88 \end{bmatrix}$	$\begin{bmatrix} 1.93 & -1.20 \\ -1.20 & 1.44 \end{bmatrix}$

Figure 4.6: The results of the EM clustering algorithm applied to the data depicted in Figure 4.4.

■

As an alternative to the EM algorithm, one can of course use continuous multiextremal optimization algorithms to directly optimize the log-likelihood function $l(\boldsymbol{\theta} \mid \tau) = \ln g(\tau \mid \boldsymbol{\theta})$ in (4.33) over the set Θ of all possible $\boldsymbol{\theta}$. This is done for example in [15], demonstrating superior results to EM when there are few data points. Closer investigation of the likelihood function reveals that there is a hidden problem with any maximum likelihood approach for clustering if Θ is chosen as large as possible — i.e., any mixture distribution is possible. To demonstrate this problem, consider Figure 4.7, depicting the probability density function, $g(\cdot \mid \boldsymbol{\theta})$ of a mixture of two Gaussian distributions, where $\boldsymbol{\theta} = [w, \mu_1, \sigma_1^2, \mu_2, \sigma_2^2]^\top$ is the vector of parameters for the mixture distribution. The log-likelihood function is given by $l(\boldsymbol{\theta} \mid \tau) = \sum_{i=1}^{4} \ln g(x_i \mid \boldsymbol{\theta})$, where $x_1, \ldots, x_4$ are the data (indicated by dots in the figure).

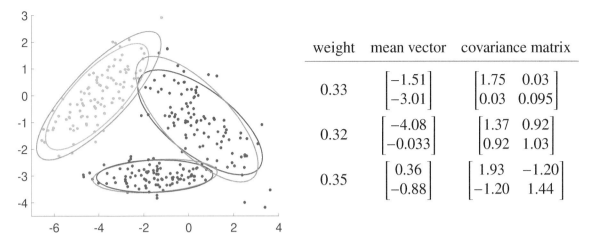

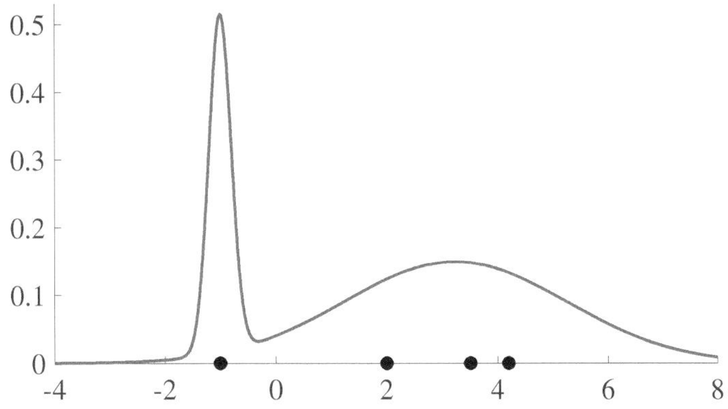

Figure 4.7: Mixture of two Gaussian distributions.

It is clear that by fixing the mixing constant w at 0.25 (say) and centering the first cluster at x_1, one can obtain an arbitrarily large likelihood value by taking the variance of the first cluster to be arbitrarily small. Similarly, for higher dimensional data, by choosing "point" or "line" clusters, or in general "degenerate" clusters, one can make the value of the likelihood infinite. This is a manifestation of the familiar overfitting problem for the training loss that we already encountered in Chapter 2. Thus, the unconstrained maximization of the log-likelihood function is an ill-posed problem, irrespective of the choice of the optimization algorithm!

Two possible solutions to this "overfitting" problem are:

1. Restrict the parameter set Θ in such a way that degenerate clusters (sometimes called spurious clusters) are not allowed.
2. Run the given algorithm and if the solution is degenerate, discard it and run the algorithm afresh. Keep restarting the algorithm until a non-degenerate solution is obtained.

The first approach is usually applied to multiextremal optimization algorithms and the second is used for the EM algorithm.

4.6 Clustering via Vector Quantization

In the previous section we introduced clustering via mixture models, as a form of parametric density estimation (as opposed to the nonparametric density estimation in Section 4.4). The clusters were modeled in a natural way via the latent variables and the EM algorithm provided a convenient way to assign the cluster members. In this section we consider a more heuristic approach to clustering by ignoring the distributional properties of the data. The resulting algorithms tend to scale better with the number of samples n and the dimensionality d.

In mathematical terms, we consider the following clustering (also called data segmentation) problem. Given a collection $\tau = \{\boldsymbol{x}_1, \dots, \boldsymbol{x}_n\}$ of data points in some d-dimensional space $\mathcal{X}$, divide this data set into K clusters (groups) such that some loss function is minimized. A convenient way to determine these clusters is to first divide up the entire space $\mathcal{X}$, using some distance function $\text{dist}(\cdot, \cdot)$ on this space. A standard choice is the Euclidean (or L_2) distance:

$$\text{dist}(\boldsymbol{x}, \boldsymbol{x}') = \|\boldsymbol{x} - \boldsymbol{x}'\| = \sqrt{\sum_{i=1}^{d} (x_i - x_i')^2}.$$

MANHATTAN DISTANCE

Other commonly used distance measures on $\mathbb{R}^d$ include the *Manhattan distance*:

$$\sum_{i=1}^{d} |x_i - x_i'|$$

MAXIMUM DISTANCE

and the *maximum distance*:

$$\max_{i=1,\dots,d} |x_i - x_i'|.$$

HAMMING DISTANCE

On the set of strings of length d, an often-used distance measure is the *Hamming distance*:

$$\sum_{i=1}^{d} \mathbb{1}\{x_i \neq x_i'\},$$

that is, the number of mismatched characters. For example, the Hamming distance between 010101 and 011010 is 4.

We can partition the space $\mathcal{X}$ into regions as follows: First, we choose K points $\boldsymbol{c}_1, \ldots, \boldsymbol{c}_K$ called *cluster centers* or *source vectors*. For each $k = 1, \ldots, K$, let

SOURCE VECTORS

$$\mathcal{R}_k = \{\boldsymbol{x} \in \mathcal{X} : \text{dist}(\boldsymbol{x}, \boldsymbol{c}_k) \leqslant \text{dist}(\boldsymbol{x}, \boldsymbol{c}_i) \text{ for all } i \neq k\}$$

be the set of points in $\mathcal{X}$ that lie closer to $\boldsymbol{c}_k$ than any other center. The regions or *cells* $\{\mathcal{R}_k\}$ divide the space $\mathcal{X}$ into what is called a *Voronoi diagram* or a *Voronoi tessellation*. Figure 4.8 shows a Voronoi tessellation of the plane into ten regions, using the Euclidean distance. Note that here the boundaries between the Voronoi cells are straight line segments. In particular, if cell $\mathcal{R}_i$ and $\mathcal{R}_j$ share a border, then a point on this border must satisfy $\|\boldsymbol{x} - \boldsymbol{c}_i\| = \|\boldsymbol{x} - \boldsymbol{c}_j\|$; that is, it must lie on the line that passes through the point $(\boldsymbol{c}_j + \boldsymbol{c}_i)/2$ (that is, the midway point of the line segment between $\boldsymbol{c}_i$ and $\boldsymbol{c}_j$) and be perpendicular to $\boldsymbol{c}_j - \boldsymbol{c}_i$.

VORONOI TESSELLATION

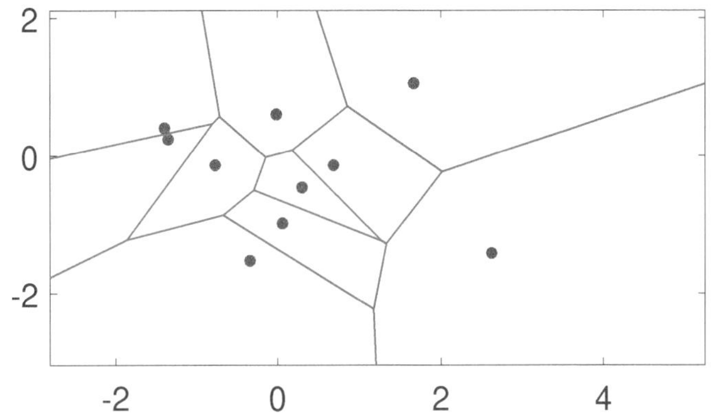

Figure 4.8: A Voronoi tessellation of the plane into ten cells, determined by the (red) centers.

Once the centers (and thus the cells $\{\mathcal{R}_k\}$) are chosen, the points in τ can be clustered according to their nearest center. Points on the boundary have to be treated separately. This is a moot point for continuous data, as generally no data points will lie exactly on the boundary.

The main remaining issue is how to choose the centers so as to cluster the data in some optimal way. In terms of our (unsupervised) learning framework, we wish to approximate a vector $\boldsymbol{x}$ via one of $\boldsymbol{c}_1, \ldots, \boldsymbol{c}_K$, using a piecewise constant vector-valued function

$$\boldsymbol{g}(\boldsymbol{x} \,|\, \mathbf{C}) := \sum_{k=1}^{K} \boldsymbol{c}_k \, \mathbb{1}\{\boldsymbol{x} \in \mathcal{R}_k\},$$

where $\mathbf{C}$ is the $d \times K$ matrix $[\boldsymbol{c}_1, \ldots, \boldsymbol{c}_K]$. Thus, $\boldsymbol{g}(\boldsymbol{x} \,|\, \mathbf{C}) = \boldsymbol{c}_k$ when $\boldsymbol{x}$ falls in region $\mathcal{R}_k$ (we ignore ties). Within this class $\mathcal{G}$ of functions, parameterized by $\mathbf{C}$, our aim is to minimize the training loss. In particular, for the squared-error loss, $\text{Loss}(\boldsymbol{x}, \boldsymbol{x}') = \|\boldsymbol{x} - \boldsymbol{x}'\|^2$, the training loss is

$$\ell_{\tau_n}(\boldsymbol{g}(\cdot \,|\, \mathbf{C})) = \frac{1}{n} \sum_{i=1}^{n} \|\boldsymbol{x}_i - \boldsymbol{g}(\boldsymbol{x}_i \,|\, \mathbf{C})\|^2 = \frac{1}{n} \sum_{k=1}^{K} \sum_{\boldsymbol{x} \in \mathcal{R}_k \cap \tau_n} \|\boldsymbol{x} - \boldsymbol{c}_k\|^2. \tag{4.40}$$

Thus, the training loss minimizes the average squared distance between the centers. This framework also combines both the encoding and decoding steps in *vector quantization*

VECTOR QUANTIZATION

[125]. Namely, we wish to "quantize" or "encode" the vectors in τ in such a way that each vector is represented by one of K source vectors $\boldsymbol{c}_1, \ldots, \boldsymbol{c}_K$, such that the loss (4.40) of this representation is minimized.

Most well-known clustering and vector quantization methods update the vector of centers, starting from some initial choice and using iterative (typically gradient-based) procedures. It is important to realize that in this case (4.40) is seen as a function of the centers, where each point $\boldsymbol{x}$ is assigned to the nearest center, thus determining the clusters. It is well known that this type of problem — optimization with respect to the centers — is highly multiextremal and, depending on the initial clusters, gradient-based procedures tend to converge to a *local minimum* rather than a global minimum.

4.6.1 K-Means

One of the simplest methods for clustering is the K-means method. It is an iterative method where, starting from an initial guess for the centers, new centers are formed by taking sample means of the current points in each cluster. The new centers are thus the *centroids* of the points in each cell. Although there exist many different varieties of the K-means algorithm, they are all essentially of the following form:

CENTROIDS

Algorithm 4.6.1: K-Means

input: Collection of points $\tau = \{\boldsymbol{x}_1, \ldots, \boldsymbol{x}_n\}$, number of clusters K, initial centers $\boldsymbol{c}_1, \ldots, \boldsymbol{c}_K$.
output: Cluster centers and cells (regions).
1 **while** a stopping criterion is not met **do**
2 $\quad \mathcal{R}_1, \ldots, \mathcal{R}_K \leftarrow \emptyset$ (empty sets).
3 $\quad$ **for** $i = 1$ **to** n **do**
4 $\quad\quad \boldsymbol{d} \leftarrow [\text{dist}(\boldsymbol{x}_i, \boldsymbol{c}_1), \ldots, \text{dist}(\boldsymbol{x}_i, \boldsymbol{c}_K)]$ // distances to centers
5 $\quad\quad k \leftarrow \text{argmin}_j\, d_j$
6 $\quad\quad \mathcal{R}_k \leftarrow \mathcal{R}_k \cup \{\boldsymbol{x}_i\}$ // assign $\boldsymbol{x}_i$ to cluster k
7 $\quad$ **for** $k = 1$ **to** K **do**
8 $\quad\quad \boldsymbol{c}_k \leftarrow \dfrac{\sum_{\boldsymbol{x} \in \mathcal{R}_k} \boldsymbol{x}}{|\mathcal{R}_k|}$ // compute the new center as a centroid of points
9 **return** $\{\boldsymbol{c}_k\}, \{\mathcal{R}_k\}$

Thus, at each iteration, for a given choice of centers, each point in τ is assigned to its nearest center. After all points have been assigned, the centers are recomputed as the centroids of all the points in the current cluster (Line 8). A typical stopping criterion is to stop when the centers no longer change very much. As the algorithm is quite sensitive to the choice of the initial centers, it is prudent to try multiple starting values, e.g., chosen randomly from the bounding box of the data points.

We can see the K-means method as a deterministic (or "hard") version of the probabilistic (or "soft") EM algorithm as follows. Suppose in the EM algorithm we have Gaussian mixtures with a fixed covariance matrix $\boldsymbol{\Sigma}_k = \sigma^2 \mathbf{I}_d$, $k = 1, \ldots, K$, where σ^2 should be thought of as being infinitesimally small. Consider iteration t of the EM algorithm. Having obtained the expectation vectors $\boldsymbol{\mu}_k^{(t-1)}$ and weights $w_k^{(t-1)}$, $k = 1, \ldots, K$, each point $\boldsymbol{x}_i$ is assigned a cluster label Z_i according to the probabilities $p_i^{(t)}(k)$, $k = 1, \ldots, K$ given in (4.36).

But for $\sigma^2 \to 0$ the probability distribution $\{p_i^{(t)}(k)\}$ becomes degenerate, putting all its probability mass on $\text{argmin}_k \|\boldsymbol{x}_i - \boldsymbol{\mu}_k\|^2$. This corresponds to the K-means rule of assigning $\boldsymbol{x}_i$ to its nearest cluster center. Moreover, in the M-step (4.38) each cluster center $\boldsymbol{\mu}_k^{(t)}$ is now updated according to the average of the $\{\boldsymbol{x}_i\}$ that have been assigned to cluster k. We thus obtain the same deterministic updating rule as in K-means.

■ **Example 4.6 (K-means Clustering)** We cluster the data from Figure 4.4 via K-means, using the Python implementation below. Note that the data points are stored as a 300×2 matrix `Xmat`. We take the same starting centers as in the EM example: $\boldsymbol{c}_1 = [-2, -3]^\top, \boldsymbol{c}_2 = [-4, 1]^\top$, and $\boldsymbol{c}_3 = [0, -1]^\top$. Note also that *squared* Euclidean distances are used in the computations, as these are slightly faster to compute than Euclidean distances (as no square root computations are required) while yielding exactly the same cluster center evaluations.

Kmeans.py

```
import numpy as np
Xmat = np.genfromtxt('clusterdata.csv', delimiter=',')
K = 3
n, D = Xmat.shape
c  = np.array([[-2.0,-4,0],[-3,1,-1]])  #initialize centers
cold = np.zeros(c.shape)
dist2 = np.zeros((K,n))
while np.abs(c - cold).sum() > 0.001:
   cold = c.copy()
   for i in range(0,K): #compute the squared distances
        dist2[i,:] = np.sum((Xmat - c[:,i].T)**2, 1)

   label = np.argmin(dist2,0) #assign the points to nearest centroid
   minvals = np.amin(dist2,0)
   for i in range(0,K): # recompute the centroids
      c[:,i] = np.mean(Xmat[np.where(label == i),:],1).reshape(1,2)

print('Loss = {:3.3f}'.format(minvals.mean()))
```

```
Loss = 2.288
```

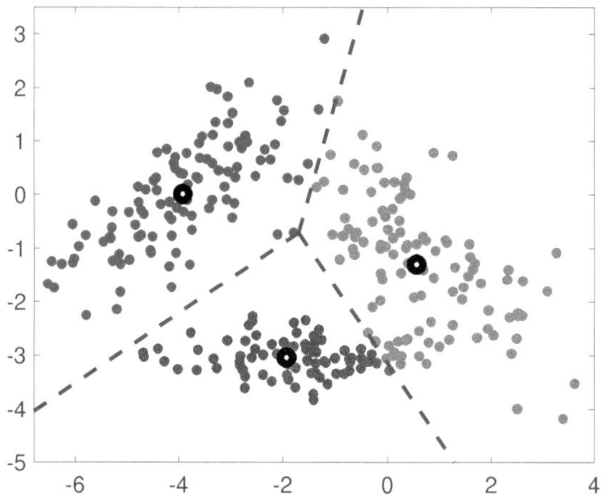

Figure 4.9: Results of the K-means algorithm applied to the data in Figure 4.4. The thick black circles are the centroids and the dotted lines define the cell boundaries.

We found the cluster centers $\mathbf{c}_1 = [-1.9286, -3.0416]^\top, \mathbf{c}_2 = [-3.9237, 0.0131]^\top$, and $\mathbf{c}_3 = [0.5611, -1.2980]^\top$, giving the clustering depicted in Figure 4.9. The corresponding loss (4.40) was found to be 2.288. ■

4.6.2 Clustering via Continuous Multiextremal Optimization

As already mentioned, the exact minimization of the loss function (4.40) is difficult to accomplish via standard local search methods such as gradient descent, as the function is highly multimodal. However, nothing is preventing us from using global optimization methods such as the CE or SCO methods discussed in Sections 3.4.2 and 3.4.3. ☞ 100

■ **Example 4.7 (Clustering via CE)** We take the same data set as in Example 4.6 and cluster the points via minimization of the loss (4.40) using the CE method. The Python code below is very similar to the code in Example 3.16, except that now we are dealing with a six-dimensional optimization problem. The loss function is implemented in the function `Scluster`, which essentially reuses the squared distance computation of the K-means code in Example 4.6. The CE program typically converges to a loss of 2.287, corresponding to the (global) minimizers $\mathbf{c}_1 = [-1.9286, -3.0416]^\top, \mathbf{c}_2 = [-3.8681, 0.0456]^\top$, and $\mathbf{c}_3 = [0.5880, -1.3526]^\top$, which slightly differs from the local minimizers for the K-means algorithm. ☞ 101

clustCE.py

```python
import numpy as np
np.set_printoptions(precision=4)

Xmat = np.genfromtxt('clusterdata.csv', delimiter=',')
K = 3
n, D = Xmat.shape

def Scluster(c):
    n, D = Xmat.shape
    dist2 = np.zeros((K,n))
    cc = c.reshape(D,K)
    for i in range(0,K):
        dist2[i,:] = np.sum((Xmat - cc[:,i].T)**2, 1)
    minvals = np.amin(dist2,0)
    return minvals.mean()

numvar = K*D
mu = np.zeros(numvar)  #initialize centers
sigma = np.ones(numvar)*2
rho = 0.1
N = 500; Nel = int(N*rho); eps = 0.001

func = Scluster
best_trj = np.array(numvar)
best_perf = np.Inf
trj = np.zeros(shape=(N,numvar))

while(np.max(sigma)>eps):
   for i in range(0,numvar):
```

```
        trj[:,i] = (np.random.randn(N,1)*sigma[i]+ mu[i]).reshape(N,)
    S = np.zeros(N)
    for i in range(0,N):
        S[i] = func(trj[i])

    sortedids = np.argsort(S) # from smallest to largest
    S_sorted = S[sortedids]
    best_trj = np.array(n)
    best_perf = np.Inf
    eliteids = sortedids[range(0,Nel)]
    eliteTrj = trj[eliteids,:]
    mu = np.mean(eliteTrj,axis=0)
    sigma = np.std(eliteTrj,axis=0)

    if(best_perf>S_sorted[0]):
        best_perf = S_sorted[0]
        best_trj = trj[sortedids[0]]

print(best_perf)
print(best_trj.reshape(2,3))
```

```
2.2874901831572947
[[-3.9238 -1.8477  0.5895]
 [ 0.0134 -3.0292 -1.2442]]
```

■

4.7 Hierarchical Clustering

It is sometimes useful to determine data clusters in a hierarchical manner; an example is the construction of evolutionary relationships between animal species. Establishing a hierarchy of clusters can be done in a bottom-up or a top-down manner. In the bottom-up approach, also called *agglomerative clustering*, the data points are merged in larger and larger clusters until all the points have been merged into a single cluster. In the top-down or *divisive clustering* approach, the data set is divided up into smaller and smaller clusters. The left panel of Figure 4.10 depicts a hierarchy of clusters.

AGGLOMERATIVE CLUSTERING

DIVISIVE CLUSTERING

In Figure 4.10, each cluster is given a cluster identifier. At the lowest level are clusters comprised of the original data points (identifiers $1, \ldots, 8$). The union of clusters 1 and 2 form a cluster with identifier 9, and the union of 3 and 4 form a cluster with identifier 10. In turn the union of clusters 9 and 10 constitutes cluster 12, and so on.

The right panel of Figure 4.10 shows a convenient way to visualize cluster hierarchies using a *dendrogram* (from the Greek *dendro* for tree). A dendrogram not only summarizes how clusters are merged or split, but also shows the distance between clusters, here on the vertical axis. The horizontal axis shows which cluster each data point (label) belongs to.

DENDROGRAM

Many different types of hierarchical clustering can be performed, depending on how the distance is defined between two data points and between two clusters. Denote the data set by $\mathcal{X} = \{\boldsymbol{x}_i, i = 1, \ldots, n\}$. As in Section 4.6, let $\text{dist}(\boldsymbol{x}_i, \boldsymbol{x}_j)$ be the distance between data points $\boldsymbol{x}_i$ and $\boldsymbol{x}_j$. The default choice is the Euclidean distance $\text{dist}(\boldsymbol{x}_i, \boldsymbol{x}_j) = \|\boldsymbol{x}_i - \boldsymbol{x}_j\|$.

Let $\mathcal{I}$ and $\mathcal{J}$ be two disjoint subsets of $\{1, \ldots, n\}$. These sets correspond to two disjoint subsets (that is, clusters) of $\mathcal{X}$: $\{\boldsymbol{x}_i, i = \mathcal{I}\}$ and $\{\boldsymbol{x}_j, j = \mathcal{J}\}$. We denote the distance between

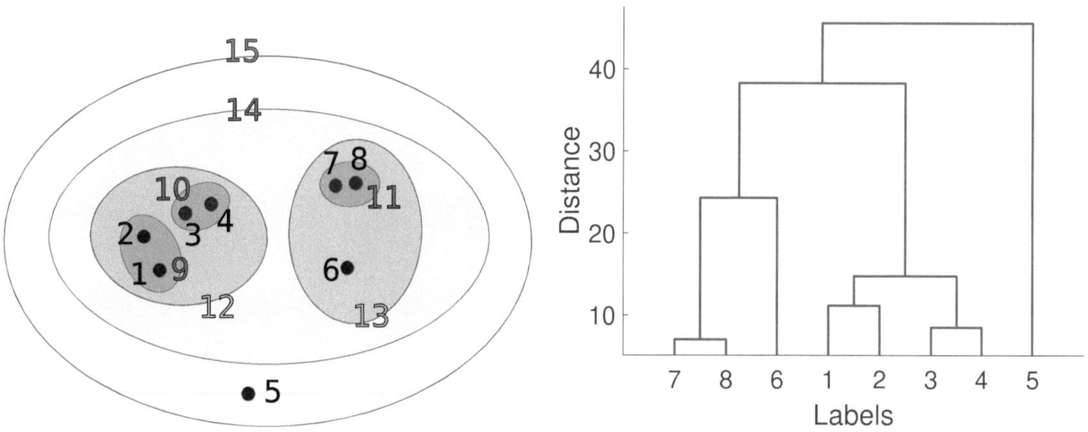

Figure 4.10: Left: a cluster hierarchy of 15 clusters. Right: the corresponding dendrogram.

LINKAGE

these two clusters by $d(\mathcal{I}, \mathcal{J})$. By specifying the function d, we indicate how the clusters are linked. For this reason it is also referred to as the *linkage* criterion. We give a number of examples:

- **Single linkage.** The closest distance between the clusters.

$$d_{\min}(\mathcal{I}, \mathcal{J}) := \min_{i \in \mathcal{I}, j \in \mathcal{J}} \operatorname{dist}(\boldsymbol{x}_i, \boldsymbol{x}_j).$$

- **Complete linkage.** The furthest distance between the clusters.

$$d_{\max}(\mathcal{I}, \mathcal{J}) := \max_{i \in \mathcal{I}, j \in \mathcal{J}} \operatorname{dist}(\boldsymbol{x}_i, \boldsymbol{x}_j).$$

- **Group average.** The mean distance between the clusters. Note that this depends on the cluster sizes.

$$d_{\text{avg}}(\mathcal{I}, \mathcal{J}) := \frac{1}{|\mathcal{I}|\,|\mathcal{J}|} \sum_{i \in \mathcal{I}} \sum_{j \in \mathcal{J}} \operatorname{dist}(\boldsymbol{x}_i, \boldsymbol{x}_j).$$

For these linkage criteria, $\mathcal{X}$ is usually assumed to be $\mathbb{R}^d$ with the Euclidean distance.

WARD'S LINKAGE

Another notable measure for the distance between clusters is *Ward's minimum variance linkage* criterion. Here, the distance between clusters is expressed as the additional amount of "variance" (expressed in terms of the sum of squares) that would be introduced if the two clusters were merged. More precisely, for any set $\mathcal{K}$ of indices (labels) let $\overline{\boldsymbol{x}}_{\mathcal{K}} = \sum_{k \in \mathcal{K}} \boldsymbol{x}_k / |\mathcal{K}|$ denote its corresponding cluster mean. Then

$$d_{\text{Ward}}(\mathcal{I}, \mathcal{J}) := \sum_{k \in \mathcal{I} \cup \mathcal{J}} \|\boldsymbol{x}_k - \overline{\boldsymbol{x}}_{\mathcal{I} \cup \mathcal{J}}\|^2 - \left(\sum_{i \in \mathcal{I}} \|\boldsymbol{x}_i - \overline{\boldsymbol{x}}_{\mathcal{I}}\|^2 + \sum_{j \in \mathcal{J}} \|\boldsymbol{x}_j - \overline{\boldsymbol{x}}_{\mathcal{J}}\|^2 \right). \tag{4.41}$$

It can be shown (see Exercise 8) that the Ward linkage depends only on the cluster means and the cluster sizes for $\mathcal{I}$ and $\mathcal{J}$:

$$d_{\text{Ward}}(\mathcal{I}, \mathcal{J}) = \frac{|\mathcal{I}|\,|\mathcal{J}|}{|\mathcal{I}| + |\mathcal{J}|} \|\overline{\boldsymbol{x}}_{\mathcal{I}} - \overline{\boldsymbol{x}}_{\mathcal{J}}\|^2.$$

> In software implementations, the Ward linkage function is often rescaled by multiplying it by a factor of 2. In this way, the distance between one-point clusters $\{\boldsymbol{x}_i\}$ and $\{\boldsymbol{x}_j\}$ is the *squared* Euclidean distance $\|\boldsymbol{x}_i - \boldsymbol{x}_j\|^2$.

Having chosen a distance on $\mathcal{X}$ and a linkage criterion, a general agglomerative clustering algorithm proceeds in the following "greedy" manner.

Algorithm 4.7.1: Greedy Agglomerative Clustering

input: Distance function dist, linkage function d, number of clusters K.
output: The label sets for the tree.

Initialize the set of cluster identifiers: $\mathcal{I} = \{1, \ldots, n\}$.
Initialize the corresponding label sets: $\mathcal{L}_i = \{i\}, i \in \mathcal{I}$.
Initialize a distance matrix $\mathbf{D} = [d_{ij}]$ with $d_{ij} = d(\{i\}, \{j\})$.
for $k = n + 1$ **to** $2n - K$ **do**
    Find i and $j > i$ in $\mathcal{I}$ such that d_{ij} is minimal.
    Create a new label set $\mathcal{L}_k := \mathcal{L}_i \cup \mathcal{L}_j$.
    Add the new identifier k to $\mathcal{I}$ and remove the old identifiers i and j from $\mathcal{I}$.
    Update the distance matrix $\mathbf{D}$ with respect to the identifiers i, j, and k.
return $\mathcal{L}_i, i = 1, \ldots, 2n - K$

Initially, the distance matrix $\mathbf{D}$ contains the (linkage) distances between the one-point clusters containing one of the data points $\boldsymbol{x}_1, \ldots, \boldsymbol{x}_n$, and hence with identifiers $1, \ldots, n$. Finding the shortest distance amounts to a table lookup in $\mathbf{D}$. When the closest clusters are found, they are merged into a new cluster, and a new identifier k (the smallest positive integer that has not yet been used as an identifier) is assigned to this cluster. The old identifiers i and j are removed from the cluster identifier set $\mathcal{I}$. The matrix $\mathbf{D}$ is then updated by adding a k-th column and row that contain the distances between k and any $m \in \mathcal{I}$. This updating step could be computationally quite costly if the cluster sizes are large and the linkage distance between the clusters depends on all points within the clusters. Fortunately, for many linkage functions, the matrix $\mathbf{D}$ can be updated in an efficient manner.

Suppose that at some stage in the algorithm, clusters $\mathcal{I}$ and $\mathcal{J}$, with identifiers i and j, respectively, are merged into a cluster $\mathcal{K} = \mathcal{I} \cup \mathcal{J}$ with identifier k. Let $\mathcal{M}$, with identifier m, be a previously assigned cluster. An update rule of the linkage distance d_{km} between $\mathcal{K}$ and $\mathcal{M}$ is called a *Lance–Williams* update if it can be written in the form LANCE–WILLIAMS

$$d_{km} = \alpha\, d_{im} + \beta\, d_{jm} + \gamma\, d_{ij} + \delta\, |d_{im} - d_{jm}|,$$

where $\alpha, \ldots, \delta$ depend only on simple characteristics of the clusters involved, such as the number of elements within the clusters. Table 4.2 shows the update constants for a number of common linkage functions. For example, for single linkage, d_{im} is the minimal distance between $\mathcal{I}$ and $\mathcal{M}$, and d_{jm} is the minimal distance between $\mathcal{J}$ and $\mathcal{M}$. The smallest of these is the minimal distance between $\mathcal{K}$ and $\mathcal{M}$. That is, $d_{km} = \min\{d_{im}, d_{jm}\} = d_{im}/2 + d_{jm}/2 - |d_{im} - d_{jm}|/2$.

Table 4.2: Constants for the Lance–Williams update rule for various linkage functions, with n_i, n_j, n_m denoting the number of elements in the corresponding clusters.

Linkage	α	β	γ	δ
Single	$1/2$	$1/2$	0	$-1/2$
Complete	$1/2$	$1/2$	0	$1/2$
Group avg.	$\frac{n_i}{n_i + n_j}$	$\frac{n_j}{n_i + n_j}$	0	0
Ward	$\frac{n_i + n_m}{n_i + n_j + n_m}$	$\frac{n_j + n_m}{n_i + n_j + n_m}$	$\frac{-n_m}{n_i + n_j + n_m}$	0

LINKAGE MATRIX

In practice, Algorithm 4.7.1 is run until a single cluster is obtained. Instead of returning the label sets of all $2n - 1$ clusters, a *linkage matrix* is returned that contains the same information. At the end of each iteration (Line 8) the linkage matrix stores the merged labels i and j, as well as the (minimal) distance d_{ij}. Other information such as the number of elements in the merged cluster can also be stored. Dendrograms and cluster labels can be directly constructed from the linkage matrix. In the following example, the linkage matrix is returned by the method **agg_cluster**.

■ **Example 4.8 (Agglomerative Hierarchical Clustering)** The Python code below gives a basic implementation of Algorithm 4.7.1 using the Ward linkage function. The methods **fcluster** and **dendrogram** from the scipy module can be used to identify the labels in a cluster and to draw the corresponding dendrogram.

AggCluster.py

```
import numpy as np
from scipy.spatial.distance import cdist

def update_distances(D,i,j, sizes): # distances for merged cluster
    n = D.shape[0]
    d = np.inf * np.ones(n+1)
    for k in range(n): # Update distances
       d[k] = ((sizes[i]+sizes[k])*D[i,k] +
       (sizes[j]+sizes[k])*D[j,k] -
       sizes[k]*D[i,j])/(sizes[i] + sizes[j] + sizes[k])

    infs =  np.inf * np.ones(n) # array of infinity
    D[i,:],D[:,i],D[j,:],D[:,j] =  infs,infs,infs,infs # deactivate
    new_D = np.inf * np.ones((n+1,n+1))
    new_D[0:n,0:n] = D # copy old matrix into new_D
    new_D[-1,:], new_D[:,-1] = d,d # add new row and column
    return new_D

def agg_cluster(X):
    n = X.shape[0]
    sizes = np.ones(n)
    D = cdist(X, X,metric = 'sqeuclidean') # initialize dist. matrix
       .
    np.fill_diagonal(D, np.inf * np.ones(D.shape[0]))
    Z = np.zeros((n-1,4))  #linkage matrix encodes hierarchy tree
    for t in range(n-1):
```

```
        i,j = np.unravel_index(D.argmin(), D.shape) # minimizer pair
        sizes = np.append(sizes, sizes[i] + sizes[j])
        Z[t,:]=np.array([i, j, np.sqrt(D[i,j]), sizes[-1]])
        D = update_distances(D, i,j, sizes)  # update distance matr.
    return Z

import scipy.cluster.hierarchy as h

X = np.genfromtxt('clusterdata.csv',delimiter=',') # read the data
Z = agg_cluster(X)  # form the linkage matrix

h.dendrogram(Z) # SciPy can produce a dendrogram from Z
# fcluster function assigns cluster ids to all points based on Z
cl = h.fcluster(Z, criterion = 'maxclust', t=3)

import matplotlib.pyplot as plt
plt.figure(2), plt.clf()
cols = ['red','green','blue']
colors = [cols[i-1] for i in cl]
plt.scatter(X[:,0], X[:,1],c=colors)
plt.show()
```

Note that the distance matrix is initialized with the squared Euclidean distance, so that the Ward linkage is rescaled by a factor of 2. Also, note that the linkage matrix stores the square root of the minimal cluster distances rather than the distances themselves. We leave it as an exercise to check that by using these modifications the results agree with the `linkage` method from `scipy`; see Exercise 9. ■

In contrast to the bottom-up (agglomerative) approach to hierarchical clustering, the divisive approach starts with one cluster, which is divided into two clusters that are as "dissimilar" as possible, which can then be further divided, and so on. We can use the same linkage criteria as for agglomerative clustering to divide a parent cluster into two child clusters by *maximizing* the distance between the child clusters. Although it is a natural to try to group together data by separating dissimilar ones as far as possible, the implementation of this idea tends to scale poorly with n. The problem is related to the well-known *max-cut problem*: given an $n \times n$ matrix of positive costs $c_{ij}, i, j \in \{1, \dots, n\}$, partition the index set $\mathcal{I} = \{1, \dots, n\}$ into two subsets $\mathcal{J}$ and $\mathcal{K}$ such that the total cost across the sets, that is, MAX-CUT PROBLEM

$$\sum_{j \in \mathcal{J}} \sum_{k \in \mathcal{K}} d_{jk},$$

is maximal. If instead we maximize according to the *average* distance, we obtain the group average linkage criterion.

■ **Example 4.9 (Divisive Clustering via CE)** The following Python code is used to divide a small data set (of size 300) into two parts according to maximal group average linkage. It uses a short cross-entropy algorithm similar to the one presented in Example 3.19. Given a vector of probabilities $\{p_i, i = 1, \dots, n\}$, the algorithm generates an $n \times n$ matrix of Bernoulli random variables with success probability p_i for column i. For each row, the 0s and 1s divide the index set into two clusters, and the corresponding average linkage ☞ 110 distance is computed. The matrix is then sorted row-wise according to these distances. Finally, the probabilities $\{p_i\}$ are updated according to the mean values of the best 10% rows. The process is repeated until the $\{p_i\}$ degenerate to a binary vector. This then presents the (approximate) solution.

clustCE2.py

```
import numpy as np
from numpy import genfromtxt
from scipy.spatial.distance import squareform
from scipy.spatial.distance import pdist
import matplotlib.pyplot as plt


def S(x,D):
   V1 = np.where(x==0)[0] # {V1,V2} is the partition
   V2 = np.where(x==1)[0]
   tmp = D[V1]
   tmp = tmp[:,V2]
   return np.mean(tmp) # the size of the cut


def maxcut(D,N,eps,rho,alpha):
   n = D.shape[1]
   Ne = int(rho*N)
   p = 1/2*np.ones(n)
   p[0] = 1.0
   while (np.max(np.minimum(p,np.subtract(1,p))) > eps):
       x = np.array(np.random.uniform(0,1,(N,n))<=p, dtype=np.int64)
       sx = np.zeros(N)
       for i in range(N):
           sx[i] = S(x[i],D)

       sortSX = np.flip(np.argsort(sx))
       #print("gamma = ",sx[sortSX[Ne-1]], " best=",sx[sortSX[0]])
       elIds = sortSX[0:Ne]
       elites = x[elIds]
       pnew = np.mean(elites, axis=0)
       p = alpha*pnew + (1.0-alpha)*p

   return np.round(p)

Xmat = genfromtxt('clusterdata.csv', delimiter=',')
n = Xmat.shape[0]
D = squareform(pdist(Xmat))
N = 1000
eps = 10**-2
rho = 0.1
alpha = 0.9

# CE
pout = maxcut(D,N,eps,rho, alpha);

cutval = S(pout,D)
```

```
print("cutvalue ",cutval)

#plot
V1 = np.where(pout==0)[0]
xblue = Xmat[V1]
V2 = np.where(pout==1)[0]
xred = Xmat[V2]
plt.scatter(xblue[:,0],xblue[:,1], c="blue")
plt.scatter(xred[:,0],xred[:,1], c="red")
```

```
cutvalue  4.625207676517948
```

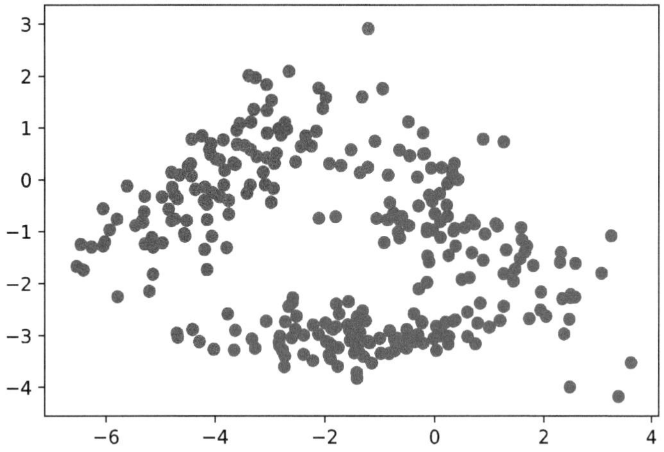

Figure 4.11: Division of the data in Figure 4.4 into two clusters, via the cross-entropy method.

■

4.8 Principal Component Analysis (PCA)

The main idea of *principal component analysis* (PCA) is to reduce the dimensionality of a data set consisting of many variables. PCA is a *feature reduction* (or *feature extraction*) mechanism, that helps us to handle high-dimensional data with more features than is convenient to interpret. PRINCIPAL COMPONENT ANALYSIS

4.8.1 Motivation: Principal Axes of an Ellipsoid

Consider a d-dimensional normal distribution with mean vector $\mathbf{0}$ and covariance matrix $\boldsymbol{\Sigma}$. The corresponding pdf (see (2.33)) is ☞ 45

$$f(\boldsymbol{x}) = \frac{1}{\sqrt{(2\pi)^n\,|\boldsymbol{\Sigma}|}}\,\mathrm{e}^{-\frac{1}{2}\,\boldsymbol{x}^\top\boldsymbol{\Sigma}^{-1}\boldsymbol{x}}, \quad \boldsymbol{x} \in \mathbb{R}^d.$$

If we were to draw many iid samples from this pdf, the points would roughly have an *ellipsoid* pattern, as illustrated in Figure 3.1, and correspond to the contours of f: sets of ☞ 71 points $\boldsymbol{x}$ such that $\boldsymbol{x}^\top\boldsymbol{\Sigma}^{-1}\boldsymbol{x} = c$, for some $c \geqslant 0$. In particular, consider the ellipsoid

$$\boldsymbol{x}^\top\boldsymbol{\Sigma}^{-1}\boldsymbol{x} = 1, \quad \boldsymbol{x} \in \mathbb{R}^d. \tag{4.42}$$

☞ 375
☞ 368

PRINCIPAL AXES

SINGULAR VALUE DECOMPOSITION

☞ 380

Let $\boldsymbol{\Sigma} = \mathbf{B}\mathbf{B}^\top$, where $\mathbf{B}$ is for example the (lower) Cholesky matrix. Then, as explained in Example A.5, the ellipsoid (4.42) can also be viewed as the linear transformation of d-dimensional unit sphere via matrix $\mathbf{B}$. Moreover, the *principal axes* of the ellipsoid can be found via a *singular value decomposition* (SVD) of $\mathbf{B}$ (or $\boldsymbol{\Sigma}$); see Section A.6.5 and Example A.8. In particular, suppose that an SVD of $\mathbf{B}$ is

$$\mathbf{B} = \mathbf{U}\mathbf{D}\mathbf{V}^\top \qquad \text{(note that an SVD of } \boldsymbol{\Sigma} \text{ is then } \mathbf{U}\mathbf{D}^2\mathbf{U}^\top\text{).}$$

PRINCIPAL COMPONENTS

The columns of the matrix $\mathbf{UD}$ correspond to the principal axes of the ellipsoid, and the relative magnitudes of the axes are given by the elements of the diagonal matrix $\mathbf{D}$. If some of these magnitudes are small compared to the others, a reduction in the dimension of the space may be achieved by *projecting* each point $\boldsymbol{x} \in \mathbb{R}^d$ onto the subspace spanned by the main (say $k \ll d$) columns of $\mathbf{U}$ — the so-called *principal components*. Suppose without loss of generality that the first k principal components are given by the first k columns of $\mathbf{U}$, and let $\mathbf{U}_k$ be the corresponding $d \times k$ matrix.

With respect to the standard basis $\{\boldsymbol{e}_i\}$, the vector $\boldsymbol{x} = x_1\boldsymbol{e}_1 + \cdots + x_d\boldsymbol{e}_d$ is represented by the d-dimensional vector $[x_1, \ldots, x_d]^\top$. With respect to the orthonormal basis $\{\boldsymbol{u}_i\}$ formed by the columns of matrix $\mathbf{U}$, the representation of $\boldsymbol{x}$ is $\mathbf{U}^\top\boldsymbol{x}$. Similarly, the projection of any point $\boldsymbol{x}$ onto the subspace spanned by the first k principal vectors is represented by the k-dimensional vector $\mathbf{U}_k^\top\boldsymbol{x}$, with respect to the orthonormal basis formed by the columns of $\mathbf{U}_k$. So, the idea is that if a point $\boldsymbol{x}$ lies close to its projection $\mathbf{U}_k\mathbf{U}_k^\top\boldsymbol{x}$, we may represent it via k numbers instead of d, using the combined features given by the k principal components. See Section A.4 for a review of projections and orthonormal bases.

☞ 364

■ **Example 4.10 (Principal Components)** Consider the matrix

$$\boldsymbol{\Sigma} = \begin{bmatrix} 14 & 8 & 3 \\ 8 & 5 & 2 \\ 3 & 2 & 1 \end{bmatrix},$$

which can be written as $\boldsymbol{\Sigma} = \mathbf{B}\mathbf{B}^\top$, with

$$\mathbf{B} = \begin{bmatrix} 1 & 2 & 3 \\ 0 & 1 & 2 \\ 0 & 0 & 1 \end{bmatrix}.$$

Figure 4.12 depicts the ellipsoid $\boldsymbol{x}^\top\boldsymbol{\Sigma}^{-1}\boldsymbol{x} = 1$, which can be obtained by linearly transforming the points on the unit sphere by means of the matrix $\mathbf{B}$. The principal axes and sizes of the ellipsoid are found through a singular value decomposition $\mathbf{B} = \mathbf{U}\mathbf{D}\mathbf{V}^\top$, where $\mathbf{U}$ and $\mathbf{D}$ are

$$\mathbf{U} = \begin{bmatrix} 0.8460 & 0.4828 & 0.2261 \\ 0.4973 & -0.5618 & -0.6611 \\ 0.1922 & -0.6718 & 0.7154 \end{bmatrix} \quad \text{and} \quad \mathbf{D} = \begin{bmatrix} 4.4027 & 0 & 0 \\ 0 & 0.7187 & 0 \\ 0 & 0 & 0.3160 \end{bmatrix}.$$

The columns of $\mathbf{U}$ show the directions of the principal axes of the ellipsoid, and the diagonal elements of $\mathbf{D}$ indicate the relative magnitudes of the principal axes. We see that the first principal component is given by the first column of $\mathbf{U}$, and the second principal component by the second column of $\mathbf{U}$.

The projection of the point $\boldsymbol{x} = [1.052, 0.6648, 0.2271]^\top$ onto the 1-dimensional space spanned by the first principal component $\boldsymbol{u}_1 = [0.8460, 0.4972, 0.1922]^\top$ is $\boldsymbol{z} = \boldsymbol{u}_1\boldsymbol{u}_1^\top\boldsymbol{x} = [1.0696, 0.6287, 0.2429]^\top$. With respect to the basis vector $\boldsymbol{u}_1$, $\boldsymbol{z}$ is represented by the number $\boldsymbol{u}_1^\top\boldsymbol{z} = 1.2643$. That is, $\boldsymbol{z} = 1.2643\boldsymbol{u}_1$.

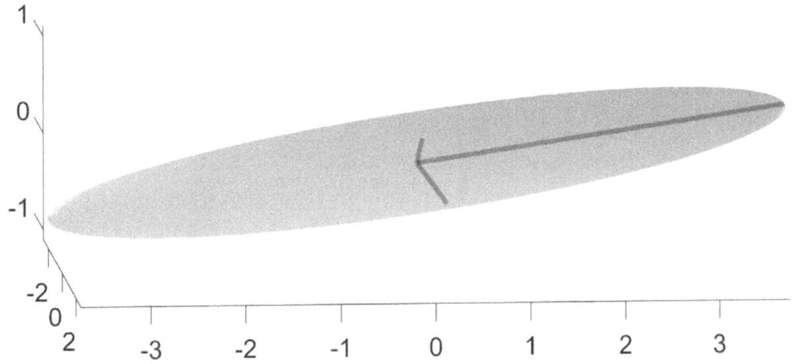

Figure 4.12: A "surfboard" ellipsoid where one principal axis is significantly larger than the other two.

■

4.8.2 PCA and Singular Value Decomposition (SVD)

In the setting above, we did not consider any data set drawn from a multivariate pdf f. The whole analysis rested on linear algebra. In *principal component analysis* (PCA) we start with data $\boldsymbol{x}_1, \ldots, \boldsymbol{x}_n$, where each $\boldsymbol{x}$ is d-dimensional. PCA does not require assumptions how the data were obtained, but to make the link with the previous section, we can think of the data as iid draws from a multivariate normal pdf. PRINCIPAL COMPONENT ANALYSIS

Let us collect the data in a matrix $\mathbf{X}$ in the usual way; that is, ☞ 43

$$\mathbf{X} = \begin{bmatrix} x_{11} & x_{12} & \ldots & x_{1d} \\ x_{21} & x_{22} & \ldots & x_{2d} \\ \vdots & \vdots & \vdots & \vdots \\ x_{n1} & x_{n2} & \ldots & x_{nd} \end{bmatrix} = \begin{bmatrix} \boldsymbol{x}_1^\top \\ \boldsymbol{x}_2^\top \\ \vdots \\ \boldsymbol{x}_n^\top \end{bmatrix}.$$

The matrix $\mathbf{X}$ will be the PCA's input. Under this setting, the data consists of points in d-dimensional space, and our goal is to present the data using n feature vectors of dimension $k < d$.

In accordance with the previous section, we assume that underlying distribution of the data has expectation vector $\mathbf{0}$. In practice, this means that before PCA is applied, the data needs to be *centered* by subtracting the *column* mean in every column:

$$x'_{ij} = x_{ij} - \overline{x}_j,$$

where $\overline{x}_j = \frac{1}{n}\sum_{i=1}^n x_{ij}$.

We assume from now on that the data comes from a general d-dimensional distribution with mean vector $\mathbf{0}$ and some covariance matrix $\mathbf{\Sigma}$. The covariance matrix $\mathbf{\Sigma}$ is by definition equal to the expectation of the random matrix $\boldsymbol{X}\boldsymbol{X}^\top$, and can be estimated from the data $\boldsymbol{x}_1, \ldots, \boldsymbol{x}_n$ via the sample average

$$\widehat{\mathbf{\Sigma}} = \frac{1}{n}\sum_{i=1}^n \boldsymbol{x}_i\boldsymbol{x}_i^\top = \frac{1}{n}\mathbf{X}^\top\mathbf{X}.$$

As $\widehat{\mathbf{\Sigma}}$ is a covariance matrix, we may conduct the same analysis for $\widehat{\mathbf{\Sigma}}$ as we did for $\mathbf{\Sigma}$ in the previous section. Specifically, suppose $\widehat{\mathbf{\Sigma}} = \mathbf{U}\mathbf{D}^2\mathbf{U}^\top$ is an SVD of $\widehat{\mathbf{\Sigma}}$ and let $\mathbf{U}_k$ be the matrix whose columns are the k principal components; that is, the k columns of $\mathbf{U}$ corresponding to the largest diagonal elements in $\mathbf{D}^2$. Note that we have used $\mathbf{D}^2$ instead of $\mathbf{D}$ to be compatible with the previous section. The transformation $\boldsymbol{z}_i = \mathbf{U}_k\mathbf{U}_k^\top\boldsymbol{x}_i$ maps each vector $\boldsymbol{x}_i \in \mathbb{R}^d$ (thus, with d features) to a vector $\boldsymbol{z}_i \in \mathbb{R}^d$ lying in the subspace spanned by the columns of $\mathbf{U}_k$. With respect to this basis, the point $\boldsymbol{z}_i$ has representation $\boldsymbol{z}_i = \mathbf{U}_k^\top(\mathbf{U}_k\mathbf{U}_k^\top\boldsymbol{x}_i) = \mathbf{U}_k^\top\boldsymbol{x}_i \in \mathbb{R}^k$ (thus with k features). The corresponding covariance matrix of the $\boldsymbol{z}_i, i = 1, \ldots, n$ is diagonal. The diagonal elements $\{d_{\ell\ell}\}$ of $\mathbf{D}$ can be interpreted as standard deviations of the data in the directions of the principal components. The quantity $v = \sum_{\ell=1} d_{\ell\ell}^2$ (that is, the trace of $\mathbf{D}^2$) is thus a measure for the amount of variance in the data. The proportion $d_{\ell\ell}^2/v$ indicates how much of the variance in the data is explained by the ℓ-th principal component.

Another way to look at PCA is by considering the question: How can we best project the data onto a k-dimensional subspace in such a way that the total squared distance between the projected points and the original points is minimal? From Section A.4, we know that any orthogonal projection to a k-dimensional subspace $\mathcal{V}_k$ can be represented by a matrix $\mathbf{U}_k\mathbf{U}_k^\top$, where $\mathbf{U}_k = [\boldsymbol{u}_1, \ldots, \boldsymbol{u}_k]$ and the $\{\boldsymbol{u}_\ell, \ell = 1, \ldots, k\}$ are orthogonal vectors of length 1 that span $\mathcal{V}_k$. The above question can thus be formulated as the minimization program:

☞ 364

$$\min_{\boldsymbol{u}_1,\ldots,\boldsymbol{u}_k} \sum_{i=1}^n \|\boldsymbol{x}_i - \mathbf{U}_k\mathbf{U}_k^\top\boldsymbol{x}_i\|^2. \tag{4.43}$$

Now observe that

$$\begin{aligned}
\frac{1}{n}\sum_{i=1}^n \|\boldsymbol{x}_i - \mathbf{U}_k\mathbf{U}_k^\top\boldsymbol{x}_i\|^2 &= \frac{1}{n}\sum_{i=1}^n (\boldsymbol{x}_i^\top - \boldsymbol{x}_i^\top\mathbf{U}_k\mathbf{U}_k^\top)(\boldsymbol{x}_i - \mathbf{U}_k\mathbf{U}_k^\top\boldsymbol{x}_i) \\
&= \underbrace{\frac{1}{n}\sum_{i=1}^n \|\boldsymbol{x}_i\|^2}_{c} - \frac{1}{n}\sum_{i=1}^n \boldsymbol{x}_i^\top\mathbf{U}_k\mathbf{U}_k^\top\boldsymbol{x}_i = c - \frac{1}{n}\sum_{i=1}^n\sum_{\ell=1}^k \mathrm{tr}(\boldsymbol{x}_i^\top\boldsymbol{u}_\ell\boldsymbol{u}_\ell^\top\boldsymbol{x}_i) \\
&= c - \frac{1}{n}\sum_{\ell=1}^k\sum_{i=1}^n \boldsymbol{u}_\ell^\top\boldsymbol{x}_i\boldsymbol{x}_i^\top\boldsymbol{u}_\ell = c - \sum_{\ell=1}^k \boldsymbol{u}_\ell^\top\widehat{\mathbf{\Sigma}}\boldsymbol{u}_\ell,
\end{aligned}$$

☞ 359

where we have used the cyclic property of a trace (Theorem A.1) and the fact that $\mathbf{U}_k\mathbf{U}_k^\top$ can be written as $\sum_{\ell=1}^k \boldsymbol{u}_\ell\boldsymbol{u}_\ell^\top$. It follows that the minimization problem(4.43) is equivalent to the maximization problem

$$\max_{\boldsymbol{u}_1,\ldots,\boldsymbol{u}_k} \sum_{\ell=1}^k \boldsymbol{u}_\ell^\top\widehat{\mathbf{\Sigma}}\boldsymbol{u}_\ell. \tag{4.44}$$

This maximum can be at most $\sum_{\ell=1}^{k} d_{\ell\ell}^2$ and is attained precisely when $\boldsymbol{u}_1, \ldots, \boldsymbol{u}_k$ are the first k principal components of $\widehat{\boldsymbol{\Sigma}}$.

■ **Example 4.11 (Singular Value Decomposition)** The following data set consists of independent samples from the three-dimensional Gaussian distribution with mean vector $\mathbf{0}$ and covariance matrix $\boldsymbol{\Sigma}$ given in Example 4.10:

$$\mathbf{X} = \begin{bmatrix} 3.1209 & 1.7438 & 0.5479 \\ -2.6628 & -1.5310 & -0.2763 \\ 3.7284 & 3.0648 & 1.8451 \\ 0.4203 & 0.3553 & 0.4268 \\ -0.7155 & -0.6871 & -0.1414 \\ 5.8728 & 4.0180 & 1.4541 \\ 4.8163 & 2.4799 & 0.5637 \\ 2.6948 & 1.2384 & 0.1533 \\ -1.1376 & -0.4677 & -0.2219 \\ -1.2452 & -0.9942 & -0.4449 \end{bmatrix}.$$

After replacing $\mathbf{X}$ with its centered version, an SVD $\mathbf{U}\mathbf{D}^2\mathbf{U}^\top$ of $\widehat{\boldsymbol{\Sigma}} = \mathbf{X}^\top\mathbf{X}/n$ yields the principal component matrix $\mathbf{U}$ and diagonal matrix $\mathbf{D}$:

$$\mathbf{U} = \begin{bmatrix} -0.8277 & 0.4613 & 0.3195 \\ -0.5300 & -0.4556 & -0.7152 \\ -0.1843 & -0.7613 & 0.6216 \end{bmatrix} \quad \text{and} \quad \mathbf{D} = \begin{bmatrix} 3.3424 & 0 & 0 \\ 0 & 0.4778 & 0 \\ 0 & 0 & 0.1038 \end{bmatrix}.$$

We also observe that, apart from the sign of the first column, the principal component matrix $\mathbf{U}$ is similar to that in Example 4.10. Likewise for the matrix $\mathbf{D}$. We see that 97.90% of the total variance is explained by the first principal component. Figure 4.13 shows the projection of the centered data onto the subspace spanned by this principal component.

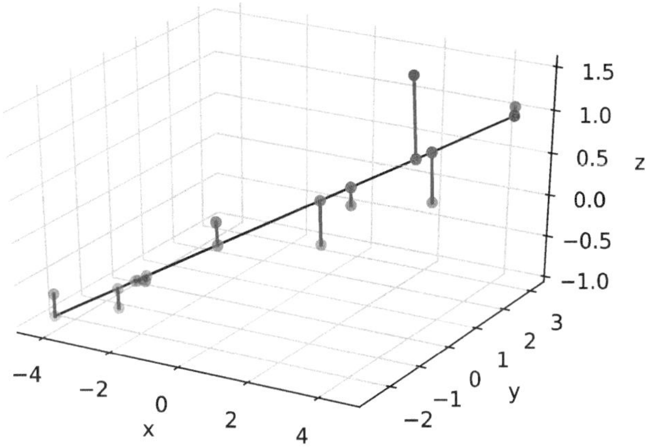

Figure 4.13: Data from the "surfboard" pdf is projected onto the subspace spanned by the largest principal component.

The following Python code was used.

PCAdat.py

```
import numpy as np
X = np.genfromtxt('pcadat.csv', delimiter=',')
n = X.shape[0]

X = X - X.mean(axis=0)
G = X.T @ X
U, _ , _ = np.linalg.svd(G/n)

# projected points
Y = X @ np.outer(U[:,0],U[:,0])

import matplotlib.pyplot as plt
from mpl_toolkits.mplot3d import Axes3D

fig = plt.figure()
ax = fig.add_subplot(111, projection='3d')
ax.w_xaxis.set_pane_color((0, 0, 0, 0))
ax.plot(Y[:,0], Y[:,1], Y[:,2], c='k', linewidth=1)
ax.scatter(X[:,0], X[:,1], X[:,2], c='b')
ax.scatter(Y[:,0], Y[:,1], Y[:,2], c='r')

for i in range(n):
    ax.plot([X[i,0], Y[i,0]], [X[i,1],Y[i,1]], [X[i,2],Y[i,2]], 'b')

ax.set_xlabel('x')
ax.set_ylabel('y')
ax.set_zlabel('z')
plt.show()
```

■

☞ 2 Next is an application of PCA to Fisher's famous iris data set, already mentioned in Section 1.1, and Exercise 1.5.

■ **Example 4.12 (PCA for the Iris Data Set)** The iris data set contains measurements on four features of the iris plant: sepal length and width, and petal length and width, for a total of 150 specimens. The full data set also contains the species name, but for the purpose of this example we ignore it.

☞ 17 Figure 1.9 shows that there is a significant correlation between the different features. Can we perhaps describe the data using fewer features by taking certain linear combinations of the original features? To investigate this, let us perform a PCA, first centering the data. The following Python code implements the PCA. It is assumed that a CSV file `irisX.csv` has been made that contains the iris data set (without the species information).

PCAiris.py

```
import seaborn as sns, numpy as np
np.set_printoptions(precision=4)

X = np.genfromtxt('IrisX.csv',delimiter=',')
n = X.shape[0]
```

```
X = X - np.mean(X, axis=0)

[U,D2,UT]= np.linalg.svd((X.T @ X)/n)
print('U = \n', U); print('\n diag(D^2) = ', D2)

z =  U[:,0].T @ X.T

sns.kdeplot(z, bw=0.15)
```

```
U =
 [[-0.3614 -0.6566  0.582   0.3155]
 [ 0.0845 -0.7302 -0.5979 -0.3197]
 [-0.8567  0.1734 -0.0762 -0.4798]
 [-0.3583  0.0755 -0.5458  0.7537]]

 diag(D^2) =  [4.2001 0.2411 0.0777 0.0237]
```

The output above shows the principal component matrix (which we called $\mathbf{U}$) as well as the diagonal of matrix $\mathbf{D}^2$. We see that a large proportion of the variance, $4.2001/(4.2001 + 0.2411 + 0.0777 + 0.0237) = 92.46\%$, is explained by the first principal component. Thus, it makes sense to transform each data point $\boldsymbol{x} \in \mathbb{R}^4$ to $\boldsymbol{u}_1^\top \boldsymbol{x} \in \mathbb{R}$. Figure 4.14 shows the kernel density estimate of the transformed data. Interestingly, we see two modes, indicating at least two clusters in the data.

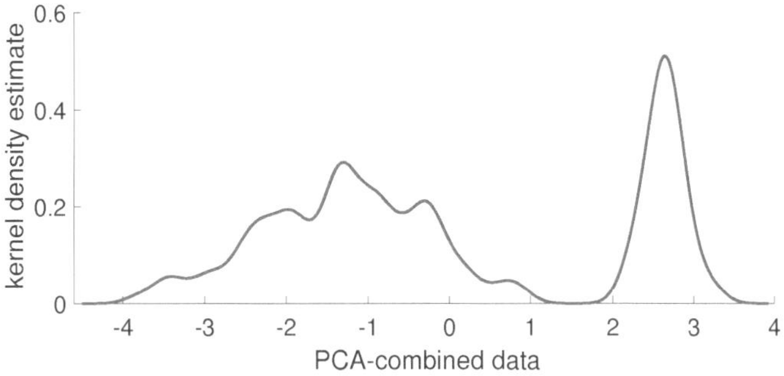

Figure 4.14: Kernel density estimate of the PCA-combined iris data.

■

Further Reading

Various information-theoretic measures to quantify uncertainty, including the Shannon entropy and Kullback–Leibler divergence, may be found in [28]. The Fisher information, the prominent information measure in statistics, is discussed in detail in [78]. Akaike's information criterion appeared in [2]. The EM algorithm was introduced in [31] and [85] gives an in-depth treatment. Convergence proofs for the EM algorithm may be found in [19, 128]. A classical reference on kernel density estimation is [113], and [14] is the main reference for the theta kernel density estimator. Theory and applications on finite mixture models may be found in [86]. For more details on clustering applications and algorithms as well as references on data compression, vector quantization, and pattern recognition, we refer to [1, 35, 107, 125]. A useful modification of the K-means algorithm is the *fuzzy K-means* algorithm; see, e.g., [9]. A popular way to choose the starting positions in K-means is given by the K-means++ heuristic, introduced in [4].

Exercises

1. This exercise is to show that the Fisher information matrix $\mathbf{F}(\boldsymbol{\theta})$ in (4.8) is equal to the matrix $\mathbf{H}(\boldsymbol{\theta})$ in (4.9), in the special case where $f = g(\cdot \mid \boldsymbol{\theta})$, and under the assumption that integration and differentiation orders can be interchanged.

(a) Let $\boldsymbol{h}$ be a vector-valued function and k a real-valued function. Prove the following *quotient rule for differentiation*:

QUOTIENT RULE FOR DIFFERENTIATION

$$\frac{\partial[\boldsymbol{h}(\boldsymbol{\theta})/k(\boldsymbol{\theta})]}{\partial \boldsymbol{\theta}} = \frac{1}{k(\boldsymbol{\theta})}\frac{\partial \boldsymbol{h}(\boldsymbol{\theta})}{\partial \boldsymbol{\theta}} - \frac{1}{k^2(\boldsymbol{\theta})}\frac{\partial k(\boldsymbol{\theta})}{\partial \boldsymbol{\theta}}\boldsymbol{h}(\boldsymbol{\theta})^\top. \tag{4.45}$$

(b) Now take $\boldsymbol{h}(\boldsymbol{\theta}) = \frac{\partial g(\boldsymbol{X} \mid \boldsymbol{\theta})}{\partial \boldsymbol{\theta}}$ and $k(\boldsymbol{\theta}) = g(\boldsymbol{X} \mid \boldsymbol{\theta})$ in (4.45) and take expectations with respect to $\mathbb{E}_{\boldsymbol{\theta}}$ on both sides to show that

$$-\mathbf{H}(\boldsymbol{\theta}) = \underbrace{\mathbb{E}_{\boldsymbol{\theta}}\left[\frac{1}{g(\boldsymbol{X} \mid \boldsymbol{\theta})}\frac{\partial \frac{\partial g(\boldsymbol{X} \mid \boldsymbol{\theta})}{\partial \boldsymbol{\theta}}}{\partial \boldsymbol{\theta}}\right]}_{\mathbf{A}} - \mathbf{F}(\boldsymbol{\theta}).$$

(c) Finally show that $\mathbf{A}$ is the zero matrix.

2. Plot the mixture of $\mathcal{N}(0, 1)$, $\mathcal{U}(0, 1)$, and $\mathsf{Exp}(1)$ distributions, with weights $w_1 = w_2 = w_3 = 1/3$.

3. Denote the pdfs in Exercise 2 by f_1, f_2, f_3, respectively. Suppose that X is simulated via the two-step procedure: First, draw Z from $\{1, 2, 3\}$, then draw X from f_Z. How likely is it that the outcome $x = 0.5$ of X has come from the uniform pdf f_2?

4. Simulate an iid training set of size 100 from the $\mathsf{Gamma}(2.3, 0.5)$ distribution, and implement the Fisher scoring method in Example 4.1 to find the maximum likelihood estimate. Plot the true and approximate pdfs.

5. Let $\mathcal{T} = \{\boldsymbol{X}_1, \dots, \boldsymbol{X}_n\}$ be iid data from a pdf $g(\boldsymbol{x} \mid \boldsymbol{\theta})$ with Fisher matrix $\mathbf{F}(\boldsymbol{\theta})$. Explain why, under the conditions where (4.7) holds,

$$\boldsymbol{S}_{\mathcal{T}}(\boldsymbol{\theta}) := \frac{1}{n}\sum_{i=1}^{n} \boldsymbol{S}(\boldsymbol{X}_i \mid \boldsymbol{\theta})$$

for large n has approximately a multivariate normal distribution with expectation vector $\mathbf{0}$ and covariance matrix $\mathbf{F}(\boldsymbol{\theta})/n$.

6. Figure 4.15 shows a Gaussian KDE with bandwidth $\sigma = 0.2$ on the points $-0.5, 0, 0.2, 0.9$, and 1.5. Reproduce the plot in Python. Using the same bandwidth, plot also the KDE for the same data, but now with $\phi(z) = 1/2, z \in [-1, 1]$.

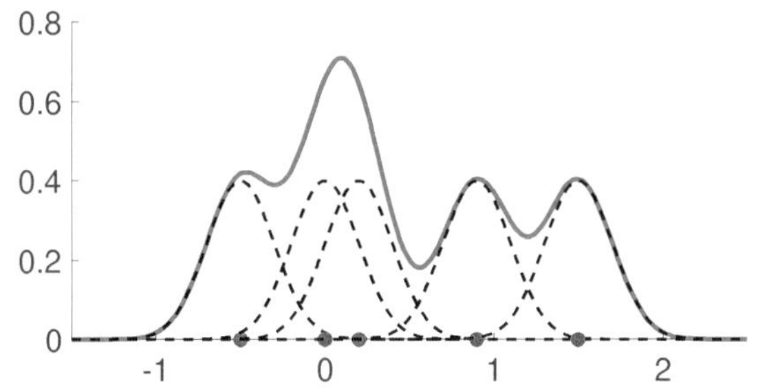

Figure 4.15: The Gaussian KDE (solid line) is the equally weighted mixture of normal pdfs centered around the data and with standard deviation $\sigma = 0.2$ (dashed).

7. For fixed x', the Gaussian kernel function

$$f(x \mid t) := \frac{1}{\sqrt{2\pi t}} \mathrm{e}^{-\frac{1}{2}\frac{(x-x')^2}{t}}$$

is the solution to Fourier's *heat equation*

$$\frac{\partial}{\partial t} f(x \mid t) = \frac{1}{2}\frac{\partial^2}{\partial x^2} f(x \mid t), \quad x \in \mathbb{R},\ t > 0,$$

with initial condition $f(x \mid 0) = \delta(x - x')$ (the Dirac function at x'). Show this. As a consequence, the Gaussian KDE is the solution to the same heat equation, but now with initial condition $f(x \mid 0) = n^{-1}\sum_{i=1}^{n} \delta(x - x_i)$. This was the motivation for the theta KDE [14], which is a solution to the same heat equation but now on a *bounded* interval.

8. Show that the Ward linkage given in (4.41) is equal to

$$d_{\text{Ward}}(\mathcal{I}, \mathcal{J}) = \frac{|\mathcal{I}|\,|\mathcal{J}|}{|\mathcal{I}| + |\mathcal{J}|} \|\overline{\boldsymbol{x}}_{\mathcal{I}} - \overline{\boldsymbol{x}}_{\mathcal{J}}\|^2.$$

9. Carry out the agglomerative hierarchical clustering of Example 4.8 via the `linkage` method from `scipy.cluster.hierarchy`. Show that the linkage matrices are the same. Give a scatterplot of the data, color coded into $K = 3$ clusters.

10. Suppose that we have the data $\tau_n = \{x_1, \ldots, x_n\}$ in $\mathbb{R}$ and decide to train the two-component Gaussian mixture model

$$g(x \mid \boldsymbol{\theta}) = w_1 \frac{1}{\sqrt{2\pi\sigma_1^2}} \exp\left(-\frac{(x-\mu_1)^2}{2\sigma_1^2}\right) + w_2 \frac{1}{\sqrt{2\pi\sigma_2^2}} \exp\left(-\frac{(x-\mu_2)^2}{2\sigma_2^2}\right),$$

where the parameter vector $\boldsymbol{\theta} = [\mu_1, \mu_2, \sigma_1, \sigma_2, w_1, w_2]^\top$ belongs to the set

$$\Theta = \{\boldsymbol{\theta} : w_1 + w_2 = 1, w_1 \in [0, 1], \mu_i \in \mathbb{R}, \sigma_i > 0,\ \forall i\}.$$

Suppose that the training is via the maximum likelihood in (2.28). Show that

$$\sup_{\boldsymbol{\theta} \in \Theta} \frac{1}{n} \sum_{i=1}^{n} \ln g(x_i \mid \boldsymbol{\theta}) = \infty.$$

In other words, find a sequence of values for $\boldsymbol{\theta} \in \Theta$ such that the likelihood grows without bound. How can we restrict the set Θ to ensure that the likelihood remains bounded?

11. A d-dimensional normal random vector $\boldsymbol{X} \sim \mathcal{N}(\boldsymbol{\mu}, \boldsymbol{\Sigma})$ can be defined via an affine transformation, $\boldsymbol{X} = \boldsymbol{\mu} + \boldsymbol{\Sigma}^{1/2}\boldsymbol{Z}$, of a standard normal random vector $\boldsymbol{Z} \sim \mathcal{N}(\boldsymbol{0}, \mathbf{I}_d)$, where $\boldsymbol{\Sigma}^{1/2}(\boldsymbol{\Sigma}^{1/2})^\top = \Sigma$. In a similar way, we can define a d-dimensional Student random vector $\boldsymbol{X} \sim \mathsf{t}_\alpha(\boldsymbol{\mu}, \boldsymbol{\Sigma})$ via a transformation

$$\boldsymbol{X} = \boldsymbol{\mu} + \frac{1}{\sqrt{S}}\boldsymbol{\Sigma}^{1/2}\boldsymbol{Z}, \tag{4.46}$$

where, $\boldsymbol{Z} \sim \mathcal{N}(\boldsymbol{0}, \mathbf{I}_d)$ and $S \sim \mathsf{Gamma}(\frac{\alpha}{2}, \frac{\alpha}{2})$ are independent, $\alpha > 0$, and $\boldsymbol{\Sigma}^{1/2}(\boldsymbol{\Sigma}^{1/2})^\top = \Sigma$. Note that we obtain the multivariate normal distribution as a limiting case for $\alpha \to \infty$.

(a) Show that the density of the $\mathsf{t}_\alpha(\boldsymbol{0}, \mathbf{I}_d)$ distribution is given by

$$t_\alpha(\boldsymbol{x}) := \frac{\Gamma((\alpha + d)/2)}{(\pi\alpha)^{d/2}\Gamma(\alpha/2)}\left(1 + \frac{1}{\alpha}\|\boldsymbol{x}\|^2\right)^{-\frac{\alpha+d}{2}}.$$

☞ 435 By the transformation rule (C.23), it follows that the density of $\boldsymbol{X} \sim \mathsf{t}_\alpha(\boldsymbol{\mu}, \boldsymbol{\Sigma})$ is given by $t_{\alpha,\boldsymbol{\Sigma}}(\boldsymbol{x} - \boldsymbol{\mu})$, where

$$t_{\alpha,\boldsymbol{\Sigma}}(\boldsymbol{x}) := \frac{1}{|\boldsymbol{\Sigma}^{1/2}|}t_\alpha(\boldsymbol{\Sigma}^{-1/2}\boldsymbol{x}).$$

[Hint: conditional on $S = s$, $\boldsymbol{X}$ has a $\mathcal{N}(\boldsymbol{0}, \mathbf{I}_d/s)$ distribution.]

(b) We wish to fit a $\mathbf{t}_\nu(\boldsymbol{\mu}, \Sigma)$ distribution to given data $\tau = \{\boldsymbol{x}_1, \ldots, \boldsymbol{x}_n\}$ in $\mathbb{R}^d$ via the EM method. We use the representation (4.46) and augment the data with the vector $\boldsymbol{S} = [S_1, \ldots, S_n]^\top$ of hidden variables. Show that the complete-data likelihood is given by

$$g(\tau, \boldsymbol{s} \mid \boldsymbol{\theta}) = \prod_i \frac{(\alpha/2)^{\alpha/2} s_i^{(\alpha+d)/2-1}\exp(-\frac{s_i}{2}\alpha - \frac{s_i}{2}\|\boldsymbol{\Sigma}^{-1/2}(\boldsymbol{x}_i - \boldsymbol{\mu})\|^2)}{\Gamma(\alpha/2)(2\pi)^{d/2}|\boldsymbol{\Sigma}^{1/2}|}. \tag{4.47}$$

(c) Show that, as a consequence, conditional on the data τ and parameter $\boldsymbol{\theta}$, the hidden data are mutually independent, and

$$(S_i \mid \tau, \boldsymbol{\theta}) \sim \mathsf{Gamma}\left(\frac{\alpha + d}{2}, \frac{\alpha + \|\boldsymbol{\Sigma}^{-1/2}(\boldsymbol{x}_i - \boldsymbol{\mu})\|^2}{2}\right), \quad i = 1, \ldots, n.$$

(d) At iteration t of the EM algorithm, let $g^{(t)}(\boldsymbol{s}) = g(\boldsymbol{s} \mid \tau, \boldsymbol{\theta}^{(t-1)})$ be the density of the missing data, given the observed data τ and the current parameter guess $\boldsymbol{\theta}^{(t-1)}$. Verify that the expected complete-data log-likelihood is given by:

$$\begin{aligned}\mathbb{E}_{g^{(t)}} \ln g(\tau, \boldsymbol{S} \mid \boldsymbol{\theta}) = &\frac{n\alpha}{2}\ln\frac{\alpha}{2} - \frac{nd}{2}\ln(2\pi) - n\ln\Gamma\left(\frac{\alpha}{2}\right) - \frac{n}{2}\ln|\boldsymbol{\Sigma}| \\ &+ \frac{\alpha + d - 2}{2}\sum_{i=1}^n \mathbb{E}_{g^{(t)}} \ln S_i - \sum_{i=1}^n \frac{\alpha + \|\boldsymbol{\Sigma}^{-1/2}(\boldsymbol{x}_i - \boldsymbol{\mu})\|^2}{2}\mathbb{E}_{g^{(t)}} S_i.\end{aligned}$$

Show that

$$\mathbb{E}_{g^{(t)}} S_i = \frac{\alpha^{(t-1)} + d}{\alpha^{(t-1)} + \|(\boldsymbol{\Sigma}^{(t-1)})^{-1/2}(\boldsymbol{x}_i - \boldsymbol{\mu}^{(t-1)})\|^2} =: w_i^{(t-1)}$$

$$\mathbb{E}_{g^{(t)}} \ln S_i = \psi\left(\frac{\alpha^{(t-1)} + d}{2}\right) - \ln\left(\frac{\alpha^{(t-1)} + d}{2}\right) + \ln w_i^{(t-1)},$$

where $\psi := (\ln\Gamma)'$ is *digamma* function.

(e) Finally, show that in the M-step of the EM algorithm $\boldsymbol{\theta}^{(t)}$ is updated from $\boldsymbol{\theta}^{(t-1)}$ as follows:

$$\boldsymbol{\mu}^{(t)} = \frac{\sum_{i=1}^n w_i^{(t-1)} \boldsymbol{x}_i}{\sum_{i=1}^n w_i^{(t-1)}}$$
$$\boldsymbol{\Sigma}^{(t)} = \frac{1}{n}\sum_{i=1}^n w_i^{(t-1)}(\boldsymbol{x}_i - \boldsymbol{\mu}^{(t)})(\boldsymbol{x}_i - \boldsymbol{\mu}^{(t)})^\top,$$

and $\alpha^{(t)}$ is defined implicitly through the solution of the nonlinear equation:

$$\ln\left(\frac{\alpha}{2}\right) - \psi\left(\frac{\alpha}{2}\right) + \psi\left(\frac{\alpha^{(t)}+d}{2}\right) - \ln\left(\frac{\alpha^{(t)}+d}{2}\right) + 1 + \frac{\sum_{i=1}^n \left(\ln(w_i^{(t-1)}) - w_i^{(t-1)}\right)}{n} = 0.$$

12. A generalization of both the gamma and inverse-gamma distribution is the *generalized inverse-gamma distribution*, which has density

GENERALIZED INVERSE-GAMMA DISTRIBUTION

$$f(s) = \frac{(a/b)^{p/2}}{2K_p(\sqrt{ab})} s^{p-1} \mathrm{e}^{-\frac{1}{2}(as+b/s)}, \quad a, b, s > 0, \; p \in \mathbb{R}, \tag{4.48}$$

where K_p is the *modified Bessel function of the second kind*, which can be defined as the integral

MODIFIED BESSEL FUNCTION OF THE SECOND KIND

$$K_p(x) = \int_0^\infty \mathrm{e}^{-x\cosh(t)} \cosh(pt)\,\mathrm{d}t, \quad x > 0, \; p \in \mathbb{R}. \tag{4.49}$$

We write $S \sim \mathsf{GIG}(a, b, p)$ to denote that S has a pdf of the form (4.48). The function K_p has many interesting properties. Special cases include

$$K_{1/2}(x) = \sqrt{\frac{x\pi}{2}}\,\mathrm{e}^{-x}\frac{1}{x}$$
$$K_{3/2}(x) = \sqrt{\frac{x\pi}{2}}\,\mathrm{e}^{-x}\left(\frac{1}{x} + \frac{1}{x^2}\right)$$
$$K_{5/2}(x) = \sqrt{\frac{x\pi}{2}}\,\mathrm{e}^{-x}\left(\frac{1}{x} + \frac{3}{x^2} + \frac{3}{x^3}\right).$$

More generally, K_p satisfies the recursion

$$K_{p+1}(x) = K_{p-1}(x) + \frac{2p}{x}K_p(x). \tag{4.50}$$

(a) Using the change of variables $\mathrm{e}^z = s\sqrt{a/b}$, show that

$$\int_0^\infty s^{p-1}\mathrm{e}^{-\frac{1}{2}(as+b/s)}\,\mathrm{d}s = 2K_p(\sqrt{ab})(b/a)^{p/2}.$$

(b) Let $S \sim \mathsf{GIG}(a, b, p)$. Show that

$$\mathbb{E}S = \frac{\sqrt{b}\,K_{p+1}(\sqrt{ab})}{\sqrt{a}\,K_p(\sqrt{ab})} \tag{4.51}$$

and

$$\mathbb{E}S^{-1} = \frac{\sqrt{a}\,K_{p+1}(\sqrt{ab})}{\sqrt{b}\,K_p(\sqrt{ab})} - \frac{2p}{b}. \tag{4.52}$$

SCALE-MIXTURE

13. In Exercise 11 we viewed the multivariate Student t_α distribution as a *scale-mixture* of the $\mathcal{N}(\mathbf{0}, \mathbf{I}_d)$ distribution. In this exercise, we consider a similar transformation, but now $\boldsymbol{\Sigma}^{1/2}\boldsymbol{Z} \sim \mathcal{N}(\mathbf{0}, \boldsymbol{\Sigma})$ is not divided but is *multiplied* by $\sqrt{S}$, with $S \sim \mathsf{Gamma}(\alpha/2, \alpha/2)$:

$$\boldsymbol{X} = \boldsymbol{\mu} + \sqrt{S}\,\boldsymbol{\Sigma}^{1/2}\,\boldsymbol{Z}, \tag{4.53}$$

where S and $\boldsymbol{Z}$ are independent and $\alpha > 0$.

BESSEL DISTRIBUTION

(a) Show, using Exercise 12, that for $\boldsymbol{\Sigma}^{1/2} = \mathbf{I}_d$ and $\boldsymbol{\mu} = \mathbf{0}$, the random vector $\boldsymbol{X}$ has a d-dimensional *Bessel distribution*, with density:

$$\kappa_\alpha(\boldsymbol{x}) := \frac{2^{1-(\alpha+d)/2}\alpha^{(\alpha+d)/4}\,\|\boldsymbol{x}\|^{(\alpha-d)/2}}{\pi^{d/2}\Gamma(\alpha/2)} K_{(\alpha-d)/2}\left(\|\boldsymbol{x}\|\,\sqrt{\alpha}\right), \quad \boldsymbol{x} \in \mathbb{R}^d,$$

where K_p is the modified Bessel function of the second kind given in (4.49). We write $\boldsymbol{X} \sim \mathsf{Bessel}_\alpha(\mathbf{0}, \mathbf{I}_d)$. A random vector $\boldsymbol{X}$ is said to have a $\mathsf{Bessel}_\alpha(\boldsymbol{\mu}, \boldsymbol{\Sigma})$ distribution if it can be written in the form (4.53). By the transformation rule (C.23), its density is given by $\frac{1}{\sqrt{|\boldsymbol{\Sigma}|}}\kappa_\alpha(\boldsymbol{\Sigma}^{-1/2}(\boldsymbol{x} - \boldsymbol{\mu}))$. Special instances of the Bessel pdf include:

$$\begin{aligned}
\kappa_2(x) &= \frac{\exp(-\sqrt{2}\,|x|)}{\sqrt{2}} \\
\kappa_4(x) &= \frac{1 + 2\,|x|}{2}\exp(-2\,|x|) \\
\kappa_4(x_1, x_2, x_3) &= \frac{1}{\pi}\exp\left(-2\sqrt{x_1^2 + x_2^2 + x_3^2}\right) \\
\kappa_{d+1}(\boldsymbol{x}) &= \frac{((d+1)/2)^{d/2}\sqrt{\pi}}{(2\pi)^{d/2}\Gamma((d+1)/2)}\exp\left(-\sqrt{d+1}\,\|\boldsymbol{x}\|\right), \quad \boldsymbol{x} \in \mathbb{R}^d.
\end{aligned}$$

Note that k_2 is the (scaled) pdf of the double-exponential or *Laplace* distribution.

(b) Given the data $\tau = \{\boldsymbol{x}_1, \ldots, \boldsymbol{x}_n\}$ in $\mathbb{R}^d$, we wish to fit a Bessel pdf to the data by employing the EM algorithm, augmenting the data with the vector $\boldsymbol{S} = [S_1, \ldots, S_n]^\top$ of missing data. We assume that α is known and $\alpha > d$. Show that conditional on τ (and given $\boldsymbol{\theta}$), the missing data vector $\boldsymbol{S}$ has independent components, with $S_i \sim \mathsf{GIG}(\alpha, b_i, (\alpha - d)/2)$, with $b_i := \|\Sigma^{-1/2}(\boldsymbol{x}_i - \boldsymbol{\mu})\|^2$, $i = 1, \ldots, n$.

(c) At iteration t of the EM algorithm, let $g^{(t)}(\boldsymbol{s}) = g(\boldsymbol{s} \,|\, \tau, \boldsymbol{\theta}^{(t-1)})$ be the density of the missing data, given the observed data τ and the current parameter guess $\boldsymbol{\theta}^{(t-1)}$. Show that the expected complete-data log-likelihood is given by:

$$Q^{(t)}(\boldsymbol{\theta}) := \mathbb{E}_{g^{(t)}} \ln g(\tau, \boldsymbol{S} \,|\, \boldsymbol{\theta}) = -\frac{1}{2}\sum_{i=1}^n b_i(\boldsymbol{\theta})\, w_i^{(t-1)} + \text{constant}, \tag{4.54}$$

where $b_i(\boldsymbol{\theta}) = \|\Sigma^{-1/2}(\boldsymbol{x}_i - \boldsymbol{\mu})\|^2$ and

$$w_i^{(t-1)} := \frac{\sqrt{\alpha}\,K_{(\alpha-d+2)/2}\left(\sqrt{\alpha\, b_i(\boldsymbol{\theta}^{(t-1)})}\right)}{\sqrt{b_i(\boldsymbol{\theta}^{(t-1)})}\,K_{(\alpha-d)/2}\left(\sqrt{\alpha\, b_i(\boldsymbol{\theta}^{(t-1)})}\right)} - \frac{\alpha - d}{b_i(\boldsymbol{\theta}^{(t-1)})}, \quad i = 1, \ldots, n.$$

(d) From (4.54) derive the M-step of the EM algorithm. That is, show how $\boldsymbol{\theta}^{(t)}$ is updated from $\boldsymbol{\theta}^{(t-1)}$.

14. Consider the ellipsoid $E = \{\boldsymbol{x} \in \mathbb{R}^d : \boldsymbol{x}\Sigma^{-1}\boldsymbol{x} = 1\}$ in (4.42). Let $\mathbf{U}\mathbf{D}^2\mathbf{U}^\top$ be an SVD of $\boldsymbol{\Sigma}$. Show that the linear transformation $\boldsymbol{x} \mapsto \mathbf{U}^\top\mathbf{D}^{-1}\boldsymbol{x}$ maps the points on E onto the unit sphere $\{z \in \mathbb{R}^d : \|z\| = 1\}$.

15. Figure 4.13 shows how the centered "surfboard" data are projected onto the first column of the principal component matrix $\mathbf{U}$. Suppose we project the data instead onto the plane spanned by the first *two* columns of $\mathbf{U}$. What are a and b in the representation $ax_1 + bx_2 = x_3$ of this plane?

16. Figure 4.14 suggests that we can assign each feature vector $\boldsymbol{x}$ in the `iris` data set to one of two clusters, based on the value of $\boldsymbol{u}_1^\top\boldsymbol{x}$, where $\boldsymbol{u}_1$ is the first principal component. Plot the sepal lengths against petal lengths and color the points for which $\boldsymbol{u}_1^\top\boldsymbol{x} < 1.5$ differently to points for which $\boldsymbol{u}_1^\top\boldsymbol{x} \geqslant 1.5$. To which species of iris do these clusters correspond?

CHAPTER 5

REGRESSION

Many supervised learning techniques can be gathered under the name "regression". The purpose of this chapter is to explain the mathematical ideas behind regression models and their practical aspects. We analyze the fundamental linear model in detail, and also discuss nonlinear and generalized linear models.

5.1 Introduction

Francis Galton observed in an article in 1889 that the heights of adult offspring are, on the whole, more "average" than the heights of their parents. Galton interpreted this as a degenerative phenomenon, using the term "regression" to indicate this "return to mediocrity". Nowadays, *regression* refers to a broad class of supervised learning techniques where the aim is to predict a quantitative response (output) variable y via a function $g(\boldsymbol{x})$ of an explanatory (input) vector $\boldsymbol{x} = [x_1, \ldots, x_p]^\top$, consisting of p features, each of which can be continuous or discrete. For instance, regression could be used to predict the birth weight of a baby (the response variable) from the weight of the mother, her socio-economic status, and her smoking habits (the explanatory variables). REGRESSION

Let us recapitulate the framework of supervised learning established in Chapter 2. The aim is to find a prediction function g that best guesses[1] what the random output Y will be for a random input vector $\boldsymbol{X}$. The joint pdf $f(\boldsymbol{x}, y)$ of $\boldsymbol{X}$ and Y is unknown, but a training set $\tau = \{(\boldsymbol{x}_1, y_1), \ldots, (\boldsymbol{x}_n, y_n)\}$ is available, which is thought of as the outcome of a random training set $\mathcal{T} = \{(\boldsymbol{X}_1, Y_1), \ldots, (\boldsymbol{X}_n, Y_n)\}$ of iid copies of $(\boldsymbol{X}, Y)$. Once we have selected a loss function $\text{Loss}(y, \widehat{y})$, such as the *squared-error loss* ☞ 19 SQUARED-ERROR LOSS

$$\text{Loss}(y, \widehat{y}) = (y - \widehat{y})^2, \tag{5.1}$$

then the "best" prediction function g is defined as the one that minimizes the *risk* $\ell(g) = \mathbb{E}\,\text{Loss}(Y, g(\boldsymbol{X}))$. We saw in Section 2.2 that for the squared-error loss this optimal prediction function is the conditional expectation RISK

$$g^*(\boldsymbol{x}) = \mathbb{E}[Y \mid \boldsymbol{X} = \boldsymbol{x}].$$

[1] Recall the mnemonic use of "g" for "guess"

As the squared-error loss is the most widely-used loss function for regression, we will adopt this loss function in most of this chapter.

The optimal prediction function g^* has to be learned from the training set τ by minimizing the training loss

$$\ell_\tau(g) = \frac{1}{n}\sum_{i=1}^{n}(y_i - g(\boldsymbol{x}_i))^2 \tag{5.2}$$

over a suitable class of functions $\mathcal{G}$. Note that in the above definition, the training set τ is assumed to be fixed. For a random training set $\mathcal{T}$, we will write the training loss as $\ell_{\mathcal{T}}(g)$. The function $g_\tau^{\mathcal{G}}$ that minimizes the training loss is the function we use for prediction — the so-called *learner*. When the function class $\mathcal{G}$ is clear from the context, we drop the superscript in the notation.

LEARNER

☞ 21

As we already saw in (2.2), conditional on $\boldsymbol{X} = \boldsymbol{x}$, the response Y can be written as

$$Y = g^*(\boldsymbol{x}) + \varepsilon(\boldsymbol{x}),$$

where $\mathbb{E}\,\varepsilon(\boldsymbol{x}) = 0$. This motivates a standard modeling assumption in supervised learning, in which the responses $Y_1, \dots, Y_n$, conditional on the explanatory variables $\boldsymbol{X}_1 = \boldsymbol{x}_1, \dots, \boldsymbol{X}_n = \boldsymbol{x}_n$, are assumed to be of the form

$$Y_i = g(\boldsymbol{x}_i) + \varepsilon_i, \quad i = 1, \dots, n,$$

where the $\{\varepsilon_i\}$ are independent with $\mathbb{E}\,\varepsilon_i = 0$ and $\mathbb{V}\mathrm{ar}\,\varepsilon_i = \sigma^2$ for some function $g \in \mathcal{G}$ and variance σ^2. The above model is usually further specified by assuming that g is completely known up to an unknown parameter vector; that is,

$$Y_i = g(\boldsymbol{x}_i \,|\, \boldsymbol{\beta}) + \varepsilon_i, \quad i = 1, \dots, n. \tag{5.3}$$

While the model (5.3) is described *conditional* on the explanatory variables, it will be convenient to make one further model simplification, and view (5.3) as if the $\{\boldsymbol{x}_i\}$ were *fixed*, while the $\{Y_i\}$ are random.

> ⚠ For the remainder of this chapter, we assume that the training feature vectors $\{\boldsymbol{x}_i\}$ are fixed and only the responses are random; that is, $\mathcal{T} = \{(\boldsymbol{x}_1, Y_1), \dots, (\boldsymbol{x}_n, Y_n)\}$.

The advantage of the model (5.3) is that the problem of estimating the *function* g from the training data is reduced to the (much simpler) problem of estimating the *parameter vector* $\boldsymbol{\beta}$. An obvious disadvantage is that functions of the form $g(\cdot \,|\, \boldsymbol{\beta})$ may not accurately approximate the true unknown g^*. The remainder of this chapter deals with the analysis of models of the form (5.3). In the important case where the function $g(\cdot \,|\, \boldsymbol{\beta})$ is *linear*, the analysis proceeds through the class of linear models. If, in addition, the error terms $\{\varepsilon_i\}$ are assumed to be *Gaussian*, this analysis can be carried out using the rich theory of normal linear models.

5.2 Linear Regression

The most basic regression model involves a linear relationship between the response and a single explanatory variable. In particular, we have measurements $(x_1, y_1), \ldots, (x_n, y_n)$ that lie approximately on a straight line, as in Figure 5.1.

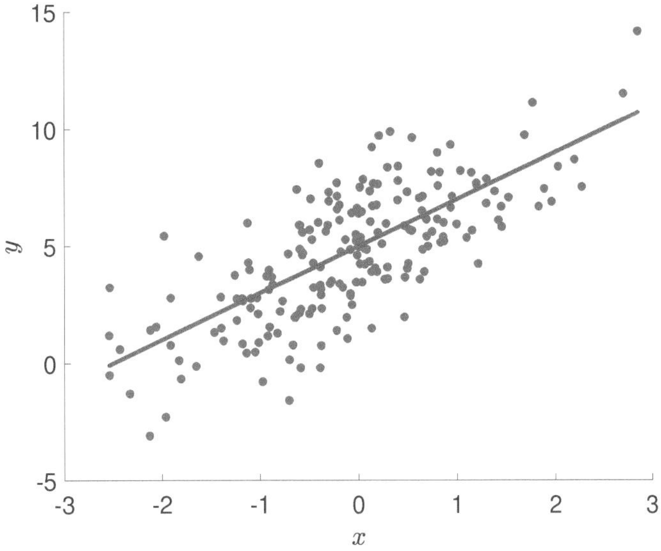

Figure 5.1: Data from a simple linear regression model.

Following the general scheme captured in (5.3), a simple model for these data is that the $\{x_i\}$ are fixed and variables $\{Y_i\}$ are random such that

$$Y_i = \beta_0 + \beta_1 x_i + \varepsilon_i, \quad i = 1, \ldots, n, \tag{5.4}$$

for certain *unknown* parameters β_0 and β_1. The $\{\varepsilon_i\}$ are assumed to be independent with expectation 0 and unknown variance σ^2. The unknown line

$$y = \underbrace{\beta_0 + \beta_1 x}_{g(x \mid \boldsymbol{\beta})} \tag{5.5}$$

is called the *regression line*. Thus, we view the responses as random variables that would lie exactly on the regression line, were it not for some "disturbance" or "error" term represented by the $\{\varepsilon_i\}$. The extent of the disturbance is modeled by the parameter σ^2. The model in (5.4) is called *simple linear regression*. This model can easily be extended to incorporate more than one explanatory variable, as follows.

REGRESSION LINE

SIMPLE LINEAR REGRESSION MODEL

Definition 5.1: Multiple Linear Regression Model

In a *multiple linear regression model* the response Y depends on a d-dimensional explanatory vector $\boldsymbol{x} = [x_1, \ldots, x_d]^\top$, via the linear relationship

$$Y = \beta_0 + \beta_1 x_1 + \cdots + \beta_d x_d + \varepsilon, \tag{5.6}$$

where $\mathbb{E}\,\varepsilon = 0$ and $\mathbb{V}\mathrm{ar}\,\varepsilon = \sigma^2$.

MULTIPLE LINEAR REGRESSION MODEL

Thus, the data lie approximately on a d-dimensional affine hyperplane

$$y = \underbrace{\beta_0 + \beta_1 x_1 + \cdots + \beta_d x_d}_{g(\boldsymbol{x} \mid \boldsymbol{\beta})},$$

where we define $\boldsymbol{\beta} = [\beta_0, \beta_1, \ldots, \beta_d]^\top$. The function $g(\boldsymbol{x} \mid \boldsymbol{\beta})$ is linear in $\boldsymbol{\beta}$, but not linear in the feature vector $\boldsymbol{x}$, due to the constant β_0. However, *augmenting* the feature space with the constant 1, the mapping $[1, \boldsymbol{x}^\top]^\top \mapsto g(\boldsymbol{x} \mid \boldsymbol{\beta}) := [1, \boldsymbol{x}^\top]\boldsymbol{\beta}$ becomes linear in the feature space and so (5.6) becomes a *linear model* (see Section 2.1). Most software packages for regression include 1 as a feature by default.

☞ 43

Note that in (5.6) we only specified the model for a single pair $(\boldsymbol{x}, Y)$. The model for the training set $\mathcal{T} = \{(\boldsymbol{x}_1, Y_1), \ldots, (\boldsymbol{x}_n, Y_n)\}$ is simply that each Y_i satisfies (5.6) (with $\boldsymbol{x} = \boldsymbol{x}_i$) and that the $\{Y_i\}$ are independent. Setting $\boldsymbol{Y} = [Y_1, \ldots, Y_n]^\top$, we can write the multiple linear regression model for the training data compactly as

$$\boldsymbol{Y} = \mathbf{X}\boldsymbol{\beta} + \boldsymbol{\varepsilon}, \tag{5.7}$$

MODEL MATRIX

where $\boldsymbol{\varepsilon} = [\varepsilon_1, \ldots, \varepsilon_n]^\top$ is a vector of iid copies of ε and $\mathbf{X}$ is the *model matrix* given by

$$\mathbf{X} = \begin{bmatrix} 1 & x_{11} & x_{12} & \cdots & x_{1d} \\ 1 & x_{21} & x_{22} & \cdots & x_{2d} \\ \vdots & \vdots & \vdots & \vdots & \vdots \\ 1 & x_{n1} & x_{n2} & \cdots & x_{nd} \end{bmatrix} = \begin{bmatrix} 1 & \boldsymbol{x}_1^\top \\ 1 & \boldsymbol{x}_2^\top \\ \vdots & \vdots \\ 1 & \boldsymbol{x}_n^\top \end{bmatrix}.$$

■ **Example 5.1 (Multiple Linear Regression Model)** Figure 5.2 depicts a realization of the multiple linear regression model

$$Y_i = x_{i1} + x_{i2} + \varepsilon_i, \quad i = 1, \ldots, 100,$$

where $\varepsilon_1, \ldots, \varepsilon_{100} \sim_{\text{iid}} \mathcal{N}(0, 1/16)$. The fixed feature vectors (vectors of explanatory variables) $\boldsymbol{x}_i = [x_{i1}, x_{i2}]^\top$, $i = 1, \ldots, 100$ lie in the unit square.

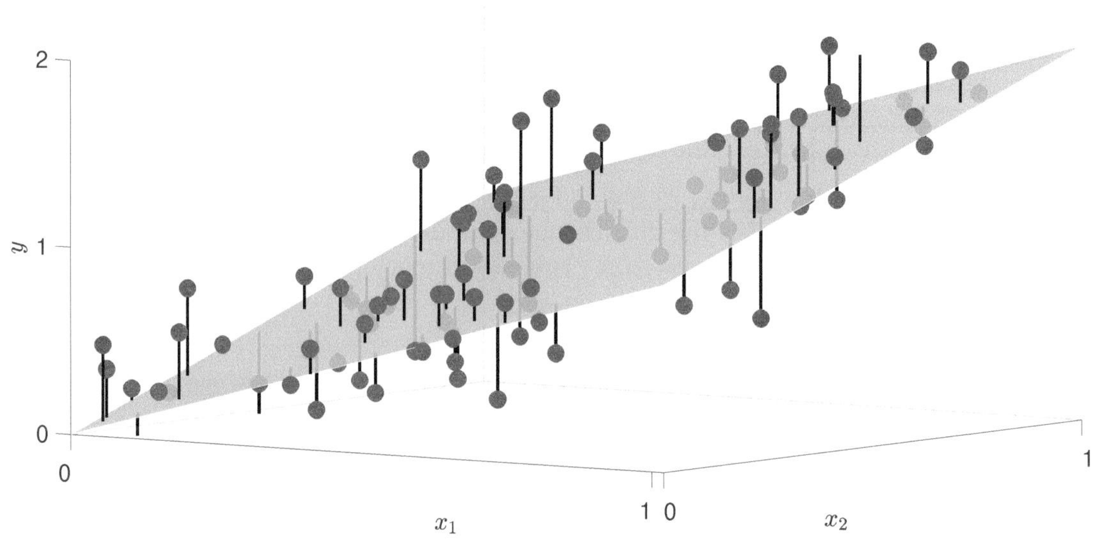

Figure 5.2: Data from a multiple linear regression model.

■

5.3 Analysis via Linear Models

Analysis of data from a linear regression model is greatly simplified through the linear model representation (5.7). In this section we present the main ideas for parameter estimation and model selection for a general linear model of the form

$$\boldsymbol{Y} = \mathbf{X}\boldsymbol{\beta} + \boldsymbol{\varepsilon}, \tag{5.8}$$

where $\mathbf{X}$ is an $n \times p$ matrix, $\boldsymbol{\beta} = [\beta_1, \ldots, \beta_p]^\top$ a vector of p parameters, and $\boldsymbol{\varepsilon} = [\varepsilon_1, \ldots, \varepsilon_n]^\top$ an n-dimensional vector of independent error terms, with $\mathbb{E}\,\varepsilon_i = 0$ and $\mathbb{V}\mathrm{ar}\,\varepsilon_i = \sigma^2$, $i = 1, \ldots, n$. Note that the model matrix $\mathbf{X}$ is assumed to be fixed, and $\boldsymbol{Y}$ and $\boldsymbol{\varepsilon}$ are random. A specific outcome of $\boldsymbol{Y}$ is denoted by $\boldsymbol{y}$ (in accordance with the notation in Section 2.8). ☞ 46

> Note that the multiple linear regression model in (5.7) was defined using a different parameterization; in particular, there we used $\boldsymbol{\beta} = [\beta_0, \beta_1, \ldots, \beta_d]^\top$. So, when applying the results in the present section to such models, be aware that $p = d + 1$. Also, in this section a feature vector $\boldsymbol{x}$ includes the constant 1, so that $\mathbf{X}^\top = [\boldsymbol{x}_1, \ldots, \boldsymbol{x}_n]$.

5.3.1 Parameter Estimation

The linear model $\boldsymbol{Y} = \mathbf{X}\boldsymbol{\beta} + \boldsymbol{\varepsilon}$ contains two unknown parameters, $\boldsymbol{\beta}$ and σ^2, which have to be estimated from the training data τ. To estimate $\boldsymbol{\beta}$, we can repeat exactly the same reasoning used in our recurring polynomial regression Example 2.1 as follows. For a linear prediction function $g(\boldsymbol{x}) = \boldsymbol{x}^\top\boldsymbol{\beta}$, the (squared-error) training loss can be written as ☞ 26

$$\ell_\tau(g) = \frac{1}{n}\|\boldsymbol{y} - \mathbf{X}\boldsymbol{\beta}\|^2,$$

and the optimal learner g_τ minimizes this quantity, leading to the least-squares estimate $\widehat{\boldsymbol{\beta}}$, which satisfies the normal equations

$$\mathbf{X}^\top\mathbf{X}\boldsymbol{\beta} = \mathbf{X}^\top\boldsymbol{y}. \tag{5.9}$$

The corresponding training loss can be taken as an estimate of σ^2; that is,

$$\widehat{\sigma^2} = \frac{1}{n}\|\boldsymbol{y} - \mathbf{X}\widehat{\boldsymbol{\beta}}\|^2. \tag{5.10}$$

To justify the latter, note that σ^2 is the second moment of the model errors $\varepsilon_i, i = 1, \ldots, n$, in (5.8) and could be estimated via the method of moments (see Section C.12.1) using the sample average $n^{-1}\sum_i \varepsilon_i^2 = \|\boldsymbol{\varepsilon}\|^2/n = \|\boldsymbol{Y} - \mathbf{X}\boldsymbol{\beta}\|^2/n$, if $\boldsymbol{\beta}$ were known. By replacing $\boldsymbol{\beta}$ with its estimator, we arrive at (5.10). Note that no distributional properties of the $\{\varepsilon_i\}$ were used other than $\mathbb{E}\,\varepsilon_i = 0$ and $\mathbb{V}\mathrm{ar}\,\varepsilon_i = \sigma^2, i = 1, \ldots, n$. The vector $\boldsymbol{e} := \boldsymbol{y} - \mathbf{X}\widehat{\boldsymbol{\beta}}$ is called the vector of *residuals* and approximates the (unknown) vector of model errors $\boldsymbol{\varepsilon}$. The quantity $\|\boldsymbol{e}\|^2 = \sum_{i=1}^n e_i^2$ is called the *residual sum of squares* (RSS). Dividing the RSS by $n - p$ gives an unbiased estimate of σ^2, which we call the estimated *residual squared error* (RSE); see Exercise 12. ☞ 457

RESIDUALS

RESIDUAL SUM OF SQUARES

RESIDUAL SQUARED ERROR

In terms of the notation given in the summary Table 2.1 for supervised learning, we ☞ 25 thus have:

1. The (observed) training data is $\tau = \{\mathbf{X}, \boldsymbol{y}\}$.

2. The function class $\mathcal{G}$ is the class of linear functions of $\boldsymbol{x}$; that is $\mathcal{G} = \{g(\cdot\,|\,\boldsymbol{\beta}) : \boldsymbol{x} \mapsto \boldsymbol{x}^\top\boldsymbol{\beta},\ \boldsymbol{\beta} \in \mathbb{R}^p\}$.

3. The (squared-error) training loss is $\ell_\tau(g(\cdot\,|\,\boldsymbol{\beta})) = \|\boldsymbol{y} - \mathbf{X}\boldsymbol{\beta}\|^2/n$.

4. The learner g_τ is given by $g_\tau(\boldsymbol{x}) = \boldsymbol{x}^\top\widehat{\boldsymbol{\beta}}$, where $\widehat{\boldsymbol{\beta}} = \operatorname{argmin}_{\boldsymbol{\beta}\in\mathbb{R}^p} \|\boldsymbol{y} - \mathbf{X}\boldsymbol{\beta}\|^2$.

5. The minimal training loss is $\ell_\tau(g_\tau) = \|\boldsymbol{y} - \mathbf{X}\widehat{\boldsymbol{\beta}}\|^2/n = \widehat{\sigma^2}$.

5.3.2 Model Selection and Prediction

Even if we restrict the learner to be a linear function, there is still the issue of which explanatory variables (features) to include. While including too few features may result in large *approximation error* (underfitting), including too many may result in large *statistical error* ☞ 31 (overfitting). As discussed in Section 2.4, we need to select the features which provide the best tradeoff between the approximation and statistical errors, so that the (expected) generalization risk of the learner is minimized. Depending on how the (expected) generalization risk is estimated, there are a number of strategies for feature selection:

1. Use *test data* $\tau' = (\mathbf{X}', \boldsymbol{y}')$ that are obtained independently from the training data τ, ☞ 24 to estimate the generalization risk $\mathbb{E}\,\|Y - g_\tau(\boldsymbol{X})\|^2$ via the test loss (2.7). Then choose the collection of features that minimizes the test loss. When there is an abundance of data, part of the data can be reserved as test data, while the remaining data is used as training data.

2. When there is a limited amount of data, we can use *cross-validation* to estimate the expected generalization risk $\mathbb{E}\,\|Y - g_{\mathcal{T}}(\boldsymbol{X})\|^2$ (where $\mathcal{T}$ is a random training set), as ☞ 37 explained in Section 2.5.2. This is then minimized over the set of possible choices for the explanatory variables.

3. When one has to choose between many potential explanatory variables, techniques such as *regularized least-squares* and *lasso regression* become important. Such methods offer another approach to model selection, via the regularization (or homotopy) paths. ☞ 216 This will be the topic of Section 6.2 in the next chapter.

4. Rather than using computer-intensive techniques, such as the ones above, one can use *theoretical* estimates of the expected generalization risk, such as the in-sample ☞ 35 risk, AIC, and BIC, as in Section 2.5, and minimize this to determine a good set of explanatory variables.

5. All of the above approaches do not assume any distributional properties of the error terms $\{\varepsilon_i\}$ in the linear model, other than that they are independent with expectation 0 and variance σ^2. If, however, they are assumed to have a *normal* (Gaussian) distribution, (that is, $\{\varepsilon_i\} \sim_{\text{iid}} \mathcal{N}(0, \sigma^2)$), then the inclusion and exclusion of variables can

be decided by means of *hypotheses tests*. This is the classical approach to model selection, and will be discussed in Section 5.4. As a consequence of the central limit theorem, one can use the same approach when the error terms are not necessarily normal, provided their variance is finite and the sample size n is large.

6. Finally, when using a Bayesian approach, comparison of two models can be achieved by computing their so-called *Bayes factor* (see Section 2.9).

All of the above strategies can be thought of as specifications of a simple rule formulated by William of Occam, which can be interpreted as:

When presented with competing models, choose the simplest one that explains the data.

This age-old principle, known as *Occam's razor*, is mirrored in a famous quote of Einstein: OCCAM'S RAZOR

Everything should be made as simple as possible, but not simpler.

In linear regression, the number of parameters or predictors is usually a reasonable measure of the simplicity of the model.

5.3.3 Cross-Validation and Predictive Residual Sum of Squares

We start by considering the n-fold cross-validation, also called *leave-one-out cross-validation*, for the linear model (5.8). We partition the data into n data sets, leaving out precisely one observation per data set, which we then predict based on the $n - 1$ remaining observations; see Section 2.5.2 for the general case. Let $\widehat{y}_{-i}$ denote the prediction for the i-th observation using all the data except y_i. The error in the prediction, $y_i - \widehat{y}_{-i}$, is called a *predicted residual* — in contrast to an ordinary residual, $e_i = y_i - \widehat{y}_i$, which is the difference between an observation and its fitted value $\widehat{y}_i = g_\tau(\boldsymbol{x}_i)$ obtained using the whole sample. In this way, we obtain the collection of predicted residuals $\{y_i - \widehat{y}_{-i}\}_{i=1}^n$ and summarize them through the *predicted residual sum of squares* (*PRESS*):

LEAVE-ONE-OUT CROSS-VALIDATION

☞ 37

PREDICTED RESIDUAL

PRESS

$$\text{PRESS} = \sum_{i=1}^{n} (y_i - \widehat{y}_{-i})^2.$$

Dividing the PRESS by n gives an estimate of the expected generalization risk.

In general, computing the PRESS is computationally intensive as it involves training and predicting n separate times. For linear models, however, the predicted residuals can be calculated quickly using only the ordinary residuals and the projection matrix $\mathbf{P} = \mathbf{X}\mathbf{X}^+$ onto the linear space spanned by the columns of the model matrix $\mathbf{X}$ (see (2.13)). The i-th diagonal element $\mathbf{P}_{ii}$ of the projection matrix is called the i-th *leverage*, and it can be shown that $0 \leqslant \mathbf{P}_{ii} \leqslant 1$ (see Exercise 10).

☞ 171

☞ 28

LEVERAGE

Theorem 5.1: PRESS for Linear Models

Consider the linear model (5.8), where the $n \times p$ model matrix $\mathbf{X}$ is of full rank. Given an outcome $\boldsymbol{y} = [y_1, \ldots, y_n]^\top$ of $\boldsymbol{Y}$, the fitted values can be obtained as $\widehat{\boldsymbol{y}} = \mathbf{P}\boldsymbol{y}$, where $\mathbf{P} = \mathbf{X}\mathbf{X}^+ = \mathbf{X}(\mathbf{X}^\top\mathbf{X})^{-1}\mathbf{X}^\top$ is the projection matrix. If the leverage value $p_i := \mathbf{P}_{ii} \neq 1$ for all $i = 1, \ldots, n$, then the predicted residual sum of squares can be written as

$$\text{PRESS} = \sum_{i=1}^{n} \left(\frac{e_i}{1 - p_i} \right)^2,$$

where $e_i = y_i - \widehat{y}_i = y_i - (\mathbf{X}\widehat{\boldsymbol{\beta}})_i$ is the i-th residual.

Proof: It suffices to show that the i-th predicted residual can be written as $y_i - \widehat{y}_{-i} = e_i/(1 - p_i)$. Let $\mathbf{X}_{-i}$ denote the model matrix $\mathbf{X}$ with the i-th row, $\boldsymbol{x}_i^\top$, removed, and define $\boldsymbol{y}_{-i}$ similarly. Then, the least-squares estimate for $\boldsymbol{\beta}$ using all but the i-th observation is $\widehat{\boldsymbol{\beta}}_{-i} = (\mathbf{X}_{-i}^\top\mathbf{X}_{-i})^{-1}\mathbf{X}_{-i}^\top\boldsymbol{y}_{-i}$. Writing $\mathbf{X}^\top\mathbf{X} = \mathbf{X}_{-i}^\top\mathbf{X}_{-i} + \boldsymbol{x}_i\boldsymbol{x}_i^\top$, we have by the Sherman–Morrison
☞ 373 formula

$$(\mathbf{X}_{-i}^\top\mathbf{X}_{-i})^{-1} = (\mathbf{X}^\top\mathbf{X})^{-1} + \frac{(\mathbf{X}^\top\mathbf{X})^{-1}\boldsymbol{x}_i\boldsymbol{x}_i^\top(\mathbf{X}^\top\mathbf{X})^{-1}}{1 - \boldsymbol{x}_i^\top(\mathbf{X}^\top\mathbf{X})^{-1}\boldsymbol{x}_i},$$

where $\boldsymbol{x}_i^\top(\mathbf{X}^\top\mathbf{X})^{-1}\boldsymbol{x}_i = p_i < 1$. Also, $\mathbf{X}_{-i}^\top\boldsymbol{y}_{-i} = \mathbf{X}^\top\boldsymbol{y} - \boldsymbol{x}_i y_i$. Combining all these identities, we have

$$\begin{aligned}
\widehat{\boldsymbol{\beta}}_{-i} &= (\mathbf{X}_{-i}^\top\mathbf{X}_{-i})^{-1}\mathbf{X}_{-i}^\top\boldsymbol{y}_{-i} \\
&= \left((\mathbf{X}^\top\mathbf{X})^{-1} + \frac{(\mathbf{X}^\top\mathbf{X})^{-1}\boldsymbol{x}_i\boldsymbol{x}_i^\top(\mathbf{X}^\top\mathbf{X})^{-1}}{1 - p_i} \right) (\mathbf{X}^\top\boldsymbol{y} - \boldsymbol{x}_i y_i) \\
&= \widehat{\boldsymbol{\beta}} + \frac{(\mathbf{X}^\top\mathbf{X})^{-1}\boldsymbol{x}_i\boldsymbol{x}_i^\top\widehat{\boldsymbol{\beta}}}{1 - p_i} - (\mathbf{X}^\top\mathbf{X})^{-1}\boldsymbol{x}_i y_i - \frac{(\mathbf{X}^\top\mathbf{X})^{-1}\boldsymbol{x}_i p_i y_i}{1 - p_i} \\
&= \widehat{\boldsymbol{\beta}} + \frac{(\mathbf{X}^\top\mathbf{X})^{-1}\boldsymbol{x}_i\boldsymbol{x}_i^\top\widehat{\boldsymbol{\beta}}}{1 - p_i} - \frac{(\mathbf{X}^\top\mathbf{X})^{-1}\boldsymbol{x}_i y_i}{1 - p_i} \\
&= \widehat{\boldsymbol{\beta}} - \frac{(\mathbf{X}^\top\mathbf{X})^{-1}\boldsymbol{x}_i(y_i - \boldsymbol{x}_i^\top\widehat{\boldsymbol{\beta}})}{1 - p_i} = \widehat{\boldsymbol{\beta}} - \frac{(\mathbf{X}^\top\mathbf{X})^{-1}\boldsymbol{x}_i e_i}{1 - p_i}.
\end{aligned}$$

It follows that the predicted value for the i-th observation is given by

$$\widehat{y}_{-i} = \boldsymbol{x}_i^\top\widehat{\boldsymbol{\beta}}_{-i} = \boldsymbol{x}_i^\top\widehat{\boldsymbol{\beta}} - \frac{\boldsymbol{x}_i^\top(\mathbf{X}^\top\mathbf{X})^{-1}\boldsymbol{x}_i e_i}{1 - p_i} = \widehat{y}_i - \frac{p_i e_i}{1 - p_i}.$$

Hence, $y_i - \widehat{y}_{-i} = e_i + p_i e_i/(1 - p_i) = e_i/(1 - p_i)$. □

☞ 26 ■ **Example 5.2 (Polynomial Regression (cont.))** We return to Example 2.1, where we estimated the generalization risk for various polynomial prediction functions using independent validation data. Instead, let us estimate the expected generalization risk via cross-validation (thus using only the training set) and apply Theorem 5.1 to compute the PRESS.
☞ 174

polyregpress.py

```
import numpy as np
import matplotlib.pyplot as plt

def generate_data(beta , sig, n):
    u = np.random.rand(n, 1)
    y = u ** np.arange(0, 4) @ beta.reshape(4,1) + (
                                sig * np.random.randn(n, 1))
    return u, y

np.random.seed(12)
beta = np.array([[10.0, -140, 400, -250]]).T;
sig=5; n = 10**2;
u,y = generate_data(beta,sig,n)

X = np.ones((n, 1))
K = 12 #maximum number of parameters
press = np.zeros(K+1)
for k in range(1,K):
    if k > 1:
        X = np.hstack((X, u**(k-1))) # add column to matrix
    P = X @ np.linalg.pinv(X) # projection matrix
    e = y - P @ y

    press[k] = np.sum((e/(1-np.diag(P).reshape(n,1)))**2)

plt.plot(press[1:K]/n)
```

The PRESS values divided by $n = 100$ for the constant, linear, quadratic, cubic, and quartic order polynomial regression models are, respectively, 152.487, 56.249, 51.606, 30.999, and 31.634. Hence, the cubic polynomial regression model has the lowest PRESS, indicating that it has the best predictive performance. ■

5.3.4 In-Sample Risk and Akaike Information Criterion

In Section 2.5.1 we introduced the *in-sample risk* as a measure for the accuracy of the prediction function. To recapitulate, given a fixed data set τ with associated response vector $\boldsymbol{y}$ and $n \times p$ matrix of explanatory variables $\mathbf{X}$, the in-sample risk of a prediction function g is defined as ☞ 35

$$\ell_{\text{in}}(g) := \mathbb{E}_{\mathbf{X}} \operatorname{Loss}(Y, g(\boldsymbol{X})), \tag{5.11}$$

where $\mathbb{E}_{\mathbf{X}}$ signifies that the expectation is taken under a different probability model, in which $\boldsymbol{X}$ takes the values $\boldsymbol{x}_1, \dots, \boldsymbol{x}_n$ with equal probability, and given $\boldsymbol{X} = \boldsymbol{x}_i$ the random variable Y is drawn from the conditional pdf $f(y \,|\, \boldsymbol{x}_i)$. The difference between the in-sample risk and the training loss is called the *optimism*. For the squared-error loss, Theorem 2.2 expresses the expected optimism of a learner $g_{\mathcal{T}}$ as two times the average covariance between the predicted values and the responses. ☞ 36

If the conditional variance of the error $Y - g^*(\boldsymbol{X})$ given $\boldsymbol{X} = \boldsymbol{x}$ does not depend on $\boldsymbol{x}$, then the expected in-sample risk of a learner g_τ, averaged over all training sets, has a simple expression:

Theorem 5.2: Expected In-Sample Risk for Linear Models

Let $\mathbf{X}$ be the model matrix for a linear model, of dimension $n \times p$. If $\mathbb{V}\text{ar}[Y - g^*(\boldsymbol{X}) \,|\, \boldsymbol{X} = \boldsymbol{x}] =: v^2$ does not depend on $\boldsymbol{x}$, then the expected in-sample risk (with respect to the squared-error loss) for a random learner $g_{\mathcal{T}}$ is given by

$$\mathbb{E}_{\mathbf{X}}\, \ell_{\text{in}}(g_{\mathcal{T}}) = \mathbb{E}_{\mathbf{X}}\, \ell_{\mathcal{T}}(g_{\mathcal{T}}) + \frac{2\ell^* p}{n}, \tag{5.12}$$

where ℓ^* is the irreducible risk.

Proof: The expected optimism is, by definition, $\mathbb{E}_{\mathbf{X}}[\ell_{\text{in}}(g_{\mathcal{T}}) - \ell_{\mathcal{T}}(g_{\mathcal{T}})]$ which, for the squared-error loss, is equal to $2\ell^* p/n$, using exactly the same reasoning as in Example 2.3. Note that here $\ell^* = v^2$. □

Equation (5.12) is the basis of the following model comparison heuristic: Estimate the irreducible risk $\ell^* = v^2$ via $\widehat{v^2}$, using a model with relatively high complexity. Then choose the linear model with the lowest value of

$$\|\boldsymbol{y} - \mathbf{X}\widehat{\boldsymbol{\beta}}\|^2 + 2\,\widehat{v^2}p. \tag{5.13}$$

☞ 122

We can also use the Akaike information criterion (AIC) as a heuristic for model comparison. We discussed the AIC in the unsupervised learning setting in Section 4.2, but the arguments used there can also be applied to the supervised case, under the in-sample model for the data. In particular, let $\boldsymbol{Z} = (\boldsymbol{X}, Y)$. We wish to predict the joint density

$$f(\boldsymbol{z}) = f(\boldsymbol{x}, y) := \frac{1}{n}\sum_{i=1}^{n} \mathbb{1}_{\{\boldsymbol{x}=\boldsymbol{x}_i\}}\, f(y \,|\, \boldsymbol{x}_i),$$

using a prediction function $g(\boldsymbol{z} \,|\, \boldsymbol{\theta})$ from a family $\mathcal{G} := \{g(\boldsymbol{z} \,|\, \boldsymbol{\theta}), \boldsymbol{\theta} \in \mathbb{R}^q\}$, where

$$g(\boldsymbol{z} \,|\, \boldsymbol{\theta}) = g(\boldsymbol{x}, y \,|\, \boldsymbol{\theta}) := \frac{1}{n}\sum_{i=1}^{n} \mathbb{1}_{\{\boldsymbol{x}=\boldsymbol{x}_i\}}\, g_i(y \,|\, \boldsymbol{\theta}).$$

Note that q is the number of parameters (typically larger than p for a linear model with a $n \times p$ design matrix).

Following Section 4.2, the in-sample cross-entropy risk in this case is

$$r(\boldsymbol{\theta}) := -\mathbb{E}_{\mathbf{X}} \ln g(\boldsymbol{Z} \,|\, \boldsymbol{\theta}),$$

and to approximate the optimal parameter $\boldsymbol{\theta}^*$ we minimize the corresponding training loss

$$r_{\tau_n}(\boldsymbol{\theta}) := -\frac{1}{n}\sum_{j=1}^{n} \ln g(\boldsymbol{z}_j \,|\, \boldsymbol{\theta}).$$

The optimal parameter $\widehat{\boldsymbol{\theta}}_n$ for the training loss is thus found by minimizing

$$-\frac{1}{n}\sum_{j=1}^{n}\Big(-\ln n + \ln g_j(y_j \,|\, \boldsymbol{\theta})\Big).$$

That is, it is the maximum likelihood estimate of $\boldsymbol{\theta}$:

$$\widehat{\boldsymbol{\theta}}_n = \underset{\boldsymbol{\theta}}{\operatorname{argmax}} \sum_{i=1}^{n} \ln g_i(y_i \,|\, \boldsymbol{\theta}).$$

Under the assumption that $f = g(\cdot \,|\, \boldsymbol{\theta}^*)$ for some parameter $\boldsymbol{\theta}^*$, we have from Theorem 4.1 that the estimated in-sample generalization risk can be approximated as ☞ 125

$$\mathbb{E}_{\mathbf{X}}\, r(\widehat{\boldsymbol{\theta}}_n) \approx r_{\mathcal{T}_n}(\widehat{\boldsymbol{\theta}}_n) + \frac{q}{n} = \ln n - \frac{1}{n} \sum_{j=1}^{n} \ln g_j(y_j \,|\, \widehat{\boldsymbol{\theta}}_n) + \frac{q}{n}.$$

This leads to the heuristic of selecting the learner $g(\cdot \,|\, \widehat{\boldsymbol{\theta}}_n)$ with the smallest value of the AIC:

$$-2 \sum_{i=1}^{n} \ln g_i(y_i \,|\, \widehat{\boldsymbol{\theta}}_n) + 2q. \tag{5.14}$$

■ **Example 5.3 (Normal Linear Model)** For the normal linear model $Y \sim \mathcal{N}(\boldsymbol{x}^\top \boldsymbol{\beta}, \sigma^2)$ (see (2.34)), with a p-dimensional vector $\boldsymbol{\beta}$, we have ☞ 46

$$g_i(y_i \,|\, \underbrace{\boldsymbol{\beta}, \sigma^2}_{=\boldsymbol{\theta}}) = \frac{1}{\sqrt{2\pi\sigma^2}} \exp\left(-\frac{1}{2} \frac{(y_i - \boldsymbol{x}_i^\top \boldsymbol{\beta})^2}{\sigma^2} \right), \quad i = 1, \ldots, n,$$

so that the AIC is

$$n \ln(2\pi) + n \ln \widehat{\sigma^2} + \frac{\|\boldsymbol{y} - \mathbf{X}\widehat{\boldsymbol{\beta}}\|^2}{\widehat{\sigma^2}} + 2q, \tag{5.15}$$

where $(\widehat{\boldsymbol{\beta}}, \widehat{\sigma^2})$ is the maximum likelihood estimate and $q = p+1$ is the number of parameters (including σ^2). For model comparison we may remove the $n \ln(2\pi)$ term if all the models are normal linear models. ■

⚠ Certain software packages report the AIC without the $n \ln \widehat{\sigma^2}$ term in (5.15). This may lead to sub-optimal model selection if normal models are compared with non-normal ones.

5.3.5 Categorical Features

Suppose that, as described in Chapter 1, the data is given in the form of a spreadsheet or data frame with n rows and $p + 1$ columns, where the first element of row i is the response variable y_i, and the remaining p elements form the vector of explanatory variables $\boldsymbol{x}_i^\top$. When all the explanatory variables (features, predictors) are *quantitative*, then the model matrix $\mathbf{X}$ can be directly read off from the data frame as the $n \times p$ matrix with rows $\boldsymbol{x}_i^\top, i = 1, \ldots, n$.

However, when some explanatory variables are *qualitative* (categorical), such a one-to-one correspondence between data frame and model matrix no longer holds. The solution is to include *indicator* or *dummy* variables.

FACTORIAL EXPERIMENTS

FACTORS

LEVELS

Linear models with continuous responses and categorical explanatory variables often arise in *factorial experiments*. These are controlled statistical experiments in which the aim is to assess how a response variable is affected by one or more *factors* tested at several *levels*. A typical example is an agricultural experiment where one wishes to investigate how the yield of a food crop depends on factors such as location, pesticide, and fertilizer.

■ **Example 5.4 (Crop Yield)** The data in Table 5.1 lists the yield of a food crop for four different crop treatments (e.g., strengths of fertilizer) on four different blocks (plots).

Table 5.1: Crop yield for different treatments and blocks.

	Treatment			
Block	1	2	3	4
1	9.2988	9.4978	9.7604	10.1025
2	8.2111	8.3387	8.5018	8.1942
3	9.0688	9.1284	9.3484	9.5086
4	8.2552	7.8999	8.4859	8.9485

The corresponding data frame, given in Table 5.2, has 16 rows and 3 columns: one column for the crop yield (the response variable), one column for the Treatment, with levels 1, 2, 3, 4, and one column for the Block, also with levels 1, 2, 3, 4. The values 1, 2, 3, and 4 have no quantitative meaning (it does not make sense to take their average, for example) — they merely identify the category of the treatment or block.

Table 5.2: Crop yield data organized as a data frame in standard format.

Yield	Treatment	Block
9.2988	1	1
8.2111	1	2
9.0688	1	3
8.2552	1	4
9.4978	2	1
8.3387	2	2
⋮	⋮	⋮
9.5086	4	3
8.9485	4	4

■

INDICATOR FEATURE

In general, suppose there are r factor (categorical) variables $u_1, \ldots, u_r$, where the j-th factor has p_j mutually exclusive levels, denoted by $1, \ldots, p_j$. In order to include these categorical variables in a linear model, a common approach is to introduce an *indicator feature* $x_{jk} = \mathbb{1}\{u_j = k\}$ for each factor j at level k. Thus, $x_{jk} = 1$ if the value of factor j is k and 0 otherwise. Since $\sum_k \mathbb{1}\{u_j = k\} = 1$, it suffices to consider only $p_j - 1$ of these indicator features for each factor j (this prevents the model matrix from being rank deficient). For a single response Y, the feature vector $\boldsymbol{x}^\top$ is thus a row vector of binary variables that indicates which levels were observed for each factor. The model assumption is that Y depends in a linear way on the indicator features, apart from an error term. That is,

$$Y = \beta_0 + \sum_{j=1}^{r} \sum_{k=2}^{p_j} \beta_{jk} \underbrace{\mathbb{1}\{u_j = k\}}_{x_{jk}} + \varepsilon,$$

where we have omitted one indicator feature (corresponding to level 1) for each factor j. For independent responses $Y_1, \ldots, Y_n$, where each Y_i corresponds to the factor values $u_{i1}, \ldots, u_{ir}$, let $x_{ijk} = \mathbb{1}\{u_{ij} = k\}$. Then, the linear model for the data becomes

$$Y_i = \beta_0 + \sum_{j=1}^{r} \sum_{k=2}^{p_j} \beta_{jk} x_{ijk} + \varepsilon_i, \tag{5.16}$$

where the $\{\varepsilon_i\}$ are independent with expectation 0 and some variance σ^2. By gathering the β_0 and $\{\beta_{jk}\}$ into a vector $\boldsymbol{\beta}$, and the $\{x_{ijk}\}$ into a matrix $\mathbf{X}$, we have again a linear model of the form (5.8). The model matrix $\mathbf{X}$ has n rows and $1 + \sum_{j=1}^{r}(p_j - 1)$ columns. Using the above convention that the β_{j1} parameters are subsumed in the parameter β_0 (corresponding to the "constant" feature), we can interpret β_0 as a baseline response when using the explanatory vector $\boldsymbol{x}^\top$ for which $x_{j1} = 1$ for all factors $j = 1, \ldots, r$. The other parameters $\{\beta_{jk}\}$ can be viewed as *incremental effects* relative to this baseline effect. For example, β_{12} describes by how much the response is expected to change if level 2 is used instead of level 1 for factor 1. INCREMENTAL EFFECTS

Example 5.5 (Crop Yield (cont.)) In Example 5.4, the linear model (5.16) has eight parameters: $\beta_0, \beta_{12}, \beta_{13}, \beta_{14}, \beta_{22}, \beta_{23}, \beta_{24}$, and σ^2. The model matrix $\mathbf{X}$ depends on how the crop yields are organized in a vector $\boldsymbol{y}$ and on the ordering of the factors. Let us order $\boldsymbol{y}$ column-wise from Table 5.1, as in $\boldsymbol{y} = [9.2988, 8.2111, 9.0688, 8.2552, 9.4978, \ldots, 8.9485]^\top$, and let Treatment be Factor 1 and Block be Factor 2. Then we can write (5.16) as

$$\boldsymbol{Y} = \underbrace{\begin{bmatrix} \mathbf{1} & \mathbf{0} & \mathbf{0} & \mathbf{0} & \mathbf{C} \\ \mathbf{1} & \mathbf{1} & \mathbf{0} & \mathbf{0} & \mathbf{C} \\ \mathbf{1} & \mathbf{0} & \mathbf{1} & \mathbf{0} & \mathbf{C} \\ \mathbf{1} & \mathbf{0} & \mathbf{0} & \mathbf{1} & \mathbf{C} \end{bmatrix}}_{\mathbf{X}} \underbrace{\begin{bmatrix} \beta_0 \\ \beta_{12} \\ \beta_{13} \\ \beta_{14} \\ \beta_{22} \\ \beta_{23} \\ \beta_{24} \end{bmatrix}}_{\boldsymbol{\beta}} + \boldsymbol{\varepsilon}, \quad \text{where} \quad \mathbf{C} = \begin{bmatrix} 0 & 0 & 0 \\ 1 & 0 & 0 \\ 0 & 1 & 0 \\ 0 & 0 & 1 \end{bmatrix},$$

and with $\mathbf{1} = [1, 1, 1, 1]^\top$ and $\mathbf{0} = [0, 0, 0, 0]^\top$. Estimation of $\boldsymbol{\beta}$ and σ^2, model selection, and prediction can now be carried out in the usual manner for linear models. ■

In the context of factorial experiments, the model matrix is often called the *design matrix*, as it specifies the design of the experiment; e.g., how many replications are taken for each combination of factor levels. The model (5.16) can be extended by adding products of indicator variables as new features. Such features are called *interaction* terms. DESIGN MATRIX INTERACTION

5.3.6 Nested Models

NESTED MODELS

Let $\mathbf{X}$ be a $n \times p$ model matrix of the form $\mathbf{X} = [\mathbf{X}_1, \mathbf{X}_2]$, where $\mathbf{X}_1$ and $\mathbf{X}_2$ are model matrices of dimension $n \times k$ and $n \times (p-k)$, respectively. The linear models $\boldsymbol{Y} = \mathbf{X}_1\boldsymbol{\beta}_1 + \boldsymbol{\varepsilon}$ and $\boldsymbol{Y} = \mathbf{X}_2\boldsymbol{\beta}_2 + \boldsymbol{\varepsilon}$ are said to be *nested within* the linear model $\boldsymbol{Y} = \mathbf{X}\boldsymbol{\beta} + \boldsymbol{\varepsilon}$. This simply means that certain features in $\mathbf{X}$ are ignored in each of the first two models. Note that $\boldsymbol{\beta}, \boldsymbol{\beta}_1$, and $\boldsymbol{\beta}_2$ are parameter vectors of dimension p, k, and $p-k$, respectively. In what follows, we assume that $n \geqslant p$ and that all model matrices are full-rank.

Suppose we wish to assess whether to use the full model matrix $\mathbf{X}$ or the reduced model matrix $\mathbf{X}_1$. Let $\widehat{\boldsymbol{\beta}}$ be the estimate of $\boldsymbol{\beta}$ under the full model (that is, obtained via (5.9)), and let $\widehat{\boldsymbol{\beta}_1}$ denote the estimate of $\boldsymbol{\beta}_1$ for the reduced model. Let $\boldsymbol{Y}^{(2)} = \mathbf{X}\widehat{\boldsymbol{\beta}}$ be the projection of $\boldsymbol{Y}$ onto the space $\mathrm{Span}(\mathbf{X})$ spanned by the columns of $\mathbf{X}$; and let $\boldsymbol{Y}^{(1)} = \mathbf{X}_1\widehat{\boldsymbol{\beta}_1}$ be the projection of $\boldsymbol{Y}$ onto the space $\mathrm{Span}(\mathbf{X}_1)$ spanned by the columns of $\mathbf{X}_1$ only; see Figure 5.3. In order to decide whether the features in $\mathbf{X}_2$ are needed, we may compare the estimated error terms of the two models, as calculated by (5.10); that is, by the residual sum of squares divided by the number of observations n. If the outcome of this comparison is that there is little difference between the model error for the full and reduced model, then it is appropriate to adopt the reduced model, as it has fewer parameters than the full model, while explaining the data just as well. The comparison is thus between the squared norms $\|\boldsymbol{Y} - \boldsymbol{Y}^{(2)}\|^2$ and $\|\boldsymbol{Y} - \boldsymbol{Y}^{(1)}\|^2$. Because of the nested nature of the linear models, $\mathrm{Span}(\mathbf{X}_1)$ is a subspace of $\mathrm{Span}(\mathbf{X})$ and, consequently, the orthogonal projection of $\boldsymbol{Y}^{(2)}$ onto $\mathrm{Span}(\mathbf{X}_1)$ is the same as the orthogonal projection of $\boldsymbol{Y}$ onto $\mathrm{Span}(\mathbf{X}_1)$; that is, $\boldsymbol{Y}^{(1)}$. By Pythagoras' theorem, we thus have the decomposition $\|\boldsymbol{Y}^{(2)} - \boldsymbol{Y}^{(1)}\|^2 + \|\boldsymbol{Y} - \boldsymbol{Y}^{(2)}\|^2 = \|\boldsymbol{Y} - \boldsymbol{Y}^{(1)}\|^2$. This is also illustrated in Figure 5.3.

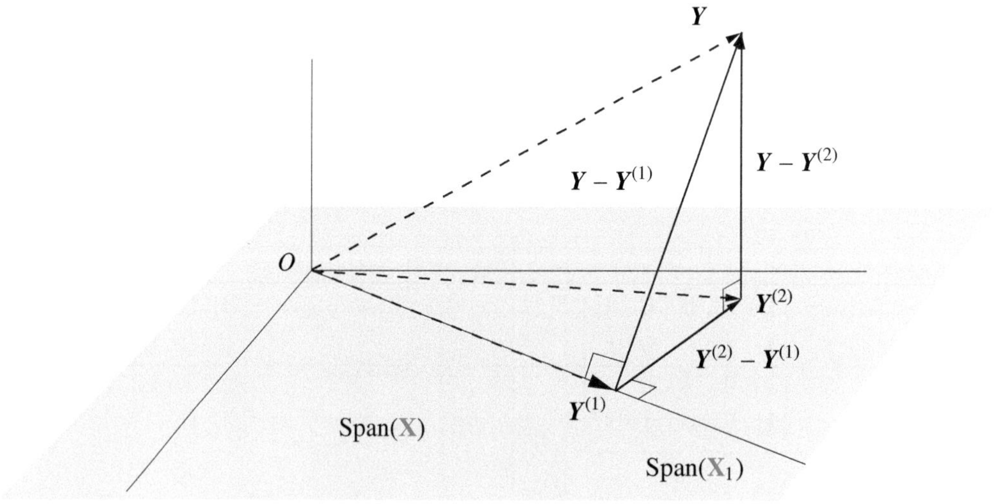

Figure 5.3: The residual sum of squares for the full model corresponds to $\|\boldsymbol{Y} - \boldsymbol{Y}^{(2)}\|^2$ and for the reduced model it is $\|\boldsymbol{Y} - \boldsymbol{Y}^{(1)}\|^2$. By Pythagoras's theorem, the difference is $\|\boldsymbol{Y}^{(2)} - \boldsymbol{Y}^{(1)}\|^2$.

The above decomposition can be generalized to more than two model matrices. Suppose that the model matrix can be decomposed into d submatrices: $\mathbf{X} = [\mathbf{X}_1, \mathbf{X}_2, \ldots, \mathbf{X}_d]$, where the matrix $\mathbf{X}_i$ has p_i columns and n rows, $i = 1, \ldots, d$. Thus, the number of columns[2]

[2]As always, we assume the columns are linearly independent.

in the full model matrix is $p = p_1 + \cdots + p_d$. This creates an increasing sequence of "nested" model matrices: $\mathbf{X}_1, [\mathbf{X}_1, \mathbf{X}_2], \ldots, [\mathbf{X}_1, \mathbf{X}_2, \ldots, \mathbf{X}_d]$, from (say) the baseline normal model matrix $\mathbf{X}_1 = \mathbf{1}$ to the full model matrix $\mathbf{X}$. Think of each model matrix corresponding to specific variables in the model.

We follow a similar projection procedure as in Figure 5.3: First project $\boldsymbol{Y}$ onto Span($\mathbf{X}$) to yield the vector $\boldsymbol{Y}^{(d)}$, then project $\boldsymbol{Y}^{(d)}$ onto Span($[\mathbf{X}_1, \ldots, \mathbf{X}_{d-1}]$) to obtain $\boldsymbol{Y}^{(d-1)}$, and so on, until $\boldsymbol{Y}^{(2)}$ is projected onto Span($\mathbf{X}_1$) to yield $\boldsymbol{Y}^{(1)} = \overline{Y}\mathbf{1}$ (in the case that $\mathbf{X}_1 = \mathbf{1}$).

By applying Pythagoras' theorem, the total sum of squares can be decomposed as

$$\underbrace{\|\boldsymbol{Y} - \boldsymbol{Y}^{(1)}\|^2}_{\text{df}=n-p_1} = \underbrace{\|\boldsymbol{Y} - \boldsymbol{Y}^{(d)}\|^2}_{\text{df}=n-p} + \underbrace{\|\boldsymbol{Y}^{(d)} - \boldsymbol{Y}^{(d-1)}\|^2}_{\text{df}=p_d} + \cdots + \underbrace{\|\boldsymbol{Y}^{(2)} - \boldsymbol{Y}^{(1)}\|^2}_{\text{df}=p_2}. \tag{5.17}$$

Software packages typically report the sums of squares as well as the corresponding *degrees of freedom* (df): $n - p, p_d, \ldots, p_2$.

DEGREES OF FREEDOM

5.3.7 Coefficient of Determination

To assess how a linear model $\boldsymbol{Y} = \mathbf{X}\boldsymbol{\beta} + \boldsymbol{\varepsilon}$ compares to the default model $\boldsymbol{Y} = \beta_0\mathbf{1} + \boldsymbol{\varepsilon}$, we can compare the variance of the original data, estimated via $\sum_i (Y_i - \overline{Y})^2/n = \|\boldsymbol{Y} - \overline{Y}\mathbf{1}\|^2/n$, with the variance of the fitted data; estimated via $\sum_i (\widehat{Y}_i - \overline{Y})^2/n = \|\widehat{\boldsymbol{Y}} - \overline{Y}\mathbf{1}\|^2/n$, where $\widehat{\boldsymbol{Y}} = \mathbf{X}\widehat{\boldsymbol{\beta}}$. The sum $\sum_i (Y_i - \overline{Y})^2/n = \|\boldsymbol{Y} - \overline{Y}\mathbf{1}\|^2$ is sometimes called the *total sum of squares* (TSS), and the quantity

TOTAL SUM OF SQUARES

$$R^2 = \frac{\|\widehat{\boldsymbol{Y}} - \overline{Y}\mathbf{1}\|^2}{\|\boldsymbol{Y} - \overline{Y}\mathbf{1}\|^2} \tag{5.18}$$

is called the *coefficient of determination* of the linear model. In the notation of Figure 5.3, $\widehat{\boldsymbol{Y}} = \boldsymbol{Y}^{(2)}$ and $\overline{Y}\mathbf{1} = \boldsymbol{Y}^{(1)}$, so that

COEFFICIENT OF DETERMINATION

$$R^2 = \frac{\|\boldsymbol{Y}^{(2)} - \boldsymbol{Y}^{(1)}\|^2}{\|\boldsymbol{Y} - \boldsymbol{Y}^{(1)}\|^2} = \frac{\|\boldsymbol{Y} - \boldsymbol{Y}^{(1)}\|^2 - \|\boldsymbol{Y} - \boldsymbol{Y}^{(2)}\|^2}{\|\boldsymbol{Y} - \boldsymbol{Y}^{(1)}\|^2} = \frac{\text{TSS} - \text{RSS}}{\text{TSS}}.$$

Note that R^2 lies between 0 and 1. An R^2 value close to 1 indicates that a large proportion of the variance in the data has been explained by the model.

Many software packages also give the *adjusted coefficient of determination*, or simply the adjusted R^2, defined by

ADJUSTED COEFFICIENT OF DETERMINATION

$$R^2_{\text{adjusted}} = 1 - (1 - R^2)\frac{n-1}{n-p}.$$

The regular R^2 is always *non-decreasing* in the number of parameters (see Exercise 15), but this may not indicate better predictive power. The adjusted R^2 compensates for this increase by decreasing the regular R^2 as the number of variables increases. This heuristic adjustment can make it easier to compare the quality of two competing models.

5.4 Inference for Normal Linear Models

So far we have not assumed any distribution for the random vector of errors $\boldsymbol{\varepsilon} = [\varepsilon_1, \ldots, \varepsilon_n]^\top$ in a linear model $\boldsymbol{Y} = \mathbf{X}\boldsymbol{\beta} + \boldsymbol{\varepsilon}$. When the error terms $\{\varepsilon_i\}$ are assumed to be normally distributed (that is, $\{\varepsilon_i\} \sim_{\text{iid}} \mathcal{N}(0, \sigma^2)$), whole new avenues open up for inference ☞ 46 on linear models. In Section 2.8 we already saw that for such *normal linear models*, estimation of $\boldsymbol{\beta}$ and σ^2 can be carried out via maximum likelihood methods, yielding the same estimators from (5.9) and (5.10).

The following theorem lists the properties of these estimators. In particular, it shows that $\widehat{\boldsymbol{\beta}}$ and $\widehat{\sigma^2} n/(n-p)$ are independent and unbiased estimators of $\boldsymbol{\beta}$ and σ^2, respectively.

Theorem 5.3: Properties of the Estimators for a Normal Linear Model

Consider the linear model $\boldsymbol{Y} = \mathbf{X}\boldsymbol{\beta} + \boldsymbol{\varepsilon}$, with $\boldsymbol{\varepsilon} \sim \mathcal{N}(\mathbf{0}, \sigma^2 \mathbf{I}_n)$, where $\boldsymbol{\beta}$ is a p-dimensional vector of parameters and σ^2 a dispersion parameter. The following results hold.

1. The maximum likelihood estimators $\widehat{\boldsymbol{\beta}}$ and $\widehat{\sigma^2}$ are independent.
2. $\widehat{\boldsymbol{\beta}} \sim \mathcal{N}(\boldsymbol{\beta}, \sigma^2(\mathbf{X}^\top\mathbf{X})^+)$.
3. $n\widehat{\sigma^2}/\sigma^2 \sim \chi^2_{n-p}$, where $p = \text{rank}(\mathbf{X})$.

☞ 362 *Proof:* Using the pseudo-inverse (Definition A.2), we can write the random vector $\widehat{\boldsymbol{\beta}}$ as $\mathbf{X}^+\boldsymbol{Y}$, which is a linear transformation of a normal random vector. Consequently, $\widehat{\boldsymbol{\beta}}$ has a ☞ 437 multivariate normal distribution; see Theorem C.6. The mean vector and covariance matrix follow from the same theorem:

$$\mathbb{E}\widehat{\boldsymbol{\beta}} = \mathbf{X}^+\, \mathbb{E}\boldsymbol{Y} = \mathbf{X}^+\mathbf{X}\boldsymbol{\beta} = \boldsymbol{\beta}$$

and

$$\mathbb{C}\text{ov}(\widehat{\boldsymbol{\beta}}) = \mathbf{X}^+\sigma^2\mathbf{I}_n(\mathbf{X}^+)^\top = \sigma^2(\mathbf{X}^\top\mathbf{X})^+.$$

To show that $\widehat{\boldsymbol{\beta}}$ and $\widehat{\sigma^2}$ are independent, define $\boldsymbol{Y}^{(2)} = \mathbf{X}\widehat{\boldsymbol{\beta}}$. Note that $\boldsymbol{Y}/\sigma$ has a $\mathcal{N}(\boldsymbol{\mu}, \mathbf{I}_n)$ distribution, with expectation vector $\boldsymbol{\mu} = \mathbf{X}\boldsymbol{\beta}/\sigma$. A direct application of Theorem C.10 ☞ 440 now shows that $(\boldsymbol{Y} - \boldsymbol{Y}^{(2)})/\sigma$ is independent of $\boldsymbol{Y}^{(2)}/\sigma$. Since $\widehat{\boldsymbol{\beta}} = \mathbf{X}^+\mathbf{X}\widehat{\boldsymbol{\beta}} = \mathbf{X}^+\boldsymbol{Y}^{(2)}$ and $\widehat{\sigma^2} = \|\boldsymbol{Y} - \boldsymbol{Y}^{(2)}\|^2/n$, it follows that $\widehat{\sigma^2}$ is independent of $\widehat{\boldsymbol{\beta}}$. Finally, by the same theorem, the random variable $\|\boldsymbol{Y} - \boldsymbol{Y}^{(2)}\|^2/\sigma^2$ has a χ^2_{n-p} distribution, as $\boldsymbol{Y}^{(2)}$ has the same expectation vector as $\boldsymbol{Y}$. □

As a corollary, we see that each estimator $\widehat{\beta_i}$ of β_i has a normal distribution with expectation β_i and variance $\sigma^2\boldsymbol{u}_i^\top\mathbf{X}^+(\mathbf{X}^+)^\top\boldsymbol{u}_i = \sigma^2\|\boldsymbol{u}_i^\top\mathbf{X}^+\|^2$, where $\boldsymbol{u}_i = [0, \ldots, 0, 1, 0, \ldots, 0]^\top$ is the i-th unit vector; in other words, the variance is $\sigma^2[(\mathbf{X}^\top\mathbf{X})^+]_{ii}$.

It is of interest to test whether certain regression parameters β_i are 0 or not, since if $\beta_i = 0$, the i-th explanatory variable has no direct effect on the expected response and so could be removed from the model. A standard procedure is to conduct a hypothesis test ☞ 460 (see Section C.14 for a review of hypothesis testing) to test the null hypothesis $H_0 : \beta_i = 0$

against the alternative $H_1 : \beta_i \neq 0$, using the test statistic

$$T = \frac{\widehat{\beta}_i / \|\boldsymbol{u}_i^\top \mathbf{X}^+\|}{\sqrt{\mathrm{RSE}}}, \tag{5.19}$$

where RSE is the residual squared error; that is $\mathrm{RSE} = \mathrm{RSS}/(n-p)$. This test statistic has a t_{n-p} distribution under H_0. To see this, write $T = Z/\sqrt{V/(n-p)}$, with

$$Z = \frac{\widehat{\beta}_i}{\sigma \|\boldsymbol{u}_i^\top \mathbf{X}^+\|} \quad \text{and} \quad V = n\,\widehat{\sigma^2}/\sigma^2.$$

Then, by Theorem 5.3, $Z \sim \mathcal{N}(0,1)$ under H_0, $V \sim \chi^2_{n-p}$, and Z and V are independent. The result now follows directly from Corollary C.1. ☞ 441

5.4.1 Comparing Two Normal Linear Models

Suppose we have the following normal linear model for data $\boldsymbol{Y} = [Y_1, \ldots, Y_n]^\top$:

$$\boldsymbol{Y} = \underbrace{\mathbf{X}_1\boldsymbol{\beta}_1 + \mathbf{X}_2\boldsymbol{\beta}_2}_{\mathbf{X}\boldsymbol{\beta}} + \boldsymbol{\varepsilon}, \quad \boldsymbol{\varepsilon} \sim \mathcal{N}(\mathbf{0}, \sigma^2 \mathbf{I}_n), \tag{5.20}$$

where $\boldsymbol{\beta}_1$ and $\boldsymbol{\beta}_2$ are unknown vectors of dimension k and $p-k$, respectively; and $\mathbf{X}_1$ and $\mathbf{X}_2$ are full-rank model matrices of dimensions $n \times k$ and $n \times (p-k)$, respectively. Above we implicitly defined $\mathbf{X} = [\mathbf{X}_1, \mathbf{X}_2]$ and $\boldsymbol{\beta}^\top = [\boldsymbol{\beta}_1^\top, \boldsymbol{\beta}_2^\top]$. Suppose we wish to test the hypothesis $H_0 : \boldsymbol{\beta}_2 = \mathbf{0}$ against $H_1 : \boldsymbol{\beta}_2 \neq \mathbf{0}$. Following Section 5.3.6, the idea is to compare the residual sum of squares for both models, expressed as $\|\boldsymbol{Y} - \boldsymbol{Y}^{(2)}\|^2$ and $\|\boldsymbol{Y} - \boldsymbol{Y}^{(1)}\|^2$. Using Pythagoras' theorem we saw that $\|\boldsymbol{Y} - \boldsymbol{Y}^{(2)}\|^2 - \|\boldsymbol{Y} - \boldsymbol{Y}^{(1)}\|^2 = \|\boldsymbol{Y}^{(2)} - \boldsymbol{Y}^{(1)}\|^2$, and so it makes sense to base the decision whether to retain or reject H_0 on the basis of the quotient of $\|\boldsymbol{Y}^{(2)} - \boldsymbol{Y}^{(1)}\|^2$ and $\|\boldsymbol{Y} - \boldsymbol{Y}^{(2)}\|^2$. This leads to the following test statistics.

Theorem 5.4: Test Statistic for Comparing Two Normal Linear Models

For the model (5.20), let $\boldsymbol{Y}^{(2)}$ and $\boldsymbol{Y}^{(1)}$ be the projections of $\boldsymbol{Y}$ onto the space spanned by the p columns of $\mathbf{X}$ and the k columns of $\mathbf{X}_1$, respectively. Then under $H_0 : \boldsymbol{\beta}_2 = \mathbf{0}$ the test statistic

$$T = \frac{\|\boldsymbol{Y}^{(2)} - \boldsymbol{Y}^{(1)}\|^2/(p-k)}{\|\boldsymbol{Y} - \boldsymbol{Y}^{(2)}\|^2/(n-p)} \tag{5.21}$$

has an $\mathsf{F}(p-k, n-p)$ distribution.

Proof: Define $\boldsymbol{X} := \boldsymbol{Y}/\sigma$ with expectation $\boldsymbol{\mu} := \mathbf{X}\boldsymbol{\beta}/\sigma$, and $\boldsymbol{X}_j := \boldsymbol{Y}^{(j)}/\sigma$ with expectation $\boldsymbol{\mu}_j$, $j = k, p$. Note that $\boldsymbol{\mu}_p = \boldsymbol{\mu}$ and, under H_0, $\boldsymbol{\mu}_k = \boldsymbol{\mu}_p$. We can directly apply Theorem C.10 to find that $\|\boldsymbol{Y} - \boldsymbol{Y}^{(2)}\|^2/\sigma^2 = \|\boldsymbol{X} - \boldsymbol{X}_p\|^2 \sim \chi^2_{n-p}$ and, under H_0, $\|\boldsymbol{Y}^{(2)} - \boldsymbol{Y}^{(1)}\|^2/\sigma^2 = \|\boldsymbol{X}_p - \boldsymbol{X}_k\|^2 \sim \chi^2_{p-k}$. Moreover, these random variables are independent of each other. The proof is completed by applying Theorem C.11. □ ☞ 440

Note that H_0 is rejected for large values of T. The testing procedure thus proceeds as follows:

1. Compute the outcome, t say, of the test statistic T in (5.21).
2. Evaluate the P-value $\mathbb{P}(T \geqslant t)$, with $T \sim \mathsf{F}(p-k, n-p)$.
3. Reject H_0 if this P-value is too small, say less than 0.05.

For nested models $[\mathbf{X}_1, \mathbf{X}_2, \ldots, \mathbf{X}_i]$, $i = 1, 2, \ldots, d$, as in Section 5.3.6, the F test statistic
☞ 183
in Theorem 5.4 can now be used to test whether certain $\mathbf{X}_i$ are needed or not. In particular, software packages will report the outcomes of

$$F_i = \frac{\|\boldsymbol{Y}^{(i)} - \boldsymbol{Y}^{(i-1)}\|^2 / p_i}{\|\boldsymbol{Y} - \boldsymbol{Y}^{(d)}\|^2 / (n-p)}, \tag{5.22}$$

in the order $i = 2, 3, \ldots, d$. Under the null hypothesis that $\boldsymbol{Y}^{(i)}$ and $\boldsymbol{Y}^{(i-1)}$ have the same expectation (that is, adding $\mathbf{X}_i$ to $\mathbf{X}_{i-1}$ has no additional effect on reducing the approximation error), the test statistic F_i has an $\mathsf{F}(p_i, n-p)$ distribution, and the corresponding P-values quantify the strength of the decision to include an additional variable in the model or not. This procedure is called *analysis of variance* (ANOVA).

ANALYSIS OF VARIANCE

⚠ Note that the output of an ANOVA table depends on the order in which the variables are considered.

■ **Example 5.6 (Crop Yield (cont.))** We continue Examples 5.4 and 5.5. Decompose the linear model as

$$\boldsymbol{Y} = \underbrace{\begin{bmatrix} \mathbf{1} \\ \mathbf{1} \\ \mathbf{1} \\ \mathbf{1} \end{bmatrix}}_{\mathbf{X}_1} \underbrace{\beta_0}_{\boldsymbol{\beta}_1} + \underbrace{\begin{bmatrix} \mathbf{0} & \mathbf{0} & \mathbf{0} \\ \mathbf{1} & \mathbf{0} & \mathbf{0} \\ \mathbf{0} & \mathbf{1} & \mathbf{0} \\ \mathbf{0} & \mathbf{0} & \mathbf{1} \end{bmatrix}}_{\mathbf{X}_2} \underbrace{\begin{bmatrix} \beta_{12} \\ \beta_{13} \\ \beta_{14} \end{bmatrix}}_{\boldsymbol{\beta}_2} + \underbrace{\begin{bmatrix} \mathbf{C} \\ \mathbf{C} \\ \mathbf{C} \\ \mathbf{C} \end{bmatrix}}_{\mathbf{X}_3} \underbrace{\begin{bmatrix} \beta_{22} \\ \beta_{23} \\ \beta_{24} \end{bmatrix}}_{\boldsymbol{\beta}_3} + \boldsymbol{\varepsilon}.$$

Is the crop yield dependent on treatment levels as well as blocks? We first test whether we can remove Block as a factor in the model against it playing a significant role in explaining the crop yields. Specifically, we test $\boldsymbol{\beta}_3 = \mathbf{0}$ versus $\boldsymbol{\beta}_3 \neq \mathbf{0}$ using Theorem 5.4. Now the vector $\boldsymbol{Y}^{(2)}$ is the projection of $\boldsymbol{Y}$ onto the $(p = 7)$-dimensional space spanned by the columns of $\mathbf{X} = [\mathbf{X}_1, \mathbf{X}_2, \mathbf{X}_3]$; and $\boldsymbol{Y}^{(1)}$ is the projection of $\boldsymbol{Y}$ onto the $(k = 4)$-dimensional space spanned by the columns of $\mathbf{X}_{12} := [\mathbf{X}_1, \mathbf{X}_2]$. The test statistic, T_{12} say, under H_0 has an $\mathsf{F}(3, 9)$ distribution.

The Python code below calculates the outcome of the test statistic T_{12} and the corresponding P-value. We find $t_{12} = 34.9998$, which gives a P-value 2.73×10^{-5}. This shows that the block effects are extremely important for explaining the data.

Using the extended model (including the block effects), we can test whether $\boldsymbol{\beta}_2 = \mathbf{0}$ or not; that is, whether the treatments have a significant effect on the crop yield in the presence of the Block factor. This is done in the last six lines of the code below. The outcome of the test statistic is 4.4878, with a P-value of 0.0346. By including the block effects, we effectively reduce the uncertainty in the model and are able to more accurately assess the effects of the treatments, to conclude that the treatment seems to have an effect on the crop yield. A closer look at the data shows that within each block (row) the crop yield roughly increases with the treatment level.

crop.py

```
import numpy as np
from scipy.stats import f
from numpy.linalg import lstsq, norm

yy = np.array([9.2988, 9.4978, 9.7604, 10.1025,
      8.2111, 8.3387, 8.5018,  8.1942,
      9.0688, 9.1284, 9.3484,  9.5086,
      8.2552, 7.8999, 8.4859,  8.9485]).reshape(4,4).T

nrow, ncol = yy.shape[0], yy.shape[1]
n = nrow * ncol
y = yy.reshape(16,)
X_1 = np.ones((n,1))

KM = np.kron(np.eye(ncol),np.ones((nrow,1)))
KM[:,0]
X_2 = KM[:,1:ncol]
IM = np.eye(nrow)
C = IM[:,1:nrow]

X_3 = np.vstack((C, C))
X_3 = np.vstack((X_3, C))
X_3 = np.vstack((X_3, C))

X = np.hstack((X_1,X_2))
X = np.hstack((X,X_3))

p = X.shape[1] #number of parameters in full model
betahat = lstsq(X, y,rcond=None)[0]  #estimate under the full model

ym = X @ betahat

X_12 = np.hstack((X_1, X_2)) #omitting the block effect
k = X_12.shape[1] #number of parameters in reduced model
betahat_12 = lstsq(X_12, y,rcond=None)[0]
y_12 = X_12 @ betahat_12
T_12=(n-p)/(p-k)*(norm(y-y_12)**2 -  norm(y-ym)**2)/norm(y-ym)**2
pval_12 = 1 - f.cdf(T_12,p-k,n-p)

X_13 = np.hstack((X_1, X_3)) #omitting the treatment effect
k = X_13.shape[1] #number of parameters in reduced model
betahat_13 = lstsq(X_13, y,rcond=None)[0]
y_13 = X_13 @ betahat_13
T_13=(n-p)/(p-k)*(norm(y-y_13)**2 - norm(y-ym)**2)/norm(y-ym)**2
pval_13 = 1 - f.cdf(T_13,p-k,n-p)
```

■

5.4.2 Confidence and Prediction Intervals

As in all supervised learning settings, linear regression is most useful when we wish to predict how a new response variable will behave on the basis of a new explanatory vector $\boldsymbol{x}$. For example, it may be difficult to measure the response variable, but by knowing the estimated regression line and the value for $\boldsymbol{x}$, we will have a reasonably good idea what Y or the expected value of Y is going to be.

Thus, consider a new $\boldsymbol{x}$ and let $Y \sim \mathcal{N}(\boldsymbol{x}^\top\boldsymbol{\beta}, \sigma^2)$, with $\boldsymbol{\beta}$ and σ^2 unknown. First we are going to look at the *expected* value of Y, that is $\mathbb{E}Y = \boldsymbol{x}^\top\boldsymbol{\beta}$. Since $\boldsymbol{\beta}$ is unknown, we do not know $\mathbb{E}Y$ either. However, we can estimate it via the estimator $\widehat{Y} = \boldsymbol{x}^\top\widehat{\boldsymbol{\beta}}$, where $\widehat{\boldsymbol{\beta}} \sim \mathcal{N}(\boldsymbol{\beta}, \sigma^2(\mathbf{X}^\top\mathbf{X})^+)$, by Theorem 5.3. Being linear in the components of $\boldsymbol{\beta}$, $\widehat{Y}$ therefore has a normal distribution with expectation $\boldsymbol{x}^\top\boldsymbol{\beta}$ and variance $\sigma^2\|\boldsymbol{x}^\top\mathbf{X}^+\|^2$. Let $Z \sim \mathcal{N}(0, 1)$ be the standardized version of $\widehat{Y}$ and $V = \|\boldsymbol{Y} - \mathbf{X}\widehat{\boldsymbol{\beta}}\|^2/\sigma^2 \sim \chi^2_{n-p}$. Then the random variable

$$T := \frac{(\boldsymbol{x}^\top\widehat{\boldsymbol{\beta}} - \boldsymbol{x}^\top\boldsymbol{\beta}) \,/\, \|\boldsymbol{x}^\top\mathbf{X}^+\|}{\|\boldsymbol{Y} - \mathbf{X}\widehat{\boldsymbol{\beta}}\| \,/\, \sqrt{(n-p)}} = \frac{Z}{\sqrt{V/(n-p)}} \tag{5.23}$$

☞ 441

CONFIDENCE INTERVAL

has, by Corollary C.1, a t_{n-p} distribution. After rearranging the identity $\mathbb{P}(|T| \leqslant t_{n-p;1-\alpha/2}) = 1 - \alpha$, where $t_{n-p;1-\alpha/2}$ is the $(1 - \alpha/2)$ quantile of the t_{n-p} distribution, we arrive at the stochastic *confidence interval*

$$\boldsymbol{x}^\top\widehat{\boldsymbol{\beta}} \pm t_{n-p;1-\alpha/2}\sqrt{\text{RSE}}\,\|\boldsymbol{x}^\top\mathbf{X}^+\|, \tag{5.24}$$

where we have identified $\|\boldsymbol{Y} - \mathbf{X}\widehat{\boldsymbol{\beta}}\|^2/(n-p)$ with RSE. This confidence interval quantifies the uncertainty in the learner (regression surface).

PREDICTION INTERVAL

A *prediction interval* for a new response Y is different from a confidence interval for $\mathbb{E}Y$. Here the idea is to construct an interval such that Y lies in this interval with a certain guaranteed probability. Note that now we have *two* sources of variation:

1. $Y \sim \mathcal{N}(\boldsymbol{x}^\top\boldsymbol{\beta}, \sigma^2)$ itself is a random variable.

2. Estimating $\boldsymbol{x}^\top\boldsymbol{\beta}$ via $\widehat{Y}$ brings another source of variation.

We can construct a $(1 - \alpha)$ prediction interval, by finding two random bounds such that the random variable Y lies between these bounds with probability $1 - \alpha$. We can reason as follows. Firstly, note that $Y \sim \mathcal{N}(\boldsymbol{x}^\top\boldsymbol{\beta}, \sigma^2)$ and $\widehat{Y} \sim \mathcal{N}(\boldsymbol{x}^\top\boldsymbol{\beta}, \sigma^2\|\boldsymbol{x}^\top\mathbf{X}^+\|^2)$ are independent. It follows that $Y - \widehat{Y}$ has a normal distribution with expectation 0 and variance

$$\sigma^2(1 + \|\boldsymbol{x}^\top\mathbf{X}^+\|^2). \tag{5.25}$$

Secondly, letting $Z \sim \mathcal{N}(0, 1)$ be the standardized version of $Y - \widehat{Y}$, and repeating the steps used for the construction of the confidence interval (5.24), we arrive at the prediction interval

$$\boldsymbol{x}^\top\widehat{\boldsymbol{\beta}} \pm t_{n-p;1-\alpha/2}\,\sqrt{\text{RSE}}\,\sqrt{1 + \|\boldsymbol{x}^\top\mathbf{X}^+\|^2}. \tag{5.26}$$

This prediction interval captures the uncertainty from an as-yet-unobserved response as well as the uncertainty in the parameters of the regression model itself.

■ **Example 5.7 (Confidence Limits in Simple Linear Regression)** The following program draws $n = 100$ samples from a simple linear regression model with parameters $\boldsymbol{\beta} = [6, 13]^\top$ and $\sigma = 2$, where the x-coordinates are evenly spaced on the interval $[0, 1]$. The parameters are estimated in the third block of the code. Estimates for $\boldsymbol{\beta}$ and σ are $[6.03, 13.09]^\top$ and $\widehat{\sigma} = 1.60$, respectively. The program then proceeds by calculating the 95% numeric confidence and prediction intervals for various values of the explanatory variable. Figure 5.4 shows the results.

confpred.py

```
import numpy as np
import matplotlib.pyplot as plt
from scipy.stats import t
from  numpy.linalg import inv, lstsq, norm
np.random.seed(123)

n = 100
x = np.linspace(0.01,1,100).reshape(n,1)
# parameters
beta = np.array([6,13])
sigma = 2
Xmat = np.hstack((np.ones((n,1)), x)) #design matrix
y = Xmat @ beta + sigma*np.random.randn(n)

# solve the normal equations
betahat = lstsq(Xmat, y,rcond=None)[0]
# estimate for sigma
sqMSE = norm(y - Xmat @ betahat)/np.sqrt(n-2)

tquant = t.ppf(0.975,n-2) # 0.975 quantile
ucl = np.zeros(n) #upper conf. limits
lcl = np.zeros(n) #lower conf. limits
upl = np.zeros(n)
lpl = np.zeros(n)
rl = np.zeros(n)  # (true) regression line
u = 0

for i in range(n):
    u = u + 1/n;
    xvec = np.array([1,u])
    sqc = np.sqrt(xvec.T @ inv(Xmat.T @ Xmat) @ xvec)
    sqp = np.sqrt(1 + xvec.T @ inv(Xmat.T @ Xmat) @ xvec)
    rl[i] = xvec.T @ beta;
    ucl[i] = xvec.T @ betahat + tquant*sqMSE*sqc;
    lcl[i] = xvec.T @ betahat - tquant*sqMSE*sqc;
    upl[i] = xvec.T @ betahat + tquant*sqMSE*sqp;
    lpl[i] = xvec.T @ betahat - tquant*sqMSE*sqp;

plt.plot(x,y, '.')
plt.plot(x,rl,'b')
plt.plot(x,ucl,'k:')
plt.plot(x,lcl,'k:')
plt.plot(x,upl,'r--')
plt.plot(x,lpl,'r--')
```

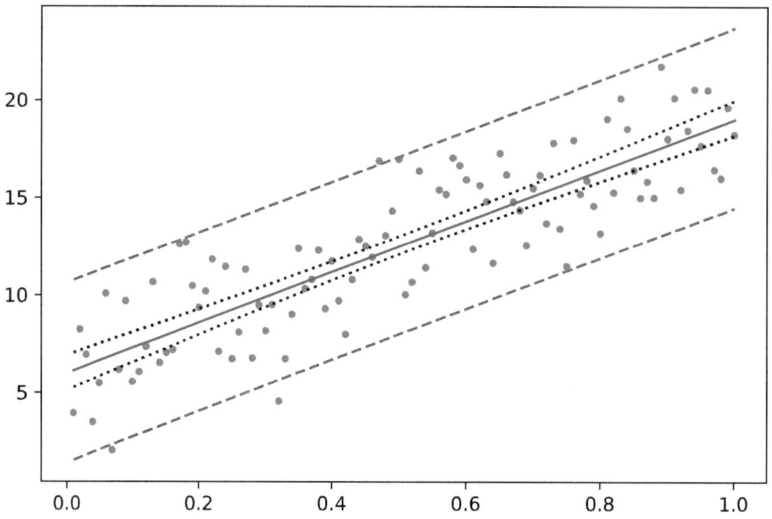

Figure 5.4: The true regression line (blue, solid) and the upper and lower 95% prediction curves (red, dashed) and confidence curves (dotted). ■

5.5 Nonlinear Regression Models

So far we have been mostly dealing with linear regression models, in which the prediction function is of the form $g(\boldsymbol{x} \mid \boldsymbol{\beta}) = \boldsymbol{x}^\top \boldsymbol{\beta}$. In this section we discuss some strategies for handling general prediction functions $g(\boldsymbol{x} \mid \boldsymbol{\beta})$, where the functional form is known up to an unknown parameter vector $\boldsymbol{\beta}$. So the regression model becomes

$$Y_i = g(\boldsymbol{x}_i \mid \boldsymbol{\beta}) + \varepsilon_i, \quad i = 1, \ldots, n, \tag{5.27}$$

where $\varepsilon_1, \ldots, \varepsilon_n$ are independent with expectation 0 and unknown variance σ^2. The model can be further specified by assuming that the error terms have a normal distribution.

Table 5.3 gives some common examples of nonlinear prediction functions for data taking values in $\mathbb{R}$.

Table 5.3: Common nonlinear prediction functions for one-dimensional data.

Name	$g(x \mid \boldsymbol{\beta})$	$\boldsymbol{\beta}$
Exponential	$a\,\mathrm{e}^{bx}$	a, b
Power law	$a\,x^b$	a, b
Logistic	$(1 + \mathrm{e}^{a+bx})^{-1}$	a, b
Weibull	$1 - \exp(-x^b/a)$	a, b
Polynomial	$\sum_{k=0}^{p-1} \beta_k x^k$	p, $\{\beta_k\}_{k=0}^{p-1}$

The logistic and polynomial prediction functions in Table 5.3 can be readily generalized to higher dimensions. For example, for $\boldsymbol{x} \in \mathbb{R}^2$ a general second-order polynomial prediction function is of the form

$$g(\boldsymbol{x} \mid \boldsymbol{\beta}) = \beta_0 + \beta_1 x_1 + \beta_2 x_2 + \beta_{11} x_1^2 + \beta_{22} x_2^2 + \beta_{12} x_1 x_2. \tag{5.28}$$

This function can be viewed as a second-order approximation to a general smooth prediction function $g(x_1, x_2)$; see also Exercise 4. Polynomial regression models are also called *response surface* models. The generalization of the above logistic prediction to $\mathbb{R}^d$ is

RESPONSE SURFACE MODEL

$$g(\boldsymbol{x} \mid \boldsymbol{\beta}) = (1 + \mathrm{e}^{-\boldsymbol{x}^\top \boldsymbol{\beta}})^{-1}. \tag{5.29}$$

This function will make its appearance in Section 5.7 and later on in Chapters 7 and 9.

The first strategy for performing regression with nonlinear prediction functions is to extend the feature space to obtain a simpler (ideally linear) prediction function in the extended feature space. We already saw an application of this strategy in Example 2.1 for the polynomial regression model, where the original feature u was extended to the feature vector $\boldsymbol{x} = [1, u, u^2, \ldots, u^{p-1}]^\top$, yielding a linear prediction function. In a similar way, the right-hand side of the polynomial prediction function in (5.28) can be viewed as a linear function of the extended feature vector $\boldsymbol{\phi}(\boldsymbol{x}) = [1, x_1, x_2, x_1^2, x_2^2, x_1 x_2]^\top$. The function $\boldsymbol{\phi}$ is called a *feature map*.

☞ 26

FEATURE MAP

The second strategy is to transform the response variable y and possibly also the explanatory variable $\boldsymbol{x}$ such that the transformed variables $\widetilde{y}, \widetilde{\boldsymbol{x}}$ are related in a simpler (ideally linear) way. For example, for the exponential prediction function $y = a\,\mathrm{e}^{-bx}$, we have $\ln y = \ln a - bx$, which is a linear relation between $\ln y$ and $[1, x]^\top$.

■ **Example 5.8 (Chlorine)** Table 5.4 lists the free chlorine concentration (in mg per liter) in a swimming pool, recorded every 8 hours for 4 days. A simple chemistry-based model for the chlorine concentration y as a function of time t is $y = a\,\mathrm{e}^{-bt}$, where a is the initial concentration and $b > 0$ is the reaction rate.

Table 5.4: Chlorine concentration (in mg/L) as a function of time (hours).

Hours	Concentration	Hours	Concentration
0	1.0056	56	0.3293
8	0.8497	64	0.2617
16	0.6682	72	0.2460
24	0.6056	80	0.1839
32	0.4735	88	0.1867
40	0.4745	96	0.1688
48	0.3563		

The exponential relationship $y = a\,\mathrm{e}^{-bt}$ suggests that a log transformation of y will result in a *linear* relationship between $\ln y$ and the feature vector $[1, t]^\top$. Thus, if for some given data $(t_1, y_1), \ldots, (t_n, y_n)$, we plot $(t_1, \ln y_1), \ldots, (t_n, \ln y_n)$, these points should approximately lie on a straight line, and hence the simple linear regression model applies. The left panel of Figure 5.5 illustrates that the transformed data indeed lie approximately on a straight line. The estimated regression line is also drawn here. The intercept and slope are $\beta_0 = -0.0555$ and $\beta_1 = -0.0190$ here. The original (non-transformed) data is shown in the right panel of Figure 5.5, along with the fitted curve $y = \widehat{a}\,\mathrm{e}^{-\widehat{b}t}$, where $\widehat{a} = \exp(\widehat{\beta}_0) = 0.9461$ and $\widehat{b} = -\widehat{\beta}_1 = 0.0190$.

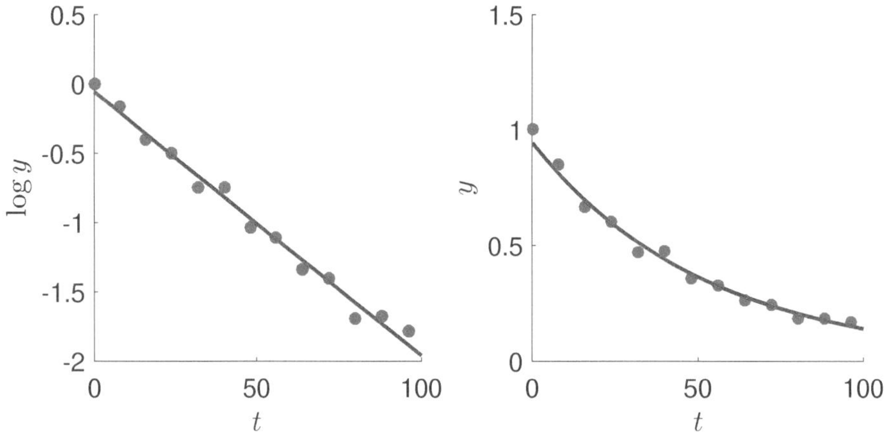

Figure 5.5: The chlorine concentration seems to have an exponential decay. ■

Recall that for a general regression problem the learner $g_\tau(\boldsymbol{x})$ for a given training set τ is obtained by minimizing the training (squared-error) loss

$$\ell_\tau(g(\cdot\,|\,\boldsymbol{\beta})) = \frac{1}{n}\sum_{i=1}^{n}(y_i - g(\boldsymbol{x}_i\,|\,\boldsymbol{\beta}))^2. \tag{5.30}$$

The third strategy for regression with nonlinear prediction functions is to directly minimize (5.30) by any means possible, as illustrated in the next example.

■ **Example 5.9 (Hougen Function)** In [7] the reaction rate y of a certain chemical reaction is posited to depend on three input variables: quantities of hydrogen x_1, n-pentane x_2, and isopentane x_3. The functional relationship is given by the *Hougen* function:

$$y = \frac{\beta_1\, x_2 - x_3/\beta_5}{1 + \beta_2\, x_1 + \beta_3\, x_2 + \beta_4\, x_3},$$

where $\beta_1, \ldots, \beta_5$ are the unknown parameters. The objective is to estimate the model parameters $\{\beta_i\}$ from the data, as given in Table 5.5.

Table 5.5: Data for the Hougen function.

x_1	x_2	x_3	y	x_1	x_2	x_3	y
470	300	10	8.55	470	190	65	4.35
285	80	10	3.79	100	300	54	13.00
470	300	120	4.82	100	300	120	8.50
470	80	120	0.02	100	80	120	0.05
470	80	10	2.75	285	300	10	11.32
100	190	10	14.39	285	190	120	3.13
100	80	65	2.54				

The estimation is carried out via the least-squares method. The objective function to minimize is thus

$$\ell_\tau(g(\cdot\,|\,\boldsymbol{\beta})) = \frac{1}{13}\sum_{i=1}^{13}\left(y_i - \frac{\beta_1\, x_{i2} - x_{i3}/\beta_5}{1 + \beta_2\, x_{i1} + \beta_3\, x_{i2} + \beta_4\, x_{i3}}\right)^2, \tag{5.31}$$

where the $\{y_i\}$ and $\{x_{ij}\}$ are given in Table 5.5.

This is a highly nonlinear optimization problem, for which standard nonlinear least-squares methods do not work well. Instead, one can use global optimization methods such as CE and SCO (see Sections 3.4.2 and 3.4.3). Using the CE method, we found the minimal value 0.02299 for the objective function, which is attained at ☞ 416 ☞ 100

$$\widehat{\boldsymbol{\beta}} = [1.2526,\ 0.0628,\ 0.0400,\ 0.1124,\ 1.1914]^\top.$$

■

5.6 Linear Models in Python

In this section we describe how to define and analyze linear models using Python and the data science module `statsmodels`. We encourage the reader to regularly refer back to the theory in the preceding sections of this chapter, so as to avoid using Python merely as a black box without understanding the underlying principles. To run the code start by importing the following code snippet:

```
import matplotlib.pyplot as plt
import pandas as pd
import statsmodels.api as sm
from statsmodels.formula.api import ols
```

5.6.1 Modeling

Although specifying a normal[3] linear model in Python is relatively easy, it requires some subtlety. The main thing to realize is that Python treats quantitative and qualitative (that is, categorical) explanatory variables differently. In `statsmodels`, ordinary least-squares linear models are specified via the function **`ols`** (short for ordinary least-squares). The main argument of this function is a formula of the form

$$\texttt{y} \sim \texttt{x1} + \texttt{x2} + \cdots + \texttt{xd}, \tag{5.32}$$

where `y` is the name of the response variable and `x1`, ..., `xd` are the names of the explanatory variables. If all variables are *quantitative*, this describes the linear model

$$Y_i = \beta_0 + \beta_1 x_{i1} + \beta_2 x_{i2} + \cdots + \beta_d x_{id} + \varepsilon_i, \quad i = 1, \dots, n, \tag{5.33}$$

where x_{ij} is the j-th explanatory variable for the i-th observation and the errors ε_i are independent normal random variables such that $\mathbb{E}\varepsilon_i = 0$ and $\mathbb{V}\mathrm{ar}\,\varepsilon_i = \sigma^2$. Or, in matrix form: $\boldsymbol{Y} = \mathbf{X}\boldsymbol{\beta} + \boldsymbol{\varepsilon}$, with

$$\boldsymbol{Y} = \begin{bmatrix} Y_1 \\ \vdots \\ Y_n \end{bmatrix}, \quad \mathbf{X} = \begin{bmatrix} 1 & x_{11} & \cdots & x_{1d} \\ 1 & x_{21} & \cdots & x_{2d} \\ \vdots & \vdots & \ddots & \vdots \\ 1 & x_{n1} & \cdots & x_{nd} \end{bmatrix}, \quad \boldsymbol{\beta} = \begin{bmatrix} \beta_0 \\ \vdots \\ \beta_d \end{bmatrix}, \text{ and } \boldsymbol{\varepsilon} = \begin{bmatrix} \varepsilon_1 \\ \vdots \\ \varepsilon_n \end{bmatrix}.$$

[3]For the rest of this section, we assume all linear models to be normal.

Thus, the first column is always taken as an "intercept" parameter, unless otherwise specified. To remove the intercept term, add `-1` to the `ols` formula, as in `ols('y~x-1')`.

For any linear model, the model matrix can be retrieved via the construction:

```
model_matrix = pd.DataFrame(model.exog,columns=model.exog_names)
```

Let us look at some examples of linear models. In the first model the variables `x1` and `x2` are both considered (by Python) to be quantitative.

```
myData = pd.DataFrame({'y' : [10,9,4,2,4,9],
     'x1' : [7.4,1.2,3.1,4.8,2.8,6.5],
     'x2' : [1,1,2,2,3,3]})
mod = ols("y~x1+x2", data=myData)
mod_matrix = pd.DataFrame(mod.exog,columns=mod.exog_names)
print(mod_matrix)
```

```
   Intercept   x1   x2
0        1.0  7.4  1.0
1        1.0  1.2  1.0
2        1.0  3.1  2.0
3        1.0  4.8  2.0
4        1.0  2.8  3.0
5        1.0  6.5  3.0
```

Suppose the second variable is actually qualitative; e.g., it represents a color, and the levels 1, 2, and 3 stand for red, blue, and green. We can account for such a categorical variable by using the **astype** method to redefine the data type (see Section 1.2). ☞ 3

```
myData['x2'] = myData['x2'].astype('category')
```

Alternatively, a categorical variable can be specified in the model formula by wrapping it with `C()`. Observe how this changes the model matrix.

```
mod2 = ols("y~x1+C(x2)", data=myData)
mod2_matrix = pd.DataFrame(mod2.exog,columns=mod2.exog_names)
print(mod2_matrix)
```

```
   Intercept  C(x2)[T.2]  C(x2)[T.3]   x1
0        1.0         0.0         0.0  7.4
1        1.0         0.0         0.0  1.2
2        1.0         1.0         0.0  3.1
3        1.0         1.0         0.0  4.8
4        1.0         0.0         1.0  2.8
5        1.0         0.0         1.0  6.5
```

Thus, if a `statsmodels` formula of the form (5.32) contains factor (qualitative) variables, the model is no longer of the form (5.33), but contains indicator variables for each level of the factor variable, except the first level.

For the case above, the corresponding linear model is

$$Y_i = \beta_0 + \beta_1 x_{i1} + \alpha_2\, \mathbb{1}\{x_{i2} = 2\} + \alpha_3\, \mathbb{1}\{x_{i2} = 3\} + \varepsilon_i, \quad i = 1, \ldots, 6, \tag{5.34}$$

where we have used parameters α_2 and α_3 to correspond to the indicator features of the qualitative variable. The parameter α_2 describes how much the response is expected to change if the factor x_2 switches from level 1 to 2. A similar interpretation holds for α_3. Such parameters can thus be viewed as incremental effects.

INTERACTION

It is also possible to model *interaction* between two variables. For two continuous variables, this simply adds the products of the original features to the model matrix. Adding interaction terms in Python is achieved by replacing "+" in the formula with "*", as the following example illustrates.

```
mod3 = ols("y~x1*C(x2)", data=myData)
mod3_matrix = pd.DataFrame(mod3.exog,columns=mod3.exog_names)
print(mod3_matrix)
```

```
   Intercept  C(x2)[T.2]  C(x2)[T.3]   x1  x1:C(x2)[T.2]  x1:C(x2)[T.3]
0        1.0         0.0         0.0  7.4            0.0            0.0
1        1.0         0.0         0.0  1.2            0.0            0.0
2        1.0         1.0         0.0  3.1            3.1            0.0
3        1.0         1.0         0.0  4.8            4.8            0.0
4        1.0         0.0         1.0  2.8            0.0            2.8
5        1.0         0.0         1.0  6.5            0.0            6.5
```

5.6.2 Analysis

Let us consider some easy linear regression models by using the student survey data set survey.csv from the book's GitHub site, which contains measurements such as height, weight, sex, etc., from a survey conducted among $n = 100$ university students. Suppose we wish to investigate the relation between the shoe size (explanatory variable) and the height (response variable) of a person. First, we load the data and draw a scatterplot of the points (height versus shoe size); see Figure 5.6 (without the fitted line).

```
survey = pd.read_csv('survey.csv')
plt.scatter(survey.shoe, survey.height)
plt.xlabel("Shoe size")
plt.ylabel("Height")
```

We observe a slight increase in the height as the shoe size increases, although this relationship is not very distinct. We analyze the data through the simple linear regression model $Y_i = \beta_0 + \beta_1 x_i + \varepsilon_i, i = 1, \ldots, n$. In statsmodels this is performed via the **ols** method as follows: ☞ 169

```
model = ols("height~shoe", data=survey) # define the model
fit = model.fit()  #fit the model defined above
b0, b1 = fit.params
print(fit.params)
```

```
Intercept    145.777570
shoe           1.004803
dtype: float64
```

The above output gives the least-squares estimates of β_0 and β_1. For this example, we have $\widehat{\beta}_0 = 145.778$ and $\widehat{\beta}_1 = 1.005$. Figure 5.6, which includes the regression line, was obtained as follows:

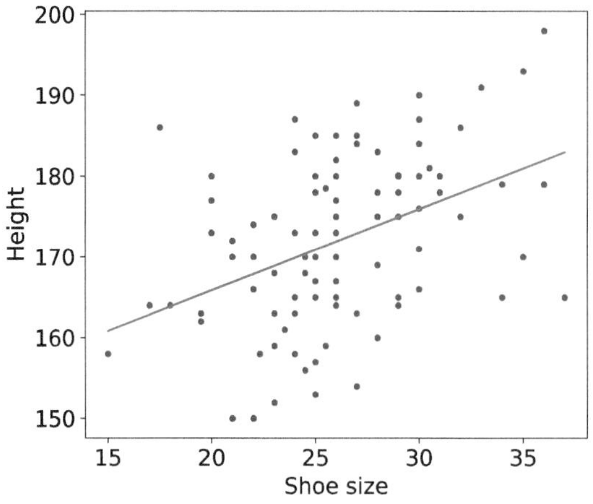

Figure 5.6: Scatterplot of height (cm) against shoe size (cm), with the fitted line.

```
plt.plot(survey.shoe, b0 + b1*survey.shoe)
plt.scatter(survey.shoe, survey.height)
plt.xlabel("Shoe size")
plt.ylabel("Height")
```

Although `ols` performs a complete analysis of the linear model, not all its calculations need to be presented. A summary of the results can be obtained with the method `summary`.

```
print(fit.summary())
```

```
Dep. Variable:              height   R-squared:                   0.178
Model:                         OLS   Adj. R-squared:              0.170
Method:              Least Squares   F-statistic:                 21.28
No. Observations:              100   Prob (F-statistic):       1.20e-05
Df Residuals:                   98   Log-Likelihood:            -363.88
Df Model:                        1   AIC:                         731.8
Covariance Type:         nonrobust   BIC:                         737.0
=====================================================================
              coef    std err          t      P>|t|    [0.025   0.975]
---------------------------------------------------------------------
Intercept  145.7776    5.763     25.296     0.000   134.341   157.214
shoe         1.0048    0.218      4.613     0.000     0.573     1.437
=====================================================================
Omnibus:                     1.958   Durbin-Watson:               1.772
Prob(Omnibus):               0.376   Jarque-Bera (JB):            1.459
Skew:                       -0.072   Prob(JB):                    0.482
Kurtosis:                    2.426   Cond. No.                     164.
```

The main output items are the following:

- `coef`: Estimates of the parameters of the regression line.
- `std error`: Standard deviations of the estimators of the regression line. These are the square roots of the variances of the $\{\widehat{\beta}_i\}$ obtained in (5.25). ☞ 186

- `t`: Realization of Student's test statistics associated with the hypotheses $H_0 : \beta_i = 0$ and $H_1 : \beta_i \neq 0$, $i = 0, 1$. In particular, the outcome of T in (5.19). ☞ 183

- `P>|t|`: P-value of Student's test (two-sided test).

- `[0.025 0.975]`: 95% confidence intervals for the parameters.

- `R-Squared`: Coefficient of determination R^2 (percentage of variation explained by the regression), as defined in (5.18). ☞ 181

- `Adj. R-Squared`: adjusted R^2 (explained in Section 5.3.7).

- `F-statistic`: Realization of the F test statistic (5.21) associated with testing the full model against the default model. The associated degrees of freedom (`Df Model` = 1 and `Df Residuals` = $n-2$) are given, as is the P-value: `Prob (F-statistic)`. ☞ 183

- `AIC`: The AIC number in (5.15); that is, minus two times the log-likelihood plus two times the number of model parameters (which is 3 here). ☞ 177

You can access all the numerical values as they are attributes of the `fit` object. First check which names are available, as in:

```
dir(fit)
```

Then access the values via the dot construction. For example, the following extracts the P-value for the slope.

```
fit.pvalues[1]
```

```
1.1994e-05
```

The results show strong evidence for a linear relationship between shoe size and height (or, more accurately, strong evidence that the slope of the regression line is not zero), as the P-value for the corresponding test is very small ($1.2 \cdot 10^{-5}$). The estimate of the slope indicates that the difference between the average height of students whose shoe size is different by one cm is 1.0048 cm.

Only 17.84% of the variability of student height is explained by the shoe size. We therefore need to add other explanatory variables to the model (multiple linear regression) to increase the model's predictive power.

5.6.3 Analysis of Variance (ANOVA)

We continue the student survey example of the previous section, but now add an extra variable, and also consider an analysis of variance of the model. Instead of "explaining" the student height via their shoe size, we include `weight` as an explanatory variable. The corresponding `ols` formula for this model is

$$\texttt{height}\sim\texttt{shoe} + \texttt{weight},$$

meaning that each random height, denoted by Height, satisfies

$$\texttt{Height} = \beta_0 + \beta_1 \texttt{shoe} + \beta_2 \texttt{weight} + \varepsilon,$$

where ε is a normally distributed error term with mean 0 and variance σ^2. Thus, the model has 4 parameters. Before analyzing the model we present a scatterplot of all pairs of variables, using **scatter_matrix**.

```
model = ols("height~shoe+weight", data=survey)
fit = model.fit()
axes = pd.plotting.scatter_matrix(
          survey[['height','shoe','weight']])
plt.show()
```

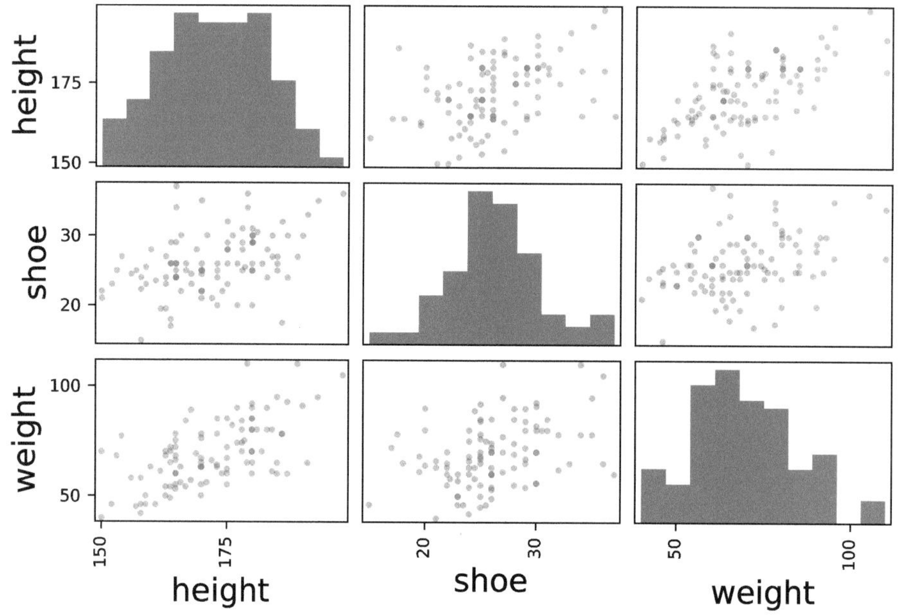

Figure 5.7: Scatterplot of all pairs of variables: height (cm), shoe (cm), and weight (kg).

As for the simple linear regression model in the previous section, we can analyze the model using the **summary** method (below we have omitted some output):

```
fit.summary()
```

```
Dep. Variable:              height   R-squared:                   0.430
Model:                         OLS   Adj. R-squared:              0.418
Method:              Least Squares   F-statistic:                 36.61
No. Observations:              100   Prob (F-statistic):       1.43e-12
Df Residuals:                   97   Log-Likelihood:             -345.58
Df Model:                        2   AIC:                         697.2
                                     BIC:                         705.0
======================================================================
              coef   std err        t     P>|t|     [0.025     0.975]
----------------------------------------------------------------------
Intercept 132.2677     5.247   25.207     0.000    121.853    142.682
```

```
shoe          0.5304     0.196      2.703     0.008      0.141      0.920
weight        0.3744     0.057      6.546     0.000      0.261      0.488
```

The `F-statistic` is used to test whether the full model (here with two explanatory variables) is better at "explaining" the height than the default model. The corresponding null hypothesis is $H_0 : \beta_1 = \beta_2 = 0$. The assertion of interest is H_1: at least one of the coefficients β_j $(j = 1, 2)$ is significantly different from zero. Given the result of this test (P-value $= 1.429 \cdot 10^{-12}$), we can conclude that at least one of the explanatory variables is associated with height. The individual Student tests indicate that:

- shoe size is linearly associated with student height, after adjusting for weight, with P-value 0.0081. At the same weight, an increase of one cm in shoe size corresponds to an increase of 0.53 cm in average student height;
- weight is linearly associated with student height, after adjusting for shoe size (the P-value is actually $2.82 \cdot 10^{-09}$; the reported value of 0.000 should be read as "less than 0.001"). At the same shoe size, an increase of one kg in weight corresponds to an increase of 0.3744 cm in average student height.

Further understanding is extracted from the model by conducting an analysis of variance. The standard `statsmodels` function is **`anova_lm`**. The input to this function is the fit object (obtained from `model.fit()`) and the output is a `DataFrame` object.

```
table = sm.stats.anova_lm(fit)
print(table)
```

```
            df       sum_sq      mean_sq          F        PR(>F)
shoe       1.0  1840.467359  1840.467359  30.371310  2.938651e-07
weight     1.0  2596.275747  2596.275747  42.843626  2.816065e-09
Residual  97.0  5878.091294    60.598879        NaN           NaN
```

The meaning of the columns is as follows.

- `df` : The degrees of freedom of the variables, according to the sum of squares decomposition (5.17). As both `shoe` and `weight` are quantitative variables, their degrees of freedom are both 1 (each corresponding to a single column in the overall model matrix). The degrees of freedom for the residuals is $n - p = 100 - 3 = 97$. ☞ 181
- `sum_sq`: The sum of squares according to (5.17). The total sum of squares is the sum of all the entries in this column. The residual error in the model that cannot be explained by the variables is RSS ≈ 5878.
- `mean_sq`: The sum of squares divided by their degrees of freedom. Note that the residual square error RSE $= \text{RSS}/(n - p) = 60.6$ is an unbiased estimate of the model variance σ^2; see Section 5.4. ☞ 182
- `F`: These are the outcomes of the test statistic (5.22). ☞ 184
- `PR(>F)`: These are the P-values corresponding to the test statistic in the preceding column and are computed using an `F` distribution whose degrees of freedom are given in the `df` column.

The ANOVA table indicates that the shoe variable explains a reasonable amount of the variation in the model, as evidenced by a sum of squares contribution of 1840 out of 1840+2596+5878 = 10314 and a very small P-value. After shoe is included in the model, it turns out that the weight variable explains even more of the remaining variability, with an even smaller P-value. The remaining sum of squares (5878) is 57% of the total sum of squares, yielding a 43% reduction, in accordance with the R^2 value reported in the summary for the ols method. As mentioned in Section 5.4.1, the *order* in which the ANOVA is conducted is important. To illustrate this, consider the output of the following commands.

```
model = ols("height~weight+shoe", data=survey)
fit = model.fit()
table = sm.stats.anova_lm(fit)
print(table)
```

```
            df      sum_sq      mean_sq         F        PR(>F)
weight     1.0  3993.860167  3993.860167  65.906502  1.503553e-12
shoe       1.0   442.882938   442.882938   7.308434  8.104688e-03
Residual  97.0  5878.091294    60.598879        NaN           NaN
```

We see that weight as a single model variable explains much more of the variability than shoe did. If we now also include shoe, we only obtain a small (but according to the P-value still significant) reduction in the model variability.

5.6.4 Confidence and Prediction Intervals

In statsmodels a method for computing confidence or prediction intervals from a dictionary of explanatory variables is get_prediction. It simply executes formula (5.24) or
☞ 186 (5.26). A simpler version is predict, which only returns the predicted value.

Continuing the student survey example, suppose we wish to predict the height of a person with shoe size 30 cm and weight 75 kg. Confidence and prediction intervals can be obtained as given in the code below. The new explanatory variable is entered as a dictionary. Notice that the 95% prediction interval (for the corresponding random response) is much wider than the 95% confidence interval (for the expectation of the random response).

```
x = {'shoe': [30.0], 'weight': [75.0]}  # new input (dictionary)
pred = fit.get_prediction(x)
pred.summary_frame(alpha=0.05).unstack()
```

```
mean           0    176.261722    # predicted value
mean_se        0      1.054015
mean_ci_lower  0    174.169795    # lower bound for CI
mean_ci_upper  0    178.353650    # upper bound for CI
obs_ci_lower   0    160.670610    # lower bound for PI
obs_ci_upper   0    191.852835    # upper bound for PI
dtype: float64
```

5.6.5 Model Validation

We can perform an analysis of residuals to examine whether the underlying assumptions of the (normal) linear regression model are verified. Various plots of the residuals can be used to inspect whether the assumptions on the errors $\{\varepsilon_i\}$ are satisfied. Figure 5.8 gives two such plots. The first is a scatterplot of the residuals $\{e_i\}$ against the fitted values $\widehat{y}_i$. When the model assumptions are valid, the residuals, as approximations of the model error, should behave approximately as iid normal random variables for each of the fitted values, with a constant variance. In this case we see no strong aberrant structure in this plot. The residuals are fairly evenly spread and symmetrical about the $y = 0$ line (not shown). The second plot is a quantile–quantile (or qq) plot. This is a useful way to check for normality of the error terms, by plotting the sample quantiles of the residuals against the theoretical quantiles of the standard normal distribution. Under the model assumptions, the points should lie approximately on a straight line. For the current case there does not seem to be an extreme departure from normality. Drawing a histogram or density plot of the residuals will also help to verify the normality assumption. The following code was used.

```
plt.plot(fit.fittedvalues,fit.resid,'.')
plt.xlabel("fitted values")
plt.ylabel("residuals")
sm.qqplot(fit.resid)
```

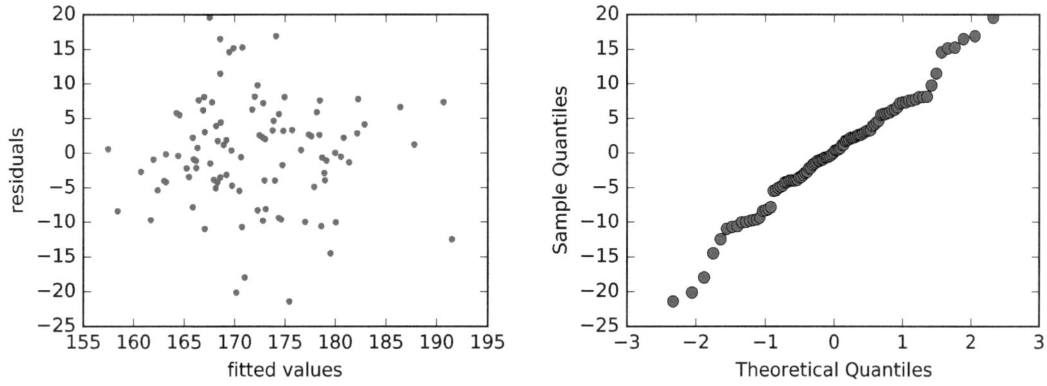

Figure 5.8: Left: residuals against fitted values. Right: a qq plot of the residuals. Neither shows clear evidence against the model assumptions of constant variance and normality.

5.6.6 Variable Selection

Among the large number of possible explanatory variables, we wish to select those which best explain the observed responses. By eliminating redundant explanatory variables, we reduce the statistical error without increasing the approximation error, and thus reduce the (expected) generalization risk of the learner.

In this section, we briefly present two methods for variable selection. They are illustrated on a few variables from the data set `birthwt` discussed in Section 1.5.3.2. The data ☞ 13
set contains information on the birth weights (masses) of babies, as well as various characteristics of the mother, such as whether she smokes, her age, etc. We wish to explain the child's weight at birth using various characteristics of the mother, her family history, and her behavior during pregnancy. The response variable is weight at birth (quantitative variable `bwt`, expressed in grams); the explanatory variables are given below.

The data can be obtained as explained in Section 1.5.3.2, or from `statsmodels` in the following way:

```
bwt = sm.datasets.get_rdataset("birthwt","MASS").data
```

Here is some information about the explanatory variables that we will investigate.

```
age:   mother's age in years
lwt:   mother's weight in lbs
race:  mother's race (1 = white, 2 = black, 3 = other)
smoke: smoking status during pregnancy (0 = no, 1 = yes)
ptl:   no. of previous premature labors
ht:    history of hypertension (0 = no, 1 = yes)
ui:    presence of uterine irritability (0 = no, 1 = yes)
ftv:   no. of physician visits during first trimester
bwt:   birth weight in grams
```

We can see the structure of the variables via `bwt.info()`. Check yourself that all variables are defined as *quantitative* (`int64`). However, the variables `race`, `smoke`, `ht`, and `ui` should really be interpreted as *qualitative* (factors). To fix this, we could redefine them with the method `astype`, similar to what we did in Chapter 1. Alternatively, we could use the `C()` construction in a `statsmodels` formula to let the program know that certain variables are factors. We will use the latter approach.

For *binary* features it does not matter whether the variables are interpreted as factorial or numerical as the numerical and summary results are identical.

We consider the explanatory variables `lwt`, `age`, `ui`, `smoke`, `ht`, and two recoded binary variables `ftv1` and `ptl1`. We define `ftv1` = 1 if there was at least one visit to a physician, and `ftv1` = 0 otherwise. Similarly, we define `ptl1` = 1 if there is at least one preterm birth in the family history, and `ptl1` = 0 otherwise.

```
ftv1 = (bwt['ftv']>=1).astype(int)
ptl1 = (bwt['ptl']>=1).astype(int)
```

5.6.6.1 Forward Selection and Backward Elimination

FORWARD SELECTION

The *forward selection* method is an iterative method for variable selection. In the first iteration we consider which feature `f1` is the most significant in terms of its P-value in the models `bwt~f1`, with `f1` $\in$ {`lwt`, `age`, . . .}. This feature is then selected into the model. In the second iteration, the feature `f2` that has the smallest P-value in the models `bwt~f1+f2` is selected, where `f2` $\neq$ `f1`, and so on. Usually only features are selected that have a P-value of at most 0.05. The following Python program automates this procedure. Instead of selecting on the P-value one could select on the AIC or BIC value.

forwardselection.py

```
import statsmodels.api as sm
from statsmodels.formula.api import ols

bwt = sm.datasets.get_rdataset("birthwt","MASS").data
ftv1 = (bwt['ftv']>=1).astype(int)
ptl1 = (bwt['ptl']>=1).astype(int)

remaining_features = {'lwt', 'age', 'C(ui)', 'smoke',
                      'C(ht)', 'ftv1', 'ptl1'}
selected_features = []
while remaining_features:
  PF = [] #list of (P value, feature)
  for f in remaining_features:
    temp = selected_features + [f] #temporary list of features
    formula = 'bwt~' + '+'.join(temp)
    fit = ols(formula,data=bwt).fit()
    pval= fit.pvalues[-1]
    if pval < 0.05:
      PF.append((pval,f))
  if PF: #if not empty
     PF.sort(reverse=True)
     (best_pval, best_f) = PF.pop()
     remaining_features.remove(best_f)
     print('feature {} with P-value = {:.2E}'.
           format(best_f, best_pval))
     selected_features.append(best_f)
  else:
     break
```

```
feature C(ui) with P-value = 7.52E-05
feature C(ht) with P-value = 1.08E-02
feature lwt with P-value = 6.01E-03
feature smoke with P-value = 7.27E-03
```

In *backward elimination* we start with the complete model (all features included) and at each step, we remove the variable with the highest P-value, as long as it is not significant (greater than 0.05). We leave it as an exercise to verify that the order in which the features are removed is: `age`, `ftv1`, and `ptl1`. In this case, forward selection and backward elimination result in the same model, but this need not be the case in general. BACKWARD ELIMINATION

This way of model selection has the advantage of being easy to use and of treating the question of variable selection in a systematic manner. The main drawback is that variables are included or deleted based on purely statistical criteria, without taking into account the aim of the study. This usually leads to a model which may be satisfactory from a statistical point of view, but in which the variables are not necessarily the most relevant when it comes to understanding and interpreting the data in the study.

Of course, we can choose to investigate any combination of features, not just the ones suggested by the above variable selection methods. For example, let us see if the mother's weight, her age, her race, and whether she smokes explain the baby's birthweight.

```
formula = 'bwt~lwt+age+C(race)+ smoke'
bwt_model = ols(formula, data=bwt).fit()
print(bwt_model.summary())
```

```
                    OLS Regression Results
======================================================================
Dep. Variable: bwt              R-squared: 0.148
Model: OLS                      Adj. R-squared: 0.125
Method: Least Squares           F-statistic: 6.373
No. Observations: 189           Prob (F-statistic):   1.76e-05
Df Residuals: 183               Log-Likelihood: -1498.4
Df Model: 5                     AIC: 3009.
                                BIC: 3028.
======================================================================
                  coef    std err       t     P>|t|  [0.025     0.975]
----------------------------------------------------------------------
Intercept     2839.4334   321.435    8.834    0.000  2205.239   3473.628
C(race)[T.2]  -510.5015   157.077   -3.250    0.001  -820.416   -200.587
C(race)[T.3]  -398.6439   119.579   -3.334    0.001  -634.575   -162.713
smoke         -401.7205   109.241   -3.677    0.000  -617.254   -186.187
lwt              3.9999     1.738    2.301    0.022     0.571      7.429
age             -1.9478     9.820   -0.198    0.843   -21.323     17.427
======================================================================
Omnibus: 3.916                    Durbin-Watson: 0.458
Prob(Omnibus): 0.141              Jarque-Bera (JB): 3.718
Skew: -0.343                      Prob(JB): 0.156
Kurtosis: 3.038                   Cond. No. 899.
```

Given the result of Fisher's global test given by `Prob (F-Statistic)` in the summary (P-value = 1.76×10^{-5}), we can conclude that at least one of the explanatory variables is associated with child weight at birth, after adjusting for the other variables. The individual Student tests indicate that:

- the mother's weight is linearly associated with child weight, after adjusting for age, race, and smoking status (P-value = 0.022). At the same age, race, and smoking status, an increase of one pound in the mother's weight corresponds to an increase of 4 g in the average child weight at birth;

- the age of the mother is not significantly linearly associated with child weight at birth, when mother weight, race, and smoking status are already taken into account (P-value = 0.843);

- weight at birth is significantly lower for a child born to a mother who smokes, compared to children born to non-smoking mothers of the same age, race, and weight, with a P-value of 0.00031 (to see this, inspect `bwt_model.pvalues`). At the same age, race, and mother weight, the child's weight at birth is 401.720 g less for a smoking mother than for a non-smoking mother;

- regarding the interpretation of the variable `race`, we note that the first level of this categorical variable corresponds to white mothers. The estimate of −510.501 g for `C(race)[T.2]` represents the difference in the child's birth weight between black mothers and white mothers (reference group), and this result is significantly different from zero (P-value = 0.001) in a model adjusted for the mother's weight, age, and smoking status.

5.6.6.2 Interaction

We can also include interaction terms in the model. Let us see whether there is any interaction effect between smoke and age via the model

$$\texttt{Bwt} = \beta_0 + \beta_1 \texttt{age} + \beta_2 \texttt{smoke} + \beta_3 \texttt{age} \times \texttt{smoke} + \varepsilon.$$

In Python this can be done as follows (below we have removed some output):

```
formula = 'bwt~age*smoke'
bwt_model = ols(formula, data=bwt).fit()
print(bwt_model.summary())
```

```
                    OLS Regression Results
==========================================================================
Dep. Variable: bwt               R-squared: 0.069
Model: OLS                       Adj. R-squared: 0.054
Method: Least Squares            F-statistic: 4.577
No. Observations: 189            Prob (F-statistic):   0.00407
Df Residuals: 183                Log-Likelihood: -1506.8
Df Model: 5                      AIC: 3009.
                                 BIC: 3028.
==========================================================================
                coef    std err       t       P>|t|    [0.025    0.975]
--------------------------------------------------------------------------
Intercept     2406.1   292.190    8.235    0.000   1829.6    2982.5
smoke          798.2   484.342    1.648    0.101   -157.4    1753.7
age             27.7    12.149    2.283    0.024      3.8      51.7
age:smoke      -46.6    20.447   -2.278    0.024    -86.9      -6.2
```

We observe that the estimate for β_3 (−46.6) is significantly different from zero (P-value = 0.024). We therefore conclude that the effect of the mother's age on the child's weight depends on the smoking status of the mother. The results on association between mother age and child weight must therefore be presented separately for the smoking and the non-smoking group. For non-smoking mothers (smoke = 0), the mean child weight at birth increases on average by 27.7 grams for each year of the mother's age. This is statistically significant, as can be seen from the 95% confidence intervals for the parameters (which does not contain zero):

```
bwt_model.conf_int()
```

```
                     0             1
Intercept  1829.605754   2982.510194
age           3.762780     51.699977
smoke      -157.368023   1753.717779
age:smoke   -86.911405     -6.232425
```

Similarly, for smoking mothers, there seems to be a decrease in birthweight, $\widehat{\beta_1} + \widehat{\beta_3} = 27.7 - 46.6 = -18.9$, but this is not statistically significant; see Exercise 6.

5.7 Generalized Linear Models

The normal linear model in Section 2.8 deals with continuous response variables — such as height and crop yield — and continuous or discrete explanatory variables. Given the feature vectors $\{\boldsymbol{x}_i\}$, the responses $\{Y_i\}$ are independent of each other, and each has a normal distribution with mean $\boldsymbol{x}_i^\top \boldsymbol{\beta}$, where $\boldsymbol{x}_i^\top$ is the i-th row of the model matrix $\mathbf{X}$. Generalized linear models allow for arbitrary response distributions, including *discrete* ones.

Definition 5.2: Generalized Linear Model

GENERALIZED LINEAR MODEL

In a *generalized linear model* (GLM) the expected response for a given feature vector $\boldsymbol{x} = [x_1, \ldots, x_p]^\top$ is of the form

$$\mathbb{E}[Y \mid \boldsymbol{X} = \boldsymbol{x}] = h(\boldsymbol{x}^\top \boldsymbol{\beta}) \tag{5.35}$$

ACTIVATION FUNCTION

for some function h, which is called the *activation function*. The distribution of Y (for a given $\boldsymbol{x}$) may depend on additional *dispersion* parameters that model the randomness in the data that is not explained by $\boldsymbol{x}$.

LINK FUNCTION

The *inverse* of function h is called the *link function*. As for the linear model, (5.35) is a model for a single pair $(\boldsymbol{x}, Y)$. Using the model simplification introduced at the end of Section 5.1, the corresponding model for a whole training set $\mathcal{T} = \{(\boldsymbol{x}_i, Y_i)\}$ is that the $\{\boldsymbol{x}_i\}$ are fixed and that the $\{Y_i\}$ are independent; each Y_i satisfying (5.35) with $\boldsymbol{x} = \boldsymbol{x}_i$. Writing $\boldsymbol{Y} = [Y_1, \ldots, Y_n]^\top$ and defining $\boldsymbol{h}$ as the multivalued function with components h, we have

$$\mathbb{E}_{\mathbf{X}} \boldsymbol{Y} = \boldsymbol{h}(\mathbf{X}\boldsymbol{\beta}),$$

where $\mathbf{X}$ is the (model) matrix with rows $\boldsymbol{x}_1^\top, \ldots, \boldsymbol{x}_n^\top$. A common assumption is that $Y_1, \ldots, Y_n$ come from the same family of distributions, e.g., normal, Bernoulli, or Poisson. The central focus is the parameter vector $\boldsymbol{\beta}$, which summarizes how the matrix of explanatory variables $\mathbf{X}$ affects the response vector $\boldsymbol{Y}$. The class of generalized linear models can encompass a wide variety of models. Obviously the normal linear model (2.34) is a generalized linear model, with $\mathbb{E}[Y \mid \boldsymbol{X} = \boldsymbol{x}] = \boldsymbol{x}^\top \boldsymbol{\beta}$, so that h is the identity function. In this case, $Y \sim \mathcal{N}(\boldsymbol{x}^\top \boldsymbol{\beta}, \sigma^2)$, $i = 1, \ldots, n$, where σ^2 is a dispersion parameter.

LOGISTIC REGRESSION

LOGISTIC DISTRIBUTION

■ **Example 5.10 (Logistic Regression)** In a *logistic regression* or *logit model*, we assume that the response variables $Y_1, \ldots, Y_n$ are independent and distributed according to $Y_i \sim \mathsf{Ber}(h(\boldsymbol{x}_i^\top \boldsymbol{\beta}))$, where h here is defined as the cdf of the *logistic distribution*:

$$h(x) = \frac{1}{1 + \mathrm{e}^{-x}}.$$

Large values of $\boldsymbol{x}_i^\top \boldsymbol{\beta}$ thus lead to a high probability that $Y_i = 1$, and small (negative) values of $\boldsymbol{x}_i^\top \boldsymbol{\beta}$ cause Y_i to be 0 with high probability. Estimation of the parameter vector $\boldsymbol{\beta}$ from the observed data is not as straightforward as for the ordinary linear model, but can be accomplished via the minimization of a suitable training loss, as explained below.

As the $\{Y_i\}$ are independent, the pdf of $\boldsymbol{Y} = [Y_1, \ldots, Y_n]^\top$ is

$$g(\boldsymbol{y} \mid \boldsymbol{\beta}, \mathbf{X}) = \prod_{i=1}^{n} [h(\boldsymbol{x}_i^\top \boldsymbol{\beta})]^{y_i} [1 - h(\boldsymbol{x}_i^\top \boldsymbol{\beta})]^{1-y_i}.$$

Maximizing the log-likelihood $\ln g(\boldsymbol{y} \,|\, \boldsymbol{\beta}, \mathbf{X})$ with respect to $\boldsymbol{\beta}$ gives the maximum likelihood estimator of $\boldsymbol{\beta}$. In a supervised learning framework, this is equivalent to minimizing:

$$\begin{aligned} -\frac{1}{n} \ln g(\boldsymbol{y} \,|\, \boldsymbol{\beta}, \mathbf{X}) &= -\frac{1}{n} \sum_{i=1}^{n} \ln g(y_i \,|\, \boldsymbol{\beta}, \boldsymbol{x}_i) \\ &= -\frac{1}{n} \sum_{i=1}^{n} \left[y_i \ln h(\boldsymbol{x}_i^\top \boldsymbol{\beta}) + (1 - y_i) \ln(1 - h(\boldsymbol{x}_i^\top \boldsymbol{\beta})) \right]. \end{aligned} \tag{5.36}$$

By comparing (5.36) with (4.4), we see that we can interpret (5.36) as the *cross-entropy training loss* associated with comparing a true conditional pdf $f(y \,|\, \boldsymbol{x})$ with an approximation pdf $g(y \,|\, \boldsymbol{\beta}, \boldsymbol{x})$ via the loss function ☞ 123

$$\text{Loss}(f(y \,|\, \boldsymbol{x}), g(y \,|\, \boldsymbol{\beta}, \boldsymbol{x})) := -\ln g(y \,|\, \boldsymbol{\beta}, \boldsymbol{x}) = -y \ln h(\boldsymbol{x}^\top \boldsymbol{\beta}) - (1 - y) \ln(1 - h(\boldsymbol{x}^\top \boldsymbol{\beta})).$$

Minimizing (5.36) in terms of $\boldsymbol{\beta}$ actually constitutes a *convex* optimization problem. Since $\ln h(\boldsymbol{x}^\top \boldsymbol{\beta}) = -\ln(1 + \mathrm{e}^{-\boldsymbol{x}^\top \boldsymbol{\beta}})$ and $\ln(1 - h(\boldsymbol{x}^\top \boldsymbol{\beta})) = -\boldsymbol{x}^\top \boldsymbol{\beta} - \ln(1 + \mathrm{e}^{-\boldsymbol{x}^\top \boldsymbol{\beta}})$, the cross-entropy training loss (5.36) can be rewritten as

$$r_\tau(\boldsymbol{\beta}) := \frac{1}{n} \sum_{i=1}^{n} \left[(1 - y_i) \boldsymbol{x}_i^\top \boldsymbol{\beta} + \ln\left(1 + \mathrm{e}^{-\boldsymbol{x}_i^\top \boldsymbol{\beta}}\right) \right].$$

We leave it as Exercise 7 to show that the gradient $\nabla r_\tau(\boldsymbol{\beta})$ and Hessian $\mathbf{H}(\boldsymbol{\beta})$ of $r_\tau(\boldsymbol{\beta})$ are given by

$$\nabla r_\tau(\boldsymbol{\beta}) = \frac{1}{n} \sum_{i=1}^{n} (\mu_i - y_i)\, \boldsymbol{x}_i \tag{5.37}$$

and

$$\mathbf{H}(\boldsymbol{\beta}) = \frac{1}{n} \sum_{i=1}^{n} \mu_i (1 - \mu_i)\, \boldsymbol{x}_i \boldsymbol{x}_i^\top, \tag{5.38}$$

respectively, where $\mu_i := h(\boldsymbol{x}_i^\top \boldsymbol{\beta})$.

Notice that $\mathbf{H}(\boldsymbol{\beta})$ is a positive semidefinite matrix for all values of $\boldsymbol{\beta}$, implying the convexity of $r_\tau(\boldsymbol{\beta})$. Consequently, we can find an optimal $\boldsymbol{\beta}$ efficiently; e.g., via Newton's method. Specifically, given an initial value $\boldsymbol{\beta}_0$, for $t = 1, 2, \ldots$, iteratively compute ☞ 405 ☞ 411

$$\boldsymbol{\beta}_t = \boldsymbol{\beta}_{t-1} - \mathbf{H}^{-1}(\boldsymbol{\beta}_{t-1}) \nabla r_\tau(\boldsymbol{\beta}_{t-1}), \tag{5.39}$$

until the sequence $\boldsymbol{\beta}_0, \boldsymbol{\beta}_1, \boldsymbol{\beta}_2, \ldots$ is deemed to have converged, using some pre-fixed convergence criterion.

Figure 5.9 shows the outcomes of 100 independent Bernoulli random variables, where each success probability, $(1 + \exp(-(\beta_0 + \beta_1 x)))^{-1}$, depends on x and $\beta_0 = -3$, $\beta_1 = 10$. The true logistic curve is also shown (dashed line). The minimum training loss curve (red line) is obtained via the Newton scheme (5.39), giving estimates $\widehat{\beta_0} = -2.66$ and $\widehat{\beta_1} = 10.08$. The Python code is given below.

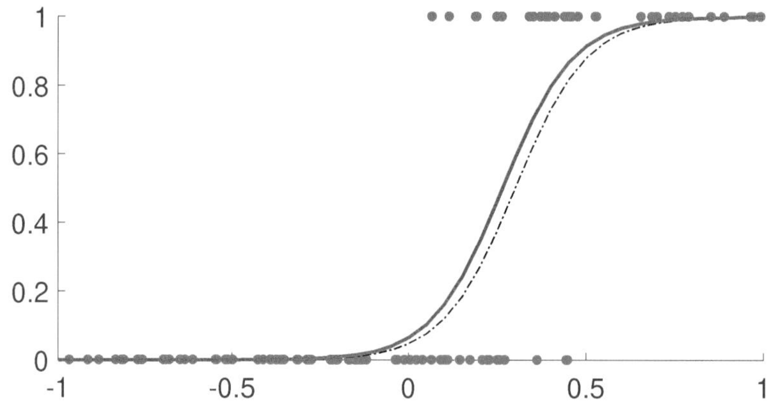

Figure 5.9: Logistic regression data (blue dots), fitted curve (red), and true curve (black dashed).

logreg1d.py

```
import numpy as np
import matplotlib.pyplot as plt
from  numpy.linalg import lstsq

n = 100                                          # sample size
x = (2*np.random.rand(n)-1).reshape(n,1)  # explanatory variables
beta = np.array([-3, 10])
Xmat = np.hstack((np.ones((n,1)), x))
p = 1/(1 + np.exp(-Xmat @ beta))
y = np.random.binomial(1,p,n)                    # response variables

# initial guess
betat = lstsq((Xmat.T @ Xmat),Xmat.T @ y, rcond=None)[0]

grad = np.array([2,1])                                 # gradient

while (np.sum(np.abs(grad)) > 1e-5) :       # stopping criteria
    mu = 1/(1+np.exp(-Xmat @ betat))
    # gradient
    delta = (mu - y).reshape(n,1)
    grad = np.sum(np.multiply( np.hstack((delta,delta)),Xmat), axis
        =0).T
    # Hessian
    H = Xmat.T @ np.diag(np.multiply(mu,(1-mu))) @ Xmat
    betat = betat - lstsq(H,grad,rcond=None)[0]
    print(betat)

plt.plot(x,y, '.') # plot data

xx = np.linspace(-1,1,40).reshape(40,1)
XXmat = np.hstack( (np.ones((len(xx),1)), xx))
yy = 1/(1 + np.exp(-XXmat @ beta))
plt.plot(xx,yy,'r-')                            #true logistic curve
yy = 1/(1 + np.exp(-XXmat @ betat));
plt.plot(xx,yy,'k--')
```

Further Reading

An excellent overview of regression is provided in [33] and an accessible mathematical treatment of *linear* regression models can be found in [108]. For extensions to *nonlinear* regression we refer the reader to [7]. A practical introduction to multilevel/hierarchical models is given in [47]. For further discussion on regression with discrete responses (classification) we refer to Chapter 7 and the further reading therein. On the important question of how to handle missing data, the classic reference is [80] (see also [85]) and a modern applied reference is [120]. ☞ 253

Exercises

1. Following his mentor Francis Galton, the mathematician/statistician Karl Pearson conducted comprehensive studies comparing hereditary traits between members of the same family. Figure 5.10 depicts the measurements of the heights of 1078 fathers and their adult sons (one son per father). The data is available from the book's GitHub site as `pearson.csv`.

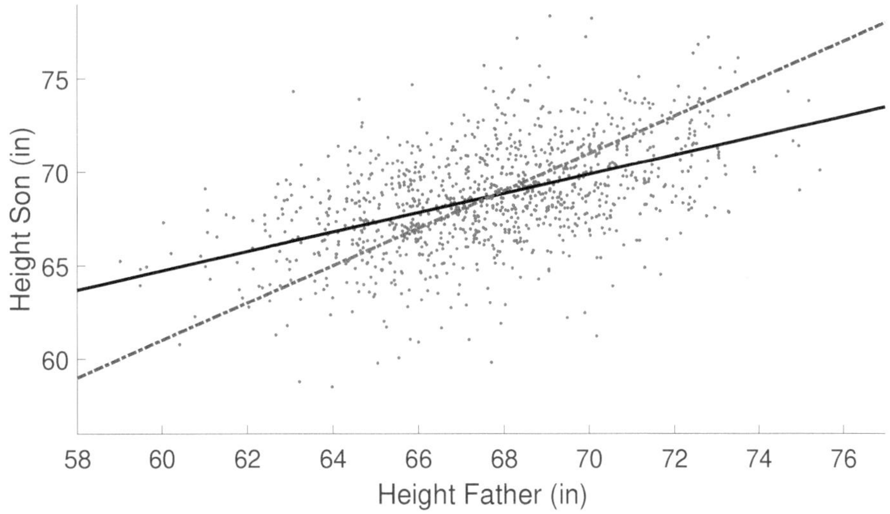

Figure 5.10: A scatterplot of heights from Pearson's data.

(a) Show that sons are on average 1 inch taller than the fathers.

(b) We could try to "explain" the height of the son by taking the height of his father and adding 1 inch. The prediction line $y = x + 1$ (red dashed) is given Figure 5.10. The black solid line is the fitted regression line. This line has a slope less than 1, and demonstrates Galton's "regression" to the average. Find the intercept and slope of the fitted regression line.

2. For the simple linear regression model, show that the values for $\widehat{\beta}_1$ and $\widehat{\beta}_0$ that solve the

equations (5.9) are:

$$\widehat{\beta}_1 = \frac{\sum_{i=1}^{n}(x_i - \overline{x})(y_i - \overline{y})}{\sum_{i=1}^{n}(x_i - \overline{x})^2} \tag{5.40}$$

$$\widehat{\beta}_0 = \overline{y} - \widehat{\beta}_1 \overline{x}, \tag{5.41}$$

provided that not all x_i are the same.

3. Edwin Hubble discovered that the universe is expanding. If v is a galaxy's recession velocity (relative to any other galaxy) and d is its distance (from that same galaxy), Hubble's law states that

$$v = Hd,$$

where H is known as Hubble's constant. The following are distance (in millions of light-years) and velocity (thousands of miles per second) measurements made on five galactic clusters.

distance	68	137	315	405	700
velocity	2.4	4.7	12.0	14.4	26.0

State the regression model and estimate H.

4. The multiple linear regression model (5.6) can be viewed as a first-order approximation of the general model

$$Y = g(\boldsymbol{x}) + \varepsilon, \tag{5.42}$$

where $\mathbb{E}\varepsilon = 0$, $\mathbb{V}\mathrm{ar}\,\varepsilon = \sigma^2$, and $g(\boldsymbol{x})$ is some known or unknown function of a d-dimensional vector $\boldsymbol{x}$ of explanatory variables. To see this, replace $g(\boldsymbol{x})$ with its first-order Taylor approximation around some point $\boldsymbol{x}_0$ and write this as $\beta_0 + \boldsymbol{x}^\top \boldsymbol{\beta}$. Express β_0 and $\boldsymbol{\beta}$ in terms of g and $\boldsymbol{x}_0$.

5. Table 5.6 shows data from an agricultural experiment where crop yield was measured for two levels of pesticide and three levels of fertilizer. There are three responses for each combination.

Table 5.6: Crop yields for pesticide and fertilizer combinations.

	Fertilizer		
Pesticide	Low	Medium	High
No	3.23, 3.20, 3.16	2.99, 2.85, 2.77	5.72, 5.77, 5.62
Yes	6.78, 6.73, 6.79	9.07, 9.09, 8.86	8.12, 8.04, 8.31

(a) Organize the data in standard form, where each row corresponds to a single measurement and the columns correspond to the response variable and the two factor variables.

(b) Let Y_{ijk} be the response for the k-th replication at level i for factor 1 and level j for factor 2. To assess which factors best explain the response variable, we use the ANOVA model

$$Y_{ijk} = \mu + \alpha_i + \beta_j + \gamma_{ij} + \varepsilon_{ijk}, \tag{5.43}$$

where $\sum_i \alpha_i = \sum_j \beta_j = \sum_i \gamma_{ij} = \sum_j \gamma_{ij} = 0$. Define $\boldsymbol{\beta} = [\mu, \alpha_1, \alpha_2, \beta_1, \beta_2, \beta_3, \gamma_{11}, \gamma_{12}, \gamma_{13}, \gamma_{21}, \gamma_{22}, \gamma_{23}]^\top$. Give the corresponding 18×12 model matrix.

(c) Note that the parameters are linearly dependent in this case. For example, $\alpha_2 = -\alpha_1$ and $\gamma_{13} = -(\gamma_{11} + \gamma_{12})$. To retain only 6 linearly independent variables consider the 6-dimensional parameter vector $\widetilde{\boldsymbol{\beta}} = [\mu, \alpha_1, \beta_1, \beta_2, \gamma_{11}, \gamma_{12}]^\top$. Find the matrix $\mathbf{M}$ such that $\mathbf{M}\widetilde{\boldsymbol{\beta}} = \boldsymbol{\beta}$.

(d) Give the model matrix corresponding to $\widetilde{\boldsymbol{\beta}}$.

6. Show that for the birthweight data in Section 5.6.6.2 there is no significant decrease in birthweight for smoking mothers. [Hint: create a new variable `nonsmoke` = 1−`smoke`, which reverses the encoding for the smoking and non-smoking mothers. Then, the parameter $\beta_1 + \beta_3$ in the original model is the same as the parameter β_1 in the model

$$\texttt{Bwt} = \beta_0 + \beta_1 \texttt{age} + \beta_2 \texttt{nonsmoke} + \beta_3 \texttt{age} \times \texttt{nonsmoke} + \varepsilon.$$

Now find a 95% for β_3 and see if it contains zero.]

7. Prove (5.37) and (5.38).

8. In the *Tobit regression* model with normally distributed errors, the response is modeled as: TOBIT REGRESSION

$$Y_i = \begin{cases} Z_i, & \text{if } u_i < Z_i \\ u_i, & \text{if } Z_i \leqslant u_i \end{cases}, \qquad \mathbf{Z} \sim \mathcal{N}(\mathbf{X}\boldsymbol{\beta}, \sigma^2 \mathbf{I}_n),$$

where the model matrix $\mathbf{X}$ and the thresholds $u_1, \ldots, u_n$ are given. Typically, $u_i = 0, i = 1, \ldots, n$. Suppose we wish to estimate $\boldsymbol{\theta} := (\boldsymbol{\beta}, \sigma^2)$ via the Expectation–Maximization method, similar to the censored data Example 4.2. Let $\boldsymbol{y} = [y_1, \ldots, y_n]^\top$ be the vector of observed data. ☞ 130

(a) Show that the likelihood of $\boldsymbol{y}$ is:

$$g(\boldsymbol{y} \,|\, \boldsymbol{\theta}) = \prod_{i:y_i > u_i} \varphi_{\sigma^2}(y_i - \boldsymbol{x}_i^\top \boldsymbol{\beta}) \times \prod_{i:y_i = u_i} \Phi((u_i - \boldsymbol{x}_i^\top \boldsymbol{\beta})/\sigma),$$

where Φ is the cdf of the $\mathcal{N}(0, 1)$ distribution and φ_{σ^2} the pdf of the $\mathcal{N}(0, \sigma^2)$ distribution.

(b) Let $\overline{\boldsymbol{y}}$ and $\underline{\boldsymbol{y}}$ be vectors that collect all $y_i > u_i$ and $y_i = u_i$, respectively. Denote the corresponding matrix of predictors by $\overline{\mathbf{X}}$ and $\underline{\mathbf{X}}$, respectively. For each observation $y_i = u_i$ introduce a latent variable z_i and collect these into a vector $\boldsymbol{z}$. For the same indices i collect the corresponding u_i into a vector $\boldsymbol{c}$. Show that the complete-data likelihood is given by

$$g(\boldsymbol{y}, \boldsymbol{z} \,|\, \boldsymbol{\theta}) = \frac{1}{(2\pi\sigma^2)^{n/2}} \exp\left(-\frac{\|\overline{\boldsymbol{y}} - \overline{\mathbf{X}}\boldsymbol{\beta}\|^2}{2\sigma^2} - \frac{\|\boldsymbol{z} - \underline{\mathbf{X}}\boldsymbol{\beta}\|^2}{2\sigma^2} \right) \mathbb{1}\{\boldsymbol{z} \leqslant \boldsymbol{c}\}.$$

(c) For the E-step, show that, for a fixed $\boldsymbol{\theta}$,

$$g(\boldsymbol{z} \mid \boldsymbol{y}, \boldsymbol{\theta}) = \prod_i g(z_i \mid \boldsymbol{y}, \boldsymbol{\theta}),$$

where each $g(z_i \mid \boldsymbol{y}, \boldsymbol{\theta})$ is the pdf of the $\mathcal{N}((\underline{\mathbf{X}}\boldsymbol{\beta})_i, \sigma^2)$ distribution, truncated to the interval $(-\infty, c_i]$.

(d) For the M-step, compute the expectation of the complete log-likelihood

$$-\frac{n}{2}\ln \sigma^2 - \frac{n}{2}\ln(2\pi) - \frac{\|\overline{\boldsymbol{y}} - \overline{\mathbf{X}}\boldsymbol{\beta}\|^2}{2\sigma^2} - \frac{\mathbb{E}\|\boldsymbol{Z} - \underline{\mathbf{X}}\boldsymbol{\beta}\|^2}{2\sigma^2}.$$

Then, derive the formulas for $\boldsymbol{\beta}$ and σ^2 that maximize the expectation of the complete log-likelihood.

9. Dowload data set `WomenWage.csv` from the book's website. This data set is a tidied-up version of the women's wages data set from [91]. The first column of the data (`hours`) is the response variable Y. It shows the hours spent in the labor force by married women in the 1970s. We want to understand what factors determine the participation rate of women in the labor force. The predictor variables are:

Table 5.7: Features for the women's wage data set.

Feature	**Description**
`kidslt6`	Number of children younger than 6 years.
`kidsge6`	Number of children older than 6 years.
`age`	Age of the married woman.
`educ`	Number of years of formal education.
`exper`	Number of years of "work experience".
`nwifeinc`	Non-wife income, that is, the income of the husband.
`expersq`	The square of `exper`, to capture any nonlinear relationships.

We observe that some of the responses are $Y = 0$, that is, some women did not participate in the labor force. For this reason, we model the data using the Tobit regression model, in which the response Y is given as:

$$Y_i = \begin{cases} Z_i, & \text{if } Z_i > 0 \\ 0, & \text{if } Z_i \leqslant 0 \end{cases}, \qquad \boldsymbol{Z} \sim \mathcal{N}(\mathbf{X}\boldsymbol{\beta}, \sigma^2 \mathbf{I}_n).$$

With $\boldsymbol{\theta} = (\boldsymbol{\beta}, \sigma^2)$, the likelihood of the data $\boldsymbol{y} = [y_1, \ldots, y_n]^\top$ is:

$$g(\boldsymbol{y} \mid \boldsymbol{\theta}) = \textstyle\prod_{i:y_i>0} \varphi_{\sigma^2}(y_i - \boldsymbol{x}_i^\top \boldsymbol{\beta}) \times \prod_{i:y_i=0} \Phi((u_i - \boldsymbol{x}_i^\top \boldsymbol{\beta})/\sigma),$$

where Φ is the standard normal cdf. In Exercise 8, we derived the EM algorithm for maximizing the log-likelihood.

(a) Write down the EM algorithm in pseudo code as it applies to this Tobit regression.

(b) Implement the EM algorithm pseudo code in Python. Comment on which factor you think is important in determining the labor participation rate of women living in the USA in the 1970s.

10. Let $\mathbf{P}$ be a projection matrix. Show that the diagonal elements of $\mathbf{P}$ all lie in the interval $[0, 1]$. In particular, for $\mathbf{P} = \mathbf{X}\mathbf{X}^+$ in Theorem 5.1, the leverage value $p_i := \mathbf{P}_{ii}$ satisfies $0 \leqslant p_i \leqslant 1$ for all i.

11. Consider the linear model $\boldsymbol{Y} = \mathbf{X}\boldsymbol{\beta} + \boldsymbol{\varepsilon}$ in (5.8), with $\mathbf{X}$ being the $n \times p$ model matrix and $\boldsymbol{\varepsilon}$ having expectation vector $\mathbf{0}$ and covariance matrix $\sigma^2\mathbf{I}_n$. Suppose that $\widehat{\boldsymbol{\beta}}_{-i}$ is the least-squares estimate obtained by omitting the i-th observation, Y_i; that is,

$$\widehat{\boldsymbol{\beta}}_{-i} = \underset{\boldsymbol{\beta}}{\operatorname{argmin}} \sum_{j \neq i} (Y_j - \boldsymbol{x}_j^\top \boldsymbol{\beta})^2,$$

where $\boldsymbol{x}_j^\top$ is the j-th row of $\mathbf{X}$. Let $\widehat{Y}_{-i} = \boldsymbol{x}_i^\top \widehat{\boldsymbol{\beta}}_{-i}$ be the corresponding fitted value at $\boldsymbol{x}_i$. Also, define $\boldsymbol{B}_i$ as the least-squares estimator of $\boldsymbol{\beta}$ based on the response data

$$\boldsymbol{Y}^{(i)} := [Y_1, \ldots, Y_{i-1}, \widehat{Y}_{-i}, Y_{i+1}, \ldots, Y_n]^\top.$$

(a) Prove that $\widehat{\boldsymbol{\beta}}_{-i} = \boldsymbol{B}_i$; that is, the linear model obtained from fitting all responses except the i-th is the same as the one obtained from fitting the data $\boldsymbol{Y}^{(i)}$.

(b) Use the previous result to verify that

$$Y_i - \widehat{Y}_{-i} = (Y_i - \widehat{Y}_i)/(1 - \mathbf{P}_{ii}),$$

where $\mathbf{P} = \mathbf{X}\mathbf{X}^+$ is the projection matrix onto the columns of $\mathbf{X}$. Hence, deduce the PRESS formula in Theorem 5.1. ☞ 174

12. Take the linear model $\boldsymbol{Y} = \mathbf{X}\boldsymbol{\beta} + \boldsymbol{\varepsilon}$,where $\mathbf{X}$ is an $n \times p$ model matrix, $\boldsymbol{\varepsilon} = \mathbf{0}$, and $\mathbb{C}\mathrm{ov}(\boldsymbol{\varepsilon}) = \sigma^2\mathbf{I}_n$. Let $\mathbf{P} = \mathbf{X}\mathbf{X}^+$ be the projection matrix onto the columns of $\mathbf{X}$.

(a) Using the properties of the pseudo-inverse (see Definition A.2), show that $\mathbf{P}\mathbf{P}^\top = \mathbf{P}$. ☞ 362

(b) Let $\boldsymbol{E} = \boldsymbol{Y} - \widehat{\boldsymbol{Y}}$ be the (random) vector of residuals, where $\widehat{\boldsymbol{Y}} = \mathbf{P}\boldsymbol{Y}$. Show that the i-th residual has a normal distribution with expectation 0 and variance $\sigma^2(1 - \mathbf{P}_{ii})$ (that is, σ^2 times 1 minus the i-th leverage).

(c) Show that σ^2 can be unbiasedly estimated via

$$S^2 := \frac{1}{n-p}\|\boldsymbol{Y} - \widehat{\boldsymbol{Y}}\|^2 = \frac{1}{n-p}\|\boldsymbol{Y} - \mathbf{X}\widehat{\boldsymbol{\beta}}\|^2. \tag{5.44}$$

[Hint: use the cyclic property of the trace as in Example 2.3.]

13. Consider a normal linear model $\boldsymbol{Y} = \mathbf{X}\boldsymbol{\beta} + \boldsymbol{\varepsilon}$, where $\mathbf{X}$ is an $n \times p$ model matrix and $\boldsymbol{\varepsilon} \sim \mathcal{N}(\mathbf{0}, \sigma^2\mathbf{I}_n)$. Exercise 12 shows that for any such model the i-th standardized residual $E_i/(\sigma\sqrt{1 - \mathbf{P}_{ii}})$ has a standard normal distribution. This motivates the use of the leverage $\mathbf{P}_{ii}$ to assess whether the i-th observation is an outlier depending on the size of the i-th residual relative to $\sqrt{1 - \mathbf{P}_{ii}}$. A more robust approach is to include an estimate for σ using

STUDENTIZED RESIDUAL

all data except the i-th observation. This gives rise to the *studentized residual* T_i, defined as

$$T_i := \frac{E_i}{S_{-i}\sqrt{1-\mathbf{P}_{ii}}},$$

where S_{-i} is an estimate of σ obtained by fitting all the observations except the i-th and $E_i = Y_i - \widehat{Y}_i$ is the i-th (random) residual. Exercise 12 shows that we can take, for example,

$$S_{-i}^2 = \frac{1}{n-1-p}\|\boldsymbol{Y}_{-i} - \mathbf{X}_{-i}\widehat{\boldsymbol{\beta}}_{-i}\|^2, \tag{5.45}$$

where $\mathbf{X}_{-i}$ is the model matrix $\mathbf{X}$ with the i-th row removed, is an unbiased estimator of σ^2. We wish to compute S_{-i}^2 efficiently, using S^2 in (5.44), as the latter will typically be available once we have fitted the linear model. To this end, define $\boldsymbol{u}_i$ as the i-th unit vector $[0,\ldots,0,1,0,\ldots,0]^\top$, and let

$$\boldsymbol{Y}^{(i)} := \boldsymbol{Y} - (Y_i - \widehat{Y}_{-i})\boldsymbol{u}_i = \boldsymbol{Y} - \frac{E_i}{1-\mathbf{P}_{ii}}\boldsymbol{u}_i,$$

where we have used the fact that $Y_i - \widehat{Y}_{-i} = E_i/(1-\mathbf{P}_{ii})$, as derived in the proof of Theorem 5.1. Now apply Exercise 11 to prove that

$$S_{-i}^2 = \frac{(n-p)\,S^2 - E_i^2/(1-\mathbf{P}_{ii})}{n-p-1}.$$

COOK'S DISTANCE

14. Using the notation from Exercises 11–13, *Cook's distance* for observation i is defined as

$$D_i := \frac{\|\widehat{\boldsymbol{Y}} - \widehat{\boldsymbol{Y}}^{(i)}\|^2}{p\,S^2}.$$

It measures the change in the fitted values when the i-th observation is removed, relative to the residual variance of the model (estimated via S^2).

By using similar arguments as those in Exercise 13, show that

$$D_i = \frac{\mathbf{P}_{ii}\,E_i^2}{(1-\mathbf{P}_{ii})^2 p\,S^2}.$$

It follows that there is no need to "omit and refit" the linear model in order to compute Cook's distance for the i-th response.

15. Prove that if we add an additional feature to the general linear model, then R^2, the coefficient of determination, is necessarily non-decreasing in value and hence cannot be used to compare models with different numbers of predictors.

16. Let $\boldsymbol{X} := [X_1,\ldots,X_n]^\top$ and $\boldsymbol{\mu} := [\mu_1,\ldots,\mu_n]^\top$. In the fundamental Theorem C.9, we use the fact that if $X_i \sim \mathcal{N}(\mu_i, 1)$, $i = 1,\ldots,n$ are independent, then $\|\boldsymbol{X}\|^2$ has (per definition) a noncentral χ_n^2 distribution. Show that $\|\boldsymbol{X}\|^2$ has moment generating function

$$\frac{\mathrm{e}^{t\|\boldsymbol{\mu}\|^2/(1-2t)}}{(1-2t)^{n/2}}, \quad t < 1/2,$$

and so the distribution of $\|\boldsymbol{X}\|^2$ depends on $\boldsymbol{\mu}$ only through the norm $\|\boldsymbol{\mu}\|$.

17. Carry out a logistic regression analysis on a (partial) *wine* data set classification problem. The data can be loaded using the following code.

```
from sklearn import datasets
import numpy as np
data = datasets.load_wine()
X = data.data[:, [9,10]]
y = np.array(data.target==1,dtype=np.uint)
X = np.append(np.ones(len(X)).reshape(-1,1),X,axis=1)
```

The model matrix has three features, including the constant feature. Instead of using Newton's method (5.39) to estimate $\boldsymbol{\beta}$, implement a simple gradient descent procedure

$$\boldsymbol{\beta}_t = \boldsymbol{\beta}_{t-1} - \alpha \nabla r_\tau(\boldsymbol{\beta}_{t-1}),$$

with learning rate $\alpha = 0.0001$, and run it for 10^6 steps. Your procedure should deliver three coefficients; one for the intercept and the rest for the explanatory variables. Solve the same problem using the `Logit` method of `statsmodels.api` and compare the results.

18. Consider again Example 5.10, where we train the learner via the Newton iteration (5.39). If $\mathbf{X}^\top := [\boldsymbol{x}_1, \ldots, \boldsymbol{x}_n]$ defines the matrix of predictors and $\boldsymbol{\mu}_t := \boldsymbol{h}(\mathbf{X}\boldsymbol{\beta}_t)$, then the gradient (5.37) and Hessian (5.38) for Newton's method can be written as: ☞ 205

$$\nabla r_\tau(\boldsymbol{\beta}_t) = \frac{1}{n}\mathbf{X}^\top(\boldsymbol{\mu}_t - \boldsymbol{y}) \quad \text{and} \quad \mathbf{H}(\boldsymbol{\beta}_t) = \frac{1}{n}\mathbf{X}^\top \mathbf{D}_t \mathbf{X},$$

where $\mathbf{D}_t := \mathrm{diag}(\boldsymbol{\mu}_t \odot (\mathbf{1} - \boldsymbol{\mu}_t))$ is a diagonal matrix. Show that the Newton iteration (5.39) can be written as the *iterative reweighted least-squares* method:

ITERATIVE REWEIGHTED LEAST SQUARES

$$\boldsymbol{\beta}_t = \underset{\boldsymbol{\beta}}{\mathrm{argmin}}\, (\widetilde{\boldsymbol{y}}_{t-1} - \mathbf{X}\boldsymbol{\beta})^\top \mathbf{D}_{t-1} (\widetilde{\boldsymbol{y}}_{t-1} - \mathbf{X}\boldsymbol{\beta}),$$

where $\widetilde{\boldsymbol{y}}_{t-1} := \mathbf{X}\boldsymbol{\beta}_{t-1} + \mathbf{D}_{t-1}^{-1}(\boldsymbol{y} - \boldsymbol{\mu}_{t-1})$ is the so-called *adjusted response*. [Hint: use the fact that $(\mathbf{M}^\top\mathbf{M})^{-1}\mathbf{M}^\top \boldsymbol{z}$ is the minimizer of $\|\mathbf{M}\boldsymbol{\beta} - \boldsymbol{z}\|^2$.]

19. In *multi-output linear regression*, the response variable is a real-valued vector of dimension, say, m. Similar to (5.8), the model can be written in matrix notation:

MULTI-OUTPUT LINEAR REGRESSION

$$\mathbf{Y} = \mathbf{X}\mathbf{B} + \begin{bmatrix} \boldsymbol{\varepsilon}_1^\top \\ \vdots \\ \boldsymbol{\varepsilon}_n^\top \end{bmatrix},$$

where:

- $\mathbf{Y}$ is an $n \times m$ matrix of n independent responses (stored as row vectors of length m);
- $\mathbf{X}$ is the usual $n \times p$ model matrix;
- $\mathbf{B}$ is an $p \times m$ matrix of model parameters;
- $\boldsymbol{\varepsilon}_1, \ldots, \boldsymbol{\varepsilon}_n \in \mathbb{R}^m$ are independent error terms with $\mathbb{E}\,\boldsymbol{\varepsilon} = \mathbf{0}$ and $\mathbb{E}\,\boldsymbol{\varepsilon}\boldsymbol{\varepsilon}^\top = \boldsymbol{\Sigma}$.

We wish to learn the matrix parameters $\mathbf{B}$ and $\mathbf{\Sigma}$ from the training set $\{\mathbf{Y}, \mathbf{X}\}$. To this end, consider minimizing the training loss:

$$\frac{1}{n}\mathrm{tr}\left((\mathbf{Y} - \mathbf{X}\mathbf{B})\,\mathbf{\Sigma}^{-1}\,(\mathbf{Y} - \mathbf{X}\mathbf{B})^\top\right),$$

☞ 359 where tr(·) is the trace of a matrix.

(a) Show that the minimizer of the training loss, denoted $\widehat{\mathbf{B}}$, satisfies the normal equations:

$$\mathbf{X}^\top \mathbf{X}\,\widehat{\mathbf{B}} = \mathbf{X}^\top \mathbf{Y}.$$

(b) Noting that

$$(\mathbf{Y} - \mathbf{X}\mathbf{B})^\top(\mathbf{Y} - \mathbf{X}\mathbf{B}) = \sum_{i=1}^{n} \boldsymbol{\varepsilon}_i \boldsymbol{\varepsilon}_i^\top,$$

explain why

$$\widehat{\mathbf{\Sigma}} := \frac{(\mathbf{Y} - \mathbf{X}\widehat{\mathbf{B}})^\top(\mathbf{Y} - \mathbf{X}\widehat{\mathbf{B}})}{n}$$

is a method-of-moments estimator of $\mathbf{\Sigma}$, just like the one given in (5.10).

CHAPTER 6

REGULARIZATION AND KERNEL METHODS

The purpose of this chapter is to familiarize the reader with two central concepts in modern data science and machine learning: regularization and kernel methods. Regularization provides a natural way to guard against overfitting and kernel methods offer a broad generalization of linear models. Here, we discuss regularized regression (ridge, lasso) as a bridge to the fundamentals of kernel methods. We introduce reproducing kernel Hilbert spaces and show that selecting the best prediction function in such spaces is in fact a finite-dimensional optimization problem. Applications to spline fitting, Gaussian process regression, and kernel PCA are given.

6.1 Introduction

In this chapter we return to the supervised learning setting of Chapter 5 (regression) and expand its scope. Given training data $\tau = \{(\boldsymbol{x}_1, y_1), \ldots, (\boldsymbol{x}_n, y_n)\}$, we wish to find a prediction function (the learner) g_τ that minimizes the (squared-error) training loss

$$\ell_\tau(g) = \frac{1}{n}\sum_{i=1}^{n}(y_i - g(\boldsymbol{x}_i))^2$$

within a class of functions $\mathcal{G}$. As noted in Chapter 2, if $\mathcal{G}$ is the set of all possible functions then choosing *any* function g with the property that $g(\boldsymbol{x}_i) = y_i$ for all i will give zero training loss, but will likely have poor generalization performance (that is, suffer from overfitting).

Recall from Theorem 2.1 that the best possible prediction function (over all g) for the squared-error *risk* $\mathbb{E}(Y - g(\boldsymbol{X}))^2$ is given by $g^*(\boldsymbol{x}) = \mathbb{E}[Y \mid \boldsymbol{X} = \boldsymbol{x}]$. The class $\mathcal{G}$ should be simple enough to permit theoretical understanding and analysis but, at the same time, rich enough to contain the optimal function g^* (or a function close to g^*). This ideal can be realized by taking $\mathcal{G}$ to be a *Hilbert space* (i.e., a complete inner product space) of functions; see Appendix A.7. ☞ 21

HILBERT SPACE ☞ 386

Many of the classes of functions that we have encountered so far are in fact Hilbert spaces. In particular, the set $\mathcal{G}$ of *linear* functions on $\mathbb{R}^p$ is a Hilbert space. To see this, identify with each element $\boldsymbol{\beta} \in \mathbb{R}^p$ the linear function $g_{\boldsymbol{\beta}} : \boldsymbol{x} \mapsto \boldsymbol{x}^\top \boldsymbol{\beta}$ and define the inner product on $\mathcal{G}$ as $\langle g_{\boldsymbol{\beta}}, g_{\boldsymbol{\gamma}} \rangle := \boldsymbol{\beta}^\top \boldsymbol{\gamma}$. In this way, $\mathcal{G}$ behaves in exactly the same way as (is isomorphic to) the space $\mathbb{R}^p$ equipped with the Euclidean inner product (dot product). The latter is a Hilbert space, because it is *complete* with respect to the Euclidean norm. See Exercise 12 for a further discussion.

☞ 362

COMPLETE VECTOR SPACE

☞ 26

Let us now turn to our "running" polynomial regression Example 2.1, where the feature vector $\boldsymbol{x} = [1, u, u^2, \ldots, u^{p-1}]^\top =: \boldsymbol{\phi}(u)$ is itself a vector-valued function of another feature u. Then, the space of functions $h_{\boldsymbol{\beta}} : u \mapsto \boldsymbol{\phi}(u)^\top \boldsymbol{\beta}$ is a Hilbert space, through the identification $h_{\boldsymbol{\beta}} \equiv \boldsymbol{\beta}$. In fact, this is true for *any* feature mapping $\boldsymbol{\phi} : u \mapsto [\phi_1(u), \ldots, \phi_p(u)]^\top$.

FEATURE MAPS

This can be further generalized by considering *feature maps* $u \mapsto \kappa_u$, where each κ_u is a real-valued *function* $v \mapsto \kappa_u(v)$ on the feature space. As we shall soon see (in Section 6.3), functions of the form $u \mapsto \sum_{i=1}^{\infty} \beta_i \kappa_{v_i}(u)$ live in a Hilbert space of functions called a *reproducing kernel Hilbert space* (RKHS). In Section 6.3 we introduce the notion of a RKHS formally, give specific examples, including the linear and Gaussian kernels, and derive various useful properties, the most important of which is the representer Theorem 6.6. Applications of such spaces include the *smoothing splines* (Section 6.6), Gaussian process regression (Section 6.7), kernel PCA (Section 6.8), and *support vector machines* for classification (Section 7.7).

RKHS

☞ 235

☞ 271

REGULARIZATION

The RKHS formalism also makes it easier to treat the important topic of *regularization*. The aim of regularization is to improve the predictive performance of the best learner in some class of functions $\mathcal{G}$ by adding a penalty term to the training loss that penalizes learners that tend to overfit the data. In the next section we introduce the main ideas behind regularization, which then segues into a discussion of kernel methods in the subsequent sections.

6.2 Regularization

Let $\mathcal{G}$ be the Hilbert space of functions over which we search for the minimizer, g_τ, of the training loss $\ell_\tau(g)$. Often, the Hilbert space $\mathcal{G}$ is rich enough so that we can find a learner g_τ within $\mathcal{G}$ such that the training loss is zero or close to zero. Consequently, if the space of functions $\mathcal{G}$ is sufficiently rich, we run the risk of overfitting. One way to avoid overfitting is to restrict attention to a subset of the space $\mathcal{G}$ by introducing a non-negative functional $J : \mathcal{G} \to \mathbb{R}_+$ which penalizes complex models (functions). In particular, we want to find functions $g \in \mathcal{G}$ such that $J(g) < c$ for some "regularization" constant $c > 0$. Thus we can formulate the quintessential supervised learning problem as:

$$\min \{\ell_\tau(g) \ : \ g \in \mathcal{G}, J(g) < c\}, \tag{6.1}$$

the solution (argmin) of which is our learner. When this optimization problem is convex, it can be solved by first obtaining the Lagrangian dual function

$$\mathcal{L}^*(\lambda) := \min_{g \in \mathcal{G}} \{\ell_\tau(g) + \lambda(J(g) - c)\},$$

☞ 409

and then maximizing $\mathcal{L}^*(\lambda)$ with respect to $\lambda \geqslant 0$; see Section B.2.3.

RIDGE REGRESSION

In order to introduce the overall ideas of kernel methods and regularization, we will proceed by exploring (6.1) in the special case of *ridge regression*, with the following running example.

■ **Example 6.1 (Ridge Regression)** *Ridge regression* is simply linear regression with a squared-norm penalty functional (also called a regularization function, or *regularizer*). Suppose we have a training set $\tau = \{(\boldsymbol{x}_i, y_i), i = 1, \ldots, n\}$, with each $\boldsymbol{x}_i \in \mathbb{R}^p$ and we use a squared-norm penalty with *regularization parameter* $\gamma > 0$. Then, the problem is to solve

REGULARIZER

REGULARIZATION PARAMETER

$$\min_{g \in \mathcal{G}} \frac{1}{n} \sum_{i=1}^{n} (y_i - g(\boldsymbol{x}_i))^2 + \gamma \, \|g\|^2, \tag{6.2}$$

where $\mathcal{G}$ is the Hilbert space of linear functions on $\mathbb{R}^p$. As explained in Section 6.1, we can identify each $g \in \mathcal{G}$ with a vector $\boldsymbol{\beta} \in \mathbb{R}^p$ and, consequently, $\|g\|^2 = \langle \boldsymbol{\beta}, \boldsymbol{\beta} \rangle = \|\boldsymbol{\beta}\|^2$. The above *functional* optimization problem is thus equivalent to the *parametric* optimization problem

$$\min_{\boldsymbol{\beta} \in \mathbb{R}^p} \frac{1}{n} \sum_{i=1}^{n} (y_i - \boldsymbol{x}_i^\top \boldsymbol{\beta})^2 + \gamma \, \|\boldsymbol{\beta}\|^2, \tag{6.3}$$

which, in the notation of Chapter 5, further simplifies to

$$\min_{\boldsymbol{\beta} \in \mathbb{R}^p} \frac{1}{n} \| \boldsymbol{y} - \mathbf{X}\boldsymbol{\beta} \|^2 + \gamma \, \|\boldsymbol{\beta}\|^2. \tag{6.4}$$

In other words, the solution to (6.2) is of the form $\boldsymbol{x} \mapsto \boldsymbol{x}^\top \boldsymbol{\beta}^*$, where $\boldsymbol{\beta}^*$ solves (6.3) (or equivalently (6.4)). Observe that as $\gamma \to \infty$, the regularization term becomes dominant and consequently the optimal g becomes identically zero.

The optimization problem in (6.4) is convex, and by multiplying by the constant $n/2$ and setting the gradient equal to zero, we obtain

$$\mathbf{X}^\top (\mathbf{X}\boldsymbol{\beta} - \boldsymbol{y}) + n\,\gamma\,\boldsymbol{\beta} = \mathbf{0}. \tag{6.5}$$

If $\gamma = 0$ these are simply the *normal equations*, albeit written in a slightly different form. If the matrix $\mathbf{X}^\top \mathbf{X} + n\,\gamma \mathbf{I}_p$ is invertible (which is the case for any $\gamma > 0$; see Exercise 13), then the solution to these modified normal equations is

☞ 28

$$\widehat{\boldsymbol{\beta}} = (\mathbf{X}^\top \mathbf{X} + n\,\gamma \mathbf{I}_p)^{-1} \mathbf{X}^\top \boldsymbol{y}.$$

■

When using regularization with respect to some Hilbert space $\mathcal{G}$, it is sometimes useful to decompose $\mathcal{G}$ into two orthogonal subspaces, $\mathcal{H}$ and $\mathcal{C}$ say, such that every $g \in \mathcal{G}$ can be uniquely written as $g = h + c$, with $h \in \mathcal{H}$, $c \in \mathcal{C}$, and $\langle h, c \rangle = 0$. Such a $\mathcal{G}$ is said to be the *direct sum* of $\mathcal{C}$ and $\mathcal{H}$, and we write $\mathcal{G} = \mathcal{H} \oplus \mathcal{C}$. Decompositions of this form become useful when functions in $\mathcal{H}$ are penalized but functions in $\mathcal{C}$ are not. We illustrate this decomposition with the ridge regression example where one of the features is a constant term, which we do not wish to penalize.

DIRECT SUM

■ **Example 6.2 (Ridge Regression (cont.))** Suppose one of the features in Example 6.1 is the constant 1, which we do not wish to penalize. The reason for this is to ensure that when $\gamma \to \infty$, the optimal g becomes the "constant" model, $g(\boldsymbol{x}) = \beta_0$, rather than the "zero" model, $g(\boldsymbol{x}) = 0$. Let us alter the notation slightly by considering the feature vectors to be of the form $\widetilde{\boldsymbol{x}} = [1, \boldsymbol{x}^\top]^\top$, where $\boldsymbol{x} = [x_1, \ldots, x_p]^\top$. We thus have $p + 1$ features, rather than p. Let $\mathcal{G}$ be the space of linear functions of $\widetilde{\boldsymbol{x}}$. Each linear function g of $\widetilde{\boldsymbol{x}}$ can be written as $g : \widetilde{\boldsymbol{x}} \mapsto \beta_0 + \boldsymbol{x}^\top \boldsymbol{\beta}$, which is the sum of the constant function $c : \widetilde{\boldsymbol{x}} \mapsto \beta_0$ and $h : \widetilde{\boldsymbol{x}} \mapsto \boldsymbol{x}^\top \boldsymbol{\beta}$. Moreover, the two functions are orthogonal with respect to the inner product on $\mathcal{G} : \langle c, h \rangle = [\beta_0, \boldsymbol{0}^\top][0, \boldsymbol{\beta}^\top]^\top = 0$, where $\boldsymbol{0}$ is a column vector of zeros.

As subspaces of $\mathcal{G}$, both $\mathcal{C}$ and $\mathcal{H}$ are again Hilbert spaces, and their inner products and norms follow directly from the inner product on $\mathcal{G}$. For example, each function $h : \widetilde{\boldsymbol{x}} \mapsto \boldsymbol{x}^\top \boldsymbol{\beta}$ in $\mathcal{H}$ has norm $\|h\|_{\mathcal{H}} = \|\boldsymbol{\beta}\|$, and the constant function $c : \widetilde{\boldsymbol{x}} \mapsto \beta_0$ in $\mathcal{C}$ has norm $|\beta_0|$.

The modification of the regularized optimization problem (6.2) where the constant term is not penalized can now be written as

$$\min_{g \in \mathcal{H} \oplus \mathcal{C}} \frac{1}{n} \sum_{i=1}^{n} (y_i - g(\widetilde{\boldsymbol{x}}_i))^2 + \gamma \, \|g\|_{\mathcal{H}}^2, \tag{6.6}$$

which further simplifies to

$$\min_{\beta_0, \boldsymbol{\beta}} \frac{1}{n} \, \| \boldsymbol{y} - \beta_0 \boldsymbol{1} - \mathbf{X}\boldsymbol{\beta} \|^2 + \gamma \, \|\boldsymbol{\beta}\|^2, \tag{6.7}$$

where $\boldsymbol{1}$ is the $n \times 1$ vector of 1s. Observe that, in this case, as $\gamma \to \infty$ the optimal g tends to the sample mean $\overline{y}$ of the $\{y_i\}$; that is, we obtain the "default" regression model, without explanatory variables. Again, this is a convex optimization problem, and the solution follows from

$$\mathbf{X}^\top (\beta_0 \boldsymbol{1} + \mathbf{X}\boldsymbol{\beta} - \boldsymbol{y}) + n\,\gamma\,\boldsymbol{\beta} = \boldsymbol{0}, \tag{6.8}$$

with

$$n\beta_0 = \boldsymbol{1}^\top (\boldsymbol{y} - \mathbf{X}\boldsymbol{\beta}). \tag{6.9}$$

This results in solving for $\boldsymbol{\beta}$ from

$$(\mathbf{X}^\top \mathbf{X} - n^{-1} \mathbf{X}^\top \boldsymbol{1}\boldsymbol{1}^\top \mathbf{X} + n\,\gamma\,\mathbf{I}_p)\boldsymbol{\beta} = (\mathbf{X}^\top - n^{-1} \mathbf{X}^\top \boldsymbol{1}\boldsymbol{1}^\top)\boldsymbol{y}, \tag{6.10}$$

and determining β_0 from (6.9).

As a precursor to the kernel methods in the following sections, let us assume that $n \geqslant p$ and that $\mathbf{X}$ has full (column) rank p. Then any vector $\boldsymbol{\beta} \in \mathbb{R}^p$ can be written as a linear combination of the feature vectors $\{\boldsymbol{x}_i\}$; that is, as linear combinations of the columns of the matrix $\mathbf{X}^\top$. In particular, let $\boldsymbol{\beta} = \mathbf{X}^\top \boldsymbol{\alpha}$, where $\boldsymbol{\alpha} = [\alpha_1, \ldots, \alpha_n]^\top \in \mathbb{R}^n$. In this case (6.10) reduces to

$$(\mathbf{X}\mathbf{X}^\top - n^{-1} \boldsymbol{1}\boldsymbol{1}^\top \mathbf{X}\mathbf{X}^\top + n\,\gamma\,\mathbf{I}_n)\boldsymbol{\alpha} = (\mathbf{I}_n - n^{-1} \boldsymbol{1}\boldsymbol{1}^\top)\boldsymbol{y}.$$

Assuming invertibility of $(\mathbf{X}\mathbf{X}^\top - n^{-1} \boldsymbol{1}\boldsymbol{1}^\top \mathbf{X}\mathbf{X}^\top + n\,\gamma\,\mathbf{I}_n)$, we have the solution

$$\widehat{\boldsymbol{\alpha}} = (\mathbf{X}\mathbf{X}^\top - n^{-1} \boldsymbol{1}\boldsymbol{1}^\top \mathbf{X}\mathbf{X}^\top + n\,\gamma\,\mathbf{I}_n)^{-1} (\mathbf{I}_n - n^{-1} \boldsymbol{1}\boldsymbol{1}^\top)\boldsymbol{y},$$

which depends on the training feature vectors $\{\boldsymbol{x}_i\}$ only through the $n \times n$ matrix of inner products: $\mathbf{X}\mathbf{X}^\top = [\langle \boldsymbol{x}_i, \boldsymbol{x}_j \rangle]$. This matrix is called the *Gram matrix* of the $\{\boldsymbol{x}_i\}$. From (6.9), the solution for the constant term is $\widehat{\beta}_0 = n^{-1} \boldsymbol{1}^\top (\boldsymbol{y} - \mathbf{X}\mathbf{X}^\top \widehat{\boldsymbol{\alpha}})$. It follows that the learner is a linear combination of inner products $\{\langle \boldsymbol{x}_i, \boldsymbol{x} \rangle\}$ plus a constant:

GRAM MATRIX

$$g_\tau(\widetilde{\boldsymbol{x}}) = \widehat{\beta}_0 + \boldsymbol{x}^\top \mathbf{X}^\top \widehat{\boldsymbol{\alpha}} = \widehat{\beta}_0 + \sum_{i=1}^{n} \widehat{\alpha}_i \, \langle \boldsymbol{x}_i, \boldsymbol{x} \rangle,$$

where the coefficients $\widehat{\beta}_0$ and $\widehat{\alpha}_i$ only depend on the inner products $\{\langle \boldsymbol{x}_i, \boldsymbol{x}_j \rangle\}$. We will see shortly that the representer Theorem 6.6 generalizes this result to a broad class of regularized optimization problems. ■

☞ 232

We illustrate in Figure 6.1 how the solutions of the ridge regression problems appearing in Examples 6.1 and 6.2 are qualitatively affected by the regularization parameter γ for a simple linear regression model. The data was generated from the model $y_i = -1.5 + 0.5x_i + \varepsilon_i$, $i = 1, \dots, 100$, where each x_i is drawn independently and uniformly from the interval $[0, 10]$ and each ε_i is drawn independently from the standard normal distribution.

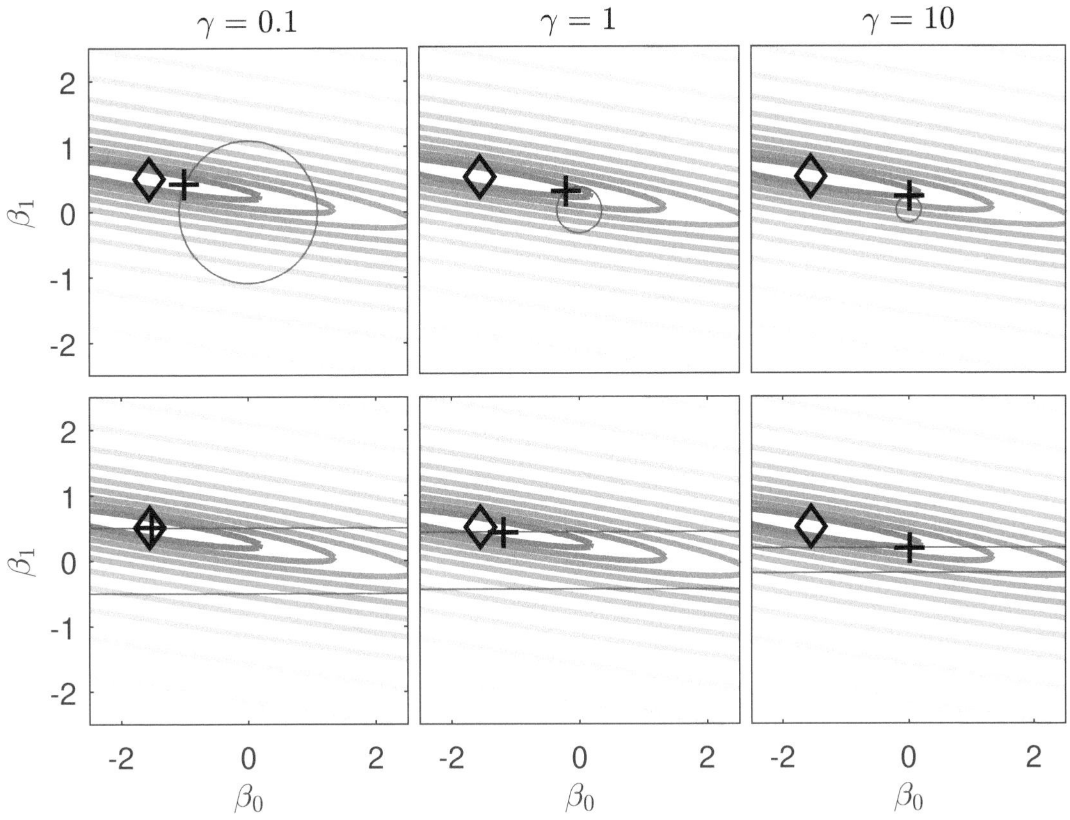

Figure 6.1: Ridge regression solutions for a simple linear regression problem. Each panel shows contours of the loss function (log scale) and the effect of the regularization parameter $\gamma \in \{0.1, 1, 10\}$, appearing in (6.4) and (6.7). Top row: both terms are penalized. Bottom row: only the non-constant term is penalized. Penalized (plus) and unpenalized (diamond) solutions are shown in each case.

The contours are those of the squared-error loss (actually the logarithm thereof), which is minimized with respect to the model parameters β_0 and β_1. The diamonds all represent the same minimizer of this loss. The plusses show each minimizer $[\beta_0^*, \beta_1^*]^\top$ of the regularized minimization problems (6.4) and (6.7) for three choices of the regularization parameter γ. For the top three panels the regularization involves both β_0 and β_1, through the squared norm $\beta_0^2 + \beta_1^2$. The circles show the points that have the same squared norm as the optimal solution. For the bottom three panels only β_1 is regularized; there, horizontal lines indicate vectors $[\beta_0, \beta_1]^\top$ for which $|\beta_1| = |\beta_1^*|$.

The problem of ridge regression discussed in Example 6.2 boils down to solving a problem of the form in (6.7), involving a squared 2-norm penalty $\|\boldsymbol{\beta}\|^2$. A natural question to ask is whether we can replace the squared 2-norm penalty by a different penalty term. Replacing it with a 1-norm gives the *lasso* (least absolute shrinkage and selection operator). The lasso equivalent of the ridge regression problem (6.7) is thus:

☞ 410
LASSO

$$\min_{\beta_0, \boldsymbol{\beta}} \frac{1}{n} \| \boldsymbol{y} - \beta_0 \mathbf{1} - \mathbf{X}\boldsymbol{\beta} \|^2 + \gamma \|\boldsymbol{\beta}\|_1, \tag{6.11}$$

where $\|\boldsymbol{\beta}\|_1 = \sum_{i=1}^{p} |\beta_i|$.

This is again a convex optimization problem. Unlike ridge regression, the lasso generally does not have an explicit solution, and so numerical methods must be used to solve it. Note that the problem (6.11) is of the form

$$\begin{aligned} \min_{\boldsymbol{x}, \boldsymbol{z}} \quad & f(\boldsymbol{x}) + g(\boldsymbol{z}) \\ \text{subject to} \quad & \mathbf{A}\boldsymbol{x} + \mathbf{B}\boldsymbol{z} = \boldsymbol{c}, \end{aligned} \tag{6.12}$$

with $\boldsymbol{x} := [\beta_0, \boldsymbol{\beta}^\top]^\top$, $\boldsymbol{z} := \boldsymbol{\beta}$, $\mathbf{A} := [\mathbf{0}_p, \mathbf{I}_p]$, $\mathbf{B} := -\mathbf{I}_p$, and $\boldsymbol{c} := \mathbf{0}_p$ (vector of zeros), and convex functions $f(\boldsymbol{x}) := \frac{1}{n} \| \boldsymbol{y} - [\mathbf{1}_n, \mathbf{X}]\, \boldsymbol{x} \|^2$ and $g(\boldsymbol{z}) := \gamma \|\boldsymbol{z}\|_1$. There exist efficient algorithms for solving such problems, including the *alternating direction method of multipliers* (ADMM) [17]. We refer to Example **??** for details on this algorithm.

☞ 418

We repeat the examples from Figure 6.1, but now using lasso regression and taking the square roots of the previous regularization parameters. The results are displayed in Figure 6.2.

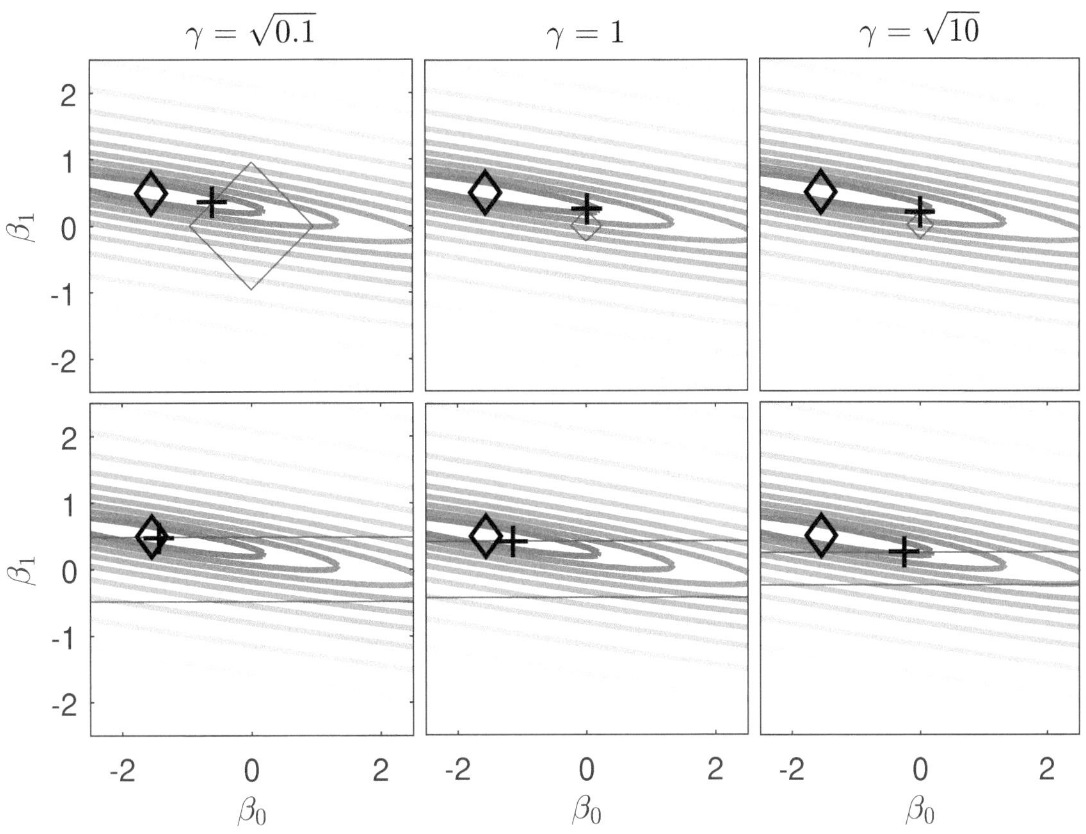

Figure 6.2: Lasso regression solutions. Compare with Figure 6.1.

One advantage of using the lasso regularization is that the resulting optimal parameter vector often has several components that are exactly 0. For example, in the top middle and right panels of Figure 6.2, the optimal solution lies exactly at a corner point of the square $\{[\beta_0, \beta_1]^\top : |\beta_0| + |\beta_1| = |\beta_0^*| + |\beta_1^*|\}$; in this case $\beta_0^* = 0$. For statistical models with many parameters, the lasso can provide a methodology for model selection. Namely, as the regularization parameter increases (or, equivalently, as the L_1 norm of the optimal solution decreases), the solution vector will have fewer and fewer non-zero parameters. By plotting the values of the parameters for each γ or L_1 one obtains the so-called *regularization paths* (also called *homotopy paths* or *coefficient profiles*) for the variables. Inspection of such paths may help assess which of the model parameters are relevant to explain the variability in the observed responses $\{y_i\}$.

REGULARIZATION PATHS

■ **Example 6.3 (Regularization Paths)** Figure 6.3 shows the regularization paths for $p =$ 60 coefficients from a multiple linear regression model ☞ 169

$$Y_i = \sum_{j=1}^{60} \beta_j x_{ij} + \varepsilon_i, \quad i = 1, \ldots, 150,$$

where $\beta_j = 1$ for $j = 1, \ldots, 10$ and $\beta_j = 0$ for $j = 11, \ldots, 60$. The error terms $\{\varepsilon_i\}$ are independent and standard normal. The explanatory variables $\{x_{ij}\}$ were independently generated from a standard normal distribution. As it is clear from the figure, the estimates of the 10 non-zero coefficients are first selected, as the L_1 norm of the solutions increases. By the time the L_1 norm reaches around 4, all 10 variables for which $\beta_j = 1$ have been correctly identified and the remaining 50 parameters are estimated as exactly 0. Only after the L_1 norm reaches around 8, will these "spurious" parameters be estimated to be non-zero. For this example, the regularization parameter γ varied from 10^{-4} to 10.

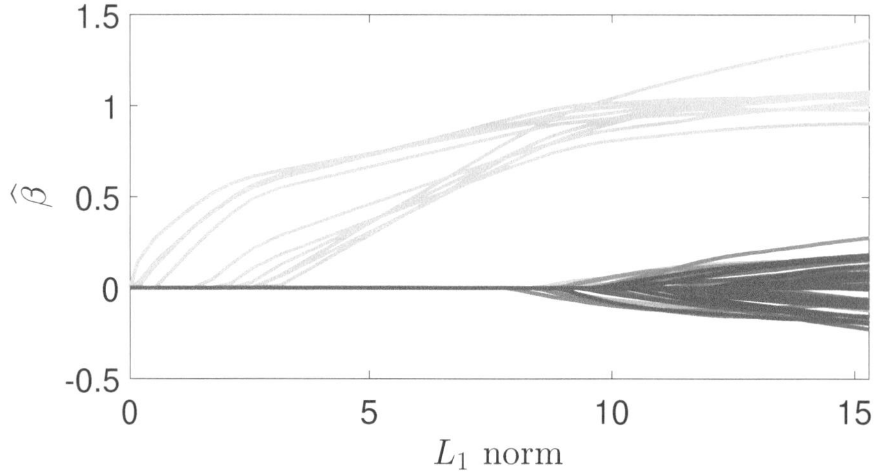

Figure 6.3: Regularization paths for lasso regression solutions as a function of the L_1 norm of the solutions.

■

6.3 Reproducing Kernel Hilbert Spaces

In this section, we formalize the idea outlined at the end of Section 6.1 of extending finite dimensional feature maps to those that are *functions* by introducing a special type of Hilbert space of functions known as a *reproducing kernel Hilbert space* (RKHS). Although the theory extends naturally to Hilbert spaces of complex-valued functions, we restrict attention to Hilbert spaces of real-valued functions here.

To evaluate the loss of a learner g in some class of functions $\mathcal{G}$, we do not need to explicitly construct g — rather, it is only required that we can evaluate g at all the feature vectors $\boldsymbol{x}_1, \ldots, \boldsymbol{x}_n$ of the training set. A defining property of an RKHS is that function evaluation at a point $\boldsymbol{x}$ can be performed by simply taking the inner product of g with some feature function $\kappa_{\boldsymbol{x}}$ associated with $\boldsymbol{x}$. We will see that this property becomes particularly useful in light of the representer theorem (see Section 6.5), which states that the learner g itself can be represented as a linear combination of the set of feature functions $\{\kappa_{\boldsymbol{x}_i}, i = 1, \ldots, n\}$. Consequently, we can evaluate a learner g at the feature vectors $\{\boldsymbol{x}_i\}$ by taking linear combinations of terms of the form $\kappa(\boldsymbol{x}_i, \boldsymbol{x}_j) = \langle \kappa_{\boldsymbol{x}_i}, \kappa_{\boldsymbol{x}_j} \rangle_{\mathcal{G}}$. Collecting these inner products into a matrix $\mathbf{K} = [\kappa(\boldsymbol{x}_i, \boldsymbol{x}_j), i, j = 1, \ldots, n]$ (the Gram matrix of the $\{\kappa_{\boldsymbol{x}_i}\}$), we will see that the feature vectors $\{\boldsymbol{x}_i\}$ only enter the loss minimization problem through $\mathbf{K}$.

☞ 231

Definition 6.1: Reproducing Kernel Hilbert Space

For a non-empty set $\mathcal{X}$, a Hilbert space $\mathcal{G}$ of functions $g : \mathcal{X} \to \mathbb{R}$ with inner product $\langle \cdot, \cdot \rangle_{\mathcal{G}}$ is called a *reproducing kernel Hilbert space* (RKHS) with *reproducing kernel* $\kappa : \mathcal{X} \times \mathcal{X} \to \mathbb{R}$ if:

1. for every $\boldsymbol{x} \in \mathcal{X}$, $\kappa_{\boldsymbol{x}} := \kappa(\boldsymbol{x}, \cdot)$ is in $\mathcal{G}$,
2. $\kappa(\boldsymbol{x}, \boldsymbol{x}) < \infty$ for all $\boldsymbol{x} \in \mathcal{X}$,
3. for every $\boldsymbol{x} \in \mathcal{X}$ and $g \in \mathcal{G}$, $g(\boldsymbol{x}) = \langle g, \kappa_{\boldsymbol{x}} \rangle_{\mathcal{G}}$.

REPRODUCING KERNEL HILBERT SPACE

The reproducing kernel of a Hilbert space of functions, if it exists, is unique; see Exercise 2. The main (third) condition in Definition 6.1 is known as the *reproducing property*. This property allows us to evaluate any function $g \in \mathcal{G}$ at a point $\boldsymbol{x} \in \mathcal{X}$ by taking the inner product of g and $\kappa_{\boldsymbol{x}}$; as such, $\kappa_{\boldsymbol{x}}$ is called the *representer of evaluation*. Further, by taking $g = \kappa_{\boldsymbol{x}'}$ and applying the reproducing property, we have $\langle \kappa_{\boldsymbol{x}'}, \kappa_{\boldsymbol{x}} \rangle_{\mathcal{G}} = \kappa(\boldsymbol{x}', \boldsymbol{x})$, and so by symmetry of the inner product it follows that $\kappa(\boldsymbol{x}, \boldsymbol{x}') = \kappa(\boldsymbol{x}', \boldsymbol{x})$. As a consequence, reproducing kernels are necessarily *symmetric* functions. Moreover, a reproducing kernel κ is a *positive semidefinite* function, meaning that for every $n \geqslant 1$ and every choice of $\alpha_1, \ldots, \alpha_n \in \mathbb{R}$ and $\boldsymbol{x}_1, \ldots, \boldsymbol{x}_n \in \mathcal{X}$, it holds that

REPRODUCING PROPERTY

POSITIVE SEMIDEFINITE

$$\sum_{i=1}^{n} \sum_{j=1}^{n} \alpha_i \, \kappa(\boldsymbol{x}_i, \boldsymbol{x}_j) \, \alpha_j \geqslant 0. \tag{6.13}$$

In other words, *every* Gram matrix $\mathbf{K}$ associated with κ is a positive semidefinite matrix; that is $\boldsymbol{\alpha}^\top \mathbf{K} \boldsymbol{\alpha} \geqslant 0$ for all $\boldsymbol{\alpha}$. The proof is addressed in Exercise 1.

The following theorem gives an alternative characterization of an RKHS. The proof uses the Riesz representation Theorem A.17. Also note that in the theorem below we could have replaced the word "bounded" with "continuous", as the two are equivalent for linear functionals; see Theorem A.16.

☞ 392

Theorem 6.1: Continuous Evaluation Functionals Characterize a RKHS

An RKHS $\mathcal{G}$ on a set $\mathcal{X}$ is a Hilbert space in which every *evaluation functional* $\delta_{\boldsymbol{x}} : g \mapsto g(\boldsymbol{x})$ is bounded. Conversely, a Hilbert space $\mathcal{G}$ of functions $\mathcal{X} \to \mathbb{R}$ for which every evaluation functional is bounded is an RKHS.

EVALUATION FUNCTIONAL

Proof: Note that, since evaluation functionals $\delta_{\boldsymbol{x}}$ are linear operators, showing boundedness is equivalent to showing continuity. Given an RKHS with reproducing kernel κ, suppose that we have a sequence $g_n \in \mathcal{G}$ converging to $g \in \mathcal{G}$, that is $\|g_n - g\|_{\mathcal{G}} \to 0$. We apply the Cauchy–Schwarz inequality (Theorem A.15) and the reproducing property of κ to find that for every $\boldsymbol{x} \in \mathcal{X}$ and any n:

☞ 391

$$\begin{aligned}|\delta_{\boldsymbol{x}} g_n - \delta_{\boldsymbol{x}} g| = |g_n(\boldsymbol{x}) - g(\boldsymbol{x})| &= |\langle g_n - g, \kappa_{\boldsymbol{x}} \rangle_{\mathcal{G}}| \leqslant \|g_n - g\|_{\mathcal{G}} \, \|\kappa_{\boldsymbol{x}}\|_{\mathcal{G}} = \|g_n - g\|_{\mathcal{G}} \sqrt{\langle \kappa_{\boldsymbol{x}}, \kappa_{\boldsymbol{x}} \rangle_{\mathcal{G}}} \\ &= \|g_n - g\|_{\mathcal{G}} \sqrt{\kappa(\boldsymbol{x}, \boldsymbol{x})}.\end{aligned}$$

Noting that $\sqrt{\kappa(\boldsymbol{x}, \boldsymbol{x})} < \infty$ by definition for every $\boldsymbol{x} \in \mathcal{X}$, and that $\|g_n - g\|_{\mathcal{G}} \to 0$ as $n \to \infty$, we have shown continuity of $\delta_{\boldsymbol{x}}$, that is $|\delta_{\boldsymbol{x}} g_n - \delta_{\boldsymbol{x}} g| \to 0$ as $n \to \infty$ for every $\boldsymbol{x} \in \mathcal{X}$.

Conversely, suppose that evaluation functionals are bounded. Then from the Riesz representation Theorem A.17, there exists some $g_{\delta_x} \in \mathcal{G}$ such that $\delta_x g = \langle g, g_{\delta_x} \rangle_{\mathcal{G}}$ for all $g \in \mathcal{G}$ — the *representer* of evaluation. If we define $\kappa(\boldsymbol{x}, \boldsymbol{x}') = g_{\delta_x}(\boldsymbol{x}')$ for all $\boldsymbol{x}, \boldsymbol{x}' \in \mathcal{X}$, then $\kappa_{\boldsymbol{x}} := \kappa(\boldsymbol{x}, \cdot) = g_{\delta_x}$ is an element of $\mathcal{G}$ for every $\boldsymbol{x} \in \mathcal{X}$ and $\langle g, \kappa_{\boldsymbol{x}} \rangle_{\mathcal{G}} = \delta_{\boldsymbol{x}} g = g(\boldsymbol{x})$, so that the reproducing property in Definition 6.1 is verified. □

The fact that an RKHS has continuous evaluation functionals means that if two functions $g, h \in \mathcal{G}$ are "close" with respect to $\|\cdot\|_{\mathcal{G}}$, then their evaluations $g(\boldsymbol{x}), h(\boldsymbol{x})$ are close *for every* $\boldsymbol{x} \in \mathcal{X}$. Formally, convergence in $\|\cdot\|_{\mathcal{G}}$ norm implies pointwise convergence for all $\boldsymbol{x} \in \mathcal{X}$.

The following theorem shows that any finite function $\kappa : \mathcal{X} \times \mathcal{X} \to \mathbb{R}$ can serve as a reproducing kernel as long as it is finite, symmetric, and positive semidefinite. The corresponding (unique!) RKHS $\mathcal{G}$ is the completion of the set of all functions of the form $\sum_{i=1}^{n} \alpha_i \kappa_{\boldsymbol{x}_i}$ where $\alpha_i \in \mathbb{R}$ for all $i = 1, \ldots, n$.

Theorem 6.2: Moore–Aronszajn

Given a non-empty set $\mathcal{X}$ and any finite symmetric positive semidefinite function $\kappa : \mathcal{X} \times \mathcal{X} \to \mathbb{R}$, there exists an RKHS $\mathcal{G}$ of functions $g : \mathcal{X} \to \mathbb{R}$ with reproducing kernel κ. Moreover, $\mathcal{G}$ is unique.

Proof: (Sketch) As the proof of uniqueness is treated in Exercise 2, the objective is to prove existence. The idea is to construct a pre-RKHS $\mathcal{G}_0$ from the given function κ that has the essential structure and then to extend $\mathcal{G}_0$ to an RKHS $\mathcal{G}$.

In particular, define $\mathcal{G}_0$ as the set of finite linear combinations of functions $\kappa_{\boldsymbol{x}}$, $\boldsymbol{x} \in \mathcal{X}$:

$$\mathcal{G}_0 := \left\{ g = \sum_{i=1}^{n} \alpha_i \kappa_{\boldsymbol{x}_i} \;\middle|\; \boldsymbol{x}_1, \ldots, \boldsymbol{x}_n \in \mathcal{X},\ \alpha_i \in \mathbb{R},\ n \in \mathbb{N} \right\}.$$

Define on $\mathcal{G}_0$ the following inner product:

$$\langle f, g \rangle_{\mathcal{G}_0} := \left\langle \sum_{i=1}^{n} \alpha_i \kappa_{\boldsymbol{x}_i}, \sum_{j=1}^{m} \beta_j \kappa_{\boldsymbol{x}'_j} \right\rangle_{\mathcal{G}_0} := \sum_{i=1}^{n} \sum_{j=1}^{m} \alpha_i \beta_j \kappa(\boldsymbol{x}_i, \boldsymbol{x}'_j).$$

Then $\mathcal{G}_0$ is an inner product space. In fact, $\mathcal{G}_0$ has the essential structure we require, namely that (i) evaluation functionals are bounded/continuous (Exercise 4) and (ii) Cauchy sequences in $\mathcal{G}_0$ that converge pointwise also converge in norm (see Exercise 5).

We then enlarge $\mathcal{G}_0$ to the set $\mathcal{G}$ of all functions $g : \mathcal{X} \to \mathbb{R}$ for which there exists a Cauchy sequence in $\mathcal{G}_0$ converging pointwise to g and define an inner product on $\mathcal{G}$ as the limit

$$\langle f, g \rangle_{\mathcal{G}} := \lim_{n \to \infty} \langle f_n, g_n \rangle_{\mathcal{G}_0}, \tag{6.14}$$

where $f_n \to f$ and $g_n \to g$. To show that $\mathcal{G}$ is an RKHS it remains to be shown that (1) this inner product is well defined; (2) evaluation functionals remain bounded; and (3) the space $\mathcal{G}$ is complete. A detailed proof is established in Exercises 6 and 7. □

6.4 Construction of Reproducing Kernels

In this section we describe various ways to construct a reproducing kernel $\kappa : \mathcal{X} \times \mathcal{X} \to \mathbb{R}$ for some feature space $\mathcal{X}$. Recall that κ needs to be a finite, symmetric, and positive semidefinite function (that is, it satisfies (6.13)). In view of Theorem 6.2, specifying the space $\mathcal{X}$ and a reproducing kernel $\kappa : \mathcal{X} \times \mathcal{X} \to \mathbb{R}$ corresponds to *uniquely* specifying an RKHS.

6.4.1 Reproducing Kernels via Feature Mapping

Perhaps the most fundamental way to construct a reproducing kernel κ is via a feature map $\boldsymbol{\phi} : \mathcal{X} \to \mathbb{R}^p$. We define $\kappa(\boldsymbol{x}, \boldsymbol{x}') := \langle \boldsymbol{\phi}(\boldsymbol{x}), \boldsymbol{\phi}(\boldsymbol{x}') \rangle$, where $\langle\ ,\ \rangle$ denotes the Euclidean inner product. The function is clearly finite and symmetric. To verify that κ is positive semidefinite, let $\boldsymbol{\Phi}$ be the matrix with rows $\boldsymbol{\phi}(\boldsymbol{x}_1)^\top, \ldots, \boldsymbol{\phi}(\boldsymbol{x}_n)^\top$ and let $\boldsymbol{\alpha} = [\alpha_1, \ldots, \alpha_n]^\top \in \mathbb{R}^n$. Then,

$$\sum_{i=1}^n \sum_{j=1}^n \alpha_i\, \kappa(\boldsymbol{x}_i, \boldsymbol{x}_j)\, \alpha_j = \sum_{i=1}^n \sum_{j=1}^n \alpha_i\, \boldsymbol{\phi}^\top(\boldsymbol{x}_i)\, \boldsymbol{\phi}(\boldsymbol{x}_j)\, \alpha_j = \boldsymbol{\alpha}^\top \boldsymbol{\Phi}\boldsymbol{\Phi}^\top \boldsymbol{\alpha} = \|\boldsymbol{\Phi}^\top \boldsymbol{\alpha}\|^2 \geqslant 0.$$

■ **Example 6.4 (Linear Kernel)** Taking the identity feature map $\boldsymbol{\phi}(\boldsymbol{x}) = \boldsymbol{x}$ on $\mathcal{X} = \mathbb{R}^p$, gives the *linear kernel* LINEAR KERNEL

$$\kappa(\boldsymbol{x}, \boldsymbol{x}') = \langle \boldsymbol{x}, \boldsymbol{x}' \rangle = \boldsymbol{x}^\top \boldsymbol{x}'.$$

As can be seen from the proof of Theorem 6.2, the RKHS of functions corresponding to the linear kernel is the space of *linear* functions on $\mathbb{R}^p$. This space is isomorphic to $\mathbb{R}^p$ itself, as discussed in the introduction (see also Exercise 12). ■

It is natural to wonder whether a given kernel function corresponds uniquely to a feature map. The answer is no, as we shall see by way of example.

■ **Example 6.5 (Feature Maps and Kernel Functions)** Let $\mathcal{X} = \mathbb{R}$ and consider feature maps $\phi_1 : \mathcal{X} \to \mathbb{R}$ and $\boldsymbol{\phi}_2 : \mathcal{X} \to \mathbb{R}^2$, with $\phi_1(x) := x$ and $\boldsymbol{\phi}_2(x) := [x, x]^\top / \sqrt{2}$. Then

$$\kappa_{\phi_1}(x, x') = \langle \phi_1(x), \phi_1(x') \rangle = x x',$$

but also

$$\kappa_{\boldsymbol{\phi}_2}(x, x') = \langle \boldsymbol{\phi}_2(x), \boldsymbol{\phi}_2(x') \rangle = x x'.$$

Thus, we arrive at the same kernel function defined for the same underlying set $\mathcal{X}$ via two different feature maps. ■

6.4.2 Kernels from Characteristic Functions

Another way to construct reproducing kernels on $\mathcal{X} = \mathbb{R}^p$ makes use of the properties of *characteristic functions*. In particular, we have the following result. We leave its proof as Exercise 10. ☞ 443

Theorem 6.3: Reproducing Kernel from a Characteristic Function

Let $X \sim \mu$ be an $\mathbb{R}^p$-valued random vector that is symmetric about the origin (that is, X and $-X$ are identically distributed), and let ψ be its characteristic function: $\psi(t) = \mathbb{E}\, e^{i t^\top X} = \int e^{i t^\top x}\, \mu(dx)$ for $t \in \mathbb{R}^p$. Then $\kappa(x, x') := \psi(x - x')$ is a valid reproducing kernel on $\mathbb{R}^p$.

■ **Example 6.6 (Gaussian Kernel)** The multivariate normal distribution with mean vector $\mathbf{0}$ and covariance matrix $b^2\, \mathbf{I}_p$ is clearly symmetric around the origin. Its characteristic function is

$$\psi(t) = \exp\left(-\frac{1}{2} b^2 \|t\|^2\right), \qquad t \in \mathbb{R}^p.$$

GAUSSIAN KERNEL

Taking $b^2 = 1/\sigma^2$, this gives the popular *Gaussian kernel* on $\mathbb{R}^p$:

$$\kappa(x, x') = \exp\left(-\frac{1}{2} \frac{\|x - x'\|^2}{\sigma^2}\right). \tag{6.15}$$

BANDWIDTH

RADIAL BASIS FUNCTION (RBF) KERNEL

The parameter σ is sometimes called the *bandwidth*. Note that in the machine learning literature, the Gaussian kernel is sometimes referred to as "the" *radial basis function (rbf) kernel*.[1]

From the proof of Theorem 6.2, we see that the RKHS $\mathcal{G}$ determined by the Gaussian kernel κ is the space of pointwise limits of functions of the form

$$g(x) = \sum_{i=1}^{n} \alpha_i \exp\left(-\frac{1}{2} \frac{\|x - x_i\|^2}{\sigma^2}\right).$$

We can think of each point x_i having a feature κ_{x_i} that is a scaled multivariate Gaussian pdf centered at x_i. ■

■ **Example 6.7 (Sinc Kernel)** The characteristic function of a Uniform$[-1, 1]$ random variable (which is symmetric around 0) is $\psi(t) = \operatorname{sinc}(t) := \sin(t)/t$, so $\kappa(x, x') = \operatorname{sinc}(x - x')$ is a valid kernel. ■

☞ 131

Inspired by kernel density estimation (Section 4.4), we may be tempted to use the pdf of a random variable that is symmetric about the origin to construct a reproducing kernel. However, doing so will not work in general, as the next example illustrates.

■ **Example 6.8 (Uniform pdf Does not Construct a Valid Reproducing Kernel)** Take the function $\psi(t) = \frac{1}{2}\mathbb{1}\{|t| \leqslant 1\}$, which is the pdf of $X \sim$ Uniform$[-1, 1]$. Unfortunately, the function $\kappa(x, x') = \psi(x - x')$ is not positive semidefinite, as can be seen for example by constructing the matrix $\mathbf{A} = [\kappa(t_i, t_j), i, j = 1, 2, 3]$ for the points $t_1 = 0$, $t_2 = 0.75$, and $t_3 = 1.5$ as follows:

$$\mathbf{A} = \begin{pmatrix} \psi(0) & \psi(-0.75) & \psi(-1.5) \\ \psi(0.75) & \psi(0) & \psi(-0.75) \\ \psi(1.5) & \psi(0.75) & \psi(0) \end{pmatrix} = \begin{pmatrix} 0.5 & 0.5 & 0 \\ 0.5 & 0.5 & 0.5 \\ 0 & 0.5 & 0.5 \end{pmatrix}.$$

The eigenvalues of $\mathbf{A}$ are $\{1/2 - \sqrt{1/2}, 1/2, 1/2 + \sqrt{1/2}\} \approx \{-0.2071, 0.5, 1.2071\}$ and so by Theorem A.9, $\mathbf{A}$ is not a positive semidefinite matrix, since it has a negative eigenvalue. Consequently, κ is not a valid reproducing kernel. ■ ☞ 369

One of the reasons why the Gaussian kernel (6.15) is popular is that it enjoys the *universal approximation property* [88]: the space of functions spanned by the Gaussian kernel is dense in the space of continuous functions with support $\mathcal{Z} \subset \mathbb{R}^p$. Naturally, this is a desirable property especially if there is little prior knowledge about the properties of g^*. However, note that *every* function g in the RKHS $\mathcal{G}$ associated with a Gaussian kernel κ is infinitely differentiable. Moreover, a Gaussian RKHS does not contain non-zero constant functions. Indeed, if $A \subset \mathcal{Z}$ is non-empty and open, then the only function of the form $g(\boldsymbol{x}) = c\,\mathbb{1}\{\boldsymbol{x} \in A\}$ contained in $\mathcal{G}$ is the zero function ($c = 0$).

UNIVERSAL APPROXIMATION PROPERTY

Consequently, if it is known that g is differentiable only to a certain order, one may prefer the *Matérn kernel* with parameters $\nu, \sigma > 0$:

MATÉRN KERNEL

$$\kappa_\nu(\boldsymbol{x}, \boldsymbol{x}') = \frac{2^{1-\nu}}{\Gamma(\nu)} \left(\sqrt{2\nu}\,\|\boldsymbol{x} - \boldsymbol{x}'\|/\sigma\right)^\nu K_\nu\left(\sqrt{2\nu}\,\|\boldsymbol{x} - \boldsymbol{x}'\|/\sigma\right), \tag{6.16}$$

which gives functions that are (weakly) differentiable to order $\lfloor \nu \rfloor$ (but not necessarily to order $\lceil \nu \rceil$). Here, K_ν denotes the modified Bessel function of the second kind; see (4.49). The particular form of the Matérn kernel appearing in (6.16) ensures that $\lim_{\nu\to\infty} \kappa_\nu(\boldsymbol{x}, \boldsymbol{x}') = \kappa(\boldsymbol{x}, \boldsymbol{x}')$, where κ is the Gaussian kernel appearing in (6.15). ☞ 163

We remark that Sobolev spaces are closely related to the Matérn kernel. Up to constants (which scale the unit ball in the space), in dimension p and for a parameter $s > p/2$, these spaces can be identified with $\psi(\boldsymbol{t}) = \frac{2^{1-s}}{\Gamma(s)} \|\boldsymbol{t}\|^{s-p/2} K_{p/2-s}(\|\boldsymbol{t}\|)$, which in turn can be viewed as the characteristic function corresponding to the (radially symmetric) multivariate Student's t distribution with s degrees of freedom: that is, with pdf $f(\boldsymbol{x}) \propto (1 + \|\boldsymbol{x}\|^2)^{-s}$. ☞ 162

6.4.3 Reproducing Kernels Using Orthonormal Features

We have seen in Sections 6.4.1 and 6.4.2 how to construct reproducing kernels from feature maps and characteristic functions. Another way to construct kernels on a space $\mathcal{X}$ is to work directly from the function class $L^2(\mathcal{X};\mu)$; that is, the set of square-integrable[2] functions on $\mathcal{X}$ with respect to μ; see also Definition A.4. For simplicity, in what follows, we will consider μ to be the Lebesgue measure, and will simply write $L^2(\mathcal{X})$ rather than $L^2(\mathcal{X};\mu)$. We will also assume that $\mathcal{X} \subseteq \mathbb{R}^p$. ☞ 387

Let $\{\xi_1, \xi_2, \ldots\}$ be an orthonormal basis of $L^2(\mathcal{X})$ and let $c_1, c_2, \ldots$ be a sequence of positive numbers. As discussed in Section 6.4.1, the kernel corresponding to a feature map $\boldsymbol{\phi} : \mathcal{X} \to \mathbb{R}^p$ is $\kappa(\boldsymbol{x}, \boldsymbol{x}') = \boldsymbol{\phi}(\boldsymbol{x})^\top \boldsymbol{\phi}(\boldsymbol{x}') = \sum_{i=1}^p \phi_i(\boldsymbol{x})\,\phi_i(\boldsymbol{x}')$. Now consider a (possibly infinite) sequence of feature functions $\phi_i = c_i\,\xi_i$, $i = 1, 2, \ldots$ and define

$$\kappa(\boldsymbol{x}, \boldsymbol{x}') := \sum_{i \geqslant 1} \phi_i(\boldsymbol{x})\,\phi_i(\boldsymbol{x}') = \sum_{i \geqslant 1} \lambda_i\,\xi_i(\boldsymbol{x})\,\xi_i(\boldsymbol{x}'), \tag{6.17}$$

[1] The term radial basis function is sometimes used more generally to mean kernels of the form $\kappa(\boldsymbol{x}, \boldsymbol{x}') = f(\|\boldsymbol{x} - \boldsymbol{x}'\|)$ for some function $f : \mathbb{R} \to \mathbb{R}$.

[2] A function $f : \mathcal{X} \to \mathbb{R}$ is said to be square-integrable if $\int f^2(\boldsymbol{x})\,\mu(\mathrm{d}\boldsymbol{x}) < \infty$, where μ is a measure on $\mathcal{X}$.

where $\lambda_i = c_i^2, i = 1, 2, \ldots$. This is well-defined as long as $\sum_{i\geqslant 1} \lambda_i < \infty$, which we assume from now on. Let $\mathcal{H}$ be the linear space of functions of the form $f = \sum_{i\geqslant 1} \alpha_i \xi_i$, where $\sum_{i\geqslant 1} \alpha_i^2/\lambda_i < \infty$. As every function $f \in L^2(\mathcal{X})$ can be represented as $f = \sum_{i\geqslant 1} \langle f, \xi_i \rangle \xi_i$, we see that $\mathcal{H}$ is a linear subspace of $L^2(\mathcal{X})$. On $\mathcal{H}$ define the inner product

$$\langle f, g \rangle_{\mathcal{H}} := \sum_{i\geqslant 1} \frac{\langle f, \xi_i \rangle \langle g, \xi_i \rangle}{\lambda_i}.$$

With this inner product, the squared norm of $f = \sum_{i\geqslant 1} \alpha_i \xi_i$ is $\|f\|_{\mathcal{H}}^2 = \sum_{i\geqslant 1} \alpha_i^2/\lambda_i < \infty$. We show that $\mathcal{H}$ is actually an RKHS with kernel κ by verifying the conditions of Definition 6.1. First,

$$\kappa_{\boldsymbol{x}} = \sum_{i\geqslant 1} \lambda_i \, \xi_i(\boldsymbol{x}) \, \xi_i \in \mathcal{H},$$

as $\sum_i \lambda_i < \infty$ by assumption, and so κ is finite. Second, the reproducing property holds. Namely, let $f = \sum_{i\geqslant 1} \alpha_i \xi_i$. Then,

$$\langle \kappa_{\boldsymbol{x}}, f \rangle_{\mathcal{H}} = \sum_{i\geqslant 1} \frac{\langle \kappa_{\boldsymbol{x}}, \xi_i \rangle \langle f, \xi_i \rangle}{\lambda_i} = \sum_{i\geqslant 1} \frac{\lambda_i \, \xi_i(\boldsymbol{x}) \, \alpha_i}{\lambda_i} = \sum_{i\geqslant 1} \alpha_i \xi_i(\boldsymbol{x}) = f(\boldsymbol{x}).$$

The discussion above demonstrates that kernels can be constructed via (6.17). In fact, (under mild conditions) any given reproducing kernel κ can be written in the form (6.17), where this series representation enjoys desirable convergence properties. This result is known as Mercer's theorem, and is given below. We leave the full proof including the precise conditions to, e.g., [40], but the main idea is that a reproducing kernel κ can be thought of as a generalization of a positive semidefinite matrix $\mathbf{K}$, and can also be written in spectral form (see also Section A.6.5). In particular, by Theorem A.9, we can write $\mathbf{K} = \mathbf{V}\mathbf{D}\mathbf{V}^\top$, where $\mathbf{V}$ is a matrix of orthonormal eigenvectors $[\boldsymbol{v}_\ell]$ and $\mathbf{D}$ the diagonal matrix of the (positive) eigenvalues $[\lambda_\ell]$; that is,

☞ 369

$$\mathbf{K}(i, j) = \sum_{\ell\geqslant 1} \lambda_\ell \, v_\ell(i) \, v_\ell(j).$$

In (6.18) below, $\boldsymbol{x}, \boldsymbol{x}'$ play the role of i, j, and ξ_ℓ plays the role of $\boldsymbol{v}_\ell$.

Theorem 6.4: Mercer

Let $\kappa : \mathcal{X} \times \mathcal{X} \to \mathbb{R}$ be a reproducing kernel for a compact set $\mathcal{X} \subset \mathbb{R}^p$. Then (under mild conditions) there exists a countable sequence of non-negative numbers $\{\lambda_\ell\}$ decreasing to zero and functions $\{\xi_\ell\}$ orthonormal in $L^2(\mathcal{X})$ such that

$$\kappa(\boldsymbol{x}, \boldsymbol{x}') = \sum_{\ell\geqslant 1} \lambda_\ell \, \xi_\ell(\boldsymbol{x}) \, \xi_\ell(\boldsymbol{x}'), \qquad \text{for all } \boldsymbol{x}, \boldsymbol{x}' \in \mathcal{X}, \tag{6.18}$$

where (6.18) converges absolutely and uniformly on $\mathcal{X} \times \mathcal{X}$.

Further, if $\lambda_\ell > 0$, then (λ_ℓ, ξ_ℓ) is an (eigenvalue, eigenfunction) pair for the integral operator $K : L^2(\mathcal{X}) \to L^2(\mathcal{X})$ defined by $[Kf](\boldsymbol{x}) := \int_{\mathcal{X}} \kappa(\boldsymbol{x}, \boldsymbol{y}) f(\boldsymbol{y}) \, \mathrm{d}\boldsymbol{y}$ for $\boldsymbol{x} \in \mathcal{X}$.

Theorem 6.4 holds if (i) the kernel κ is continuous on $\mathcal{X} \times \mathcal{X}$, (ii) the function $\widetilde{\kappa}(\boldsymbol{x}) := \kappa(\boldsymbol{x}, \boldsymbol{x})$ defined for $\boldsymbol{x} \in \mathcal{X}$ is integrable. Extensions of Theorem 6.4 to more general spaces $\mathcal{X}$ and measures μ hold; see, e.g., [115] or [40].

The key importance of Theorem 6.4 lies in the fact that the series representation (6.18) converges absolutely and uniformly on $\mathcal{X}\times\mathcal{X}$. The uniform convergence is a much stronger condition than pointwise convergence, and means for instance that properties of the sequence of partial sums, such as continuity and integrability, are transferred to the limit.

■ **Example 6.9 (Mercer)** Suppose $\mathcal{X} = [-1, 1]$ and the kernel is $\kappa(x, x') = 1 + xx'$ which corresponds to the RKHS $\mathcal{G}$ of affine functions from $\mathcal{X} \to \mathbb{R}$. To find the (eigenvalue, eigenfunction) pairs for the integral operator appearing in Theorem 6.4, we need to find numbers $\{\lambda_\ell\}$ and orthonormal functions $\{\xi_\ell(x)\}$ that solve

$$\int_{-1}^{1} (1 + xx')\,\xi_\ell(x')\,\mathrm{d}x' = \lambda_\ell\,\xi_\ell(x), \quad \text{for all } x \in [-1, 1].$$

Consider first a constant function $\xi_1(x) = c$. Then, for all $x \in [-1, 1]$, we have that $2c = \lambda_1 c$, and the normalization condition requires that $\int_{-1}^{1} c^2\,\mathrm{d}x = 1$. Together, these give $\lambda_1 = 2$ and $c = \pm 1/\sqrt{2}$. Next, consider an affine function $\xi_2(x) = a + bx$. Orthogonality requires that

$$\int_{-1}^{1} c(a + bx)\,\mathrm{d}x = 0,$$

which implies $a = 0$ (since $c \neq 0$). Moreover, the normalization condition then requires

$$\int_{-1}^{1} b^2x^2\,\mathrm{d}x = 1,$$

or, equivalently, $2b^2/3 = 1$, implying $b = \pm\sqrt{3/2}$. Finally, the integral equation reads

$$\int_{-1}^{1} (1 + xx')\,bx'\,\mathrm{d}x' = \lambda_2\,bx \iff \frac{2bx}{3} = \lambda_2 bx,$$

implying that $\lambda_2 = 2/3$. We take the positive solutions (i.e., $c > 0$ and $b > 0$), and note that

$$\lambda_1\,\xi_1(x)\,\xi_1(x') + \lambda_2\,\xi_2(x)\,\xi_2(x') = 2\frac{1}{\sqrt{2}}\frac{1}{\sqrt{2}} + \frac{2}{3}\frac{\sqrt{3}}{\sqrt{2}}x\frac{\sqrt{3}}{\sqrt{2}}x' = 1 + xx' = \kappa(x, x'),$$

and so we have found the decomposition appearing in (6.18). As an aside, observe that ξ_1 and ξ_2 are orthonormal versions of the first two Legendre polynomials. The corresponding ☞ 389
feature map can be explicitly identified as $\phi_1(x) = \sqrt{\lambda_1}\,\xi_1(x) = 1$ and $\phi_2(x) = \sqrt{\lambda_2}\,\xi_2(x) = x$. ■

6.4.4 Kernels from Kernels

The following theorem lists some useful properties for constructing reproducing kernels from existing reproducing kernels.

Theorem 6.5: Rules for Constructing Kernels from Other Kernels

1. If $\kappa : \mathbb{R}^p \times \mathbb{R}^p \to \mathbb{R}$ is a reproducing kernel and $\boldsymbol{\phi} : \mathcal{X} \to \mathbb{R}^p$ is a function, then $\kappa(\boldsymbol{\phi}(\boldsymbol{x}), \boldsymbol{\phi}(\boldsymbol{x}'))$ is a reproducing kernel from $\mathcal{X} \times \mathcal{X} \to \mathbb{R}$.

2. If $\kappa : \mathcal{X} \times \mathcal{X} \to \mathbb{R}$ is a reproducing kernel and $f : \mathcal{X} \to \mathbb{R}_+$ is a function, then $f(\boldsymbol{x})\kappa(\boldsymbol{x}, \boldsymbol{x}')f(\boldsymbol{x}')$ is also a reproducing kernel from $\mathcal{X} \times \mathcal{X} \to \mathbb{R}$.

3. If κ_1 and κ_2 are reproducing kernels from $\mathcal{X} \times \mathcal{X} \to \mathbb{R}$, then so is their sum $\kappa_1 + \kappa_2$.

4. If κ_1 and κ_2 are reproducing kernels from $\mathcal{X} \times \mathcal{X} \to \mathbb{R}$, then so is their product $\kappa_1 \kappa_2$.

5. If κ_1 and κ_2 are reproducing kernels from $\mathcal{X} \times \mathcal{X} \to \mathbb{R}$ and $\mathcal{Y} \times \mathcal{Y} \to \mathbb{R}$ respectively, then $\kappa_+((\boldsymbol{x}, \boldsymbol{y}), (\boldsymbol{x}', \boldsymbol{y}')) := \kappa_1(\boldsymbol{x}, \boldsymbol{x}') + \kappa_2(\boldsymbol{y}, \boldsymbol{y}')$ and $\kappa_\times((\boldsymbol{x}, \boldsymbol{y}), (\boldsymbol{x}', \boldsymbol{y}')) := \kappa_1(\boldsymbol{x}, \boldsymbol{x}')\kappa_2(\boldsymbol{y}, \boldsymbol{y}')$ are reproducing kernels from $(\mathcal{X} \times \mathcal{Y}) \times (\mathcal{X} \times \mathcal{Y}) \to \mathbb{R}$.

Proof: For Rules 1, 2, and 3 it is easy to verify that the resulting function is finite, symmetric, and positive semidefinite, and so is a valid reproducing kernel by Theorem 6.2. For example, for Rule 1 we have $\sum_{i=1}^n \sum_{j=1}^n \alpha_i \, \kappa(\boldsymbol{y}_i, \boldsymbol{y}_j)\alpha_j \geqslant 0$ for every choice of $\{\alpha_i\}_{i=1}^n$ and $\{\boldsymbol{y}_i\}_{i=1}^n \in \mathbb{R}^p$, since κ is a reproducing kernel. In particular, it holds true for $\boldsymbol{y}_i = \boldsymbol{\phi}(\boldsymbol{x}_i)$, $i = 1, \ldots, n$. Rule 4 is easy to show for kernels κ_1, κ_2 that admit a representation of the form (6.17), since

$$\begin{aligned}
\kappa_1(\boldsymbol{x}, \boldsymbol{x}')\, \kappa_2(\boldsymbol{x}, \boldsymbol{x}') &= \left(\sum_{i \geqslant 1} \phi_i^{(1)}(\boldsymbol{x})\, \phi_i^{(1)}(\boldsymbol{x}')\right)\left(\sum_{j \geqslant 1} \phi_j^{(2)}(\boldsymbol{x})\, \phi_j^{(2)}(\boldsymbol{x}')\right) \\
&= \sum_{i,j \geqslant 1} \phi_i^{(1)}(\boldsymbol{x})\, \phi_j^{(2)}(\boldsymbol{x})\, \phi_i^{(1)}(\boldsymbol{x}')\, \phi_j^{(2)}(\boldsymbol{x}') \\
&= \sum_{k \geqslant 1} \phi_k(\boldsymbol{x})\, \phi_k(\boldsymbol{x}') =: \kappa(\boldsymbol{x}, \boldsymbol{x}'),
\end{aligned}$$

showing that $\kappa = \kappa_1 \kappa_2$ also admits a representation of the form (6.17), where the new (possibly infinite) sequence of features (ϕ_k) is identified in a one-to-one way with the sequence $(\phi_i^{(1)} \phi_j^{(2)})$. We leave the proof of rule 5 as an exercise (Exercise 8). □

■ **Example 6.10 (Polynomial Kernel)** Consider $\boldsymbol{x}, \boldsymbol{x}' \in \mathbb{R}^2$ with

$$\kappa(\boldsymbol{x}, \boldsymbol{x}') = (1 + \langle \boldsymbol{x}, \boldsymbol{x}' \rangle)^2,$$

POLYNOMIAL KERNEL

where $\langle \boldsymbol{x}, \boldsymbol{x}' \rangle = \boldsymbol{x}^\top \boldsymbol{x}'$. This is an example of a *polynomial kernel*. Combining the fact that sums and products of kernels are again kernels (rules 3 and 4 of Theorem 6.5), we find that, since $\langle \boldsymbol{x}, \boldsymbol{x}' \rangle$ and the constant function 1 are kernels, so are $1 + \langle \boldsymbol{x}, \boldsymbol{x}' \rangle$ and $(1 + \langle \boldsymbol{x}, \boldsymbol{x}' \rangle)^2$. By writing

$$\begin{aligned}
\kappa(\boldsymbol{x}, \boldsymbol{x}') &= (1 + x_1 x_1' + x_2 x_2')^2 \\
&= 1 + 2x_1 x_1' + 2x_2 x_2' + 2x_1 x_2 x_1' x_2' + (x_1 x_1')^2 + (x_2 x_2')^2,
\end{aligned}$$

we see that $\kappa(\boldsymbol{x}, \boldsymbol{x}')$ can be written as the inner product in $\mathbb{R}^6$ of the two feature vectors $\boldsymbol{\phi}(\boldsymbol{x})$ and $\boldsymbol{\phi}(\boldsymbol{x}')$, where the feature map $\boldsymbol{\phi} : \mathbb{R}^2 \to \mathbb{R}^6$ can be explicitly identified as

$$\boldsymbol{\phi}(\boldsymbol{x}) = [1,\ \sqrt{2}x_1,\ \sqrt{2}x_2,\ \sqrt{2}x_1x_2, x_1^2, x_2^2]^\top.$$

Thus, the RKHS determined by κ can be explicitly identified with the space of functions $\boldsymbol{x} \mapsto \boldsymbol{\phi}(\boldsymbol{x})^\top \boldsymbol{\beta}$ for some $\boldsymbol{\beta} \in \mathbb{R}^6$. ■

In the above example we could explicitly identify the feature map. However, in general a feature map need not be explicitly available. Using a particular reproducing kernel corresponds to using an *implicit* (possibly infinite dimensional!) feature map that never needs to be explicitly computed.

6.5 Representer Theorem

Recall the setting discussed at the beginning of this chapter: we are given training data $\tau = \{(\boldsymbol{x}_i, y_i)\}_{i=1}^n$ and a loss function that measures the fit to the data, and we wish to find a function g that minimizes the training loss, with the addition of a regularization term, as described in Section 6.2. To do this, we assume first that the class $\mathcal{G}$ of prediction functions can be decomposed as the direct sum of an RKHS $\mathcal{H}$, defined by a kernel function $\kappa : \mathcal{X} \times \mathcal{X} \to \mathbb{R}$, and another linear space of real-valued functions $\mathcal{H}_0$ on $\mathcal{X}$; that is,

$$\mathcal{G} = \mathcal{H} \oplus \mathcal{H}_0,$$

meaning that any element $g \in \mathcal{G}$ can be written as $g = h + h_0$, with $h \in \mathcal{H}$ and $h_0 \in \mathcal{H}_0$. In minimizing the training loss we wish to penalize the h term of g but not the h_0 term. Specifically, the aim is to solve the functional optimization problem

$$\min_{g \in \mathcal{H} \oplus \mathcal{H}_0} \frac{1}{n} \sum_{i=1}^{n} \text{Loss}(y_i, g(\boldsymbol{x}_i)) + \gamma \, \|g\|_{\mathcal{H}}^2. \tag{6.19}$$

Here, we use a slight abuse of notation: $\|g\|_{\mathcal{H}}$ means $\|h\|_{\mathcal{H}}$ if $g = h + h_0$, as above. In this way, we can view $\mathcal{H}_0$ as the null space of the functional $g \mapsto \|g\|_{\mathcal{H}}$. This null space may be empty, but typically has a small dimension m; for example it could be the one-dimensional space of constant functions, as in Example 6.2. ☞ 217

■ **Example 6.11 (Null Space)** Consider again the setting of Example 6.2, for which we have feature vectors $\widetilde{\boldsymbol{x}} = [1, \boldsymbol{x}^\top]^\top$ and $\mathcal{G}$ consists of functions of the form $g : \widetilde{\boldsymbol{x}} \mapsto \beta_0 + \boldsymbol{x}^\top \boldsymbol{\beta}$. Each function g can be decomposed as $g = h + h_0$, where $h : \widetilde{\boldsymbol{x}} \mapsto \boldsymbol{x}^\top \boldsymbol{\beta}$, and $h_0 : \widetilde{\boldsymbol{x}} \mapsto \beta_0$.

Given $g \in \mathcal{G}$, we have $\|g\|_{\mathcal{H}} = \|\boldsymbol{\beta}\|$, and so the null space $\mathcal{H}_0$ of the functional $g \mapsto \|g\|_{\mathcal{H}}$ (that is, the set of all functions $g \in \mathcal{G}$ for which $\|g\|_{\mathcal{H}} = 0$) is the set of constant functions here, which has dimension $m = 1$. ■

Regularization favors elements in $\mathcal{H}_0$ and penalizes large elements in $\mathcal{H}$. As the regularization parameter γ varies between zero and infinity, solutions to (6.19) vary from "complex" ($g \in \mathcal{H} \oplus \mathcal{H}_0$) to "simple" ($g \in \mathcal{H}_0$).

A key reason why RKHSs are so useful is the following. By choosing $\mathcal{H}$ to be an RKHS in (6.19) this *functional* optimization problem effectively becomes a *parametric* optimization problem. The reason is that any solution to (6.19) can be represented as a finite-dimensional linear combination of kernel functions, evaluated at the training sample. This is known as the *kernel trick*.

KERNEL TRICK

Theorem 6.6: Representer Theorem

The solution to the penalized optimization problem (6.19) is of the form

$$g(\boldsymbol{x}) = \sum_{i=1}^{n} \alpha_i \, \kappa(\boldsymbol{x}_i, \boldsymbol{x}) + \sum_{j=1}^{m} \eta_j \, q_j(\boldsymbol{x}), \tag{6.20}$$

where $\{q_1, \ldots, q_m\}$ is a basis of $\mathcal{H}_0$.

Proof: Let $\mathcal{F} = \operatorname{Span}\{\kappa_{\boldsymbol{x}_i}, i = 1, \ldots, n\}$. Clearly, $\mathcal{F} \subseteq \mathcal{H}$. Then, the Hilbert space $\mathcal{H}$ can be represented as $\mathcal{H} = \mathcal{F} \oplus \mathcal{F}^\perp$, where $\mathcal{F}^\perp$ is the orthogonal complement of $\mathcal{F}$. In other words, $\mathcal{F}^\perp$ is the class of functions

$$\{f^\perp \in \mathcal{H} : \langle f^\perp, f \rangle_{\mathcal{H}} = 0, \; f \in \mathcal{F}\} \equiv \{f^\perp : \langle f^\perp, \kappa_{\boldsymbol{x}_i} \rangle_{\mathcal{H}} = 0, \; \forall i\}.$$

It follows, by the reproducing kernel property, that for all $f^\perp \in \mathcal{F}^\perp$:

$$f^\perp(\boldsymbol{x}_i) = \langle f^\perp, \kappa_{\boldsymbol{x}_i} \rangle_{\mathcal{H}} = 0, \quad i = 1, \ldots, n.$$

Now, take any $g \in \mathcal{H} \oplus \mathcal{H}_0$, and write it as $g = f + f^\perp + h_0$, with $f \in \mathcal{F}$, $f^\perp \in \mathcal{F}^\perp$, and $h_0 \in \mathcal{H}_0$. By the definition of the null space $\mathcal{H}_0$, we have $\|g\|_{\mathcal{H}}^2 = \|f + f^\perp\|_{\mathcal{H}}^2$. Moreover, by Pythagoras' theorem, the latter is equal to $\|f\|_{\mathcal{H}}^2 + \|f^\perp\|_{\mathcal{H}}^2$. It follows that

$$\begin{aligned}
\frac{1}{n}\sum_{i=1}^{n} \operatorname{Loss}(y_i, g(\boldsymbol{x}_i)) + \gamma \|g\|_{\mathcal{H}}^2 &= \frac{1}{n}\sum_{i=1}^{n} \operatorname{Loss}(y_i, f(\boldsymbol{x}_i) + h_0(\boldsymbol{x}_i)) + \gamma \left(\|f\|_{\mathcal{H}}^2 + \|f^\perp\|_{\mathcal{H}}^2\right) \\
&\geqslant \frac{1}{n}\sum_{i=1}^{n} \operatorname{Loss}(y_i, f(\boldsymbol{x}_i) + h_0(\boldsymbol{x}_i)) + \gamma \|f\|_{\mathcal{H}}^2.
\end{aligned}$$

Since we can obtain equality by taking $f^\perp = 0$, this implies that the minimizer of the penalized optimization problem (6.19) lies in the subspace $\mathcal{F} \oplus \mathcal{H}_0$ of $\mathcal{G} = \mathcal{H} \oplus \mathcal{H}_0$, and hence is of the form (6.20). □

Substituting the representation (6.20) of g into (6.19) gives the finite-dimensional optimization problem:

$$\min_{\boldsymbol{\alpha} \in \mathbb{R}^n, \boldsymbol{\eta} \in \mathbb{R}^m} \frac{1}{n}\sum_{i=1}^{n} \operatorname{Loss}(y_i, (\mathbf{K}\boldsymbol{\alpha} + \mathbf{Q}\boldsymbol{\eta})_i) + \gamma \, \boldsymbol{\alpha}^\top \mathbf{K} \boldsymbol{\alpha}, \tag{6.21}$$

where

- $\mathbf{K}$ is the $n \times n$ (Gram) matrix with entries $[\kappa(\boldsymbol{x}_i, \boldsymbol{x}_j), i = 1, \ldots, n, \; j = 1, \ldots, n]$.
- $\mathbf{Q}$ is the $n \times m$ matrix with entries $[q_j(\boldsymbol{x}_i), i = 1, \ldots, n, \; j = 1, \ldots, m]$.

In particular, for the squared-error loss we have

$$\min_{\boldsymbol{\alpha}\in\mathbb{R}^n,\,\boldsymbol{\eta}\in\mathbb{R}^m} \frac{1}{n}\left\|\boldsymbol{y} - (\mathbf{K}\boldsymbol{\alpha} + \mathbf{Q}\boldsymbol{\eta})\right\|^2 + \gamma\,\boldsymbol{\alpha}^\top\mathbf{K}\boldsymbol{\alpha}. \tag{6.22}$$

This is a convex optimization problem, and its solution is found by differentiating (6.22) with respect to $\boldsymbol{\alpha}$ and $\boldsymbol{\eta}$ and equating to zero, leading to the following system of $(n + m)$ linear equations:

$$\begin{bmatrix} \mathbf{K}\mathbf{K}^\top + n\gamma\mathbf{K} & \mathbf{K}\mathbf{Q} \\ \mathbf{Q}^\top\mathbf{K}^\top & \mathbf{Q}^\top\mathbf{Q}\end{bmatrix}\begin{bmatrix}\boldsymbol{\alpha} \\ \boldsymbol{\eta}\end{bmatrix} = \begin{bmatrix}\mathbf{K}^\top \\ \mathbf{Q}^\top\end{bmatrix}\boldsymbol{y}. \tag{6.23}$$

As long as $\mathbf{Q}$ is of full column rank, the minimizing function is unique.

■ **Example 6.12 (Ridge Regression (cont.))** We return to Example 6.2 and identify that $\mathcal{H}$ is the RKHS with linear kernel function $\kappa(\boldsymbol{x}, \boldsymbol{x}') = \boldsymbol{x}^\top\boldsymbol{x}'$ and $C = \mathcal{H}_0$ is the linear space of constant functions. In this case, $\mathcal{H}_0$ is spanned by the function $q_1 \equiv 1$. Moreover, $\mathbf{K} = \mathbf{X}\mathbf{X}^\top$ and $\mathbf{Q} = \mathbf{1}$.

If we appeal to the representer theorem directly, then the problem in (6.6) becomes, as a result of (6.21):

$$\min_{\boldsymbol{\alpha},\eta_0} \frac{1}{n}\left\|\boldsymbol{y} - \eta_0\mathbf{1} - \mathbf{X}\mathbf{X}^\top\boldsymbol{\alpha}\right\|^2 + \gamma\,\|\mathbf{X}^\top\boldsymbol{\alpha}\|^2.$$

This is a convex optimization problem, and so the solution follows by taking derivatives and setting them to zero. This gives the equations

$$\mathbf{X}\mathbf{X}^\top((\mathbf{X}\mathbf{X}^\top + n\gamma\mathbf{I}_n)\boldsymbol{\alpha} + \eta_0\mathbf{1} - \boldsymbol{y}) = \mathbf{0},$$

and

$$n\,\eta_0 = \mathbf{1}^\top(\boldsymbol{y} - \mathbf{X}\mathbf{X}^\top\boldsymbol{\alpha}).$$

Note that these are equivalent to (6.8) and (6.9) (once again assuming that $n \geqslant p$ and $\mathbf{X}$ has full rank p). Equivalently, the solution is found by solving (6.23):

$$\begin{bmatrix} \mathbf{X}\mathbf{X}^\top\mathbf{X}\mathbf{X}^\top + n\gamma\mathbf{X}\mathbf{X}^\top & \mathbf{X}\mathbf{X}^\top\mathbf{1} \\ \mathbf{1}^\top\mathbf{X}\mathbf{X}^\top & n\end{bmatrix}\begin{bmatrix}\boldsymbol{\alpha} \\ \eta_0\end{bmatrix} = \begin{bmatrix}\mathbf{X}\mathbf{X}^\top \\ \mathbf{1}^\top\end{bmatrix}\boldsymbol{y}.$$

This is a system of $(n + 1)$ linear equations, and is typically of much larger dimension than the $(p + 1)$ linear equations given by (6.8) and (6.9). As such, one may question the practicality of reformulating the problem in this way. However, the benefit of this formulation is that the problem can be expressed entirely through the Gram matrix $\mathbf{K}$, without having to explicitly compute the feature vectors — in turn permitting the (implicit) use of infinite dimensional feature spaces. ■

■ **Example 6.13 (Estimating the Peaks Function)** Figure 6.4 shows the surface plot of the *peaks* function:

$$f(x_1, x_2) = 3(1 - x_1)^2 \mathrm{e}^{-x_1^2-(x_2+1)^2} - 10\left(\frac{x_1}{5} - x_1^3 - x_2^5\right)\mathrm{e}^{-x_1^2-x_2^2} - \frac{1}{3}\mathrm{e}^{-(x_1+1)^2-x_2^2}. \tag{6.24}$$

The goal is to learn the function $y = f(\boldsymbol{x})$ based on a small set of training data (pairs of $(\boldsymbol{x}, y)$ values). The red dots in the figure represent data $\tau = \{(\boldsymbol{x}_i, y_i)\}_{i=1}^{20}$, where $y_i = f(\boldsymbol{x}_i)$ and the $\{\boldsymbol{x}_i\}$ have been chosen in a *quasi-random* way, using *Hammersley points* (with bases 2 QUASI-RANDOM and 3) on the square $[-3, 3]^2$. Quasi-random point sets have better space-filling properties than either a regular grid of points or a set of pseudo-random points. We refer to [71] for details. Note that there is no observation noise in this particular problem.

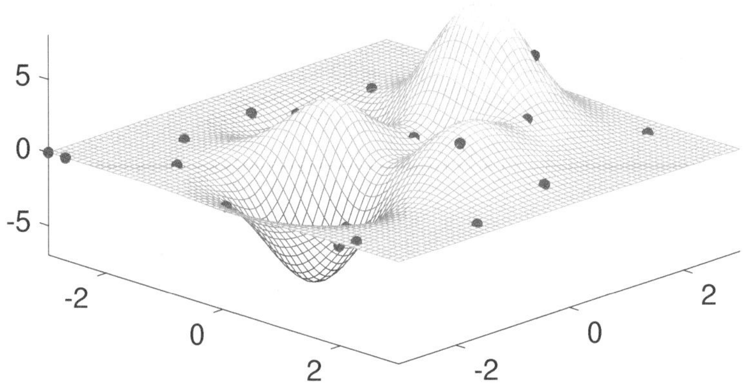

Figure 6.4: Peaks function sampled at 20 Hammersley points.

The purpose of this example is to illustrate how, using the small data set of size $n = 20$, the entire *peaks* function can be approximated well using kernel methods. In particular, we use the Gaussian kernel (6.15) on $\mathbb{R}^2$, and denote by $\mathcal{H}$ the unique RKHS corresponding to this kernel. We omit the regularization term in (6.19), and thus our objective is to find the solution to

$$\min_{g \in \mathcal{H}} \frac{1}{n} \sum_{i=1}^{n} (y_i - g(\boldsymbol{x}_i))^2.$$

By the representer theorem, the optimal function is of the form

$$g(\boldsymbol{x}) = \sum_{i=1}^{n} \alpha_i \exp\left(-\frac{1}{2}\frac{\|\boldsymbol{x} - \boldsymbol{x}_i\|^2}{\sigma^2}\right),$$

where $\boldsymbol{\alpha} := [\alpha_1, \ldots, \alpha_n]^\top$ is, by (6.23), the solution to the set of linear equations $\mathbf{K}\mathbf{K}^\top \boldsymbol{\alpha} = \mathbf{K}\boldsymbol{y}$.

Note that we are performing regression over the class of functions $\mathcal{H}$ with an implicit feature space. Due to the representer theorem, the solution to this problem coincides with the solution to the linear regression problem for which the i-th feature (for $i = 1, \ldots, n$) is chosen to be the vector $[\kappa(\boldsymbol{x}_1, \boldsymbol{x}_i), \ldots, \kappa(\boldsymbol{x}_n, \boldsymbol{x}_i)]^\top$.

The following code performs these calculations and gives the contour plots of g and the peaks functions, shown in Figure 6.5. We see that the two are quite close. Code for the generation of Hammersley points is available from the book's GitHub site as `genham.py`.

peakskernel.py

```
from genham import hammersley
import numpy as np
import matplotlib.pyplot as plt
from mpl_toolkits.mplot3d import Axes3D
```

```
from matplotlib import cm
from numpy.linalg import norm

import numpy as np
def peaks(x,y):
    z =  (3*(1-x)**2 * np.exp(-(x**2) - (y+1)**2)
         - 10*(x/5 - x**3 - y**5) * np.exp(-x**2 - y**2)
         - 1/3 * np.exp(-(x+1)**2 - y**2))
    return(z)

n = 20
x = -3 + 6*hammersley([2,3],n)
z = peaks(x[:,0],x[:,1])
xx, yy = np.mgrid[-3:3:150j,-3:3:150j]
zz = peaks(xx,yy)
plt.contour(xx,yy,zz,levels=50)

fig=plt.figure()
ax = fig.add_subplot(111,projection='3d')
ax.plot_surface(xx,yy,zz,rstride=1,cstride=1,color='c',alpha=0.3,
   linewidth=0)
ax.scatter(x[:,0],x[:,1],z,color='k',s=20)
plt.show()

sig2 = 0.3 # kernel parameter
def k(x,u):
    return(np.exp(-0.5*norm(x- u)**2/sig2))
K = np.zeros((n,n))
for i in range(n):
    for j in range(n):
        K[i,j] = k(x[i,:],x[j])
alpha = np.linalg.solve(K@K.T, K@z)



N, = xx.flatten().shape
Kx = np.zeros((n,N))
for i in range(n):
    for j in range(N):
        Kx[i,j] = k(x[i,:],np.array([xx.flatten()[j],yy.flatten()[j
           ]]))

g = Kx.T @ alpha
dim = np.sqrt(N).astype(int)
yhat = g.reshape(dim,dim)
plt.contour(xx,yy,yhat,levels=50)
```

■

6.6 Smoothing Cubic Splines

A striking application of kernel methods is to fitting "well-behaved" functions to data. Key examples of "well-behaved" functions are those that do not have large second-

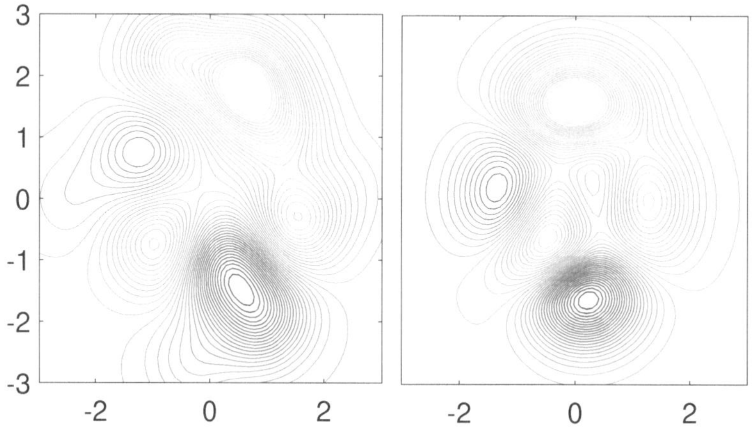

Figure 6.5: Contour plots for the prediction function g (left) and the *peaks* function given in (6.24) (right).

order derivatives. Consider functions $g : [0, 1] \to \mathbb{R}$ that are twice differentiable and define $\|g''\|^2 := \int_0^1 (g''(x))^2 \, dx$ as a measure of the size of the second derivative.

■ **Example 6.14 (Behavior of $\|g''\|^2$)** Intuitively, the larger $\|g''\|^2$ is, the more "wiggly" the function g will be. As an explicit example, consider $g(x) = \sin(\omega x)$ for $x \in [0, 1]$, where ω is a free parameter. We can explicitly compute $g''(x) = -\omega^2 \sin(\omega x)$, and consequently

$$\|g''\|^2 = \int_0^1 \omega^4 \sin^2(\omega x) \, dx = \frac{\omega^4}{2}(1 - \text{sinc}(2\omega)).$$

As $|\omega| \to \infty$, the frequency of g increases and we have $\|g''\|^2 \to \infty$. ■

Now, in the context of data fitting, consider the following penalized least-squares optimization problem on $[0, 1]$:

$$\min_{g \in \mathcal{G}} \frac{1}{n} \sum_{i=1}^{n} (y_i - g(x_i))^2 + \gamma \|g''\|^2, \tag{6.25}$$

where we will specify $\mathcal{G}$ in what follows. In order to apply the kernel machinery, we want to write this in the form (6.19), for some RKHS $\mathcal{H}$ and null space $\mathcal{H}_0$. Clearly, the norm on $\mathcal{H}$ should be of the form $\|g\|_{\mathcal{H}} = \|g''\|$ and should be well-defined (i.e., finite and ensuring g and g' are absolutely continuous). This suggests that we take

$$\mathcal{H} = \{g \in L^2[0, 1] : \|g''\| < \infty, g, g' \text{ absolutely continuous}, \; g(0) = g'(0) = 0\},$$

with inner product

$$\langle f, g \rangle_{\mathcal{H}} := \int_0^1 f''(x) \, g''(x) \, dx.$$

One rationale for imposing the boundary conditions $g(0) = g'(0) = 0$ is as follows: when expanding g about the point $x = 0$, Taylor's theorem (with integral remainder term) states that

$$g(x) = g(0) + g'(0) \, x + \int_0^x g''(s) \, (x - s) \, ds.$$

Imposing the condition that $g(0) = g'(0) = 0$ for functions in $\mathcal{H}$ will ensure that $\mathcal{G} = \mathcal{H} \oplus \mathcal{H}_0$ where the null space $\mathcal{H}_0$ contains only linear functions, as we will see.

To see that this $\mathcal{H}$ is in fact an RKHS, we derive its reproducing kernel. Using integration by parts (or directly from the Taylor expansion above), write

$$g(x) = \int_0^x g'(s)\,\mathrm{d}s = \int_0^x g''(s)\,(x-s)\,\mathrm{d}s = \int_0^1 g''(s)\,(x-s)_+\,\mathrm{d}s.$$

If κ is a kernel, then by the reproducing property it must hold that

$$g(x) = \langle g, \kappa_x \rangle_{\mathcal{H}} = \int_0^1 g''(s)\,\kappa_x''(s)\,\mathrm{d}s,$$

so that κ must satisfy $\frac{\partial^2}{\partial s^2}\kappa(x,s) = (x-s)_+$, where $y_+ := \max\{y, 0\}$. Therefore, noting that $\kappa(x,u) = \langle \kappa_x, \kappa_u \rangle_{\mathcal{H}}$, we have (see Exercise 15)

$$\kappa(x,u) = \int_0^1 \frac{\partial^2 \kappa(x,s)}{\partial s^2}\frac{\partial^2 \kappa(u,s)}{\partial s^2}\,\mathrm{d}s = \frac{\max\{x,u\}\min\{x,u\}^2}{2} - \frac{\min\{x,u\}^3}{6}.$$

The last expression is a cubic function with quadratic and cubic terms that misses the constant and linear monomials. This is not surprising considering the Taylor's theorem interpretation of a function $g \in \mathcal{H}$. If we now take $\mathcal{H}_0$ as the space of functions of the following form (having zero second derivative):

$$h_0 = \eta_1 + \eta_2\, x, \quad x \in [0,1],$$

then (6.25) is exactly of the form (6.19).

As a consequence of the representer Theorem 6.6, the optimal solution to (6.25) is a linear combination of piecewise cubic functions:

$$g(x) = \eta_1 + \eta_2\, x + \sum_{i=1}^{n} \alpha_i\, \kappa(x_i, x). \tag{6.26}$$

Such a function is called a *cubic spline* with *n knots* (with one knot at each data point x_i) — so called, because the piecewise cubic function between knots is required to be "tied together" at the knots. The parameters $\boldsymbol{\alpha}, \boldsymbol{\eta}$ are determined from (6.21) for instance by solving (6.23) with matrices $\mathbf{K} = [\kappa(x_i, x_j)]_{i,j=1}^n$ and $\mathbf{Q}$ with i-th row of the form $[1, x_i]$ for $i = 1, \ldots, n$. CUBIC SPLINE

■ **Example 6.15 (Smoothing Spline)** Figure 6.6 shows various cubic smoothing splines for the data $(0.05, 0.4), (0.2, 0.2), (0.5, 0.6), (0.75, 0.7), (1, 1)$. In the figure, we use the re-parameterization $r = 1/(1 + n\,\gamma)$ for the smoothing parameter. Thus $r \in [0, 1]$, where $r = 0$ means an infinite penalty for curvature (leading to the ordinary linear regression solution) and $r = 1$ does not penalize curvature at all and leads to a perfect fit via the so-called *natural spline*. Of course the latter will generally lead to overfitting. For r from 0 up to 0.8 the solutions will be close to the simple linear regression line, while only for r very close to 1, the shape of the curve changes significantly.

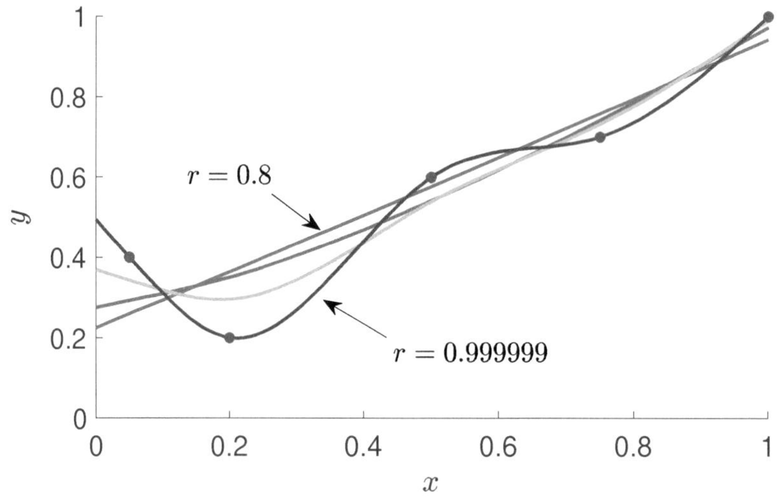

Figure 6.6: Various cubic smoothing splines for smoothing parameter $r = 1/(1 + n\gamma) \in \{0.8, 0.99, 0.999, 0.999999\}$. For $r = 1$, the natural spline through the data points is obtained; for $r = 0$, the simple linear regression line is found.

The following code first computes the matrices **K** and **Q**, and then solves the linear system (6.23). Finally, the smoothing curve is determined via (6.26), for selected points, and then plotted. Note that the code plots only a single curve corresponding to the specified value of p.

smoothspline.py

```
import matplotlib.pyplot as plt
import numpy as np

x = np.array([[0.05, 0.2, 0.5, 0.75, 1.]]).T
y = np.array([[0.4, 0.2, 0.6, 0.7, 1.]]).T

n = x.shape[0]
r = 0.999
ngamma = (1-r)/r

k = lambda x1, x2 : (1/2)* np.max((x1,x2)) * np.min((x1,x2)) ** 2 \
                              - ((1/6)* np.min((x1,x2))**3)
K = np.zeros((n,n))
for i in range(n):
    for j in range(n):
        K[i,j] = k(x[i], x[j])

Q = np.hstack((np.ones((n,1)), x))

m1 = np.hstack((K @ K.T + (ngamma * K), K @ Q))
m2 = np.hstack((Q.T @ K.T, Q.T @ Q))
M = np.vstack((m1,m2))

c = np.vstack((K, Q.T)) @ y

ad = np.linalg.solve(M,c)
```

```
# plot the curve
xx = np.arange(0,1+0.01,0.01).reshape(-1,1)

g = np.zeros_like(xx)
Qx = np.hstack((np.ones_like(xx), xx))
g = np.zeros_like(xx)
N = np.shape(xx)[0]

Kx = np.zeros((n,N))
for i in range(n):
    for j in range(N):
        Kx[i,j] = k(x[i], xx[j])

g = g + np.hstack((Kx.T, Qx)) @ ad

plt.ylim((0,1.15))
plt.plot(xx, g, label = 'r = {}'.format(r), linewidth = 2)
plt.plot(x,y, 'b.', markersize=15)
plt.xlabel('$x$')
plt.ylabel('$y$')
plt.legend()
```

■

6.7 Gaussian Process Regression

Another application of the kernel machinery is to Gaussian process regression. A *Gaussian process* (GP) on a space $\mathcal{X}$ is a stochastic process $\{Z_{\boldsymbol{x}}, \boldsymbol{x} \in \mathcal{X}\}$ where, for any choice of indices $\boldsymbol{x}_1, \ldots, \boldsymbol{x}_n$, the vector $[Z_{\boldsymbol{x}_1}, \ldots Z_{\boldsymbol{x}_n}]^\top$ has a multivariate Gaussian distribution. As such, the distribution of a GP is completely specified by its mean and covariance functions $\mu : \mathcal{X} \to \mathbb{R}$ and $\kappa : \mathcal{X} \times \mathcal{X} \to \mathbb{R}$, respectively. The covariance function is a finite positive semidefinite function, and hence, in view of Theorem 6.2, can be viewed as a reproducing kernel on $\mathcal{X}$. GAUSSIAN PROCESS

As for ordinary regression, the objective of GP regression is to learn a regression function g that predicts a response $y = g(\boldsymbol{x})$ for each feature vector $\boldsymbol{x}$. This is done in a Bayesian fashion, by establishing (1) a prior pdf for g and (2) the likelihood of the data, for a given g. From these two we then derive, via Bayes' formula, the posterior distribution of g given the data. We refer to Section 2.9 for the general Bayesian framework. ☞ 168 ☞ 47

A simple Bayesian model for GP regression is as follows. First, the prior distribution of g is taken to be the distribution of a GP with some known mean function μ and covariance function (that is, kernel) κ. Most often μ is taken to be a constant, and for simplicity of exposition, we take it to be 0. The Gaussian kernel (6.15) is often used for the covariance function. For radial basis function kernels (including the Gaussian kernel), points that are closer will be more highly correlated or "similar" [97], independent of translations in space.

Second, similar to standard regression, we view the observed feature vectors $\boldsymbol{x}_1, \ldots, \boldsymbol{x}_n$ as fixed and the responses $y_1, \ldots, y_n$ as outcomes of random variables $Y_1, \ldots, Y_n$. Specifically, given g, we model the $\{Y_i\}$ as

$$Y_i = g(\boldsymbol{x}_i) + \varepsilon_i, \quad i = 1, \ldots, n, \tag{6.27}$$

where $\{\varepsilon_i\} \overset{\text{iid}}{\sim} \mathcal{N}(0, \sigma^2)$. To simplify the analysis, let us assume that σ^2 is known, so no prior needs to be specified for σ^2. Let $\boldsymbol{g} = [g(\boldsymbol{x}_1), \ldots, g(\boldsymbol{x}_n)]^\top$ be the (unknown) vector of regression values. Placing a GP prior on the function g is equivalent to placing a multivariate Gaussian prior on the vector $\boldsymbol{g}$:

$$\boldsymbol{g} \sim \mathcal{N}(\boldsymbol{0}, \mathbf{K}), \tag{6.28}$$

where the covariance matrix $\mathbf{K}$ of $\boldsymbol{g}$ is a Gram matrix (implicitly associated with a feature map through the kernel κ), given by:

$$\mathbf{K} = \begin{bmatrix} \kappa(\boldsymbol{x}_1, \boldsymbol{x}_1) & \kappa(\boldsymbol{x}_1, \boldsymbol{x}_2) & \ldots & \kappa(\boldsymbol{x}_1, \boldsymbol{x}_n) \\ \kappa(\boldsymbol{x}_2, \boldsymbol{x}_1) & \kappa(\boldsymbol{x}_2, \boldsymbol{x}_2) & \ldots & \kappa(\boldsymbol{x}_2, \boldsymbol{x}_n) \\ \vdots & \vdots & \ddots & \vdots \\ \kappa(\boldsymbol{x}_n, \boldsymbol{x}_1) & \kappa(\boldsymbol{x}_n, \boldsymbol{x}_2) & \ldots & \kappa(\boldsymbol{x}_n, \boldsymbol{x}_n) \end{bmatrix}. \tag{6.29}$$

The likelihood of our data given $\boldsymbol{g}$, denoted $p(\boldsymbol{y} \,|\, \boldsymbol{g})$, is obtained directly from the model (6.27):

$$(\boldsymbol{Y} \,|\, \boldsymbol{g}) \sim \mathcal{N}(\boldsymbol{g}, \sigma^2 \mathbf{I}_n). \tag{6.30}$$

Solving this Bayesian problem involves deriving the posterior distribution of $(\boldsymbol{g} \,|\, \boldsymbol{Y})$. To do so, we first note that since $\boldsymbol{Y}$ has covariance matrix $\mathbf{K} + \sigma^2 \mathbf{I}_n$ (which can be seen from (6.27)), the joint distribution of $\boldsymbol{Y}$ and $\boldsymbol{g}$ is again normal, with mean $\boldsymbol{0}$ and covariance matrix:

$$\mathbf{K}_{y,g} = \begin{bmatrix} \mathbf{K} + \sigma^2 \mathbf{I}_n & \mathbf{K} \\ \mathbf{K} & \mathbf{K} \end{bmatrix}. \tag{6.31}$$

☞ 438 The posterior can then be found by conditioning on $\boldsymbol{Y} = \boldsymbol{y}$, via Theorem C.8, giving

$$(\boldsymbol{g} \,|\, \boldsymbol{y}) \sim \mathcal{N}\left(\mathbf{K}^\top (\mathbf{K} + \sigma^2 \mathbf{I}_n)^{-1} \boldsymbol{y},\ \mathbf{K} - \mathbf{K}^\top (\mathbf{K} + \sigma^2 \mathbf{I}_n)^{-1} \mathbf{K}\right).$$

This only gives information about g at the observed points $\boldsymbol{x}_1, \ldots, \boldsymbol{x}_n$. It is more interesting to consider the posterior predictive distribution of $\widetilde{g} := g(\widetilde{\boldsymbol{x}})$ for a new input $\widetilde{\boldsymbol{x}}$. We can find the corresponding posterior predictive pdf $p(\widetilde{g} \,|\, \boldsymbol{y})$ by integrating out the joint posterior pdf $p(\widetilde{g}, \boldsymbol{g} \,|\, \boldsymbol{y})$, which is equivalent to taking the expectation of $p(\widetilde{g} \,|\, \boldsymbol{g})$ when $\boldsymbol{g}$ is distributed according to the posterior pdf $p(\boldsymbol{g} \,|\, \boldsymbol{y})$; that is,

$$p(\widetilde{g} \,|\, \boldsymbol{y}) = \int p(\widetilde{g} \,|\, \boldsymbol{g})\, p(\boldsymbol{g} \,|\, \boldsymbol{y})\, \mathrm{d}\boldsymbol{g}.$$

To do so more easily than direct evaluation via the above integral representation of $p(\widetilde{g} \,|\, \boldsymbol{y})$, we can begin with the joint distribution of $[\boldsymbol{y}^\top, \widetilde{g}]^\top$, which is multivariate normal with mean $\boldsymbol{0}$ and covariance matrix

$$\widetilde{\mathbf{K}} = \begin{bmatrix} \mathbf{K} + \sigma^2 \mathbf{I}_n & \boldsymbol{\kappa} \\ \boldsymbol{\kappa}^\top & \kappa(\widetilde{\boldsymbol{x}}, \widetilde{\boldsymbol{x}}) \end{bmatrix}, \tag{6.32}$$

where $\boldsymbol{\kappa} = [\kappa(\widetilde{\boldsymbol{x}}, \boldsymbol{x}_1), \ldots, \kappa(\widetilde{\boldsymbol{x}}, \boldsymbol{x}_n)]^\top$. It now follows, again by using Theorem C.8, that $(\widetilde{g} \,|\, \boldsymbol{y})$ has a normal distribution with mean and variance given respectively by

$$\mu(\widetilde{\boldsymbol{x}}) = \boldsymbol{\kappa}^\top (\mathbf{K} + \sigma^2 \mathbf{I}_n)^{-1} \boldsymbol{y} \tag{6.33}$$

and

$$\sigma^2(\widetilde{\boldsymbol{x}}) = \kappa(\widetilde{\boldsymbol{x}}, \widetilde{\boldsymbol{x}}) - \boldsymbol{\kappa}^\top (\mathbf{K} + \sigma^2 \mathbf{I}_n)^{-1} \boldsymbol{\kappa}. \tag{6.34}$$

PREDICTIVE

These are sometimes called the *predictive* mean and variance. It is important to note that we are predicting the *expected* response $\mathbb{E}\widetilde{Y} = g(\widetilde{\boldsymbol{x}})$ here, and not the actual response $\widetilde{Y}$.

■ **Example 6.16 (GP Regression)** Suppose the regression function is

$$g(x) = 2\sin(2\pi x), \quad x \in [0, 1].$$

We use GP regression to estimate g, using a Gaussian kernel of the form (6.15) with bandwidth parameter 0.2. The explanatory variables $x_1, \ldots, x_{30}$ were drawn uniformly on the interval $[0, 1]$, and the responses were obtained from (6.27), with noise level $\sigma = 0.5$. Figure 6.7 shows 10 samples from the prior distribution for g as well as the data points and the true sinusoidal regression function g.

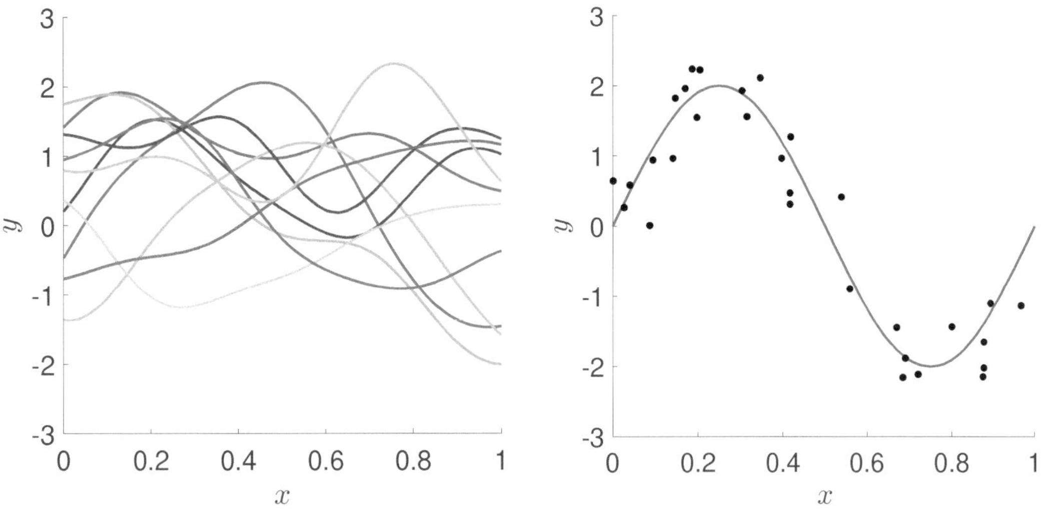

Figure 6.7: Left: samples drawn from the GP prior distribution. Right: the true regression function with the data points.

Again assuming that the variance σ^2, is known, the predictive distribution as determined by (6.33) and (6.34) is shown in Figure 6.8 for bandwidth 0.2 (left) and 0.02 (right). Clearly, decreasing the bandwidth leads to the covariance between points x and x' decreasing at a faster rate with respect to the squared distance $\|x - x'\|^2$, leading to a predictive mean that is less smooth. ■

In the above exposition, we have taken the mean function for the prior distribution of g to be identically zero. If instead we have a general mean function m and write $\boldsymbol{m} = [m(\boldsymbol{x}_1), \ldots, m(\boldsymbol{x}_n)]^\top$ then the predictive variance (6.34) remains unchanged, and the predictive mean (6.33) is modified to read

$$\mu(\widetilde{\boldsymbol{x}}) = m(\widetilde{\boldsymbol{x}}) + \boldsymbol{\kappa}^\top (\mathbf{K} + \sigma^2 \mathbf{I}_n)^{-1} (\boldsymbol{y} - \boldsymbol{m}). \tag{6.35}$$

Typically, the variance σ^2 appearing in (6.27) is not known, and the kernel κ itself depends on several parameters — for instance a Gaussian kernel (6.15) with an unknown bandwidth parameter. In the Bayesian framework, one typically specifies a hierarchical model by introducing a prior $p(\boldsymbol{\theta})$ for the vector $\boldsymbol{\theta}$ of such *hyperparameters*. Now, the GP prior $(g \mid \boldsymbol{\theta})$ (equivalently, specifying $p(\boldsymbol{g} \mid \boldsymbol{\theta})$) and the model for the likelihood of the data given $\boldsymbol{Y} | \boldsymbol{g}, \boldsymbol{\theta}$, namely $p(\boldsymbol{y} \mid \boldsymbol{g}, \boldsymbol{\theta})$, are both dependent on $\boldsymbol{\theta}$. The posterior distribution of $(\boldsymbol{g} \mid \boldsymbol{y}, \boldsymbol{\theta})$ is as before. HYPERPARAMETERS

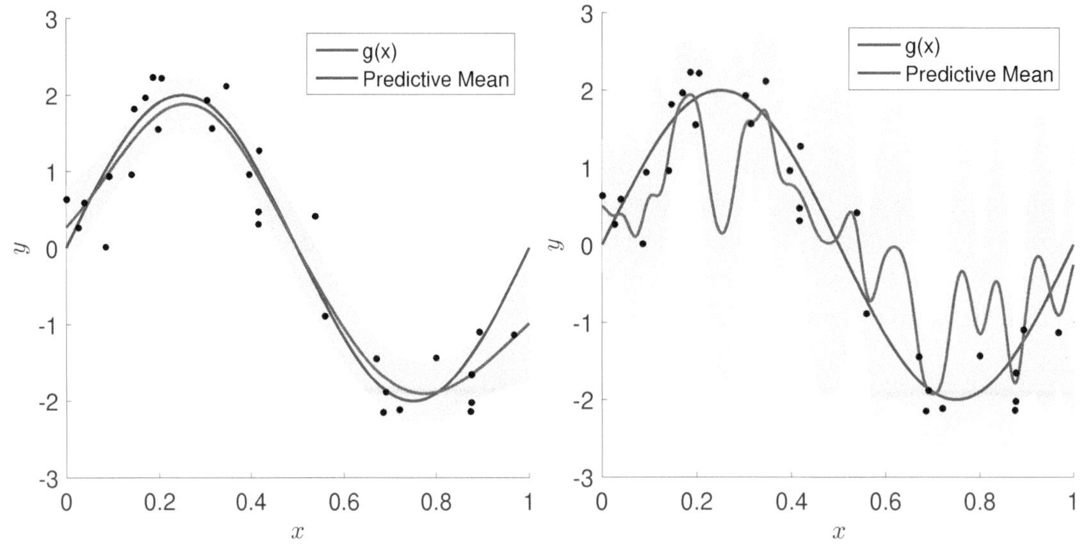

Figure 6.8: GP regression of synthetic data set with bandwidth 0.2 (left) and 0.02 (right). The black dots represent the data and the blue curve is the latent function $g(x) = 2\sin(2\pi x)$. The red curve is the mean of the GP predictive distribution given by (6.33), and the shaded region is the 95% confidence band, corresponding to the predictive variance given in (6.34).

One approach to setting the hyperparameter $\boldsymbol{\theta}$ is to determine its posterior $p(\boldsymbol{\theta} \,|\, \mathbf{y})$ and obtain a point estimate, for instance via its maximum a posteriori estimate. However, this can be a computationally demanding exercise. What is frequently done in practice is to consider instead the *marginal likelihood* $p(\mathbf{y} \,|\, \boldsymbol{\theta})$ and maximize this with respect to $\boldsymbol{\theta}$. This procedure is called *empirical Bayes*.

EMPIRICAL BAYES

Considering again the mean function m to be identically zero, from (6.31), we have that $(\boldsymbol{Y} \,|\, \boldsymbol{\theta})$ is multivariate normal with mean $\mathbf{0}$ and covariance matrix $\mathbf{K}_y = \mathbf{K} + \sigma^2 \mathbf{I}_n$, immediately giving an expression for the marginal log-likelihood:

$$\ln p(\mathbf{y} \,|\, \boldsymbol{\theta}) = -\frac{n}{2}\ln(2\pi) - \frac{1}{2}\ln|\det(\mathbf{K}_y)| - \frac{1}{2}\mathbf{y}^\top \mathbf{K}_y^{-1}\mathbf{y}. \tag{6.36}$$

We notice that only the second and third terms in (6.36) depend on $\boldsymbol{\theta}$. Considering a partial derivative of (6.36) with respect to a single element θ of the hyperparameter vector $\boldsymbol{\theta}$ yields

$$\frac{\partial}{\partial\theta}\ln p(\mathbf{y} \,|\, \boldsymbol{\theta}) = -\frac{1}{2}\mathrm{tr}\left(\mathbf{K}_y^{-1}\left[\frac{\partial}{\partial\theta}\mathbf{K}_y\right]\right) + \frac{1}{2}\mathbf{y}^\top \mathbf{K}_y^{-1}\left[\frac{\partial}{\partial\theta}\mathbf{K}_y\right]\mathbf{K}_y^{-1}\mathbf{y}, \tag{6.37}$$

where $\left[\frac{\partial}{\partial\theta}\mathbf{K}_y\right]$ is the element-wise derivative of matrix K_y with respect to θ. If these partial derivatives can be computed for each hyperparameter θ, gradient information could be used when maximizing (6.36).

■ **Example 6.17 (GP Regression (cont.))** Continuing Example 6.16, we plot in Figure 6.9 the marginal log-likelihood as a function of the noise level σ and bandwidth parameter.

The maximum is attained for a bandwidth parameter around 0.20 and $\sigma \approx 0.44$, which is very close to the left panel of Figure 6.8 for the case where σ was assumed to be known (and equal to 0.5). We note here that the marginal log-likelihood is extremely flat, perhaps owing to the small number of points. ■

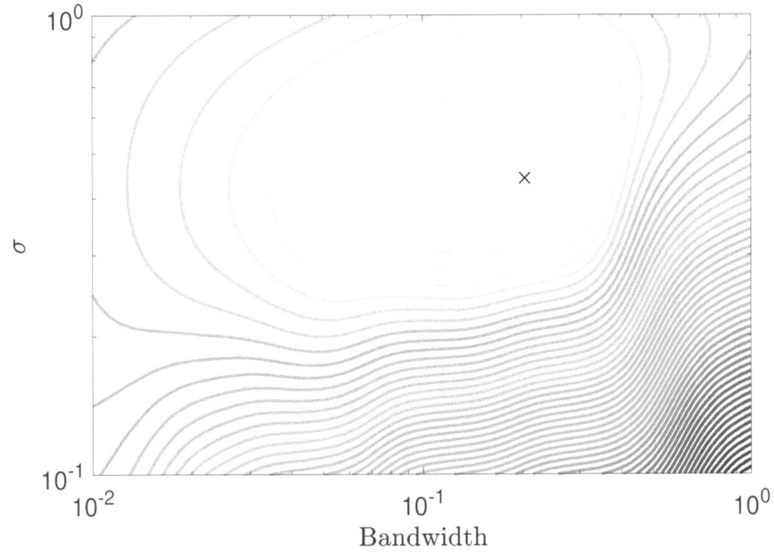

Figure 6.9: Contours of the marginal log-likelihood for the GP regression example. The maximum is denoted by a cross.

6.8 Kernel PCA

In its basic form, kernel PCA (principal component analysis) can be thought of as PCA in feature space. The main motivation for PCA introduced in Section 4.8 was as a dimensionality reduction technique. There, the analysis rested on an SVD of the matrix $\widehat{\mathbf{\Sigma}} = \frac{1}{n}\mathbf{X}^\top\mathbf{X}$, where the data in $\mathbf{X}$ was first centered via $x'_{i,j} = x_{i,j} - \overline{x}_j$ where $\overline{x}_i = \frac{1}{n}\sum_{i=1}^n x_{i,j}$. ☞ 153

What we shall do is to first re-cast the problem in terms of the Gram matrix $\mathbf{K} = \mathbf{X}\mathbf{X}^\top = [\langle \boldsymbol{x}_i, \boldsymbol{x}_j \rangle]$ (note the different order of $\mathbf{X}$ and $\mathbf{X}^\top$), and subsequently replace the inner product $\langle \boldsymbol{x}, \boldsymbol{x}' \rangle$ with $\kappa(\boldsymbol{x}, \boldsymbol{x}')$ for a general reproducing kernel κ. To make the link, let us start with an SVD of $\mathbf{X}^\top$:

$$\mathbf{X}^\top = \mathbf{U}\mathbf{D}\mathbf{V}^\top. \tag{6.38}$$

The dimensions of $\mathbf{X}^\top$, $\mathbf{U}$, $\mathbf{D}$, and $\mathbf{V}$ are $d \times n$, $d \times d$, $d \times n$, and $n \times n$, respectively. Then an SVD of $\mathbf{X}^\top\mathbf{X}$ is

$$\mathbf{X}^\top\mathbf{X} = (\mathbf{U}\mathbf{D}\mathbf{V}^\top)(\mathbf{U}\mathbf{D}\mathbf{V}^\top)^\top = \mathbf{U}(\mathbf{D}\mathbf{D}^\top)\mathbf{U}^\top$$

and an SVD of $\mathbf{K}$ is

$$\mathbf{K} = (\mathbf{U}\mathbf{D}\mathbf{V}^\top)^\top(\mathbf{U}\mathbf{D}\mathbf{V}^\top) = \mathbf{V}(\mathbf{D}^\top\mathbf{D})\mathbf{V}^\top.$$

Let $\lambda_1 \geqslant \cdots \geqslant \lambda_r > 0$ denote the non-zero eigenvalues of $\mathbf{X}^\top\mathbf{X}$ (or, equivalently, of $\mathbf{K}$) and denote the corresponding $r \times r$ diagonal matrix by $\mathbf{\Lambda}$. Without loss of generality we can assume that the eigenvector of $\mathbf{X}^\top\mathbf{X}$ corresponding to λ_k is the k-th column of $\mathbf{U}$ and that the k-th column of $\mathbf{V}$ is an eigenvector of $\mathbf{K}$. Similar to Section 4.8, let $\mathbf{U}_k$ and $\mathbf{V}_k$ contain the first k columns of $\mathbf{U}$ and $\mathbf{V}$, respectively, and let $\mathbf{\Lambda}_k$ be the corresponding $k\times k$ submatrix of $\mathbf{\Lambda}$, $k = 1, \ldots, r$. ☞ 153

By the SVD (6.38), we have $\mathbf{X}^\top\mathbf{V}_k = \mathbf{U}\mathbf{D}\mathbf{V}^\top\mathbf{V}_k = \mathbf{U}_k\mathbf{\Lambda}_k^{1/2}$. Next, consider the projection of a point $\boldsymbol{x}$ onto the k-dimensional linear space spanned by the columns of $\mathbf{U}_k$ — the first k principal components. We saw in Section 4.8 that this projection simply is the linear mapping $\boldsymbol{x} \mapsto \mathbf{U}_k^\top\boldsymbol{x}$. Using the fact that $\mathbf{U}_k = \mathbf{X}^\top\mathbf{V}_k\mathbf{\Lambda}^{-1/2}$, we find that $\boldsymbol{x}$ is projected to a point z given by

$$z = \mathbf{\Lambda}_k^{-1/2} \mathbf{V}_k^\top \mathbf{X} \boldsymbol{x} = \mathbf{\Lambda}_k^{-1/2} \mathbf{V}_k^\top \boldsymbol{\kappa}_{\boldsymbol{x}},$$

where we have (suggestively) defined $\boldsymbol{\kappa}_{\boldsymbol{x}} := [\langle \boldsymbol{x}_1, \boldsymbol{x} \rangle, \ldots, \langle \boldsymbol{x}_n, \boldsymbol{x} \rangle]^\top$. The important point is that z is completely determined by the vector of inner products $\boldsymbol{\kappa}_{\boldsymbol{x}}$ and the k principal eigenvalues and (right) eigenvectors of the Gram matrix $\mathbf{K}$. Note that each component z_m of z is of the form

$$z_m = \sum_{i=1}^{n} \alpha_{m,i}\, \kappa(\boldsymbol{x}_i, \boldsymbol{x}), \quad m = 1, \ldots, k. \tag{6.39}$$

The preceding discussion assumed centering of the columns of $\mathbf{X}$. Consider now an uncentered data matrix $\widetilde{\mathbf{X}}$. Then the centered data can be written as $\mathbf{X} = \widetilde{\mathbf{X}} - \frac{1}{n}\mathbf{E}_n \widetilde{\mathbf{X}}$, where $\mathbf{E}_n$ is the $n \times n$ matrix of ones. Consequently,

$$\mathbf{X}\mathbf{X}^\top = \widetilde{\mathbf{X}}\widetilde{\mathbf{X}}^\top - \frac{1}{n}\mathbf{E}_n \widetilde{\mathbf{X}}\widetilde{\mathbf{X}}^\top - \frac{1}{n}\widetilde{\mathbf{X}}\widetilde{\mathbf{X}}^\top \mathbf{E}_n + \frac{1}{n^2}\mathbf{E}_n \widetilde{\mathbf{X}}\widetilde{\mathbf{X}}^\top \mathbf{E}_n,$$

or, more compactly, $\mathbf{X}\mathbf{X}^\top = \mathbf{H}\widetilde{\mathbf{X}}\widetilde{\mathbf{X}}^\top \mathbf{H}$, where $\mathbf{H} = \mathbf{I}_n - \frac{1}{n}\mathbf{1}_n \mathbf{1}_n^\top$, $\mathbf{I}_n$ is the $n \times n$ identity matrix, and $\mathbf{1}_n$ is the $n \times 1$ vector of ones.

To generalize to the kernel setting, we replace $\widetilde{\mathbf{X}}\widetilde{\mathbf{X}}^\top$ by $\mathbf{K} = [\kappa(\boldsymbol{x}_i, \boldsymbol{x}_j), i, j = 1, \ldots, n]$ and set $\boldsymbol{\kappa}_{\boldsymbol{x}} = [\kappa(\boldsymbol{x}_1, \boldsymbol{x}), \ldots, \kappa(\boldsymbol{x}_n, \boldsymbol{x})]^\top$, so that $\mathbf{\Lambda}_k$ is the diagonal matrix of the k largest eigenvalues of $\mathbf{HKH}$ and $\mathbf{V}_k$ is the corresponding matrix of eigenvectors. Note that the "usual" PCA is recovered when we use the linear kernel $\kappa(\boldsymbol{x}, \boldsymbol{y}) = \boldsymbol{x}^\top \boldsymbol{y}$. However, instead of having only kernels that are explicitly inner products of feature vectors, we are now permitted to implicitly use *infinite* feature maps (functions) by using kernels.

■ **Example 6.18 (Kernel PCA)** We simulated 200 points, $\boldsymbol{x}_1, \ldots, \boldsymbol{x}_{200}$, from the uniform distribution on the set $B_1 \cup (B_4 \cap B_3^c)$, where $B_r := \{(x, y) \in \mathbb{R}^2 : x^2 + y^2 \leqslant r^2\}$ (disk with radius r). We apply kernel PCA with Gaussian kernel $\kappa(\boldsymbol{x}, \boldsymbol{x}') = \exp\left(-\|\boldsymbol{x} - \boldsymbol{x}'\|^2\right)$ and compute the functions $z_m(\boldsymbol{x}), m = 1, \ldots, 9$ in (6.39). Their density plots are shown in Figure 6.10. The data points are superimposed in each plot. From this we see that the principal components identify the radial structure present in the data. Finally, Figure 6.11 shows the projections $[z_1(\boldsymbol{x}_i), z_2(\boldsymbol{x}_i)]^\top, i = 1, \ldots, 200$ of the original data points onto the first two principal components. We see that the projected points can be separated by a straight line, whereas this is not possible for the original data; see also, Example 7.6 for a related problem.

☞ 274

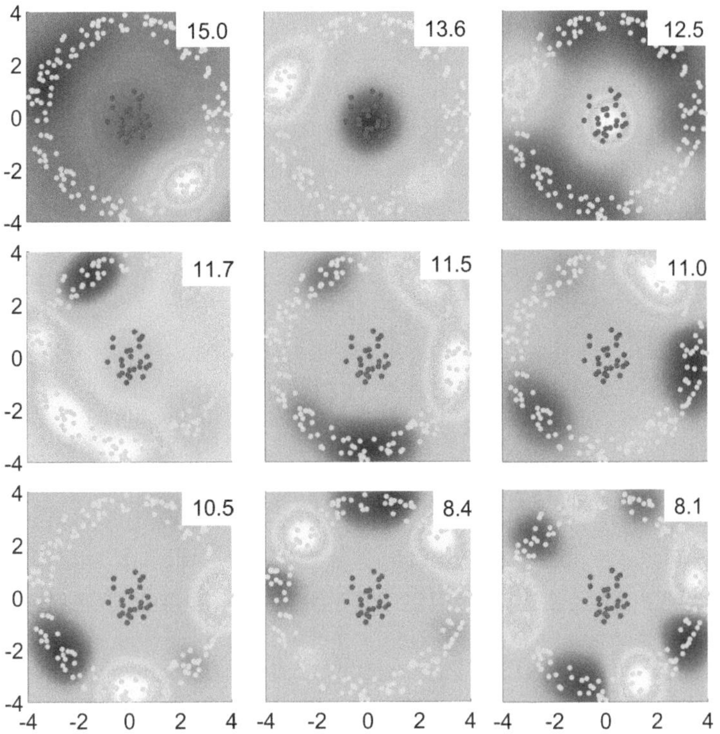

Figure 6.10: First nine eigenfunctions using a Gaussian kernel for the two-dimensional data set formed by the red and cyan points.

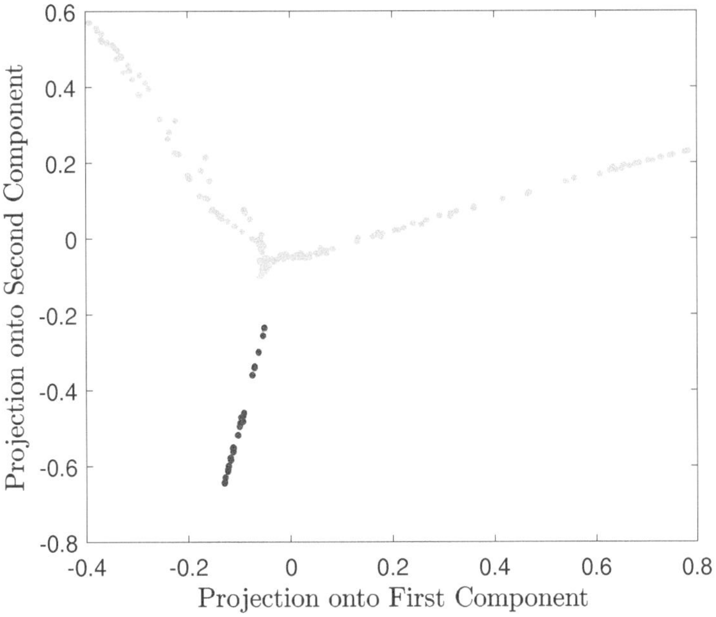

Figure 6.11: Projection of the data onto the first two principal components. Observe that already the projections of the inner and outer points are well separated.

■

Further Reading

For a good overview of the ridge regression and the lasso, we refer the reader to [36, 56]. For overviews of the theory of RKHS we refer to [3, 115, 126], and for in-depth background on splines and their connection to RKHSs we refer to [123]. For further details on GP regression we refer to [97] and for kernel PCA in particular we refer to [12, 92]. Finally, many facts about kernels and their corresponding RKHSs can be found in [115].

Exercises

1. Let $\mathcal{G}$ be an RKHS with reproducing kernel κ. Show that κ is a positive semidefinite function.

2. Show that a reproducing kernel, if it exists, is unique.

3. Let $\mathcal{G}$ be a Hilbert space of functions $g : \mathcal{X} \to \mathbb{R}$. Recall that the *evaluation functional* is the map $\delta_{\boldsymbol{x}} : g \mapsto g(\boldsymbol{x})$ for a given $\boldsymbol{x} \in \mathcal{X}$. Show that evaluation functionals are linear operators.

4. Let $\mathcal{G}_0$ be the pre-RKHS $\mathcal{G}_0$ constructed in the proof of Theorem 6.2. Thus, $g \in \mathcal{G}_0$ is of the form $g = \sum_{i=1}^{n} \alpha_i \kappa_{\boldsymbol{x}_i}$ and

$$\langle g, \kappa_{\boldsymbol{x}} \rangle_{\mathcal{G}_0} = \sum_{i=1}^{n} \alpha_i \langle \kappa_{\boldsymbol{x}_i}, \kappa_{\boldsymbol{x}} \rangle_{\mathcal{G}_0} = \sum_{i=1}^{n} \alpha_i \kappa(\boldsymbol{x}_i, \boldsymbol{x}) = g(\boldsymbol{x}).$$

   Therefore, we may write the evaluation functional of $g \in \mathcal{G}_0$ at $\boldsymbol{x}$ as $\delta_{\boldsymbol{x}} g := \langle g, \kappa_{\boldsymbol{x}} \rangle_{\mathcal{G}_0}$. Show that $\delta_{\boldsymbol{x}}$ is bounded on $\mathcal{G}_0$ for every $\boldsymbol{x}$; that is, $|\delta_{\boldsymbol{x}} f| < \gamma \|f\|_{\mathcal{G}_0}$, for some $\gamma < \infty$.

5. Continuing Exercise 4, let (f_n) be a Cauchy sequence in $\mathcal{G}_0$ such that $|f_n(\boldsymbol{x})| \to 0$ for all $\boldsymbol{x}$. Show that $\|f_n\|_{\mathcal{G}_0} \to 0$.

6. Continuing Exercises 5 and 4, to show that the inner product (6.14) is well defined, a number of facts have to be checked.

   (a) Verify that the limit converges.

   (b) Verify that the limit is independent of the Cauchy sequences used.

   (c) Verify that the properties of an inner product are satisfied. The only non-trivial property to verify is that $\langle f, f \rangle_{\mathcal{G}} = 0$ if and only if $f = 0$.

7. Exercises 4–6 show that $\mathcal{G}$ defined in the proof of Theorem 6.2 is an inner product space. It remains to prove that $\mathcal{G}$ is an RKHS. This requires us to prove that the inner product space $\mathcal{G}$ is complete (and thus Hilbert), and that its evaluation functionals are bounded and hence continuous (see Theorem A.16). This is done in a number of steps. ☞ 391

   (a) Show that $\mathcal{G}_0$ is dense in $\mathcal{G}$ in the sense that every $f \in \mathcal{G}$ is a limit point (with respect to the norm on $\mathcal{G}$) of a Cauchy sequence (f_n) in $\mathcal{G}_0$.

(b) Show that every evaluation functional δ_x on $\mathcal{G}$ is continuous at the 0 function. That is,

$$\forall \varepsilon > 0 : \exists \delta > 0 : \forall f \in \mathcal{G} : \|f\|_{\mathcal{G}} < \delta \Rightarrow |f(\boldsymbol{x})| < \varepsilon. \quad (6.40)$$

Continuity of δ_x at all functions $g \in \mathcal{G}$ then follows automatically from linearity.

(c) Show that $\mathcal{G}$ is complete; that is, every Cauchy sequence $(f_n) \in \mathcal{G}$ converges in the norm $\|\cdot\|_{\mathcal{G}}$.

8. If κ_1 and κ_2 are kernels on $\mathcal{X}$ and $\mathcal{Y}$, then $\kappa_+((\boldsymbol{x},\boldsymbol{y}),(\boldsymbol{x}',\boldsymbol{y}')) := \kappa_1(\boldsymbol{x},\boldsymbol{x}') + \kappa_2(\boldsymbol{y},\boldsymbol{y}')$ and $\kappa_\times((\boldsymbol{x},\boldsymbol{y}),(\boldsymbol{x}',\boldsymbol{y}')) := \kappa_1(\boldsymbol{x},\boldsymbol{x}')\kappa_2(\boldsymbol{y},\boldsymbol{y}')$ are kernels on the Cartesian product $\mathcal{X}\times\mathcal{Y}$. Prove this.

9. An RKHS enjoys the following desirable smoothness property: if (g_n) is a sequence belonging to RKHS $\mathcal{G}$ on $\mathcal{X}$, and $\|g_n - g\|_{\mathcal{G}} \to 0$, then $g(\boldsymbol{x}) = \lim_n g_n(\boldsymbol{x})$ for all $\boldsymbol{x} \in \mathcal{X}$. Prove this, using Cauchy–Schwarz.

10. Let $\boldsymbol{X}$ be an $\mathbb{R}^d$-valued random variable that is symmetric about the origin (that is, $\boldsymbol{X}$ and $(-\boldsymbol{X})$ are identically distributed). Denote by μ is its distribution and $\psi(\boldsymbol{t}) = \mathbb{E}\,\mathrm{e}^{\mathrm{i}\boldsymbol{t}^\top \boldsymbol{X}} = \int \mathrm{e}^{\mathrm{i}\boldsymbol{t}^\top \boldsymbol{x}}\,\mu(\mathrm{d}\boldsymbol{x})$ for $\boldsymbol{t} \in \mathbb{R}^d$ is its characteristic function. Verify that $\kappa(\boldsymbol{x},\boldsymbol{x}') = \psi(\boldsymbol{x}-\boldsymbol{x}')$ is a real-valued positive semidefinite function.

11. Suppose an RKHS $\mathcal{G}$ of functions from $\mathcal{X} \to \mathbb{R}$ (with kernel κ) is invariant under a group $\mathcal{T}$ of transformations $T : \mathcal{X} \to \mathcal{X}$; that is, for all $f, g \in \mathcal{G}$ and $T \in \mathcal{T}$, we have (i) $f \circ T \in \mathcal{G}$ and (ii) $\langle f \circ T, g \circ T\rangle_{\mathcal{G}} = \langle f, g\rangle_{\mathcal{G}}$. Show that $\kappa(T\boldsymbol{x}, T\boldsymbol{x}') = \kappa(\boldsymbol{x},\boldsymbol{x}')$ for all $\boldsymbol{x},\boldsymbol{x}' \in \mathcal{X}$ and $T \in \mathcal{T}$.

12. Given two Hilbert spaces $\mathcal{H}$ and $\mathcal{G}$, we call a mapping $A : \mathcal{H} \to \mathcal{G}$ a *Hilbert space isomorphism* if it is

HILBERT SPACE ISOMORPHISM

(i) a linear map; that is, $A(af + bg) = aA(f) + bA(g)$ for any $f, g \in \mathcal{H}$ and $a, b \in \mathbb{R}$.

(ii) a surjective map; and

(iii) an isometry; that is, for all $f, g \in \mathcal{H}$, it holds that $\langle f, g\rangle_{\mathcal{H}} = \langle Af, Ag\rangle_{\mathcal{G}}$.

Let $\mathcal{H} = \mathbb{R}^p$ (equipped with the usual Euclidean inner product) and construct its (continuous) *dual space* $\mathcal{G}$, consisting of all continuous linear functions from $\mathbb{R}^p$ to $\mathbb{R}$, as follows: (a) For each $\boldsymbol{\beta} \in \mathbb{R}^p$, define $g_{\boldsymbol{\beta}} : \mathbb{R}^p \to \mathbb{R}$ via $g_{\boldsymbol{\beta}}(\boldsymbol{x}) = \langle \boldsymbol{\beta}, \boldsymbol{x}\rangle = \boldsymbol{\beta}^\top \boldsymbol{x}$, for all $\boldsymbol{x} \in \mathbb{R}^p$. (b) Equip $\mathcal{G}$ with the inner product $\langle g_{\boldsymbol{\beta}}, g_{\boldsymbol{\gamma}}\rangle_{\mathcal{G}} := \boldsymbol{\beta}^\top \boldsymbol{\gamma}$.

Show that $A : \mathcal{H} \to \mathcal{G}$ defined by $A(\boldsymbol{\beta}) = g_{\boldsymbol{\beta}}$ for $\boldsymbol{\beta} \in \mathbb{R}^p$ is a Hilbert space isomorphism.

13. Let $\mathbf{X}$ be an $n \times p$ model matrix. Show that $\mathbf{X}^\top\mathbf{X} + n\gamma\mathbf{I}_p$ for $\gamma > 0$ is invertible.

14. As Example 6.8 clearly illustrates, the pdf of a random variable that is symmetric about the origin is not in general a valid reproducing kernel. Take two such iid random variables X and X' with common pdf f, and define $Z = X + X'$. Denote by ψ_Z and f_Z the characteristic function and pdf of Z, respectively.

Show that if ψ_Z is in $L^1(\mathbb{R})$, f_Z is a positive semidefinite function. Use this to show that $\kappa(x, x') = f_Z(x - x') = \mathbb{1}\{|x - x'| \leqslant 2\}(1 - |x - x'|/2)$ is a valid reproducing kernel.

15. For the smoothing cubic spline of Section 6.6, show that $\kappa(x, u) = \frac{\max\{x,u\}\min\{x,u\}^2}{2} - \frac{\min\{x,u\}^3}{6}$.

16. Let $\mathbf{X}$ be an $n \times p$ model matrix and let $\boldsymbol{u} \in \mathbb{R}^p$ be the unit-length vector with k-th entry equal to one ($u_k = \|\boldsymbol{u}\| = 1$). Suppose that the k-th column of $\mathbf{X}$ is $\boldsymbol{v}$ and that it is replaced with a new predictor $\boldsymbol{w}$, so that we obtain the new model matrix:

$$\widetilde{\mathbf{X}} = \mathbf{X} + (\boldsymbol{w} - \boldsymbol{v})\boldsymbol{u}^\top.$$

(a) Denoting

$$\boldsymbol{\delta} := \mathbf{X}^\top(\boldsymbol{w} - \boldsymbol{v}) + \frac{\|\boldsymbol{w} - \boldsymbol{v}\|^2}{2}\boldsymbol{u},$$

show that

$$\widetilde{\mathbf{X}}^\top\widetilde{\mathbf{X}} = \mathbf{X}^\top\mathbf{X} + \boldsymbol{u}\boldsymbol{\delta}^\top + \boldsymbol{\delta}\boldsymbol{u}^\top = \mathbf{X}^\top\mathbf{X} + \frac{(\boldsymbol{u} + \boldsymbol{\delta})(\boldsymbol{u} + \boldsymbol{\delta})^\top}{2} - \frac{(\boldsymbol{u} - \boldsymbol{\delta})(\boldsymbol{u} - \boldsymbol{\delta})^\top}{2}.$$

In other words, $\widetilde{\mathbf{X}}^\top\widetilde{\mathbf{X}}$ differs from $\mathbf{X}^\top\mathbf{X}$ by a symmetric matrix of rank two.

(b) Suppose that $\mathbf{B} := (\mathbf{X}^\top\mathbf{X} + n\gamma\mathbf{I}_p)^{-1}$ is already computed. Explain how the Sherman–Morrison formulas in Theorem A.10 can be applied twice to compute the inverse and log-determinant of the matrix $\widetilde{\mathbf{X}}^\top\widetilde{\mathbf{X}} + n\gamma\mathbf{I}_p$ in $O((n + p)p)$ computing time, rather than the usual $O((n + p^2)p)$ computing time.[3]

☞ 373

(c) Write a Python program for updating a matrix $\mathbf{B} = (\mathbf{X}^\top\mathbf{X} + n\gamma\mathbf{I}_p)^{-1}$ when we change the k-th column of $\mathbf{X}$, as shown in the following pseudo-code.

Algorithm 6.8.1: Updating via Sherman–Morrison Formula

input: Matrices $\mathbf{X}$ and $\mathbf{B}$, index k, and replacement $\boldsymbol{w}$ for the k-th column of $\mathbf{X}$.
output: Updated matrices $\mathbf{X}$ and $\mathbf{B}$.

Set $\boldsymbol{v} \in \mathbb{R}^n$ to be the k-th column of $\mathbf{X}$.
Set $\boldsymbol{u} \in \mathbb{R}^p$ to be the unit-length vector such that $u_k = \|\boldsymbol{u}\| = 1$.
$\mathbf{B} \leftarrow \mathbf{B} - \dfrac{\mathbf{B}\boldsymbol{u}\boldsymbol{\delta}^\top\mathbf{B}}{1 + \boldsymbol{\delta}^\top\mathbf{B}\boldsymbol{u}}$
$\mathbf{B} \leftarrow \mathbf{B} - \dfrac{\mathbf{B}\boldsymbol{\delta}\boldsymbol{u}^\top\mathbf{B}}{1 + \boldsymbol{u}^\top\mathbf{B}\boldsymbol{\delta}}$
Update the k-th column of $\mathbf{X}$ with $\boldsymbol{w}$.
return $\mathbf{X}, \mathbf{B}$

17. Use Algorithm 6.8.1 from Exercise 16 to write Python code that computes the ridge regression coefficient $\boldsymbol{\beta}$ in (6.5) and use it to replicate the results on Figure 6.1. The following pseudo-code (with running cost of $O((n+p)p^2)$) may help with the writing of the Python code.

☞ 217

[3]This Sherman–Morrison updating is not always numerically stable. A more numerically stable method will perform two consecutive rank-one updates of the Cholesky decomposition of $\mathbf{X}^\top\mathbf{X} + n\gamma\mathbf{I}_p$.

Algorithm 6.8.2: Ridge Regression Coefficients via Sherman–Morrison Formula

input: Training set $\{\mathbf{X}, \boldsymbol{y}\}$ and regularization parameter $\gamma > 0$.
output: Solution $\widehat{\boldsymbol{\beta}} = (n\,\gamma\,\mathbf{I}_p + \mathbf{X}^\top\mathbf{X})^{-1}\mathbf{X}^\top\boldsymbol{y}$.

```
Set A to be an n × p matrix of zeros and B ← (n γ I_p)^{-1}.
for j = 1, ..., p do
    Set w to be the j-th column of X.
    Update {A, B} via Algorithm 6.8.1 with inputs {A, B, j, w}.
β̂ ← B(X^T y)
return β̂
```

18. Consider Example 2.10 with $\mathbf{D} = \mathrm{diag}(\lambda_1, \ldots, \lambda_p)$ for some nonnegative vector $\boldsymbol{\lambda} \in \mathbb{R}^p$, so that twice the negative logarithm of the *model evidence* can be written as ☞ 55

$$-2\ln g(\boldsymbol{y}) = l(\boldsymbol{\lambda}) := n\ln[\boldsymbol{y}^\top(\mathbf{I} - \mathbf{X}\boldsymbol{\Sigma}\mathbf{X}^\top)\boldsymbol{y}] + \ln|\mathbf{D}| - \ln|\boldsymbol{\Sigma}| + c,$$

where c is a constant that depends only on n.

(a) Use the *Woodbury identities* (A.15) and (A.16) to show that ☞ 373

$$\begin{aligned}\mathbf{I} - \mathbf{X}\boldsymbol{\Sigma}\mathbf{X}^\top &= (\mathbf{I} + \mathbf{X}\mathbf{D}\mathbf{X}^\top)^{-1}\\ \ln|\mathbf{D}| - \ln|\boldsymbol{\Sigma}| &= \ln|\mathbf{I} + \mathbf{X}\mathbf{D}\mathbf{X}^\top|.\end{aligned}$$

Deduce that $l(\boldsymbol{\lambda}) = n\ln[\boldsymbol{y}^\top\mathbf{C}\boldsymbol{y}] - \ln|\mathbf{C}| + c$, where $\mathbf{C} := (\mathbf{I} + \mathbf{X}\mathbf{D}\mathbf{X}^\top)^{-1}$.

(b) Let $[\boldsymbol{v}_1, \ldots, \boldsymbol{v}_p] := \mathbf{X}$ denote the p columns/predictors of $\mathbf{X}$. Show that

$$\mathbf{C}^{-1} = \mathbf{I} + \sum_{k=1}^{p} \lambda_k \boldsymbol{v}_k \boldsymbol{v}_k^\top.$$

Explain why setting $\lambda_k = 0$ has the effect of excluding the k-th predictor from the regression model. How can this observation be used for model selection?

(c) Prove the following formulas for the gradient and Hessian elements of $l(\boldsymbol{\lambda})$:

$$\begin{aligned}\frac{\partial l}{\partial \lambda_i} &= \boldsymbol{v}_i^\top\mathbf{C}\boldsymbol{v}_i - n\frac{(\boldsymbol{v}_i^\top\mathbf{C}\boldsymbol{y})^2}{\boldsymbol{y}^\top\mathbf{C}\boldsymbol{y}}\\ \frac{\partial^2 l}{\partial \lambda_i\,\partial \lambda_j} &= (n-1)(\boldsymbol{v}_i^\top\mathbf{C}\boldsymbol{v}_j)^2 - n\left[\boldsymbol{v}_i^\top\mathbf{C}\boldsymbol{v}_j - \frac{(\boldsymbol{v}_i^\top\mathbf{C}\boldsymbol{y})(\boldsymbol{v}_j^\top\mathbf{C}\boldsymbol{y})}{\boldsymbol{y}^\top\mathbf{C}\boldsymbol{y}}\right]^2.\end{aligned} \tag{6.41}$$

(d) One method to determine which predictors in $\mathbf{X}$ are important is to compute

$$\boldsymbol{\lambda}^* := \underset{\boldsymbol{\lambda} \geqslant \mathbf{0}}{\mathrm{argmin}}\; l(\boldsymbol{\lambda})$$

using, for example, the interior-point minimization Algorithm B.4.1 with gradient and Hessian computed from (6.41). Write Python code to compute $\boldsymbol{\lambda}^*$ and use it to select the best polynomial model in Example 2.10. ☞ 421

☞ 55

19. (Exercise 18 continued.) Consider again Example 2.10 with $\mathbf{D} = \text{diag}(\lambda_1, \ldots, \lambda_p)$ for some nonnegative model-selection parameter $\boldsymbol{\lambda} \in \mathbb{R}^p$. A Bayesian choice for $\boldsymbol{\lambda}$ is the maximizer of the marginal likelihood $g(\mathbf{y} \mid \boldsymbol{\lambda})$; that is,

$$\boldsymbol{\lambda}^* = \underset{\boldsymbol{\lambda} \geqslant \mathbf{0}}{\text{argmax}} \iint g(\boldsymbol{\beta}, \sigma^2, \mathbf{y} \mid \boldsymbol{\lambda}) \, \mathrm{d}\boldsymbol{\beta} \, \mathrm{d}\sigma^2,$$

where

$$\ln g(\boldsymbol{\beta}, \sigma^2, \mathbf{y} \mid \boldsymbol{\lambda}) = -\frac{\|\mathbf{y} - \mathbf{X}\boldsymbol{\beta}\|^2 + \boldsymbol{\beta}^\top \mathbf{D}^{-1} \boldsymbol{\beta}}{2\sigma^2} - \frac{1}{2} \ln |\mathbf{D}| - \frac{n+p}{2} \ln(2\pi\sigma^2) - \ln \sigma^2.$$

☞ 128

To maximize $g(\mathbf{y} \mid \boldsymbol{\lambda})$, one can use the *EM algorithm* with $\boldsymbol{\beta}$ and σ^2 acting as *latent variables* in the *complete-data log-likelihood* $\ln g(\boldsymbol{\beta}, \sigma^2, \mathbf{y} \mid \boldsymbol{\lambda})$. Define

$$\begin{aligned} \boldsymbol{\Sigma} &:= (\mathbf{D}^{-1} + \mathbf{X}^\top \mathbf{X})^{-1} \\ \overline{\boldsymbol{\beta}} &:= \boldsymbol{\Sigma} \mathbf{X}^\top \mathbf{y} \\ \widehat{\sigma^2} &:= \left(\|\mathbf{y}\|^2 - \mathbf{y}^\top \mathbf{X} \overline{\boldsymbol{\beta}} \right) / n. \end{aligned} \tag{6.42}$$

(a) Show that the conditional density of the latent variables $\boldsymbol{\beta}$ and σ^2 is such that

$$\begin{aligned} \left(\sigma^{-2} \,\middle|\, \boldsymbol{\lambda}, \mathbf{y} \right) &\sim \mathsf{Gamma}\left(\frac{n}{2}, \frac{n}{2} \widehat{\sigma^2} \right) \\ \left(\boldsymbol{\beta} \,\middle|\, \boldsymbol{\lambda}, \sigma^2, \mathbf{y} \right) &\sim \mathcal{N}\left(\overline{\boldsymbol{\beta}}, \sigma^2 \boldsymbol{\Sigma} \right). \end{aligned}$$

☞ 432

(b) Use Theorem C.2 to show that the expected complete-data log-likelihood is

$$-\frac{\overline{\boldsymbol{\beta}}^\top \mathbf{D}^{-1} \overline{\boldsymbol{\beta}}}{2\widehat{\sigma^2}} - \frac{\text{tr}(\mathbf{D}^{-1}\boldsymbol{\Sigma}) + \ln |\mathbf{D}|}{2} + c_1,$$

where c_1 is a constant that does not depend on $\boldsymbol{\lambda}$.

☞ 361

(c) Use Theorem A.2 to simplify the expected complete-data log-likelihood and to show that it is maximized at $\lambda_i = \Sigma_{ii} + (\overline{\beta}_i / \widehat{\sigma})^2$ for $i = 1, \ldots, p$. Hence, deduce the following E and M steps in the EM algorithm:

E-step. Given $\boldsymbol{\lambda}$, update $(\boldsymbol{\Sigma}, \overline{\boldsymbol{\beta}}, \widehat{\sigma^2})$ via the formulas (6.42).

M-step. Given $(\boldsymbol{\Sigma}, \overline{\boldsymbol{\beta}}, \widehat{\sigma^2})$, update $\boldsymbol{\lambda}$ via $\lambda_i = \Sigma_{ii} + (\overline{\beta}_i / \widehat{\sigma})^2$, $i = 1, \ldots, p$.

(d) Write Python code to compute $\boldsymbol{\lambda}^*$ via the EM algorithm, and use it to select the best polynomial model in Example 2.10. A possible stopping criterion is to terminate the EM iterations when

$$\ln g(\mathbf{y} \mid \boldsymbol{\lambda}_{t+1}) - \ln g(\mathbf{y} \mid \boldsymbol{\lambda}_t) < \varepsilon$$

for some small $\varepsilon > 0$, where the marginal log-likelihood is

$$\ln g(\mathbf{y} \mid \boldsymbol{\lambda}) = -\frac{n}{2} \ln(n\pi\widehat{\sigma^2}) - \frac{1}{2} \ln |\mathbf{D}| + \frac{1}{2} \ln |\boldsymbol{\Sigma}| + \ln \Gamma(n/2).$$

20. In this exercise we explore how the *early stopping* of the *gradient descent* iterations (see Example B.10), ☞ 414

$$\boldsymbol{x}_{t+1} = \boldsymbol{x}_t - \alpha \nabla f(\boldsymbol{x}_t), \quad t = 0, 1, \ldots,$$

is (approximately) equivalent to the global minimization of $f(\boldsymbol{x}) + \frac{1}{2}\gamma\|\boldsymbol{x}\|^2$ for certain values of the *ridge regularization* parameter $\gamma > 0$ (see Example 6.1). We illustrate the *early stopping* idea on the quadratic function $f(\boldsymbol{x}) = \frac{1}{2}(\boldsymbol{x} - \boldsymbol{\mu})^\top \mathbf{H}(\boldsymbol{x} - \boldsymbol{\mu})$, where $\mathbf{H} \in \mathbb{R}^{n \times n}$ is a symmetric positive-definite (Hessian) matrix with eigenvalues $\{\lambda_k\}_{k=1}^n$. EARLY STOPPING

(a) Verify that for a symmetric matrix $\mathbf{A} \in \mathbb{R}^n$ such that $\mathbf{I} - \mathbf{A}$ is invertible, we have

$$\mathbf{I} + \mathbf{A} + \cdots + \mathbf{A}^{t-1} = (\mathbf{I} - \mathbf{A}^t)(\mathbf{I} - \mathbf{A})^{-1}.$$

(b) Let $\mathbf{H} = \mathbf{Q}\boldsymbol{\Lambda}\mathbf{Q}^\top$ be the diagonalization of $\mathbf{H}$ as per Theorem A.8. If $\boldsymbol{x}_0 = \mathbf{0}$, show that the formula for $\boldsymbol{x}_t$ is ☞ 368

$$\boldsymbol{x}_t = \boldsymbol{\mu} - \mathbf{Q}(\mathbf{I} - \alpha\boldsymbol{\Lambda})^t \mathbf{Q}^\top \boldsymbol{\mu}.$$

Hence, deduce that a necessary condition for $\boldsymbol{x}_t$ to converge is $\alpha < 2/\max_k \lambda_k$.

(c) Show that the minimizer of $f(\boldsymbol{x}) + \frac{1}{2}\gamma\|\boldsymbol{x}\|^2$ can be written as

$$\boldsymbol{x}^* = \boldsymbol{\mu} - \mathbf{Q}(\mathbf{I} + \gamma^{-1}\boldsymbol{\Lambda})^{-1}\mathbf{Q}^\top \boldsymbol{\mu}.$$

(d) For a fixed value of t, let the learning rate $\alpha \downarrow 0$. Using part (b) and (c), show that if $\gamma \simeq 1/(t\,\alpha)$ as $\alpha \downarrow 0$, then $\boldsymbol{x}_t \simeq \boldsymbol{x}^*$. In other words, $\boldsymbol{x}_t$ is approximately equal to $\boldsymbol{x}^*$ for small α, provided that γ is inversely proportional to $t\,\alpha$.

CHAPTER 7

CLASSIFICATION

The purpose of this chapter is to explain the mathematical ideas behind well-known classification techniques such as the naïve Bayes method, linear and quadratic discriminant analysis, logistic/softmax classification, the K-nearest neighbors method, and support vector machines.

7.1 Introduction

Classification methods are supervised learning methods in which a categorical *response* variable Y takes one of c possible values (for example whether a person is sick or healthy), which is to be predicted from a vector $\boldsymbol{X}$ of *explanatory* variables (for example, the blood pressure, age, and smoking status of the person), using a *prediction function* g. In this sense, g classifies the input $\boldsymbol{X}$ into one of the classes, say in the set $\{0, \ldots, c-1\}$. For this reason, we will call g a *classification function* or simply *classifier*. As with any supervised learning technique (see Section 2.3), the goal is to minimize the expected loss or *risk* CLASSIFIER

$$\ell(g) = \mathbb{E}\,\text{Loss}(Y, g(\boldsymbol{X})) \tag{7.1}$$

for some loss function, $\text{Loss}(y, \widehat{y})$, that quantifies the impact of classifying a response y via $\widehat{y} = g(\boldsymbol{x})$. The natural loss function is the *zero–one* (also written 0–1) or *indicator loss*: $\text{Loss}(y, \widehat{y}) := \mathbb{1}\{y \neq \widehat{y}\}$; that is, there is no loss for a correct classification ($y = \widehat{y}$) and a unit loss for a misclassification ($y \neq \widehat{y}$). In this case the optimal classifier g^* is given in the following theorem. INDICATOR LOSS

Theorem 7.1: Optimal classifier

For the loss function $\text{Loss}(y, \widehat{y}) = \mathbb{1}\{y \neq \widehat{y}\}$, an optimal classification function is

$$g^*(\boldsymbol{x}) = \underset{y \in \{0,\ldots,c-1\}}{\text{argmax}}\ \mathbb{P}[Y = y \,|\, \boldsymbol{X} = \boldsymbol{x}]. \tag{7.2}$$

Proof: The goal is to minimize $\ell(g) = \mathbb{E}\,\mathbb{1}\{Y \neq g(\boldsymbol{X})\}$ over all functions g taking values in $\{0, \ldots, c-1\}$. Conditioning on $\boldsymbol{X}$ gives, by the tower property, $\ell(g) = \mathbb{E}\,(\mathbb{P}[Y \neq g(\boldsymbol{X}) \,|\, \boldsymbol{X}\,])$, ☞ 433
and so minimizing $\ell(g)$ with respect to g can be accomplished by *maximizing* $\mathbb{P}[Y =$

$g(\boldsymbol{x}) \mid \boldsymbol{X} = \boldsymbol{x}]$ with respect to $g(\boldsymbol{x})$, for every fixed $\boldsymbol{x}$. In other words, take $g(\boldsymbol{x})$ to be equal to the class label y for which $\mathbb{P}[Y = y \mid \boldsymbol{X} = \boldsymbol{x}]$ is maximal. □

The formulation (7.2) allows for "ties", when there is an equal probability between optimal classes for a feature vector $\boldsymbol{x}$. Assigning one of these tied classes arbitrarily (or randomly) to $\boldsymbol{x}$ does not affect the loss function and so we assume for simplicity that $g^*(\boldsymbol{x})$ is always a scalar value.

☞ 21

Note that, as was the case for the regression (see, e.g., Theorem 2.1), the optimal prediction function depends on the conditional pdf $f(y \mid \boldsymbol{x}) = \mathbb{P}[Y = y \mid \boldsymbol{X} = \boldsymbol{x}]$. However, since we assign $\boldsymbol{x}$ to class y if $f(y \mid \boldsymbol{x}) \geqslant f(z \mid \boldsymbol{x})$ for all z, we do not need to learn the entire surface of the function $f(y \mid \boldsymbol{x})$; we only need to estimate it well enough near the decision boundary $\{\boldsymbol{x} : f(y \mid \boldsymbol{x}) = f(z \mid \boldsymbol{x})\}$ for any choice of classes y and z. This is because the assignment (7.2) divides the feature space into c regions, $\mathcal{R}_y = \{\boldsymbol{x} : f(y \mid \boldsymbol{x}) = \max_z f(z \mid \boldsymbol{x})\}$, $y = 0, \ldots, c-1$.

Recall that for any supervised learning problem the smallest possible expected loss (that is, the irreducible risk) is given by $\ell^* = \ell(g^*)$. For the indicator loss, the irreducible risk is equal to $\mathbb{P}[Y \neq g^*(\boldsymbol{X})]$. This smallest possible probability of misclassification is often called the *Bayes error rate*.

BAYES ERROR RATE

For a given training set τ, a classifier is often derived from a *pre-classifier* g_τ, which is a prediction function (learner) that can take any real value, rather than only values in the set of class labels. A typical situation is the case of binary classification with labels -1 and 1, where the prediction function g_τ is a function taking values in the interval $[-1, 1]$ and the actual classifier is given by $\text{sign}(g_\tau)$. It will be clear from the context whether a prediction function g_τ should be interpreted as a classifier or pre-classifier.

The indicator loss function may not always be the most appropriate choice of loss function for a given classification problem. For example, when diagnosing an illness, the mistake in misclassifying a person as being sick when in fact the person is healthy may be less serious than classifying the person as healthy when in fact the person is sick. In Section 7.2 we consider various classification metrics.

There are many ways to fit a classifier to a training set $\tau = \{(\boldsymbol{x}_1, y_1), \ldots, (\boldsymbol{x}_n, y_n)\}$. The approach taken in Section 7.3 is to use a Bayesian framework for classification. Here the conditional pdf $f(y \mid \boldsymbol{x})$ is viewed as a posterior pdf $f(y \mid \boldsymbol{x}) \propto f(\boldsymbol{x} \mid y) f(y)$ for a given class prior $f(y)$ and likelihood $f(\boldsymbol{x} \mid y)$. Section 7.4 discusses linear and quadratic discriminant analysis for classification, which assumes that the class of approximating functions for the conditional pdf $f(\boldsymbol{x} \mid y)$ is a parametric class $\mathcal{G}$ of Gaussian densities. As a result of this choice of $\mathcal{G}$, the marginal $f(\boldsymbol{x})$ is approximated via a Gaussian mixture density.

In contrast, in the logistic or soft-max classification in Section 7.5, the conditional pdf $f(y \mid \boldsymbol{x})$ is approximated using a more flexible class of approximating functions. As a result of this, the approximation to the marginal density $f(\boldsymbol{x})$ does not belong to a simple parametric class (such as a Gaussian mixture). As in unsupervised learning, the cross-entropy loss is the most common choice for training the learner.

The K-nearest neighbors method, discussed in Section 7.6, is yet another approach to classification that makes minimal assumptions on the class $\mathcal{G}$. Here the aim is to directly estimate the conditional pdf $f(y \mid \boldsymbol{x})$ from the training data, using only feature vectors in the neighborhood of $\boldsymbol{x}$. In Section 7.7 we explain the support vector methodology for classification; this is based on the same Reproducing Kernel Hilbert Space ideas that proved successful for regression analysis in Section 6.3. Finally, a versatile way to do both classification and regression is to use classification and regression trees. This is the topic of Chapter 8. Neural networks (Chapter 9) provide yet another way to perform classification.

☞ 222

☞ 289

☞ 325

7.2 Classification Metrics

The effectiveness of a classifier g is, theoretically, measured in terms of the risk (7.1), which depends on the loss function used. Fitting a classifier to iid training data $\tau = \{(\boldsymbol{x}_i, y_i)\}_{i=1}^n$ is established by minimizing the *training loss*

$$\ell_\tau(g) = \frac{1}{n} \sum_{i=1}^{n} \mathrm{Loss}(y_i, g(\boldsymbol{x}_i)) \tag{7.3}$$

over some class of functions $\mathcal{G}$. As the training loss is often a poor estimator of the risk, the risk is usually estimated as in (7.3), using instead a test set $\tau' = \{(\boldsymbol{x}'_i, y'_i)\}_{i=1}^{n'}\}$ that is independent of the training set, as explained in Section 2.3. To measure the performance of a classifier on a training or test set, it is convenient to introduce the notion of a *loss matrix*. Consider a classification problem with classifier g, loss function Loss, and classes $0, \ldots, c-1$. If an input feature vector $\boldsymbol{x}$ is classified as $\widehat{y} = g(\boldsymbol{x})$ when the observed class is y, the loss incurred is, by definition, $\mathrm{Loss}(y, \widehat{y})$. Consequently, we may identify the loss function with a matrix $\mathbf{L} = [\mathrm{Loss}(j,k), j,k \in \{0,\ldots,c-1\}]$. For the indicator loss function, the matrix $\mathbf{L}$ has 0s on the diagonal and 1s everywhere else. Another useful matrix is the *confusion matrix*, denoted by $\mathbf{M}$, where the (j,k)-th element of $\mathbf{M}$ counts the number of times that, for the training or test data, the actual (observed) class is j whereas the predicted class is k. Table 7.1 shows the confusion matrix of some Dog/Cat/Possum classifier.

☞ 23

LOSS MATRIX

CONFUSION MATRIX

Table 7.1: Confusion matrix for three classes.

	Predicted		
Actual	Dog	Cat	Possum
Dog	30	2	6
Cat	8	22	15
Possum	7	4	41

We can now express the classifier performance (7.3) in terms of $\mathbf{L}$ and $\mathbf{M}$ as

$$\frac{1}{n} \sum_{j,k} [\mathbf{L} \odot \mathbf{M}]_{jk}, \tag{7.4}$$

where $\mathbf{L} \odot \mathbf{M}$ is the elementwise product of $\mathbf{L}$ and $\mathbf{M}$. Note that for the indicator loss, (7.4) is simply $1 - \mathrm{tr}(\mathbf{M})/n$, and is called the *misclassification error*. The expression (7.4) makes it clear that both the counts and the loss are important in determining the performance of a classifier.

MISCLASSIFICATION ERROR

☞ 461

TRUE POSITIVE

TRUE NEGATIVE

FALSE POSITIVE

FALSE NEGATIVE

In the spirit of Table C.4 for hypothesis testing, it is sometimes useful to divide the elements of a confusion matrix into four groups. The diagonal elements are the *true positive* counts; that is, the numbers of correct classifications for each class. The true positive counts for the Dog, Cat, and Possum classes in Table 7.1 are 30, 22, and 41, respectively. Similarly, the *true negative* count for a class is the sum of all matrix elements that do not belong to the row or the column of this particular class. For the Dog class it is $22 + 15 + 4 + 41 = 82$. The *false positive* count for a class is the sum of the corresponding column elements without the diagonal element. For the Dog class it is $8 + 7 = 15$. Finally, the *false negative* count for a specific class, can be calculated by summing over the corresponding row elements (again, without counting the diagonal element). For the Dog class it is $2 + 6 = 8$.

In terms of the elements of the confusion matrix, we have the following counts for class $j = 0, \dots, c - 1$:

$$
\begin{aligned}
&\text{True positive} && \mathrm{tp}_j = \mathbf{M}_{jj}, && \\
&\text{False positive} && \mathrm{fp}_j = \sum_{k \neq j} \mathbf{M}_{kj}, && \text{(column sum)} \\
&\text{False negative} && \mathrm{fn}_j = \sum_{k \neq j} \mathbf{M}_{jk}, && \text{(row sum)} \\
&\text{True negative} && \mathrm{tn}_j = n - \mathrm{fn}_j - \mathrm{fp}_j - \mathrm{tp}_j. &&
\end{aligned}
$$

Note that in the binary classification case ($c = 2$), and using the indicator loss function, the misclassification error (7.4) can be written as

$$
\mathrm{error}_j = \frac{\mathrm{fp}_j + \mathrm{fn}_j}{n}. \tag{7.5}
$$

This does not depend on which of the two classes is considered, as $\mathrm{fp}_0 + \mathrm{fn}_0 = \mathrm{fp}_1 + \mathrm{fn}_1$.

ACCURACY

Similarly, the *accuracy* measures the fraction of correctly classified objects:

$$
\mathrm{accuracy}_j = 1 - \mathrm{error}_j = \frac{\mathrm{tp}_j + \mathrm{tn}_j}{n}. \tag{7.6}
$$

In some cases, classification error (or accuracy) alone is not sufficient to adequately describe the effectiveness of a classifier. As an example, consider the following two classification problems based on a fingerprint detection system:

1. Identification of authorized personnel in a top-secret military facility.

2. Identification to get an online discount for some retail chain.

Both problems are binary classification problems. However, a false positive in the first problem is extremely dangerous, while a false positive in the second problem will make a customer happy. Let us examine a classifier in the top-secret facility. The corresponding confusion matrix is given in Table 7.2.

Table 7.2: Confusion matrix for authorized personnel classification.

	Predicted	
Actual	authorized	non-authorized
authorized	100	400
non-authorized	50	100,000

From (7.6), we conclude that the accuracy of classification is equal to

$$\text{accuracy} = \frac{\text{tp} + \text{tn}}{\text{tp} + \text{tn} + \text{fp} + \text{fn}} = \frac{100 + 100,000}{100 + 100,000 + 50 + 400} \approx 99.55\%.$$

However, we can see that in this particular case, accuracy is a problematic metric, since the algorithm allowed 50 non-authorized personnel to enter the facility. One way to deal with this issue is to modify the loss function to give a much higher loss to non-authorized access. Thus, instead of an (indicator) loss matrix, we could for example take the loss matrix

$$\mathbf{L} = \begin{pmatrix} 0 & 1 \\ 1000 & 0 \end{pmatrix}.$$

An alternative approach is to keep the indicator loss function and consider additional classification metrics. Below we give a list of commonly used metrics. For simplicity we call an object whose actual class is j a "j-object".

- The *precision* (also called *positive predictive value*) is the fraction of all objects classified as j that are actually j-objects. Specifically, PRECISION

$$\text{precision}_j = \frac{\text{tp}_j}{\text{tp}_j + \text{fp}_j}.$$

- The *recall* (also called *sensitivity*) is the fraction of all j-objects that are correctly classified as such. That is, RECALL

$$\text{recall}_j = \frac{\text{tp}_j}{\text{tp}_j + \text{fn}_j}.$$

- The *specificity* measures the fraction of all non-j-objects that are correctly classified as such. Specifically, SPECIFICITY

$$\text{specificity}_j = \frac{\text{tn}_j}{\text{fp}_j + \text{tn}_j}.$$

- The F_β *score* is a combination of the precision and the recall and is used as a single measurement for a classifier's performance. The F_β score is given by F_β SCORE

$$F_{\beta,j} = \frac{(\beta^2 + 1)\,\text{tp}_j}{(\beta^2 + 1)\,\text{tp}_j + \beta^2\,\text{fn}_j + \text{fp}_j}.$$

  For $\beta = 0$ we obtain the precision and for $\beta \to \infty$ we obtain the recall.

The particular choice of metric is clearly application dependent. For example, in the classification of authorized personnel in a top-secret military facility, suppose we have two classifiers. The first (Classifier 1) has a confusion matrix given in Table 7.2, and the second (Classifier 2) has a confusion matrix given in Table 7.3. Various metrics for these two classifiers are show in Table 7.4. In this case we prefer Classifier 1, which has a much higher precision.

Table 7.3: Confusion matrix for authorized personnel classification, using a different classifier (Classifier 2).

	Predicted	
Actual	Authorized	Non-Authorized
authorized	50	10
non-authorized	450	100,040

Table 7.4: Comparing the metrics for the confusion matrices in Tables 7.2 and 7.3.

Metric	Classifier 1	Classifier 2
accuracy	9.955×10^{-1}	9.954×10^{-1}
precision	$\mathbf{6.667 \times 10^{-1}}$	$\mathbf{1.000 \times 10^{-1}}$
recall	2.000×10^{-1}	8.333×10^{-1}
specificity	9.995×10^{-1}	9.955×10^{-1}
F_1	3.077×10^{-1}	1.786×10^{-1}

■ **Remark 7.1 (Multilabel and Hierarchical Classification)** In standard classification the classes are assumed to be mutually exclusive. For example a satellite image could be classified as "cloudy", "clear", or "foggy". In *multilabel classification* the classes (often called labels) do not have to be mutually exclusive. In this case the response is a subset $\mathcal{Y}$ of some collection of labels $\{0, \ldots, c-1\}$. Equivalently, the response can be viewed as a binary vector of length c, where the y-th element is 1 if the response belongs to label y and 0 otherwise. Again, consider the satellite image example and add two labels, such as "road" and "river" to the previous three labels. Clearly, an image can contain both a road and a river. In addition, the image can be clear, cloudy, or foggy.

MULTILABEL CLASSIFICATION

In *hierarchical classification* a hierarchical relation between classes/labels is taken into account during the classification process. Usually, the relations are modeled via a tree or a directed acyclic graph. A visual comparison between the hierarchical and non-hierarchical (flat) classification tasks for satellite image data is presented in Figure 7.1.

HIERARCHICAL CLASSIFICATION

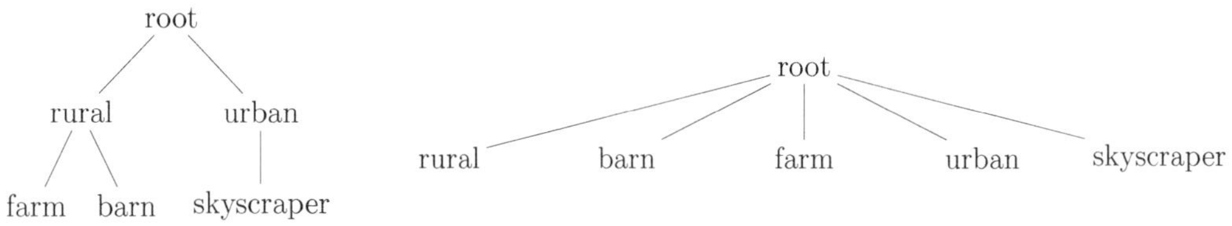

Figure 7.1: Hierarchical (left) and non-hierarchical (right) classification schemes. Barns and farms are common in rural areas, while skyscrapers are generally located in cities. While this relation can be clearly observed in the hierarchical model scheme, the connection is missing in the non-hierarchical design.

■

In multilabel classification, both the prediction $\widehat{\mathcal{Y}} := g(\boldsymbol{x})$ and the true response $\mathcal{Y}$ are *subsets* of the label set $\{0, \ldots, c-1\}$. A reasonable metric is the so-called *exact match ratio*,

EXACT MATCH RATIO

defined as

$$\text{exact match ratio} = \frac{\sum_{i=1}^n \mathbb{1}\{\widehat{\mathcal{Y}}_i = \mathcal{Y}_i\}}{n}.$$

The exact match ratio is rather stringent, as it requires a full match. In order to consider partial correctness, the following metrics could be used instead.

- The *accuracy* is defined as the ratio of correctly predicted labels and the total number of predicted and actual labels. The formula is given by

$$\text{accuracy} = \frac{\sum_{i=1}^n |\mathcal{Y}_i \cap \widehat{\mathcal{Y}}_i|}{\sum_{i=1}^n |\mathcal{Y}_i \cup \widehat{\mathcal{Y}}_i|}.$$

- The *precision* is defined as the ratio of correctly predicted labels and the total number of predicted labels. Specifically,

$$\text{precision} = \frac{\sum_{i=1}^n |\mathcal{Y}_i \cap \widehat{\mathcal{Y}}_i|}{\sum_{i=1}^n |\widehat{\mathcal{Y}}_i|}. \tag{7.7}$$

- The *recall* is defined as the ratio of correctly predicted labels and the total number of actual labels. Specifically,

$$\text{recall} = \frac{\sum_{i=1}^n |\mathcal{Y}_i \cap \widehat{\mathcal{Y}}_i|}{\sum_{i=1}^n |\mathcal{Y}_i|}. \tag{7.8}$$

- The *Hamming loss* counts the average number of incorrect predictions for all classes, calculated as

$$\text{Hamming} = \frac{1}{n\,c}\sum_{i=1}^{n}\sum_{y=0}^{c-1} \mathbb{1}\{y \in \widehat{\mathcal{Y}}_i\}\,\mathbb{1}\{y \notin \mathcal{Y}_i\} + \mathbb{1}\{y \notin \widehat{\mathcal{Y}}_i\}\,\mathbb{1}\{y \in \mathcal{Y}_i\}.$$

7.3 Classification via Bayes' Rule

We saw from Theorem 7.1 that the optimal classifier for classes $0, \ldots, c-1$ divides the feature space into c regions, depending on $f(y \,|\, \boldsymbol{x})$: the conditional pdf of the response Y given the feature vector $\boldsymbol{X} = \boldsymbol{x}$. In particular, if $f(y \,|\, \boldsymbol{x}) > f(z \,|\, \boldsymbol{x})$ for all $z \neq y$, the feature vector $\boldsymbol{x}$ is classified as y. Classifying feature vectors on the basis of their conditional class probabilities is a natural thing to do, especially in a Bayesian learning context; see Section 2.9 for an overview of Bayesian terminology and usage. Specifically, the conditional probability $f(y \,|\, \boldsymbol{x})$ is interpreted as a *posterior* probability, of the form ☞ 47

$$f(y \,|\, \boldsymbol{x}) \propto f(\boldsymbol{x} \,|\, y) f(y), \tag{7.9}$$

where $f(\boldsymbol{x} \,|\, y)$ is the *likelihood* of obtaining feature vector $\boldsymbol{x}$ from class y and $f(y)$ is the *prior* probability[1] of class y. By making various modeling assumptions about the prior

[1]Here we have used the Bayesian notation convention of "overloading" the notation f.

(e.g., all classes are *a priori* equally likely) and the likelihood function, one obtains the posterior pdf via Bayes' formula (7.9). A class $\widehat{y}$ is then assigned to a feature vector $\boldsymbol{x}$ according to the highest posterior probability; that is, we classify according to the *Bayes optimal decision rule*:

BAYES OPTIMAL DECISION RULE

$$\widehat{y} = \underset{y}{\operatorname{argmax}}\, f(y \,|\, \boldsymbol{x}), \tag{7.10}$$

which is exactly (7.2). Since the discrete density $f(y \,|\, \boldsymbol{x})$, $y = 0, \ldots, c-1$ is usually not known, the aim is to approximate it well with a function $g(y \,|\, \boldsymbol{x})$ from some class of functions $\mathcal{G}$. Note that in this context, $g(\cdot \,|\, \boldsymbol{x})$ refers to a discrete density (a probability mass function) for a given $\boldsymbol{x}$.

Suppose a feature vector $\boldsymbol{x} = [x_1, \ldots, x_p]^\top$ of p features has to be classified into one of the classes $0, \ldots, c-1$. For example, the classes could be different people and the features could be various facial measurements, such as the width of the eyes divided by the distance between the eyes, or the ratio of the nose height and mouth width. In the *naïve Bayes* method, the class of approximating functions $\mathcal{G}$ is chosen such that $g(\boldsymbol{x} \,|\, y) = g(x_1 \,|\, y) \cdots g(x_p \,|\, y)$, that is, conditional on the label, all features are independent. Assuming a uniform prior for y, the posterior pdf can thus be written as

NAÏVE BAYES

$$g(y \,|\, \boldsymbol{x}) \propto \prod_{j=1}^{p} g(x_j \,|\, y),$$

where the marginal pdfs $g(x_j \,|\, y)$, $j = 1, \ldots, p$ belong to a given class of approximating functions $\mathcal{G}$. To classify $\boldsymbol{x}$, simply take the y that maximizes the unnormalized posterior pdf.

For instance, suppose that the approximating class $\mathcal{G}$ is such that $(X_j \,|\, y) \sim \mathcal{N}(\mu_{yj}, \sigma^2)$, $y = 0, \ldots, c-1$, $j = 1, \ldots, p$. The corresponding posterior pdf is then

$$g(y \,|\, \boldsymbol{\theta}, \boldsymbol{x}) \propto \exp\left(-\frac{1}{2}\sum_{j=1}^{p}\frac{(x_j - \mu_{yj})^2}{\sigma^2}\right) = \exp\left(-\frac{1}{2}\frac{\|\boldsymbol{x} - \boldsymbol{\mu}_y\|^2}{\sigma^2}\right),$$

where $\boldsymbol{\mu}_y := [\mu_{y1}, \ldots, \mu_{yp}]^\top$ and $\boldsymbol{\theta} := \{\boldsymbol{\mu}_0, \ldots, \boldsymbol{\mu}_{c-1}, \sigma^2\}$ collects all model parameters. The probability $g(y \,|\, \boldsymbol{\theta}, \boldsymbol{x})$ is maximal when $\|\boldsymbol{x} - \boldsymbol{\mu}_y\|$ is minimal. Thus $\widehat{y} = \operatorname{argmin}_y \|\boldsymbol{x} - \boldsymbol{\mu}_y\|$ is the classifier that maximizes the posterior probability. That is, classify $\boldsymbol{x}$ as y when $\boldsymbol{\mu}_y$ is closest to $\boldsymbol{x}$ in Euclidean distance. Of course, the parameters (here, the $\{\boldsymbol{\mu}_y\}$ and σ^2) are unknown and have to be estimated from the training data.

We can extend the above idea to the case where also the variance σ^2 depends on the class y and feature j, as in the next example.

■ **Example 7.1 (Naïve Bayes Classification)** Table 7.5 lists the means μ and standard deviations σ of $p = 3$ normally distributed features, for $c = 4$ different classes. How should a feature vector $\boldsymbol{x} = [1.67, 2.00, 4.23]^\top$ be classified? The posterior pdf is

$$g(y \,|\, \boldsymbol{\theta}, \boldsymbol{x}) \propto (\sigma_{y1}\sigma_{y2}\sigma_{y3})^{-1} \exp\left(-\frac{1}{2}\sum_{j=1}^{3}\frac{(x_j - \mu_{yj})^2}{\sigma_{yj}^2}\right),$$

where $\boldsymbol{\theta} := \{\boldsymbol{\sigma}_j, \boldsymbol{\mu}_j\}_{j=0}^{c-1}$ again collects all model parameters. The (unscaled) values for $g(y \,|\, \boldsymbol{\theta}, \boldsymbol{x})$, $y = 0, 1, 2, 3$ are 53.5, 0.24, 8.37, and 3.5×10^{-6}, respectively. Hence, the feature vector should be classified as 0. The code follows.

Table 7.5: Feature parameters.

	Feature 1		Feature 2		Feature 3	
Class	μ	σ	μ	σ	μ	σ
0	1.6	0.1	2.4	0.5	4.3	0.2
1	1.5	0.2	2.9	0.6	6.1	0.9
2	1.8	0.3	2.5	0.3	4.2	0.3
3	1.1	0.2	3.1	0.7	5.6	0.3

naiveBayes.py

```
import numpy as np
x = np.array([1.67,2,4.23]).reshape(1,3)
mu = np.array([1.6, 2.4, 4.3,
               1.5, 2.9, 6.1,
               1.8, 2.5, 4.2,
               1.1, 3.1, 5.6]).reshape(4,3)
sig = np.array([0.1, 0.5, 0.2,
                0.2, 0.6, 0.9,
                0.3, 0.3, 0.3,
                0.2, 0.7, 0.3]).reshape(4,3)
g = lambda y: 1/np.prod(sig[y,:]) * np.exp(
      -0.5*np.sum((x-mu[y,:])**2/sig[y,:]**2));
for y in range(0,4):
    print('{:3.2e}'.format(g(y)))
```

```
5.35e+01
2.42e-01
8.37e+00
3.53e-06
```

■

7.4 Linear and Quadratic Discriminant Analysis

The Bayesian viewpoint for classification of the previous section (not limited to naïve Bayes) leads in a natural way to the well-established technique of *discriminant analysis*. We discuss the binary classification case first, with classes 0 and 1. DISCRIMINANT ANALYSIS

We consider a class of approximating functions $\mathcal{G}$ such that, conditional on the class $y \in \{0, 1\}$, the feature vector $\boldsymbol{X} = [X_1, \ldots, X_p]^\top$ has a $\mathcal{N}(\boldsymbol{\mu}_y, \boldsymbol{\Sigma}_y)$ distribution (see (2.33)): ☞ 45

$$g(\boldsymbol{x} \,|\, \boldsymbol{\theta}, y) = \frac{1}{\sqrt{(2\pi)^p \, |\boldsymbol{\Sigma}_y|}} \, \mathrm{e}^{-\frac{1}{2}\,(\boldsymbol{x}-\boldsymbol{\mu}_y)^\top \boldsymbol{\Sigma}_y^{-1}(\boldsymbol{x}-\boldsymbol{\mu}_y)}, \quad \boldsymbol{x} \in \mathbb{R}^p, \quad y \in \{0, 1\}, \tag{7.11}$$

where $\boldsymbol{\theta} = \{\alpha_j, \boldsymbol{\mu}_j, \boldsymbol{\Sigma}_j\}_{j=0}^{c-1}$ collects all model parameters, including the probability vector $\boldsymbol{\alpha}$ (that is, $\sum_i \alpha_i = 1$ and $\alpha_i \geqslant 0$) which helps define the prior density: $g(y \,|\, \boldsymbol{\theta}) = \alpha_y, \; y \in \{0, 1\}$. Then, the posterior density is

$$g(y \,|\, \boldsymbol{\theta}, \boldsymbol{x}) \propto \alpha_y \times g(\boldsymbol{x} \,|\, \boldsymbol{\theta}, y),$$

and, according to the Bayes optimal decision rule (7.10), we classify $\boldsymbol{x}$ to come from class 0 if $\alpha_0 g(\boldsymbol{x} \mid \boldsymbol{\theta}, 0) > \alpha_1 g(\boldsymbol{x} \mid \boldsymbol{\theta}, 1)$ or, equivalently (by taking logarithms) if,

$$\ln \alpha_0 - \frac{1}{2}\ln|\boldsymbol{\Sigma}_0| - \frac{1}{2}(\boldsymbol{x}-\boldsymbol{\mu}_0)^\top \boldsymbol{\Sigma}_0^{-1}(\boldsymbol{x}-\boldsymbol{\mu}_0) > \ln \alpha_1 - \frac{1}{2}\ln|\boldsymbol{\Sigma}_1| - \frac{1}{2}(\boldsymbol{x}-\boldsymbol{\mu}_1)^\top \boldsymbol{\Sigma}_1^{-1}(\boldsymbol{x}-\boldsymbol{\mu}_1).$$

The function

$$\delta_y(\boldsymbol{x}) = \ln \alpha_y - \frac{1}{2}\ln|\boldsymbol{\Sigma}_y| - \frac{1}{2}(\boldsymbol{x}-\boldsymbol{\mu}_y)^\top \boldsymbol{\Sigma}_y^{-1}(\boldsymbol{x}-\boldsymbol{\mu}_y), \quad \boldsymbol{x} \in \mathbb{R}^p \tag{7.12}$$

QUADRATIC DISCRIMINANT FUNCTION

is called the *quadratic discriminant function* for class $y = 0, 1$. A point $\boldsymbol{x}$ is classified to class y for which $\delta_y(\boldsymbol{x})$ is largest. The function is quadratic in $\boldsymbol{x}$ and so the decision boundary $\{\boldsymbol{x} \in \mathbb{R}^p : \delta_0(\boldsymbol{x}) = \delta_1(\boldsymbol{x})\}$ is quadratic as well. An important simplification arises for the case where the assumption is made that $\boldsymbol{\Sigma}_0 = \boldsymbol{\Sigma}_1 = \boldsymbol{\Sigma}$. Now, the decision boundary is the set of $\boldsymbol{x}$ for which

$$\ln \alpha_0 - \frac{1}{2}(\boldsymbol{x}-\boldsymbol{\mu}_0)^\top \boldsymbol{\Sigma}^{-1}(\boldsymbol{x}-\boldsymbol{\mu}_0) = \ln \alpha_1 - \frac{1}{2}(\boldsymbol{x}-\boldsymbol{\mu}_1)^\top \boldsymbol{\Sigma}^{-1}(\boldsymbol{x}-\boldsymbol{\mu}_1).$$

Expanding the above expression shows that the quadratic term in $\boldsymbol{x}$ is eliminated, giving a *linear* decision boundary in $\boldsymbol{x}$:

$$\ln \alpha_0 - \frac{1}{2}\boldsymbol{\mu}_0^\top \boldsymbol{\Sigma}^{-1}\boldsymbol{\mu}_0 + \boldsymbol{x}^\top \boldsymbol{\Sigma}^{-1}\boldsymbol{\mu}_0 = \ln \alpha_1 - \frac{1}{2}\boldsymbol{\mu}_1^\top \boldsymbol{\Sigma}^{-1}\boldsymbol{\mu}_1 + \boldsymbol{x}^\top \boldsymbol{\Sigma}^{-1}\boldsymbol{\mu}_1.$$

LINEAR DISCRIMINANT FUNCTION

The corresponding *linear discriminant function* for class y is

$$\delta_y(\boldsymbol{x}) = \ln \alpha_y - \frac{1}{2}\boldsymbol{\mu}_y^\top \boldsymbol{\Sigma}^{-1}\boldsymbol{\mu}_y + \boldsymbol{x}^\top \boldsymbol{\Sigma}^{-1}\boldsymbol{\mu}_y, \quad \boldsymbol{x} \in \mathbb{R}^p. \tag{7.13}$$

■ **Example 7.2 (Linear Discriminant Analysis)** Consider the case where $\alpha_0 = \alpha_1 = 1/2$ and

$$\boldsymbol{\Sigma} = \begin{bmatrix} 2 & 0.7 \\ 0.7 & 2 \end{bmatrix}, \quad \boldsymbol{\mu}_0 = \begin{bmatrix} 0 \\ 0 \end{bmatrix}, \quad \boldsymbol{\mu}_1 = \begin{bmatrix} 2 \\ 4 \end{bmatrix}.$$

☞ 135

The distribution of $\boldsymbol{X}$ is a mixture of two bivariate normal distributions. Its pdf,

$$\frac{1}{2}g(\boldsymbol{x} \mid \boldsymbol{\theta}, y = 0) + \frac{1}{2}g(\boldsymbol{x} \mid \boldsymbol{\theta}, y = 1),$$

is depicted in Figure 7.2.

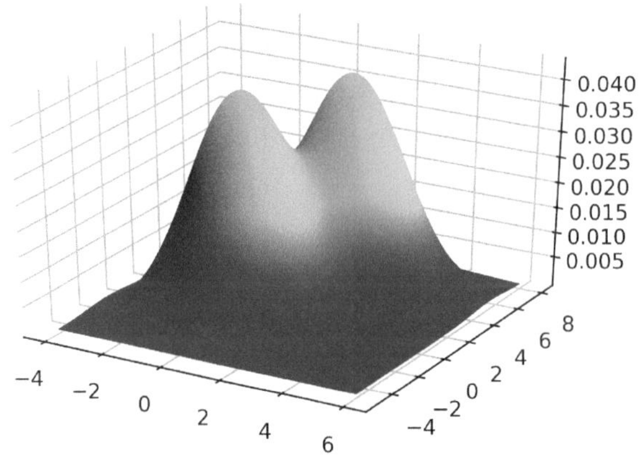

Figure 7.2: A Gaussian mixture density where the two mixture components have the same covariance matrix.

We used the following Python code to make this figure.

LDAmixture.py

```
import numpy as np, matplotlib.pyplot as plt
from scipy.stats import multivariate_normal
from mpl_toolkits.mplot3d import Axes3D
from matplotlib.colors import LightSource

mu0, mu1 = np.array([0,0]), np.array([2,4])
Sigma = np.array([[2,0.7],[0.7, 2]])
x, y = np.mgrid[-4:6:150j,-5:8:150j]
mvn0 = multivariate_normal( mu0, Sigma )
mvn1 = multivariate_normal( mu1, Sigma )

xy = np.hstack((x.reshape(-1,1),y.reshape(-1,1)))
z = 0.5*mvn0.pdf(xy).reshape(x.shape) +  0.5*mvn1.pdf(xy).reshape(x.
    shape)

fig = plt.figure()
ax = fig.gca(projection='3d')
ls = LightSource(azdeg=180, altdeg=65)
cols = ls.shade(z, plt.cm.winter)
surf = ax.plot_surface(x, y, z, rstride=1, cstride=1, linewidth=0,
                        antialiased=False, facecolors=cols)
plt.show()
```

The following Python code, which imports the previous code, draws a contour plot of the mixture density, simulates 1000 data points from the mixture density, and draws the decision boundary. To compute and display the linear decision boundary, let $[a_1, a_2]^\top = 2\mathbf{\Sigma}^{-1}(\boldsymbol{\mu}_1 - \boldsymbol{\mu}_0)$ and $b = \boldsymbol{\mu}_0^\top \mathbf{\Sigma}^{-1}\boldsymbol{\mu}_0 - \boldsymbol{\mu}_1^\top \mathbf{\Sigma}^{-1}\boldsymbol{\mu}_1$. Then, the decision boundary can be written as $a_1 x_1 + a_2 x_2 + b = 0$ or, equivalently, $x_2 = -(a_1 x_1 + b)/a_2$. We see in Figure 7.3 that the decision boundary nicely separates the two modes of the mixture density.

LDA.py

```
from LDAmixture import *
from numpy.random import rand
from numpy.linalg import inv

fig = plt.figure()
plt.contourf(x, y,z, cmap=plt.cm.Blues, alpha= 0.9,extend='both')
plt.ylim(-5.0,8.0)
plt.xlim(-4.0,6.0)
M = 1000
r = (rand(M,1) < 0.5)
for i in range(0,M):
    if r[i]:
        u = np.random.multivariate_normal(mu0,Sigma,1)
        plt.plot(u[0][0],u[0][1],'.r',alpha = 0.4)
    else:
        u = np.random.multivariate_normal(mu1,Sigma,1)
        plt.plot(u[0][0],u[0][1],'+k',alpha = 0.6)
```

```
a = 2*inv(Sigma) @ (mu1-mu0);
b = ( mu0.reshape(1,2) @ inv(Sigma) @ mu0.reshape(2,1)
    - mu1.reshape(1,2) @ inv(Sigma) @mu1.reshape(2,1) )
xx = np.linspace(-4,6,100)
yy = (-(a[0]*xx +b)/a[1])[0]
plt.plot(xx,yy,'m')
plt.show()
```

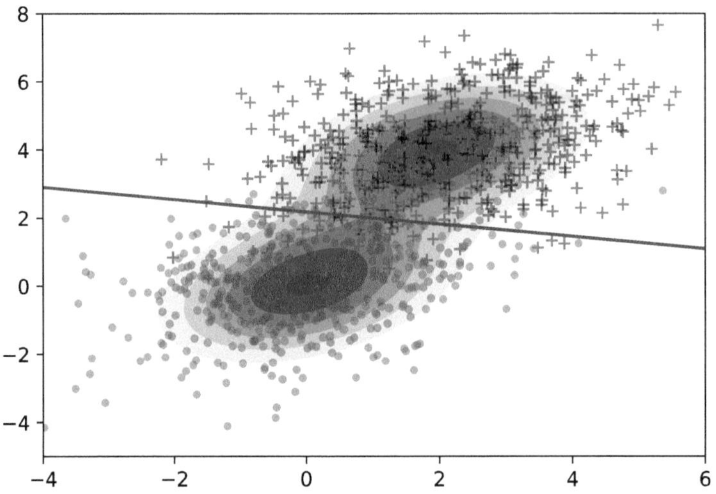

Figure 7.3: The linear discriminant boundary lies between the two modes of the mixture density and is linear.

■

To illustrate the difference between the linear and quadratic case, we specify different covariance matrices for the mixture components in the next example.

■ **Example 7.3 (Quadratic Discriminant Analysis)** As in Example 7.2 we consider a mixture of two Gaussians, but now with different covariance matrices. Figure 7.4 shows the quadratic decision boundary. The Python code follows.

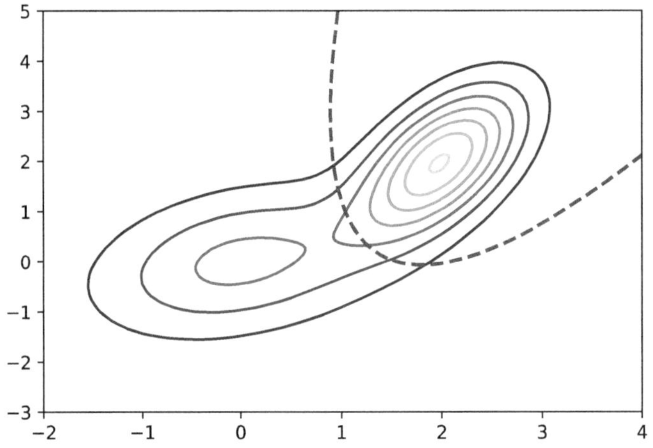

Figure 7.4: A quadratic decision boundary.

QDA.py

```
import numpy as np
import matplotlib.pyplot as plt
from scipy.stats import multivariate_normal

mu1 = np.array([0,0])
mu2 = np.array([2,2])
Sigma1 = np.array([[1,0.3],[0.3, 1]])
Sigma2 = np.array([[0.3,0.3],[0.3, 1]])
x, y = np.mgrid[-2:4:150j,-3:5:150j]
mvn1 = multivariate_normal( mu1, Sigma1 )
mvn2 = multivariate_normal( mu2, Sigma2 )

xy = np.hstack((x.reshape(-1,1),y.reshape(-1,1)))
z = ( 0.5*mvn1.pdf(xy).reshape(x.shape) +
      0.5*mvn2.pdf(xy).reshape(x.shape) )
plt.contour(x,y,z)

z1 = ( 0.5*mvn1.pdf(xy).reshape(x.shape) -
       0.5*mvn2.pdf(xy).reshape(x.shape))
plt.contour(x,y,z1, levels=[0],linestyles ='dashed',
            linewidths = 2, colors = 'm')
plt.show()
```

■

Of course, in practice the true parameter $\boldsymbol{\theta} = \{\alpha_j, \boldsymbol{\Sigma}_j, \boldsymbol{\mu}_j\}_{j=1}^c$ is not known and must be estimated from the training data — for example, by minimizing the *cross-entropy training loss* (4.4) with respect to $\boldsymbol{\theta}$: ☞ 123

$$\frac{1}{n}\sum_{i=1}^{n} \mathrm{Loss}(f(\boldsymbol{x}_i, y_i), g(\boldsymbol{x}_i, y_i \,|\, \boldsymbol{\theta})) = -\frac{1}{n}\sum_{i=1}^{n} \ln g(\boldsymbol{x}_i, y_i \,|\, \boldsymbol{\theta}),$$

where

$$\ln g(\boldsymbol{x}, y \,|\, \boldsymbol{\theta}) = \ln \alpha_y - \frac{1}{2}\ln|\boldsymbol{\Sigma}_y| - \frac{1}{2}(\boldsymbol{x} - \boldsymbol{\mu}_y)^\top \boldsymbol{\Sigma}_y^{-1}(\boldsymbol{x} - \boldsymbol{\mu}_y) - \frac{p}{2}\ln(2\pi).$$

The corresponding estimates of the model parameters (see Exercise 2) are:

$$\begin{aligned}
\widehat{\alpha}_y &= \frac{n_y}{n} \\
\widehat{\boldsymbol{\mu}}_y &= \frac{1}{n_y}\sum_{i:y_i=y} \boldsymbol{x}_i \\
\widehat{\boldsymbol{\Sigma}}_y &= \frac{1}{n_y}\sum_{i:y_i=y} (\boldsymbol{x}_i - \widehat{\boldsymbol{\mu}}_y)(\boldsymbol{x}_i - \widehat{\boldsymbol{\mu}}_y)^\top
\end{aligned} \tag{7.14}$$

for $y = 0, \ldots, c-1$, where $n_y := \sum_{i=1}^{n} \mathbb{1}\{y_i = y\}$. For the case where $\boldsymbol{\Sigma}_y = \boldsymbol{\Sigma}$ for all y, we have $\widehat{\boldsymbol{\Sigma}} = \sum_y \widehat{\alpha}_y \widehat{\boldsymbol{\Sigma}}_y$.

When $c > 2$ classes are involved, the classification procedure carries through in exactly the same way, leading to quadratic and linear discriminant functions (7.12) and (7.13) for each class. The space $\mathbb{R}^p$ now is partitioned into c regions, determined by the linear or quadratic boundaries determined by each pair of Gaussians.

For the linear discriminant case (that is, when $\boldsymbol{\Sigma}_y = \boldsymbol{\Sigma}$ for all y), it is convenient to first "whiten" or *sphere the data* as follows. Let $\mathbf{B}$ be an invertible matrix such that $\boldsymbol{\Sigma} = \mathbf{B}\mathbf{B}^\top$, obtained, for example, via the Cholesky method. We linearly transform each data point $\boldsymbol{x}$ to $\boldsymbol{x}' := \mathbf{B}^{-1}\boldsymbol{x}$ and each mean $\boldsymbol{\mu}_y$ to $\boldsymbol{\mu}'_y := \mathbf{B}^{-1}\boldsymbol{\mu}_y$, $y = 0, \dots, c-1$. Let the random vector $\boldsymbol{X}$ be distributed according to the mixture pdf

SPHERE THE DATA

☞ 375

$$g_{\boldsymbol{X}}(\boldsymbol{x} \mid \boldsymbol{\theta}) := \sum_y \alpha_y \frac{1}{\sqrt{(2\pi)^p\, |\boldsymbol{\Sigma}_y|}} \,\mathrm{e}^{-\frac{1}{2}(\boldsymbol{x}-\boldsymbol{\mu}_y)^\top \boldsymbol{\Sigma}_y^{-1}(\boldsymbol{x}-\boldsymbol{\mu}_y)}.$$

Then, by the transformation Theorem C.4, the vector $\boldsymbol{X}' = \mathbf{B}^{-1}\boldsymbol{X}$ has density

☞ 435

$$g_{\boldsymbol{X}'}(\boldsymbol{x}' \mid \boldsymbol{\theta}) = \frac{g_{\boldsymbol{X}}(\boldsymbol{x} \mid \boldsymbol{\theta})}{|\mathbf{B}^{-1}|} = \sum_{y=0}^{c-1} \frac{\alpha_y}{\sqrt{(2\pi)^p}} \,\mathrm{e}^{-\frac{1}{2}(\boldsymbol{x}-\boldsymbol{\mu}_y)^\top (\mathbf{B}\mathbf{B}^\top)^{-1}(\boldsymbol{x}-\boldsymbol{\mu}_y)}$$

$$= \sum_{y=0}^{c-1} \frac{\alpha_y}{\sqrt{(2\pi)^p}} \,\mathrm{e}^{-\frac{1}{2}(\boldsymbol{x}'-\boldsymbol{\mu}'_y)^\top (\boldsymbol{x}'-\boldsymbol{\mu}'_y)} = \sum_{y=0}^{c-1} \frac{\alpha_y}{\sqrt{(2\pi)^p}} \,\mathrm{e}^{-\frac{1}{2}\|\boldsymbol{x}'-\boldsymbol{\mu}'_y\|^2}.$$

This is the pdf of a mixture of standard p-dimensional normal distributions. The name "sphering" derives from the fact that the contours of each mixture component are perfect spheres. Classification of the transformed data is now particularly easy: classify $\boldsymbol{x}$ as $\widehat{y} := \mathrm{argmin}_y\{\, \|\boldsymbol{x}' - \boldsymbol{\mu}'_y\|^2 - 2\ln \alpha_y\}$. Note that this rule only depends on the prior probabilities and the distance from $\boldsymbol{x}'$ to the transformed means $\{\boldsymbol{\mu}'_y\}$. This procedure can lead to a significant dimensionality reduction of the data. Namely, the data can be projected onto the space spanned by the differences between the mean vectors $\{\boldsymbol{\mu}'_y\}$. When there are c classes, this is a $(c-1)$-dimensional space, as opposed to the p-dimensional space of the original data. We explain the precise ideas via an example.

■ **Example 7.4 (Classification after Data Reduction)** Consider an equal mixture of three 3-dimensional Gaussian distributions with identical covariance matrices. After sphering the data, the covariance matrices are all equal to the identity matrix. Suppose the mean vectors of the sphered data are $\boldsymbol{\mu}_1 = [2, 1, -3]^\top$, $\boldsymbol{\mu}_2 = [1, -4, 0]^\top$, and $\boldsymbol{\mu}_3 = [2, 4, 6]^\top$. The left panel of Figure 7.5 shows the 3-dimensional (sphered) data from each of the three classes.

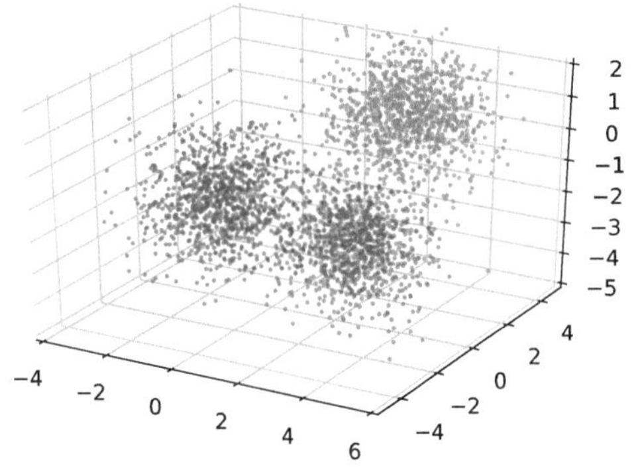

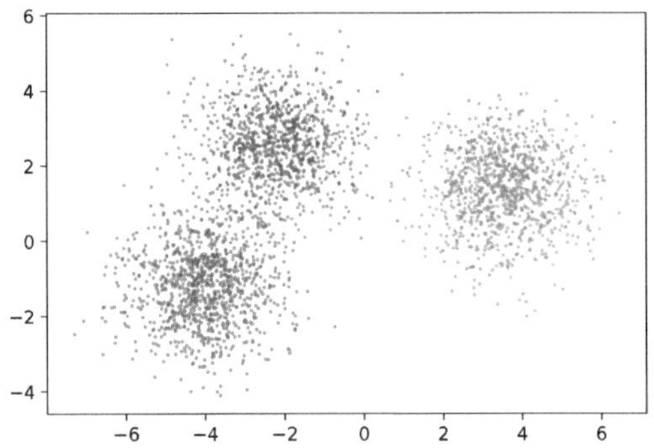

Figure 7.5: Left: original data. Right: projected data.

The data are stored in three 1000×3 matrices $\mathbf{X}_1$, $\mathbf{X}_2$, and $\mathbf{X}_3$. Here is how the data was generated and plotted.

datared.py

```
import numpy as np
from numpy.random import randn
import matplotlib.pyplot as plt
from mpl_toolkits.mplot3d import Axes3D

n=1000
mu1 = np.array([2,1,-3])
mu2 = np.array([1,-4,0])
mu3 = np.array([2,4,0])
X1 = randn(n,3) + mu1
X2 = randn(n,3) + mu2
X3 = randn(n,3) + mu3
fig = plt.figure()
ax = fig.gca(projection='3d',)
ax.plot(X1[:,0],X1[:,1],X1[:,2],'r.',alpha=0.5,markersize=2)
ax.plot(X2[:,0],X2[:,1],X2[:,2],'b.',alpha=0.5,markersize=2)
ax.plot(X3[:,0],X3[:,1],X3[:,2],'g.',alpha=0.5,markersize=2)
ax.set_xlim3d(-4,6)
ax.set_ylim3d(-5,5)
ax.set_zlim3d(-5,2)
plt.show()
```

Since we have equal mixtures, we classify each data point $\boldsymbol{x}$ according to the closest distance to $\boldsymbol{\mu}_1$, $\boldsymbol{\mu}_2$, or $\boldsymbol{\mu}_3$. We can achieve a reduction in the dimensionality of the data by *projecting* the data onto the two-dimensional affine space spanned by the $\{\boldsymbol{\mu}_i\}$; that is, all vectors are of the form

$$\boldsymbol{\mu}_1 + \beta_1(\boldsymbol{\mu}_2 - \boldsymbol{\mu}_1) + \beta_2(\boldsymbol{\mu}_3 - \boldsymbol{\mu}_1), \quad \beta_1, \beta_2 \in \mathbb{R}.$$

In fact, one may just as well project the data onto the subspace spanned by the vectors $\boldsymbol{\mu}_{21} = \boldsymbol{\mu}_2 - \boldsymbol{\mu}_1$ and $\boldsymbol{\mu}_{31} = \boldsymbol{\mu}_3 - \boldsymbol{\mu}_1$. Let $\mathbf{W} = [\boldsymbol{\mu}_{21}, \boldsymbol{\mu}_{31}]$ be the 3×2 matrix whose columns are $\boldsymbol{\mu}_{21}$ and $\boldsymbol{\mu}_{31}$. The orthogonal projection matrix onto the subspace $\mathcal{W}$ spanned by the columns of $\mathbf{W}$ is (see Theorem A.4): ☞ 364

$$\mathbf{P} = \mathbf{W}\mathbf{W}^+ = \mathbf{W}(\mathbf{W}^\top\mathbf{W})^{-1}\mathbf{W}^\top.$$

Let $\mathbf{U}\mathbf{D}\mathbf{V}^\top$ be the singular value decomposition of $\mathbf{W}$. Then $\mathbf{P}$ can also be written as

$$\mathbf{P} = \mathbf{U}\mathbf{D}(\mathbf{D}^\top\mathbf{D})^{-1}\mathbf{D}^\top\mathbf{U}^\top.$$

Note that $\mathbf{D}$ has dimension 3×2, so is not square. The first two columns of $\mathbf{U}$, say $\boldsymbol{u}_1$ and $\boldsymbol{u}_2$, form an orthonormal basis of the subspace $\mathcal{W}$. What we want to do is rotate this subspace to the $x-y$ plane, mapping $\boldsymbol{u}_1$ and $\boldsymbol{u}_2$ to $[1,0,0]^\top$ and $[0,1,0]^\top$, respectively. This is achieved via the rotation matrix $\mathbf{U}^{-1} = \mathbf{U}^\top$, giving the skewed projection matrix

$$\mathbf{R} = \mathbf{U}^\top\mathbf{P} = \mathbf{D}(\mathbf{D}^\top\mathbf{D})^{-1}\mathbf{D}^\top\mathbf{U}^\top,$$

whose 3rd row only contains zeros. Applying $\mathbf{R}$ to all the data points, and ignoring the 3rd component of the projected points (which is 0), gives the right panel of Figure 7.5. We see that the projected points are much better separated than the original ones. We have achieved dimensionality reduction of the data while retaining all the necessary information required for classification. Here is the rest of the Python code.

dataproj.py
```
from datared import *
from numpy.linalg import svd, pinv
mu21 = (mu2 - mu1).reshape(3,1)
mu31 = (mu3 - mu1).reshape(3,1)
W = np.hstack((mu21, mu31))
U,_,_ = svd(W)  # we only need U
P = W @ pinv(W)
R = U.T @ P

RX1 = (R @ X1.T).T
RX2 = (R @ X2.T).T
RX3 = (R @ X3.T).T
plt.plot(RX1[:,0],RX1[:,1],'b.',alpha=0.5,markersize=2)
plt.plot(RX2[:,0],RX2[:,1],'g.',alpha=0.5,markersize=2)
plt.plot(RX3[:,0],RX3[:,1],'r.',alpha=0.5,markersize=2)
plt.show()
```

■

7.5 Logistic Regression and Softmax Classification

☞ 204

In Example 5.10 we introduced the logistic (logit) regression model as a generalized linear model where, conditional on a p-dimensonal feature vector $\boldsymbol{x}$, the random response Y has a $\mathsf{Ber}(h(\boldsymbol{x}^\top\boldsymbol{\beta}))$ distribution with $h(u) = 1/(1+\mathrm{e}^{-u})$. The parameter $\boldsymbol{\beta}$ was then learned from the training data by maximizing the likelihood of the training responses or, equivalently, by minimizing the supervised version of the *cross-entropy training loss* (4.4):

☞ 123

$$-\frac{1}{n}\sum_{i=1}^{n} \ln g(y_i \,|\, \boldsymbol{\beta}, \boldsymbol{x}_i),$$

where $g(y = 1 \,|\, \boldsymbol{\beta}, \boldsymbol{x}) = 1/(1 + \mathrm{e}^{-\boldsymbol{x}^\top\boldsymbol{\beta}})$ and $g(y = 0 \,|\, \boldsymbol{\beta}, \boldsymbol{x}) = \mathrm{e}^{-\boldsymbol{x}^\top\boldsymbol{\beta}}/(1 + \mathrm{e}^{-\boldsymbol{x}^\top\boldsymbol{\beta}})$. In particular, we have

$$\ln \frac{g(y = 1 \,|\, \boldsymbol{\beta}, \boldsymbol{x})}{g(y = 0 \,|\, \boldsymbol{\beta}, \boldsymbol{x})} = \boldsymbol{x}^\top\boldsymbol{\beta}. \tag{7.15}$$

LOG-ODDS RATIO

In other words, the *log-odds ratio* is a linear function of the feature vector. As a consequence, the decision boundary $\{\boldsymbol{x} : g(y = 0 \,|\, \boldsymbol{\beta}, \boldsymbol{x}) = g(y = 1 \,|\, \boldsymbol{\beta}, \boldsymbol{x})\}$ is the hyperplane $\boldsymbol{x}^\top\boldsymbol{\beta} = 0$. Note that $\boldsymbol{x}$ typically includes the constant feature. If the constant feature is considered separately, that is $\boldsymbol{x} = [1, \widetilde{\boldsymbol{x}}^\top]^\top$, then the boundary is an affine hyperplane in $\widetilde{\boldsymbol{x}}$.

Suppose that training on $\tau = \{(\boldsymbol{x}_i, y_i)\}$ yields the estimate $\widehat{\boldsymbol{\beta}}$ with the corresponding learner $g_\tau(y = 1 \,|\, \boldsymbol{x}) = 1/(1 + \mathrm{e}^{-\boldsymbol{x}^\top\widehat{\boldsymbol{\beta}}})$. The learner can be used as a pre-classifier from which we obtain the classifier $\mathbb{1}\{g_\tau(y = 1 \,|\, \boldsymbol{x}) > 1/2\}$ or, equivalently,

$$\widehat{y} := \underset{j\in\{0,1\}}{\operatorname{argmax}}\, g_\tau(y = j \,|\, \boldsymbol{x}),$$

in accordance with the fundamental classification rule (7.2).

MULTI-LOGIT

The above classification methodology for the logit model can be generalized to the *multi-logit* model where the response takes values in the set $\{0, \ldots, c-1\}$. The key idea is to replace (7.15) with

$$\ln \frac{g(y = j \mid \mathbf{W}, \boldsymbol{b}, \boldsymbol{x})}{g(y = 0 \mid \mathbf{W}, \boldsymbol{b}, \boldsymbol{x})} = \boldsymbol{x}^\top \boldsymbol{\beta}_j, \quad j = 1, \ldots, c-1, \tag{7.16}$$

where the matrix $\mathbf{W} \in \mathbb{R}^{(c-1)\times(p-1)}$ and vector $\boldsymbol{b} \in \mathbb{R}^{c-1}$ reparameterize all $\boldsymbol{\beta}_j \in \mathbb{R}^p$ such that (recall $\boldsymbol{x} = [1, \widetilde{\boldsymbol{x}}^\top]^\top$):

$$\mathbf{W}\widetilde{\boldsymbol{x}} + \boldsymbol{b} = [\boldsymbol{\beta}_1, \ldots, \boldsymbol{\beta}_{c-1}]^\top \boldsymbol{x}.$$

Observe that the random response Y is assumed to have a conditional probability distribution for which the log-odds ratio with respect to class j and a "reference" class (in this case 0) is *linear*. The separating boundaries between two pairs of classes are again affine hyperplanes.

The model (7.16) completely specifies the distribution of Y, namely:

$$g(y \mid \mathbf{W}, \boldsymbol{b}, \boldsymbol{x}) = \frac{\exp(z_{y+1})}{\sum_{k=1}^{c} \exp(z_k)}, \quad y = 0, \ldots, c-1,$$

where z_1 is an arbitrary constant, say 0, corresponding to the "reference" class $y = 0$, and

$$[z_2, \ldots, z_c]^\top := \mathbf{W}\widetilde{\boldsymbol{x}} + \boldsymbol{b}.$$

Note that $g(y \mid \mathbf{W}, \boldsymbol{b}, \boldsymbol{x})$ is the $(y+1)$-st component of $\boldsymbol{a} = \text{softmax}(\boldsymbol{z})$, where

$$\text{softmax} : \boldsymbol{z} \mapsto \frac{\exp(\boldsymbol{z})}{\sum_k \exp(z_k)}$$

is the *softmax* function and $\boldsymbol{z} = [z_1, \ldots, z_c]^\top$. Finally, we can write the classifier as SOFTMAX

$$\widehat{y} = \underset{j \in \{0,\ldots,c-1\}}{\text{argmax}}\ a_{j+1}.$$

In summary, we have the sequence of mappings transforming the input $\boldsymbol{x}$ into the output $\widehat{y}$:

$$\boldsymbol{x} \to \mathbf{W}\widetilde{\boldsymbol{x}} + \boldsymbol{b} \to \text{softmax}(\boldsymbol{z}) \to \underset{j \in \{0,\ldots,c-1\}}{\text{argmax}}\ a_{j+1} \to \widehat{y}.$$

In Example 9.4 we will revisit the multi-logit model and reinterpret this sequence of mappings as a *neural network*. In the context of neural networks, $\mathbf{W}$ is called a *weight* matrix and $\boldsymbol{b}$ is called a *bias* vector. ☞ 335

The parameters $\mathbf{W}$ and $\boldsymbol{b}$ have to be learned from the training data, which involves minimization of the supervised version of the *cross-entropy training loss* (4.4): ☞ 123

$$\frac{1}{n}\sum_{i=1}^{n} \text{Loss}(f(y_i \mid \boldsymbol{x}_i), g(y_i \mid \mathbf{W}, \boldsymbol{b}, \boldsymbol{x}_i)) = -\frac{1}{n}\sum_{i=1}^{n} \ln g(y_i \mid \mathbf{W}, \boldsymbol{b}, \boldsymbol{x}_i).$$

Using the softmax function, the *cross-entropy* loss can be simplified to:

$$\text{Loss}(f(y \mid \boldsymbol{x}), g(y \mid \mathbf{W}, \boldsymbol{b}, \boldsymbol{x})) = -z_{y+1} + \ln \sum_{k=1}^{c} \exp(z_k). \tag{7.17}$$

The discussion on training is postponed until Chapter 9, where we reinterpret the multi-logit model as a neural net, which can be trained using the *limited-memory BFGS* method (Exercise 11). Note that in the binary case ($c = 2$), where there is only one vector $\boldsymbol{\beta}$ to be estimated, Example 5.10 already established that minimization of the cross-entropy training loss is equivalent to likelihood maximization. ☞ 354

7.6 K-Nearest Neighbors Classification

Let $\tau = \{(\boldsymbol{x}_i, y_i)\}_{i=1}^n$ be the training set, with $y_i \in \{0, \ldots, c-1\}$, and let $\boldsymbol{x}$ be a new feature vector. Define $\boldsymbol{x}_{(1)}, \boldsymbol{x}_{(2)}, \ldots, \boldsymbol{x}_{(n)}$ as the feature vectors ordered by closeness to $\boldsymbol{x}$ in some distance $\text{dist}(\boldsymbol{x}, \boldsymbol{x}_i)$, e.g., the Euclidean distance $\|\boldsymbol{x}-\boldsymbol{x}'\|$. Let $\tau(\boldsymbol{x}) := \{(\boldsymbol{x}_{(1)}, y_{(1)}) \ldots, (\boldsymbol{x}_{(K)}, y_{(K)})\}$ be the subset of τ that contains K feature vectors $\boldsymbol{x}_i$ that are closest to $\boldsymbol{x}$. Then the *K-nearest neighbors* classification rule classifies $\boldsymbol{x}$ according to the most frequently occurring class labels in $\tau(\boldsymbol{x})$. If two or more labels receive the same number of votes, the feature vector is classified by selecting one of these labels randomly with equal probability. For the case $K = 1$ the set $\tau(\boldsymbol{x})$ contains only one element, say $(\boldsymbol{x}', y')$, and $\boldsymbol{x}$ is classified as y'. This divides the space into n regions

K-NEAREST NEIGHBORS

$$\mathcal{R}_i = \{\boldsymbol{x} : \text{dist}(\boldsymbol{x}, \boldsymbol{x}_i) \leqslant \text{dist}(\boldsymbol{x}, \boldsymbol{x}_j), j \neq i\}, \quad i = 1, \ldots, n.$$

For a feature space $\mathbb{R}^p$ with the Euclidean distance, this gives a Voronoi tessellation of the feature space, similar to what was done for vector quantization in Section 4.6.

☞ 142

■ **Example 7.5 (Nearest Neighbor Classification)** The Python program below simulates 80 random points above and below the line $x_2 = x_1$. Points above the line $x_2 = x_1$ have label 0 and points below this line have label 1. Figure 7.6 shows the Voronoi tessellation obtained from the 1-nearest neighbor classification.

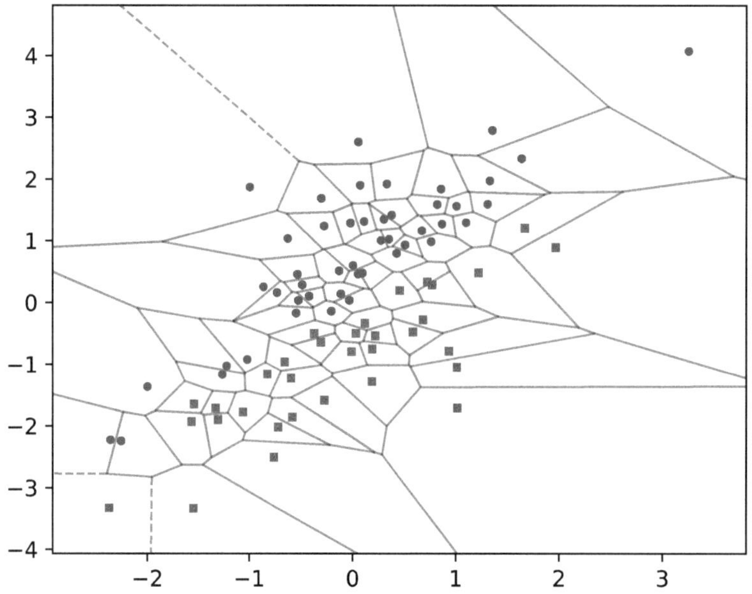

Figure 7.6: The 1-nearest neighbor algorithm divides up the space into Voronoi cells.

nearestnb.py

```
import numpy as np
from numpy.random import rand,randn
import matplotlib.pyplot as plt
from scipy.spatial import Voronoi, voronoi_plot_2d
```

```
np.random.seed(12345)
M = 80
x = randn(M,2)
y = np.zeros(M) # pre-allocate list

for i in range(M):
    if rand()<0.5:
        x[i,1], y[i] = x[i,0] + np.abs(randn()), 0
    else:
        x[i,1], y[i] = x[i,0] - np.abs(randn()), 1

vor = Voronoi(x)
plt_options = {'show_vertices':False, 'show_points':False,
               'line_alpha':0.5}
fig = voronoi_plot_2d(vor, **plt_options)
plt.plot(x[y==0,0], x[y==0,1],'bo',
         x[y==1,0], x[y==1,1],'rs', markersize=3)
```

■

7.7 Support Vector Machine

Suppose we are given the training set $\tau = \{(\boldsymbol{x}_i, y_i)\}_{i=1}^n$, where each response[2] y_i takes either the value -1 or 1, and we wish to construct a classifier taking values in $\{-1, 1\}$. As this merely involves a relabeling of the 0–1 classification problem in Section 7.1, the optimal classification function for the indicator loss, $\mathbb{1}\{y \neq \widehat{y}\}$, is, by Theorem 7.1, equal to

$$g^*(\boldsymbol{x}) = \begin{cases} 1 & \text{if} \quad \mathbb{P}[Y = 1 \mid \boldsymbol{X} = \boldsymbol{x}] \geqslant 1/2, \\ -1 & \text{if} \quad \mathbb{P}[Y = 1 \mid \boldsymbol{X} = \boldsymbol{x}] < 1/2. \end{cases}$$

It is not difficult to show, see Exercise 5, that the function g^* can be viewed as the minimizer of the risk for the *hinge loss* function, $\text{Loss}(y, \widehat{y}) = (1 - y\widehat{y})_+ := \max\{0,\ 1 - y\widehat{y}\}$, over all prediction functions g (not necessarily taking values only in the set $\{-1, 1\}$). That is, HINGE LOSS

$$g^* = \underset{g}{\operatorname{argmin}}\, \mathbb{E}\,(1 - Y\, g(\boldsymbol{X}))_+. \tag{7.18}$$

Given the training set τ, we can approximate the risk $\ell(g) = \mathbb{E}\,(1 - Y\, g(\boldsymbol{X}))_+$ with the training loss

$$\ell_\tau(g) = \frac{1}{n} \sum_{i=1}^{n} (1 - y_i\, g(\boldsymbol{x}_i))_+,$$

and minimize this over a (smaller) class of functions to obtain the optimal prediction function g_τ. Finally, as the prediction function g_τ generally is not a classifier by itself (it usually does not only take values -1 or 1), we take the classifier

$$\operatorname{sign} g_\tau(\boldsymbol{x}).$$

[2]The reason why we use responses -1 and 1 here, instead of 0 and 1, is that the notation becomes easier.

OPTIMAL DECISION BOUNDARY

Therefore, a feature vector $\boldsymbol{x}$ is classified according to 1 or -1 depending on whether $g_\tau(\boldsymbol{x}) \geqslant 0$ or < 0, respectively. The *optimal decision boundary* is given by the set of $\boldsymbol{x}$ for which $g_\tau(\boldsymbol{x}) = 0$.

Similar to the cubic smoothing spline or RKHS setting in (6.19), we can consider finding the best classifier, given the training data, via the penalized goodness-of-fit optimization:

$$\min_{g \in \mathcal{H} \oplus \mathcal{H}_0} \frac{1}{n} \sum_{i=1}^{n} [1 - y_i\, g(\boldsymbol{x}_i)]_+ + \widetilde{\gamma}\, \|g\|_{\mathcal{H}}^2,$$

for some regularization parameter $\widetilde{\gamma}$. It will be convenient to define $\gamma := 2n\widetilde{\gamma}$ and to solve the equivalent problem

$$\min_{g \in \mathcal{H} \oplus \mathcal{H}_0} \sum_{i=1}^{n} [1 - y_i\, g(\boldsymbol{x}_i)]_+ + \frac{\gamma}{2}\, \|g\|_{\mathcal{H}}^2.$$

☞ 232

We know from the Representer Theorem 6.6 that if κ is the reproducing kernel corresponding to $\mathcal{H}$, then the solution is of the form (assuming that the null space $\mathcal{H}_0$ has a constant term only):

$$g(\boldsymbol{x}) = \alpha_0 + \sum_{i=1}^{n} \alpha_i\, \kappa(\boldsymbol{x}_i, \boldsymbol{x}). \tag{7.19}$$

☞ 232

Substituting into the minimization expression yields the analogue of (6.21):

$$\min_{\boldsymbol{\alpha}, \alpha_0} \sum_{i=1}^{n} [1 - y_i(\alpha_0 + \{\mathbf{K}\boldsymbol{\alpha}\}_i)]_+ + \frac{\gamma}{2}\, \boldsymbol{\alpha}^\top \mathbf{K} \boldsymbol{\alpha}, \tag{7.20}$$

where $\mathbf{K}$ is the Gram matrix. This is a *convex* optimization problem, as it is the sum of a convex quadratic and piecewise linear term in $\boldsymbol{\alpha}$. Defining $\lambda_i := \gamma\alpha_i / y_i$, $i = 1, \ldots, n$ and $\boldsymbol{\lambda} := [\lambda_1, \ldots, \lambda_n]^\top$, we show in Exercise 10 that the optimal $\boldsymbol{\alpha}$ and α_0 in (7.20) can be obtained by solving the "dual" convex optimization problem

$$\begin{aligned} \max_{\boldsymbol{\lambda}} \quad & \sum_{i=1}^{n} \lambda_i - \frac{1}{2\gamma} \sum_{i=1}^{n} \sum_{j=1}^{n} \lambda_i \lambda_j y_i y_j\, \kappa(\boldsymbol{x}_i, \boldsymbol{x}_j) \\ \text{subject to:} \quad & \boldsymbol{\lambda}^\top \boldsymbol{y} = 0, \ \mathbf{0} \leqslant \boldsymbol{\lambda} \leqslant \mathbf{1}, \end{aligned} \tag{7.21}$$

and $\alpha_0 = y_j - \sum_{i=1} \alpha_i\, \kappa(\boldsymbol{x}_i, \boldsymbol{x}_j)$ for any j for which $\lambda_j \in (0, 1)$. In view of (7.19), the optimal prediction function (pre-classifier) g_τ is then given by

$$g_\tau(\boldsymbol{x}) = \alpha_0 + \sum_{i=1}^{n} \alpha_i\, \kappa(\boldsymbol{x}_i, \boldsymbol{x}) = \alpha_0 + \frac{1}{\gamma} \sum_{i=1}^{n} y_i \lambda_i\, \kappa(\boldsymbol{x}_i, \boldsymbol{x}). \tag{7.22}$$

To mitigate possible numerical problems in the calculation of α_0 it is customary to take an overall average:

$$\alpha_0 = \frac{1}{|\mathcal{J}|} \sum_{j \in \mathcal{J}} \left\{ y_j - \sum_{i=1}^{n} \alpha_i\, \kappa(\boldsymbol{x}_i, \boldsymbol{x}_j) \right\},$$

where $\mathcal{J} := \{j : \lambda_j \in (0, 1)\}$.

Note that, from (7.22), the optimal pre-classifier $g(\boldsymbol{x})$ and the classifier sign $g(\boldsymbol{x})$ only depend on vectors $\boldsymbol{x}_i$ for which $\lambda_i \neq 0$. These vectors are called the *support vectors* of the support vector machine. It is also important to note that the quadratic function in (7.21) depends on the regularization parameter γ. By defining $\nu_i := \lambda_i/\gamma$, $i = 1, \ldots, n$, we can rewrite (7.21) as

SUPPORT VECTORS

$$\begin{aligned} \min_{\nu} \quad & \frac{1}{2} \sum_{i,j} \nu_i \nu_j y_i y_j \kappa(\boldsymbol{x}_i, \boldsymbol{x}_j) - \sum_{i=1}^{n} \nu_i \\ \text{subject to:} \quad & \sum_{i=1}^{n} \nu_i y_i = 0, \quad 0 \leqslant \nu_i \leqslant 1/\gamma =: C, \quad i = 1, \ldots, n. \end{aligned} \tag{7.23}$$

For perfectly separable data, that is, data for which an affine plane can be drawn to perfectly separate the two classes, we may take $C = \infty$, as explained below. Otherwise, C needs to be chosen via cross-validation or a test data set, for example.

Geometric interpretation

For the linear kernel function $\kappa(\boldsymbol{x}, \boldsymbol{x}') = \boldsymbol{x}^\top \boldsymbol{x}'$, we have

$$g_\tau(\boldsymbol{x}) = \beta_0 + \boldsymbol{\beta}^\top \boldsymbol{x},$$

with $\beta_0 = \alpha_0$ and $\boldsymbol{\beta} = \gamma^{-1} \sum_{i=1}^n \lambda_i y_i \boldsymbol{x}_i = \sum_{i=1}^n \alpha_i \boldsymbol{x}_i$, and so the decision boundary is an affine plane. The situation is illustrated in Figure 7.7. The decision boundary is formed by the points $\boldsymbol{x}$ such that $g_\tau(\boldsymbol{x}) = 0$. The two sets $\{\boldsymbol{x} : g_\tau(\boldsymbol{x}) = -1\}$ and $\{\boldsymbol{x} : g_\tau(\boldsymbol{x}) = 1\}$ are called the *margins*. The distance from the points on a margin to the decision boundary is $1/\|\boldsymbol{\beta}\|$.

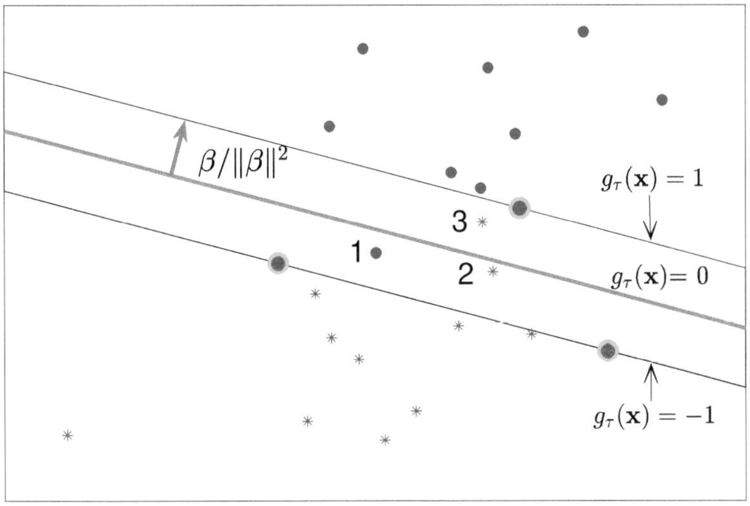

Figure 7.7: Classifying two classes (red and blue) using SVM.

Based on the "multipliers" $\{\lambda_i\}$, we can divide the training samples $\{(\boldsymbol{x}_i, y_i)\}$ into three categories (see Exercise 11):

- Points for which $\lambda_i \in (0, 1)$. These are the support vectors on the margins (green encircled in the figure) and are correctly classified.

- Points for which $\lambda_i = 1$. These points, which are also support vectors, lie strictly inside the margins (points 1, 2, and 3 in the figure). Such points may or may not be correctly classified.

- Points for which $\lambda_i = 0$. These are the non-support vectors, which all lie outside the margins. Every such point is correctly classified.

If the classes of points $\{\boldsymbol{x}_i : y_i = 1\}$ and $\{\boldsymbol{x}_i : y_i = -1\}$ are perfectly separable by some affine plane, then there will be no points strictly inside the margins, so all support vectors will lie exactly on the margins. In this case (7.20) reduces to

$$\begin{aligned} \min_{\boldsymbol{\beta},\beta_0} \; & \|\boldsymbol{\beta}\|^2 \\ \text{subject to:} \quad & y_i(\beta_0 + \boldsymbol{x}_i^\top \boldsymbol{\beta}) \geqslant 1, \; i = 1, \ldots, n, \end{aligned} \tag{7.24}$$

using the fact that $\alpha_0 = \beta_0$ and $\mathbf{K}\boldsymbol{\alpha} = \mathbf{X}\mathbf{X}^\top\boldsymbol{\alpha} = \mathbf{X}\boldsymbol{\beta}$. We may replace $\min \|\boldsymbol{\beta}\|^2$ in (7.24) with $\max 1/\|\boldsymbol{\beta}\|$, as this gives the same optimal solution. As $1/\|\boldsymbol{\beta}\|$ is equal to half the margin width, the latter optimization problem has a simple interpretation: separate the points via an affine hyperplane such that the margin width is maximized.

■ **Example 7.6 (Support Vector Machine)** The data in Figure 7.8 was uniformly generated on the unit disc. Class-1 points (blue dots) have a radius less than 1/2 (y-values 1) and class-2 points (red crosses) have a radius greater than 1/2 (y-values -1).

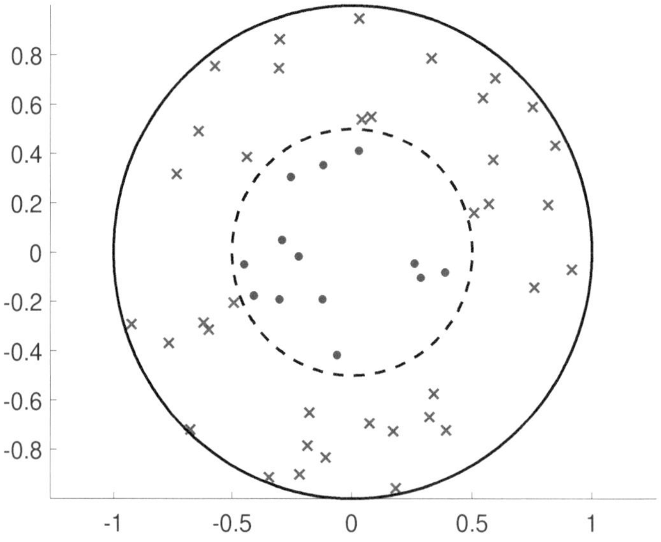

Figure 7.8: Separate the two classes.

Of course it is not possible to separate the two groups of points via a straight line in $\mathbb{R}^2$. However, it is possible to separate them in $\mathbb{R}^3$ by considering three-dimensional feature vectors $\boldsymbol{z} = [z_1, z_2, z_3]^\top = [x_1, x_2, x_1^2 + x_2^2]^\top$. For any $\boldsymbol{x} \in \mathbb{R}^2$, the corresponding feature vector $\boldsymbol{z}$ lies on a quadratic surface. In this space it is possible to separate the $\{\boldsymbol{z}_i\}$ points into two groups by means of a planar surface, as illustrated in Figure 7.9.

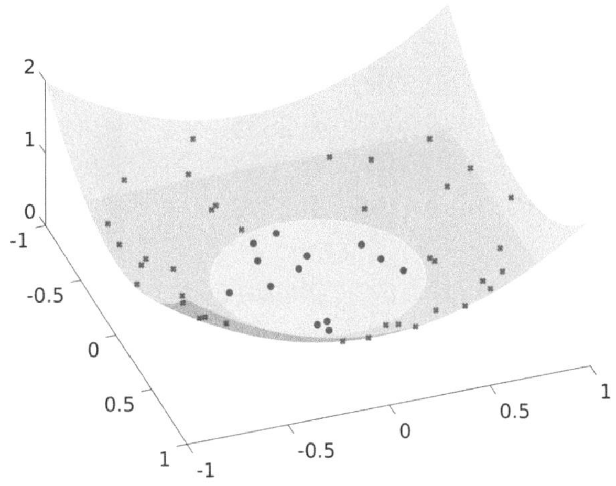

Figure 7.9: In feature space $\mathbb{R}^3$ the points can be separated by a plane.

We wish to find a separating plane in $\mathbb{R}^3$ using the transformed features. The following Python code uses the **SVC** function of the `sklearn` module to solve the quadratic optimization problem (7.23) (with $C = \infty$). The results are summarized in Table 7.6. The data is available from the book's GitHub site as `svmcirc.csv`.

svmquad.py

```
import numpy as np
from numpy import genfromtxt
from sklearn.svm import SVC

data = genfromtxt('svmcirc.csv', delimiter=',')
x = data[:,[0,1]] #vectors are rows
y = data[:,[2]].reshape(len(x),) #labels

tmp = np.sum(np.power(x,2),axis=1).reshape(len(x),1)
z = np.hstack((x,tmp))

clf = SVC(C = np.inf, kernel='linear')
clf.fit(z,y)

print("Support Vectors \n", clf.support_vectors_)
print("Support Vector Labels ",y[clf.support_])
print("Nu",clf.dual_coef_)
print("Bias",clf.intercept_)
```

```
  Support Vectors
 [[ 0.038758    0.53796     0.29090314]
 [-0.49116    -0.20563     0.28352184]
 [-0.45068    -0.04797     0.20541358]
 [-0.061107   -0.41651     0.17721465]]
Support Vector Labels  [-1. -1.  1.  1.]
Nu [[ -46.49249413 -249.01807328  265.31805855   30.19250886]]
Bias  [5.617891]
```

Table 7.6: Optimal support vector machine parameters for the $\mathbb{R}^3$ data.

$z^\top$			y	$\alpha = \nu y$
0.0388	0.5380	0.2909	−1	−46.4925
−0.4912	−0.2056	0.2835	−1	−249.0181
−0.4507	−0.0480	0.2054	1	265.3181
−0.0611	−0.4165	0.1772	1	30.1925

It follows that the normal vector of the plane is

$$\boldsymbol{\beta} = \sum_{i \in \mathcal{S}} \alpha_i \boldsymbol{z}_i = [-0.9128, 0.8917, -24.2764]^\top,$$

where $\mathcal{S}$ is the set of indices of the support vectors. We see that the plane is almost perpendicular to the z_1, z_2 plane. The bias term β_0 can also be found from the table above. In particular, for any $\boldsymbol{x}^\top$ and y in Table 7.6, we have $y - \boldsymbol{\beta}^\top \boldsymbol{z} = \beta_0 = 5.6179$.

To draw the separating boundary in $\mathbb{R}^2$ we need to project the intersection of the separating plane with the quadratic surface onto the z_1, z_2 plane. That is, we need to find all points (z_1, z_2) such that

$$5.6179 - 0.9128 z_1 + 0.8917 z_2 = 24.2764\,(z_1^2 + z_2^2). \tag{7.25}$$

This is the equation of a circle with (approximate) center $(0.019, -0.018)$ and radius 0.48, which is very close to the true circular boundary between the two groups, with center $(0, 0)$ and radius 0.5. This circle is drawn in Figure 7.10.

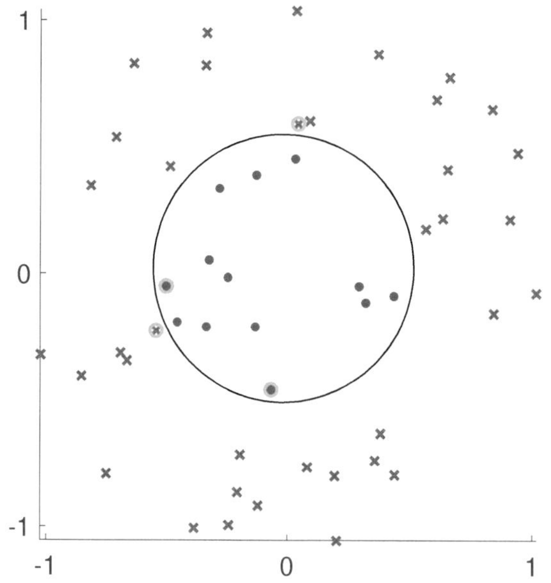

Figure 7.10: The circular decision boundary can be viewed equivalently as (a) the projection onto the x_1, x_2 plane of the intersection of the separating plane with the quadratic surface (both in $\mathbb{R}^3$), or (b) the set of points $\boldsymbol{x} = (x_1, x_2)$ for which $g_\tau(\boldsymbol{x}) = \beta_0 + \boldsymbol{\beta}^\top \boldsymbol{\phi}(\boldsymbol{x}) = 0$.

An equivalent way to derive this circular separating boundary is to consider the feature map $\boldsymbol{\phi}(\boldsymbol{x}) = [x_1, x_2, x_1^2 + x_2^2]^\top$ on $\mathbb{R}^2$, which defines a reproducing kernel

$$\kappa(\boldsymbol{x}, \boldsymbol{x}') = \boldsymbol{\phi}(\boldsymbol{x})^\top \boldsymbol{\phi}(\boldsymbol{x}'),$$

on $\mathbb{R}^2$, which in turn gives rise to a (unique) RKHS $\mathcal{H}$. The optimal prediction function (7.19) is now of the form

$$g_\tau(\boldsymbol{x}) = \alpha_0 + \frac{1}{\gamma}\sum_{i=1}^{n} y_i\,\lambda_i\,\boldsymbol{\phi}(\boldsymbol{x}_i)^\top\boldsymbol{\phi}(\boldsymbol{x}) = \beta_0 + \boldsymbol{\beta}^\top\boldsymbol{\phi}(\boldsymbol{x}), \tag{7.26}$$

where $\alpha_0 = \beta_0$ and

$$\boldsymbol{\beta} = \frac{1}{\gamma}\sum_{i=1}^{n} y_i\,\lambda_i\,\boldsymbol{\phi}(\boldsymbol{x}_i).$$

The decision boundary, $\{\boldsymbol{x} : g_\tau(\boldsymbol{x}) = 0\}$, is again a circle in $\mathbb{R}^2$. The following code determines the fitted model parameters and the decision boundary. Figure 7.10 shows the optimal decision boundary, which is identical to (7.25). The function **mykernel** specifies the custom kernel above.

svmkern.py

```
import numpy as np, matplotlib.pyplot as plt
from numpy import genfromtxt
from sklearn.svm import SVC

def mykernel(U,V):
    tmpU = np.sum(np.power(U,2),axis=1).reshape(len(U),1)
    U = np.hstack((U,tmpU))
    tmpV = np.sum(np.power(V,2),axis=1).reshape(len(V),1)
    V = np.hstack((V,tmpV))
    K = U @ V.T
    print(K.shape)
    return K

# read in the data
inp = genfromtxt('svmcirc.csv', delimiter=',')
data = inp[:,[0,1]] #vectors are rows
y = inp[:,[2]].reshape(len(data),) #labels

clf = SVC(C = np.inf, kernel=mykernel, gamma='auto') # custom kernel
# clf = SVC(C = np.inf, kernel="rbf", gamma='scale') # inbuilt

clf.fit(data,y)

print("Support Vectors \n", clf.support_vectors_)
print("Support Vector Labels ",y[clf.support_])
print("Nu ",clf.dual_coef_)
print("Bias ",clf.intercept_)

# plot
d = 0.001
x_min, x_max = -1,1
y_min, y_max = -1,1
xx, yy = np.meshgrid(np.arange(x_min, x_max, d), np.arange(y_min,
   y_max, d))
plt.plot(data[clf.support_,0],data[clf.support_,1],'go')
plt.plot(data[y==1,0],data[y==1,1],'b.')
plt.plot(data[y==-1,0],data[y==-1,1],'rx')
```

```
Z = clf.predict(np.c_[xx.ravel(), yy.ravel()])
Z = Z.reshape(xx.shape)
plt.contour(xx, yy, Z,colors ="k")
plt.show()
```

Finally, we illustrate the use of the Gaussian kernel

$$\kappa(\boldsymbol{x}, \boldsymbol{x}') = \mathrm{e}^{-c\,\|\boldsymbol{x}-\boldsymbol{x}'\|^2}, \tag{7.27}$$

where $c > 0$ is some tuning constant. This is an example of a *radial basis function kernel*, which are reproducing kernels of the form $\kappa(\boldsymbol{x}, \boldsymbol{x}') = f(\|\boldsymbol{x} - \boldsymbol{x}'\|)$, for some positive real-valued function f. Each feature vector $\boldsymbol{x}$ is now transformed to a *function* $\kappa_{\boldsymbol{x}} = \kappa(\boldsymbol{x}, \cdot)$. We can think of it as the (unnormalized) pdf of a Gaussian distribution centered around $\boldsymbol{x}$, and g_τ is a (signed) *mixture* of these pdfs, plus a constant; that is,

$$g_\tau(\boldsymbol{x}) = \alpha_0 + \sum_{i=1}^{n} \alpha_i\, \mathrm{e}^{-c\,\|\boldsymbol{x}_i-\boldsymbol{x}\|^2}.$$

Replacing in Line 2 of the previous code `mykernel` with `'rbf'` produces the SVM parameters given in Table 7.7. Figure 7.11 shows the decision boundary, which is not exactly circular, but is close to the true (circular) boundary $\{\boldsymbol{x} : \|\boldsymbol{x}\| = 1/2\}$. There are now seven support vectors, rather than the four in Figure 7.10.

Table 7.7: Optimal support vector machine parameters for the Gaussian kernel case.

$\boldsymbol{x}^\top$		y	α ($\times 10^9$)	$\boldsymbol{x}^\top$		y	α ($\times 10^9$)
0.0388	0.5380	−1	−0.0635	−0.4374	0.3854	−1	−1.4399
−0.4912	−0.2056	−1	−9.4793	0.3402	−0.5740	−1	−0.1000
0.5086	0.1576	−1	−0.5240	−0.4098	−0.1763	1	6.0662
−0.4507	−0.0480	1	5.5405				

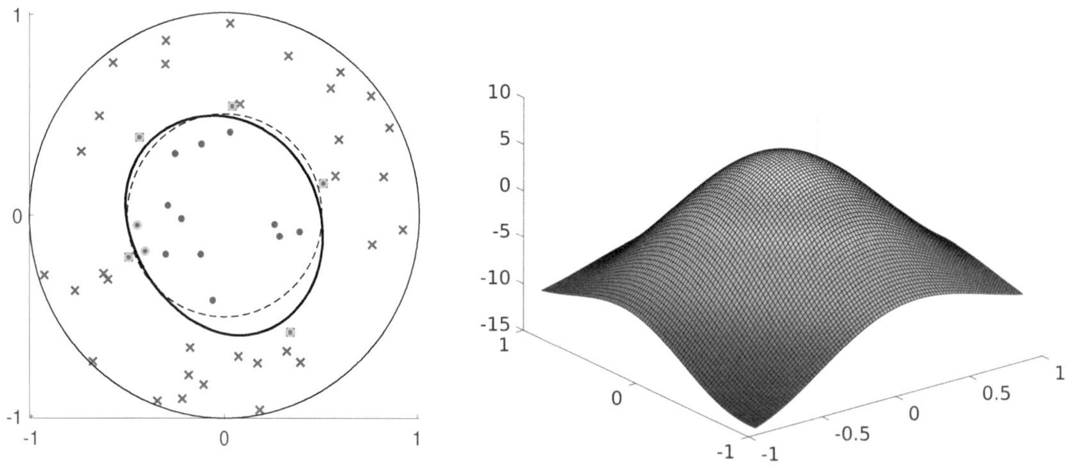

Figure 7.11: Left: The decision boundary $\{\boldsymbol{x} : g_\tau(\boldsymbol{x}) = 0\}$ is roughly circular, and separates the two classes well. There are seven support vectors, indicated by green circles. Right: The graph of g_τ is a scaled mixture of Gaussian pdfs plus a constant.

■ **Remark 7.2 (Scaling and Penalty Parameters)** When using a radial basis function in SVC in sklearn, the scaling c (7.27) can be set via the parameter gamma. Note that large values of gamma lead to highly peaked predicted functions, and small values lead to highly smoothed predicted functions. The parameter C in SVC refers $C = 1/\gamma$ in (7.23). ■

7.8 Classification with Scikit-Learn

In this section we apply several classification methods to a real-world data set, using the Python module sklearn (the package name is Scikit-Learn). Specifically, the data is obtained from UCI's Breast Cancer Wisconsin data set. This data set, first published and analyzed in [118], contains the measurements related to 569 images of 357 benign and 212 malignant breast masses. The goal is to classify a breast mass as benign or malignant based on 10 features: Radius, Texture, Perimeter, Area, Smoothness, Compactness, Concavity, Concave Points, Symmetry, and Fractal Dimension of each mass. The mean, standard error, and "worst" of these attributes were computed for each image, resulting in 30 features. For instance, feature 1 is Mean Radius, feature 11 is Radius SE, feature 21 is Worst Radius.

The following Python code reads the data, extracts the response vector and model (feature) matrix and divides the data into a training and test set.

skclass1.py

```
from numpy import genfromtxt
from sklearn.model_selection import train_test_split
url1 = "http://mlr.cs.umass.edu/ml/machine-learning-databases/"
url2 = "breast-cancer-wisconsin/"
name = "wdbc.data"
data = genfromtxt(url1 + url2 + name, delimiter=',', dtype=str)
y = data[:,1] #responses
X = data[:,2:].astype('float') #features as an ndarray matrix

X_train , X_test , y_train , y_test = train_test_split(
        X, y, test_size = 0.4, random_state = 1234)
```

To visualize the data we create a 3D scatterplot for the features mean *radius*, mean *texture*, and mean *concavity*, which correspond to the columns 0, 1, and 6 of the model matrix X. Figure 7.12 suggests that the malignant and benign breast masses could be well separated using these three features.

skclass2.py

```
from skclass1 import X, y
import matplotlib.pyplot as plt
from mpl_toolkits.mplot3d import Axes3D
import numpy as np

Bidx = np.where(y == 'B')
Midx= np.where(y == 'M')

# plot features Radius (column 0), Texture (1), Concavity (6)
```

```
fig = plt.figure()
ax = fig.gca(projection = '3d')
ax.scatter(X[Bidx,0], X[Bidx,1], X[Bidx,6],
           c='r', marker='^', label='Benign')
ax.scatter(X[Midx,0], X[Midx,1], X[Midx,6],
           c='b', marker='o', label='Malignant')
ax.legend()
ax.set_xlabel('Mean Radius')
ax.set_ylabel('Mean Texture')
ax.set_zlabel('Mean Concavity')
plt.show()
```

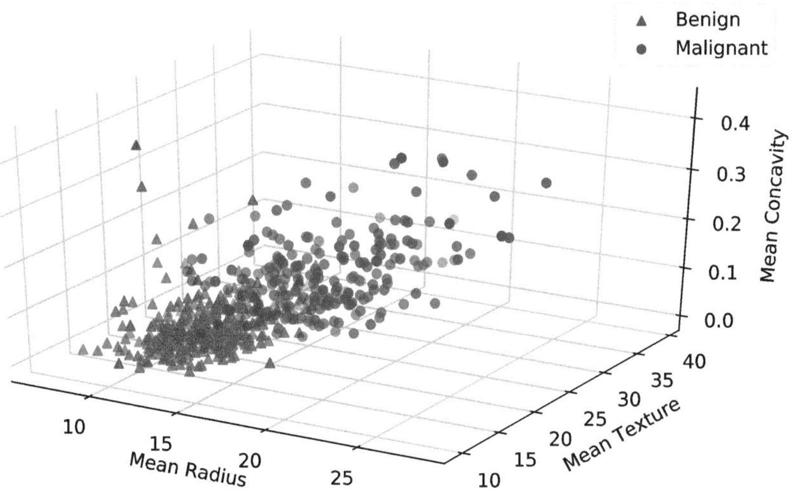

Figure 7.12: Scatterplot of three features of the benign and malignant breast masses.

The following code uses various classifiers to predict the category of breast masses (benign or malignant). In this case the training set has 341 elements and the test set has 228 elements. For each classifier the percentage of correct predictions (that is, the accuracy) in the test set is reported. We see that in this case quadratic discriminant analysis gives the highest accuracy (0.956). Exercise 18 explores the question whether this metric is the most appropriate for these data.

skclass3.py

```
from skclass1 import X_train, y_train, X_test, y_test
from sklearn.metrics import accuracy_score

import sklearn.discriminant_analysis as DA
from sklearn.naive_bayes import GaussianNB
from sklearn.neighbors import KNeighborsClassifier
from sklearn.linear_model import LogisticRegression
from sklearn.svm import SVC

names = ["Logit","NBayes", "LDA", "QDA", "KNN", "SVM"]
```

```
classifiers = [LogisticRegression(C=1e5),
               GaussianNB(),
               DA.LinearDiscriminantAnalysis(),
               DA.QuadraticDiscriminantAnalysis(),
               KNeighborsClassifier(n_neighbors=5),
               SVC(kernel='rbf', gamma = 1e-4)]

print('Name  Accuracy\n'+14*'-')
for name, clf in zip(names, classifiers):
  clf.fit(X_train, y_train)
  y_pred = clf.predict(X_test)
  print('{:6}  {:3.3f}'.format(name, accuracy_score(y_test,y_pred)))
```

```
Name  Accuracy
--------------
Logit   0.943
NBayes  0.908
LDA     0.943
QDA     0.956
KNN     0.925
SVM     0.939
```

Further Reading

An excellent source for understanding various pattern recognition techniques is the book [35] by Duda et al. Theoretical foundations of classification, including the Vapnik–Chernovenkis dimension and the fundamental theorem of learning, are discussed in [109, 121, 122]. A popular measure for characterizing the performance of a binary classifier is the *receiver operating characteristic* (ROC) curve [38]. The naïve Bayes classification paradigm can be extended to handle explanatory variable dependency via graphical models such as Bayesian networks and Markov random fields [46, 66, 69]. For a detailed discussion on Bayesian decision theory, see [8].

Exercises

1. Let $0 \leqslant w \leqslant 1$. Show that the solution to the convex optimization problem

$$\begin{aligned} &\min_{p_1,\ldots,p_n} \sum_{i=1}^{n} p_i^2 \\ &\text{subject to: } \sum_{i-1}^{n-1} p_i = w \text{ and } \sum_{i=1}^{n} p_i = 1, \end{aligned} \tag{7.28}$$

is given by $p_i = w/(n-1), i = 1, \ldots, n-1$ and $p_n = 1 - w$.

2. Derive the formulas (7.14) by minimizing the cross-entropy training loss:

$$-\frac{1}{n}\sum_{i=1}^{n} \ln g(\boldsymbol{x}_i, y_i \,|\, \boldsymbol{\theta}),$$

where $g(\boldsymbol{x}, y \mid \boldsymbol{\theta})$ is such that:

$$\ln g(\boldsymbol{x}, y \mid \boldsymbol{\theta}) = \ln \alpha_y - \frac{1}{2}\ln|\boldsymbol{\Sigma}_y| - \frac{1}{2}(\boldsymbol{x} - \boldsymbol{\mu}_y)^\top \boldsymbol{\Sigma}_y^{-1}(\boldsymbol{x} - \boldsymbol{\mu}_y) - \frac{p}{2}\ln(2\pi).$$

3. Adapt the code in Example 7.2 to plot the estimated decision boundary instead of the true one in Figure 7.3. Compare the true and estimated decision boundaries.

4. Recall from equation (7.16) that the decision boundaries of the multi-logit classifier are linear, and that the pre-classifier can be written as a conditional pdf of the form:

$$g(y \mid \mathbf{W}, \boldsymbol{b}, \boldsymbol{x}) = \frac{\exp(z_{y+1})}{\sum_{i=1}^{c}\exp(z_i)}, \quad y \in \{0, \dots, c-1\},$$

where $\boldsymbol{x}^\top = [1, \widetilde{\boldsymbol{x}}^\top]$ and $\boldsymbol{z} = \mathbf{W}\widetilde{\boldsymbol{x}} + \boldsymbol{b}$.

(a) Show that the linear discriminant pre-classifier in Section 7.4 can also be written as a conditional pdf of the form ($\boldsymbol{\theta} = \{\alpha_y, \boldsymbol{\Sigma}_y, \boldsymbol{\mu}_y\}_{y=0}^{c-1}$):

$$g(y \mid \boldsymbol{\theta}, \boldsymbol{x}) = \frac{\exp(z_{y+1})}{\sum_{i=1}^{c}\exp(z_i)}, \quad y \in \{0, \dots, c-1\},$$

where $\boldsymbol{x}^\top = [1, \widetilde{\boldsymbol{x}}^\top]$ and $\boldsymbol{z} = \mathbf{W}\widetilde{\boldsymbol{x}} + \boldsymbol{b}$. Find formulas for the corresponding $\boldsymbol{b}$ and $\mathbf{W}$ in terms of the linear discriminant parameters $\{\alpha_y, \boldsymbol{\mu}_y, \boldsymbol{\Sigma}_y\}_{y=0}^{c-1}$, where $\boldsymbol{\Sigma}_y = \boldsymbol{\Sigma}$ for all y.

(b) Explain which pre-classifier has smaller approximation error: the linear discriminant or multi-logit one? Justify your answer by proving an inequality between the two approximation errors.

5. Consider a binary classification problem where the response Y takes values in $\{-1, 1\}$. Show that optimal prediction function for the hinge loss $\text{Loss}(y, \widehat{y}) = (1 - y\widehat{y})_+ := \max\{0, 1 - y\widehat{y}\}$ is the same as the optimal prediction function g^* for the indicator loss:

$$g^*(\boldsymbol{x}) = \begin{cases} 1 & \text{if } \mathbb{P}[Y = 1 \mid \boldsymbol{X} = \boldsymbol{x}] > 1/2, \\ -1 & \text{if } \mathbb{P}[Y = 1 \mid \boldsymbol{X} = \boldsymbol{x}] < 1/2. \end{cases}$$

That is, show that

$$\mathbb{E}\,(1 - Y\,h(\boldsymbol{X}))_+ \geqslant \mathbb{E}\,(1 - Y\,g^*(\boldsymbol{X}))_+ \tag{7.29}$$

for all functions h.

☞ 158

6. In Example 4.12, we applied a principal component analysis (PCA) to the `iris` data, but refrained from classifying the flowers based on their feature vectors $\boldsymbol{x}$. Implement a 1-nearest neighbor algorithm, using a training set of 50 randomly chosen data pairs $(\boldsymbol{x}, y)$ from the `iris` data set. How many of the remaining 100 flowers are correctly classified? Now classify these entries with an off-the-shelf multi-logit classifier, e.g., such as can be found in the `sklearn` and `statsmodels` packages.

7. Figure 7.13 displays two groups of data points, given in Table 7.8. The convex hulls have also been plotted. It is possible to separate the two classes of points via a straight line.

In fact, many such lines are possible. SVM gives the best separation, in the sense that the gap (margin) between the points is maximal.

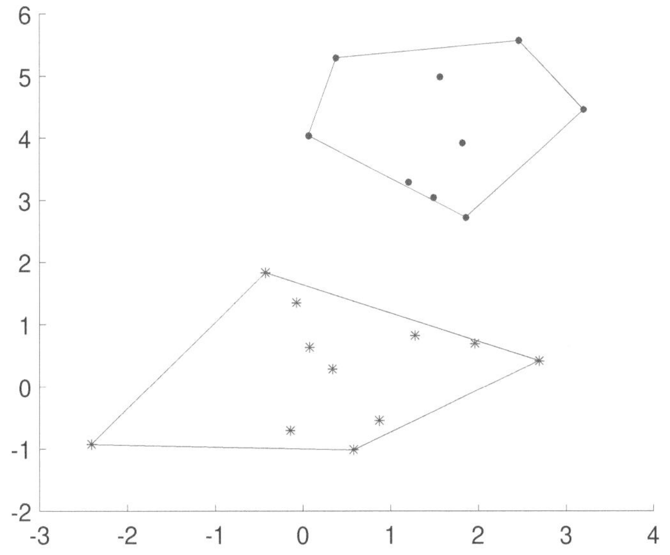

Figure 7.13: Separate the points by a straight line so that the separation between the two groups is maximal.

Table 7.8: Data for Figure 7.13.

x_1	x_2	y	x_1	x_2	y
2.4524	5.5673	−1	0.5819	−1.0156	1
1.2743	0.8265	1	1.2065	3.2984	−1
0.8773	−0.5478	1	2.6830	0.4216	1
1.4837	3.0464	−1	−0.0734	1.3457	1
0.0628	4.0415	−1	0.0787	0.6363	1
−2.4151	−0.9309	1	0.3816	5.2976	−1
1.8152	3.9202	−1	0.3386	0.2882	1
1.8557	2.7262	−1	−0.1493	−0.7095	1
−0.4239	1.8349	1	1.5554	4.9880	−1
1.9630	0.6942	1	3.2031	4.4614	−1

(a) Identify from the figure the three support vectors.

(b) For a separating boundary (line) given by $\beta_0 + \boldsymbol{\beta}^\top \boldsymbol{x} = 0$, show that the margin width is $2/\|\boldsymbol{\beta}\|$.

(c) Show that the parameters β_0 and $\boldsymbol{\beta}$ that solve the convex optimization problem (7.24) provide the maximal width between the margins.

(d) Solve (7.24) using a penalty approach; see Section B.4. In particular, minimize the ☞ 417 penalty function

$$S(\boldsymbol{\beta}, \beta_0) = \|\boldsymbol{\beta}\|^2 - C \sum_{i=1}^{n} \min\left\{(\beta_0 + \boldsymbol{\beta}^\top \boldsymbol{x}_i)\, y_i - 1,\ 0\right\}$$

for some positive penalty constant C.

(e) Find the solution the dual optimization problem (7.21) by using `sklearn`'s `SCV` method. Note that, as the two point sets are separable, the constraint $\boldsymbol{\lambda} \leqslant \mathbf{1}$ may be removed, and the value of γ can be set to 1.

8. In Example 7.6 we used the feature map $\boldsymbol{\phi}(\boldsymbol{x}) = [x_1, x_2, x_1^2 + x_2^2]^\top$ to classify the points. An easier way is to map the points into $\mathbb{R}^1$ via the feature map $\boldsymbol{\phi}(\boldsymbol{x}) = \|\boldsymbol{x}\|$ or any monotone function thereof. Translated back into $\mathbb{R}^2$ this yields a circular separating boundary. Find the radius and center of this circle, using the fact that here the sorted norms for the two groups are $\ldots, 0.4889, 0.5528, \ldots$.

9. Let $Y \in \{0, 1\}$ be a response variable and let $h(\boldsymbol{x})$ be the regression function

$$h(\boldsymbol{x}) := \mathbb{E}[Y \,|\, \boldsymbol{X} = \boldsymbol{x}] = \mathbb{P}[Y = 1 \,|\, \boldsymbol{X} = \boldsymbol{x}].$$

Recall that the Bayes classifier is $g^*(\boldsymbol{x}) = \mathbb{1}\{h(\boldsymbol{x}) > 1/2\}$. Let $g : \mathbb{R} \to \{0, 1\}$ be any other classifier function. Below, we denote all probabilities and expectations conditional on $\boldsymbol{X} = \boldsymbol{x}$ as $\mathbb{P}_{\boldsymbol{x}}[\cdot]$ and $\mathbb{E}_{\boldsymbol{x}}[\cdot]$.

(a) Show that

$$\mathbb{P}_{\boldsymbol{x}}[g(\boldsymbol{x}) \neq Y] = \overbrace{\mathbb{P}_{\boldsymbol{x}}[g^*(\boldsymbol{x}) \neq Y]}^{\text{irreducible error}} + |2h(\boldsymbol{x}) - 1|\, \mathbb{1}\{g(\boldsymbol{x}) \neq g^*(\boldsymbol{x})\}.$$

Hence, deduce that for a learner $g_{\mathcal{T}}$ constructed from a training set $\mathcal{T}$, we have

$$\mathbb{E}[\mathbb{P}_{\boldsymbol{x}}[g_{\mathcal{T}}(\boldsymbol{x}) \neq Y \,|\, \mathcal{T}]] = \mathbb{P}_{\boldsymbol{x}}[g^*(\boldsymbol{x}) \neq Y] + |2h(\boldsymbol{x}) - 1|\, \mathbb{P}[g_{\mathcal{T}}(\boldsymbol{x}) \neq g^*(\boldsymbol{x})],$$

where the first expectation and last probability operations are with respect to $\mathcal{T}$.

(b) Using the previous result, deduce that for the unconditional error (that is, we no longer condition on $\boldsymbol{X} = \boldsymbol{x}$), we have

$$\mathbb{P}[g^*(\boldsymbol{X}) \neq Y] \leqslant \mathbb{P}[g_{\mathcal{T}}(\boldsymbol{X}) \neq Y].$$

(c) Show that, if $g_{\mathcal{T}} := \mathbb{1}\{h_{\mathcal{T}}(\boldsymbol{x}) > 1/2\}$ is a classifier function such that as $n \to \infty$

$$h_{\mathcal{T}}(\boldsymbol{x}) \xrightarrow{\text{d}} Z \sim \mathcal{N}(\mu(\boldsymbol{x}), \sigma^2(\boldsymbol{x}))$$

for some mean and variance functions $\mu(\boldsymbol{x})$ and $\sigma^2(\boldsymbol{x})$, respectively, then

$$\mathbb{P}_{\boldsymbol{x}}[g_{\mathcal{T}}(\boldsymbol{x}) \neq g^*(\boldsymbol{x})] \longrightarrow \Phi\left(\frac{\operatorname{sign}(1 - 2h(\boldsymbol{x}))(2\mu(\boldsymbol{x}) - 1)}{2\sigma(\boldsymbol{x})}\right),$$

where Φ is the cdf of a standard normal random variable.

10. The purpose of this exercise is to derive the dual program (7.21) from the primal program (7.20). The starting point is to introduce a vector of auxiliary variables $\boldsymbol{\xi} := [\xi_1, \ldots, \xi_n]^\top$ and write the primal program as

$$\begin{aligned} \min_{\alpha, \alpha_0, \boldsymbol{\xi}} \; & \sum_{i=1}^n \xi_i + \frac{\gamma}{2} \boldsymbol{\alpha}^\top \mathbf{K} \boldsymbol{\alpha} \\ \text{subject to:} \quad & \boldsymbol{\xi} \geqslant \mathbf{0}, \\ & y_i(\alpha_0 + \{\mathbf{K}\boldsymbol{\alpha}\}_i) \geqslant 1 - \xi_i, \; i = 1, \ldots, n. \end{aligned} \tag{7.30}$$

(a) Apply the Lagrangian optimization theory from Section B.2.2 to obtain the Lagrangian function $\mathcal{L}(\{\alpha_0, \boldsymbol{\alpha}, \boldsymbol{\xi}\}, \{\boldsymbol{\lambda}, \boldsymbol{\mu}\})$, where $\boldsymbol{\mu}$ and $\boldsymbol{\lambda}$ are the Lagrange multipliers corresponding to the first and second inequality constraints, respectively. ☞ 408

(b) Show that the Karush–Kuhn–Tucker (see Theorem B.2) conditions for optimizing $\mathcal{L}$ are: ☞ 409

$$\begin{aligned} \boldsymbol{\lambda}^\top \boldsymbol{y} &= 0 \\ \boldsymbol{\alpha} &= \boldsymbol{y} \odot \boldsymbol{\lambda}/\gamma \\ \mathbf{0} &\leqslant \boldsymbol{\lambda} \leqslant \mathbf{1} \\ (\mathbf{1} - \boldsymbol{\lambda}) \odot \boldsymbol{\xi} = \mathbf{0}, &\quad \lambda_i\,(y_i g(\boldsymbol{x}_i) - 1 + \xi_i) = 0, \; i = 1, \ldots, n \\ \boldsymbol{\xi} \geqslant \mathbf{0}, &\quad y_i g(\boldsymbol{x}_i) - 1 + \xi_i \geqslant 0, \; i = 1, \ldots, n. \end{aligned} \tag{7.31}$$

Here $\odot$ stands for componentwise multiplication; e.g., $\boldsymbol{y} \odot \boldsymbol{\lambda} = [y_1\lambda_1, \ldots, y_n\lambda_n]^\top$, and we have abbreviated $\alpha_0 + \{\mathbf{K}\boldsymbol{\alpha}\}_i$ to $g(\boldsymbol{x}_i)$, in view of (7.19). [Hint: one of the KKT conditions is $\boldsymbol{\lambda} = \mathbf{1} - \boldsymbol{\mu}$; thus we can eliminate $\boldsymbol{\mu}$.]

(c) Using the KKT conditions (7.31), reduce the Lagrange dual function $\mathcal{L}^*(\boldsymbol{\lambda}) := \min_{\alpha_0, \alpha, \xi} \mathcal{L}(\{\alpha_0, \boldsymbol{\alpha}, \boldsymbol{\xi}\}, \{\boldsymbol{\lambda}, \mathbf{1} - \boldsymbol{\lambda}\})$ to

$$\mathcal{L}^*(\boldsymbol{\lambda}) = \sum_{i=1}^n \lambda_i - \frac{1}{2\gamma} \sum_{i=1}^n \sum_{j=1}^n \lambda_i \lambda_j y_i y_j\, \kappa(\boldsymbol{x}_i, \boldsymbol{x}_j). \tag{7.32}$$

(d) As a consequence of (7.19) and (a)–(c), show that the optimal prediction function g_τ is given by

$$g_\tau(\boldsymbol{x}) = \alpha_0 + \frac{1}{\gamma} \sum_{i=1}^n y_i \lambda_i\, \kappa(\boldsymbol{x}_i, \boldsymbol{x}), \tag{7.33}$$

where $\boldsymbol{\lambda}$ is the solution to

$$\begin{aligned} \max_{\boldsymbol{\lambda}} \quad & \mathcal{L}^*(\boldsymbol{\lambda}) \\ \text{subject to:} \quad & \boldsymbol{\lambda}^\top \boldsymbol{y} = 0, \; \mathbf{0} \leqslant \boldsymbol{\lambda} \leqslant \mathbf{1}, \end{aligned} \tag{7.34}$$

and $\alpha_0 = y_j - \frac{1}{\gamma} \sum_{i=1}^n y_i \lambda_i\, \kappa(\boldsymbol{x}_i, \boldsymbol{x}_j)$ for any j such that $\lambda_j \in (0, 1)$.

11. Consider SVM classification as illustrated in Figure 7.7. The goal of this exercise is to classify the training points $\{(\boldsymbol{x}_i, y_i)\}$ based on the value of the multipliers $\{\lambda_i\}$ in Exercise 10. Let ξ_i be the auxiliary variable in Exercise 10, $i = 1, \ldots, n$.

(a) For $\lambda_i \in (0, 1)$ show that $(\boldsymbol{x}_i, y_i)$ lies exactly on the decision border.

(b) For $\lambda_i = 1$, show that $(\boldsymbol{x}_i, y_i)$ lies strictly inside the margins.

(c) Show that for $\lambda_i = 0$ the point $(\boldsymbol{x}_i, y_i)$ lies outside the margins and is correctly classified.

12. A well-known data set is the MNIST handwritten digit database, containing many thousands of digitalized numbers (from 0 to 9), each described by a 28×28 matrix of gray scales. A similar but much smaller data set is described in [63]. Here, each handwritten digit is summarized by a 8×8 matrix with integer entries from 0 (white) to 15 (black). Figure 7.14 shows the first 50 digitized images. The data set can be accessed with Python using the `sklearn` package as follows.

```
from sklearn import datasets
digits = datasets.load_digits()
x_digits = digits.data     # explanatory variables
y_digits = digits.target   # responses
```

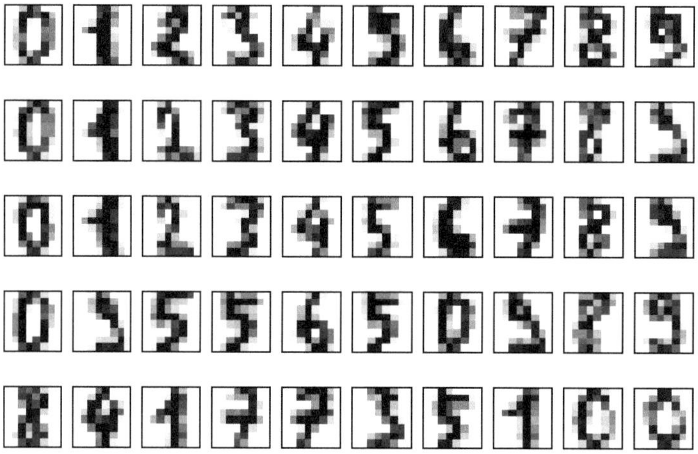

Figure 7.14: Classify the digitized images.

(a) Divide the data into a 75% training set and 25% test set.

(b) Compare the effectiveness of the K-nearest neighbors and naïve Bayes method to classify the data.

(c) Assess which K to use in the K-nearest neighbors classification.

13. Download the `winequality-red.csv` data set from UCI's wine-quality website. The response here is the wine quality (from 0 to 10) as specified by a wine "expert" and the explanatory variables are various characteristics such as acidity and sugar content. Use the **SVC** classifier of `sklearn.svm` with a linear kernel and penalty parameter $\mathsf{C} = 1$ (see Remark 7.2) to fit the data. Use the method **cross_val_score** from

`sklearn.model_selection` to obtain a five-fold cross-validation score as an estimate of the probability that the predicted class matches the expert's class.

14. Consider the credit approval data set `crx.data` from UCI's credit approval website. The data set is concerned with credit card applications. The last column in the data set indicates whether the application is approved (+) or not (−). With the view of preserving data privacy, all 15 explanatory variables were anonymized. Note that some explanatory variables are continuous and some are categorical.

(a) Load and prepare the data for analysis with `sklearn`. First, eliminate data rows with missing values. Next, encode categorical explanatory variables using a **`OneHotEncoder`** object from `sklearn.preprocessing` to create a model matrix `X` with indicator variables for the categorical variables, as described in Section 5.3.5. ☞ 177

(b) The model matrix should contain 653 rows and 46 columns. The response variable should be a 0/1 variable (reject/approve). We will consider several classification algorithms and test their performance (using a zero-one loss) via ten-fold cross validation.

   i. Write a function which takes 3 parameters: `X`, $\boldsymbol{y}$, and a model, and returns the ten-fold cross-validation estimate of the expected generalization risk.

   ii. Consider the following `sklearn` classifiers: **`KNeighborsClassifier`** ($k = 5$), **`LogisticRegression`**, and **`MPLClassifier`** (multilayer perceptron). Use the function from (i) to identify the best performing classifier.

15. Consider a synthetic data set that was generated in the following fashion. The explanatory variable follows a standard normal distribution. The response label is 0 if the explanatory variable is between the 0.95 and 0.05 quantiles of the standard normal distribution, and 1, otherwise. The data set was generated using the following code.

```
import numpy as np
import scipy.stats
# generate data

np.random.seed(12345)
N = 100
X = np.random.randn(N)
q = scipy.stats.norm.ppf(0.95)
y = np.zeros(N)
y[X>=q] = 1
y[X<=-q] = 1
X = X.reshape(-1,1)
```

Compare the K-nearest neighbors classifier with $K = 5$ and logistic regression classifier. Without computation, which classifier is likely to be better for these data? Verify your answer by coding both classifiers and printing the corresponding training 0–1 loss.

16. Consider the `digits` data set from Exercise 12. In this exercise, we would like to train a binary classifier for the identification of digit 8.

(a) Divide the data such that the first 1000 rows are used as the training set and the rest are used as the test set.

(b) Train the `LogisticRegression` classifier from the `sklearn.linear_model` package.

(c) "Train" a naïve classifier that always returns 0. That is, the naïve classifier identifies each instance as being not 8.

(d) Compare the zero-one test losses of the logistic regression and the naïve classifiers.

(e) Find the confusion matrix, the precision, and the recall of the logistic regression classifier.

(f) Find the fraction of eights that are correctly detected by the logistic regression classifier.

17. Repeat Exercise 16 with the original MNIST data set. Use the first 60,000 rows as the train set and the remaining 10,000 rows as the test set. The original data set can be obtained using the following code.

```
from sklearn.datasets import fetch_openml

X, y = fetch_openml('mnist_784', version=1, return_X_y=True)
```

☞ 279
☞ 255

18. For the breast cancer data in Section 7.8, investigate and discuss whether *accuracy* is the relevant metric to use or if other metrics discussed in Section 7.2 are more appropriate.

CHAPTER 8

DECISION TREES AND ENSEMBLE METHODS

Statistical learning methods based on decision trees have gained tremendous popularity due to their simplicity, intuitive representation, and predictive accuracy. This chapter gives an introduction to the construction and use of such trees. We also discuss two key ensemble methods, namely bootstrap aggregation and boosting, which can further improve the efficiency of decision trees and other learning methods.

8.1 Introduction

Tree-based methods provide a simple, intuitive, and powerful mechanism for both regression and classification. The main idea is to divide a (potentially complicated) feature space $\mathcal{X}$ into smaller regions and fit a simple prediction function to each region. For example, in a regression setting, one could take the mean of the training responses associated with the training features that fall in that specific region. In the classification setting, a commonly used prediction function takes the majority vote among the corresponding response variables. We start with a simple classification example.

■ **Example 8.1 (Decision Tree for Classification)** The left panel of Figure 8.1 shows a training set of 15 two-dimensional points (features) falling into two classes (red and blue). How should the new feature vector (black point) be classified?

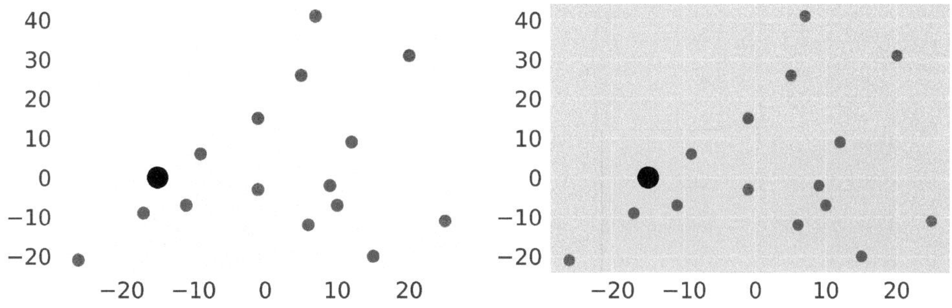

Figure 8.1: Left: training data and a new feature. Right: a partition of the feature space.

It is not possible to linearly separate the training set, but we can partition the feature space $\mathcal{X} = \mathbb{R}^2$ into rectangular regions and assign a class (color) to each region, as shown in the right panel of Figure 8.1. Points in these regions are classified accordingly as blue or red. The partition thus defines a classifier (prediction function) g that assigns to each feature vector $\boldsymbol{x}$ a class "red" or "blue". For example, for $\boldsymbol{x} = [-15, 0]^\top$ (solid black point), $g(\boldsymbol{x})$ = "blue", since it belongs to a blue region of the feature space.

DECISION TREE

Both the classification procedure and the partitioning of the feature space can be conveniently represented by a binary *decision tree*. This is a tree where each node v corresponds to a region (subset) $\mathcal{R}_v$ of the feature space $\mathcal{X}$ — the root node corresponding to the feature space itself.

Each internal node v contains a logical condition that divides $\mathcal{R}_v$ into two disjoint subregions. The leaf nodes (the terminal nodes of the tree) are not subdivided, and their corresponding regions form a partition of $\mathcal{X}$, as they are disjoint and their union is $\mathcal{X}$. Associated with each leaf node w is also a regional prediction function g^w on $\mathcal{R}_w$.

The partitioning of Figure 8.1 was obtained from the decision tree shown in Figure 8.2. As an illustration of the decision procedure, consider again the input $\boldsymbol{x} = [x_1, x_2]^\top = [-15, 0]^\top$. The classification process starts from the tree root, which contains the condition $x_2 \leqslant 12.0$. As the second component of $\boldsymbol{x}$ is 0, the root condition is satisfied and we proceed to the left child, which contains the condition $x_1 \leqslant -20.5$. The next step is similar. As $0 > -20.5$, the condition is not satisfied and we proceed to the right child. Such an evaluation of logical conditions along the tree path will eventually bring us to a leaf node and its associated region. In this case the process terminates in a leaf that corresponds to the left blue region in the right-hand panel of Figure 8.1.

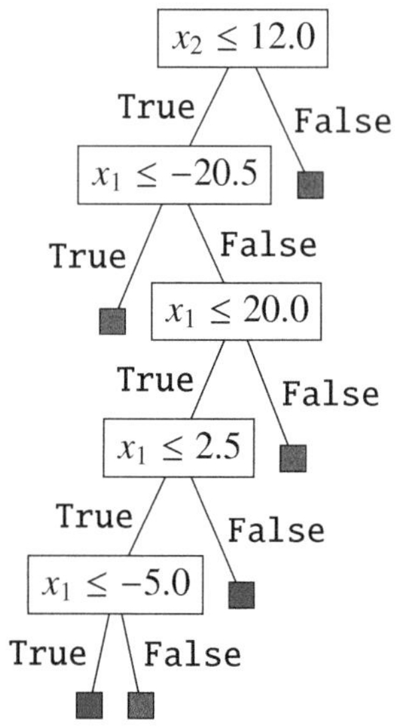

Figure 8.2: The decision-tree that corresponds to the partition in Figure 8.1. ■

More generally, a binary tree $\mathbb{T}$ will partition the feature space $\mathcal{X}$ into as many regions as there are leaf nodes. Denote the set of leaf nodes by $\mathcal{W}$. The overall prediction function g that corresponds to the tree can then be written as

$$g(\boldsymbol{x}) = \sum_{w \in \mathcal{W}} g^w(\boldsymbol{x}) \, \mathbb{1}\{\boldsymbol{x} \in \mathcal{R}_w\}, \tag{8.1}$$

REGIONAL PREDICTION FUNCTIONS

where $\mathbb{1}$ denotes the indicator function. The representation (8.1) is very general and depends on (1) how the regions $\{\mathcal{R}_w\}$ are constructed via the logical conditions in the decision tree, as well as (2) how the *regional prediction functions* of the leaf nodes are defined. Simple logical conditions of the form $x_j \leqslant \xi$ split a Euclidean feature space into rectangles aligned with the axes. For example, Figure 8.2 partitions the feature space into six rectangles: two blue and four red rectangles.

In a classification setting, the regional prediction function g^w corresponding to a leaf node w takes values in the set of possible class labels. In most cases, as in Example 8.1, it is taken to be constant on the corresponding region $\mathcal{R}_w$. In a regression setting, g^w is real-valued and also usually takes only one value. That is, every feature vector in $\mathcal{R}_w$ leads to the same predicted value. Of course, different regions will usually have different predicted values.

Constructing a tree with a training set $\tau = \{(\boldsymbol{x}_i, y_i)\}\}_{i=1}^n$ amounts to minimizing the training loss

$$\ell_\tau(g) = \frac{1}{n}\sum_{i=1}^n \text{Loss}(y_i, g(\boldsymbol{x}_i)) \tag{8.2}$$

for some loss function; see Chapter 2. With g of the form (8.1), we can write ☞ 19

$$\ell_\tau(g) = \frac{1}{n}\sum_{i=1}^n \text{Loss}(y_i, g(\boldsymbol{x}_i)) = \frac{1}{n}\sum_{i=1}^n \sum_{w\in\mathcal{W}} \mathbb{1}\{\boldsymbol{x}_i \in \mathcal{R}_w\} \text{Loss}(y_i, g(\boldsymbol{x}_i)) \tag{8.3}$$

$$= \sum_{w\in\mathcal{W}} \underbrace{\frac{1}{n}\sum_{i=1}^n \mathbb{1}\{\boldsymbol{x}_i \in \mathcal{R}_w\} \text{Loss}(y_i, g^w(\boldsymbol{x}_i)),}_{(*)} \tag{8.4}$$

where $(*)$ is the contribution by the regional prediction function g^w to the overall training loss. In the case where all $\{\boldsymbol{x}_i\}$ are different, finding a decision tree $\mathbb{T}$ that gives a zero squared-error or zero–one training loss is easy, see Exercise 1, but such an "overfitted" tree will have poor predictive behavior, expressed in terms of the generalization risk. Instead we consider a restricted class of decision trees and aim to minimize the training loss within that class. It is common to use a top-down greedy approach, which can only achieve an approximate minimization of the training loss.

8.2 Top-Down Construction of Decision Trees

Let $\tau = \{(\boldsymbol{x}_i, y_i)\}_{i=1}^n$ be the training set. The key to constructing a binary decision tree $\mathbb{T}$ is to specify a *splitting rule* for each node v, which can be defined as a logical function $s : \mathcal{X} \to \{\texttt{False}, \texttt{True}\}$ or, equivalently, a binary function $s : \mathcal{X} \to \{0, 1\}$. For example, in the decision tree of Figure 8.2 the root node has splitting rule $\boldsymbol{x} \mapsto \mathbb{1}\{x_2 \leqslant 12.0\}$, in correspondence with the logical condition $\{x_2 \leqslant 12.0\}$. During the construction of the tree, each node v is associated with a specific region $\mathcal{R}_v \subseteq \mathcal{X}$ and therefore also the training subset $\{(\boldsymbol{x}, y) \in \tau : \boldsymbol{x} \in \mathcal{R}_v\} \subseteq \tau$. Using a splitting rule s, we can divide any subset σ of the training set τ into two sets: SPLITTING RULE

$$\sigma_\text{T} := \{(\boldsymbol{x}, y) \in \sigma \ :\ s(\boldsymbol{x}) = \texttt{True}\} \text{ and } \sigma_\text{F} := \{(\boldsymbol{x}, y) \in \sigma \ :\ s(\boldsymbol{x}) = \texttt{False}\}. \tag{8.5}$$

Starting from an empty tree and the initial data set τ, a generic decision tree construction takes the form of the recursive Algorithm 8.2.1. Here we use the notation $\mathbb{T}_v$ for a subtree of $\mathbb{T}$ starting from node v. The final tree $\mathbb{T}$ is thus obtained via $\mathbb{T} = \texttt{Construct\_Subtree}(v_0, \tau)$, where v_0 is the root of the tree.

Algorithm 8.2.1: `Construct_Subtree`

Input: A node v and a subset of the training data: $\sigma \subseteq \tau$.
Output: A (sub) decision tree $\mathbb{T}_v$.

1 **if** termination criterion is met **then** `// v is a leaf node`
2 Train a regional prediction function g^v using the training data σ.
3 **else** `// split the node`
4 Find the best splitting rule s_v for node v.
5 Create successors v_T and v_F of v.
6 $\sigma_\mathrm{T} \leftarrow \{(\boldsymbol{x}, y) \in \sigma \; : \; s_v(\boldsymbol{x}) = \texttt{True}\}$
7 $\sigma_\mathrm{F} \leftarrow \{(\boldsymbol{x}, y) \in \sigma \; : \; s_v(\boldsymbol{x}) = \texttt{False}\}$
8 $\mathbb{T}_{v_\mathrm{T}} \leftarrow$ `Construct_Subtree` $(v_\mathrm{T}, \sigma_\mathrm{T})$ `// left branch`
9 $\mathbb{T}_{v_\mathrm{F}} \leftarrow$ `Construct_Subtree` $(v_\mathrm{F}, \sigma_\mathrm{F})$ `// right branch`
10 **return** $\mathbb{T}_v$

The splitting rule s_v divides the region $\mathcal{R}_v$ into two disjoint parts, say $\mathcal{R}_{v_\mathrm{T}}$ and $\mathcal{R}_{v_\mathrm{F}}$. The corresponding prediction functions, g^T and g^F, satisfy

$$g^v(\boldsymbol{x}) = g^\mathrm{T}(\boldsymbol{x})\, \mathbb{1}\{\boldsymbol{x} \in \mathcal{R}_{v_\mathrm{T}}\} + g^\mathrm{F}(\boldsymbol{x})\, \mathbb{1}\{\boldsymbol{x} \in \mathcal{R}_{v_\mathrm{F}}\}, \quad \boldsymbol{x} \in \mathcal{R}_v.$$

In order to implement the procedure described in Algorithm 8.2.1, we need to address the construction of the regional prediction functions g^v at the leaves (Line 2), the specification of the splitting rule (Line 4), and the termination criterion (Line 1). These important aspects are detailed in the following Sections 8.2.1, 8.2.2, and 8.2.3, respectively.

8.2.1 Regional Prediction Functions

In general, there is no restriction on how to choose the prediction function g^w for a leaf node $v = w$ in Line 2 of Algorithm 8.2.1. In principle we can train any model from the data; e.g., via linear regression. However, in practice very simple prediction functions are used. Below, we detail a popular choice for classification, as well as one for regression.

1. In the *classification* setting with class labels $0, \ldots, c-1$, the regional prediction function g^w for leaf node w is usually chosen to be *constant* and equal to the most common class label of the training data in the associated region $\mathcal{R}_w$ (ties can be broken randomly). More precisely, let n_w be the number of feature vectors in region $\mathcal{R}_w$ and let

$$p_z^w = \frac{1}{n_w} \sum_{\{(\boldsymbol{x},y)\in\tau \,:\, \boldsymbol{x}\in\mathcal{R}_w\}} \mathbb{1}_{\{y=z\}},$$

be the proportion of feature vectors in $\mathcal{R}_w$ that have class label $z = 0, \ldots, c-1$. The regional prediction function for node w is chosen to be the constant

$$g^w(\boldsymbol{x}) = \underset{z\in\{0,\ldots,c-1\}}{\operatorname{argmax}}\; p_z^w. \tag{8.6}$$

2. In the *regression* setting, g^w is usually chosen as the mean response in the region; that is,

$$g^w(\boldsymbol{x}) = \overline{y}_{\mathcal{R}_w} := \frac{1}{n_w} \sum_{\{(\boldsymbol{x},y)\in\tau \,:\, \boldsymbol{x}\in\mathcal{R}_w\}} y, \tag{8.7}$$

where n_w is again the number of feature vectors in $\mathcal{R}_w$. It is not difficult to show that $g^w(\boldsymbol{x}) = \overline{y}_{\mathcal{R}_w}$ minimizes the squared-error loss with respect to all constant functions, in the region $\mathcal{R}_w$; see Exercise 2.

8.2.2 Splitting Rules

In Line 4 in Algorithm 8.2.1, we divide region $\mathcal{R}_v$ into two sets, using a splitting rule (function) s_v. Consequently, the data set σ associated with node v (that is, the subset of the original data set τ whose feature vectors lie in $\mathcal{R}_v$), is also split — into σ_T and σ_F. What is the benefit of such a split in terms of a reduction in the training loss? If v were set to a leaf node, its contribution to the training loss would be (see (8.4)):

$$\frac{1}{n}\sum_{i=1}^{n} \mathbb{1}_{\{(\boldsymbol{x},y)\in\sigma\}} \operatorname{Loss}(y_i, g^v(\boldsymbol{x}_i)). \tag{8.8}$$

If v were to be split instead, its contribution to the overall training loss would be:

$$\frac{1}{n}\sum_{i=1}^{n} \mathbb{1}_{\{(\boldsymbol{x},y)\in\sigma_\text{T}\}} \operatorname{Loss}(y_i, g^\text{T}(\boldsymbol{x}_i)) + \frac{1}{n}\sum_{i=1}^{n} \mathbb{1}_{\{(\boldsymbol{x},y)\in\sigma_\text{F}\}} \operatorname{Loss}(y_i, g^\text{F}(\boldsymbol{x}_i)), \tag{8.9}$$

where g^T and g^F are the prediction functions belonging to the child nodes v_T and v_F. A *greedy* heuristic is to pretend that the tree construction algorithm immediately terminates after the split, in which case v_T and v_F are leaf nodes, and g^T and g^F are readily evaluated — e.g., as in Section 8.2.1. Note that for any splitting rule the contribution (8.8) is always greater than or equal to (8.9). It therefore makes sense to choose the splitting rule such that (8.9) is minimized. Moreover, the termination criterion may involve comparing (8.9) with (8.8). If their difference is too small it may not be worth further splitting the feature space.

As an example, suppose the feature space is $\mathcal{X} = \mathbb{R}^p$ and we consider splitting rules of the form

$$s(\boldsymbol{x}) = \mathbb{1}\{x_j \leqslant \xi\}, \tag{8.10}$$

for some $1 \leqslant j \leqslant p$ and $\xi \in \mathbb{R}$, where we identify 0 with `False` and 1 with `True`. Due to the computational and interpretative simplicity, such binary splitting rules are implemented in many software packages and are considered to be the *de facto* standard. As we have seen, these rules divide up the feature space into rectangles, as in Figure 8.1. It is natural to ask how j and ξ should be chosen so as to minimize (8.9). For a regression problem, using a squared-error loss and a constant regional prediction function as in (8.7), the sum (8.9) is given by

$$\frac{1}{n}\sum_{(\boldsymbol{x},y)\in\tau:x_j\leqslant\xi} \left(y - \overline{y}_\text{T}\right)^2 + \frac{1}{n}\sum_{(\boldsymbol{x},y)\in\tau:x_j>\xi} \left(y - \overline{y}_\text{F}\right)^2, \tag{8.11}$$

where $\overline{y}_\text{T}$ and $\overline{y}_\text{F}$ are the average responses for the σ_T and σ_F data, respectively. Let $\{x_{j,k}\}_{k=1}^m$ denote the possible values of x_j, $j = 1,\ldots,p$ within the training subset σ (with $m \leqslant n$ elements). Note that, for a fixed j, (8.11) is a piecewise constant function of ξ, and that its minimal value is attained at some value $x_{j,k}$. As a consequence, to minimize (8.11) over all j and ξ, it suffices to evaluate (8.11) for each of the $m \times p$ values $x_{j,k}$ and then take the minimizing pair $(j, x_{j,k})$.

For a classification problem, using the indicator loss and a constant regional prediction function as in (8.6), the aim is to choose a splitting rule that minimizes

$$\frac{1}{n}\sum_{(\boldsymbol{x},y)\in\sigma_{\mathrm{T}}} \mathbb{1}\{y \neq y^*_{\mathrm{T}}\} + \frac{1}{n}\sum_{(\boldsymbol{x},y)\in\sigma_{\mathrm{F}}} \mathbb{1}\{y \neq y^*_{\mathrm{F}}\}, \tag{8.12}$$

where $y^*_{\mathrm{T}} = g^{\mathrm{T}}(\boldsymbol{x})$ is the most prevalent class (majority vote) in the data set σ_{T} and y^*_{F} is the most prevalent class in σ_{F}. If the feature space is $\mathcal{X} = \mathbb{R}^p$ and the splitting rules are of the form (8.10), then the optimal splitting rule can be obtained in the same way as described above for the regression case; the only difference is that (8.11) is replaced with (8.12).

We can view the minimization of (8.12) as minimizing a weighted average of "impurities" of nodes σ_{T} and σ_{F}. Namely, for an arbitrary training subset $\sigma \subseteq \tau$, if y^* is the most prevalent label, then

$$\frac{1}{|\sigma|}\sum_{(\boldsymbol{x},y)\in\sigma} \mathbb{1}\{y \neq y^*\} = 1 - \frac{1}{|\sigma|}\sum_{(\boldsymbol{x},y)\in\sigma} \mathbb{1}\{y = y^*\} = 1 - p_{y^*} = 1 - \max_{z\in\{0,\dots,c-1\}} p_z,$$

where p_z is the proportion of data points in σ that have class label z, $z = 0, \dots, c-1$. The quantity

$$1 - \max_{z\in\{0,\dots,c-1\}} p_z$$

MISCLASSIFICATION IMPURITY

measures the diversity of the labels in σ and is called the *misclassification impurity*. Consequently, (8.12) is the weighted sum of the misclassification impurities of σ_{T} and σ_{F}, with weights by $|\sigma_{\mathrm{T}}|/n$ and $|\sigma_{\mathrm{F}}|/n$, respectively. Note that the misclassification impurity only depends on the label proportions rather than on the individual responses. Instead of using the misclassification impurity to decide if and how to split a data set σ, we can use other impurity measures that only depend on the label proportions. Two popular choices are the *entropy impurity*:

ENTROPY IMPURITY

$$-\sum_{z=0}^{c-1} p_z \log_2(p_z)$$

GINI IMPURITY

and the *Gini impurity*:

$$\frac{1}{2}\left(1 - \sum_{z=0}^{c-1} p_z^2\right).$$

All of these impurities are maximal when the label proportions are equal to $1/c$. Typical shapes of the above impurity measures are illustrated in Figure 8.3 for the two-label case, with class probabilities p and $1-p$. We see here the similarity of the different impurity measures. Note that impurities can be arbitrarily scaled, and so using $\ln(p_z) = \log_2(p_z)\ln(2)$ instead of $\log_2(p_z)$ above gives an equivalent entropy impurity.

8.2.3 Termination Criterion

When building a tree, one can define various types of termination conditions. For example, we might stop when the number of data points in the tree node (the size of the input σ set in Algorithm 8.2.1) is less than or equal to some predefined number. Or we might choose the maximal depth of the tree in advance. Another possibility is to stop when there is no

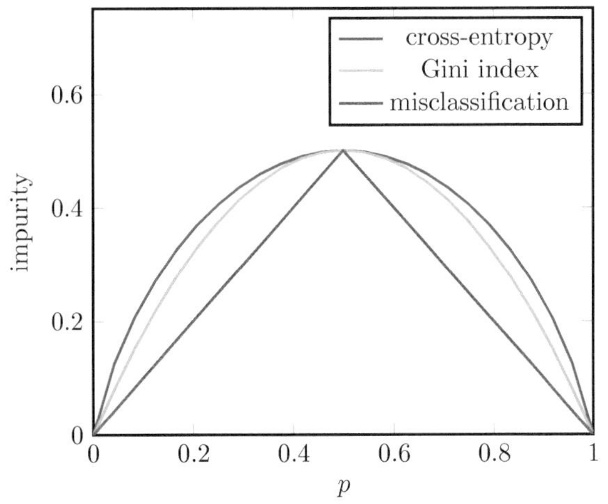

Figure 8.3: Entropy, Gini, and misclassification impurities for binary classification, with class frequencies $p_1 = p$ and $p_2 = 1 - p$. The entropy impurity was normalized (divided by 2), to ensure that all impurity measures attain the same maximum value of $1/2$ at $p = 1/2$.

significant advantage, in terms of training loss, to split regions. Ultimately, the quality of a tree is determined by its predictive performance (generalization risk) and the termination condition should aim to strike a balance between minimizing the approximation error and minimizing the statistical error, as discussed in Section 2.4. ☞ 31

■ **Example 8.2 (Fixed Tree Depth)** To illustrate how the tree depth impacts on the generalization risk, consider Figure 8.4, which shows the typical behavior of the cross-validation loss as a function of the tree depth. Recall that the cross-validation loss is an estimate of the expected generalization risk. Complicated (deep) trees tend to overfit the training data by producing many divisions of the feature space. As we have seen, this overfitting problem is typical of all learning methods; see Chapter 2 and in particular Example 2.1. To conclude, increasing the maximal depth does not necessarily result in better performance. ☞ 26

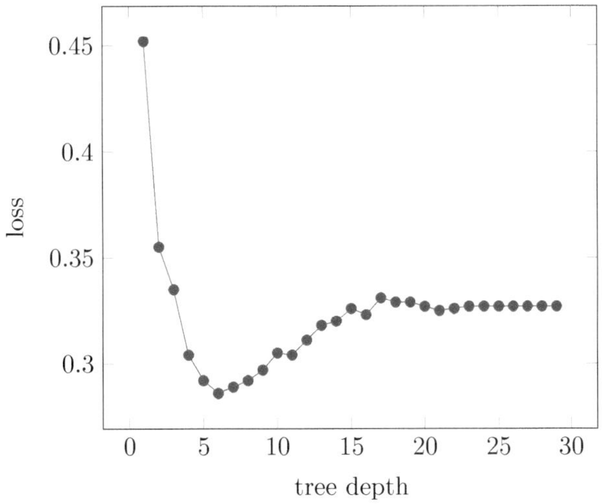

Figure 8.4: The ten-fold cross-validation loss as a function of the maximal tree depth for a classification problem. The optimal maximal tree depth is here 6.

☞ 492

To create Figure 8.4 we used[1] the Python method **make_blobs** from the sklearn module to produce a training set of size $n = 5000$ with ten-dimensional feature vectors (thus, $p = 10$ and $\mathcal{X} = \mathbb{R}^{10}$), each of which is classified into one of $c = 3$ classes. The full code is given below.

TreeDepthCV.py

```
import numpy as np
from sklearn.datasets import make_blobs
from sklearn.model_selection import cross_val_score
from sklearn.tree import DecisionTreeClassifier
from sklearn.metrics import zero_one_loss
import matplotlib.pyplot as plt

def ZeroOneScore(clf, X, y):
    y_pred = clf.predict(X)
    return zero_one_loss(y, y_pred)

# Construct the training set
X, y =  make_blobs(n_samples=5000, n_features=10, centers=3,
                           random_state=10, cluster_std=10)

# construct a decision tree classifier
clf = DecisionTreeClassifier(random_state=0)

# Cross-validation loss as a function of tree depth (1 to 30)
xdepthlist = []
cvlist = []
tree_depth = range(1,30)
for d in tree_depth:
    xdepthlist.append(d)
    clf.max_depth=d
    cv = np.mean(cross_val_score(clf, X, y, cv=10, scoring=
        ZeroOneScore))
    cvlist.append(cv)

plt.xlabel('tree depth', fontsize=18, color='black')
plt.ylabel('loss', fontsize=18, color='black')
plt.plot(xdepthlist, cvlist,'-*' , linewidth=0.5)
```

■

The code above relies heavily on sklearn and hides the implementation details. To show how decision trees are actually constructed using the previous theory, we proceed with a very basic implementation.

8.2.4 Basic Implementation

In this section we implement a regression tree, step by step. To run the program, amalgamate the code snippets below into one file, in the order presented. First, we import various packages and define a function to generate the training and test data.

[1]The data used for Figure 8.1 was produced in a similar way.

BasicTree.py

```
import numpy as np
from sklearn.datasets import make_friedman1
from sklearn.model_selection import train_test_split

def makedata():
   n_points = 500 # number of samples

   X, y =  make_friedman1(n_samples=n_points, n_features=5,
                          noise=1.0, random_state=100)
   return train_test_split(X, y, test_size=0.5, random_state=3)
```

The "main" method calls the `makedata` method, uses the training data to build a regression tree, and then predicts the responses of the test set and reports the mean squared-error loss.

```
def main():
  X_train, X_test, y_train, y_test = makedata()
  maxdepth = 10 # maximum tree depth
  # Create tree root at depth 0
  treeRoot = TNode(0, X_train,y_train)

  # Build the regression tree with maximal depth equal to max_depth
  Construct_Subtree(treeRoot, maxdepth)

  # Predict
  y_hat = np.zeros(len(X_test))
  for i in range(len(X_test)):
     y_hat[i] = Predict(X_test[i],treeRoot)

  MSE = np.mean(np.power(y_hat - y_test,2))
  print("Basic tree: tree loss = ",  MSE)
```

The next step is to specify a tree node as a Python class. Each node has a number of attributes, including the features and the response data (`X` and `y`) and the depth at which the node is placed in the tree. The root node has depth 0. Each node w can calculate its contribution to the squared-error training loss $\sum_{i=1}^{n} \mathbb{1}\{\boldsymbol{x}_i \in \mathcal{R}^w\}(y_i - g^w(\boldsymbol{x}_i))^2$. Note that we have omitted the constant $1/n$ term when training the tree, which simply scales the loss (8.2).

```
class TNode:
   def __init__(self, depth, X, y):
      self.depth = depth
      self.X = X   # matrix of features
      self.y = y   # vector of response variables
      # initialize optimal split parameters
      self.j = None
      self.xi = None
      # initialize children to be None
      self.left = None
      self.right = None
      # initialize the regional predictor
```

```
        self.g = None

    def CalculateLoss(self):
        if(len(self.y)==0):
            return 0

        return np.sum(np.power(self.y - self.y.mean(),2))
```

The function below implements the training (tree-building) Algorithm 8.2.1.

```
def Construct_Subtree(node, max_depth):
    if(node.depth == max_depth or len(node.y) == 1):
        node.g  = node.y.mean()
    else:
        j, xi = CalculateOptimalSplit(node)
        node.j = j
        node.xi = xi
        Xt, yt, Xf, yf = DataSplit(node.X, node.y, j, xi)

        if(len(yt)>0):
            node.left = TNode(node.depth+1,Xt,yt)
            Construct_Subtree(node.left, max_depth)

        if(len(yf)>0):
            node.right = TNode(node.depth+1, Xf,yf)
            Construct_Subtree(node.right, max_depth)

    return node
```

This requires an implementation of the **CalculateOptimalSplit** function. To start, we implement a function **DataSplit** that splits the data according to $s(\boldsymbol{x}) = \mathbb{1}\{x_j \leqslant \xi\}$.

```
def DataSplit(X,y,j,xi):
    ids = X[:,j]<=xi
    Xt  = X[ids == True,:]
    Xf  = X[ids == False,:]
    yt  = y[ids == True]
    yf  = y[ids == False]
    return Xt, yt, Xf, yf
```

The **CalculateOptimalSplit** method runs through the possible splitting thresholds ξ from the set $\{x_{j,k}\}$ and finds the optimal split.

```
def CalculateOptimalSplit(node):
    X = node.X
    y = node.y
    best_var = 0
    best_xi = X[0,best_var]
    best_split_val = node.CalculateLoss()

    m, n  = X.shape

    for j in range(0,n):
```

```
        for i in range(0,m):
            xi = X[i,j]
            Xt, yt, Xf, yf = DataSplit(X,y,j,xi)
            tmpt = TNode(0, Xt, yt)
            tmpf = TNode(0, Xf, yf)
            loss_t = tmpt.CalculateLoss()
            loss_f = tmpf.CalculateLoss()
            curr_val =  loss_t + loss_f
            if (curr_val < best_split_val):
                best_split_val = curr_val
                best_var = j
                best_xi = xi
    return best_var,  best_xi
```

Finally, we implement the recursive method for prediction.

```
def Predict(X,node):
    if(node.right == None and node.left != None):
        return Predict(X,node.left)

    if(node.right != None and node.left == None):
        return Predict(X,node.right)

    if(node.right == None and node.left == None):
        return node.g
    else:
        if(X[node.j] <= node.xi):
            return Predict(X,node.left)
        else:
            return Predict(X,node.right)
```

Running the **main** function defined above gives a similar[2] result to what one would achieve with the sklearn package, using the **DecisionTreeRegressor** method.

```
main()  # run the main program

# compare with sklearn
from sklearn.tree import DecisionTreeRegressor

X_train, X_test, y_train, y_test = makedata() # use the same data
regTree = DecisionTreeRegressor(max_depth = 10, random_state=0)
regTree.fit(X_train,y_train)
y_hat = regTree.predict(X_test)
MSE2 = np.mean(np.power(y_hat - y_test,2))
print("DecisionTreeRegressor: tree loss = ",  MSE2)
```

```
Basic tree: tree loss =  9.067077996170276
DecisionTreeRegressor: tree loss =  10.197991295531748
```

[2]After establishing a best split $\xi = x_{j,k}$, sklearn assigns the corresponding feature vector randomly to one of the two child nodes, rather than to the True child.

8.3 Additional Considerations

8.3.1 Binary Versus Non-Binary Trees

While it is possible to split a tree node into more than two groups (multiway splits), it generally produces inferior results compared to the simple binary split. The major reason is that multiway splits can lead to too many nodes near the tree root that have only a few data points, thus leaving insufficient data for later splits. As multiway splits can be represented as several binary splits, the latter is preferred [55].

8.3.2 Data Preprocessing

☞ 153

Sometimes, it can be beneficial to preprocess the data prior to the tree construction. For example, PCA can be used with a view to identify the most important dimensions, which in turn will lead to simpler and possibly more informative splitting rules in the internal nodes.

8.3.3 Alternative Splitting Rules

We restricted our attention to splitting rules of the type $s(\boldsymbol{x}) = \mathbb{1}\{x_j \leqslant \xi\}$, where $j \in \{1, \ldots, p\}$ and $\xi \in \mathbb{R}$. These types of rules may not always result in a simple partition of the feature space, as illustrated by the binary data in Figure 8.5. In this case, the feature space could have been partitioned into just two regions, separated by a straight line.

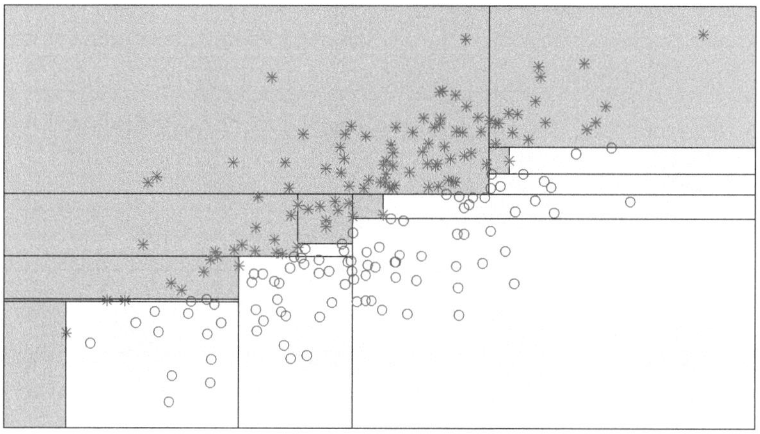

Figure 8.5: The two groups of points can here be separated by a straight line. Instead, the classification tree divides up the space into many rectangles, leading to an unnecessarily complicated classification procedure.

☞ 261

In this case many classification methods discussed in Chapter 7, such as linear discriminant analysis (Section 7.4), will work very well, whereas the classification tree is rather elaborate, dividing the feature set into too many regions. An obvious remedy is to use splitting rules of the form

$$s(\boldsymbol{x}) = \mathbb{1}\{\boldsymbol{a}^\top \boldsymbol{x} \leqslant \xi\}.$$

In some cases, such as the one just discussed, it may be useful to use a splitting rule that involves several variables, as opposed to a single one. The decision regarding the split type clearly depends on the problem domain. For example, for logical (binary) variables our domain knowledge may indicate that a different behavior is expected when both x_i and x_j $(i \neq j)$ are `True`. In this case, we will naturally introduce a decision rule of the form:

$$s(\boldsymbol{x}) = \mathbb{1}\{x_i = \texttt{True and } x_j = \texttt{True}\}.$$

8.3.4 Categorical Variables

When an explanatory variable is categorical with labels (levels) say $\{1, \ldots, k\}$, the splitting rule is generally defined via a partition of the label set $\{1, \ldots, k\}$ into two subsets. Specifically, let L and R be a partition of $\{1, \ldots, k\}$. Then, the splitting rule is defined via

$$s(\boldsymbol{x}) = \mathbb{1}\{x_j \in L\}.$$

For the general supervised learning case, finding the optimal partition in the sense of minimal loss requires one to consider 2^k subsets of $\{1, \ldots, k\}$. Consequently, finding a good splitting rule for categorical variables can be challenging when the number of labels p is large.

8.3.5 Missing Values

Missing data is present in many real-life problems. Generally, when working with incomplete feature vectors, where one or more values are missing, it is typical to either completely delete the feature vector from the data (which may distort the data) or to impute (guess) its missing values from the available data; see e.g., [120]. Tree methods, however, allow an elegant approach for handling missing data. Specifically, in the general case, the missing data problem can be handled via *surrogate* splitting rules [20].

> When dealing with categorical (factor) features, we can introduce an additional category "missing" for the absent data.

The main idea of surrogate rules is as follows. First, we construct a decision (regression or a classification) tree via Algorithm 8.2.1. During this construction process, the solution of the optimization problem (8.9) is calculated only over the observations that are not missing a particular variable. Suppose that a tree node v has a splitting rule $s^*(\boldsymbol{x}) = \mathbb{1}\{x_{j^*} \leqslant \xi^*\}$ for some $1 \leqslant j^* \leqslant p$ and threshold ξ^*.

For the node v we can introduce a set of alternative splitting rules that resemble the original splitting rule, sometimes called the *primary* splitting rule, using different variables and thresholds. Namely, we look for a binary splitting rule $s(\boldsymbol{x} \mid j, \xi)$, $j \neq j^*$ such that the data split introduced by s will be *similar* to the original data split from s^*. The similarity is generally measured via a binary misclassification loss, where the true classes of observations are determined by the primary splitting rule and the surrogate splitting rules serve as classifiers. Consider, for example, the data in Table 8.1 and suppose that the primary splitting rule at node v is $\mathbb{1}\{\text{Age} \leqslant 25\}$. That is, the five data points are split such that the left and the right child of v contains two and three data points, respectively. Next, the following surrogate splitting rules can be considered:

1. $\mathbb{1}\{\text{Salary} \leqslant 1500\}$, and
2. $\mathbb{1}\{\text{Height} \leqslant 173\}$.

Table 8.1: Example data with three variables (Age, Height, and Salary).

Id	Age	Height	Salary
1	20	173	1000
2	25	168	1500
3	38	191	1700
4	49	170	1900
5	62	182	2000

The $\mathbb{1}\{\text{Salary} \leqslant 1500\}$ surrogate rule completely mimics the primary rule, in the sense that the data splits induced by these rules are identical. Namely, both rules partition the data into two sets (by Id) $\{1, 2\}$ and $\{3, 4, 5\}$. On the other hand, the $\mathbb{1}\{\text{Height} \leqslant 173\}$ rule is less similar to the primary rule, since it causes the different partition $\{1, 2, 4\}$ and $\{3, 5\}$.

It is up to the user to define the number of surrogate rules for each tree node. As soon as these surrogate rules are available, we can use them to handle a new data point, even if the main rule cannot be applied due to a missing value of the primary variable x_{j^*}. Specifically, if the observation is missing the primary split variable, we apply the first (best) surrogate rule. If the first surrogate variable is also missing, we apply the second best surrogate rule, and so on.

8.4 Controlling the Tree Shape

Eventually, we are interested in getting the right-size tree. Namely, a tree that shows good generalization properties. It was already discussed in Section 8.2.3 (Figure 8.4) that shallow trees tend to underfit and deep trees tend to overfit the data. Basically, a shallow tree does not produce a sufficient number of splits and a deep tree will produce many partitions and thus many leaf nodes. If we grow the tree to a sufficient depth, each training sample will occupy a separate leaf and we will observe a zero loss with respect to the training data. The above phenomenon is illustrated in Figure 8.6, which presents the cross-validation loss and the training loss as a function of the tree depth.

In order to overcome the under- and the overfitting problem, Breiman et al. [20] examined the possibility of stopping the tree from growing as soon as the decrease in loss due to a split of node v, as expressed in the difference of (8.8) and (8.9), is smaller than some predefined parameter $\delta \in \mathbb{R}$. Under this setting, the tree construction process will terminate when no leaf node can be split such that the contribution to the training loss after this split is greater than δ.

The authors found that this approach was unsatisfactory. Specifically, it was noted that a very small δ leads to an excessive amount of splitting and thus causes overfitting. Increasing δ did not work either. The problem is that the nature of the proposed rule is *one-step-look-ahead*. To see this, consider a tree node for which the best possible decrease in loss is

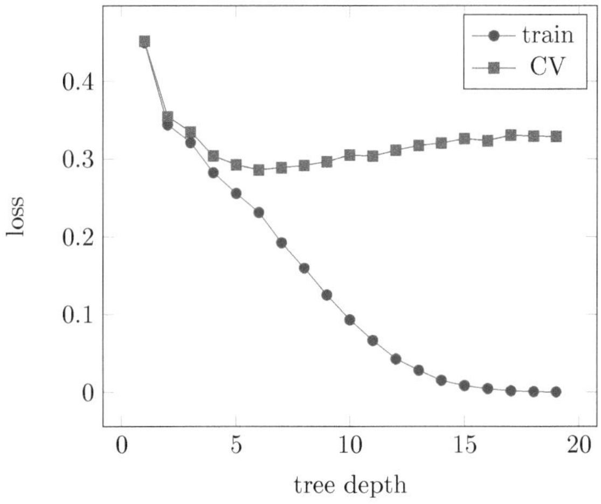

Figure 8.6: The cross-validation and the training loss as a function of the tree depth for a binary classification problem.

smaller than δ. According to the proposed procedure, this node will not be split further. This may, however, be sub-optimal, because it could happen that one of the node's *descendants*, if split, could lead to a major decrease in loss.

To address these issues, a so-called *pruning* routine can be employed. The idea is as follows. We first grow a very deep tree and then prune (remove nodes) it upwards until we reach the root node. Consequently, the pruning process causes the number of tree nodes to decrease. While the tree is being pruned, the generalization risk gradually decreases up to the point where it starts increasing again, at which point the pruning is stopped. This decreasing/increasing behavior is due to the bias–variance tradeoff (2.22). TREE PRUNING

We next describe the details. To start with, let v and v' be tree nodes. We say that v' is a *descendant* of v if there is a path down the tree, which leads from v to v'. If such a path exists, we also say that v is an *ancestor* of v'. Consider the tree in Figure 8.7.

To formally define pruning, we will require the following Definition 8.1. An example of pruning is demonstrated in Figure 8.8.

Definition 8.1: Branches and Pruning

1. A *tree branch* $\mathbb{T}_v$ of the tree $\mathbb{T}$ is a sub-tree of $\mathbb{T}$ rooted at node $v \in \mathbb{T}$. TREE BRANCH

2. The pruning of branch $\mathbb{T}_v$ from a tree $\mathbb{T}$ is performed via deletion of the entire branch $\mathbb{T}_v$ from $\mathbb{T}$ except the branch's root node v. The resulting pruned tree is denoted by $\mathbb{T} - \mathbb{T}_v$.

3. A sub-tree $\mathbb{T} - \mathbb{T}_v$ is called a pruned sub-tree of $\mathbb{T}$. We indicate this with the notation $\mathbb{T} - \mathbb{T}_v \prec \mathbb{T}$ or $\mathbb{T} \succ \mathbb{T} - \mathbb{T}_v$.

A basic decision tree pruning procedure is summarized in Algorithm 8.4.1.

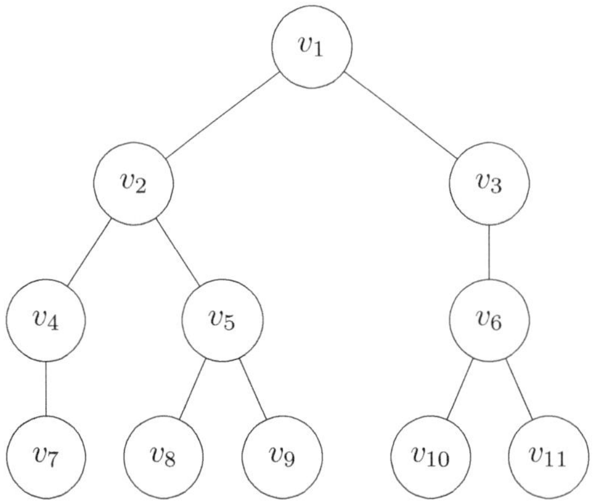

Figure 8.7: The node v_9 is a descendant of v_2, and v_2 is an ancestor of $\{v_4, v_5, v_7, v_8, v_9\}$, but v_6 is not a descendant of v_2.

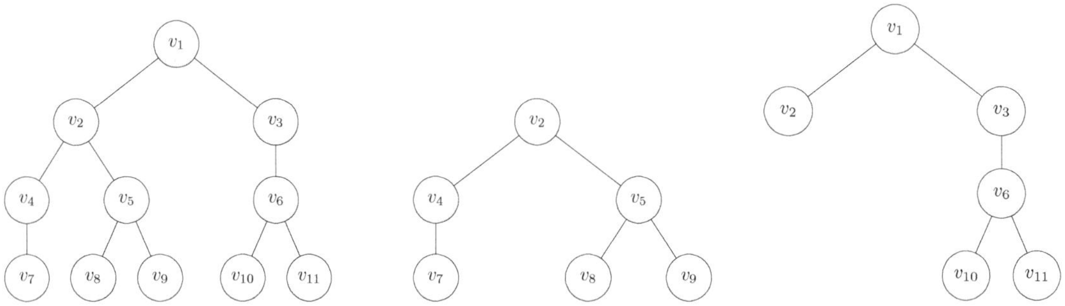

(a) $\mathbb{T}$ (b) $\mathbb{T}_{v_2}$ (c) $\mathbb{T} - \mathbb{T}_{v_2}$

Figure 8.8: The pruned tree $\mathbb{T} - \mathbb{T}_{v_2}$ in (c) is the result of pruning the $\mathbb{T}_{v_2}$ branch in (b) from the original tree $\mathbb{T}$ in (a).

Algorithm 8.4.1: Decision Tree Pruning

Input: Training set τ.

Output: Sequence of decision trees $\mathbb{T}^0 \succ \mathbb{T}^1 \succ \cdots$

Build a large decision tree $\mathbb{T}^0$ via Algorithm 8.2.1. [A possible termination criterion for that algorithm is to have some small predetermined number of data points at each terminal node of $\mathbb{T}^0$.]

$\mathbb{T}' \leftarrow \mathbb{T}^0$

$k \leftarrow 0$

while $\mathbb{T}'$ has more than one node **do**

  $k \leftarrow k + 1$

  Choose $v \in \mathbb{T}'$.

  Prune the branch rooted at v from $\mathbb{T}'$.

  $\mathbb{T}^k \leftarrow \mathbb{T}' - \mathbb{T}_v$ and $\mathbb{T}' \leftarrow \mathbb{T}^k$.

return $\mathbb{T}^0, \mathbb{T}^1, \ldots, \mathbb{T}^k$

Let $\mathbb{T}^0$ be the initial (deep) tree and let $\mathbb{T}^k$ be the tree obtained after the k-th pruning operation, for $k = 1, \ldots, K$. As soon as the sequence of trees $\mathbb{T}^0 > \mathbb{T}^1 > \cdots > \mathbb{T}^K$ is available, one can choose the best tree of $\{\mathbb{T}^k\}_{k=1}^K$ according to the smallest generalization risk. Specifically, we can split the data into training and validation sets. In this case, Algorithm 8.4.1 is executed using the training set and the generalization risks of $\{\mathbb{T}^k\}_{k=1}^K$ are estimated via the validation set.

While Algorithm 8.4.1 and the corresponding best tree selection process look appealing, there is still an important question to consider; namely, how to choose the node v and the corresponding branch $\mathbb{T}_v$ in Line 6 of the algorithm. In order to overcome this problem, Breiman proposed a method called *cost complexity pruning*, which we discuss next.

8.4.1 Cost-Complexity Pruning

Let $\mathbb{T} < \mathbb{T}^0$ be a tree obtained via pruning of a tree $\mathbb{T}^0$. Denote the set of leaf (terminal) nodes of $\mathbb{T}$ by $\mathcal{W}$. The number of leaves $|\mathcal{W}|$ is a measure for the complexity of the tree; recall that $|\mathcal{W}|$ is the number of regions $\{\mathcal{R}_w\}$ in the partition of $\mathcal{X}$. Corresponding to each tree $\mathbb{T}$ is a prediction function g, as in (8.1). In *cost-complexity pruning* the objective is to find a prediction function g (or, equivalently, tree $\mathbb{T}$) that minimizes the training loss $\ell_\tau(g)$ while taking into account the complexity of the tree. The idea is to regularize the training loss, similar to what was done in Chapter 6, by adding a penalty term for the complexity of the tree. This leads to the following definition.

COST-COMPLEXITY PRUNING

Definition 8.2: Cost-Complexity Measure

Let $\tau = \{(\boldsymbol{x}_i, y_i)\}_{i=1}^n$ be a data set and $\gamma \geqslant 0$ be a real number. For a given tree $\mathbb{T}$, the *cost-complexity measure* $C_\tau(\gamma, \mathbb{T})$ is defined as:

COST-COMPLEXITY MEASURE

$$\begin{aligned} C_\tau(\gamma, \mathbb{T}) &:= \frac{1}{n} \sum_{w \in \mathcal{W}} \left(\sum_{i=1}^n \mathbb{1}\{\boldsymbol{x}_i \in \mathcal{R}_w\} \operatorname{Loss}(y_i, g^w(\boldsymbol{x}_i)) \right) + \gamma\, |\mathcal{W}| \qquad (8.13) \\ &= \ell_\tau(g) + \gamma\, |\mathcal{W}|, \end{aligned}$$

where $\ell_\tau(g)$ is the training loss (8.2).

Small values of γ result in a small penalty for the tree complexity $|\mathcal{W}|$, and thus large trees (that fit the entire *training* data well) will minimize the measure $C_\tau(\gamma, \mathbb{T})$. In particular, for $\gamma = 0$, $\mathbb{T} = \mathbb{T}^0$ will be the minimizer of $C_\tau(\gamma, \mathbb{T})$. On the other hand, large values of γ will prefer smaller trees or, more precisely, trees with fewer leaves. For sufficiently large γ, the solution $\mathbb{T}$ will collapse to a single (root) node.

It can be shown that, for every value of γ, there exists a smallest minimizing subtree with respect to the cost-complexity measure. In practice, a suitable γ is selected via observing the performance of the learner on the validation set or by cross-validation.

These advantages and the corresponding limitations are detailed next.

8.4.2 Advantages and Limitations of Decision Trees

We list a number of advantages and disadvantages of decision trees, as compared with other supervised learning methods such as were discussed in Chapters 5, 6, and 7.

Advantages

1. The tree structure can handle both categorical and numerical features in a natural and straightforward way. Specifically, there is no need to pre-process categorical features, say via the introduction of dummy variables.

2. The final tree obtained after the training phase can be compactly stored for the purpose of making predictions for new feature vectors. The prediction process only involves a single tree traversal from the tree root to a leaf.

3. The hierarchical nature of decision trees allows for an efficient encoding of the feature's conditional information. Specifically, after an internal split of a feature x_j via the standard splitting rule (8.10), Algorithm 8.2.1 will only consider such subsets of data that were constructed based on this split, thus implicitly exploiting the corresponding conditional information from the initial split of x_j.

4. The tree structure can be easily understood and interpreted by domain experts with little statistical knowledge, since it is essentially a logical decision flow diagram.

5. The sequential decision tree growth procedure in Algorithm 8.2.1, and in particular the fact that the tree has been split using the most important features, provides an implicit step-wise variable elimination procedure. In addition, the partition of the variable space into smaller regions results in simpler prediction problems in these regions.

6. Decision trees are invariant under monotone transformations of the data. To see this, consider the (optimal) splitting rule $s(\boldsymbol{x}) = \mathbb{1}\{x_3 \leqslant 2\}$, where x_3 is a positive feature. Suppose that x_3 is transformed to $x_3' = x_3^2$. Now, the optimal splitting rule will take the form $s(\boldsymbol{x}) = \mathbb{1}\{x_3' \leqslant 4\}$.

7. In the classification setting, it is common to report not only the predicted value of a feature vector, e.g., as in (8.6), but also the respective class probabilities. Decision trees handle this task without any additional effort. Specifically, consider a new feature vector. During the estimation process, we will perform a tree traversal and the point will end up in a certain leaf w. The probability of this feature vector lying in class z can be estimated as the proportion of training points in w that are in class z.

8. As each training point is treated equally in the construction of a tree, their structure of the tree will be relatively robust to outliers. In a way, trees exhibit a similar kind of robustness as the sample median does for real-valued data.

Limitations

Despite the fact that the decision trees are extremely interpretable, the predictive accuracy is generally inferior to other established statistical learning methods. In addition, decision trees, and in particular very deep trees that were not subject to pruning, are heavily reliant on their training set. A small change in the training set can result in a dramatic change of the resulting decision tree. Their inferior predictive accuracy, however, is a direct consequence of the bias–variance tradeoff. Specifically, a decision tree model generally exhibits a high variance. To overcome the above limitations, several promising approaches such as *bagging*, *random forest*, and *boosting* are introduced below.

The bagging approach was initially introduced in the context of an ensemble of decision trees. However, both the bagging and the boosting methods can be applied to improve the accuracy of general prediction functions.

8.5 Bootstrap Aggregation

The major idea of the *bootstrap aggregation* or *bagging* method is to combine prediction functions learned from multiple data sets, with a view to improving overall prediction accuracy. Bagging is especially beneficial when dealing with predictors that tend to overfit the data, such as in decision trees, where the (unpruned) tree structure is very sensitive to small changes in the training set [37, 55]. BAGGING

To start with, consider an idealized setting for a regression tree, where we have access to B iid copies[3] $\mathcal{T}_1, \ldots, \mathcal{T}_B$ of a training set $\mathcal{T}$. Then, we can train B separate regression models (B different decision trees) using these sets, giving learners $g_{\mathcal{T}_1}, \ldots, g_{\mathcal{T}_B}$, and take their average:

$$g_{\text{avg}}(\boldsymbol{x}) = \frac{1}{B} \sum_{b=1}^{B} g_{\mathcal{T}_b}(\boldsymbol{x}). \tag{8.14}$$

By the law of large numbers, as $B \to \infty$, the average prediction function converges to the expected prediction function $g^\dagger := \mathbb{E} g_{\mathcal{T}}$. The following result shows that using $g^\dagger$ as a prediction function (if it were known) would result in an expected squared-error generalization risk that is less than or equal to the expected gencralization risk for a general prediction function $g_{\mathcal{T}}$. It thus suggests that taking an average of prediction functions may lead to a better expected squared-error generalization risk. ☞ 447 ☞ 24

Theorem 8.1: Expected Squared-Error Generalization Risk

Let $\mathcal{T}$ be a random training set and let $\boldsymbol{X}, Y$ be a random feature vector and response that are independent of $\mathcal{T}$. Then,

$$\mathbb{E}\Big(Y - g_{\mathcal{T}}(\boldsymbol{X})\Big)^2 \geqslant \mathbb{E}\Big(Y - g^\dagger(\boldsymbol{X})\Big)^2.$$

[3] In this section $\mathcal{T}_k$ means the k-th training set, not a training set of size k.

Proof: We have

$$\mathbb{E}\left[\left(Y - g_{\mathcal{T}}(\mathbf{X})\right)^2 \,\middle|\, \mathbf{X}, Y\right] \geqslant \left(\mathbb{E}[Y \mid \mathbf{X}, Y] - \mathbb{E}[g_{\mathcal{T}}(\mathbf{X}) \mid \mathbf{X}, Y]\right)^2 = \left(Y - g^{\dagger}(\mathbf{X})\right)^2,$$

where the inequality follows from $\mathbb{E}U^2 \geqslant (\mathbb{E}U)^2$ for any (conditional) expectation. Consequently, by the tower property,

☞ 433

$$\mathbb{E}\left(Y - g_{\mathcal{T}}(\mathbf{X})\right)^2 = \mathbb{E}\left[\mathbb{E}\left[(Y - g_{\mathcal{T}}(\mathbf{X}))^2 \mid \mathbf{X}, Y\right]\right] \geqslant \mathbb{E}\left(Y - g^{\dagger}(\mathbf{X})\right)^2.$$

□

Unfortunately, multiple independent data sets are rarely available. But we can substitute them by bootstrapped ones. Specifically, instead of the $\mathcal{T}_1, \ldots, \mathcal{T}_B$ sets, we can obtain random training sets $\mathcal{T}_1^*, \ldots, \mathcal{T}_B^*$ by resampling them from a *single* (fixed) training set τ, similar to Algorithm 3.2.6, and use them to train B separate models. By model averaging as in (8.14) we obtain the bootstrapped aggregated estimator or *bagged estimator* of the form:

☞ 76

BAGGED ESTIMATOR

$$g_{\text{bag}}(\mathbf{x}) = \frac{1}{B} \sum_{b=1}^{B} g_{\mathcal{T}_b^*}(\mathbf{x}). \tag{8.15}$$

Algorithm 8.5.1: Bootstrap Aggregation Sampling

Input: Training set $\tau = \{(\mathbf{x}_i, y_i)\}_{i=1}^n$ and resample size B.
Output: Bootstrapped data sets.

```
for b = 1 to B do
    T*_b ← ∅
    for i = 1 to n do
        Draw U ~ U(0, 1)
        I ← ⌈nU⌉                          // select random index
        T*_b ← T*_b ∪ {(x_I, y_I)}.
return T*_b, b = 1, ..., B.
```

■ **Remark 8.1 (Bootstrap Aggregation for Classification Problems)** Note that (8.15) is suitable for handling regression problems. However, the bagging idea can be readily extended to handle classification settings as well. For example, g_{bag} can take the majority vote among $\{g_{\mathcal{T}_b^*}\}, b = 1, \ldots, B$; that is, to accept the most frequent class among B predictors. ■

While bagging can be applied for any statistical model (such as decision trees, neural networks, linear regression, K-nearest neighbors, and so on), it is most effective for predictors that are sensitive to small changes in the training set. The reason becomes clear when we decompose the expected generalization risk as

$$\mathbb{E}\, \ell(g_{\mathcal{T}}) = \ell^* + \underbrace{\mathbb{E}\left(\mathbb{E}[g_{\mathcal{T}}(\mathbf{X}) \mid \mathbf{X}] - g^*(\mathbf{X})\right)^2}_{\text{expected squared bias}} + \underbrace{\mathbb{E}[\mathbb{V}\text{ar}[g_{\mathcal{T}}(\mathbf{X}) \mid \mathbf{X}]]}_{\text{expected variance}}, \tag{8.16}$$

similar to (2.22). Compare this with the same decomposition for the average prediction function g_{bag} in (8.14). As $\mathbb{E}g_{\text{bag}}(\boldsymbol{x}) = \mathbb{E}g_{\mathcal{T}}(\boldsymbol{x})$, we see that any possible improvement in the generalization risk must be due to the expected variance term. Averaging and bagging are thus only useful for predictors with a large expected variance, relative to the other two terms. Examples of such "unstable" predictors include decision trees, neural networks, and subset selection in linear regression [22]. On the other hand, "stable" predictors are insensitive to small data changes, an example being the K-nearest neighbors method. Note that for independent training sets $\mathcal{T}_1, \ldots, \mathcal{T}_B$ a reduction of the variance by a factor B is achieved: $\mathbb{V}\text{ar}\, g_{\text{bag}}(\boldsymbol{x}) = B^{-1}\mathbb{V}\text{ar}\, g_{\mathcal{T}}(\boldsymbol{x})$. Again, it depends on the squared bias and irreducible loss how significant this reduction is for the generalization risk. ☞ 35

■ **Remark 8.2 (Limitations of Bagging)** It is important to remember that g_{bag} is not exactly equal to g_{avg}, which in turn is not exactly $g^{\dagger}$. Specifically, g_{bag} is constructed from the bootstrap approximation of the sampling pdf f. As a consequence, for stable predictors, it can happen that g_{bag} will perform worse than $g_{\mathcal{T}}$. In addition to the deterioration of the bagging performance for stable procedures, it can also happen that $g_{\mathcal{T}}$ has already achieved a near optimal predictive accuracy given the available training data. In this case, bagging will not introduce a significant improvement. ■

The bagging process provides an opportunity to estimate the generalization risk of the bagged model without an additional test set. Specifically, recall that we obtain the $\mathcal{T}_1^*, \ldots, \mathcal{T}_B^*$ sets from a single training set τ by sampling via Algorithm 8.5.1, and use them to train B separate models. It can be shown (see Exercise 8) that, for large sample sizes, on average about a third (more precisely, a fraction $\mathrm{e}^{-1} \approx 0.37$) of the original sample points are not included in bootstrapped set $\mathcal{T}_b^*$ for $1 \leqslant b \leqslant B$. Therefore, these samples can be used for the loss estimation. These samples are called *out-of-bag* (OOB) observations. OUT-OF-BAG

Specifically, for each sample from the original data set, we calculate the OOB loss using predictors that were trained without this particular sample. The estimation procedure is summarized in Algorithm 8.5.2. Hastie et al. [55] observe that, under certain conditions, the OOB loss is almost identical to the n-fold cross-validation loss. In addition, the OOB loss can be used to determine the number of trees required. Specifically, we can train predictors until the OOB loss stops changing. Namely, decision trees are added until the OOB loss stabilizes.

Algorithm 8.5.2: Out-of-Bag Loss Estimation

Input: The original data set $\tau = \{(\boldsymbol{x}_1, y_1), \ldots, (\boldsymbol{x}_n, y_n)\}$, the bootstrapped data sets $\{\mathcal{T}_1^*, \ldots, \mathcal{T}_B^*\}$, and the trained predictors $\left\{g_{\mathcal{T}_1^*}, \ldots, g_{\mathcal{T}_B^*}\right\}$.
Output: Out-of-bag loss for the averaged model.

```
for i = 1 to n do
    C_i ← ∅                      // Indices of predictors not depending on (x_i, y_i)
    for b = 1 to B do
        if (x_i, y_i) ∉ T*_b then C_i ← C_i ∪ {b}
    Y'_i ← |C_i|^{-1} Σ_{b∈C_i} g_{T*_b}(x_i)
    L_i ← Loss(y_i, Y'_i)
L_OOB ← (1/n) Σ_{i=1}^n L_i
return L_OOB.
```

■ **Example 8.3 (Bagging for a Regression Tree)** We next proceed with a basic bagging example for a regression tree, in which we compare the decision tree estimator with the corresponding bagged estimator. We use the R^2 metric (coefficient of determination) for comparison.

BaggingExample.py

```
import numpy as np
from sklearn.datasets import make_friedman1
from sklearn.tree import DecisionTreeRegressor
from sklearn.model_selection import train_test_split
from sklearn.metrics import r2_score

np.random.seed(100)

# create regression problem
n_points = 1000 # points
x, y =  make_friedman1(n_samples=n_points, n_features=15,
                       noise=1.0, random_state=100)

# split to train/test set
x_train, x_test, y_train, y_test = \
        train_test_split(x, y, test_size=0.33, random_state=100)

# training
regTree = DecisionTreeRegressor(random_state=100)
regTree.fit(x_train,y_train)

# test
yhat = regTree.predict(x_test)

# Bagging construction
n_estimators=500
bag = np.empty((n_estimators), dtype=object)
bootstrap_ds_arr = np.empty((n_estimators), dtype=object)
for i in range(n_estimators):
    # sample bootstrapped data set
    ids = np.random.choice(range(0,len(x_train)),size=len(x_train),
                replace=True)
    x_boot = x_train[ids]
    y_boot = y_train[ids]
    bootstrap_ds_arr[i] = np.unique(ids)

    bag[i] = DecisionTreeRegressor()
    bag[i].fit(x_boot,y_boot)

# bagging prediction
yhatbag = np.zeros(len(y_test))
for i in range(n_estimators):
    yhatbag = yhatbag + bag[i].predict(x_test)

yhatbag = yhatbag/n_estimators

# out of bag loss estimation
```

```
oob_pred_arr = np.zeros(len(x_train))
for i in range(len(x_train)):
    x = x_train[i].reshape(1, -1)
    C = []
    for b in range(n_estimators):
        if(np.isin(i, bootstrap_ds_arr[b])==False):
            C.append(b)
    for pred in  bag[C]:
        oob_pred_arr[i] = oob_pred_arr[i] + (pred.predict(x)/len(C))

L_oob = r2_score(y_train, oob_pred_arr)

print("DecisionTreeRegressor R^2 score = ",r2_score(y_test, yhat),
      "\nBagging R^2 score = ", r2_score(y_test, yhatbag),
      "\nBagging OOB R^2 score = ",L_oob)
```

```
DecisionTreeRegressor R^2 score =  0.575438224929718
Bagging R^2 score =  0.7612121189201985
Bagging OOB R^2 score =  0.7758253149069059
```

The decision tree bagging improves the test-set R^2 score by about 32% (from 0.575 to 0.761). Moreover, the OOB score (0.776) is very close to the true generalization risk (0.761) of the bagged estimator. ■

The bagging procedure can be further enhanced by introducing random forests, which is discussed next.

8.6 Random Forests

In Section 8.5, we discussed the intuition behind the prediction averaging procedure. Specifically, for some feature vector $\boldsymbol{x}$ let $Z_b = g_{\mathcal{T}_b}(\boldsymbol{x}), b = 1, 2, \ldots, B$ be iid prediction values, obtained from independent training sets $\mathcal{T}_1, \ldots, \mathcal{T}_B$. Suppose that $\mathbb{V}\text{ar}\, Z_b = \sigma^2$ for all $b = 1, \ldots, B$. Then the variance of the average prediction value $\overline{Z}_B$ is equal to σ^2/B. However, if bootstrapped data sets $\{\mathcal{T}_b^*\}$ are used instead, the corresponding random variables $\{Z_b\}$ will be *correlated*. In particular, $Z_b = g_{\mathcal{T}_b^*}(\boldsymbol{x})$ for $b = 1, \ldots, B$ are identically distributed (but not independent) with some positive pairwise correlation ϱ. It then holds that (see Exercise 9)

$$\mathbb{V}\text{ar}\, \overline{Z}_B = \varrho\, \sigma^2 + \sigma^2 \frac{(1-\varrho)}{B}. \tag{8.17}$$

While the second term of (8.17) goes to zero as the number of observation B increases, the first term remains constant.

This issue is particularly relevant for bagging with decision trees. For example, consider a situation in which there exists a feature that provides a very good split of the data. Such a feature will be selected and split for every $\{g_{\mathcal{T}_b^*}\}_{b=1}^B$ at the root level and we will consequently end up with highly correlated predictions. In such a situation, prediction averaging will not introduce the desired improvement in the performance of the bagged predictor.

The major idea of random forests is to perform bagging in combination with a "decorrelation" of the trees by including only a subset of features during the tree construction. For each bootstrapped training set $\mathcal{T}_b^*$ we build a decision tree using a randomly selected subset of $m \leqslant p$ features for the splitting rules. This simple but powerful idea will decorrelate the trees, since strong predictors will have a smaller chance to be considered at the root levels.

Consequentially, we can expect to improve the predictive performance of the bagged estimator. The resulting predictor (random forest) construction is summarized in Algorithm 8.6.1.

Algorithm 8.6.1: Random Forest Construction

Input: Training set $\tau = \{(\boldsymbol{x}_i, y_i)\}_{i=1}^n$, the number of trees in the forest B, and the number $m \leqslant p$ of features to be included, where p is the total number of features in $\boldsymbol{x}$.

Output: Ensemble of trees.

1 Generate bootstrapped training sets $\{\mathcal{T}_1^*, \dots, \mathcal{T}_B^*\}$ via Algorithm 8.5.1.
2 **for** $b = 1$ **to** B **do**
3 Train a decision tree $g_{\mathcal{T}_b^*}$ via Algorithm 8.2.1, where each split is performed using m randomly selected features out of p.
4 **return** $\{g_{\mathcal{T}_b^*}\}_{b=1}^B$.

For regression problems, the output of Algorithm 8.6.1 is combined to yield the random forest prediction function:

$$g_{\mathrm{RF}}(\boldsymbol{x}) = \frac{1}{B} \sum_{b=1}^{B} g_{\mathcal{T}_b^*}(\boldsymbol{x}).$$

In the classification setting, similar to Remark 8.1, we take instead the majority vote from the $\{g_{\mathcal{T}_b^*}\}$.

■ **Example 8.4 (Random Forest for a Regression Tree)** We continue with the basic bagging Example 8.3 for a regression tree, in which we compared the decision tree estimator with the corresponding bagged estimator. Here, however, we use the random forest with $B = 500$ trees and a subset size $m = 8$. It can be seen that the random forest's R^2 score is outperforming that of the bagged estimator.

BaggingExampleRF.py

```
from sklearn.datasets import make_friedman1
from sklearn.model_selection import train_test_split
from sklearn.metrics import r2_score
from sklearn.ensemble import RandomForestRegressor

# create regression problem
n_points = 1000 # points
x, y =  make_friedman1(n_samples=n_points, n_features=15,
                       noise=1.0, random_state=100)
# split to train/test set
x_train, x_test, y_train, y_test = \
        train_test_split(x, y, test_size=0.33, random_state=100)
rf = RandomForestRegressor(n_estimators=500, oob_score = True,
```

```
    max_features=8,random_state=100)
rf.fit(x_train,y_train)
yhatrf = rf.predict(x_test)

print("RF R^2 score = ", r2_score(y_test, yhatrf),
      "\nRF OOB R^2 score = ", rf.oob_score_)
```

```
RF R^2 score =  0.8106589580845707
RF OOB R^2 score =  0.8260541058404149
```

■

■ **Remark 8.3 (The Optimal Number of Subset Features m)** The default values for m are $\lfloor p/3 \rfloor$ and $\lfloor \sqrt{p} \rfloor$ for regression and classification setting, respectively. However, the standard practice is to treat m as a hyperparameter that requires tuning, depending on the specific problem at hand [55]. ■

Note that the procedure of bagging decision trees is a special case of a random forest construction (see Exercise 11). Consequently, the OOB loss is readily available for random forests.

While the advantage of bagging in the sense of enhanced accuracy is clear, we should also consider its negative aspects and, in particular, the loss of interpretability. Specifically a random forest consists of many trees, thus making the prediction process both hard to visualize and interpret. For example, given a random forest, it is not easy to determine a subset of features that are essential for accurate prediction.

The feature importance measure intends to address this issue. The idea is as follows. Each *internal* node of a decision tree induces a certain decrease in the training loss; see (8.9). Let us denote this decrease in the training loss by $\Delta_{\text{Loss}}(v)$, where v is not a leaf node of $\mathbb{T}$. In addition, recall that for splitting rules of the type $\mathbb{1}\{x_j \leqslant \xi\}$ $(1 \leqslant j \leqslant p)$, each node v is associated with a feature x_j that determines the split. Using the above definitions, we can define the *feature importance* of x_j as

FEATURE IMPORTANCE

$$\mathcal{I}_{\mathbb{T}}(x_j) = \sum_{v \text{ internal } \in \mathbb{T}} \Delta_{\text{Loss}}(v)\ \mathbb{1}\{x_j \text{ is associated with } v\}, \quad 1 \leqslant j \leqslant p. \tag{8.18}$$

While (8.18) is defined for a single tree, it can be readily extended to random forests. Specifically, the feature importance in that case will be averaged over all trees of the forest; that is, for a forest consisting of B trees $\{\mathbb{T}_1, \ldots, \mathbb{T}_B\}$, the feature importance measure is:

$$\mathcal{I}_{\text{RF}}(x_j) = \frac{1}{B} \sum_{b=1}^{B} \mathcal{I}_{\mathbb{T}_b}(x_j), \quad 1 \leqslant j \leqslant p. \tag{8.19}$$

■ **Example 8.5 (Feature Importance)** We consider a classification problem with 15 features. The data is specifically designed to contain only 5 informative features out of 15. In the code below, we apply the random forest procedure and calculate the corresponding feature importance measures, which are summarized in Figure 8.9.

VarImportance.py

```
import numpy as np
from sklearn.datasets import make_classification
from sklearn.ensemble import RandomForestClassifier
import matplotlib.pyplot as plt, pylab

n_points = 1000 # create regression data with 1000 data points
x, y =  make_classification(n_samples=n_points, n_features=15,
  n_informative=5, n_redundant=0, n_repeated=0, random_state=100,
      shuffle=False)

rf = RandomForestClassifier(n_estimators=200, max_features="log2")
rf.fit(x,y)

importances = rf.feature_importances_
indices = np.argsort(importances)[::-1]

for f in range(15):
    print("Feature %d (%f)" % (indices[f]+1, importances[indices[f
        ]]))

std = np.std([rf.feature_importances_ for tree in rf.estimators_],
             axis=0)
f = plt.figure()
plt.bar(range(x.shape[1]), importances[indices],
        color="b", yerr=std[indices], align="center")
plt.xticks(range(x.shape[1]), indices+1)
plt.xlim([-1, x.shape[1]])
pylab.xlabel("feature index")
pylab.ylabel("importance")
plt.show()
```

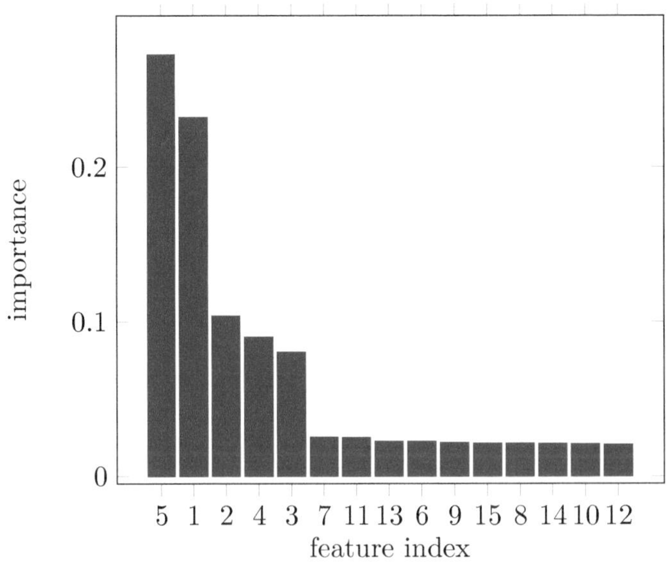

Figure 8.9: Importance measure for the 15-feature data set with only 5 informative features x_1, x_2, x_3, x_4, and x_5.

Clearly, it is hard to visualize and understand the prediction process based on 200 trees. However, Figure 8.9 shows that the features x_1, x_2, x_3, x_4, and x_5 were correctly identified as being important. ■

8.7 Boosting

Boosting is a powerful idea that aims to improve the accuracy of any learning algorithm, especially when involving *weak learners* — simple prediction functions that exhibit performance slightly better than random guessing. Shallow decision trees typically yield weak learners. WEAK LEARNERS

Originally, boosting was developed for binary classification tasks, but it can be readily extended to handle general classification and regression problems. The boosting approach has some similarity with the bagging method in the sense that boosting uses an ensemble of prediction functions. Despite this similarity, there exists a fundamental difference between these methods. Specifically, while bagging involves the fitting of prediction functions to bootstrapped data, the predicting functions in boosting are learned *sequentially*. That is, each learner uses information from previous learners.

The idea is to start with a simple model (weak learner) g_0 for the data $\tau = \{(\boldsymbol{x}_i, y_i)\}_{i=1}^n$ and then to improve or "boost" this learner to a learner $g_1 := g_0 + h_1$. Here, the function h_1 is found by minimizing the training loss for $g_0 + h_1$ over all functions h in some class of functions $\mathcal{H}$. For example, $\mathcal{H}$ could be the set of prediction functions that can be obtained via a decision tree of maximal depth 2. Given a loss function Loss, the function h_1 is thus obtained as the solution to the optimization problem

$$h_1 = \underset{h \in \mathcal{H}}{\operatorname{argmin}} \frac{1}{n} \sum_{i=1}^{n} \operatorname{Loss}\left(y_i, g_0(\boldsymbol{x}_i) + h\left(\boldsymbol{x}_i\right)\right). \tag{8.20}$$

This process can be repeated for g_1 to obtain $g_2 = g_1 + h_2$, and so on, yielding the boosted prediction function

$$g_B(\boldsymbol{x}) = g_0(\boldsymbol{x}) + \sum_{b=1}^{B} h_b(\boldsymbol{x}). \tag{8.21}$$

Instead of using the updating step $g_b = g_{b-1} + h_b$, one prefers to use the smooth updating step $g_b = g_{b-1} + \gamma h_b$, for some suitably chosen step-size parameter γ. As we shall see shortly, this helps reduce overfitting.

Boosting can be used for regression and classification problems. We start with a simple regression setting, using the squared-error loss; thus, $\operatorname{Loss}(y, \widehat{y}) = (y - \widehat{y})^2$. In this case, it is common to start with $g_0(\boldsymbol{x}) = n^{-1} \sum_{i=1}^n y_i$, and each h_b for $b = 1, \ldots, B$ is chosen as a learner for the data set τ_b of residuals corresponding to g_{b-1}. That is, $\tau_b := \left\{\left(\boldsymbol{x}_i, e_i^{(b)}\right)\right\}_{i=1}^n$, with

$$e_i^{(b)} := y_i - g_{b-1}(\boldsymbol{x}_i). \tag{8.22}$$

This leads to the following boosting procedure for regression with squared-error loss.

Algorithm 8.7.1: Regression Boosting with Squared-Error Loss

Input: Training set $\tau = \{(\boldsymbol{x}_i, y_i)\}_{i=1}^n$, the number of boosting rounds B, and a shrinkage step-size parameter γ.

Output: Boosted prediction function.

Set $g_0(\boldsymbol{x}) \leftarrow n^{-1} \sum_{i=1}^n y_i$.
for $b = 1$ **to** B **do**
  Set $e_i^{(b)} \leftarrow y_i - g_{b-1}(\boldsymbol{x}_i)$ for $i = 1, \ldots, n$, and let $\tau_b \leftarrow \left\{\left(\boldsymbol{x}_i, e_i^{(b)}\right)\right\}_{i=1}^n$.
  Fit a prediction function h_b on the training data τ_b.
  Set $g_b(\boldsymbol{x}) \leftarrow g_{b-1}(\boldsymbol{x}) + \gamma\, h_b(\boldsymbol{x})$.
return g_B.

STEP-SIZE PARAMETER γ

The *step-size parameter* γ introduced in Algorithm 8.7.1 controls the speed of the fitting process. Specifically, for small values of γ, boosting takes smaller steps towards the training loss minimization. The step-size γ is of great practical importance, since it helps the boosting algorithm to avoid overfitting. This phenomenon is demonstrated in Figure 8.10.

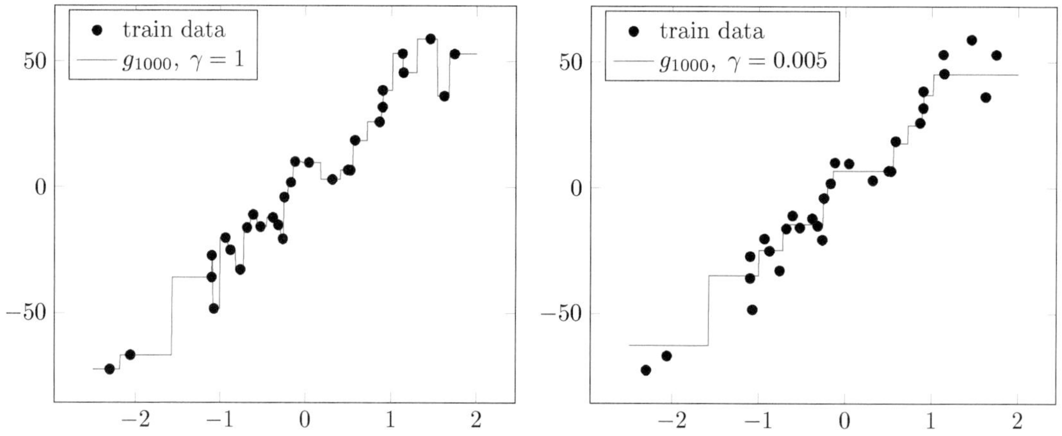

Figure 8.10: The left and the right panels show the fitted boosting regression model g_{1000} with $\gamma = 1.0$ and $\gamma = 0.005$, respectively. Note the overfitting on the left.

A very basic implementation of Algorithm 8.7.1 which reproduces Figure 8.10 is provided below.

RegressionBoosting.py

```
import numpy as np
from sklearn.tree import DecisionTreeRegressor
from sklearn.model_selection import train_test_split
from sklearn.datasets import make_regression
import matplotlib.pyplot as plt


def TrainBoost(alpha,BoostingRounds,x,y):
    g_0 = np.mean(y)
```

```
    residuals  = y-alpha*g_0

    # list of basic regressor
    g_boost = []

    for i in range(BoostingRounds):
        h_i = DecisionTreeRegressor(max_depth=1)
        h_i.fit(x,residuals)
        residuals = residuals  - alpha*h_i.predict(x)
        g_boost.append(h_i)

    return g_0, g_boost

def Predict(g_0, g_boost,alpha, x):
    yhat = alpha*g_0*np.ones(len(x))
    for j in range(len(g_boost)):
        yhat = yhat+alpha*g_boost[j].predict(x)

    return yhat


np.random.seed(1)
sz = 30

# create data set
x,y = make_regression(n_samples=sz, n_features=1, n_informative=1,
   noise=10.0)

# boosting algorithm
BoostingRounds = 1000
alphas = [1, 0.005]

for alpha in alphas:
    g_0, g_boost = TrainBoost(alpha,BoostingRounds,x,y)
    yhat = Predict(g_0, g_boost, alpha, x)

    # plot
    tmpX =  np.reshape(np.linspace(-2.5,2,1000),(1000,1))
    yhatX = Predict(g_0, g_boost, alpha, tmpX)
    f = plt.figure()
    plt.plot(x,y,'*')
    plt.plot(tmpX,yhatX)
    plt.show()
```

The parameter γ can be viewed as a step size made in the direction of the negative gradient of the squared-error training loss. To see this, note that the negative gradient

$$-\left.\frac{\partial \operatorname{Loss}(y_i, z)}{\partial z}\right|_{z=g_{b-1}(\boldsymbol{x}_i)} = -\left.\frac{\partial\,(y_i - z)^2}{\partial z}\right|_{z=g_{b-1}(\boldsymbol{x}_i)} = 2(y_i - g_{b-1}(\boldsymbol{x}_i))$$

is two times the residual $e_i^{(b)}$ given in (8.22) that is used in Algorithm 8.7.1 to fit the prediction function h_b.

In fact, one of the major advances in the theory of boosting was the recognition that one can use a similar gradient descent method for any differentiable loss function. The

GRADIENT BOOSTING

☞ 414

resulting algorithm is called *gradient boosting*. The general gradient boosting algorithm is summarized in Algorithm 8.7.2. The main idea is to mimic a gradient descent algorithm in the following sense. At each stage of the boosting procedure, we calculate a negative gradient on n training points $\boldsymbol{x}_1, \ldots, \boldsymbol{x}_n$ (Lines 3–4). Then, we fit a simple model (such as a shallow decision tree) to approximate the gradient (Line 5) for any feature $\boldsymbol{x}$. Finally, similar to the gradient descent method, we make a γ-sized step in the direction of the negative gradient (Line 6).

Algorithm 8.7.2: Gradient Boosting

Input: Training set $\tau = \{(\boldsymbol{x}_i, y_i)\}_{i=1}^n$, the number of boosting rounds B, a differentiable loss function $\text{Loss}(y, \widehat{y})$, and a gradient step-size parameter γ.

Output: Gradient boosted prediction function.

1 Set $g_0(\boldsymbol{x}) \leftarrow 0$.

2 **for** $b = 1$ **to** B **do**

3   **for** $i = 1$ **to** n **do**

4   Evaluate the negative gradient of the loss at $(\boldsymbol{x}_i, y_i)$ via

$$r_i^{(b)} \leftarrow - \left. \frac{\partial \, \text{Loss}\,(y_i, z)}{\partial z} \right|_{z = g_{b-1}(\boldsymbol{x}_i)} \qquad i = 1, \ldots, n.$$

5   Approximate the negative gradient by solving

$$h_b = \underset{h \in \mathcal{H}}{\text{argmin}} \, \frac{1}{n} \sum_{i=0}^{n} \left(r_i^{(b)} - h\,(\boldsymbol{x}_i) \right)^2. \tag{8.23}$$

6   Set $g_b(\boldsymbol{x}) \leftarrow g_{b-1}(\boldsymbol{x}) + \gamma \, h_b(\boldsymbol{x})$.

7 **return** g_B

■ **Example 8.6 (Gradient Boosting for a Regression Tree)** Let us continue with the basic bagging and random forest examples for a regression tree (Examples 8.3 and 8.4), where we compared the standard decision tree estimator with the corresponding bagging and random forest estimators. Now, we use the gradient boosting estimator from Algorithm 8.7.2, as implemented in `sklearn`. We use $\gamma = 0.1$ and perform $B = 100$ boosting rounds. As a prediction function h_b for $b = 1, \ldots, B$ we use small regression trees of depth at most 3. Note that such individual trees do not usually give good performance; that is, they are weak prediction functions. We can see that the resulting boosting prediction function gives the R^2 score equal to 0.899, which is better than R^2 scores of simple decision tree (0.5754), the bagged tree (0.761), and the random forest (0.8106).

GradientBoostingRegression.py

```
import numpy as np
from sklearn.datasets import make_friedman1
from sklearn.tree import DecisionTreeRegressor
from sklearn.model_selection import train_test_split
from sklearn.metrics import r2_score
```

```
# create regression problem
n_points = 1000 # points
x, y =  make_friedman1(n_samples=n_points, n_features=15,
                       noise=1.0, random_state=100)

# split to train/test set
x_train, x_test, y_train, y_test = \
        train_test_split(x, y, test_size=0.33, random_state=100)

# boosting sklearn
from sklearn.ensemble import GradientBoostingRegressor

breg = GradientBoostingRegressor(learning_rate=0.1,
            n_estimators=100, max_depth =3, random_state=100)
breg.fit(x_train,y_train)
yhat = breg.predict(x_test)
print("Gradient Boosting R^2 score = ",r2_score(y_test, yhat))
```

```
Gradient Boosting R^2 score =  0.8993055635639531
```

■

We proceed with the classification setting and consider the original boosting algorithm: *AdaBoost*. The inventors of the AdaBoost method considered a binary classification problem, where the response variable belongs to the $\{-1, 1\}$ set. The idea of AdaBoost is similar to the one presented in the regression setting, that is, AdaBoost fits a sequence of prediction functions $g_0, g_1 = g_0 + h_1, g_2 = g_0 + h_1 + h_2, \ldots$ with final prediction function ADABOOST

$$g_B(\boldsymbol{x}) = g_0(\boldsymbol{x}) + \sum_{b=1}^{B} h_b(\boldsymbol{x}), \tag{8.24}$$

where each function h_b is of the form $h_b(\boldsymbol{x}) = \alpha_b \, c_b(\boldsymbol{x})$, with $\alpha_b \in \mathbb{R}_+$ and where c_b is a proper (but weak) classifier in some class C. Thus, $c_b(\boldsymbol{x}) \in \{-1, 1\}$. Exactly as in (8.20), we solve at each boosting iteration the optimization problem

$$(\alpha_b, c_b) = \underset{\alpha \geqslant 0,\, c \in C}{\operatorname{argmin}} \frac{1}{n} \sum_{i=1}^{n} \operatorname{Loss}\left(y_i, g_{b-1}(\boldsymbol{x}_i) + \alpha\, c(\boldsymbol{x}_i)\right). \tag{8.25}$$

However, in this case the loss function is defined as $\operatorname{Loss}(y, \widehat{y}) = \mathrm{e}^{-y\widehat{y}}$. The algorithm starts with a simple model $g_0 := 0$ and for each successive iteration $b = 1, \ldots, B$ solves (8.25). Thus,

$$(\alpha_b, c_b) = \underset{\alpha \geqslant 0,\, c \in C}{\operatorname{argmin}} \sum_{i=1}^{n} \underbrace{\mathrm{e}^{-y_i\, g_{b-1}(\boldsymbol{x}_i)}}_{w_i^{(b)}} \mathrm{e}^{-y_i\, \alpha\, c(\boldsymbol{x}_i)} = \underset{\alpha \geqslant 0,\, c \in C}{\operatorname{argmin}} \sum_{i=1}^{n} w_i^{(b)} \, \mathrm{e}^{-y_i \alpha\, c(\boldsymbol{x}_i)},$$

where $w_i^{(b)} := \exp\{-y_i\, g_{b-1}(\boldsymbol{x}_i)\}$ does not depend on α or c. It follows that

$$\begin{aligned}(\alpha_b, c_b) &= \underset{\alpha \geqslant 0,\, c \in C}{\operatorname{argmin}}\, \mathrm{e}^{-\alpha} \sum_{i=1}^{n} w_i^{(b)} \mathbb{1}\{c(\boldsymbol{x}_i) = y_i\} + \mathrm{e}^{\alpha} \sum_{i=1}^{n} w_i^{(b)} \mathbb{1}\{c(\boldsymbol{x}_i) \neq y_i\} \\ &= \underset{\alpha \geqslant 0,\, c \in C}{\operatorname{argmin}}\, (\mathrm{e}^{\alpha} - \mathrm{e}^{-\alpha})\, \ell_\tau^{(b)}(c) + \mathrm{e}^{-\alpha},\end{aligned} \tag{8.26}$$

where

$$\ell_{\tau}^{(b)}(c) := \frac{\sum_{i=1}^{n} w_i^{(b)} \, \mathbb{1}\{c(\boldsymbol{x}_i) \neq y_i\}}{\sum_{i=1}^{n} w_i^{(b)}}$$

can be interpreted as the weighted zero–one training loss at iteration b.

For any $\alpha \geqslant 0$, the program (8.26) is minimized by a classifier $c \in C$ that minimizes this weighted training loss; that is,

$$c_b(\boldsymbol{x}) = \underset{c \in C}{\operatorname{argmin}} \, \ell_{\tau}^{(b)}. \tag{8.27}$$

Substituting (8.27) into (8.26) and solving for the optimal α gives

$$\alpha_b = \frac{1}{2} \ln \left(\frac{1 - \ell_{\tau}^{(b)}(c_b)}{\ell_{\tau}^{(b)}(c_b)} \right). \tag{8.28}$$

This gives the AdaBoost algorithm, summarized below.

Algorithm 8.7.3: AdaBoost

Input: Training set $\tau = \{(\boldsymbol{x}_i, y_i)\}_{i=1}^{n}$, and the number of boosting rounds B.
Output: AdaBoost prediction function.

```
Set g_0(x) ← 0.
for i = 1 to n do
    w_i^(1) ← 1/n
for b = 1 to B do
    Fit a classifier c_b on the training set τ by solving

        c_b = argmin_{c∈C} ℓ_τ^(b)(c) = argmin_{c∈C} ( Σ_{i=1}^n w_i^(b) 1{c(x_i) ≠ y_i} ) / ( Σ_{i=1}^n w_i^(b) ).

    Set α_b ← ½ ln( (1 − ℓ_τ^(b)(c_b)) / ℓ_τ^(b)(c_b) ).      // Update weights
    for i = 1 to n do
        w_i^(b+1) ← w_i^(b) exp{−y_i α_b c_b(x_i)}.
return g_B(x) := Σ_{b=1}^B α_b c_b(x).
```

Algorithm 8.7.3 is quite intuitive. At the first step ($b = 1$), AdaBoost assigns an equal weight $w_i^{(1)} = 1/n$ to each training sample $(\boldsymbol{x}_i, y_i)$ in the set $\tau = \{(\boldsymbol{x}_i, y_i)\}_{i=1}^{n}$. Note that, in this case, the weighted zero–one training loss is equal to the regular zero–one training loss. At each successive step $b > 1$, the weights of observations that were incorrectly classified by the previous boosting prediction function g_b are increased, and the weights of correctly classified observations are decreased. Due to the use of the weighted zero–one loss, the set of incorrectly classified training samples will receive an extra weight and thus have a better chance of being classified correctly by the next classifier c_{b+1}. As soon as the AdaBoost algorithm finds the prediction function g_B, the final classification is delivered via

$$\operatorname{sign}\left(\sum_{b=1}^{B} \alpha_b \, c_b(\boldsymbol{x}) \right).$$

> ⚠ The step-size parameter α_b found by the AdaBoost algorithm in Line 6 can be viewed as an optimal step-size in the sense of training loss minimization. However, similar to the regression setting, one can slow down the AdaBoost algorithm by setting α_b to be a fixed (small) value $\alpha_b = \gamma$. As usual, when the latter is done in practice, it is tackling the problem of overfitting.

We consider an implementation of Algorithm 8.7.3 for a binary classification problem. Specifically, during all boosting rounds, we use simple decision trees of depth 1 (also called decision tree *stumps*) as weak learners. The exponential and zero–one training losses as a function of the number of boosting rounds are presented in Figure 8.11.

STUMPS

AdaBoost.py

```
from sklearn.datasets import make_blobs
from sklearn.tree import DecisionTreeClassifier
from sklearn.model_selection import train_test_split
from sklearn.metrics import zero_one_loss
import numpy as np


def ExponentialLoss(y,yhat):
    n = len(y)
    loss = 0
    for i in range(n):
        loss = loss+np.exp(-y[i]*yhat[i])
    loss = loss/n
    return loss

# create binary classification problem
np.random.seed(100)

n_points = 100 # points
x, y =  make_blobs(n_samples=n_points, n_features=5,  centers=2,
                        cluster_std=20.0, random_state=100)
y[y==0]=-1

# AdaBoost implementation
BoostingRounds = 1000
n = len(x)
W = 1/n*np.ones(n)

Learner = []
alpha_b_arr = []

for i in range(BoostingRounds):
    clf = DecisionTreeClassifier(max_depth=1)
    clf.fit(x,y, sample_weight=W)

    Learner.append(clf)

    train_pred = clf.predict(x)
    err_b = 0
```

```
    for i in range(n):
        if(train_pred[i]!=y[i]):
            err_b = err_b+W[i]

    err_b = err_b/np.sum(W)
    alpha_b = 0.5*np.log((1-err_b)/err_b)
    alpha_b_arr.append(alpha_b)

    for i in range(n):
        W[i] = W[i]*np.exp(-y[i]*alpha_b*train_pred[i])

yhat_boost = np.zeros(len(y))

for j in range(BoostingRounds):
    yhat_boost = yhat_boost+alpha_b_arr[j]*Learner[j].predict(x)


yhat = np.zeros(n)
yhat[yhat_boost>=0] = 1
yhat[yhat_boost<0] = -1
print("AdaBoost Classifier exponential loss = ", ExponentialLoss(y,
   yhat_boost))
print("AdaBoost Classifier zero--one loss = ",zero_one_loss(y,yhat))
```

```
AdaBoost Classifier exponential loss =  0.004224013663777142
AdaBoost Classifier zero--one loss =  0.0
```

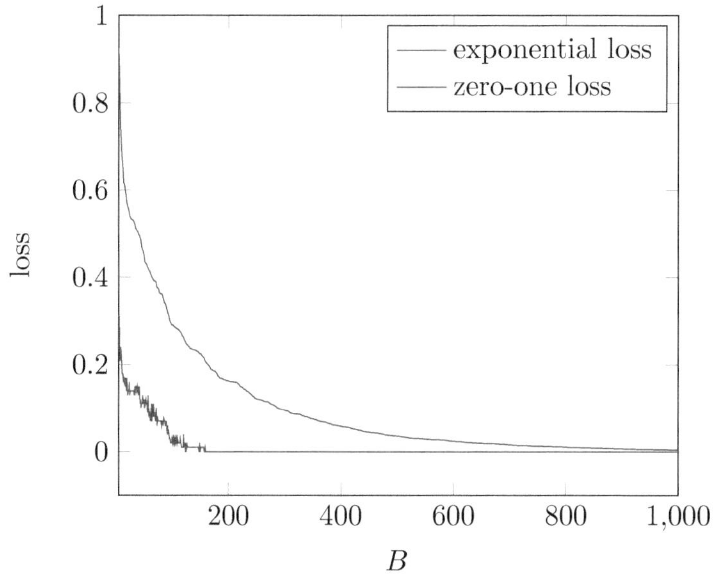

Figure 8.11: Exponential and zero–one training loss as a function of the number of boosting rounds B for a binary classification problem.

Further Reading

Breiman's book on decision trees, [20], serves as a great starting point. Some additional advances can be found in [62, 96]. From the computational point of view, there exists an efficient recursive procedure for tree pruning; see Chapters 3 and 10 in [20]. Several advantages and disadvantages of using decision trees are debated in [37, 55]. A detailed discussion on bagging and random forests can be found in [21] and [23], respectively. Freund and Schapire [44] provide the first boosting algorithm, the AdaBoost. While AdaBoost was developed in the context of the computational complexity of learning, it was later discovered by Friedman [45] that AdaBoost is a special case of an additive model. In addition, it was shown that for any differentiable loss function, there exists an efficient boosting procedure which mimics the gradient descent algorithm. The foundation of the resulting gradient boosting method is detailed in [45]. Python packages that implement gradient boosting include `XGBoost` and `LightGBM`.

Exercises

1. Show that any training set $\tau = \{(\boldsymbol{x}, y_i), i = 1, \ldots, n\}$ can be fitted via a tree with zero training loss.

2. Suppose during the construction of a decision tree we wish to specify a constant regional prediction function g^w on the region $\mathcal{R}_w$, based on the training data in $\mathcal{R}_w$, say $\{(\boldsymbol{x}_1, y_1), \ldots, (\boldsymbol{x}_k, y_k)\}$. Show that $g^w(\boldsymbol{x}) := k^{-1} \sum_{i=1}^{k} y_i$ minimizes the squared-error loss.

3. Using the program from Section 8.2.4, write a basic implementation of a decision tree for a binary classification problem. Implement the misclassification, Gini index, and entropy impurity criteria to split nodes. Compare the results.

4. Suppose in the decision tree of Example 8.1, there are 3 blue and 2 red data points in a certain tree region. Calculate the misclassification impurity, the Gini impurity, and the entropy impurity. Repeat these calculations for 2 blue and 3 red data points.

5. Consider the procedure of finding the best splitting rule for a categorical variable with k labels from Section 8.3.4. Show that one needs to consider 2^k subsets of $\{1, \ldots, k\}$ to find the optimal partition of labels.

6. Reproduce Figure 8.6 using the following classification data.

```
from sklearn.datasets import make_blobs
X, y =  make_blobs(n_samples=5000, n_features=10, centers=3,
                   random_state=10, cluster_std=10)
```

7. Prove (8.13); that is, show that

$$\sum_{w \in \mathcal{W}} \left(\sum_{i=1}^{n} \mathbb{1}\{\boldsymbol{x}_i \in \mathcal{R}_w\} \operatorname{Loss}(y_i, g^w(\boldsymbol{x}_i)) \right) = n\, \ell_\tau(g)\,.$$

8. Suppose τ is a training set with n elements and τ^*, also of size n, is obtained from τ by bootstrapping; that is, resampling with replacement. Show that for large n, τ^* does not contain a fraction of about $e^{-1} \approx 0.37$ of the points from τ.

9. Prove Equation (8.17).

10. Consider the following training/test split of the data. Construct a random forest regressor and identify the optimal subset size m in the sense of R^2 score (see Remark 8.3).

```
import numpy as np
from sklearn.datasets import make_friedman1
from sklearn.tree import DecisionTreeRegressor
from sklearn.model_selection import train_test_split
from sklearn.metrics import r2_score

# create regression problem
n_points = 1000 # points
x, y =  make_friedman1(n_samples=n_points, n_features=15,
                       noise=1.0, random_state=100)

# split to train/test set
x_train, x_test, y_train, y_test = \
        train_test_split(x, y, test_size=0.33, random_state=100)
```

11. Explain why bagging decision trees are a special case of random forests.

12. Show that (8.28) holds.

13. Consider the following classification data and module imports:

```
from sklearn.datasets import make_blobs
from sklearn.metrics import zero_one_loss
from sklearn.model_selection import train_test_split
import numpy as np
import matplotlib.pyplot as plt
from sklearn.ensemble import GradientBoostingClassifier

X_train, y_train =  make_blobs(n_samples=5000, n_features=10,
   centers=3, random_state=10, cluster_std=5)
```

Using the gradient boosting algorithm with $B = 100$ rounds, plot the training loss as a function of γ, for $\gamma = 0.1, 0.3, 0.5, 0.7, 1$. What is your conclusion regarding the relation between B and γ?

CHAPTER 9

DEEP LEARNING

In this chapter, we show how one can construct a rich class of approximating functions called neural networks. The learners belonging to the neural-network class of functions have attractive properties that have made them ubiquitous in modern machine learning applications — their training is computationally feasible and their complexity is easy to control and fine-tune.

9.1 Introduction

In Chapter 2 we described the basic supervised learning task; namely, we wish to predict a random output Y from a random input $\boldsymbol{X}$, using a prediction function $g : \boldsymbol{x} \mapsto y$ that belongs to a suitably chosen class of approximating functions $\mathcal{G}$. More generally, we may wish to predict a vector-valued output $\boldsymbol{y}$ using a prediction function $\boldsymbol{g} : \boldsymbol{x} \mapsto \boldsymbol{y}$ from class $\mathcal{G}$.

⚠ In this chapter $\boldsymbol{y}$ denotes the vector-valued output for a given input $\boldsymbol{x}$. This differs from our previous use (e.g., in Table 2.1), where $\boldsymbol{y}$ denotes a vector of scalar outputs.

In the machine learning context, the class $\mathcal{G}$ is sometimes referred to as the *hypothesis space* or the *universe of possible models*, and the *representational capacity* of a hypothesis space $\mathcal{G}$ is simply its complexity. REPRESENTATIONAL CAPACITY

Suppose that we have a class of functions $\mathcal{G}_L$, indexed by a parameter L that controls the complexity of the class, so that $\mathcal{G}_L \subset \mathcal{G}_{L+1} \subset \mathcal{G}_{L+2} \subset \cdots$. In selecting a suitable class of functions, we have to be mindful of the *approximation–estimation tradeoff*. On the one hand, the class $\mathcal{G}_L$ must be complex (rich) enough to accurately represent the optimal unknown prediction function g^*, which may require a very large L. On the other hand, the learners in the class $\mathcal{G}_L$ must be simple enough to train with small estimation error and with minimal demands on computer memory, which may necessitate a small L. ☞ 31

In balancing these competing objectives, it helps if the more complex class $\mathcal{G}_{L+1}$ is easily constructed from an already existing and simpler $\mathcal{G}_L$. The simpler class of functions $\mathcal{G}_L$ may itself be constructed by modifying an even simpler class $\mathcal{G}_{L-1}$, and so on.

A class of functions that permits such a natural hierarchical construction is the class of *neural networks*. Conceptually, a neural network with L layers is a nonlinear parametric regression model whose representational capacity can easily be controlled by L. NEURAL NETWORKS ☞ 188

Alternatively, in (9.3) we will define the output of a neural network as the repeated composition of linear and (componentwise) nonlinear functions. As we shall see, this representation of the output will provide a flexible class of nonlinear functions that can be easily differentiated. As a result, the training of learners via gradient optimization methods involves mostly standard matrix operations that can be performed very efficiently. ☞ 414

Historically, neural networks were originally intended to mimic the workings of the human brain, with the network nodes modeling neurons and the network links modeling the axons connecting neurons. For this reason, rather than using the terminology of the regression models in Chapter 5, we prefer to use a nomenclature inspired by the apparent resemblance of neural networks to structures in the human brain.

We note, however, that the attempts at building efficient machine learning algorithms by mimicking the functioning of the human brain have been as unsuccessful as the attempts at building flying aircraft by mimicking the flapping of birds' wings. Instead, many effective machine algorithms have been inspired by age-old mathematical ideas for function approximation. One such idea is the following fundamental result (see [119] for a proof).

Theorem 9.1: Kolmogorov (1957)

Every continuous function $g^* : [0,1]^p \mapsto \mathbb{R}$ with $p \geqslant 2$ can be written as

$$g^*(\boldsymbol{x}) = \sum_{j=1}^{2p+1} h_j\left(\sum_{i=1}^{p} h_{ij}(x_i)\right),$$

where $\{h_j, h_{ij}\}$ is a set of univariate continuous functions that depend on g^*.

This result tells us that any continuous high-dimensional map can be represented as the function composition of much simpler (one-dimensional) maps. The composition of the maps needed to compute the output $g^*(\boldsymbol{x})$ for a given input $\boldsymbol{x} \in \mathbb{R}^p$ are depicted in Figure 9.1, showing a directed graph or *neural network* with three layers, denoted as $l = 0, 1, 2$.

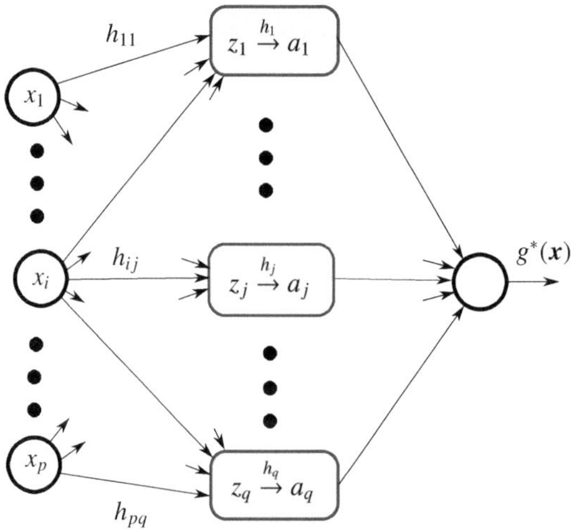

Figure 9.1: Every continuous function $g^* : [0,1]^p \mapsto \mathbb{R}$ can be represented by a neural network with one hidden layer ($l = 1$), an input layer ($l = 0$), and an output layer ($l = 2$).

In particular, each of the p components of the input $\boldsymbol{x}$ is represented as a node in the *input layer* ($l = 0$). In the *hidden layer* ($l = 1$) there are $q := 2p + 1$ nodes, each of which is associated with a pair of variables (z, a) with values

HIDDEN LAYER

$$z_j := \sum_{i=1}^{p} h_{ij}(x_i) \quad \text{and} \quad a_j := h_j(z_j).$$

A link between nodes (z_j, a_j) and x_i with weight h_{ij} signifies that the value of z_j depends on the value of x_i via the function h_{ij}. Finally, the *output layer* ($l = 2$) represents the value $g^*(\boldsymbol{x}) = \sum_{j=1}^{q} a_j$. Note that the arrows on the graph remind us that the sequence of the computations is executed from left to right, or from the input layer $l = 0$ through to the output layer $l = 2$.

In practice, we do not know the collection of functions $\{h_j, h_{ij}\}$, because they depend on the unknown g^*. In the unlikely event that g^* is linear, then all of the $(2p + 1)(p + 1)$ one-dimensional functions will be linear as well. However, in general, we should expect that each of the functions in $\{h_j, h_{ij}\}$ is nonlinear.

Unfortunately, Theorem 9.1 only asserts the existence of $\{h_j, h_{ij}\}$, and does not tell us how to construct these nonlinear functions. One way out of this predicament is to replace these $(2p + 1)(p + 1)$ unknown functions with a much larger number of known nonlinear functions called *activation functions*.[1] For example, a logistic activation function is

ACTIVATION FUNCTIONS

$$S(z) = (1 + \exp(-z))^{-1}.$$

We then hope that such a network, built from a sufficiently large number of activation functions, will have similar representational capacity as the neural network in Figure 9.1 with $(2p + 1)(p + 1)$ functions.

In general, we wish to use the simplest activation functions that will allow us to build a learner with large representational capacity and low training cost. The logistic function is merely one possible choice for an activation function from among infinite possibilities. Figure 9.2 shows a small selection of activation functions with different regularity or smoothness properties.

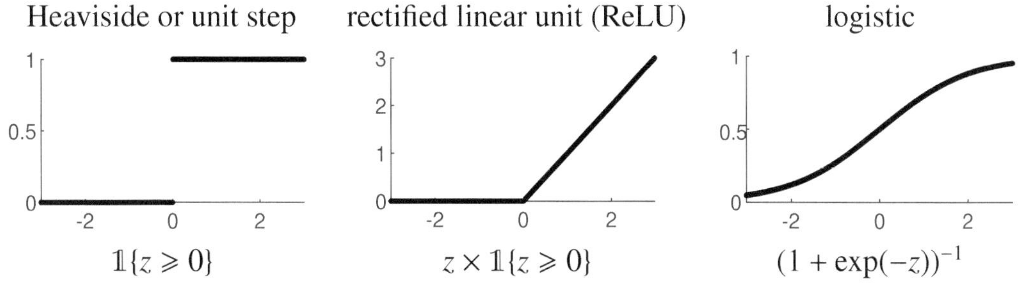

Figure 9.2: Some common activation functions $S(z)$ with their defining formulas and plots. The logistic function is an example of a *sigmoid* (that is, an S-shaped) function. Some books define the logistic function as $2S(z) - 1$ (in terms of our definition).

In addition to choosing the type and number of activation functions in a neural network, we can improve its representational capacity in another important way: introduce more hidden layers. In the next section we explore this possibility in detail.

[1]Activation functions derive their name from models of a neuron's response when exposed to chemical or electric stimuli.

9.2 Feed-Forward Neural Networks

In a neural network with $L+1$ layers, the zero or input layer ($l = 0$) encodes the input feature vector $\boldsymbol{x}$, and the last or output layer ($l = L$) encodes the (multivalued) output function $\boldsymbol{g}(\boldsymbol{x})$. The remaining layers are called *hidden layers*. Each layer has a number of nodes, say p_l nodes for layer $l = 0, \ldots, L$. In this notation, p_0 is the dimension of the input feature vector $\boldsymbol{x}$ and, for example, $p_L = 1$ signifies that $g(\boldsymbol{x})$ is a scalar output. All nodes in the hidden layers ($l = 1, \ldots, L-1$) are associated with a pair of variables (z, a), which we gather into p_l-dimensional column vectors $\boldsymbol{z}_l$ and $\boldsymbol{a}_l$. In the so-called *feed-forward* networks, the variables in any layer l are simple functions of the variables in the preceding layer $l-1$. In particular, $\boldsymbol{z}_l$ and $\boldsymbol{a}_{l-1}$ are related via the linear relation $\boldsymbol{z}_l = \mathbf{W}_l\, \boldsymbol{a}_{l-1} + \boldsymbol{b}_l$, for some *weight matrix* $\mathbf{W}_l$ and *bias vector* $\boldsymbol{b}_l$.

FEED-FORWARD

WEIGHT MATRIX

BIAS VECTOR

Within any hidden layer $l = 1, \ldots, L-1$, the components of the vectors $\boldsymbol{z}_l$ and $\boldsymbol{a}_l$ are related via $\boldsymbol{a}_l = \boldsymbol{S}_l(\boldsymbol{z}_l)$, where $\boldsymbol{S}_l : \mathbb{R}^{p_l} \mapsto \mathbb{R}^{p_l}$ is a nonlinear multivalued function. All of these multivalued functions are typically of the form

$$\boldsymbol{S}_l(\boldsymbol{z}) = [S(z_1), \ldots, S(z_{\dim(\boldsymbol{z})})]^\top, \quad l = 1, \ldots, L-1, \tag{9.1}$$

where S is an activation function common to all hidden layers. The function $\boldsymbol{S}_L : \mathbb{R}^{p_{L-1}} \mapsto \mathbb{R}^{p_L}$ in the output layer is more general and its specification depends, for example, on whether the network is used for classification or for the prediction of a continuous output Y. A four-layer ($L = 3$) network is illustrated in Figure 9.3.

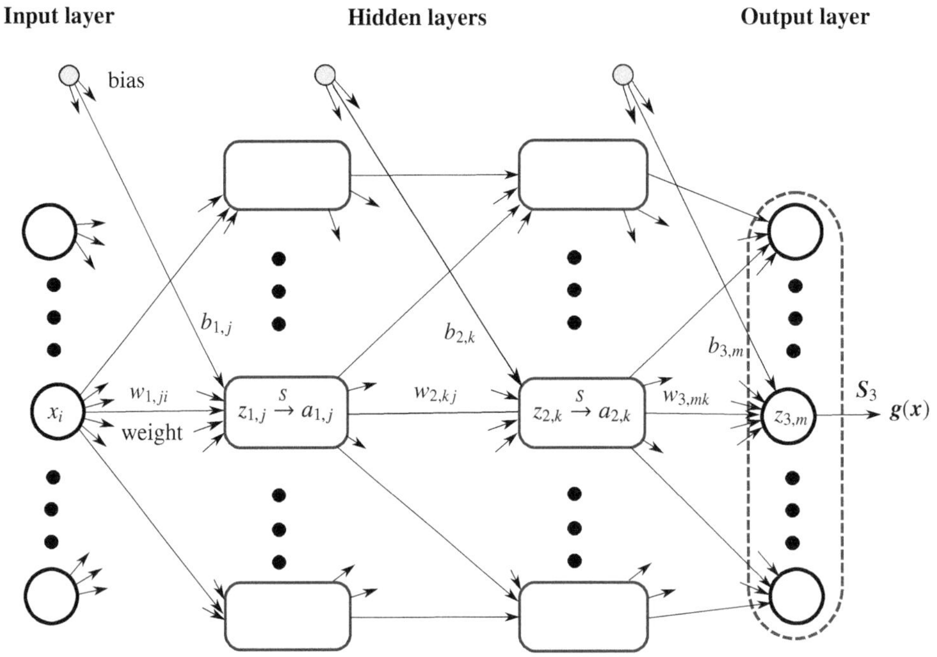

Figure 9.3: A neural network with $L = 3$: the $l = 0$ layer is the input layer, followed by two hidden layers, and the output layer. Hidden layers may have different numbers of nodes.

The output of this neural network is determined by the input vector $\boldsymbol{x}$, (nonlinear) functions $\{\boldsymbol{S}_l\}$, as well as weight matrices $\mathbf{W}_l = [w_{l,ij}]$ and bias vectors $\boldsymbol{b}_l = [b_{l,j}]$ for $l = 1, 2, 3$.

Here, the (i, j)-th element of the weight matrix $\mathbf{W}_l = [w_{l,ij}]$ is the weight that connects the j-th node in the $(l-l)$-st layer with the i-th node in the l-th layer.

NETWORK DEPTH

NETWORK WIDTH

The name given to L (the number of layers without the input layer) is the *network depth* and $\max_l p_l$ is called the *network width*. While we mostly study networks that have an equal number of nodes in the hidden layers ($p_1 = \cdots = p_{L-1}$), in general there can be different numbers of nodes in each hidden layer.

The output $\boldsymbol{g}(\boldsymbol{x})$ of a multiple-layer neural network is obtained from the input $\boldsymbol{x}$ via the following sequence of computations:

$$\begin{aligned}\underbrace{\boldsymbol{x}}_{\boldsymbol{a}_0} &\to \underbrace{\mathbf{W}_1\,\boldsymbol{a}_0 + \boldsymbol{b}_1}_{\boldsymbol{z}_1} \to \underbrace{\boldsymbol{S}_1(\boldsymbol{z}_1)}_{\boldsymbol{a}_1} \to \underbrace{\mathbf{W}_2\,\boldsymbol{a}_1 + \boldsymbol{b}_2}_{\boldsymbol{z}_2} \to \underbrace{\boldsymbol{S}_2(\boldsymbol{z}_2)}_{\boldsymbol{a}_2} \to \cdots \\ &\to \underbrace{\mathbf{W}_L\,\boldsymbol{a}_{L-1} + \boldsymbol{b}_L}_{\boldsymbol{z}_L} \to \underbrace{\boldsymbol{S}_L(\boldsymbol{z}_L)}_{\boldsymbol{a}_L} = \boldsymbol{g}(\boldsymbol{x}).\end{aligned} \tag{9.2}$$

Denoting the function $\boldsymbol{z} \mapsto \mathbf{W}_l\,\boldsymbol{z} + \boldsymbol{b}_l$ by $\boldsymbol{M}_l$, the output $\boldsymbol{g}(\boldsymbol{x})$ can thus be written as the function composition

$$\boldsymbol{g}(\boldsymbol{x}) = \boldsymbol{S}_L \circ \boldsymbol{M}_L \circ \cdots \circ \boldsymbol{S}_2 \circ \boldsymbol{M}_2 \circ \boldsymbol{S}_1 \circ \boldsymbol{M}_1(\boldsymbol{x}). \tag{9.3}$$

The algorithm for computing the output $\boldsymbol{g}(\boldsymbol{x})$ for an input $\boldsymbol{x}$ is summarized next. Note that we leave open the possibility that the activation functions $\{\boldsymbol{S}_l\}$ have different definitions for each layer. In some cases, $\boldsymbol{S}_l$ may even depend on some or all of the already computed $\boldsymbol{z}_1, \boldsymbol{z}_2, \ldots$ and $\boldsymbol{a}_1, \boldsymbol{a}_2, \ldots$.

Algorithm 9.2.1: Feed-Forward Propagation for a Neural Network

input: Feature vector $\boldsymbol{x}$; weights $\{w_{l,ij}\}$, biases $\{b_{l,i}\}$ for each layer $l = 1, \ldots, L$.
output: The value of the prediction function $\boldsymbol{g}(\boldsymbol{x})$.

$\boldsymbol{a}_0 \leftarrow \boldsymbol{x}$ `// the zero or input layer`

for $l = 1$ **to** L **do**

  Compute the hidden variable $z_{l,i}$ for each node i in layer l:

$$\boldsymbol{z}_l \leftarrow \mathbf{W}_l\,\boldsymbol{a}_{l-1} + \boldsymbol{b}_l$$

  Compute the activation function $a_{l,i}$ for each node i in layer l:

$$\boldsymbol{a}_l \leftarrow \boldsymbol{S}_l(\boldsymbol{z}_l)$$

return $\boldsymbol{g}(\boldsymbol{x}) \leftarrow \boldsymbol{a}_L$ `// the output layer`

■ **Example 9.1 (Nonlinear Multi-Output Regression)** Given the input $\boldsymbol{x} \in \mathbb{R}^{p_0}$ and an activation function $S : \mathbb{R} \mapsto \mathbb{R}$, the output $\boldsymbol{g}(\boldsymbol{x}) := [g_1(\boldsymbol{x}), \ldots, g_{p_2}(\boldsymbol{x})]^\top$ of a *nonlinear multi-output regression* model can be computed via a neural network with:

☞ 213

$$\begin{aligned} \boldsymbol{z}_1 &= \mathbf{W}_1 \boldsymbol{x} + \boldsymbol{b}_1, \quad \text{where } \mathbf{W}_1 \in \mathbb{R}^{p_1 \times p_0}, \boldsymbol{b}_1 \in \mathbb{R}^{p_1}, \\ a_{1,k} &= S(z_{1,k}), \quad k = 1, \ldots, p_1, \\ \boldsymbol{g}(\boldsymbol{x}) &= \mathbf{W}_2 \boldsymbol{a}_1 + \boldsymbol{b}_2, \quad \text{where } \mathbf{W}_2 \in \mathbb{R}^{p_2 \times p_1}, \boldsymbol{b}_2 \in \mathbb{R}^{p_2}, \end{aligned}$$

which is a neural network with one hidden layer and output function $\boldsymbol{S}_2(\boldsymbol{z}) = \boldsymbol{z}$. In the special case where $p_1 = p_2 = 1$, $\boldsymbol{b}_2 = 0, \mathbf{W}_2 = 1$, and we collect all parameters into the vector $\boldsymbol{\theta}^\top = [\boldsymbol{b}_1, \mathbf{W}_1] \in \mathbb{R}^{p_0+1}$, the neural network can be interpreted as a *generalized linear model* with $\mathbb{E}[Y \,|\, \boldsymbol{X} = \boldsymbol{x}] = h([1, \boldsymbol{x}^\top]\,\boldsymbol{\theta})$ for some activation function h. ■

☞ 204

■ **Example 9.2 (Multi-Logit Classification)** Suppose that, for a classification problem, an input $\boldsymbol{x}$ has to be classified into one of c classes, labeled $0, \ldots, c-1$. We can perform the classification via a neural network with one hidden layer, with $p_1 = c$ nodes. In particular, we have

$$\boldsymbol{z}_1 = \mathbf{W}_1 \boldsymbol{x} + \boldsymbol{b}_1, \quad \boldsymbol{a}_1 = \boldsymbol{S}_1(\boldsymbol{z}_1),$$

where $\boldsymbol{S}_1$ is the *softmax* function:

SOFTMAX

$$\text{softmax} : \boldsymbol{z} \mapsto \frac{\exp(\boldsymbol{z})}{\sum_k \exp(z_k)}.$$

For the output, we take $\boldsymbol{g}(\boldsymbol{x}) = [g_1(\boldsymbol{x}), \ldots, g_c(\boldsymbol{x})]^\top = \boldsymbol{a}_1$, which can then be used as a *pre-classifier* of $\boldsymbol{x}$. The actual classifier of $\boldsymbol{x}$ into one of the categories $0, 1, \ldots, c-1$ is then

☞ 254

$$\underset{k \in \{0,\ldots,c-1\}}{\text{argmax}} \; g_{k+1}(\boldsymbol{x}).$$

This is equivalent to the multi-logit classifier in Section 7.5. Note, however, that there we used a slightly different notation, with $\widetilde{\boldsymbol{x}}$ instead of $\boldsymbol{x}$ and we have a reference class; see Exercise 13. ■

☞ 268

> In practical implementations, the softmax function can cause numerical over- and under-flow errors when either one of the $\exp(z_k)$ happens to be extremely large or $\sum_k \exp(z_k)$ happens to be very small. In such cases we can exploit the invariance property (Exercise 1):
>
> $$\text{softmax}(\boldsymbol{z}) = \text{softmax}(\boldsymbol{z} + c \times \mathbf{1}) \quad \text{for any constant } c.$$
>
> Using this property, we can compute softmax($\boldsymbol{z}$) with greater numerical stability via softmax($\boldsymbol{z} - \max_k\{z_k\} \times \mathbf{1}$).

When neural networks are used for classification into c classes and the number of output nodes is $c - 1$, then the $g_i(\boldsymbol{x})$ may be viewed as nonlinear *discriminant functions*.

☞ 262

■ **Example 9.3 (Density Estimation)** Estimating the density f of some random feature $X \in \mathbb{R}$ is the prototypical unsupervised learning task, which we tackled in Section 4.5.2 using Gaussian mixture models. We can view a Gaussian mixture model with p_1 components and a common scale parameter $\sigma > 0$ as a neural network with two hidden layers, similar to the one on Figure 9.3. In particular, if the activation function in the first hidden layer, $\boldsymbol{S}_1$, is of the form (9.1) with $S(z) := \exp(-z^2/(2\sigma^2))/\sqrt{2\pi\sigma^2}$, then the density value $g(x)$ is computed via:

☞ 137

$$\begin{aligned} \boldsymbol{z}_1 &= \mathbf{W}_1\, x + \boldsymbol{b}_1, & \boldsymbol{a}_1 &= \boldsymbol{S}_1(\boldsymbol{z}_1), \\ \boldsymbol{z}_2 &= \mathbf{W}_2\, \boldsymbol{a}_1 + \boldsymbol{b}_2, & \boldsymbol{a}_2 &= \boldsymbol{S}_2(\boldsymbol{z}_2), \\ g(x) &= \boldsymbol{a}_1^\top \boldsymbol{a}_2, \end{aligned}$$

where $\mathbf{W}_1 = \mathbf{1}$ is a $p_1 \times 1$ column vector of ones, $\mathbf{W}_2 = \mathbf{O}$ is a $p_1 \times p_1$ matrix of zeros, and $\boldsymbol{S}_2$ is the softmax function. We identify the column vector $\boldsymbol{b}_1$ with the p_1 location parameters, $[\mu_1, \ldots, \mu_{p_1}]^\top$ of the Gaussian mixture and $\boldsymbol{b}_2 \in \mathbb{R}^{p_1}$ with the p_1 weights of the mixture. Note the unusual activation function of the output layer — it requires the value of $\boldsymbol{a}_1$ from the first hidden layer and $\boldsymbol{a}_2$ from the second hidden layer. ■

There are a number of key design characteristics of a feed-forward network. First, we need to choose the activation function(s). Second, we need to choose the loss function for the training of the network. As we shall explain in the next section, the most common choices are the ReLU activation function and the cross-entropy loss. Crucially, we need to carefully construct the *network architecture* — the number of connections among the nodes in different layers and the overall number of layers of the network.

NETWORK ARCHITECTURE

For example, if the connections from one layer to the next are pruned (called *sparse connectivity*) and the links share the same weight values $\{w_{l,ij}\}$ (called *parameter sharing*) for all $\{(i, j) : |i - j| = 0, 1, \ldots\}$, then the weight matrices will be sparse and *Toeplitz*.

☞ 381

Intuitively, the parameter sharing and sparse connectivity can speed up the training of the network, because there are fewer parameters to learn, and the Toeplitz structure permits quick computation of the matrix-vector products in Algorithm 9.2.1. An important example of such a network is the *convolution neural network* (CNN), in which some or all of the network layers encode the linear operation of *convolution*:

CONVOLUTION NEURAL NETWORK

$$\mathbf{W}_l\, \boldsymbol{a}_{l-1} = \boldsymbol{w}_l * \boldsymbol{a}_{l-1},$$

where $[\boldsymbol{x} * \boldsymbol{y}]_i := \sum_k x_k y_{i-k+1}$. As discussed in Example A.10, a convolution matrix is a special type of sparse Toeplitz matrix, and its action on a vector of learning parameters can be evaluated quickly via the *fast Fourier transform*.

☞ 382

☞ 396

CNNs are particularly suited to image processing problems, because their *convolution layers* closely mimic the neurological properties of the visual cortex. In particular, the cortex partitions the visual field into many small regions and assigns a group of neurons to every such region. Moreover, some of these groups of neurons respond only to the presence of particular features (for example, edges).

This neurological property is naturally modeled via convolution layers in the neural network. Specifically, suppose that the input image is given by an $m_1 \times m_2$ matrix of pixels. Now, define a $k \times k$ matrix (sometimes called a *kernel*, where k is generally taken to be 3 or 5). Then, the convolution layer output can be calculated using the discrete convolution of all possible $k \times k$ input matrix regions and the kernel matrix; (see Example A.10). In particular, by noting that there are $(m_1 - k + 1) \times (m_2 - k + 1)$ possible regions in the original image, we conclude that the convolution layer output size is $(m_1 - k + 1) \times (m_2 - k + 1)$. In practice, we frequently define several kernel matrices, giving an output layer of size $(m_1 - k + 1) \times (m_2 - k + 1) \times$ (the number of kernels). Figure 9.4 shows a 5×5 input image and a 2×2 kernel with a 4×4 output matrix. An example of using a CNN for image classification is given in Section 9.5.2.

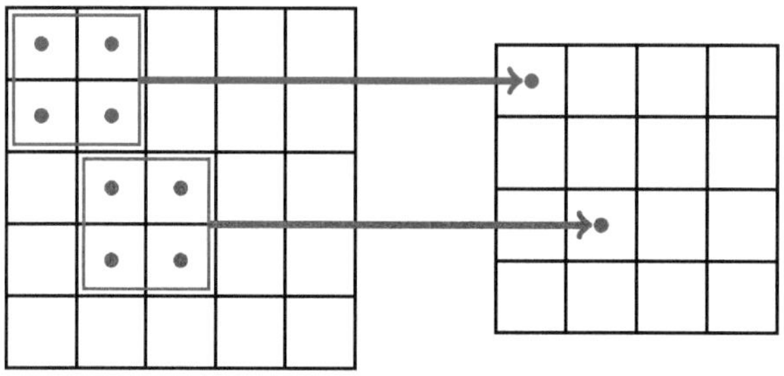

Figure 9.4: An example 5×5 input image and a 2×2 kernel. The kernel is applied to every 2×2 region of the original image.

9.3 Back-Propagation

The training of neural networks is a major challenge that requires both ingenuity and much experimentation. The algorithms for training neural networks with great depth are collectively referred to as *deep learning* methods. One of the simplest and most effective methods for training is via *steepest descent* and its variations.

DEEP LEARNING

☞ 414

Steepest descent requires computation of the gradient with respect to all bias vectors and weight matrices. Given the potentially large number of parameters (weight and bias terms) in a neural network, we need to find an efficient method to calculate this gradient.

To illustrate the nature of the gradient computations, let $\boldsymbol{\theta} = \{\mathbf{W}_l, \boldsymbol{b}_l\}$ be a column vector of length $\dim(\boldsymbol{\theta}) = \sum_{l=1}^{L}(p_{l-1}p_l + p_l)$ that collects all the weight parameters (numbering $\sum_{l=1}^{L} p_{l-1}p_l$) and bias parameters (numbering $\sum_{l=1}^{L} p_l$) of a multiple-layer network with training loss:

$$\ell_\tau(\boldsymbol{g}(\cdot \,|\, \boldsymbol{\theta})) := \frac{1}{n}\sum_{i=1}^{n} \text{Loss}(\boldsymbol{y}_i, \boldsymbol{g}(\boldsymbol{x}_i \,|\, \boldsymbol{\theta})).$$

Writing $C_i(\boldsymbol{\theta}) := \text{Loss}(\boldsymbol{y}_i, \boldsymbol{g}(\boldsymbol{x}_i \,|\, \boldsymbol{\theta}))$ for short (using C for cost), we have

$$\ell_\tau(\boldsymbol{g}(\cdot \,|\, \boldsymbol{\theta})) = \frac{1}{n}\sum_{i=1}^{n} C_i(\boldsymbol{\theta}), \tag{9.4}$$

so that obtaining the gradient of ℓ_τ requires computation of $\partial C_i/\partial \boldsymbol{\theta}$ for every i. For activation functions of the form (9.1), define $\mathbf{D}_l$ as the diagonal matrix with the vector of derivatives

$$\boldsymbol{S}'(\boldsymbol{z}) := [S'(z_{l,1}), \ldots, S'(z_{l,p_l})]^\top$$

down its main diagonal; that is,

$$\mathbf{D}_l := \text{diag}(S'(z_{l,1}), \ldots, S'(z_{l,p_l})), \quad l = 1, \ldots, L-1.$$

The following theorem provides us with the formulas needed to compute the gradient of a typical $C_i(\boldsymbol{\theta})$.

Theorem 9.2: Gradient of Training Loss

For a given (input, output) pair $(\boldsymbol{x}, \boldsymbol{y})$, let $\boldsymbol{g}(\boldsymbol{x} \mid \boldsymbol{\theta})$ be the output of Algorithm 9.2.1, and let $C(\boldsymbol{\theta}) = \text{Loss}(\boldsymbol{y}, \boldsymbol{g}(\boldsymbol{x} \mid \boldsymbol{\theta}))$ be an almost-everywhere differentiable loss function. Suppose $\{\boldsymbol{z}_l, \boldsymbol{a}_l\}_{l=1}^L$ are the vectors obtained during the feed-forward propagation ($\boldsymbol{a}_0 = \boldsymbol{x}, \boldsymbol{a}_L = \boldsymbol{g}(\boldsymbol{x} \mid \boldsymbol{\theta})$). Then, we have for $l = 1, \ldots, L$:

$$\frac{\partial C}{\partial \mathbf{W}_l} = \boldsymbol{\delta}_l \, \boldsymbol{a}_{l-1}^\top \quad \text{and} \quad \frac{\partial C}{\partial \boldsymbol{b}_l} = \boldsymbol{\delta}_l,$$

where $\boldsymbol{\delta}_l := \partial C / \partial \boldsymbol{z}_l$ is computed recursively for $l = L, \ldots, 2$:

$$\boldsymbol{\delta}_{l-1} = \mathbf{D}_{l-1} \mathbf{W}_l^\top \boldsymbol{\delta}_l \quad \text{with} \quad \boldsymbol{\delta}_L = \frac{\partial \boldsymbol{S}_L}{\partial \boldsymbol{z}_L} \frac{\partial C}{\partial \boldsymbol{g}}. \tag{9.5}$$

Proof: The scalar value C is obtained from the transitions (9.2), followed by the mapping $\boldsymbol{g}(\boldsymbol{x} \mid \boldsymbol{\theta}) \mapsto \text{Loss}(\boldsymbol{y}, \boldsymbol{g}(\boldsymbol{x} \mid \boldsymbol{\theta}))$. Using the chain rule (see Appendix B.1.2), we have ☞ 402

$$\boldsymbol{\delta}_L = \frac{\partial C}{\partial \boldsymbol{z}_L} = \frac{\partial \boldsymbol{g}(\boldsymbol{x})}{\partial \boldsymbol{z}_L} \frac{\partial C}{\partial \boldsymbol{g}(\boldsymbol{x})} = \frac{\partial \boldsymbol{S}_L}{\partial \boldsymbol{z}_L} \frac{\partial C}{\partial \boldsymbol{g}}.$$

Recall that the vector/vector derivative of a linear mapping $\boldsymbol{z} \mapsto \mathbf{W}\boldsymbol{z}$ is given by $\mathbf{W}^\top$; see (B.5). It follows that, since $\boldsymbol{z}_l = \mathbf{W}_l \, \boldsymbol{a}_{l-1} + \boldsymbol{b}_l$ and $\boldsymbol{a}_l = \boldsymbol{S}(\boldsymbol{z}_l)$, the chain rule gives ☞ 401

$$\frac{\partial \boldsymbol{z}_l}{\partial \boldsymbol{z}_{l-1}} = \frac{\partial \boldsymbol{a}_{l-1}}{\partial \boldsymbol{z}_{l-1}} \frac{\partial \boldsymbol{z}_l}{\partial \boldsymbol{a}_{l-1}} = \mathbf{D}_{l-1} \mathbf{W}_l^\top.$$

Hence, the recursive formula (9.5):

$$\boldsymbol{\delta}_{l-1} = \frac{\partial C}{\partial \boldsymbol{z}_{l-1}} = \frac{\partial \boldsymbol{z}_l}{\partial \boldsymbol{z}_{l-1}} \frac{\partial C}{\partial \boldsymbol{z}_l} = \mathbf{D}_{l-1} \mathbf{W}_l^\top \boldsymbol{\delta}_l, \quad l = L, \ldots, 3, 2.$$

Using the $\{\boldsymbol{\delta}_l\}$, we can now compute the derivatives with respect to the weight matrices and the biases. In particular, applying the "scalar/matrix" differentiation rule (B.10) to $\boldsymbol{z}_l = \mathbf{W}_l \, \boldsymbol{a}_{l-1} + \boldsymbol{b}_l$ gives:

$$\frac{\partial C}{\partial \mathbf{W}_l} = \frac{\partial C}{\partial \boldsymbol{z}_l} \frac{\partial \boldsymbol{z}_l}{\partial \mathbf{W}_l} = \boldsymbol{\delta}_l \, \boldsymbol{a}_{l-1}^\top, \quad l = 1, \ldots, L$$

and

$$\frac{\partial C}{\partial \boldsymbol{b}_l} = \frac{\partial \boldsymbol{z}_l}{\partial \boldsymbol{b}_l} \frac{\partial C}{\partial \boldsymbol{z}_l} = \boldsymbol{\delta}_l, \quad l = 1, \ldots, L.$$

□

From the theorem we can see that for each pair $(\boldsymbol{x}, \boldsymbol{y})$ in the training set, we can compute the gradient $\partial C / \partial \boldsymbol{\theta}$ in a sequential manner, by computing $\boldsymbol{\delta}_L, \ldots, \boldsymbol{\delta}_1$. This procedure is called *back-propagation*. Since back-propagation mostly involves simple matrix multiplication, it BACK-PROPAGATION can be efficiently implemented using dedicated computing hardware such as graphical processor units (GPUs) and other parallel computing architecture. Note also that many matrix computations that run in quadratic time can be replaced with linear-time componentwise multiplication. Specifically, multiplication of a vector with a diagonal matrix is equivalent to componentwise multiplication:

$$\underbrace{\mathbf{A}}_{\text{diag}(\boldsymbol{a})}\,\boldsymbol{b} = \boldsymbol{a} \odot \boldsymbol{b}.$$

Consequently, we can write $\boldsymbol{\delta}_{l-1} = \mathbf{D}_{l-1}\mathbf{W}_l^\top \boldsymbol{\delta}_l$ as: $\boldsymbol{\delta}_{l-1} = \boldsymbol{S}'(\boldsymbol{z}_{l-1}) \odot \mathbf{W}_l^\top \boldsymbol{\delta}_l$, $l = L, \ldots, 3, 2$.

We now summarize the back-propagation algorithm for the computation of a typical $\partial C/\partial \boldsymbol{\theta}$. In the following algorithm, Lines 1 to 5 are the feed-forward part of the algorithm, and Lines 7 to 10 are the back-propagation part of the algorithm.

Algorithm 9.3.1: Computing the Gradient of a Typical $C(\boldsymbol{\theta})$

input: Training example $(\boldsymbol{x}, \boldsymbol{y})$, weight matrices and bias vectors $\{\mathbf{W}_l, \boldsymbol{b}_l\}_{l=1}^L =: \boldsymbol{\theta}$, activation functions $\{\boldsymbol{S}_l\}_{l=1}^L$.

output: The derivatives with respect to all weight matrices and bias vectors.

1 $\boldsymbol{a}_0 \leftarrow \boldsymbol{x}$
2 **for** $l = 1, \ldots, L$ **do** // feed-forward
3 $\quad \boldsymbol{z}_l \leftarrow \mathbf{W}_l\, \boldsymbol{a}_{l-1} + \boldsymbol{b}_l$
4 $\quad \boldsymbol{a}_l \leftarrow \boldsymbol{S}_l(\boldsymbol{z}_l)$
5 $\boldsymbol{\delta}_L \leftarrow \frac{\partial \boldsymbol{S}_L}{\partial \boldsymbol{z}_L}\frac{\partial C}{\partial \boldsymbol{g}}$
6 $\boldsymbol{z}_0 \leftarrow \mathbf{0}$ // arbitrary assignment needed to finish the loop
7 **for** $l = L, \ldots, 1$ **do** // back-propagation
8 $\quad \frac{\partial C}{\partial \boldsymbol{b}_l} \leftarrow \boldsymbol{\delta}_l$
9 $\quad \frac{\partial C}{\partial \mathbf{W}_l} \leftarrow \boldsymbol{\delta}_l\, \boldsymbol{a}_{l-1}^\top$
10 $\quad \boldsymbol{\delta}_{l-1} \leftarrow \boldsymbol{S}'(\boldsymbol{z}_{l-1}) \odot \mathbf{W}_l^\top \boldsymbol{\delta}_l$
11 **return** $\frac{\partial C}{\partial \mathbf{W}_l}$ and $\frac{\partial C}{\partial \boldsymbol{b}_l}$ for all $l = 1, \ldots, L$ and the value $\boldsymbol{g}(\boldsymbol{x}) \leftarrow \boldsymbol{a}_L$ (if needed)

Note that for the gradient of $C(\boldsymbol{\theta})$ to exist at every point, we need the activation functions to be differentiable everywhere. This is the case, for example, for the logistic activation function in Figure 9.2. It is not the case for the ReLU function, which is differentiable everywhere, except at $z = 0$. However, in practice, the kink of the ReLU function at $z = 0$ is unlikely to trip the back-propagation algorithm, because rounding errors and the finite-precision computer arithmetic make it extremely unlikely that we will need to evaluate the ReLU at precisely $z = 0$. This is the reason why in Theorem 9.2 we merely required that $C(\boldsymbol{\theta})$ is almost-everywhere differentiable.

In spite of its kink at the origin, the ReLU has an important advantage over the logistic function. While the derivative of the logistic function decays exponentially fast to zero as we move away from the origin, a phenomenon referred to as *saturation*, the derivative of the ReLU function is always unity for positive z. Thus, for large positive z, the derivative of the logistic function does not carry any useful information, but the derivative of the ReLU can help guide a gradient optimization algorithm. The situation for the Heaviside function in Figure 9.2 is even worse, because its derivative is completely noninformative for any $z \neq 0$. In this respect, the lack of saturation of the ReLU function for $z > 0$ makes it a desirable activation function for training a network via back-propagation.

SATURATION

Finally, note that to obtain the gradient $\partial \ell_\tau / \partial \boldsymbol{\theta}$ of the training loss, we simply need to loop Algorithm 9.3.1 over all the n training examples, as follows.

Algorithm 9.3.2: Computing the Gradient of the Training Loss

input: Training set $\tau = \{(\boldsymbol{x}_i, \boldsymbol{y}_i)\}_{i=1}^n$, weight matrices and bias vectors $\{\mathbf{W}_l, \boldsymbol{b}_l\}_{l=1}^L =: \boldsymbol{\theta}$, activation functions $\{\boldsymbol{S}_l\}_{l=1}^L$.

output: The gradient of the training loss.

```
for i = 1, ..., n do                          // loop over all training examples
    Run Algorithm 9.3.1 with input (x_i, y_i) to compute { ∂C_i/∂W_l , ∂C_i/∂b_l }_{l=1}^L
return ∂C/∂W_l = (1/n) Σ_{i=1}^n ∂C_i/∂W_l and ∂C/∂b_l = (1/n) Σ_{i=1}^n ∂C_i/∂b_l for all l = 1, ..., L
```

■ **Example 9.4 (Squared-Error and Cross-Entropy Loss)** The back-propagation Algorithm 9.3.1 requires a formula for $\boldsymbol{\delta}_L$ in line 5. In particular, to execute line 5 we need to specify both a loss function and an $\boldsymbol{S}_L$ that defines the output layer: $\boldsymbol{g}(\boldsymbol{x} \,|\, \boldsymbol{\theta}) = \boldsymbol{a}_L = \boldsymbol{S}_L(\boldsymbol{z}_L)$.

For instance, in the *multi-logit* classification of inputs $\boldsymbol{x}$ into p_L categories labeled $0, 1, \ldots, (p_L - 1)$, the output layer is defined via the softmax function: ☞ 268

$$\boldsymbol{S}_L : \boldsymbol{z}_L \mapsto \frac{\exp(\boldsymbol{z}_L)}{\sum_{k=1}^{p_L} \exp(z_{L,k})}.$$

In other words, $\boldsymbol{g}(\boldsymbol{x} \,|\, \boldsymbol{\theta})$ is a probability vector such that its $(y+1)$-st component $g_{y+1}(\boldsymbol{x} \,|\, \boldsymbol{\theta}) = g(y \,|\, \boldsymbol{\theta}, \boldsymbol{x})$ is the estimate or prediction of the true conditional probability $f(y \,|\, \boldsymbol{x})$. Combining the softmax output with the *cross-entropy* loss, as was done in (7.17), yields: ☞ 269

$$\begin{aligned}
\text{Loss}(f(y \,|\, \boldsymbol{x}), g(y \,|\, \boldsymbol{\theta}, \boldsymbol{x})) &= -\ln g(y \,|\, \boldsymbol{\theta}, \boldsymbol{x}) \\
&= -\ln g_{y+1}(\boldsymbol{x} \,|\, \boldsymbol{\theta}) \\
&= -z_{y+1} + \ln \textstyle\sum_{k=1}^{p_L} \exp(z_k).
\end{aligned}$$

Hence, we obtain the vector $\boldsymbol{\delta}_L$ with components $(k = 1, \ldots, p_L)$

$$\delta_{L,k} = \frac{\partial}{\partial z_k}\left(-z_{y+1} + \ln \textstyle\sum_{k=1}^{p_L} \exp(z_k)\right) = g_k(\boldsymbol{x} \,|\, \boldsymbol{\theta}) - \mathbb{1}\{y = k - 1\}.$$

Note that we can remove a node from the final layer of the multi-logit network, because $g_1(\boldsymbol{x} \,|\, \boldsymbol{\theta})$ (which corresponds to the $y = 0$ class) can be eliminated, using the fact that $g_1(\boldsymbol{x} \,|\, \boldsymbol{\theta}) = 1 - \sum_{k=2}^{p_L} g_k(\boldsymbol{x} \,|\, \boldsymbol{\theta})$. For a numerical comparison, see Exercise 13.

As another example, in nonlinear multi-output regression (see Example 9.1), the output function $\boldsymbol{S}_L$ is typically of the form (9.1), so that $\partial \boldsymbol{S}_L / \partial \boldsymbol{z} = \text{diag}(S_L'(z_1), \ldots, S_L'(z_{p_L}))$. Combining the output $\boldsymbol{g}(\boldsymbol{x} \,|\, \boldsymbol{\theta}) = \boldsymbol{S}_L(\boldsymbol{z}_L)$ with the squared-error loss yields:

$$\text{Loss}(\boldsymbol{y}, \boldsymbol{g}(\boldsymbol{x} \,|\, \boldsymbol{\theta})) = \|\boldsymbol{y} - \boldsymbol{g}(\boldsymbol{x} \,|\, \boldsymbol{\theta})\|^2 = \sum_{j=1}^{p_L} (y_j - g_j(\boldsymbol{x} \,|\, \boldsymbol{\theta}))^2.$$

Hence, line 5 in Algorithm 9.3.1 simplifies to:

$$\boldsymbol{\delta}_L = \frac{\partial \boldsymbol{S}_L}{\partial \boldsymbol{z}} \frac{\partial C}{\partial \boldsymbol{g}} = \boldsymbol{S}_L'(\boldsymbol{z}_L) \odot 2(\boldsymbol{g}(\boldsymbol{x} \,|\, \boldsymbol{\theta}) - \boldsymbol{y}).$$

■

9.4 Methods for Training

Neural networks have been studied for a long time, yet it is only recently that there have been sufficient computational resources to train them effectively. The training of neural networks requires minimization of a training loss, $\ell_\tau(\boldsymbol{g}(\cdot\,|\,\boldsymbol{\theta})) = \frac{1}{n}\sum_{i=1}^{n} C_i(\boldsymbol{\theta})$, which is typically a difficult high-dimensional optimization problem with multiple local minima. We next consider a number of simple training methods.

> In this section, the vectors $\boldsymbol{\delta}_t$ and $\boldsymbol{g}_t$ use the notation of Section B.3.2 and should not be confused with the derivative $\boldsymbol{\delta}$ and the prediction function $\boldsymbol{g}$, respectively.

9.4.1 Steepest Descent

☞ 414

If we can compute the gradient of $\ell_\tau(\boldsymbol{g}(\cdot\,|\,\boldsymbol{\theta}))$ via back-propagation, then we can apply the steepest descent algorithm, which reads as follows. Starting from a guess $\boldsymbol{\theta}_1$, we iterate the following step until convergence:

$$\boldsymbol{\theta}_{t+1} = \boldsymbol{\theta}_t - \alpha_t \boldsymbol{u}_t, \qquad t = 1, 2, \ldots, \tag{9.6}$$

LEARNING RATE

where $\boldsymbol{u}_t := \frac{\partial \ell_\tau}{\partial \boldsymbol{\theta}}(\boldsymbol{\theta}_t)$ and α_t is the *learning rate*.

Observe that, rather than operating directly on the weights and biases, we operate instead on $\boldsymbol{\theta} := \{\mathbf{W}_l, \boldsymbol{b}_l\}_{l=1}^{L}$ — a column vector of length $\sum_{l=1}^{L}(p_{l-1}p_l + p_l)$ that stores all the weight and bias parameters. The advantage of organizing the computations in this way is that we can easily compute the learning rate α_t; for example, via the *Barzilai–Borwein* formula in (B.26).

☞ 415

Algorithm 9.4.1: Training via Steepest Descent

---

input: Training set $\tau = \{(\boldsymbol{x}_i, \boldsymbol{y}_i)\}_{i=1}^{n}$, initial weight matrices and bias vectors $\{\mathbf{W}_l, \boldsymbol{b}_l\}_{l=1}^{L} =: \boldsymbol{\theta}_1$, activation functions $\{S_l\}_{l=1}^{L}$.
output: The parameters of the trained learner.

$t \leftarrow 1$, $\boldsymbol{\delta} \leftarrow 0.1 \times \mathbf{1}$, $\boldsymbol{u}_{t-1} \leftarrow \mathbf{0}$, $\alpha \leftarrow 0.1$ `// initialization`
while stopping condition is not met **do**
  compute the gradient $\boldsymbol{u}_t = \frac{\partial \ell_\tau}{\partial \boldsymbol{\theta}}(\boldsymbol{\theta}_t)$ using Algorithm 9.3.2
  $\boldsymbol{g} \leftarrow \boldsymbol{u}_t - \boldsymbol{u}_{t-1}$
  if $\boldsymbol{\delta}^\top \boldsymbol{g} > 0$ **then** `// check if Hessian is positive-definite`
    $\alpha \leftarrow \boldsymbol{\delta}^\top \boldsymbol{g} / \|\boldsymbol{g}\|^2$ `// Barzilai-Borwein`
  else
    $\alpha \leftarrow 2 \times \alpha$ `// failing positivity, do something heuristic`
  $\boldsymbol{\delta} \leftarrow -\alpha \boldsymbol{u}_t$
  $\boldsymbol{\theta}_{t+1} \leftarrow \boldsymbol{\theta}_t + \boldsymbol{\delta}$
  $t \leftarrow t + 1$
return $\boldsymbol{\theta}_t$ as the minimizer of the training loss

---

Typically, we initialize the algorithm with small random values for $\boldsymbol{\theta}_1$, while being careful to avoid saturating the activation function. For example, in the case of the ReLU activation function, we will use small positive values to ensure that its derivative is not zero. A zero derivative of the activation function prevents the propagation of information useful for computing a good search direction.

Recall that computation of the gradient of the training loss via Algorithm 9.3.2 requires averaging over all training examples. When the size n of the training set τ_n is too large, computation of the gradient $\partial \ell_{\tau_n} / \partial \boldsymbol{\theta}$ via Algorithm 9.3.2 may be too costly. In such cases, we may employ the *stochastic gradient descent* algorithm. In this algorithm, we view the training loss as an expectation that can be approximated via Monte Carlo sampling. In particular, if K is a random variable with distribution $\mathbb{P}[K = k] = 1/n$ for $k = 1, \ldots, n$, then we can write

STOCHASTIC GRADIENT DESCENT

$$\ell_\tau(\boldsymbol{g}(\cdot \,|\, \boldsymbol{\theta})) = \frac{1}{n} \sum_{k=1}^{n} \mathrm{Loss}(\boldsymbol{y}_k, \boldsymbol{g}(\boldsymbol{x}_k \,|\, \boldsymbol{\theta})) = \mathbb{E}\, \mathrm{Loss}(\boldsymbol{y}_K, \boldsymbol{g}(\boldsymbol{x}_K \,|\, \boldsymbol{\theta})).$$

We can thus approximate $\ell_\tau(\boldsymbol{g}(\cdot \,|\, \boldsymbol{\theta}))$ via a Monte Carlo estimator using N iid copies of K:

$$\widehat{\ell_\tau}(\boldsymbol{g}(\cdot \,|\, \boldsymbol{\theta})) := \frac{1}{N} \sum_{i=1}^{N} \mathrm{Loss}(\boldsymbol{y}_{K_i}, \boldsymbol{g}(\boldsymbol{x}_{K_i} \,|\, \boldsymbol{\theta})).$$

The iid Monte Carlo sample $K_1, \ldots, K_N$ is called a *minibatch* (see also Exercise 3). Typically, $n \gg N$ so that the probability of observing ties in a minibatch of size N is negligible.

MINIBATCH

Finally, note that if the learning rate of the stochastic gradient descent algorithm satisfies the conditions in (3.30), then the stochastic gradient descent algorithm is simply a version of the *stochastic approximation* Algorithm 3.4.5.

☞ 106

9.4.2 Levenberg–Marquardt Method

Since a neural network with squared-error loss is a special type of nonlinear regression model, it is possible to train it using classical nonlinear least-squares minimization methods, such as the *Levenberg–Marquardt* algorithm.

☞ 417

For simplicity of notation, suppose that the output of the net for an input $\boldsymbol{x}$ is a *scalar* $g(\boldsymbol{x})$. For a given input parameter $\boldsymbol{\theta}$ of dimension $d = \dim(\boldsymbol{\theta})$, the Levenberg–Marquardt Algorithm B.3.3 requires computation of the following *vector* of outputs:

$$\boldsymbol{g}(\tau \,|\, \boldsymbol{\theta}) := [g(\boldsymbol{x}_1 \,|\, \boldsymbol{\theta}), \ldots, g(\boldsymbol{x}_n \,|\, \boldsymbol{\theta})]^\top,$$

as well as the $n \times d$ matrix of Jacobi, $\mathbf{G}$, of $\boldsymbol{g}$ at $\boldsymbol{\theta}$. To compute these quantities, we can again use the back-propagation Algorithm 9.3.1, as follows.

Algorithm 9.4.2: Output for Training via Levenberg–Marquardt

input: Training set $\tau = \{(\boldsymbol{x}_i, y_i)\}_{i=1}^n$, parameter $\boldsymbol{\theta}$.
output: Vector $\boldsymbol{g}(\tau \,|\, \boldsymbol{\theta})$ and matrix of Jacobi $\mathbf{G}$ for use in Algorithm B.3.3.

for $i = 1, \ldots, n$ **do** `// loop over all training examples`
  Run Algorithm 9.3.1 with input $(\boldsymbol{x}_i, y_i)$ (using $\frac{\partial C}{\partial g} = 1$ in line 5) to compute $g(\boldsymbol{x}_i \,|\, \boldsymbol{\theta})$ and $\frac{\partial g(\boldsymbol{x}_i \,|\, \boldsymbol{\theta})}{\partial \boldsymbol{\theta}}$.
$\boldsymbol{g}(\tau \,|\, \boldsymbol{\theta}) \leftarrow [g(\boldsymbol{x}_1 \,|\, \boldsymbol{\theta}), \ldots, g(\boldsymbol{x}_n \,|\, \boldsymbol{\theta})]^\top$
$\mathbf{G} \leftarrow \left[\frac{\partial g(\boldsymbol{x}_1 \,|\, \boldsymbol{\theta})}{\partial \boldsymbol{\theta}}, \cdots, \frac{\partial g(\boldsymbol{x}_n \,|\, \boldsymbol{\theta})}{\partial \boldsymbol{\theta}}\right]^\top$
return $\boldsymbol{g}(\tau \,|\, \boldsymbol{\theta})$ and $\mathbf{G}$

The Levenberg–Marquardt algorithm is not suitable for networks with a large number of parameters, because the cost of the matrix computations becomes prohibitive. For instance, obtaining the Levenberg–Marquardt search direction in (B.28) usually incurs an $O(d^3)$ cost. In addition, the Levenberg–Marquardt algorithm is applicable only when we wish to train the network using the squared-error loss. Both of these shortcomings are mitigated to an extent with the quasi-Newton or adaptive gradient methods described next.

9.4.3 Limited-Memory BFGS Method

All the methods discussed so far have been *first-order* optimization methods, that is, methods that only use the gradient vector $\boldsymbol{u}_t := \frac{\partial \ell_\tau}{\partial \boldsymbol{\theta}}(\boldsymbol{\theta}_t)$ at the current (and/or immediate past) candidate solution $\boldsymbol{\theta}_t$. In trying to design a more efficient *second-order* optimization method, we may be tempted to use Newton's method with a search direction:

☞ 412

$$-\mathbf{H}_t^{-1}\boldsymbol{u}_t,$$

where $\mathbf{H}_t$ is the $d \times d$ matrix of *second-order* partial derivatives of $\ell_\tau(\boldsymbol{g}(\cdot \mid \boldsymbol{\theta}))$ at $\boldsymbol{\theta}_t$.

There are two problems with this approach. First, while the computation of $\boldsymbol{u}_t$ via Algorithm 9.3.2 typically costs $O(d)$, the computation of $\mathbf{H}_t$ costs $O(d^2)$. Second, even if we have somehow computed $\mathbf{H}_t$ very fast, computing the search direction $\mathbf{H}_t^{-1}\,\boldsymbol{u}_t$ still incurs an $O(d^3)$ cost. Both of these considerations make Newton's method impractical for large d.

QUASI-NEWTON METHOD

☞ 413

Instead, a practical alternative is to use a *quasi-Newton method*, in which we directly aim to approximate $\mathbf{H}_t^{-1}$ via a matrix $\mathbf{C}_t$ that satisfies the *secant condition*:

$$\mathbf{C}_t\,\boldsymbol{g}_t = \boldsymbol{\delta}_t,$$

where $\boldsymbol{\delta}_t := \boldsymbol{\theta}_t - \boldsymbol{\theta}_{t-1}$ and $\boldsymbol{g}_t := \boldsymbol{u}_t - \boldsymbol{u}_{t-1}$.

An ingenious formula that generates a suitable sequence of approximating matrices $\{\mathbf{C}_t\}$ (each satisfying the secant condition) is the BFGS updating formula (B.23), which can be written as the recursion (see Exercise 9):

$$\mathbf{C}_t = \left(\mathbf{I} - \upsilon_t\,\boldsymbol{g}_t\boldsymbol{\delta}_t^\top\right)^\top \mathbf{C}_{t-1}\left(\mathbf{I} - \upsilon_t\,\boldsymbol{g}_t\boldsymbol{\delta}_t^\top\right) + \upsilon_t\,\boldsymbol{\delta}_t\boldsymbol{\delta}_t^\top, \quad \upsilon_t := (\boldsymbol{g}_t^\top\boldsymbol{\delta}_t)^{-1}. \tag{9.7}$$

This formula allows us to update $\mathbf{C}_{t-1}$ to $\mathbf{C}_t$ and then compute $\mathbf{C}_t\,\boldsymbol{u}_t$ in $O(d^2)$ time. While this quasi-Newton approach is better than the $O(d^3)$ cost of Newton's method, it may be still too costly in large-scale applications.

LIMITED MEMORY BFGS

Instead, an approximate or *limited memory BFGS* updating can be achieved in $O(d)$ time. The idea is to store a few of the most recent pairs $\{\boldsymbol{\delta}_t, \boldsymbol{g}_t\}$ in order to evaluate its action on a vector $\boldsymbol{u}_t$ without explicitly constructing and storing $\mathbf{C}_t$ in computer memory. This is possible, because updating $\mathbf{C}_0$ to $\mathbf{C}_1$ in (9.7) requires only the pair $\boldsymbol{\delta}_1, \boldsymbol{g}_1$, and similarly computing $\mathbf{C}_t$ from $\mathbf{C}_0$ only requires the history of the updates $\boldsymbol{\delta}_1, \boldsymbol{g}_1 \ldots, \boldsymbol{\delta}_t, \boldsymbol{g}_t$, which can be shown as follows.

Define the matrices $\mathbf{A}_t, \ldots, \mathbf{A}_0$ via the backward recursion ($j = 1, \ldots, t$):

$$\mathbf{A}_t := \mathbf{I}, \quad \mathbf{A}_{j-1} := \left(\mathbf{I} - \upsilon_j\,\boldsymbol{g}_j\boldsymbol{\delta}_j^\top\right)\mathbf{A}_j,$$

and observe that all matrix vector products: $\mathbf{A}_j\,\boldsymbol{u} =: \boldsymbol{q}_j$, for $j = 0, \ldots, t$ can be computed efficiently via the backward recursion starting with $\boldsymbol{q}_t = \boldsymbol{u}$:

$$\tau_j := \boldsymbol{\delta}_j^\top\boldsymbol{q}_j, \quad \boldsymbol{q}_{j-1} = \boldsymbol{q}_j - \upsilon_j\tau_j\,\boldsymbol{g}_j, \quad j = t, t-1, \ldots, 1. \tag{9.8}$$

In addition to $\{\boldsymbol{q}_j\}$, we will make use of the vectors $\{\boldsymbol{r}_j\}$ defined via the recursion:

$$\boldsymbol{r}_0 := \mathbf{C}_0\,\boldsymbol{q}_0, \quad \boldsymbol{r}_j = \boldsymbol{r}_{j-1} + \upsilon_j\left(\tau_j - \boldsymbol{g}_j^\top \boldsymbol{r}_{j-1}\right)\boldsymbol{\delta}_j, \quad j = 1,\ldots,t. \tag{9.9}$$

At the final iteration t, the BFGS updating formula (9.7) can be rewritten in the form:

$$\mathbf{C}_t = \mathbf{A}_{t-1}^\top \mathbf{C}_{t-1}\mathbf{A}_{t-1} + \upsilon_t\,\boldsymbol{\delta}_t\boldsymbol{\delta}_t^\top.$$

By iterating the recursion (9.7) backwards to $\mathbf{C}_0$, we can write:

$$\mathbf{C}_t = \mathbf{A}_0^\top \mathbf{C}_0\mathbf{A}_0 + \sum_{j=1}^{t} \upsilon_j\,\mathbf{A}_j^\top \boldsymbol{\delta}_j\boldsymbol{\delta}_j^\top \mathbf{A}_j,$$

that is, we can express $\mathbf{C}_t$ in terms of the initial $\mathbf{C}_0$ and the entire history of all BFGS values $\{\boldsymbol{\delta}_j, \boldsymbol{g}_j\}$, as claimed. Further, with the $\{\boldsymbol{q}_j, \boldsymbol{r}_j\}$ computed via (9.8) and (9.9), we can write:

$$\begin{aligned}
\mathbf{C}_t\,\boldsymbol{u} &= \mathbf{A}_0^\top \mathbf{C}_0\,\boldsymbol{q}_0 + \sum_{j=1}^{t} \upsilon_j\left(\boldsymbol{\delta}_j^\top \boldsymbol{q}_j\right)\mathbf{A}_j^\top \boldsymbol{\delta}_j \\
&= \mathbf{A}_0^\top \boldsymbol{r}_0 + \upsilon_1\tau_1\mathbf{A}_1^\top \boldsymbol{\delta}_1 + \sum_{j=2}^{t} \upsilon_j\tau_j\mathbf{A}_j^\top \boldsymbol{\delta}_j \\
&= \mathbf{A}_1^\top\left[(\mathbf{I} - \upsilon_1\boldsymbol{\delta}_1\boldsymbol{g}_1^\top)\,\boldsymbol{r}_0 + \upsilon_1\tau_1\boldsymbol{\delta}_1\right] + \sum_{j=2}^{t} \upsilon_j\tau_j\mathbf{A}_j^\top \boldsymbol{\delta}_j.
\end{aligned}$$

Hence, from the definition of the $\{\boldsymbol{r}_j\}$ in (9.9), we obtain

$$\begin{aligned}
\mathbf{C}_t\boldsymbol{u} &= \mathbf{A}_1^\top \boldsymbol{r}_1 + \sum_{j=2}^{t} \upsilon_j\tau_j\mathbf{A}_j^\top \boldsymbol{\delta}_j \\
&= \mathbf{A}_2^\top \boldsymbol{r}_2 + \sum_{j=3}^{t} \upsilon_j\tau_j\mathbf{A}_j^\top \boldsymbol{\delta}_j \\
&= \cdots = \mathbf{A}_t^\top \boldsymbol{r}_t + \mathbf{0} = \boldsymbol{r}_t.
\end{aligned}$$

Given $\mathbf{C}_0$ and the history of all recent BFGS values $\{\boldsymbol{\delta}_j, \boldsymbol{g}_j\}_{j=1}^{h}$, the computation of the quasi-Newton search direction $\boldsymbol{d} = -\mathbf{C}_h\,\boldsymbol{u}$ can be accomplished via the recursions (9.8) and (9.9) as summarized in Algorithm 9.4.3.

Note that if $\mathbf{C}_0$ is a diagonal matrix, say the identity matrix, then $\mathbf{C}_0\,\boldsymbol{q}$ is cheap to compute and the cost of running Algorithm 9.4.3 is $\mathcal{O}(h\,d)$. Thus, for a fixed length of the BFGS history, the cost of the limited-memory BFGS updating grows linearly in d, making it a viable optimization algorithm in large-scale applications.

Algorithm 9.4.3: Limited-Memory BFGS Update

input: BFGS history list $\{\boldsymbol{\delta}_j, \boldsymbol{g}_j\}_{j=1}^h$, initial $\mathbf{C}_0$, and input $\boldsymbol{u}$.

output: $\boldsymbol{d} = -\mathbf{C}_h \boldsymbol{u}$, where $\mathbf{C}_t = (\mathbf{I} - \upsilon_t \boldsymbol{\delta}_t \boldsymbol{g}_t^\top)\, \mathbf{C}_{t-1} \left(\mathbf{I} - \upsilon_t \boldsymbol{g}_t \boldsymbol{\delta}_t^\top\right) + \upsilon_t \boldsymbol{\delta}_t \boldsymbol{\delta}_t^\top$.

$\boldsymbol{q} \leftarrow \boldsymbol{u}$
for $i = h, h-1, \ldots, 1$ **do** // backward recursion to compute $\mathbf{A}_0 \boldsymbol{u}$
    $\upsilon_i \leftarrow \left(\boldsymbol{\delta}_i^\top \boldsymbol{g}_i\right)^{-1}$
    $\tau_i \leftarrow \boldsymbol{\delta}_i^\top \boldsymbol{q}$
    $\boldsymbol{q} \leftarrow \boldsymbol{q} - \upsilon_i \tau_i \boldsymbol{g}_i$
$\boldsymbol{q} \leftarrow \mathbf{C}_0 \boldsymbol{q}$ // compute $\mathbf{C}_0(\mathbf{A}_0 \boldsymbol{u})$
for $i = 1, \ldots, h$ **do** // compute recursion (9.9)
    $\boldsymbol{q} \leftarrow \boldsymbol{q} + \upsilon_i(\tau_i - \boldsymbol{g}_i^\top \boldsymbol{q})\, \boldsymbol{\delta}_i$
return $\boldsymbol{d} \leftarrow -\boldsymbol{q}$, the value of $-\mathbf{C}_h \boldsymbol{u}$

In summary, a quasi-Newton algorithm with limited-memory BFGS updating reads as follows.

Algorithm 9.4.4: Quasi-Newton Minimization with Limited-Memory BFGS

input: Training set $\tau = \{(\boldsymbol{x}_i, \boldsymbol{y}_i)\}_{i=1}^n$, initial weight matrices and bias vectors $\{\mathbf{W}_l, \boldsymbol{b}_l\}_{l=1}^L =: \boldsymbol{\theta}_1$, activation functions $\{S_l\}_{l=1}^L$, and history parameter h.

output: The parameters of the trained learner.

$t \leftarrow 1$, $\boldsymbol{\delta} \leftarrow 0.1 \times \mathbf{1}$, $\boldsymbol{u}_{t-1} \leftarrow \mathbf{0}$ // initialization
while stopping condition is not met **do**
    Compute $\ell_{\text{value}} = \ell_\tau(\boldsymbol{g}(\cdot \,|\, \boldsymbol{\theta}_t))$ and $\boldsymbol{u}_t = \frac{\partial \ell_\tau}{\partial \boldsymbol{\theta}}(\boldsymbol{\theta}_t)$ via Algorithm 9.3.2.
    $\boldsymbol{g} \leftarrow \boldsymbol{u}_t - \boldsymbol{u}_{t-1}$
    Add $(\boldsymbol{\delta}, \boldsymbol{g})$ to the BFGS history as the newest BFGS pair.
    if the number of pairs in the BFGS history is greater than h **then**
        remove the oldest pair from the BFGS history
    Compute $\boldsymbol{d}$ via Algorithm 9.4.3 using the BFGS history, $\mathbf{C}_0 = \mathbf{I}$, and $\boldsymbol{u}_t$.
    $\alpha \leftarrow 1$
    while $\ell_\tau(\boldsymbol{g}(\cdot \,|\, \boldsymbol{\theta}_t + \alpha \boldsymbol{d})) \geqslant \ell_{\text{value}} + 10^{-4} \alpha\, \boldsymbol{d}^\top \boldsymbol{u}_t$ **do**
        $\alpha \leftarrow \alpha / 1.5$ // line-search along quasi-Newton direction
    $\boldsymbol{\delta} \leftarrow \alpha \boldsymbol{d}$
    $\boldsymbol{\theta}_{t+1} \leftarrow \boldsymbol{\theta}_t + \boldsymbol{\delta}$
    $t \leftarrow t + 1$
return $\boldsymbol{\theta}_t$ as the minimizer of the training loss

9.4.4 Adaptive Gradient Methods

Recall that the limited-memory BFGS method in the previous section determines a search direction using the recent history of previously computed gradients $\{\boldsymbol{u}_t\}$ and input parameters $\{\boldsymbol{\theta}_t\}$. This is because the BFGS pairs $\{\boldsymbol{\delta}_t, \boldsymbol{g}_t\}$ can be easily constructed from the identities: $\boldsymbol{\delta}_t = \boldsymbol{\theta}_t - \boldsymbol{\theta}_{t-1}$ and $\boldsymbol{g}_t = \boldsymbol{u}_t - \boldsymbol{u}_{t-1}$. In other words, using only past gradient computations and with little extra computation, it is possible to infer some of the second-order information contained in the Hessian matrix of $\ell_\tau(\boldsymbol{\theta})$. In addition to the BFGS method, there are other ways in which we can exploit the history of past gradient computations.

One approach is to use the *normal approximation method*, in which the Hessian of ℓ_τ at $\boldsymbol{\theta}_t$ is approximated via ☞ 416

$$\widehat{\mathbf{H}}_t = \gamma \mathbf{I} + \frac{1}{h} \sum_{i=t-h+1}^{t} \boldsymbol{u}_i \boldsymbol{u}_i^\top, \tag{9.10}$$

where $\boldsymbol{u}_{t-h+1}, \ldots, \boldsymbol{u}_t$ are the h most recently computed gradients and γ is a tuning parameter (for example, $\gamma = 1/h$). The search direction is then given by

$$-\widehat{\mathbf{H}}_t^{-1} \boldsymbol{u}_t,$$

which can be computed quickly in $O(h^2\, d)$ time either using the QR decomposition (Exercises 5 and 6), or the Sherman–Morrison Algorithm A.6.1. This approach requires that we store the last h gradient vectors in memory. ☞ 375

Another approach that completely bypasses the need to invert a Hessian approximation is the *Adaptive Gradient* or *AdaGrad* method, in which we only store the diagonal of $\widehat{\mathbf{H}}_t$ and use the search direction:

ADAGRAD

$$-\mathrm{diag}(\widehat{\mathbf{H}}_t)^{-1/2} \boldsymbol{u}_t.$$

We can avoid storing any of the gradient history by instead using the slightly different search direction[2]

$$-\boldsymbol{u}_t \Big/ \sqrt{\boldsymbol{v}_t + \gamma \times \mathbf{1}},$$

where the vector $\boldsymbol{v}_t$ is updated recursively via

$$\boldsymbol{v}_t = \left(1 - \frac{1}{h}\right) \boldsymbol{v}_{t-1} + \frac{1}{h} \boldsymbol{u}_t \odot \boldsymbol{u}_t.$$

With this updating of $\boldsymbol{v}_t$, the difference between the vector $\boldsymbol{v}_t + \gamma \times \mathbf{1}$ and the diagonal of the Hessian $\widehat{\mathbf{H}}_t$ will be negligible.

A more sophisticated version of AdaGrad is the *adaptive moment estimation* or *Adam* method, in which we not only average the vectors $\{\boldsymbol{v}_t\}$, but also average the gradient vectors $\{\boldsymbol{u}_t\}$, as follows.

ADAM

Algorithm 9.4.5: Updating of Search Direction at Iteration t via *Adam*

input: $\boldsymbol{u}_t, \widehat{\boldsymbol{u}}_{t-1}, \boldsymbol{v}_{t-1}, \boldsymbol{\theta}_t$, and parameters (α, h_v, h_u), equal to, e.g., $(10^{-3}, 10^3, 10)$.
output: $\widehat{\boldsymbol{u}}_t, \boldsymbol{v}_t, \boldsymbol{\theta}_{t+1}$.

$\widehat{\boldsymbol{u}}_t \leftarrow \left(1 - \frac{1}{h_u}\right) \widehat{\boldsymbol{u}}_{t-1} + \frac{1}{h_u} \boldsymbol{u}_t$
$\boldsymbol{v}_t \leftarrow \left(1 - \frac{1}{h_v}\right) \boldsymbol{v}_{t-1} + \frac{1}{h_v} \boldsymbol{u}_t \odot \boldsymbol{u}_t$
$\boldsymbol{u}_t^* \leftarrow \widehat{\boldsymbol{u}}_t \Big/ \left(1 - (1 - h_u^{-1})^t\right)$
$\boldsymbol{v}_t^* \leftarrow \boldsymbol{v}_t \Big/ \left(1 - (1 - h_v^{-1})^t\right)$
$\boldsymbol{\theta}_{t+1} \leftarrow \boldsymbol{\theta}_t - \alpha\, \boldsymbol{u}_t^* \Big/ \left(\sqrt{\boldsymbol{v}_t^*} + 10^{-8} \times \mathbf{1}\right)$
return $\widehat{\boldsymbol{u}}_t, \boldsymbol{v}_t, \boldsymbol{\theta}_{t+1}$

[2]Here we divide two vectors componentwise.

MOMENTUM METHOD

Yet another computationally cheap approach is the *momentum method*, in which the steepest descent iteration (9.6) is modified to

$$\boldsymbol{\theta}_{t+1} = \boldsymbol{\theta}_t - \alpha_t \boldsymbol{u}_t + \gamma \boldsymbol{\delta}_t,$$

where $\boldsymbol{\delta}_t = \boldsymbol{\theta}_t - \boldsymbol{\theta}_{t-1}$ and γ is a tuning parameter. This strategy frequently performs better than the "vanilla" steepest descent method, because the search direction is less likely to change abruptly.

Numerical experience suggests that the vanilla steepest-descent Algorithm 9.4.1 and the Levenberg–Marquardt Algorithm B.3.3 are effective for networks with shallow architectures, but not for networks with deep architectures. In comparison, the stochastic gradient descent method, the limited-memory BFGS Algorithm 9.4.4, or any of the adaptive gradient methods in this section, can frequently handle networks with many hidden layers (provided that any tuning parameters and initialization values are carefully chosen via experimentation).

9.5 Examples in Python

In this section we provide two numerical examples in Python. In the first example, we train a neural network with the *stochastic gradient descent* method using the polynomial regression data from Example 2.1, and without using any specialized Python packages. ☞ 26

In the second example, we consider a realistic application of a neural network to image recognition and classification. Here we use the specialized open-source Python package `Pytorch`.

9.5.1 Simple Polynomial Regression

Consider again the polynomial regression data set depicted in Figure 2.4. We use a network with architecture

$$[p_0, p_1, p_2, p_3] = [1, 20, 20, 1].$$

In other words, we have two hidden layers with 20 neurons, resulting in a learner with a total of $\dim(\boldsymbol{\theta}) = 481$ parameters. To implement such a neural network, we first import the `numpy` and the `matplotlib` packages, then read the regression problem data and define the feed-forward neural network layers.

NeuralNetPurePython.py

```
import numpy as np
import matplotlib.pyplot as plt

#%%
# import data
data = np.genfromtxt('polyreg.csv',delimiter=',')
X = data[:,0].reshape(-1,1)
y = data[:,1].reshape(-1,1)

# Network setup
p = [X.shape[1],20,20,1] # size of layers
L = len(p)-1             # number of layers
```

Next, the **initialize** method generates random initial weight matrices and bias vectors $\{\mathbf{W}_l, \boldsymbol{b}_l\}_{l=1}^{L}$. Specifically, all parameters are initialized with values distributed according to the standard normal distribution.

```
def initialize(p, w_sig = 1):
    W, b = [[]]*len(p), [[]]*len(p)

    for l in range(1,len(p)):
       W[l]= w_sig * np.random.randn(p[l], p[l-1])
       b[l]= w_sig * np.random.randn(p[l], 1)
    return W,b

W,b = initialize(p) # initialize weight matrices and bias vectors
```

The following code implements the ReLU activation function from Figure 9.2 and the squared error loss. Note that these functions return both the function values and the corresponding gradients.

```
def RELU(z,l):  # RELU activation function: value and derivative
    if l == L: return z, np.ones_like(z)
    else:
        val = np.maximum(0,z) # RELU function element-wise
        J = np.array(z>0, dtype = float) # derivative of RELU
           element-wise
        return val, J

def loss_fn(y,g):
    return (g - y)**2, 2 * (g - y)

S = RELU
```

Next, we implement the feed-forward and backward-propagation Algorithm 9.3.1. Here, we have implemented Algorithm 9.3.2 inside the backward-propagation loop.

```
def feedforward(x,W,b):
    a, z, gr_S = [0]*(L+1), [0]*(L+1), [0]*(L+1)

    a[0] = x.reshape(-1,1)
    for l in range(1,L+1):
        z[l] = W[l] @ a[l-1] + b[l] # affine transformation
        a[l], gr_S[l] = S(z[l],l) # activation function
    return a, z, gr_S

def backward(W,b,X,y):
    n =len(y)
    delta = [0]*(L+1)
    dC_db, dC_dW = [0]*(L+1), [0]*(L+1)
    loss=0

    for i in range(n): # loop over training examples
        a, z, gr_S = feedforward(X[i,:].T, W, b)
        cost, gr_C = loss_fn(y[i], a[L]) # cost i and gradient wrt g
        loss += cost/n
```

```
        delta[L] = gr_S[L] @ gr_C

        for l in range(L,0,-1): # l = L,...,1
            dCi_dbl = delta[l]
            dCi_dWl = delta[l] @  a[l-1].T

            # ---- sum up over samples ----
            dC_db[l] = dC_db[l] + dCi_dbl/n
            dC_dW[l] = dC_dW[l] + dCi_dWl/n
            # -----------------------------

            delta[l-1] =  gr_S[l-1] * W[l].T @ delta[l]

    return dC_dW, dC_db, loss
```

As explained in Section 9.4, it is sometimes more convenient to collect all the weight matrices and bias vectors $\{\mathbf{W}_l, \boldsymbol{b}_l\}_{l=1}^L$ into a single vector $\boldsymbol{\theta}$. Consequently, we code two functions that map the weight matrices and the bias vectors into a single parameter vector, and vice versa.

```
def list2vec(W,b):
    # converts list of weight matrices and bias vectors into
    # one column vector
    b_stack = np.vstack([b[i] for i in range(1,len(b))] )
    W_stack = np.vstack(W[i].flatten().reshape(-1,1) for i in range
        (1,len(W)))
    vec = np.vstack([b_stack, W_stack])
    return vec
#%%
def vec2list(vec, p):
    # converts vector to weight matrices and bias vectors
    W, b = [[]]*len(p),[[]]*len(p)
    p_count = 0

    for l in range(1,len(p)): # construct bias vectors
        b[l] = vec[p_count:(p_count+p[l])].reshape(-1,1)
        p_count = p_count + p[l]

    for l in range(1,len(p)): # construct weight matrices
        W[l] = vec[p_count:(p_count + p[l]*p[l-1])].reshape(p[l], p[
            l-1])
        p_count = p_count + (p[l]*p[l-1])

    return W, b
```

Finally, we run the *stochastic gradient descent* for 10^4 iterations using a minibatch of size 20 and a constant learning rate of $\alpha_t = 0.005$.

```
batch_size = 20
lr = 0.005
beta = list2vec(W,b)
loss_arr = []
```

```
n = len(X)
num_epochs = 10000
print("epoch | batch loss")
print("----------------------------")
for epoch in range(1,num_epochs+1):
    batch_idx = np.random.choice(n,batch_size)
    batch_X = X[batch_idx].reshape(-1,1)
    batch_y=y[batch_idx].reshape(-1,1)
    dC_dW, dC_db, loss = backward(W,b,batch_X,batch_y)
    d_beta = list2vec(dC_dW,dC_db)
    loss_arr.append(loss.flatten()[0])
    if(epoch==1 or np.mod(epoch,1000)==0):
        print(epoch,": ",loss.flatten()[0])
    beta = beta - lr*d_beta
    W,b = vec2list(beta,p)

# calculate the loss of the entire training set
dC_dW, dC_db, loss = backward(W,b,X,y)
print("entire training set loss = ",loss.flatten()[0])
xx = np.arange(0,1,0.01)
y_preds = np.zeros_like(xx)

for i in range(len(xx)):
    a, _, _ = feedforward(xx[i],W,b)
    y_preds[i], = a[L]

plt.plot(X,y, 'r.', markersize = 4,label = 'y')
plt.plot(np.array(xx), y_preds, 'b',label = 'fit')
plt.legend()
plt.xlabel('x')
plt.ylabel('y')
plt.show()
plt.plot(np.array(loss_arr), 'b')
plt.xlabel('iteration')
plt.ylabel('Training Loss')
plt.show()
```

```
epoch | batch loss
----------------------------
1 :  158.6779278688539
1000 :  54.52430507401445
2000 :  38.346572088604965
3000 :  31.02036319180713
4000 :  22.91114276931535
5000 :  27.75810262906341
6000 :  22.296907007032928
7000 :  17.337367420038046
8000 :  19.233689945334195
9000 :  39.54261478969857
10000 :  14.754724387604416
entire training set loss =  28.904957963612727
```

The left panel of Figure 9.5 shows a trained neural network with a training loss of approximately 28.9. As seen from the right panel of Figure 9.5, the algorithm initially makes rapid progress until it settles down into a stationary regime after 400 iterations.

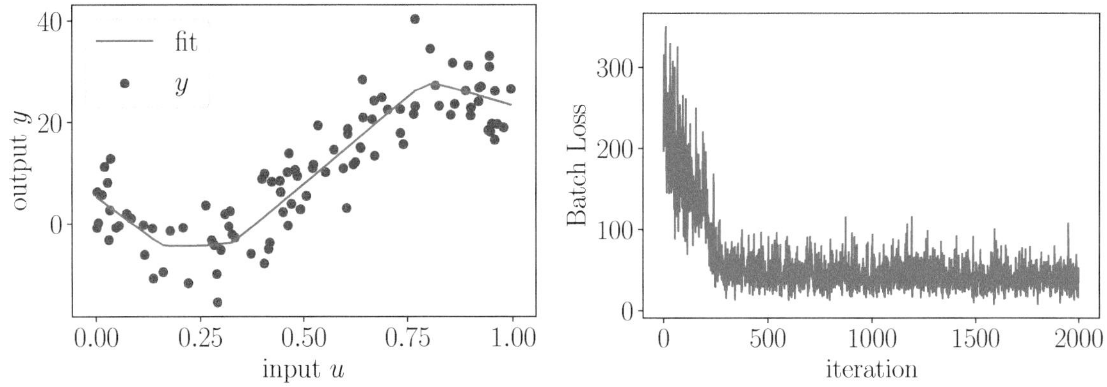

Figure 9.5: Left panel: The fitted neural network with training loss of $\ell_\tau(g_\tau) \approx 28.9$. Right panel: The evolution of the estimated loss, $\widehat{\ell_\tau}(g_\tau(\cdot \,|\, \boldsymbol{\theta}))$, over the steepest-descent iterations.

9.5.2 Image Classification

In this section, we will use the package `Pytorch`, which is an open-source machine learning library for Python. `Pytorch` can easily exploit any graphics processing unit (GPU) for accelerated computation. As an example, we consider the Fashion-MNIST data set from `https://www.kaggle.com/zalando-research/fashionmnist`. The Fashion-MNIST data set contains 28×28 gray-scale images of clothing. Our task is to classify each image according to its label. Specifically, the labels are: T-Shirt, Trouser, Pullover, Dress, Coat, Sandal, Shirt, Sneaker, Bag, and Ankle Boot. Figure 9.6 depicts a typical ankle boot in the left panel and a typical dress in the right panel. To start with, we import the required libraries and load the Fashion-MNIST data set.

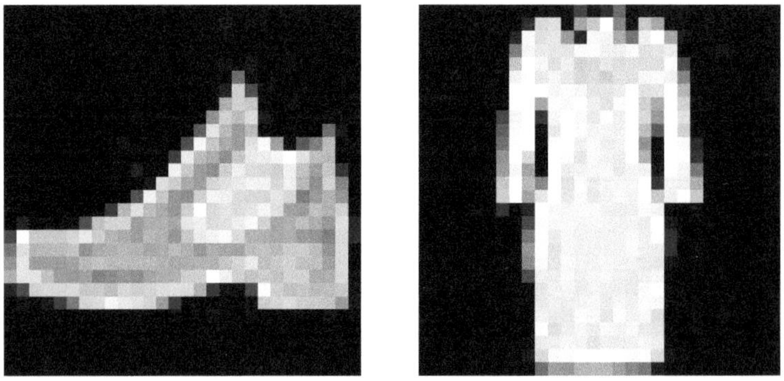

Figure 9.6: Left: an ankle boot. Right: a dress.

ImageClassificationPytorch.py

```
import torch
import torch.nn as nn
from torch.autograd import Variable
import pandas as pd
import numpy as np
```

```
import matplotlib.pyplot as plt
from torch.utils.data import Dataset, DataLoader
from PIL import Image
import torch.nn.functional as F
#################################################################
# data loader class
#################################################################
class LoadData(Dataset):
    def __init__(self, fName, transform=None):
        data = pd.read_csv(fName)
        self.X = np.array(data.iloc[:, 1:], dtype=np.uint8).reshape
           (-1, 1, 28, 28)
        self.y = np.array(data.iloc[:, 0])

    def __len__(self):
        return len(self.X)

    def __getitem__(self, idx):
        img = self.X[idx]
        lbl = self.y[idx]
        return (img, lbl)

# load the image data
train_ds = LoadData('fashionmnist/fashion-mnist_train.csv')
test_ds  = LoadData('fashionmnist/fashion-mnist_test.csv')

# set labels dictionary
labels = {0 : 'T-Shirt', 1 : 'Trouser', 2 : 'Pullover',
          3 : 'Dress', 4 : 'Coat', 5 : 'Sandal', 6 : 'Shirt',
          7 : 'Sneaker', 8 : 'Bag', 9 : 'Ankle Boot'}
```

Since an image input data is generally memory intensive, it is important to partition the data set into (mini-)batches. The code below defines a batch size of 100 images and initializes the Pytorch data loader objects. These objects will be used for efficient iteration over the data set.

```
# load the data in batches
batch_size = 100

train_loader = torch.utils.data.DataLoader(dataset=train_ds,
                                           batch_size=batch_size,
                                           shuffle=True)
test_loader = torch.utils.data.DataLoader(dataset=test_ds,
                                          batch_size=batch_size,
                                          shuffle=True)
```

Next, to define the network architecture in Pytorch all we need to do is define an instance of the `torch.nn.Module` class. Choosing a network architecture with good generalization properties can be a difficult task. Here, we use a network with two convolution layers (defined in the `cnn_layer` block), a 3×3 kernel, and three hidden layers (defined in the `flat_layer` block). Since there are ten possible output labels, the output layer has ten nodes. More specifically, the first and the second convolution layers have 16 and 32 output channels. Combining this with the definition of the 3×3 kernel, we conclude that the size of the first flat hidden layer should be:

$$\left(\overbrace{\underbrace{(28-3+1)}_{\text{first convolution layer}} -3+1}^{\text{second convolution layer}}\right)^2 \times 32 = 18432,$$

where the multiplication by 32 follows from the fact that the second convolution layer has 32 output channels. Having said that, the `flat_fts` variable determines the number of output layers of the convolution block. This number is used to define the size of the first hidden layer of the `flat_layer` block. The rest of the hidden layers have 100 neurons and we use the ReLU activation function for all layers. Finally, note that the `forward` method in the `CNN` class implements the forward pass.

```
# define the network
class CNN(nn.Module):

    def __init__(self):
        super(CNN, self).__init__()

        self.cnn_layer = nn.Sequential(
                nn.Conv2d(1, 16, kernel_size=3, stride=(1,1)),
                nn.ReLU(),
                nn.Conv2d(16, 32, kernel_size=3, stride=(1,1)),
                nn.ReLU(),
        )
        self.flat_fts = (((28-3+1)-3+1)**2)*32


        self.flat_layer = nn.Sequential(
                nn.Linear(self.flat_fts, 100),
                nn.ReLU(),
                nn.Linear(100, 100),
                nn.ReLU(),
                nn.Linear(100, 100),
                nn.ReLU(),
                nn.Linear(100, 10))

    def forward(self, x):
        out = self.cnn_layer(x)
        out = out.view(-1, self.flat_fts)
        out = self.flat_layer(out)
        return out
```

Next, we specify how the network will be trained. We choose the device type, namely, the central processing unit (CPU) or the GPU (if available), the number of training iterations (epochs), and the learning rate. Then, we create an instance of the proposed convolution network and send it to the predefined device (CPU or GPU). Note how easily one can switch between the CPU or the GPU without major changes to the code.

In addition to the specifications above, we need to choose an appropriate loss function and training algorithm. Here, we use the *cross-entropy* loss and the *Adam* adaptive gradient Algorithm 9.4.5. Once these parameters are set, the learning proceeds to evaluate the gradient of the loss function via the back-propagation algorithm.

☞ 269

```
# learning parameters
num_epochs = 50
learning_rate = 0.001

#device = torch.device ('cpu') # use this to run on CPU
device = torch.device ('cuda') # use this to run on GPU

#instance of the Conv Net
cnn = CNN()
cnn.to(device=device)

#loss function and optimizer
criterion = nn.CrossEntropyLoss()
optimizer = torch.optim.Adam(cnn.parameters(), lr=learning_rate)

# the learning loop
losses = []
for epoch in range(1,num_epochs+1):
    for i, (images, labels) in enumerate(train_loader):
        images = Variable(images.float()).to(device=device)
        labels = Variable(labels).to(device=device)

        optimizer.zero_grad()
        outputs = cnn(images)
        loss = criterion(outputs, labels)
        loss.backward()
        optimizer.step()

        losses.append(loss.item())
    if(epoch==1 or epoch % 10 == 0):
        print ("Epoch : ", epoch, ", Training Loss: ",  loss.item())

# evaluate on the test set
cnn.eval()
correct = 0
total = 0
for images, labels in test_loader:
    images = Variable(images.float()).to(device=device)
    outputs = cnn(images)
    _, predicted = torch.max(outputs.data, 1)
    total += labels.size(0)
    correct += (predicted.cpu() == labels).sum()
print("Test Accuracy of the model on the 10,000 training test images
    : ", (100 * correct.item() / total),"%")

# plot
plt.rc('text', usetex=True)
plt.rc('font', family='serif',size=20)
plt.tight_layout()

plt.plot(np.array(losses)[10:len(losses)])
plt.xlabel(r'{iteration}',fontsize=20)
plt.ylabel(r'{Batch Loss}',fontsize=20)
plt.subplots_adjust(top=0.8)
plt.show()
```

```
Epoch :  1 , Training Loss:  0.412550151348114
Epoch :  10 , Training Loss:  0.05452106520533562
Epoch :  20 , Training Loss:  0.07233225554227829
Epoch :  30 , Training Loss:  0.01696968264877796
Epoch :  40 , Training Loss:  0.0008199119474738836
Epoch :  50 , Training Loss:  0.006860652007162571
Test Accuracy of the model on the 10,000 training test images: 91.02 %
```

Finally, we evaluate the network performance using the test data set. A typical mini-batch loss as a function of iteration is shown in Figure 9.7 and the proposed neural network achieves about 91% accuracy on the test set.

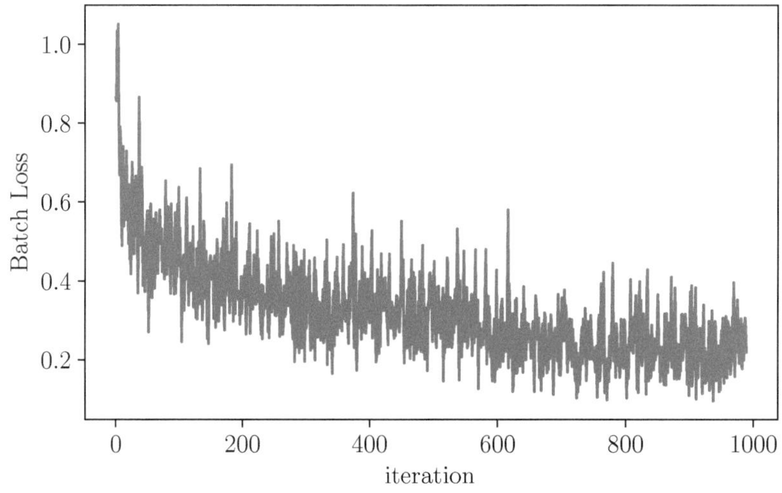

Figure 9.7: The batch loss history.

Further Reading

A popular book written by some of the pioneers of deep learning is [53]. For an excellent and gentle introduction to the intuition behind neural networks, we recommend [94]. A summary of many effective gradient descent methods for training of deep networks is given in [105]. An early resource on the limited-memory BFGS method is [81], and a more recent resource includes [13], which makes recommendations on the best choice for the length of the BFGS history (that is, the value of the parameter h).

Exercises

1. Show that the softmax function

$$\text{softmax} : \boldsymbol{z} \mapsto \frac{\exp(\boldsymbol{z})}{\sum_k \exp(z_k)}$$

satisfies the invariance property:

$$\text{softmax}(\boldsymbol{z}) = \text{softmax}(\boldsymbol{z} + c \times \mathbf{1}), \quad \text{for any constant } c.$$

2. *Projection pursuit* is a network with one hidden layer that can be written as: PROJECTION PURSUIT

$$g(\boldsymbol{x}) = S(\boldsymbol{\omega}^\top \boldsymbol{x}),$$

where S is a univariate *smoothing cubic spline*. If we use squared-error loss with $\tau_n = \{y_i, \boldsymbol{x}_i\}_{i=1}^n$, we need to minimize the training loss: ☞ 235

$$\frac{1}{n}\sum_{i=1}^{n}\left(y_i - S(\boldsymbol{\omega}^\top \boldsymbol{x}_i)\right)^2$$

with respect to $\boldsymbol{\omega}$ and all cubic smoothing splines. This training of the network is typically tackled iteratively in a manner similar to the *EM algorithm*. In particular, we iterate ($t = 1, 2, \ldots$) the following steps until convergence. ☞ 139

(a) Given the *missing data* $\boldsymbol{\omega}_t$, compute the spline S_t by training a cubic smoothing spline on $\{y_i, \boldsymbol{\omega}_t^\top \boldsymbol{x}_i\}$. The smoothing coefficient of the spline may be determined as part of this step.

(b) Given the spline function S_t, compute the next projection vector $\boldsymbol{\omega}_{t+1}$ via *iterative reweighted least squares*: ITERATIVE REWEIGHTED LEAST SQUARES

$$\boldsymbol{\omega}_{t+1} = \underset{\boldsymbol{\beta}}{\operatorname{argmin}}\, (\boldsymbol{e}_t - \mathbf{X}\boldsymbol{\beta})^\top \boldsymbol{\Sigma}_t (\boldsymbol{e}_t - \mathbf{X}\boldsymbol{\beta}), \tag{9.11}$$

where

$$e_{t,i} := \boldsymbol{\omega}_t^\top \boldsymbol{x}_i + \frac{y_i - S_t(\boldsymbol{\omega}_t^\top \boldsymbol{x}_i)}{S_t'(\boldsymbol{\omega}_t^\top \boldsymbol{x}_i)}, \quad i = 1, \ldots, n$$

is the adjusted response, and $\boldsymbol{\Sigma}_t^{1/2} = \operatorname{diag}(S_t'(\boldsymbol{\omega}_t^\top \boldsymbol{x}_1), \ldots, S_t'(\boldsymbol{\omega}_t^\top \boldsymbol{x}_n))$ is a diagonal matrix.

Apply Taylor's Theorem B.1 to the function S_t and derive the iterative reweighted least-squares optimization program (9.11). ☞ 402

3. Suppose that in the *stochastic gradient descent* method we wish to repeatedly draw minibatches of size N from τ_n, where we assume that $N \times m = n$ for some large integer m. Instead of repeatedly resampling from τ_n, an alternative is to reshuffle τ_n via a random permutation $\boldsymbol{\Pi}$ and then advance sequentially through the reshuffled training set to construct m non-overlapping minibatches. A single traversal of such a reshuffled training set is called an *epoch*. The following pseudo-code describes the procedure. ☞ 337 ☞ 115 EPOCH

Algorithm 9.5.1: Stochastic Gradient Descent with Reshuffling

input: Training set $\tau_n = \{(\boldsymbol{x}_i, \boldsymbol{y}_i)\}_{i=1}^n$, initial weight matrices and bias vectors $\{\mathbf{W}_l, \boldsymbol{b}_l\}_{l=1}^L \to \boldsymbol{\theta}_1$, activation functions $\{S_l\}_{l=1}^L$, learning rates $\{\alpha_1, \alpha_2, \ldots\}$.
output: The parameters of the trained learner.

$t \leftarrow 1$ and epoch $\leftarrow 0$
while stopping condition is not met **do**
  Draw $U_1, \ldots, U_n \overset{\text{iid}}{\sim} \mathcal{U}(0,1)$.
  Let $\boldsymbol{\Pi}$ be the permutation of $\{1, \ldots, n\}$ that satisfies $U_{\Pi_1} < \cdots < U_{\Pi_n}$.
  $(\boldsymbol{x}_i, \boldsymbol{y}_i) \leftarrow (\boldsymbol{x}_{\Pi_i}, \boldsymbol{y}_{\Pi_i})$ for $i = 1, \ldots, n$ // reshuffle τ_n
  for $j = 1, \ldots, m$ **do**
    $\widehat{\ell_\tau} \leftarrow \frac{1}{N} \sum_{i=(j-1)N+1}^{jN} \text{Loss}(\boldsymbol{y}_i, \boldsymbol{g}(\boldsymbol{x}_i \,|\, \boldsymbol{\theta}))$
    $\boldsymbol{\theta}_{t+1} \leftarrow \boldsymbol{\theta}_t - \alpha_t \frac{\partial \widehat{\ell_\tau}}{\partial \boldsymbol{\theta}}(\boldsymbol{\theta}_t)$
    $t \leftarrow t + 1$
  epoch $\leftarrow$ epoch + 1 // number of reshuffles or epochs
return $\boldsymbol{\theta}_t$ as the minimizer of the training loss

☞ 342
Write Python code that implements the stochastic gradient descent with data reshuffling, and use it to train the neural net in Section 9.5.1.

4. Denote the pdf of the $\mathcal{N}(\mathbf{0}, \boldsymbol{\Sigma})$ distribution by $\varphi_{\boldsymbol{\Sigma}}(\cdot)$, and let

$$\mathcal{D}(\boldsymbol{\mu}_0, \boldsymbol{\Sigma}_0 \,|\, \boldsymbol{\mu}_1, \boldsymbol{\Sigma}_1) = \int_{\mathbb{R}^d} \varphi_{\boldsymbol{\Sigma}_0}(\boldsymbol{x} - \boldsymbol{\mu}_0) \ln \frac{\varphi_{\boldsymbol{\Sigma}_0}(\boldsymbol{x} - \boldsymbol{\mu}_0)}{\varphi_{\boldsymbol{\Sigma}_1}(\boldsymbol{x} - \boldsymbol{\mu}_1)} \, \mathrm{d}\boldsymbol{x}$$

☞ 42
be the Kullback–Leibler divergence between the densities of the $\mathcal{N}(\boldsymbol{\mu}_0, \boldsymbol{\Sigma}_0)$ and $\mathcal{N}(\boldsymbol{\mu}_1, \boldsymbol{\Sigma}_1)$ distributions on $\mathbb{R}^d$. Show that

$$2\mathcal{D}(\boldsymbol{\mu}_0, \boldsymbol{\Sigma}_0 \,|\, \boldsymbol{\mu}_1, \boldsymbol{\Sigma}_1) = \text{tr}(\boldsymbol{\Sigma}_1^{-1}\boldsymbol{\Sigma}_0) - \ln |\boldsymbol{\Sigma}_1^{-1}\boldsymbol{\Sigma}_0| + (\boldsymbol{\mu}_1 - \boldsymbol{\mu}_0)^\top \boldsymbol{\Sigma}_1^{-1} (\boldsymbol{\mu}_1 - \boldsymbol{\mu}_0) - d.$$

Hence, deduce the formula in (B.22).

5. Suppose that we wish to compute the inverse and log-determinant of the matrix

$$\mathbf{I}_n + \mathbf{U}\mathbf{U}^\top,$$

where $\mathbf{U}$ is an $n \times h$ matrix with $h \ll n$. Show that

$$(\mathbf{I}_n + \mathbf{U}\mathbf{U}^\top)^{-1} = \mathbf{I}_n - \mathbf{Q}_n\mathbf{Q}_n^\top,$$

☞ 377
where $\mathbf{Q}_n$ contains the first n rows of the $(n+h) \times h$ matrix $\mathbf{Q}$ in the QR factorization of the $(n+h) \times h$ matrix:

$$\begin{bmatrix} \mathbf{U} \\ \mathbf{I}_h \end{bmatrix} = \mathbf{Q}\mathbf{R}.$$

In addition, show that $\ln |\mathbf{I}_n + \mathbf{U}\mathbf{U}^\top| = \sum_{i=1}^h \ln r_{ii}^2$, where $\{r_{ii}\}$ are the diagonal elements of the $h \times h$ matrix $\mathbf{R}$.

6. Suppose that

$$\mathbf{U} = [\boldsymbol{u}_0, \boldsymbol{u}_1, \ldots, \boldsymbol{u}_{h-1}],$$

where all $\boldsymbol{u} \in \mathbb{R}^n$ are column vectors and we have computed $(\mathbf{I}_n + \mathbf{U}\mathbf{U}^\top)^{-1}$ via the QR factorization method in Exercise 5. If the columns of matrix $\mathbf{U}$ are updated to

$$[\boldsymbol{u}_1, \ldots, \boldsymbol{u}_{h-1}, \boldsymbol{u}_h],$$

show that the inverse $(\mathbf{I}_n + \mathbf{U}\mathbf{U}^\top)^{-1}$ can be updated in $\mathcal{O}(h\,n)$ time (rather than computed from scratch in $\mathcal{O}(h^2\,n)$ time). Deduce that the computing cost of updating the Hessian approximation (9.10) is the same as that for the *limited-memory BFGS* Algorithm 9.4.3.

In your solution you may use the following facts from [29]. Suppose we are given the $\mathbf{Q}$ and $\mathbf{R}$ factors in the QR factorization of a matrix $\mathbf{A} \in \mathbb{R}^{n\times h}$. If a row/column is added to matrix $\mathbf{A}$, then the $\mathbf{Q}$ and $\mathbf{R}$ factors need not be recomputed from scratch (in $\mathcal{O}(h^2\,n)$ time), but can be updated efficiently in $\mathcal{O}(h\,n)$ time. Similarly, if a row/column is removed from matrix $\mathbf{A}$, then the $\mathbf{Q}$ and $\mathbf{R}$ factors can be updated in $\mathcal{O}(h^2)$ time.

7. Suppose that $\mathbf{U} \in \mathbb{R}^{n\times h}$ has its k-th column $\boldsymbol{v}$ replaced with $\boldsymbol{w}$, giving the updated $\widetilde{\mathbf{U}}$.

(a) If $\boldsymbol{e} \in \mathbb{R}^h$ denotes the unit-length vector such that $e_k = \|\boldsymbol{e}\| = 1$ and

$$\boldsymbol{r}_\pm := \frac{\sqrt{2}}{2}\mathbf{U}^\top(\boldsymbol{w} - \boldsymbol{v}) + \frac{\sqrt{2}\,\|\boldsymbol{w} - \boldsymbol{v}\|^2}{4}\boldsymbol{e} \pm \frac{\sqrt{2}}{2}\boldsymbol{e},$$

show that

$$\widetilde{\mathbf{U}}^\top\widetilde{\mathbf{U}} = \mathbf{U}^\top\mathbf{U} + \boldsymbol{r}_+\boldsymbol{r}_+^\top - \boldsymbol{r}_-\boldsymbol{r}_-^\top.$$

[Hint: You may find Exercise 16 in Chapter 6 useful.] ☞ 248

(b) Let $\mathbf{B} := (\mathbf{I}_h + \mathbf{U}^\top\mathbf{U})^{-1}$. Use the *Woodbury identity* (A.15) to show that ☞ 373

$$(\mathbf{I}_n + \widetilde{\mathbf{U}}\widetilde{\mathbf{U}}^\top)^{-1} = \mathbf{I}_n - \widetilde{\mathbf{U}}\left(\mathbf{B}^{-1} + \boldsymbol{r}_+\boldsymbol{r}_+^\top - \boldsymbol{r}_-\boldsymbol{r}_-^\top\right)^{-1}\widetilde{\mathbf{U}}^\top.$$

(c) Suppose that we have stored $\mathbf{B}$ in computer memory. Use Algorithm 6.8.1 and parts (a) and (b) to write pseudo-code that updates $(\mathbf{I}_n+\mathbf{U}\mathbf{U}^\top)^{-1}$ to $(\mathbf{I}_n+\widetilde{\mathbf{U}}\widetilde{\mathbf{U}}^\top)^{-1}$ in $\mathcal{O}((n+h)h)$ computing time.

8. Equation (9.7) gives the rank-two BFGS update of the inverse Hessian $\mathbf{C}_{t-1}$ to $\mathbf{C}_t$. Instead of using a two-rank update, we can consider a one-rank update, in which $\mathbf{C}_{t-1}$ is updated to $\mathbf{C}_t$ by the general rank-one formula:

$$\mathbf{C}_t = \mathbf{C}_{t-1} + \upsilon_t\,\boldsymbol{r}_t\boldsymbol{r}_t^\top.$$

Find values for the scalar υ_t and vector $\boldsymbol{r}_t$, such that $\mathbf{C}_t$ satisfies the secant condition $\mathbf{C}_t\boldsymbol{g}_t = \boldsymbol{\delta}_t$.

9. Show that the *BFGS formula* (B.23) can be written as:

$$\mathbf{C} \leftarrow \left(\mathbf{I} - \upsilon\boldsymbol{g}\boldsymbol{\delta}^\top\right)^\top\mathbf{C}\left(\mathbf{I} - \upsilon\boldsymbol{g}\boldsymbol{\delta}^\top\right) + \upsilon\boldsymbol{\delta}\boldsymbol{\delta}^\top,$$

where $\upsilon := (\boldsymbol{g}^\top\boldsymbol{\delta})^{-1}$.

10. Show that the *BFGS formula* (B.23) is the solution to the constrained optimization problem:

$$\mathbf{C}_{\text{BFGS}} = \underset{\mathbf{A} \text{ subject to } \mathbf{A}\boldsymbol{g} = \boldsymbol{\delta},\, \mathbf{A} = \mathbf{A}^\top}{\operatorname{argmin}} \mathcal{D}(\mathbf{0}, \mathbf{C} \,|\, \mathbf{0}, \mathbf{A}),$$

where $\mathcal{D}$ is the Kullback–Leibler discrepancy defined in (B.22). On the other hand, show that the *DFP formula* (B.24) is the solution to the constrained optimization problem:

$$\mathbf{C}_{\text{DFP}} = \underset{\mathbf{A} \text{ subject to } \mathbf{A}\boldsymbol{g} = \boldsymbol{\delta},\, \mathbf{A} = \mathbf{A}^\top}{\operatorname{argmin}} \mathcal{D}(\mathbf{0}, \mathbf{A} \,|\, \mathbf{0}, \mathbf{C}).$$

☞ 213
11. Consider again the logistic regression model in Exercise 5.18, which used *iterative reweighted least squares* for training the learner. Repeat all the computations, but this time using the *limited-memory BFGS* Algorithm 9.4.4. Which training algorithm converges faster to the optimal solution?

12. Download the `seeds_dataset.txt` data set from the book's GitHub site, which contains 210 independent examples. The categorical output (response) here is the type of wheat grain: Kama, Rosa, and Canadian (encoded as 1, 2, and 3), so that $c = 3$. The seven continuous features (explanatory variables) are measurements of the geometrical properties of the grain (area, perimeter, compactness, length, width, asymmetry coefficient, and length of kernel groove). Thus, $\boldsymbol{x} \in \mathbb{R}^7$ (which does not include the constant feature 1) and the
☞ 330
multi-logit pre-classifier in Example 9.2 can be written as $\boldsymbol{g}(\boldsymbol{x}) = \text{softmax}(\mathbf{W}\boldsymbol{x} + \boldsymbol{b})$, where $\mathbf{W} \in \mathbb{R}^{3\times 7}$ and $\boldsymbol{b} \in \mathbb{R}^3$. Implement and train this pre-classifier on the first $n = 105$ examples of the seeds data set using, for example, Algorithm 9.4.1. Use the remaining $n' = 105$ examples in the data set to estimate the generalization risk of the learner using the cross-entropy loss. [Hint: Use the cross-entropy loss formulas from Example 9.4.]

13. In Exercise 12 above, we train the multi-logit classifier using a weight matrix $\mathbf{W} \in \mathbb{R}^{3\times 7}$ and bias vector $\boldsymbol{b} \in \mathbb{R}^3$. Repeat the training of the multi-logit model, but this time keeping z_1 as an arbitrary constant (say $z_1 = 0$), and thus setting $c = 0$ to be a "reference" class. This has the effect of removing a node from the output layer of the network, giving a weight
☞ 269
matrix $\mathbf{W} \in \mathbb{R}^{2\times 7}$ and bias vector $\boldsymbol{b} \in \mathbb{R}^2$ of smaller dimensions than in (7.16).

☞ 335
14. Consider again Example 9.4, where we used a *softmax* output function $\boldsymbol{S}_L$ in conjunction with the *cross-entropy* loss: $C(\boldsymbol{\theta}) = -\ln g_{y+1}(\boldsymbol{x} \,|\, \boldsymbol{\theta})$. Find formulas for $\frac{\partial C}{\partial \boldsymbol{g}}$ and $\frac{\partial \boldsymbol{S}_L}{\partial \boldsymbol{z}_L}$. Hence, verify that:

$$\frac{\partial \boldsymbol{S}_L}{\partial \boldsymbol{z}_L}\frac{\partial C}{\partial \boldsymbol{g}} = \boldsymbol{g}(\boldsymbol{x} \,|\, \boldsymbol{\theta}) - \boldsymbol{e}_{y+1},$$

where $\boldsymbol{e}_i$ is the unit length vector with an entry of 1 in the i-th position.

☞ 414
15. Derive the formula (B.25) for a diagonal Hessian update in a quasi-Newton method for minimization. In other words, given a current minimizer $\boldsymbol{x}_t$ of $f(\boldsymbol{x})$, a diagonal matrix $\mathbf{C}$ of approximating the Hessian of f, and a gradient vector $\boldsymbol{u} = \nabla f(\boldsymbol{x}_t)$, find the solution to the constrained optimization program:

$$\min_{\mathbf{A}} \mathcal{D}(\boldsymbol{x}_t, \mathbf{C} \,|\, \boldsymbol{x}_t - \mathbf{A}\boldsymbol{u}, \mathbf{A})$$
$$\text{subject to: } \mathbf{A}\boldsymbol{g} \geqslant \boldsymbol{\delta},\ \mathbf{A} \text{ is diagonal},$$

where $\mathcal{D}$ is the Kullback–Leibler distance defined in (B.22) (see Exercise 4).

16. Consider again the Python implementation of the polynomial regression in Section 9.5.1, where the *stochastic gradient descent* was used for training.

Using the polynomial regression data set, implement and run the following four alternative training methods:

(a) the steepest-descent Algorithm 9.4.1;

(b) the Levenberg–Marquardt Algorithm B.3.3, in conjunction with Algorithm 9.4.2 for computing the matrix of Jacobi; ☞ 417

(c) the *limited-memory BFGS* Algorithm 9.4.4;

(d) the *Adam* Algorithm 9.4.5, which uses past gradient values to determine the next search direction.

For each training algorithm, using trial and error, tune any algorithmic parameters so that the network training is as fast as possible. Comment on the relative advantages and disadvantages of each training/optimization method. For example, comment on which optimization method makes rapid initial progress, but gets trapped in a suboptimal solution, and which method is slower, but more consistent in finding good optima.

17. Consider again the `Pytorch` code in Section 9.5.2. Repeat all the computations, but this time using the *momentum* method for training of the network. Comment on which method is preferable: the *momentum* or the *Adam* method?

APPENDIX A

LINEAR ALGEBRA AND FUNCTIONAL ANALYSIS

The purpose of this appendix is to review some important topics in linear algebra and functional analysis. We assume that the reader has some familiarity with matrix and vector operations, including matrix multiplication and the computation of determinants.

A.1 Vector Spaces, Bases, and Matrices

Linear algebra is the study of vector spaces and linear mappings. Vectors are, by definition, elements of some *vector space* $\mathcal{V}$ and satisfy the usual rules of addition and scalar multiplication, e.g., VECTOR SPACE

$$\text{if } \boldsymbol{x} \in \mathcal{V} \text{ and } \boldsymbol{y} \in \mathcal{V}, \text{ then } \alpha\boldsymbol{x} + \beta\boldsymbol{y} \in \mathcal{V} \text{ for all } \alpha, \beta \in \mathbb{R} \ \ (\text{or } \mathbb{C}).$$

We will be dealing mostly with vectors in the Euclidean vector space $\mathbb{R}^n$ for some n. That is, we view the points of $\mathbb{R}^n$ as objects that can be added up and multiplied with a scalar, e.g., $(x_1, x_2) + (y_1, y_2) = (x_1 + y_1, x_2 + y_2)$ for points in $\mathbb{R}^2$. Sometimes it is convenient to work with the complex vector space $\mathbb{C}^n$ instead of $\mathbb{R}^n$; see also Section A.3.

Vectors $\boldsymbol{v}_1, \ldots, \boldsymbol{v}_k$ are called *linearly independent* if none of them can be expressed as a linear combination of the others; that is, if $\alpha_1\boldsymbol{v}_1 + \cdots + \alpha_n\boldsymbol{v}_n = \boldsymbol{0}$, then it must hold that $\alpha_i = 0$ for all $i = 1, \ldots, n$. LINEARLY INDEPENDENT

Definition A.1: Basis of a Vector Space

A set of vectors $\mathcal{B} = \{\boldsymbol{v}_1, \ldots, \boldsymbol{v}_n\}$ is called a *basis* of the vector space $\mathcal{V}$ if every vector $\boldsymbol{x} \in \mathcal{V}$ can be written as a unique linear combination of the vectors in $\mathcal{B}$: BASIS

$$\boldsymbol{x} = \alpha_1 \boldsymbol{v}_1 + \cdots + \alpha_n \boldsymbol{v}_n.$$

The (possibly infinite) number n is called the *dimension* of $\mathcal{V}$. DIMENSION

Using a basis $\mathcal{B}$ of $\mathcal{V}$, we can thus represent each vector $\boldsymbol{x} \in \mathcal{V}$ as a row or column of numbers

$$[\alpha_1, \ldots, \alpha_n] \quad \text{or} \quad \begin{bmatrix} \alpha_1 \\ \vdots \\ \alpha_n \end{bmatrix}. \tag{A.1}$$

STANDARD BASIS

Typically, vectors in $\mathbb{R}^n$ are represented via the *standard basis*, consisting of unit vectors (points) $\boldsymbol{e}_1 = (1, 0, \ldots, 0), \ldots, \boldsymbol{e}_n = (0, 0, \ldots, 0, 1)$. As a consequence, any point $(x_1, \ldots, x_n) \in \mathbb{R}^n$ can be represented, using the standard basis, as a row or column vector of the form (A.1) above, with $\alpha_i = x_i, i = 1, \ldots, n$. We will also write $[x_1, x_2, \ldots, x_n]^\top$, for the corresponding column vector, where $^\top$ denotes the *transpose*.

TRANSPOSE

> To avoid confusion, we will use the convention from now on that a generic vector $\boldsymbol{x}$ is always represented via the standard basis as a *column* vector. The corresponding row vector is denoted by $\boldsymbol{x}^\top$.

MATRIX

LINEAR TRANSFORMATION

A *matrix* can be viewed as an array of m rows and n columns that defines a *linear transformation* from $\mathbb{R}^n$ to $\mathbb{R}^m$ (or for complex matrices, from $\mathbb{C}^n$ to $\mathbb{C}^m$). The matrix is said to be *square* if $m = n$. If $\boldsymbol{a}_1, \boldsymbol{a}_2, \ldots, \boldsymbol{a}_n$ are the columns of $\mathbf{A}$, that is, $\mathbf{A} = [\boldsymbol{a}_1, \boldsymbol{a}_2, \ldots, \boldsymbol{a}_n]$, and if $\boldsymbol{x} = [x_1, \ldots, x_n]^\top$, then $\mathbf{A}\boldsymbol{x} = x_1\,\boldsymbol{a}_1 + \cdots + x_n\,\boldsymbol{a}_n$. In particular, the standard basis vector $\boldsymbol{e}_k$ is mapped to the vector $\boldsymbol{a}_k, k = 1, \ldots, n$. We sometimes use the notation $\mathbf{A} = [a_{ij}]$, to denote a matrix whose (i, j)-th element is a_{ij}. When we wish to emphasize that a matrix $\mathbf{A}$ is real-valued with m rows and n columns, we write $\mathbf{A} \in \mathbb{R}^{m \times n}$. The *rank* of a matrix is the number of linearly independent rows or, equivalently, the number of linearly independent columns.

RANK

■ **Example A.1 (Linear Transformation)** Take the matrix

$$\mathbf{A} = \begin{bmatrix} 1 & 1 \\ -0.5 & -2 \end{bmatrix}.$$

It transforms the two basis vectors $[1, 0]^\top$ and $[0, 1]^\top$, shown in red and blue in the left panel of Figure A.1, to the vectors $[1, -0.5]^\top$ and $[1, -2]^\top$, shown on the right panel. Similarly, the points on the unit circle are transformed to an ellipse.

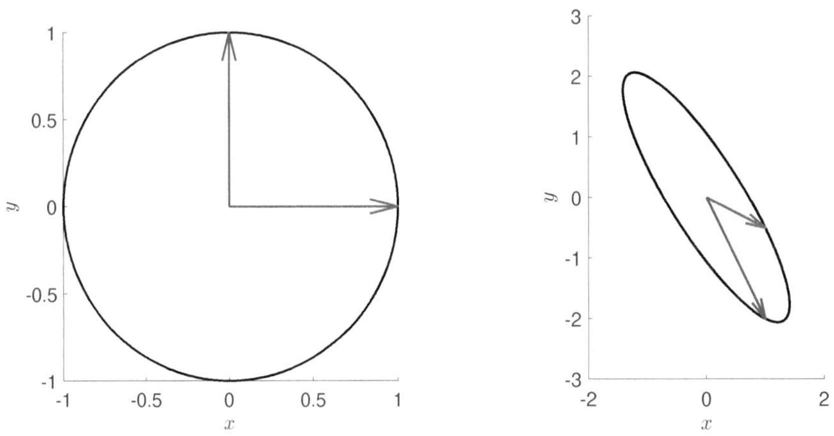

Figure A.1: A linear transformation of the unit circle. ■

Suppose $\mathbf{A} = [\boldsymbol{a}_1, \ldots, \boldsymbol{a}_n]$, where the $\mathcal{A} = \{\boldsymbol{a}_i\}$ form a basis of $\mathbb{R}^n$. Take any vector $\boldsymbol{x} = [x_1, \ldots, x_n]_{\mathcal{E}}^\top$ with respect to the standard basis $\mathcal{E}$ (we write subscript $\mathcal{E}$ to stress this). Then the representation of this vector with respect to $\mathcal{A}$ is simply

$$\boldsymbol{y} = \mathbf{A}^{-1}\boldsymbol{x},$$

where $\mathbf{A}^{-1}$ is the *inverse* of $\mathbf{A}$; that is, the matrix such that $\mathbf{A}\mathbf{A}^{-1} = \mathbf{A}^{-1}\mathbf{A} = \mathbf{I}_n$, where $\mathbf{I}_n$ is the n-dimensional identity matrix. To see this, note that $\mathbf{A}^{-1}\boldsymbol{a}_i$ gives the i-th unit vector representation, for $i = 1, \ldots, n$, and recall that each vector in $\mathbb{R}^n$ is a unique linear combination of these basis vectors. INVERSE

Example A.2 (Basis Representation) Consider the matrix

$$\mathbf{A} = \begin{bmatrix} 1 & 2 \\ 3 & 4 \end{bmatrix} \quad \text{with inverse} \quad \mathbf{A}^{-1} = \begin{bmatrix} -2 & 1 \\ 3/2 & -1/2 \end{bmatrix}. \tag{A.2}$$

The vector $\boldsymbol{x} = [1, 1]_{\mathcal{E}}^\top$ in the standard basis has representation $\boldsymbol{y} = \mathbf{A}^{-1}\boldsymbol{x} = [-1, 1]_{\mathcal{A}}^\top$ in the basis consisting of the columns of $\mathbf{A}$. Namely,

$$\mathbf{A}\boldsymbol{y} = -\begin{bmatrix} 1 \\ 3 \end{bmatrix} + \begin{bmatrix} 2 \\ 4 \end{bmatrix} = \begin{bmatrix} 1 \\ 1 \end{bmatrix}.$$

■

The *transpose* of a matrix $\mathbf{A} = [a_{ij}]$ is the matrix $\mathbf{A}^\top = [a_{ji}]$; that is, the (i, j)-th element of $\mathbf{A}^\top$ is the (j, i)-th element of $\mathbf{A}$. The *trace* of a square matrix is the sum of its diagonal elements. A useful result is the following cyclic property. TRANSPOSE TRACE

Theorem A.1: Cyclic Property

The trace is invariant under cyclic permutations: $\mathrm{tr}(\mathbf{ABC}) = \mathrm{tr}(\mathbf{BCA}) = \mathrm{tr}(\mathbf{CAB})$.

Proof: It suffices to show that $\mathrm{tr}(\mathbf{DE})$ is equal to $\mathrm{tr}(\mathbf{ED})$ for any $m \times n$ matrix $\mathbf{D} = [d_{ij}]$ and $n \times m$ matrix $\mathbf{E} = [e_{ij}]$. The diagonal elements of $\mathbf{DE}$ are $\sum_{j=1}^n d_{ij}\, e_{ji}$, $i = 1, \ldots, m$ and the diagonal elements of $\mathbf{ED}$ are $\sum_{i=1}^m e_{ji}\, d_{ij}$, $j = 1, \ldots, n$. They sum up to the same number $\sum_{i=1}^m \sum_{j=1}^n d_{ij}\, e_{ji}$. □

A square matrix has an inverse if and only if its columns (or rows) are linearly independent. This is the same as the matrix being of *full rank*; that is, its rank is equal to the number of columns. An equivalent statement is that its determinant is not zero. The *determinant* of an $n \times n$ matrix $\mathbf{A} = [a_{i,j}]$ is defined as DETERMINANT

$$\det(\mathbf{A}) := \sum_{\pi} (-1)^{\zeta(\pi)} \prod_{i=1}^{n} a_{\pi_i, i}, \tag{A.3}$$

where the sum is over all permutations $\boldsymbol{\pi} = (\pi_1, \ldots, \pi_n)$ of $(1, \ldots, n)$, and $\zeta(\boldsymbol{\pi})$ is the number of pairs (i, j) for which $i < j$ and $\pi_i > \pi_j$. For example, $\zeta(2, 3, 4, 1) = 3$ for the pairs $(1, 4), (2, 4), (3, 4)$. The determinant of a *diagonal matrix* — a matrix with only zero elements off the diagonal — is simply the product of its diagonal elements. DIAGONAL MATRIX

Geometrically, the determinant of a square matrix $\mathbf{A} = [\boldsymbol{a}_1, \ldots, \boldsymbol{a}_n]$ is the (signed) *volume* of the parallelepiped (n-dimensional parallelogram) defined by the columns $\boldsymbol{a}_1, \ldots, \boldsymbol{a}_n$; that is, the set of points $\boldsymbol{x} = \sum_{i=1}^n \alpha_i \, \boldsymbol{a}_i$, where $0 \leqslant \alpha_i \leqslant 1, i = 1, \ldots, n$.

The easiest way to compute a determinant of a general matrix is to apply simple operations to the matrix that potentially reduce its complexity (as in the number of non-zero elements, for example), while retaining its determinant:

- Adding a multiple of one column (or row) to another, does not change the determinant.
- Multiplying a column (or row) with a number multiplies the determinant by the same number.
- Swapping two rows changes the sign of the determinant.

By applying these rules repeatedly one can reduce any matrix to a diagonal matrix. It follows then that the determinant of the original matrix is equal to the product of the diagonal elements of the resulting diagonal matrix multiplied by a known constant.

■ **Example A.3 (Determinant and Volume)** Figure A.2 illustrates how the determinant of a matrix can be viewed as a signed volume, which can be computed by repeatedly applying the first rule above. Here, we wish to compute the area of red parallelogram determined by the matrix $\mathbf{A}$ given in (A.2). In particular, the corner points of the parallelogram correspond to the vectors $[0, 0]^\top, [1, 3]^\top, [2, 4]^\top$, and $[3, 7]^\top$.

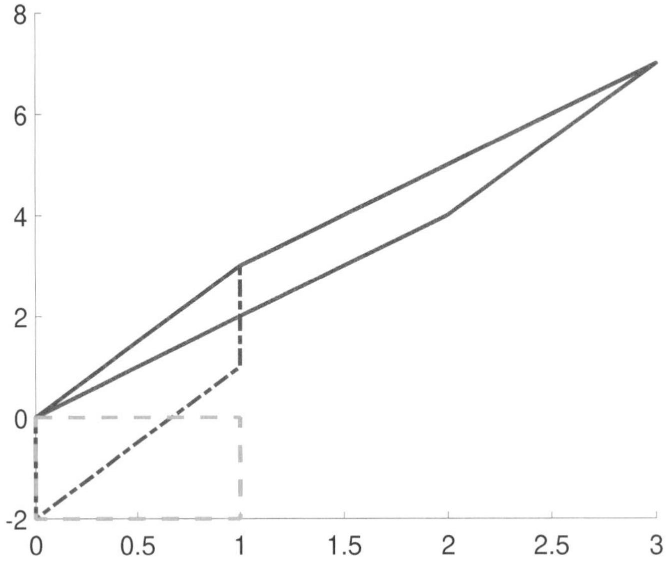

Figure A.2: The volume of the red parallelogram can be obtained by a number of shear operations that do not change the volume.

Adding -2 times the first column of $\mathbf{A}$ to the second column gives the matrix

$$\mathbf{B} = \begin{bmatrix} 1 & 0 \\ 3 & -2 \end{bmatrix},$$

corresponding to the blue parallelogram. The linear operation that transforms the red to the blue parallelogram can be thought of as a succession of two linear transformations. The first is to transform the coordinates of points on the red parallelogram (in standard basis) to the basis formed by the columns of **A**. Second, relative to this new basis, we apply the matrix **B** above. Note that the input of this matrix is with respect to the new basis, whereas the output is with respect to the standard basis. The matrix for the combined operation is now

$$\mathbf{B}\mathbf{A}^{-1} = \begin{bmatrix} 1 & 0 \\ 3 & -2 \end{bmatrix} \begin{bmatrix} -2 & 1 \\ 3/2 & -1/2 \end{bmatrix} = \begin{bmatrix} -2 & 1 \\ -9 & 4 \end{bmatrix},$$

which maps $[1,3]^\top$ to $[1,3]^\top$ (does not change) and $[2,4]^\top$ to $[0,-2]^\top$. We say that we apply a *shear* in the direction $[1,3]^\top$. The significance of such an operation is that a shear *does not alter the volume of the parallelogram*. The second (blue) parallelogram has an easier form, because one of the sides is parallel to the y-axis. By applying another shear, in the direction $[0,-2]^\top$, we can obtain a simple (green) rectangle, whose volume is 2. In matrix terms, we add 3/2 times the second column of **B** to the first column of **B**, to obtain the matrix SHEAR

$$\mathbf{C} = \begin{bmatrix} 1 & 0 \\ 0 & -2 \end{bmatrix},$$

which is a diagonal matrix, whose determinant is -2, corresponding to the volume 2 of all the parallelograms. ■

Theorem A.2 summarizes a number of useful matrix rules for the concepts that we have discussed so far. We leave the proofs, which typically involves "writing out" the equations, as an exercise for the reader; see also [116].

Theorem A.2: Useful Matrix Rules

1. $(\mathbf{AB})^\top = \mathbf{B}^\top\mathbf{A}^\top$
2. $(\mathbf{AB})^{-1} = \mathbf{B}^{-1}\mathbf{A}^{-1}$
3. $(\mathbf{A}^{-1})^\top = (\mathbf{A}^\top)^{-1} =: \mathbf{A}^{-\top}$
4. $\det(\mathbf{AB}) = \det(\mathbf{A})\det(\mathbf{B})$
5. $\boldsymbol{x}^\top\mathbf{A}\boldsymbol{x} = \mathrm{tr}(\mathbf{A}\boldsymbol{x}\boldsymbol{x}^\top)$
6. $\det(\mathbf{A}) = \prod_i a_{ii}$ if $\mathbf{A} = [a_{ij}]$ is triangular

Next, consider an $n \times p$ matrix **A** for which the matrix inverse fails to exist. That is, **A** is either non-square ($n \neq p$) or its determinant is 0. Instead of the inverse, we can use its so-called pseudo-inverse, which always exists.

MOORE–PENROSE PSEUDO-INVERSE

> **Definition A.2: Moore–Penrose Pseudo-Inverse**
>
> The *Moore–Penrose pseudo-inverse* of a real matrix $\mathbf{A} \in \mathbb{R}^{n \times p}$ is defined as the unique matrix $\mathbf{A}^+ \in \mathbb{R}^{p \times n}$ that satisfies the conditions:
>
> 1. $\mathbf{A}\mathbf{A}^+\mathbf{A} = \mathbf{A}$
> 2. $\mathbf{A}^+\mathbf{A}\mathbf{A}^+ = \mathbf{A}^+$
> 3. $(\mathbf{A}\mathbf{A}^+)^\top = \mathbf{A}\mathbf{A}^+$
> 4. $(\mathbf{A}^+\mathbf{A})^\top = \mathbf{A}^+\mathbf{A}$

We can write $\mathbf{A}^+$ explicitly in terms of $\mathbf{A}$ when $\mathbf{A}$ has a full column or row rank. For example, we always have

$$\mathbf{A}^\top\mathbf{A}\mathbf{A}^+ = \mathbf{A}^\top(\mathbf{A}\mathbf{A}^+)^\top = ((\mathbf{A}\mathbf{A}^+)\mathbf{A})^\top = (\mathbf{A})^\top = \mathbf{A}^\top. \tag{A.4}$$

LEFT PSEUDO-INVERSE

If $\mathbf{A}$ has a full column rank p, then $(\mathbf{A}^\top\mathbf{A})^{-1}$ exists, so that from (A.4) it follows that $\mathbf{A}^+ = (\mathbf{A}^\top\mathbf{A})^{-1}\mathbf{A}^\top$. This is referred to as the *left pseudo-inverse*, as $\mathbf{A}^+\mathbf{A} = \mathbf{I}_p$. Similarly, if $\mathbf{A}$ has a full row rank n, that is, $(\mathbf{A}\mathbf{A}^\top)^{-1}$ exists, then it follows from

$$\mathbf{A}^+\mathbf{A}\mathbf{A}^\top = (\mathbf{A}^+\mathbf{A})^\top\mathbf{A}^\top = (\mathbf{A}(\mathbf{A}^+\mathbf{A}))^\top = \mathbf{A}^\top$$

RIGHT PSEUDO-INVERSE

that $\mathbf{A}^+ = \mathbf{A}^\top(\mathbf{A}\mathbf{A}^\top)^{-1}$. This is the *right pseudo-inverse*, as $\mathbf{A}\mathbf{A}^+ = \mathbf{I}_n$. Finally, if $\mathbf{A}$ is of full rank and square, then $\mathbf{A}^+ = \mathbf{A}^{-1}$.

A.2 Inner Product

INNER PRODUCT

The (Euclidean) *inner product* of two real vectors $\boldsymbol{x} = [x_1, \ldots, x_n]^\top$ and $\boldsymbol{y} = [y_1, \ldots, y_n]^\top$ is defined as the number

$$\langle \boldsymbol{x}, \boldsymbol{y} \rangle = \sum_{i=1}^{n} x_i y_i = \boldsymbol{x}^\top \boldsymbol{y}.$$

Here $\boldsymbol{x}^\top\boldsymbol{y}$ is the matrix multiplication of the $(1 \times n)$ matrix $\boldsymbol{x}^\top$ and the $(n \times 1)$ matrix $\boldsymbol{y}$. The inner product induces a geometry on the linear space $\mathbb{R}^n$, allowing for the definition of length, angle, and so on. The inner product satisfies the following properties:

1. $\langle \alpha\boldsymbol{x} + \beta\boldsymbol{y}, \boldsymbol{z} \rangle = \alpha\langle \boldsymbol{x}, \boldsymbol{z} \rangle + \beta\langle \boldsymbol{y}, \boldsymbol{z} \rangle$;
2. $\langle \boldsymbol{x}, \boldsymbol{y} \rangle = \langle \boldsymbol{y}, \boldsymbol{x} \rangle$;
3. $\langle \boldsymbol{x}, \boldsymbol{x} \rangle \geqslant 0$;
4. $\langle \boldsymbol{x}, \boldsymbol{x} \rangle = 0$ if and only if $\boldsymbol{x} = \mathbf{0}$.

ORTHOGONAL

EUCLIDEAN NORM

Vectors $\boldsymbol{x}$ and $\boldsymbol{y}$ are called *perpendicular* (or *orthogonal*) if $\langle \boldsymbol{x}, \boldsymbol{y} \rangle = 0$. The *Euclidean norm* (or length) of a vector $\boldsymbol{x}$ is defined as

$$\|\boldsymbol{x}\| = \sqrt{x_1^2 + \cdots + x_n^2} = \sqrt{\langle \boldsymbol{x}, \boldsymbol{x} \rangle}.$$

If x and y are perpendicular, then *Pythagoras' theorem* holds: PYTHAGORAS' THEOREM

$$\|x + y\|^2 = \langle x + y,\ x + y\rangle = \langle x, x\rangle + 2\langle x, y\rangle + \langle y, y\rangle = \|x\|^2 + \|y\|^2. \tag{A.5}$$

A basis $\{v_1, \dots, v_n\}$ of $\mathbb{R}^n$ in which all the vectors are pairwise perpendicular and have norm 1 is called an *orthonormal* (short for orthogonal and normalized) basis. For example, the standard basis is orthonormal. ORTHONORMAL

Theorem A.3: Orthonormal Basis Representation

If $\{v_1, \dots, v_n\}$ is an orthonormal basis of $\mathbb{R}^n$, then any vector $x \in \mathbb{R}^n$ can be expressed as

$$x = \langle x, v_1\rangle v_1 + \cdots + \langle x, v_n\rangle v_n. \tag{A.6}$$

Proof: Observe that, because the $\{v_i\}$ form a basis, there exist unique $\alpha_1, \dots, \alpha_n$ such that $x = \alpha_1 v_1 + \cdots + \alpha_n v_n$. By the linearity of the inner product and the orthonormality of the $\{v_i\}$ it follows that $\langle x, v_j\rangle = \langle \sum_i \alpha_i v_i, v_j\rangle = \alpha_j$. □

An $n \times n$ matrix $\mathbf{V}$ whose columns form an orthonormal basis is called an *orthogonal matrix*.[1] Note that for an orthogonal matrix $\mathbf{V} = [v_1, \dots, v_n]$, we have ORTHOGONAL MATRIX

$$\mathbf{V}^\top \mathbf{V} = \begin{bmatrix} v_1^\top \\ v_2^\top \\ \vdots \\ v_n^\top \end{bmatrix} [v_1, v_2, \dots, v_n] = \begin{bmatrix} v_1^\top v_1 & v_1^\top v_2 & \dots & v_1^\top v_n \\ \vdots & \vdots & \vdots & \vdots \\ v_n^\top v_1 & v_n^\top v_2 & \dots & v_n^\top v_n \end{bmatrix} = \mathbf{I}_n.$$

Hence, $\mathbf{V}^{-1} = \mathbf{V}^\top$. Note also that an orthogonal transformation is *length preserving*; that is, $\mathbf{V}x$ has the same length as x. This follows from LENGTH PRESERVING

$$\|\mathbf{V}x\|^2 = \langle \mathbf{V}x, \mathbf{V}x\rangle = x^\top \mathbf{V}^\top \mathbf{V} x = x^\top x = \|x\|^2.$$

A.3 Complex Vectors and Matrices

Instead of the vector space $\mathbb{R}^n$ of n-dimensional real vectors, it is sometimes useful to consider the vector space $\mathbb{C}^n$ of n-dimensional complex vectors. In this case the *adjoint* or *conjugate transpose* operation ($*$) replaces the transpose operation ($\top$). This involves the usual transposition of the matrix or vector with the additional step that any complex number $z = x + \mathrm{i}\,y$ is replaced by its complex conjugate $\overline{z} = x - \mathrm{i}\,y$. For example, if ADJOINT

$$x = \begin{bmatrix} a_1 + \mathrm{i}\,b_1 \\ a_2 + \mathrm{i}\,b_2 \end{bmatrix} \quad \text{and} \quad \mathbf{A} = \begin{bmatrix} a_{11} + \mathrm{i}\,b_{11} & a_{12} + \mathrm{i}\,b_{12} \\ a_{21} + \mathrm{i}\,b_{21} & a_{22} + \mathrm{i}\,b_{22} \end{bmatrix},$$

then

$$x^* = [a_1 - \mathrm{i}\,b_1,\ a_2 - \mathrm{i}\,b_2] \quad \text{and} \quad \mathbf{A}^* = \begin{bmatrix} a_{11} - \mathrm{i}\,b_{11} & a_{21} - \mathrm{i}\,b_{21} \\ a_{12} - \mathrm{i}\,b_{12} & a_{22} - \mathrm{i}\,b_{22} \end{bmatrix}.$$

[1] The qualifier "orthogonal" for such matrices has been fixed by history. A better term would have been "orthonormal".

The (Euclidean) inner product of $\boldsymbol{x}$ and $\boldsymbol{y}$ (viewed as column vectors) is now defined as

$$\langle \boldsymbol{x}, \boldsymbol{y} \rangle = \boldsymbol{y}^* \boldsymbol{x} = \sum_{i=1}^{n} x_i \overline{y}_i,$$

which is no longer symmetric: $\langle \boldsymbol{x}, \boldsymbol{y} \rangle = \overline{\langle \boldsymbol{y}, \boldsymbol{x} \rangle}$. Note that this generalizes the real-valued inner product. The determinant of a complex matrix $\mathbf{A}$ is defined exactly as in (A.3). As a consequence, $\det(\mathbf{A}^*) = \overline{\det(\mathbf{A})}$.

HERMITIAN
UNITARY

A complex matrix is said to be *Hermitian* or *self-adjoint* if $\mathbf{A}^* = \mathbf{A}$, and *unitary* if $\mathbf{A}^*\mathbf{A} = \mathbf{I}$ (that is, if $\mathbf{A}^* = \mathbf{A}^{-1}$). For real matrices "Hermitian" is the same as "symmetric", and "unitary" is the same as "orthogonal".

A.4 Orthogonal Projections

Let $\{\boldsymbol{u}_1, \ldots, \boldsymbol{u}_k\}$ be a set of linearly independent vectors in $\mathbb{R}^n$. The set

$$\mathcal{V} = \text{Span}\,\{\boldsymbol{u}_1, \ldots, \boldsymbol{u}_k\} = \{\alpha_1 \boldsymbol{u}_1 + \cdots + \alpha_k \boldsymbol{u}_k,\ \alpha_1, \ldots, \alpha_k \in \mathbb{R}\},$$

LINEAR SUBSPACE
ORTHOGONAL COMPLEMENT
ORTHOGONAL PROJECTION MATRIX

is called the *linear subspace spanned by* $\{\boldsymbol{u}_1, \ldots, \boldsymbol{u}_k\}$. The *orthogonal complement* of $\mathcal{V}$, denoted by $\mathcal{V}^\perp$, is the set of all vectors $\boldsymbol{w}$ that are orthogonal to $\mathcal{V}$, in the sense that $\langle \boldsymbol{w}, \boldsymbol{v} \rangle = 0$ for all $\boldsymbol{v} \in \mathcal{V}$. The matrix $\mathbf{P}$ such that $\mathbf{P}\boldsymbol{x} = \boldsymbol{x}$, for all $\boldsymbol{x} \in \mathcal{V}$, and $\mathbf{P}\boldsymbol{x} = \mathbf{0}$, for all $\boldsymbol{x} \in \mathcal{V}^\perp$ is called the *orthogonal projection matrix* onto $\mathcal{V}$. Suppose that $\mathbf{U} = [\boldsymbol{u}_1, \ldots, \boldsymbol{u}_k]$ has full rank, in which case $\mathbf{U}^\top\mathbf{U}$ is an invertible matrix. The orthogonal projection matrix $\mathbf{P}$ onto $\mathcal{V} = \text{Span}\,\{\boldsymbol{u}_1, \ldots, \boldsymbol{u}_k\}$ is then given by

$$\mathbf{P} = \mathbf{U}(\mathbf{U}^\top\mathbf{U})^{-1}\mathbf{U}^\top.$$

Namely, since $\mathbf{PU} = \mathbf{U}$, the matrix $\mathbf{P}$ projects any vector in $\mathcal{V}$ onto itself. Moreover, $\mathbf{P}$ projects any vector in $\mathcal{V}^\perp$ onto the zero vector. Using the pseudo-inverse, it is possible to specify the projection matrix also for the case where $\mathbf{U}$ is not of full rank, leading to the following theorem.

Theorem A.4: Orthogonal Projection

Let $\mathbf{U} = [\boldsymbol{u}_1, \ldots, \boldsymbol{u}_k]$. Then, the orthogonal projection matrix $\mathbf{P}$ onto $\mathcal{V} = \text{Span}\{\boldsymbol{u}_1, \ldots, \boldsymbol{u}_k\}$ is given by

$$\mathbf{P} = \mathbf{U}\,\mathbf{U}^+, \tag{A.7}$$

where $\mathbf{U}^+$ is the (right) pseudo-inverse of $\mathbf{U}$.

Proof: By Property 1 of Definition A.2 we have $\mathbf{PU} = \mathbf{UU}^+\mathbf{U} = \mathbf{U}$, so that $\mathbf{P}$ projects any vector in $\mathcal{V}$ onto itself. Moreover, $\mathbf{P}$ projects any vector in $\mathcal{V}^\perp$ onto the zero vector. □

Note that in the special case where $\boldsymbol{u}_1, \ldots, \boldsymbol{u}_k$ above form an orthonormal basis of $\mathcal{V}$, then the projection onto $\mathcal{V}$ is very simple to describe, namely we have

$$\mathbf{P}\boldsymbol{x} = \mathbf{U}\mathbf{U}^\top\boldsymbol{x} = \sum_{i=1}^{k} \langle \boldsymbol{x}, \boldsymbol{u}_i \rangle\, \boldsymbol{u}_i. \tag{A.8}$$

For any point $x \in \mathbb{R}^n$, the point in $\mathcal{V}$ that is closest to x is its orthogonal projection $\mathbf{P}x$, as the following theorem shows.

Theorem A.5: Orthogonal Projection and Minimal Distance

Let $\{u_1, \ldots, u_k\}$ be an orthonormal basis of subspace $\mathcal{V}$ and let $\mathbf{P}$ be the orthogonal projection matrix onto $\mathcal{V}$. The solution to the minimization program

$$\min_{y \in \mathcal{V}} \|x - y\|^2$$

is $y = \mathbf{P}x$. That is, $\mathbf{P}x \in \mathcal{V}$ is closest to x.

Proof: We can write each point $y \in \mathcal{V}$ as $y = \sum_{i=1}^k \alpha_i u_i$. Consequently,

$$\|x - y\|^2 = \left\langle x - \sum_{i=1}^k \alpha_i u_i,\ x - \sum_{i=1}^k \alpha_i u_i \right\rangle = \|x\|^2 - 2\sum_{i=1}^k \alpha_i \langle x, u_i \rangle + \sum_{i=1}^k \alpha_i^2.$$

Minimizing this with respect to the $\{\alpha_i\}$ gives $\alpha_i = \langle x, u_i \rangle, i = 1, \ldots, k$. In view of (A.8), the optimal y is thus $\mathbf{P}x$. □

A.5 Eigenvalues and Eigenvectors

Let $\mathbf{A}$ be an $n \times n$ matrix. If $\mathbf{A}v = \lambda v$ for some number λ and non-zero vector v, then λ is called an *eigenvalue* of $\mathbf{A}$ with *eigenvector* v.

EIGENVALUE
EIGENVECTOR

If (λ, v) is an (eigenvalue, eigenvector) pair, the matrix $\lambda\mathbf{I} - \mathbf{A}$ maps any multiple of v to the zero vector. Consequently, the columns of $\lambda\mathbf{I} - \mathbf{A}$ are linearly *dependent*, and hence its determinant is 0. This provides a way to identify the eigenvalues, namely as the $r \leqslant n$ different roots $\lambda_1, \ldots, \lambda_r$ of the *characteristic polynomial*

CHARACTERISTIC POLYNOMIAL

$$\det(\lambda\mathbf{I} - \mathbf{A}) = (\lambda - \lambda_1)^{\alpha_1} \cdots (\lambda - \lambda_r)^{\alpha_r},$$

where $\alpha_1 + \cdots + \alpha_r = n$. The integer α_i is called the *algebraic multiplicity* of λ_i. The eigenvectors that correspond to an eigenvalue λ_i lie in the *kernel* or *null space* of the matrix $\lambda_i\mathbf{I} - \mathbf{A}$; that is, the linear space of vectors v such that $(\lambda_i\mathbf{I} - \mathbf{A})v = \mathbf{0}$. This space is called the *eigenspace* of λ_i. Its dimension, $d_i \in \{1, \ldots, n\}$, is called the *geometric multiplicity* of λ_i. It always holds that $d_i \leqslant \alpha_i$. If $\sum_i d_i = n$, then we can construct a basis for $\mathbb{R}^n$ consisting of eigenvectors, as illustrated next.

ALGEBRAIC MULTIPLICITY
NULL SPACE
GEOMETRIC MULTIPLICITY

■ **Example A.4 (Linear Transformation (cont.))** We revisit the linear transformation in Figure A.1, where

$$\mathbf{A} = \begin{bmatrix} 1 & 1 \\ -1/2 & -2 \end{bmatrix}.$$

The characteristic polynomial is $(\lambda - 1)(\lambda + 2) + 1/2$, with roots $\lambda_1 = -1/2 - \sqrt{7}/2 \approx -1.8229$ and $\lambda_2 = -1/2 + \sqrt{7}/2 \approx 0.8229$. The corresponding unit eigenvectors are $v_1 \approx [0.3339, -0.9426]^\top$ and $v_2 \approx [0.9847, -0.1744]^\top$. The eigenspace corresponding to λ_1 is

$\mathcal{V}_1 = \text{Span}\{\boldsymbol{v}_1\} = \{\beta \boldsymbol{v}_1 : \beta \in \mathbb{R}\}$ and the eigenspace corresponding to λ_2 is $\mathcal{V}_2 = \text{Span}\{\boldsymbol{v}_2\}$. The algebraic and geometric multiplicities are 1 in this case. Any pair of vectors taken from $\mathcal{V}_1$ and $\mathcal{V}_2$ forms a basis for $\mathbb{R}^2$. Figure A.3 shows how $\boldsymbol{v}_1$ and $\boldsymbol{v}_2$ are transformed to $\mathbf{A}\boldsymbol{v}_1 \in \mathcal{V}_1$ and $\mathbf{A}\boldsymbol{v}_2 \in \mathcal{V}_2$, respectively.

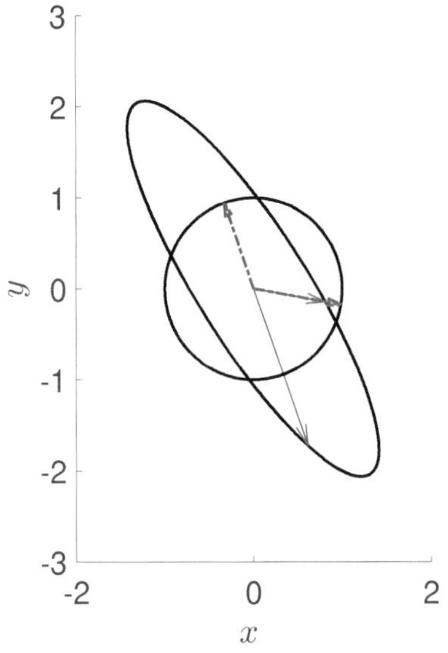

Figure A.3: The dashed arrows are the unit eigenvectors $\boldsymbol{v}_1$ (blue) and $\boldsymbol{v}_2$ (red) of matrix $\mathbf{A}$. Their transformed values $\mathbf{A}\boldsymbol{v}_1$ and $\mathbf{A}\boldsymbol{v}_2$ are indicated by solid arrows. ■

SEMI-SIMPLE
DIAGONALIZABLE

A matrix for which the algebraic and geometric multiplicities of all its eigenvalues are the same is called *semi-simple*. This is equivalent to the matrix being *diagonalizable*, meaning that there is a matrix $\mathbf{V}$ and a diagonal matrix $\mathbf{D}$ such that

$$\mathbf{A} = \mathbf{V}\mathbf{D}\mathbf{V}^{-1}.$$

EIGEN-DECOMPOSITION

To see that this so-called *eigen-decomposition* holds, suppose $\mathbf{A}$ is a semi-simple matrix with eigenvalues

$$\underbrace{\lambda_1, \ldots, \lambda_1}_{d_1}, \cdots, \underbrace{\lambda_r, \ldots, \lambda_r}_{d_r}.$$

Let $\mathbf{D}$ be the diagonal matrix whose diagonal elements are the eigenvalues of $\mathbf{A}$, and let $\mathbf{V}$ be a matrix whose columns are linearly independent eigenvectors corresponding to these eigenvalues. Then, for each (eigenvalue, eigenvector) pair $(\lambda, \boldsymbol{v})$, we have $\mathbf{A}\boldsymbol{v} = \lambda\boldsymbol{v}$. Hence, in matrix notation, we have $\mathbf{A}\mathbf{V} = \mathbf{V}\mathbf{D}$, and so $\mathbf{A} = \mathbf{V}\mathbf{D}\mathbf{V}^{-1}$.

A.5.1 Left- and Right-Eigenvectors

The eigenvector as defined in the previous section is called a *right*-eigenvector, as it lies on the right of $\mathbf{A}$ in the equation $\mathbf{A}\boldsymbol{v} = \lambda\boldsymbol{v}$.

If $\mathbf{A}$ is a complex matrix with an eigenvalue λ, then the eigenvalue's complex conjugate $\overline{\lambda}$ is an eigenvalue of $\mathbf{A}^*$. To see this, define $\mathbf{B} := \lambda\mathbf{I} - \mathbf{A}$ and $\mathbf{B}^* := \overline{\lambda}\mathbf{I} - \mathbf{A}^*$. Since λ is an eigenvalue, we have $\det(\mathbf{B}) = 0$. Applying the identity $\det(\mathbf{B}) = \overline{\det(\mathbf{B}^*)}$, we see that therefore $\det(\mathbf{B}^*) = 0$, and hence that $\overline{\lambda}$ is an eigenvalue of $\mathbf{A}^*$. Let $\boldsymbol{w}$ be an eigenvector corresponding to $\overline{\lambda}$. Then, $\mathbf{A}^*\boldsymbol{w} = \overline{\lambda}\boldsymbol{w}$ or, equivalently,

$$\boldsymbol{w}^*\mathbf{A} = \lambda \boldsymbol{w}^*.$$

For this reason, we call $\boldsymbol{w}^*$ the *left-eigenvector* of $\mathbf{A}$ for eigenvalue λ. If $\boldsymbol{v}$ is a (right-) eigenvector of $\mathbf{A}$, then its adjoint $\boldsymbol{v}^*$ is usually *not* a left-eigenvector, unless $\mathbf{A}^*\mathbf{A} = \mathbf{A}\mathbf{A}^*$ (such matrices are called *normal*; a real symmetric matrix is normal). However, the important property holds that left- and right-eigenvectors belonging to *different* eigenvalues are *orthogonal*. Namely, if $\boldsymbol{w}^*$ is a left-eigenvalue of λ_1 and $\boldsymbol{v}$ a right-eigenvalue of $\lambda_2 \neq \lambda_1$, then

LEFT-EIGENVECTOR

NORMAL MATRIX

$$\lambda_1 \boldsymbol{w}^*\boldsymbol{v} = \boldsymbol{w}^*\mathbf{A}\boldsymbol{v} = \lambda_2 \boldsymbol{w}^*\boldsymbol{v},$$

which can only be true if $\boldsymbol{w}^*\boldsymbol{v} = 0$.

Theorem A.6: Schur Triangulation

For any complex matrix $\mathbf{A}$, there exists a unitary matrix $\mathbf{U}$ such that $\mathbf{T} = \mathbf{U}^{-1}\mathbf{A}\mathbf{U}$ is upper triangular.

Proof: The proof is by induction on the dimension n of the matrix. Clearly, the statement is true for $n = 1$, as $\mathbf{A}$ is simply a complex number and we can take $\mathbf{U}$ equal to 1. Suppose that the result is true for dimension n. We wish to show that it also holds for dimension $n + 1$. Any matrix $\mathbf{A}$ always has at least one eigenvalue λ with eigenvector $\boldsymbol{v}$, normalized to have length 1. Let $\mathbf{U}$ be any unitary matrix whose first column is $\boldsymbol{v}$. Such a matrix can always be constructed[2]. As $\mathbf{U}$ is unitary, the first row of $\mathbf{U}^{-1}$ is $\boldsymbol{v}^*$, and $\mathbf{U}^{-1}\mathbf{A}\mathbf{U}$ is of the form

$$\left[\begin{array}{c} \boldsymbol{v}^* \\ \hline * \end{array}\right] \mathbf{A} \underbrace{\left[\begin{array}{c|c} \boldsymbol{v} & * \end{array}\right]}_{\mathbf{U}} = \left[\begin{array}{c|c} \lambda & * \\ \hline \mathbf{0} & \mathbf{B} \end{array}\right],$$

for some matrix $\mathbf{B}$. By the induction hypothesis, there exists a unitary matrix $\mathbf{W}$ and an upper triangular matrix $\mathbf{T}$ such that $\mathbf{W}^{-1}\mathbf{B}\mathbf{W} = \mathbf{T}$. Now, define

$$\mathbf{V} := \left[\begin{array}{c|c} 1 & \mathbf{0}^\top \\ \hline \mathbf{0} & \mathbf{W} \end{array}\right].$$

Then,

$$\mathbf{V}^{-1}\left(\mathbf{U}^{-1}\mathbf{A}\mathbf{U}\right)\mathbf{V} = \left[\begin{array}{c|c} 1 & \mathbf{0}^\top \\ \hline \mathbf{0} & \mathbf{W}^{-1} \end{array}\right]\left[\begin{array}{c|c} \lambda & * \\ \hline \mathbf{0} & \mathbf{B} \end{array}\right]\left[\begin{array}{c|c} 1 & \mathbf{0}^\top \\ \hline \mathbf{0} & \mathbf{W} \end{array}\right] = \left[\begin{array}{c|c} \lambda & * \\ \hline \mathbf{0} & \mathbf{W}^{-1}\mathbf{B}\mathbf{W} \end{array}\right] = \left[\begin{array}{c|c} \lambda & * \\ \hline \mathbf{0} & \mathbf{T} \end{array}\right],$$

which is upper triangular of dimension $n + 1$. Since $\mathbf{U}\mathbf{V}$ is unitary, this completes the induction, and hence the result is true for all n. □

The theorem above can be used to prove an important property of *Hermitian* matrices, i.e., matrices for which $\mathbf{A}^* = \mathbf{A}$.

[2] After specifying $\boldsymbol{v}$ we can complete the rest of the unitary matrix via the Gram–Schmidt procedure, for example; see Section A.6.4.

Theorem A.7: Eigenvalues of a Hermitian Matrix

Any $n \times n$ Hermitian matrix has real eigenvalues. The corresponding matrix of normalized eigenvectors is a unitary matrix.

Proof: Let $\mathbf{A}$ be a Hermitian matrix. By Theorem A.6 there exists a unitary matrix $\mathbf{U}$ such that $\mathbf{U}^{-1}\mathbf{A}\mathbf{U} = \mathbf{T}$, where $\mathbf{T}$ is upper triangular. It follows that the adjoint $(\mathbf{U}^{-1}\mathbf{A}\mathbf{U})^* = \mathbf{T}^*$ is lower triangular. However, $(\mathbf{U}^{-1}\mathbf{A}\mathbf{U})^* = \mathbf{U}^{-1}\mathbf{A}\mathbf{U}$, since $\mathbf{A}^* = \mathbf{A}$ and $\mathbf{U}^* = \mathbf{U}^{-1}$. Hence, $\mathbf{T}$ and $\mathbf{T}^*$ must be the same, which can only be the case if $\mathbf{T}$ is a *real diagonal* matrix $\mathbf{D}$. Since $\mathbf{A}\mathbf{U} = \mathbf{D}\mathbf{U}$, the diagonal elements are exactly the eigenvalues and the corresponding eigenvectors are the columns of $\mathbf{U}$. □

In particular, the eigenvalues of a real symmetric matrix are real. We can now repeat the proof of Theorem A.6 with real eigenvalues and eigenvectors, so that there exists an orthogonal matrix $\mathbf{Q}$ such that $\mathbf{Q}^{-1}\mathbf{A}\mathbf{Q} = \mathbf{Q}^\top\mathbf{A}\mathbf{Q} = \mathbf{D}$. The eigenvectors can be chosen as the columns of $\mathbf{Q}$, which form an orthonormal basis. This proves the following theorem.

Theorem A.8: Real Symmetric Matrices are Orthogonally Diagonizable

Any real symmetric matrix $\mathbf{A}$ can be written as

$$\mathbf{A} = \mathbf{Q}\mathbf{D}\mathbf{Q}^\top,$$

where $\mathbf{D}$ is the diagonal matrix of (real) eigenvalues and $\mathbf{Q}$ is an orthogonal matrix whose columns are eigenvectors of $\mathbf{A}$.

■ **Example A.5 (Real Symmetric Matrices and Ellipses)** As we have seen, linear transformations map circles into ellipses. We can use the above theory for real symmetric matrices to identify the principal axes. Consider, for example, the transformation with matrix $\mathbf{A} = [1, 1; -1/2, -2]$ in (A.1). A point $\boldsymbol{x}$ on the unit circle is mapped to a point $\boldsymbol{y} = \mathbf{A}\boldsymbol{x}$. Since for such points $\|\boldsymbol{x}\|^2 = \boldsymbol{x}^\top\boldsymbol{x} = 1$, we have that $\boldsymbol{y}$ satisfies $\boldsymbol{y}^\top(\mathbf{A}^{-1})^\top\mathbf{A}^{-1}\boldsymbol{y} = 1$, which gives the equation for the ellipse

$$\frac{17\, y_1^2}{9} + \frac{20\, y_1 y_2}{9} + \frac{8\, y_2^2}{9} = 1.$$

Let $\mathbf{Q}$ be the orthogonal matrix of eigenvectors of the symmetric matrix $(\mathbf{A}^{-1})^\top\mathbf{A}^{-1} = (\mathbf{A}\mathbf{A}^\top)^{-1}$, so $\mathbf{Q}^\top(\mathbf{A}\mathbf{A}^\top)^{-1}\mathbf{Q} = \mathbf{D}$ for some diagonal matrix $\mathbf{D}$. Taking the inverse on both sides of the previous equation, we have $\mathbf{Q}^\top\mathbf{A}\mathbf{A}^\top\mathbf{Q} = \mathbf{D}^{-1}$, which shows that $\mathbf{Q}$ is also the matrix of eigenvectors of $\mathbf{A}\mathbf{A}^\top$. These eigenvectors point precisely in the direction of the principal axes, as shown in Figure A.4. It turns out, see Section A.6.5, that the square roots of the eigenvalues of $\mathbf{A}\mathbf{A}^\top$, here approximately 2.4221 and 0.6193, correspond to the sizes of the principal axes of the ellipse, as illustrated in Figure A.4.

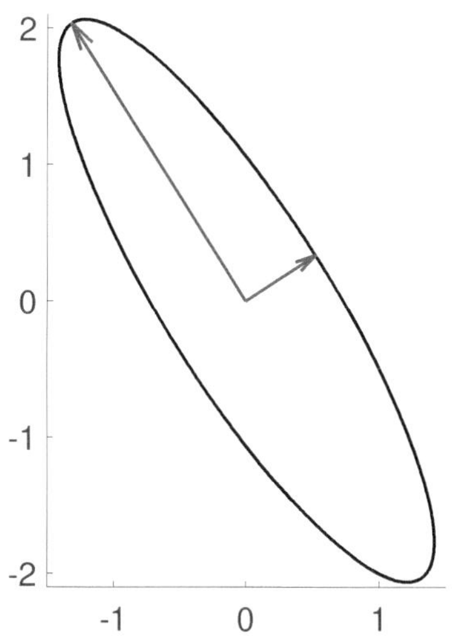

Figure A.4: The eigenvectors and eigenvalues of $\mathbf{A}\mathbf{A}^\top$ determine the principal axes of the ellipse. ■

The following definition generalizes the notion of positivity of a real variable to that of a (Hermitian) matrix, providing a crucial concept for multivariate differentiation and optimization; see Appendix B. ☞ 399

Definition A.3: Positive (Semi)Definite Matrix

A Hermitian matrix $\mathbf{A}$ is called *positive semidefinite* (we write $\mathbf{A} \succeq 0$) if $\langle \mathbf{A}\boldsymbol{x}, \boldsymbol{x}\rangle \geqslant 0$ for all $\boldsymbol{x}$. It is called *positive definite* (we write $\mathbf{A} \succ 0$) if $\langle \mathbf{A}\boldsymbol{x}, \boldsymbol{x}\rangle > 0$ for all $\boldsymbol{x} \neq \mathbf{0}$.

POSITIVE SEMIDEFINITE

The positive (semi)definiteness of a matrix can be directly related to the positivity of its eigenvalues, as follows:

Theorem A.9: Eigenvalues of a Positive Semidefinite Matrix

All eigenvalues of a positive semidefinite matrix are non-negative and all eigenvalues of a positive definite matrix are strictly positive.

Proof: Let $\mathbf{A}$ be a positive semidefinite matrix. By Theorem A.7, the eigenvalues of $\mathbf{A}$ are all *real*. Suppose λ is an eigenvalue with eigenvector $\boldsymbol{v}$. As $\mathbf{A}$ is positive semidefinite, we have

$$0 \leqslant \langle \mathbf{A}\boldsymbol{v}, \boldsymbol{v}\rangle = \lambda \langle \boldsymbol{v}, \boldsymbol{v}\rangle = \lambda \, \|\boldsymbol{v}\|^2,$$

which can only be true if $\lambda \geqslant 0$. Similarly, for a positive definite matrix, λ must be strictly greater than 0. □

Corollary A.1 Any real positive semidefinite matrix $\mathbf{A}$ can be written as

$$\mathbf{A} = \mathbf{B}\mathbf{B}^\top$$

for some real matrix $\mathbf{B}$. Conversely, for any real matrix $\mathbf{B}$, the matrix $\mathbf{B}\mathbf{B}^\top$ is positive semidefinite.

Proof: The matrix $\mathbf{A}$ is both Hermitian (by definition) and real (by assumption) and hence it is symmetric. By Theorem A.8, we can write $\mathbf{A} = \mathbf{Q}\mathbf{D}\mathbf{Q}^\top$, where $\mathbf{D}$ is the diagonal matrix of (real) eigenvalues of $\mathbf{A}$. By Theorem A.9 all eigenvalues are non-negative, and thus their square root is real-valued. Now, define $\mathbf{B} = \mathbf{Q}\sqrt{\mathbf{D}}$, where $\sqrt{\mathbf{D}}$ is defined as the diagonal matrix whose diagonal elements are the *square roots* of the eigenvalues of $\mathbf{A}$. Then, $\mathbf{B}\mathbf{B}^\top = \mathbf{Q}\sqrt{\mathbf{D}}(\sqrt{\mathbf{D}})^\top\mathbf{Q}^\top = \mathbf{Q}\mathbf{D}\mathbf{Q}^\top = \mathbf{A}$. The converse statement follows from the fact that $\boldsymbol{x}^\top\mathbf{B}\mathbf{B}^\top\boldsymbol{x} = \|\mathbf{B}^\top\boldsymbol{x}\|^2 \geqslant 0$ for all $\boldsymbol{x}$. □

A.6 Matrix Decompositions

Matrix decompositions are frequently used in linear algebra to simplify proofs, avoid numerical instability, and to speed up computations. We mention three important matrix decompositions: (P)LU, QR, and SVD.

A.6.1 (P)LU Decomposition

Every invertible matrix $\mathbf{A}$ can be written as the product of three matrices:

$$\mathbf{A} = \mathbf{P}\mathbf{L}\mathbf{U}, \tag{A.9}$$

PERMUTATION MATRIX

PLU DECOMPOSITION

where $\mathbf{L}$ is a lower triangular matrix, $\mathbf{U}$ an upper triangular matrix, and $\mathbf{P}$ a *permutation matrix*. A permutation matrix is a square matrix with a single 1 in each row and column, and zeros otherwise. The matrix product $\mathbf{P}\mathbf{B}$ simply permutes the rows of a matrix $\mathbf{B}$ and, likewise, $\mathbf{B}\mathbf{P}$ permutes its columns. A decomposition of the form (A.9) is called a *PLU decomposition*. As a permutation matrix is orthogonal, its transpose is equal to its inverse, and so we can write (A.9) as

$$\mathbf{P}^\top\mathbf{A} = \mathbf{L}\mathbf{U}.$$

The decomposition is not unique, and in many cases $\mathbf{P}$ can be taken to be the identity matrix, in which case we speak of the *LU decomposition* of $\mathbf{A}$, also called the LR for left–right (triangular) decomposition.

A PLU decomposition of an invertible $n \times n$ matrix $\mathbf{A}_0$ can be obtained recursively as follows. The first step is to swap the rows of $\mathbf{A}_0$ such that the element in the first column and first row of the pivoted matrix is as large as possible in absolute value. Write the resulting matrix as

$$\widetilde{\mathbf{P}}_0\mathbf{A}_0 = \begin{bmatrix} a_1 & \boldsymbol{b}_1^\top \\ \boldsymbol{c}_1 & \mathbf{D}_1 \end{bmatrix},$$

where $\widetilde{\mathbf{P}}_0$ is the permutation matrix that swaps the first and k-th row, where k is the row that contains the largest element in the first column. Next, add the matrix $-\boldsymbol{c}_1[1, \boldsymbol{b}_1^\top/a_1]$ to the last $n-1$ rows of $\widetilde{\mathbf{P}}_0\mathbf{A}_0$, to obtain the matrix

$$\begin{bmatrix} a_1 & \boldsymbol{b}_1^\top \\ \mathbf{0} & \mathbf{D}_1 - \boldsymbol{c}_1\boldsymbol{b}_1^\top/a_1 \end{bmatrix} =: \begin{bmatrix} a_1 & \boldsymbol{b}_1^\top \\ \mathbf{0} & \mathbf{A}_1 \end{bmatrix}.$$

In effect, we add some multiple of the first row to each of the remaining rows in order to obtain zeros in the first column, except for the first element.

We now apply the same procedure to $\mathbf{A}_1$ as we did to $\mathbf{A}_0$ and then to subsequent smaller matrices $\mathbf{A}_2, \ldots, \mathbf{A}_{n-1}$:

1. Swap the first row with the row having the maximal absolute value element in the first column.

2. Make every other element in the first column equal to 0 by adding appropriate multiples of the first row to the other rows.

Suppose that $\mathbf{A}_t$ has a PLU decomposition $\mathbf{P}_t\mathbf{L}_t\mathbf{U}_t$. Then it is easy to check that

$$\underbrace{\widetilde{\mathbf{P}}_{t-1}^\top \begin{bmatrix} 1 & \mathbf{0}^\top \\ \mathbf{0} & \mathbf{P}_t \end{bmatrix}}_{\mathbf{P}_{t-1}} \underbrace{\begin{bmatrix} 1 & \mathbf{0}^\top \\ \mathbf{P}_t^\top \boldsymbol{c}_t/a_t & \mathbf{L}_t \end{bmatrix}}_{\mathbf{L}_{t-1}} \underbrace{\begin{bmatrix} a_t & \boldsymbol{b}_t^\top \\ \mathbf{0} & \mathbf{U}_t \end{bmatrix}}_{\mathbf{U}_{t-1}} \tag{A.10}$$

is a PLU decomposition of $\mathbf{A}_{t-1}$. Since the PLU decomposition for the scalar $\mathbf{A}_{n-1}$ is trivial, by working backwards we obtain a PLU decomposition $\mathbf{P}_0\mathbf{L}_0\mathbf{U}_0$ of $\mathbf{A}$.

■ **Example A.6 (PLU Decomposition)** Take

$$\mathbf{A} = \begin{bmatrix} 0 & -1 & 7 \\ 3 & 2 & 0 \\ 1 & 1 & 1 \end{bmatrix}.$$

Our goal is to modify $\mathbf{A}$ via Steps 1 and 2 above so as to obtain an upper triangular matrix with maximal elements on the diagonal. We first swap the first and second row. Next, we add $-1/3$ times the first row to the third row and $1/3$ times the second row to the third row:

$$\begin{bmatrix} 0 & -1 & 7 \\ 3 & 2 & 0 \\ 1 & 1 & 1 \end{bmatrix} \longrightarrow \begin{bmatrix} 3 & 2 & 0 \\ 0 & -1 & 7 \\ 1 & 1 & 1 \end{bmatrix} \longrightarrow \begin{bmatrix} 3 & 2 & 0 \\ 0 & -1 & 7 \\ 0 & 1/3 & 1 \end{bmatrix} \longrightarrow \begin{bmatrix} 3 & 2 & 0 \\ 0 & -1 & 7 \\ 0 & 0 & 10/3 \end{bmatrix}.$$

The final matrix is $\mathbf{U}_0$, and in the process we have applied the permutation matrices

$$\widetilde{\mathbf{P}}_0 = \begin{bmatrix} 0 & 1 & 0 \\ 1 & 0 & 0 \\ 0 & 0 & 1 \end{bmatrix}, \quad \widetilde{\mathbf{P}}_1 = \begin{bmatrix} 1 & 0 \\ 0 & 1 \end{bmatrix}.$$

Using the recursion (A.10) we can now recover $\mathbf{P}_0$ and $\mathbf{L}_0$. Namely, at the final iteration we have $\mathbf{P}_2 = 1, \mathbf{L}_2 = 1$, and $\mathbf{U}_2 = 10/3$. And subsequently,

$$\mathbf{P}_1 = \begin{bmatrix} 1 & 0 \\ 0 & 1 \end{bmatrix}, \quad \mathbf{L}_1 = \begin{bmatrix} 1 & 0 \\ -1/3 & 1 \end{bmatrix}, \quad \mathbf{P}_0 = \begin{bmatrix} 0 & 1 & 0 \\ 1 & 0 & 0 \\ 0 & 0 & 1 \end{bmatrix}, \quad \mathbf{L}_0 = \begin{bmatrix} 1 & 0 & 0 \\ 0 & 1 & 0 \\ 1/3 & -1/3 & 1 \end{bmatrix},$$

observing that $a_1 = 3, \boldsymbol{c}_1 = [0, 1]^\top, a_2 = -1$, and $\boldsymbol{c}_2 = 1/3$. ■

PLU decompositions can be used to solve large systems of linear equations of the form $\mathbf{A}\boldsymbol{x} = \boldsymbol{b}$ efficiently, especially when such an equation has to be solved for many different $\boldsymbol{b}$. This is done by first decomposing $\mathbf{A}$ into $\mathbf{PLU}$, and then solving two triangular systems:

1. $\mathbf{L}\boldsymbol{y} = \mathbf{P}^\top\boldsymbol{b}$.

2. $\mathbf{U}\boldsymbol{x} = \boldsymbol{y}$.

FORWARD SUBSTITUTION

BACKWARD SUBSTITUTION

The first equation can be solved efficiently via *forward substitution*, and the second via *backward substitution*, as illustrated in the following example.

■ **Example A.7 (Solving Linear Equations with an LU Decomposition)** Let $\mathbf{A} = \mathbf{PLU}$ be the same as in Example A.6. We wish to solve $\mathbf{A}\boldsymbol{x} = [1, 2, 3]^\top$. First, solving

$$\begin{bmatrix} 1 & 0 & 0 \\ 0 & 1 & 0 \\ 1/3 & -1/3 & 1 \end{bmatrix} \begin{bmatrix} y_1 \\ y_2 \\ y_3 \end{bmatrix} = \begin{bmatrix} 2 \\ 1 \\ 3 \end{bmatrix}$$

gives, $y_1 = 2$, $y_2 = 1$ and $y_3 = 3 - 2/3 + 1/3 = 8/3$, by forward substitution. Next,

$$\begin{bmatrix} 3 & 2 & 0 \\ 0 & -1 & 7 \\ 0 & 0 & 10/3 \end{bmatrix} \begin{bmatrix} x_1 \\ x_2 \\ x_3 \end{bmatrix} = \begin{bmatrix} 2 \\ 1 \\ 8/3 \end{bmatrix}$$

gives $x_3 = 4/5, x_2 = -1 + 28/5 = 23/5$, and $x_1 = 2(1 - 23/5)/3 = -12/5$, so $\boldsymbol{x} = [-12, 23, 4]^\top/5$. ■

A.6.2 Woodbury Identity

LU (or more generally PLU) decompositions can also be applied to block matrices. A starting point is the following LU decomposition for a general 2×2 matrix:

$$\begin{bmatrix} a & b \\ c & d \end{bmatrix} = \begin{bmatrix} a & 0 \\ c & d - bc/a \end{bmatrix} \begin{bmatrix} 1 & b/a \\ 0 & 1 \end{bmatrix},$$

which holds as long as $a \neq 0$; this can be seen by simply writing out the matrix product. The block matrix generalization for matrices $\mathbf{A} \in \mathbb{R}^{n\times n}, \mathbf{B} \in \mathbb{R}^{n\times k}, \mathbf{C} \in \mathbb{R}^{k\times n}, \mathbf{D} \in \mathbb{R}^{k\times k}$ is

$$\boldsymbol{\Sigma} := \begin{bmatrix} \mathbf{A} & \mathbf{B} \\ \mathbf{C} & \mathbf{D} \end{bmatrix} = \begin{bmatrix} \mathbf{A} & \mathbf{O}_{n\times k} \\ \mathbf{C} & \mathbf{D} - \mathbf{C}\mathbf{A}^{-1}\mathbf{B} \end{bmatrix} \begin{bmatrix} \mathbf{I}_n & \mathbf{A}^{-1}\mathbf{B} \\ \mathbf{O}_{k\times n} & \mathbf{I}_k \end{bmatrix}, \tag{A.11}$$

provided that $\mathbf{A}$ is invertible (again, write out the block matrix product). Here, we use the notation $\mathbf{O}_{p\times q}$ to denote the $p \times q$ matrix of zeros. We can further rewrite this as:

$$\boldsymbol{\Sigma} = \begin{bmatrix} \mathbf{I}_n & \mathbf{O}_{n\times k} \\ \mathbf{C}\mathbf{A}^{-1} & \mathbf{I}_k \end{bmatrix} \begin{bmatrix} \mathbf{A} & \mathbf{O}_{n\times k} \\ \mathbf{O}_{k\times n} & \mathbf{D} - \mathbf{C}\mathbf{A}^{-1}\mathbf{B} \end{bmatrix} \begin{bmatrix} \mathbf{I}_n & \mathbf{A}^{-1}\mathbf{B} \\ \mathbf{O}_{k\times n} & \mathbf{I}_k \end{bmatrix}.$$

Thus, inverting both sides, we obtain

$$\boldsymbol{\Sigma}^{-1} = \begin{bmatrix} \mathbf{I}_n & \mathbf{A}^{-1}\mathbf{B} \\ \mathbf{O}_{k\times n} & \mathbf{I}_k \end{bmatrix}^{-1} \begin{bmatrix} \mathbf{A} & \mathbf{O}_{n\times k} \\ \mathbf{O}_{k\times n} & \mathbf{D} - \mathbf{C}\mathbf{A}^{-1}\mathbf{B} \end{bmatrix}^{-1} \begin{bmatrix} \mathbf{I}_n & \mathbf{O}_{n\times k} \\ \mathbf{C}\mathbf{A}^{-1} & \mathbf{I}_k \end{bmatrix}^{-1}.$$

Inversion of the above block matrices gives (again write out)

$$\boldsymbol{\Sigma}^{-1} = \begin{bmatrix} \mathbf{I}_n & -\mathbf{A}^{-1}\mathbf{B} \\ \mathbf{O}_{k\times n} & \mathbf{I}_k \end{bmatrix} \begin{bmatrix} \mathbf{A}^{-1} & \mathbf{O}_{n\times k} \\ \mathbf{O}_{k\times n} & (\mathbf{D} - \mathbf{C}\mathbf{A}^{-1}\mathbf{B})^{-1} \end{bmatrix} \begin{bmatrix} \mathbf{I}_n & \mathbf{O}_{n\times k} \\ -\mathbf{C}\mathbf{A}^{-1} & \mathbf{I}_k \end{bmatrix}. \tag{A.12}$$

Assuming that $\mathbf{D}$ is invertible, we could also perform a block UL (as opposed to LU) decomposition:

$$\boldsymbol{\Sigma} = \begin{bmatrix} \mathbf{A} - \mathbf{B}\mathbf{D}^{-1}\mathbf{C} & \mathbf{B} \\ \mathbf{O}_{k\times n} & \mathbf{D} \end{bmatrix} \begin{bmatrix} \mathbf{I}_n & \mathbf{O}_{n\times k} \\ \mathbf{D}^{-1}\mathbf{C} & \mathbf{I}_k \end{bmatrix}, \tag{A.13}$$

which, after a similar calculation as the one above, yields

$$\boldsymbol{\Sigma}^{-1} = \begin{bmatrix} \mathbf{I}_n & \mathbf{O}_{n\times k} \\ -\mathbf{D}^{-1}\mathbf{C} & \mathbf{I}_k \end{bmatrix} \begin{bmatrix} (\mathbf{A} - \mathbf{B}\mathbf{D}^{-1}\mathbf{C})^{-1} & \mathbf{O}_{n\times k} \\ \mathbf{O}_{k\times n} & \mathbf{D}^{-1} \end{bmatrix} \begin{bmatrix} \mathbf{I}_n & -\mathbf{B}\mathbf{D}^{-1} \\ \mathbf{O}_{k\times n} & \mathbf{I}_k \end{bmatrix}. \tag{A.14}$$

The upper-left block of $\boldsymbol{\Sigma}^{-1}$ from (A.14) must be the same as the upper-left block of $\boldsymbol{\Sigma}^{-1}$ from (A.12), leading to the *Woodbury identity*:

WOODBURY IDENTITY

$$(\mathbf{A} - \mathbf{B}\mathbf{D}^{-1}\mathbf{C})^{-1} = \mathbf{A}^{-1} + \mathbf{A}^{-1}\mathbf{B}(\mathbf{D} - \mathbf{C}\mathbf{A}^{-1}\mathbf{B})^{-1}\mathbf{C}\mathbf{A}^{-1}. \tag{A.15}$$

From (A.11) and the fact that the determinant of a product is the product of the determinants, we see that $\det(\boldsymbol{\Sigma}) = \det(\mathbf{A})\det(\mathbf{D} - \mathbf{C}\mathbf{A}^{-1}\mathbf{B})$. Similarly, from (A.13) we have $\det(\boldsymbol{\Sigma}) = \det(\mathbf{A} - \mathbf{B}\mathbf{D}^{-1}\mathbf{C})\det(\mathbf{D})$, leading to the identity

$$\det(\mathbf{A} - \mathbf{B}\mathbf{D}^{-1}\mathbf{C})\det(\mathbf{D}) = \det(\mathbf{A})\det(\mathbf{D} - \mathbf{C}\mathbf{A}^{-1}\mathbf{B}). \tag{A.16}$$

The following special cases of (A.16) and (A.15) are of particular importance.

Theorem A.10: Sherman–Morrison Formula

Suppose that $\mathbf{A} \in \mathbb{R}^{n\times n}$ is invertible and $\boldsymbol{x}, \boldsymbol{y} \in \mathbb{R}^n$. Then,

$$\det(\mathbf{A} + \boldsymbol{x}\boldsymbol{y}^\top) = \det(\mathbf{A})(1 + \boldsymbol{y}^\top\mathbf{A}^{-1}\boldsymbol{x}).$$

If in addition $\boldsymbol{y}^\top\mathbf{A}^{-1}\boldsymbol{x} \neq -1$, then the *Sherman–Morrison formula* holds:

SHERMAN–MORRISON FORMULA

$$(\mathbf{A} + \boldsymbol{x}\boldsymbol{y}^\top)^{-1} = \mathbf{A}^{-1} - \frac{\mathbf{A}^{-1}\boldsymbol{x}\boldsymbol{y}^\top\mathbf{A}^{-1}}{1 + \boldsymbol{y}^\top\mathbf{A}^{-1}\boldsymbol{x}}.$$

Proof: Take $\mathbf{B} = \boldsymbol{x}$, $\mathbf{C} = -\boldsymbol{y}^\top$, and $\mathbf{D} = 1$ in (A.16) and (A.15). □

One important application of the Sherman–Morrison formula is in the efficient solution of the linear system $\mathbf{A}\boldsymbol{x} = \boldsymbol{b}$, where $\mathbf{A}$ is an $n \times n$ matrix of the form:

$$\mathbf{A} = \mathbf{A}_0 + \sum_{j=1}^{p} \boldsymbol{a}_j\boldsymbol{a}_j^\top$$

for some column vectors $\boldsymbol{a}_1, \ldots, \boldsymbol{a}_p \in \mathbb{R}^n$ and $n\times n$ diagonal (or otherwise easily invertible) matrix $\mathbf{A}_0$. Such linear systems arise, for example, in the context of *ridge regression* and optimization.

☞ 217
☞ 416

To see how the Sherman–Morrison formula can be exploited, define the matrices $\mathbf{A}_0, \ldots, \mathbf{A}_p$ via the recursion:

$$\mathbf{A}_k = \mathbf{A}_{k-1} + \boldsymbol{a}_k\boldsymbol{a}_k^\top, \quad k = 1, \ldots, p.$$

Application of Theorem A.10 for $k = 1, \ldots, p$ yields the identities:[3]

$$\mathbf{A}_k^{-1} = \mathbf{A}_{k-1}^{-1} - \frac{\mathbf{A}_{k-1}^{-1}\boldsymbol{a}_k\boldsymbol{a}_k^\top\mathbf{A}_{k-1}^{-1}}{1 + \boldsymbol{a}_k^\top\mathbf{A}_{k-1}^{-1}\boldsymbol{a}_k}$$

$$|\mathbf{A}_k| = |\mathbf{A}_{k-1}| \times \left(1 + \boldsymbol{a}_k^\top\mathbf{A}_{k-1}^{-1}\boldsymbol{a}_k\right).$$

[3] Here $|\mathbf{A}|$ is a shorthand notation for $\det(\mathbf{A})$.

Therefore, by evolving the recursive relationships up until $k = p$, we obtain:

$$\mathbf{A}_p^{-1} = \mathbf{A}_0^{-1} - \sum_{j=1}^{p} \frac{\mathbf{A}_{j-1}^{-1}\boldsymbol{a}_j\boldsymbol{a}_j^\top \mathbf{A}_{j-1}^{-1}}{1 + \boldsymbol{a}_j^\top \mathbf{A}_{j-1}^{-1}\boldsymbol{a}_j}$$

$$|\mathbf{A}_p| = |\mathbf{A}_0| \times \prod_{j=1}^{p} \left(1 + \boldsymbol{a}_j^\top \mathbf{A}_{j-1}^{-1}\boldsymbol{a}_j\right).$$

These expressions will allow us to easily compute $\mathbf{A}^{-1} = \mathbf{A}_p^{-1}$ and $|\mathbf{A}| = |\mathbf{A}_p|$ provided the following quantities are available:

$$\boldsymbol{c}_{k,j} := \mathbf{A}_{k-1}^{-1}\boldsymbol{a}_j, \quad k = 1, \ldots, p-1, \quad j = k+1, \ldots, p.$$

Since, by Theorem A.10, we can write:

$$\mathbf{A}_{k-1}^{-1}\boldsymbol{a}_j = \mathbf{A}_{k-2}^{-1}\boldsymbol{a}_j - \frac{\mathbf{A}_{k-2}^{-1}\boldsymbol{a}_{k-1}\boldsymbol{a}_{k-1}^\top \mathbf{A}_{k-2}^{-1}}{1 + \boldsymbol{a}_{k-1}^\top \mathbf{A}_{k-2}^{-1}\boldsymbol{a}_{k-1}}\boldsymbol{a}_j,$$

the quantities $\{\boldsymbol{c}_{k,j}\}$ can be computed from the recursion:

$$\begin{aligned} \boldsymbol{c}_{1,j} &= \mathbf{A}_0^{-1}\boldsymbol{a}_j, \quad j = 1, \ldots, p \\ \boldsymbol{c}_{k,j} &= \boldsymbol{c}_{k-1,j} - \frac{\boldsymbol{a}_{k-1}^\top \boldsymbol{c}_{k-1,j}}{1 + \boldsymbol{a}_{k-1}^\top \boldsymbol{c}_{k-1,k-1}}\boldsymbol{c}_{k-1,k-1}, \quad k = 2, \ldots, p, \quad j = k, \ldots, p. \end{aligned} \tag{A.17}$$

Observe that this recursive computation takes $O(p^2 n)$ time and that once $\{\boldsymbol{c}_{k,j}\}$ are available, we can express $\mathbf{A}^{-1}$ and $|\mathbf{A}|$ as:

$$\mathbf{A}^{-1} = \mathbf{A}_0^{-1} - \sum_{j=1}^{p} \frac{\boldsymbol{c}_{j,j}\,\boldsymbol{c}_{j,j}^\top}{1 + \boldsymbol{a}_j^\top \boldsymbol{c}_{j,j}}$$

$$|\mathbf{A}| = |\mathbf{A}_0| \times \prod_{j=1}^{p} \left(1 + \boldsymbol{a}_j^\top \boldsymbol{c}_{j,j}\right).$$

In summary, we have proved the following.

Theorem A.11: Sherman–Morrison Recursion

The inverse and determinant of the $n \times n$ matrix $\mathbf{A} = \mathbf{A}_0 + \sum_{k=1}^{p} \boldsymbol{a}_k\boldsymbol{a}_k^\top$ are given respectively by:

$$\mathbf{A}^{-1} = \mathbf{A}_0^{-1} - \mathbf{C}\mathbf{D}^{-1}\mathbf{C}^\top$$
$$\det(\mathbf{A}) = \det(\mathbf{A}_0)\det(\mathbf{D}),$$

where $\mathbf{C} \in \mathbb{R}^{n\times p}$ and $\mathbf{D} \in \mathbb{R}^{p\times p}$ are the matrices

$$\mathbf{C} := \left[\boldsymbol{c}_{1,1}, \ldots, \boldsymbol{c}_{p,p}\right], \quad \mathbf{D} := \operatorname{diag}\left(1 + \boldsymbol{a}_1^\top \boldsymbol{c}_{1,1}, \cdots, 1 + \boldsymbol{a}_p^\top \boldsymbol{c}_{p,p}\right),$$

and all the $\{\boldsymbol{c}_{j,k}\}$ are computed from the recursion (A.17) in $O(p^2 n)$ time.

As a consequence of Theorem A.11, the solution to the linear system $\mathbf{A}\boldsymbol{x} = \boldsymbol{b}$ can be computed in $O(p^2 n)$ time via:

$$\boldsymbol{x} = \mathbf{A}_0^{-1}\boldsymbol{b} - \mathbf{C}\mathbf{D}^{-1}[\mathbf{C}^\top \boldsymbol{b}].$$

If $n > p$, the Sherman–Morrison recursion can frequently be much faster than the $O(n^3)$ direct solution via the LU decomposition method in Section A.6.1. ☞ 370

In summary, the following algorithm computes the matrices $\mathbf{C}$ and $\mathbf{D}$ in Theorem A.11 via the recursion (A.17).

Algorithm A.6.1: Sherman–Morrison Recursion

input: Easily invertible matrix $\mathbf{A}_0$ and column vectors $\boldsymbol{a}_1, \ldots, \boldsymbol{a}_p$.

output: Matrices $\mathbf{C}$ and $\mathbf{D}$ such that $\mathbf{C}\mathbf{D}^{-1}\mathbf{C}^\top = \mathbf{A}_0^{-1} - \left(\mathbf{A}_0 + \sum_j \boldsymbol{a}_j \boldsymbol{a}_j^\top\right)^{-1}$.

```
c_k ← A_0^{-1} a_k for k = 1, ..., p (assuming A_0 is diagonal or easily invertible matrix)
for k = 1, ..., p − 1 do
    d_k ← 1 + a_k^T c_k
    for j = k + 1, ..., p do
        c_j ← c_j − (a_k^T c_j / d_k) c_k
d_p ← 1 + a_p^T c_p
C ← [c_1, ..., c_p]
D ← diag(d_1, ..., d_p)
return C and D
```

Finally, note that if $\mathbf{A}_0$ is a diagonal matrix and we only store the diagonal elements of $\mathbf{D}$ and $\mathbf{A}_0$ (as opposed to storing the full matrices $\mathbf{D}$ and $\mathbf{A}_0$), then the storage or memory requirements of Algorithm A.6.1 are only $O(p\,n)$.

A.6.3 Cholesky Decomposition

If $\mathbf{A}$ is a real-valued positive definite matrix (and therefore symmetric), e.g., a covariance matrix, then an LU decomposition can be achieved with matrices $\mathbf{L}$ and $\mathbf{U} = \mathbf{L}^\top$.

Theorem A.12: Cholesky Decomposition

A real-valued positive definite matrix $\mathbf{A} = [a_{ij}] \in \mathbb{R}^{n \times n}$ can be decomposed as

$$\mathbf{A} = \mathbf{L}\mathbf{L}^\top,$$

where the real $n \times n$ lower triangular matrix $\mathbf{L} = [l_{kj}]$ satisfies the recursive formula

$$l_{kj} = \frac{a_{kj} - \sum_{i=1}^{j-1} l_{ji}\, l_{ki}}{\sqrt{a_{jj} - \sum_{i=1}^{j-1} l_{ji}^2}}, \quad \text{where} \quad \sum_{i=1}^{0} l_{ji}\, l_{ki} := 0 \tag{A.18}$$

for $k = 1, \ldots, n$ and $j = 1, \ldots, k$.

Proof: The proof is by inductive construction. For $k = 1, \dots, n$, let $\mathbf{A}_k$ be the left-upper $k \times k$ submatrix of $\mathbf{A} = \mathbf{A}_n$. With $\boldsymbol{e}_1 := [1, 0, \dots, 0]^\top$, we have $\mathbf{A}_1 = a_{11} = \boldsymbol{e}_1^\top \mathbf{A} \boldsymbol{e}_1 > 0$ by the positive-definiteness of $\mathbf{A}$. It follows that $l_{11} = \sqrt{a_{11}}$. Suppose that $\mathbf{A}_{k-1}$ has a Cholesky factorization $\mathbf{L}_{k-1}\mathbf{L}_{k-1}^\top$ with $\mathbf{L}_{k-1}$ having strictly positive diagonal elements, we can construct a Cholesky factorization of $\mathbf{A}_k$ as follows. First write

$$\mathbf{A}_k = \begin{bmatrix} \mathbf{L}_{k-1}\mathbf{L}_{k-1}^\top & \boldsymbol{a}_{k-1} \\ \boldsymbol{a}_{k-1}^\top & a_{kk} \end{bmatrix}$$

and propose $\mathbf{L}_k$ to be of the form

$$\mathbf{L}_k = \begin{bmatrix} \mathbf{L}_{k-1} & \mathbf{0} \\ \boldsymbol{l}_{k-1}^\top & l_{kk} \end{bmatrix}$$

for some vector $\boldsymbol{l}_{k-1} \in \mathbb{R}^{k-1}$ and scalar l_{kk}, for which it must hold that

$$\begin{bmatrix} \mathbf{L}_{k-1}\mathbf{L}_{k-1}^\top & \boldsymbol{a}_{k-1} \\ \boldsymbol{a}_{k-1}^\top & a_{kk} \end{bmatrix} = \begin{bmatrix} \mathbf{L}_{k-1} & \mathbf{0} \\ \boldsymbol{l}_{k-1}^\top & l_{kk} \end{bmatrix} \begin{bmatrix} \mathbf{L}_{k-1}^\top & \boldsymbol{l}_{k-1} \\ \mathbf{0}^\top & l_{kk} \end{bmatrix}.$$

To establish that such an $\boldsymbol{l}_{k-1}$ and l_{kk} exist, we must verify that the set of equations

$$\begin{aligned} \mathbf{L}_{k-1}\boldsymbol{l}_{k-1} &= \boldsymbol{a}_{k-1} \\ \boldsymbol{l}_{k-1}^\top \boldsymbol{l}_{k-1} + l_{kk}^2 &= a_{kk} \end{aligned} \tag{A.19}$$

has a solution. The system $\mathbf{L}_{k-1}\boldsymbol{l}_{k-1} = \boldsymbol{a}_{k-1}$ has a unique solution, because (by assumption) $\mathbf{L}_{k-1}$ is lower diagonal with strictly positive entries down the main diagonal and we can solve for $\boldsymbol{l}_{k-1}$ using forward substitution: $\boldsymbol{l}_{k-1} = \mathbf{L}_{k-1}^{-1}\boldsymbol{a}_{k-1}$. We can solve the second equation as $l_{kk} = \sqrt{a_{kk} - \|\boldsymbol{l}_k\|^2}$, provided that the term within the square root is positive. We demonstrate this using the fact that $\mathbf{A}$ is a positive definite matrix. In particular, for $\boldsymbol{x} \in \mathbb{R}^n$ of the form $[\boldsymbol{x}_1^\top, x_2, \mathbf{0}^\top]^\top$, where $\boldsymbol{x}_1$ is a non-zero $(k-1)$-dimensional vector and x_2 a non-zero number, we have

$$0 < \boldsymbol{x}^\top \mathbf{A} \boldsymbol{x} = [\boldsymbol{x}_1^\top, x_2] \begin{bmatrix} \mathbf{L}_{k-1}\mathbf{L}_{k-1}^\top & \boldsymbol{a}_{k-1} \\ \boldsymbol{a}_{k-1}^\top & a_{kk} \end{bmatrix} \begin{bmatrix} \boldsymbol{x}_1 \\ x_2 \end{bmatrix} = \|\mathbf{L}_{k-1}^\top \boldsymbol{x}_1\|^2 + 2\boldsymbol{x}_1^\top \boldsymbol{a}_{k-1} x_2 + a_{kk} x_2^2.$$

Now take $\boldsymbol{x}_1 = -x_2 \mathbf{L}_{k-1}^{-\top} \boldsymbol{l}_{k-1}$ to obtain $0 < \boldsymbol{x}^\top \mathbf{A} \boldsymbol{x} = x_2^2 (a_{kk} - \|\boldsymbol{l}_{k-1}\|^2)$. Therefore, (A.19) can be uniquely solved. As we have already solved it for $k = 1$, we can solve it for any $k = 1, \dots, n$, leading to the recursive formula (A.18) and Algorithm A.6.2 below. □

An implementation of Cholesky's decomposition that uses the notation in the proof of Theorem A.6.3 is the following algorithm, whose running cost is $\mathcal{O}(n^3)$.

Algorithm A.6.2: Cholesky Decomposition

input: Positive-definite $n \times n$ matrix $\mathbf{A}_n$ with entries $\{a_{ij}\}$.
output: Lower triangular $\mathbf{L}_n$ such that $\mathbf{L}_n \mathbf{L}_n^\top = \mathbf{A}_n$.

$\mathbf{L}_1 \leftarrow \sqrt{a_{11}}$
for $k = 2, \dots, n$ **do**
  $\boldsymbol{a}_{k-1} \leftarrow [a_{1k}, \dots, a_{k-1,k}]^\top$
  $\boldsymbol{l}_{k-1} \leftarrow \mathbf{L}_{k-1}^{-1} \boldsymbol{a}_{k-1}$ (computed in $\mathcal{O}(k^2)$ time via forward substitution)
  $l_{kk} \leftarrow \sqrt{a_{kk} - \boldsymbol{l}_{k-1}^\top \boldsymbol{l}_{k-1}}$
  $\mathbf{L}_k \leftarrow \begin{bmatrix} \mathbf{L}_{k-1} & \mathbf{0} \\ \boldsymbol{l}_{k-1}^\top & l_{kk} \end{bmatrix}$
return $\mathbf{L}_n$

A.6.4 QR Decomposition and the Gram–Schmidt Procedure

Let $\mathbf{A}$ be an $n \times p$ matrix, where $p \leqslant n$. Then, there exists a matrix $\mathbf{Q} \in \mathbb{R}^{n \times p}$ satisfying $\mathbf{Q}^\top \mathbf{Q} = \mathbf{I}_p$, and an upper triangular matrix $\mathbf{R} \in \mathbb{R}^{p \times p}$, such that

$$\mathbf{A} = \mathbf{QR}.$$

This is the *QR decomposition* for real-valued matrices. When $\mathbf{A}$ has full column rank, such a decomposition can be obtained via the *Gram–Schmidt* procedure, which constructs an *orthonormal basis* $\{\boldsymbol{u}_1, \ldots, \boldsymbol{u}_p\}$ of the column space of $\mathbf{A}$ spanned by $\{\boldsymbol{a}_1, \ldots, \boldsymbol{a}_p\}$, in the following way (see also Figure A.5):

GRAM–SCHMIDT

1. Take $\boldsymbol{u}_1 = \boldsymbol{a}_1 / \|\boldsymbol{a}_1\|$.
2. Let $\boldsymbol{p}_1$ be the projection of $\boldsymbol{a}_2$ onto Span $\{\boldsymbol{u}_1\}$. That is, $\boldsymbol{p}_1 = \langle \boldsymbol{u}_1, \boldsymbol{a}_2 \rangle \boldsymbol{u}_1$. Now take $\boldsymbol{u}_2 = (\boldsymbol{a}_2 - \boldsymbol{p}_1)/\|\boldsymbol{a}_2 - \boldsymbol{p}_1\|$. This vector is perpendicular to $\boldsymbol{u}_1$ and has unit length.
3. Let $\boldsymbol{p}_2$ be the projection of $\boldsymbol{a}_3$ onto Span $\{\boldsymbol{u}_1, \boldsymbol{u}_2\}$. That is, $\boldsymbol{p}_2 = \langle \boldsymbol{u}_1, \boldsymbol{a}_3 \rangle \boldsymbol{u}_1 + \langle \boldsymbol{u}_2, \boldsymbol{a}_3 \rangle \boldsymbol{u}_2$. Now take $\boldsymbol{u}_3 = (\boldsymbol{a}_3 - \boldsymbol{p}_2)/\|\boldsymbol{a}_3 - \boldsymbol{p}_2\|$. This vector is perpendicular to both $\boldsymbol{u}_1$ and $\boldsymbol{u}_2$ and has unit length.
4. Continue this process to obtain $\boldsymbol{u}_4, \ldots, \boldsymbol{u}_p$.

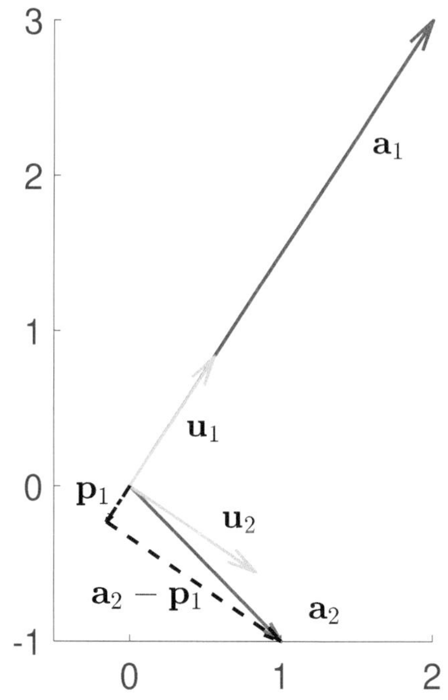

Figure A.5: Illustration of the Gram–Schmidt procedure.

At the end of the procedure, a set $\{\boldsymbol{u}_1, \ldots, \boldsymbol{u}_p\}$ of p orthonormal vectors are obtained. Consequently, as a result of Theorem A.3,

$$\boldsymbol{a}_j = \sum_{i=1}^{j} \underbrace{\langle \boldsymbol{a}_j, \boldsymbol{u}_i \rangle}_{r_{ij}} \boldsymbol{u}_i, \quad j = 1, \ldots, p,$$

for some numbers r_{ij}, $j = 1, \ldots, i$, $i = 1, \ldots, p$. Denoting the corresponding upper triangular matrix $[r_{ij}]$ by $\mathbf{R}$, we have in matrix notation:

$$\mathbf{QR} = [\boldsymbol{u}_1, \ldots, \boldsymbol{u}_p] \begin{bmatrix} r_{11} & r_{12} & r_{13} & \ldots & r_{1p} \\ 0 & r_{22} & r_{23} & \ldots & r_{2p} \\ \vdots & 0 & \ddots & \ddots & \vdots \\ 0 & 0 & 0 & \ldots & r_{pp} \end{bmatrix} = [\boldsymbol{a}_1, \ldots, \boldsymbol{a}_p] = \mathbf{A},$$

which yields a QR decomposition. The QR decomposition can be used to efficiently solve least-squares problems; this will be shown shortly. It can also be used to calculate the determinant of the matrix $\mathbf{A}$, whenever $\mathbf{A}$ is square. Namely, $\det(\mathbf{A}) = \det(\mathbf{Q})\det(\mathbf{R}) = \det(\mathbf{R})$; and since $\mathbf{R}$ is triangular, its determinant is the product of its diagonal elements. There exist various improvements of the Gram–Schmidt process (for example, the *Householder transformation* [52]) that not only improve the numerical stability of the QR decomposition, but also can be applied even when $\mathbf{A}$ is not full rank.

An important application of the QR decomposition is found in solving the least-squares problem in $O(p^2\, n)$ time:

$$\min_{\boldsymbol{\beta} \in \mathbb{R}^p} \|\mathbf{X}\boldsymbol{\beta} - \boldsymbol{y}\|^2$$

for some $\mathbf{X} \in \mathbb{R}^{n \times p}$ (model) matrix. Using the defining properties of the pseudo-inverse in
☞ 362 Definition A.2, one can show that $\|\mathbf{X}\mathbf{X}^+\boldsymbol{y} - \boldsymbol{y}\|^2 \leqslant \|\mathbf{X}\boldsymbol{\beta} - \boldsymbol{y}\|^2$ for any $\boldsymbol{\beta}$. In other words, $\widehat{\boldsymbol{\beta}} := \mathbf{X}^+\boldsymbol{y}$ minimizes $\|\mathbf{X}\boldsymbol{\beta} - \boldsymbol{y}\|$. If we have the QR decomposition $\mathbf{X} = \mathbf{QR}$, then a numerically stable way to calculate $\widehat{\boldsymbol{\beta}}$ with an $O(p^2\, n)$ cost is via

$$\widehat{\boldsymbol{\beta}} = (\mathbf{QR})^+\boldsymbol{y} = \mathbf{R}^+\mathbf{Q}^+\boldsymbol{y} = \mathbf{R}^+\mathbf{Q}^\top\boldsymbol{y}.$$

If $\mathbf{X}$ has full column rank, then $\mathbf{R}^+ = \mathbf{R}^{-1}$.

Note that while the QR decomposition is the method of choice for solving the ordinary
☞ 374 least-squares regression problem, the *Sherman–Morrison recursion* is the method of choice
☞ 217 for solving the regularized least-squares (or ridge) regression problem.

A.6.5 Singular Value Decomposition

SINGULAR VALUE DECOMPOSITION

One of the most useful matrix decompositions is the *singular value decomposition* (SVD).

Theorem A.13: Singular Value Decomposition

Any (complex) matrix $m \times n$ matrix $\mathbf{A}$ admits a unique decomposition

$$\mathbf{A} = \mathbf{U}\boldsymbol{\Sigma}\mathbf{V}^*,$$

where $\mathbf{U}$ and $\mathbf{V}$ are unitary matrices of dimension m and n, respectively, and $\boldsymbol{\Sigma}$ is a real $m \times n$ diagonal matrix. If $\mathbf{A}$ is real, then $\mathbf{U}$ and $\mathbf{V}$ are both orthogonal matrices.

Proof: Without loss of generality we can assume that $m \geqslant n$ (otherwise consider the transpose of $\mathbf{A}$). Then $\mathbf{A}^*\mathbf{A}$ is a positive semidefinite Hermitian matrix, because $\langle \mathbf{A}^*\mathbf{A}\boldsymbol{v}, \boldsymbol{v}\rangle = \boldsymbol{v}^*\mathbf{A}^*\mathbf{A}\boldsymbol{v} = \|\mathbf{A}\boldsymbol{v}\|^2 \geqslant 0$ for all $\boldsymbol{v}$. Hence, $\mathbf{A}^*\mathbf{A}$ has non-negative real eigenvalues, $\lambda_1 \geqslant \lambda_2 \geqslant$

$\cdots \geqslant \lambda_n \geqslant 0$. By Theorem A.7 the matrix $\mathbf{V} = [\boldsymbol{v}_1, \ldots, \boldsymbol{v}_n]$ of right-eigenvectors is a unitary matrix. Define the i-th *singular value* as $\sigma_i = \sqrt{\lambda_i}$, $i = 1, \ldots, n$ and suppose $\lambda_1, \ldots, \lambda_r$ are all greater than 0, and $\lambda_{r+1}, \ldots, \lambda_n = 0$. In particular, $\mathbf{A}\boldsymbol{v}_i = \mathbf{0}$ for $i = r+1, \ldots, n$. Let $\boldsymbol{u}_i = \mathbf{A}\boldsymbol{v}_i/\sigma_i$, $i = 1, \ldots, r$. Then, for $i, j \leqslant r$,

SINGULAR VALUE

$$\langle \boldsymbol{u}_i, \boldsymbol{u}_j \rangle = \boldsymbol{u}_j^* \boldsymbol{u}_i = \frac{\boldsymbol{v}_j^* \mathbf{A}^* \mathbf{A} \boldsymbol{v}_i}{\sigma_i \sigma_j} = \frac{\lambda_i \, \mathbb{1}\{i = j\}}{\sigma_i \sigma_j} = \mathbb{1}\{i = j\}.$$

We can extend $\boldsymbol{u}_1, \ldots, \boldsymbol{u}_r$ to an orthonormal basis $\{\boldsymbol{u}_1, \ldots, \boldsymbol{u}_m\}$ of $\mathbb{C}^m$ (e.g., using the Gram–Schmidt procedure). Let $\mathbf{U} = [\boldsymbol{u}_1, \ldots, \boldsymbol{u}_n]$ be the corresponding unitary matrix. Defining $\mathbf{\Sigma}$ to be the $m \times n$ diagonal matrix with diagonal $(\sigma_1, \ldots, \sigma_r, 0, \ldots, 0)$, we have,

$$\mathbf{U\Sigma} = [\mathbf{A}\boldsymbol{v}_1, \ldots, \mathbf{A}\boldsymbol{v}_r, \mathbf{0}, \ldots, \mathbf{0}] = \mathbf{AV},$$

and hence $\mathbf{A} = \mathbf{U\Sigma V}^*$. □

Note that

$$\mathbf{AA}^* = \mathbf{U\Sigma V}^*\mathbf{V\Sigma}^*\mathbf{U}^* = \mathbf{U\Sigma\Sigma}^\top\mathbf{U}^* \quad \text{and} \quad \mathbf{A}^*\mathbf{A} = \mathbf{V\Sigma}^*\mathbf{U}^*\mathbf{U\Sigma V}^* = \mathbf{V\Sigma}^\top\mathbf{\Sigma V}^*.$$

So, $\mathbf{U}$ is a unitary matrix whose columns are eigenvectors of $\mathbf{AA}^*$ and $\mathbf{V}$ is a unitary matrix whose columns are eigenvectors of $\mathbf{A}^*\mathbf{A}$.

The SVD makes it possible to write the matrix $\mathbf{A}$ as a sum of rank-1 matrices, weighted by the singular values $\{\sigma_i\}$:

$$\mathbf{A} = \begin{bmatrix} \boldsymbol{u}_1, \boldsymbol{u}_2, \ldots, \boldsymbol{u}_m \end{bmatrix} \begin{bmatrix} \sigma_1 & 0 & \ldots & \ldots & 0 \\ 0 & \ddots & 0 & \ldots & 0 \\ 0 & \ldots & \sigma_r & \ldots & 0 \\ 0 & \ldots & \ldots & 0 \ldots & 0 \\ 0 & \ldots & \ldots & \ldots \ddots & 0 \end{bmatrix} \begin{bmatrix} \boldsymbol{v}_1^* \\ \boldsymbol{v}_2^* \\ \vdots \\ \boldsymbol{v}_n^* \end{bmatrix} = \sum_{i=1}^{r} \sigma_i \, \boldsymbol{u}_i \, \boldsymbol{v}_i^*, \tag{A.20}$$

which is called the *dyade* or *spectral representation* of $\mathbf{A}$.

SPECTRAL REPRESENTATION

For real-valued matrices, the SVD has a nice geometric interpretation, illustrated in Figure A.6. The linear mapping defined by matrix $\mathbf{A}$ can be thought of as a succession of three linear operations: (1) an orthogonal transformation (i.e., a rotation with a possible flipping of some axes), corresponding to matrix $\mathbf{V}^\top$, followed by (2) a simple scaling of the unit vectors, corresponding to $\mathbf{\Sigma}$, followed by (3) another orthogonal transformation, corresponding to $\mathbf{U}$.

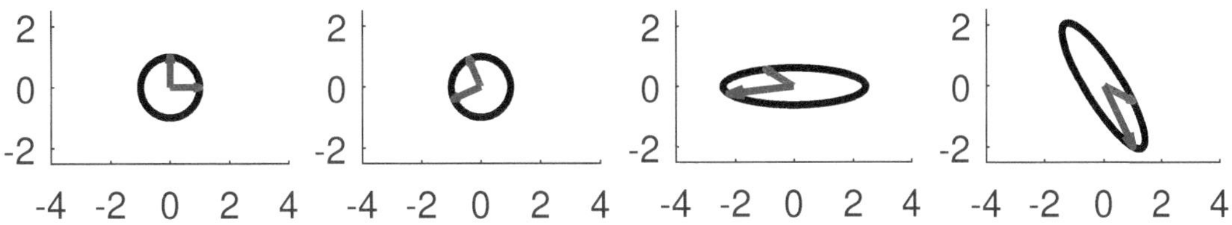

Figure A.6: The figure shows how the unit circle and unit vectors (first panel) are first rotated (second panel), then scaled (third panel), and finally rotated and flipped.

■ **Example A.8 (Ellipses)** We continue Example A.5. Using the `svd` method of the module `numpy.linalg`, we obtain the following SVD matrices for matrix $\mathbf{A}$:

$$\mathbf{U} = \begin{bmatrix} -0.5430 & 0.8398 \\ 0.8398 & 0.5430 \end{bmatrix}, \quad \mathbf{\Sigma} = \begin{bmatrix} 2.4221 & 0 \\ 0 & 0.6193 \end{bmatrix}, \quad \text{and} \quad \mathbf{V} = \begin{bmatrix} -0.3975 & 0.9176 \\ -0.9176 & -0.3975 \end{bmatrix}.$$

☞ 369 Figure A.4 shows the columns of the matrix $\mathbf{U\Sigma}$ as the two principal axes of the ellipse that is obtained by applying matrix $\mathbf{A}$ to the points of the unit circle. ■

A practical method to compute the pseudo-inverse of a real-valued matrix $\mathbf{A}$ is via the singular value decomposition $\mathbf{A} = \mathbf{U\Sigma V}^\top$, where $\mathbf{\Sigma}$ is the diagonal matrix collecting all the positive singular values, say $\sigma_1, \ldots, \sigma_r$, as in Theorem A.13. In this case, $\mathbf{A}^+ = \mathbf{V\Sigma}^+\mathbf{U}^\top$, where $\mathbf{\Sigma}^+$ is the $n \times m$ diagonal (pseudo-inverse) matrix:

$$\mathbf{\Sigma}^+ = \begin{bmatrix} \sigma_1^{-1} & 0 & \ldots & & \ldots & 0 \\ 0 & \ddots & 0 & & \ldots & 0 \\ 0 & \ldots & \sigma_r^{-1} & & \ldots & 0 \\ 0 & \ldots & \ldots & 0 & \ldots & 0 \\ 0 & \ldots & \ldots & \ldots & \ddots & 0 \end{bmatrix}.$$

We conclude with a typical application of the pseudo-inverse for a least-squares optimization problem from data science.

■ **Example A.9 (Rank-Deficient Least Squares)** Given is an $n \times p$ data matrix

$$\mathbf{X} = \begin{bmatrix} x_{11} & x_{12} & \cdots & x_{1p} \\ x_{21} & x_{22} & \cdots & x_{2p} \\ \vdots & \vdots & \vdots & \vdots \\ x_{n1} & x_{n2} & \cdots & x_{np} \end{bmatrix}.$$

It is assumed that the matrix is of full row rank (all rows of $\mathbf{X}$ are linearly independent) and that the number of rows is less than the number of columns: $n < p$. Under this setting, any solution to the equation $\mathbf{X}\boldsymbol{\beta} = \boldsymbol{y}$ provides a perfect fit to the data and minimizes (to 0) the least-squares problem

$$\widehat{\boldsymbol{\beta}} = \underset{\boldsymbol{\beta} \in \mathbb{R}^p}{\operatorname{argmin}} \, \|\mathbf{X}\boldsymbol{\beta} - \boldsymbol{y}\|^2. \tag{A.21}$$

In particular, if $\boldsymbol{\beta}^*$ minimizes $\|\mathbf{X}\boldsymbol{\beta} - \boldsymbol{y}\|^2$ then so does $\boldsymbol{\beta}^* + \boldsymbol{u}$ for all $\boldsymbol{u}$ in the null space $\mathcal{N}_{\mathbf{X}} := \{\boldsymbol{u} : \mathbf{X}\boldsymbol{u} = \mathbf{0}\}$, which has dimension $p - n$. To cope with the non-uniqueness of solutions, a possible approach is to solve instead the following optimization problem:

$$\begin{aligned} &\underset{\boldsymbol{\beta} \in \mathbb{R}^p}{\text{minimize}} \quad \boldsymbol{\beta}^\top \boldsymbol{\beta} \\ &\text{subject to} \quad \mathbf{X}\boldsymbol{\beta} - \boldsymbol{y} = \mathbf{0}. \end{aligned}$$

That is, we are interested in a solution $\boldsymbol{\beta}$ with the smallest squared norm (or, equivalently, the smallest norm). The solution can be obtained via Lagrange's method (see Section B.2.2). Specifically, set $\mathcal{L}(\boldsymbol{\beta}, \boldsymbol{\lambda}) = \boldsymbol{\beta}^\top\boldsymbol{\beta} - \boldsymbol{\lambda}^\top(\mathbf{X}\boldsymbol{\beta} - \boldsymbol{y})$, and solve ☞ 408

$$\nabla_{\boldsymbol{\beta}} \mathcal{L}(\boldsymbol{\beta}, \boldsymbol{\lambda}) = 2\boldsymbol{\beta} - \mathbf{X}^\top \boldsymbol{\lambda} = \mathbf{0}, \tag{A.22}$$

and

$$\nabla_\lambda \mathcal{L}(\boldsymbol{\beta}, \boldsymbol{\lambda}) = \mathbf{X}\boldsymbol{\beta} - \boldsymbol{y} = \mathbf{0}. \tag{A.23}$$

From (A.22) we get $\boldsymbol{\beta} = \mathbf{X}^\top \boldsymbol{\lambda}/2$. By substituting it in (A.23), we arrive at $\boldsymbol{\lambda} = 2(\mathbf{X}\mathbf{X}^\top)^{-1}\boldsymbol{y}$, and hence $\boldsymbol{\beta}$ is given by

$$\boldsymbol{\beta} = \frac{\mathbf{X}^\top \boldsymbol{\lambda}}{2} = \frac{\mathbf{X}^\top 2(\mathbf{X}\mathbf{X}^\top)^{-1}\boldsymbol{y}}{2} = \mathbf{X}^\top(\mathbf{X}\mathbf{X}^\top)^{-1}\boldsymbol{y} = \mathbf{X}^+\boldsymbol{y}.$$

An example Python code is given below.

svdexample.py

```
from numpy import diag, zeros,vstack
from numpy.random import rand, seed
from numpy.linalg import svd, pinv
seed(12345)
n = 5
p = 8
X = rand(n,p)
y = rand(n,1)
U,S,VT = svd(X)
SI  = diag(1/S)
# compute pseudo inverse
pseudo_inv = VT.T @ vstack((SI, zeros((p-n,n)))) @ U.T
b = pseudo_inv @ y
#b = pinv(X) @ y   #remove comment for the built-in pseudo inverse
print(X @ b - y)
```

```
[[5.55111512e-16]
 [1.11022302e-16]
 [5.55111512e-16]
 [8.60422844e-16]
 [2.22044605e-16]]
```

■

A.6.6 Solving Structured Matrix Equations

For a general matrix $\mathbf{A} \in \mathbb{C}^{n\times n}$, performing matrix–vector multiplications takes $O(n^2)$ operations; and solving linear systems $\mathbf{A}\boldsymbol{x} = \boldsymbol{b}$, and carrying out LU decompositions takes $O(n^3)$ operations. However, when $\mathbf{A}$ is *sparse* (i.e., has relatively few non-zero elements) or has a special structure, the computational complexity for these operations can often be reduced. Matrices $\mathbf{A}$ that are "structured" in this way often satisfy a *Sylvester equation*, of the form

SYLVESTER EQUATION

$$\mathbf{M}_1\mathbf{A} - \mathbf{A}\mathbf{M}_2^* = \mathbf{G}_1\mathbf{G}_2^*, \tag{A.24}$$

where $\mathbf{M}_i \in \mathbb{C}^{n\times n}, i = 1,2$ are sparse matrices and $\mathbf{G}_i \in \mathbb{C}^{n\times r}$, $i = 1,2$ are matrices of rank $r \ll n$. The elements of $\mathbf{A}$ must be easy to recover from these matrices, e.g., with $O(1)$ operations. A typical example is a (square) *Toeplitz matrix*, which has the following structure:

TOEPLITZ MATRIX

$$\mathbf{A} = \begin{bmatrix} a_0 & a_{-1} & \cdots & a_{-(n-2)} & a_{-(n-1)} \\ a_1 & a_0 & a_{-1} & & a_{-(n-2)} \\ \vdots & a_1 & a_0 & \ddots & \vdots \\ a_{n-2} & & \ddots & \ddots & a_{-1} \\ a_{n-1} & a_{n-2} & \cdots & a_1 & a_0 \end{bmatrix}.$$

A general square Toeplitz matrix $\mathbf{A}$ is completely determined by the $2n-1$ elements along its first row and column. If $\mathbf{A}$ is also Hermitian (i.e., $\mathbf{A}^* = \mathbf{A}$), then clearly it is determined by only n elements. If we define the matrices:

$$\mathbf{M}_1 = \begin{bmatrix} 0 & 0 & \cdots & 0 & 1 \\ 1 & 0 & 0 & & 0 \\ \vdots & 1 & 0 & \ddots & \vdots \\ 0 & & \ddots & \ddots & 0 \\ 0 & 0 & \cdots & 1 & 0 \end{bmatrix} \quad \text{and} \quad \mathbf{M}_2 = \begin{bmatrix} 0 & 1 & \cdots & 0 & 0 \\ 0 & 0 & 1 & & 0 \\ \vdots & 0 & 0 & \ddots & \vdots \\ 0 & & \ddots & \ddots & 1 \\ -1 & 0 & \cdots & 0 & 0 \end{bmatrix},$$

then (A.24) is satisfied with

$$\mathbf{G}_1\mathbf{G}_2^* := \begin{bmatrix} 1 & 0 \\ 0 & a_1 + a_{-(n-1)} \\ 0 & a_2 + a_{-(n-2)} \\ \vdots & \vdots \\ 0 & a_{n-1} + a_{-1} \end{bmatrix} \begin{bmatrix} a_{n-1} - a_{-1} & a_{n-2} - a_{-2} & \ldots & a_1 - a_{-(n-1)} & 2a_0 \\ 0 & 0 & \ldots & 0 & 1 \end{bmatrix}$$

$$= \begin{bmatrix} a_{n-1} - a_{-1} & a_{n-2} - a_{-2} & \ldots & a_1 - a_{-(n-1)} & 2a_0 \\ 0 & 0 & \ldots & 0 & a_1 + a_{-(n-1)} \\ \vdots & \vdots & \ldots & \vdots & a_2 + a_{-(n-2)} \\ \vdots & \vdots & \ldots & \vdots & \vdots \\ 0 & 0 & \ldots & 0 & a_{n-1} + a_{-1} \end{bmatrix},$$

which has rank $r \leqslant 2$.

■ **Example A.10 (Discrete Convolution of Vectors)** The convolution of two vectors can be represented as multiplication of one of the vectors by a Toeplitz matrix. Suppose $\boldsymbol{a} = [a_1, \ldots, a_n]^\top$ and $\boldsymbol{b} = [b_1, \ldots, b_n]^\top$ are two complex-valued vectors. Then, their *convolution* is defined as the vector $\boldsymbol{a} * \boldsymbol{b}$ with i-th element

CONVOLUTION

$$[\boldsymbol{a} * \boldsymbol{b}]_i = \sum_{k=1}^{n} a_k\, b_{i-k+1}, \quad i = 1, \ldots, n,$$

where $b_j := 0$ for $j \leqslant 0$. It is easy to verify that the convolution can be written as

$$\boldsymbol{a} * \boldsymbol{b} = \mathbf{A}\boldsymbol{b},$$

where, denoting the d-dimensional column vector of zeros by $\mathbf{0}_d$, we have that

$$\mathbf{A} = \begin{bmatrix} \boldsymbol{a} & 0 & & & \\ \mathbf{0}_{n-1} & \boldsymbol{a} & \ddots & & \\ & \mathbf{0}_{n-2} & \ddots & \mathbf{0}_{n-2} & \\ & & \ddots & \boldsymbol{a} & \mathbf{0}_{n-1} \\ & & & 0 & \boldsymbol{a} \end{bmatrix}.$$

Clearly, the matrix $\mathbf{A}$ is a (sparse) Toeplitz matrix. ■

A *circulant matrix* is a special Toeplitz matrix which is obtained from a vector $\boldsymbol{c}$ by circularly permuting its indices as follows: CIRCULANT MATRIX

$$\mathbf{C} = \begin{bmatrix} c_0 & c_{n-1} & \dots & c_2 & c_1 \\ c_1 & c_0 & c_{n-1} & & c_2 \\ \vdots & c_1 & c_0 & \ddots & \vdots \\ c_{n-2} & & \ddots & \ddots & c_{n-1} \\ c_{n-1} & c_{n-2} & \dots & c_1 & c_0 \end{bmatrix}. \tag{A.25}$$

Note that $\mathbf{C}$ is completely determined by the n elements of its first column, $\boldsymbol{c}$.

To illustrate how structured matrices allow for faster matrix computations, consider solving the $n \times n$ linear system:

$$\mathbf{A}_n \boldsymbol{x}_n = \boldsymbol{a}_n$$

for $\boldsymbol{x}_n = [x_1, \dots, x_n]^\top$, where $\boldsymbol{a}_n = [a_1, \dots, a_n]^\top$, and

$$\mathbf{A}_n := \begin{bmatrix} 1 & a_1 & \dots & a_{n-2} & a_{n-1} \\ a_1 & 1 & \ddots & & a_{n-2} \\ \vdots & \ddots & \ddots & \ddots & \vdots \\ a_{n-2} & & \ddots & \ddots & a_1 \\ a_{n-1} & a_{n-2} & \cdots & a_1 & 1 \end{bmatrix} \tag{A.26}$$

is a real-valued symmetric positive-definite Toeplitz matrix (so that it is invertible). Note that the entries of $\mathbf{A}_n$ are completely determined by the right-hand side of the linear equation: vector $\boldsymbol{a}_n$. As we shall see shortly in Example A.11, the solution to the more general linear equation $\mathbf{A}_n \boldsymbol{x}_n = \boldsymbol{b}_n$, where $\boldsymbol{b}_n$ is arbitrary, can be efficiently computed using the solution to this specific system $\mathbf{A}_n \boldsymbol{x}_n = \boldsymbol{a}_n$, obtained via a special recursive algorithm (Algorithm A.6.3 below).

For every $k = 1, \dots, n$ the $k \times k$ Toeplitz matrix $\mathbf{A}_k$ satisfies

$$\mathbf{A}_k = \mathbf{P}_k \mathbf{A}_k \mathbf{P}_k,$$

where $\mathbf{P}_k$ is a permutation matrix that "flips" the order of elements — rows when pre-multiplying and columns when post-multiplying. For example,

$$\begin{bmatrix} 1 & 2 & 3 & 4 & 5 \\ 6 & 7 & 8 & 9 & 10 \end{bmatrix} \mathbf{P}_5 = \begin{bmatrix} 5 & 4 & 3 & 2 & 1 \\ 10 & 9 & 8 & 7 & 6 \end{bmatrix}, \quad \text{where} \quad \mathbf{P}_5 = \begin{bmatrix} 0 & 0 & 0 & 0 & 1 \\ 0 & 0 & 0 & 1 & 0 \\ 0 & 0 & 1 & 0 & 0 \\ 0 & 1 & 0 & 0 & 0 \\ 1 & 0 & 0 & 0 & 0 \end{bmatrix}.$$

Clearly, $\mathbf{P}_k = \mathbf{P}_k^\top$ and $\mathbf{P}_k\mathbf{P}_k = \mathbf{I}_k$ hold, so that in fact $\mathbf{P}_k$ is an orthogonal matrix.

We can solve the $n \times n$ linear system $\mathbf{A}_n\boldsymbol{x}_n = \boldsymbol{a}_n$ in $O(n^2)$ time recursively, as follows. Assume that we have somehow solved for the upper $k \times k$ block $\mathbf{A}_k\,\boldsymbol{x}_k = \boldsymbol{a}_k$ and now we wish to solve for the $(k+1) \times (k+1)$ block:

$$\mathbf{A}_{k+1}\,\boldsymbol{x}_{k+1} = \boldsymbol{a}_{k+1} \quad \Longleftrightarrow \quad \begin{bmatrix} \mathbf{A}_k & \mathbf{P}_k\,\boldsymbol{a}_k \\ \boldsymbol{a}_k^\top\,\mathbf{P}_k & 1 \end{bmatrix} \begin{bmatrix} \boldsymbol{z} \\ \alpha \end{bmatrix} = \begin{bmatrix} \boldsymbol{a}_k \\ a_{k+1} \end{bmatrix}.$$

Therefore,

$$\begin{aligned} \alpha &= a_{k+1} - \boldsymbol{a}_k^\top\,\mathbf{P}_k\,\boldsymbol{z} \\ \mathbf{A}_k\,\boldsymbol{z} &= \boldsymbol{a}_k - \alpha\,\mathbf{P}_k\,\boldsymbol{a}_k. \end{aligned}$$

Since $\mathbf{A}_k^{-1}\,\mathbf{P}_k = \mathbf{P}_k\,\mathbf{A}_k^{-1}$, the second equation above simplifies to

$$\begin{aligned} \boldsymbol{z} &= \mathbf{A}_k^{-1}\,\boldsymbol{a}_k - \alpha\,\mathbf{A}_k^{-1}\,\mathbf{P}_k\,\boldsymbol{a}_k \\ &= \boldsymbol{x}_k - \alpha\,\mathbf{P}_k\,\boldsymbol{x}_k. \end{aligned}$$

Substituting $\boldsymbol{z} = \boldsymbol{x}_k - \alpha\,\mathbf{P}_k\,\boldsymbol{x}_k$ into $\alpha = a_{k+1} - \boldsymbol{a}_k^\top\,\mathbf{P}_k\,\boldsymbol{z}$ and solving for α yields:

$$\alpha = \frac{a_{k+1} - \boldsymbol{a}_k^\top\,\mathbf{P}_k\,\boldsymbol{x}_k}{1 - \boldsymbol{a}_k^\top\,\boldsymbol{x}_k}.$$

Finally, with the value of α computed above, we have

$$\boldsymbol{x}_{k+1} = \begin{bmatrix} \boldsymbol{x}_k - \alpha\,\mathbf{P}_k\,\boldsymbol{x}_k \\ \alpha \end{bmatrix}.$$

LEVINSON–DURBIN

This gives the following *Levinson–Durbin* recursive algorithm for solving $\mathbf{A}_n\,\boldsymbol{x}_n = \boldsymbol{a}_n$.

Algorithm A.6.3: Levinson–Durbin Recursion for Solving $\mathbf{A}_n\,\boldsymbol{x}_n = \boldsymbol{a}_n$

input: First row $[1, a_1, \ldots, a_{n-1}] = [1, \boldsymbol{a}_{n-1}^\top]$ of matrix $\mathbf{A}_n$.
output: Solution $\boldsymbol{x}_n = \mathbf{A}_n^{-1}\,\boldsymbol{a}_n$.

$\boldsymbol{x}_1 \leftarrow a_1$
for $k = 1, \ldots, n-1$ **do**
  $\beta_k \leftarrow 1 - \boldsymbol{a}_k^\top\,\boldsymbol{x}_k$
  $\check{\boldsymbol{x}} \leftarrow [x_{k,k}, x_{k,k-1}, \ldots, x_{k,1}]^\top$
  $\alpha \leftarrow (a_{k+1} - \boldsymbol{a}_k^\top\,\check{\boldsymbol{x}})/\beta_k$
  $\boldsymbol{x}_{k+1} \leftarrow \begin{bmatrix} \boldsymbol{x}_k - \alpha\,\check{\boldsymbol{x}} \\ \alpha \end{bmatrix}$
return $\boldsymbol{x}_n$

In the algorithm above, we have identified $\boldsymbol{x}_k = [x_{k,1}, x_{k,2}, \ldots, x_{k,k}]^\top$. The advantage of the Levinson–Durbin algorithm is that its running cost is $O(n^2)$, instead of the usual $O(n^3)$.

Using the $\{\boldsymbol{x}_k, \beta_k\}$ computed in Algorithm A.6.3, we construct the following lower triangular matrix recursively, setting $\mathbf{L}_1 = 1$ and

$$\mathbf{L}_{k+1} = \begin{bmatrix} \mathbf{L}_k & \mathbf{0}_k \\ -(\mathbf{P}_k\boldsymbol{x}_k)^\top & 1 \end{bmatrix}, \qquad k = 1, \ldots, n-1. \tag{A.27}$$

Then, we have the following factorization of $\mathbf{A}_n$.

Theorem A.14: Diagonalization of Toeplitz Correlation Matrix $\mathbf{A}_n$

For a real-valued symmetric positive-definite Toeplitz matrix $\mathbf{A}_n$ of the form (A.26), we have

$$\mathbf{L}_n\,\mathbf{A}_n\,\mathbf{L}_n^\top = \mathbf{D}_n,$$

where $\mathbf{L}_n$ is a the lower diagonal matrix (A.27) and $\mathbf{D}_n := \mathrm{diag}(1,\beta_1,\ldots,\beta_{n-1})$ is a diagonal matrix.

Proof: We give a proof by induction. Obviously, $\mathbf{L}_1\mathbf{A}_1\mathbf{L}_1^\top = 1\cdot 1\cdot 1 = 1 = \mathbf{D}_1$ is true. Next, assume that the factorization $\mathbf{L}_k\mathbf{A}_k\mathbf{L}_k^\top = \mathbf{D}_k$ holds for a given k. Observe that

$$\mathbf{L}_{k+1}\mathbf{A}_{k+1} = \begin{bmatrix} \mathbf{L}_k & \mathbf{0}_k \\ -(\mathbf{P}_k\boldsymbol{x}_k)^\top & 1 \end{bmatrix}\begin{bmatrix} \mathbf{A}_k & \mathbf{P}_k\,\boldsymbol{a}_k \\ \boldsymbol{a}_k^\top\,\mathbf{P}_k & 1 \end{bmatrix} = \begin{bmatrix} \mathbf{L}_k\mathbf{A}_k\,, & \mathbf{L}_k\mathbf{P}_k\,\boldsymbol{a}_k \\ -(\mathbf{P}_k\boldsymbol{x}_k)^\top\,\mathbf{A}_k + \boldsymbol{a}_k^\top\,\mathbf{P}_k\,, & -(\mathbf{P}_k\boldsymbol{x}_k)^\top\,\mathbf{P}_k\,\boldsymbol{a}_k + 1 \end{bmatrix}.$$

It is straightforward to verify that $[-(\mathbf{P}_k\boldsymbol{x}_k)^\top\,\mathbf{A}_k + \boldsymbol{a}_k^\top\,\mathbf{P}_k, -(\mathbf{P}_k\boldsymbol{x}_k)^\top\,\mathbf{P}_k\,\boldsymbol{a}_k + 1] = [\mathbf{0}_k^\top, \beta_k]$, yielding the recursion

$$\mathbf{L}_{k+1}\mathbf{A}_{k+1} = \begin{bmatrix} \mathbf{L}_k\mathbf{A}_k & \mathbf{L}_k\mathbf{P}_k\,\boldsymbol{a}_k \\ \mathbf{0}_k^\top & \beta_k \end{bmatrix}.$$

Secondly, observe that

$$\mathbf{L}_{k+1}\mathbf{A}_{k+1}\mathbf{L}_{k+1}^\top = \begin{bmatrix} \mathbf{L}_k\mathbf{A}_k & \mathbf{L}_k\mathbf{P}_k\,\boldsymbol{a}_k \\ \mathbf{0}_k^\top & \beta_k \end{bmatrix}\begin{bmatrix} \mathbf{L}_k^\top & -\mathbf{P}_k\boldsymbol{x}_k \\ \mathbf{0}_k^\top & 1 \end{bmatrix} = \begin{bmatrix} \mathbf{L}_k\mathbf{A}_k\mathbf{L}_k^\top\,, & -\mathbf{L}_k\mathbf{A}_k\mathbf{P}_k\boldsymbol{x}_k + \mathbf{L}_k\mathbf{P}_k\boldsymbol{a}_k \\ \mathbf{0}_k^\top\,, & \beta_k \end{bmatrix}.$$

By noting that $\mathbf{A}_k\mathbf{P}_k\boldsymbol{x}_k = \mathbf{P}_k\mathbf{P}_k\mathbf{A}_k\mathbf{P}_k\boldsymbol{x}_k = \mathbf{P}_k\mathbf{A}_k\boldsymbol{x}_k = \mathbf{P}_k\boldsymbol{a}_k$, we obtain:

$$\mathbf{L}_{k+1}\mathbf{A}_{k+1}\mathbf{L}_{k+1}^\top = \begin{bmatrix} \mathbf{L}_k\mathbf{A}_k\mathbf{L}_k^\top & \mathbf{0}_k \\ \mathbf{0}_k^\top & \beta_k \end{bmatrix}.$$

Hence, the result follows by induction. □

■ **Example A.11 (Solving $\mathbf{A}_n\boldsymbol{x}_n = \boldsymbol{b}_n$ in $O(n^2)$ Time)** One application of the factorization in Theorem A.14 is in the fast solution of a linear system $\mathbf{A}_n\boldsymbol{x}_n = \boldsymbol{b}_n$, where the right-hand side is an arbitrary vector $\boldsymbol{b}_n$. Since the solution $\boldsymbol{x}_n$ can be written as

$$\boldsymbol{x}_n = \mathbf{A}_n^{-1}\boldsymbol{b}_n = \mathbf{L}_n^\top\mathbf{D}_n^{-1}\mathbf{L}_n\boldsymbol{b}_n,$$

we can compute $\boldsymbol{x}_n$ in $O(n^2)$ time, as follows.

Algorithm A.6.4: Solving $\mathbf{A}_n\,\boldsymbol{x}_n = \boldsymbol{b}_n$ for a General Right-Hand Side

input: First row $[1, \boldsymbol{a}_{n-1}^\top]$ of matrix $\mathbf{A}_n$ and right-hand side $\boldsymbol{b}_n$.
output: Solution $\boldsymbol{x}_n = \mathbf{A}_n^{-1}\,\boldsymbol{b}_n$.
Compute $\mathbf{L}_n$ in (A.27) and the numbers $\beta_1,\ldots,\beta_{n-1}$ via Algorithm A.6.3.
$[x_1,\ldots,x_n]^\top \leftarrow \mathbf{L}_n\boldsymbol{b}_n$ (computed in $O(n^2)$ time)
$x_i \leftarrow x_i/\beta_{i-1}$ for $i = 2,\ldots,n$ (computed in $O(n)$ time)
$[x_1,\ldots,x_n] \leftarrow [x_1,\ldots,x_n]\,\mathbf{L}_n$ (computed in $O(n^2)$ time)
return $\boldsymbol{x}_n \leftarrow [x_1,\ldots,x_n]^\top$

Note that it is possible to avoid the explicit construction of the lower triangular matrix in (A.27) via the following modification of Algorithm A.6.3, which only stores an extra vector $\boldsymbol{y}$ at each recursive step of the Levinson–Durbin algorithm.

Algorithm A.6.5: Solving $\mathbf{A}_n\, \boldsymbol{x}_n = \boldsymbol{b}_n$ with $O(n)$ Memory Cost

input: First row $[1, \boldsymbol{a}_{n-1}^\top]$ of matrix $\mathbf{A}_n$ and right-hand side $\boldsymbol{b}_n$.
output: Solution $\boldsymbol{x}_n = \mathbf{A}_n^{-1}\, \boldsymbol{b}_n$.

$\boldsymbol{x} \leftarrow b_1$
$\boldsymbol{y} \leftarrow a_1$
for $k = 1, \ldots, n-1$ **do**
  $\breve{\boldsymbol{x}} \leftarrow [x_k, x_{k-1}, \ldots, x_1]$
  $\breve{\boldsymbol{y}} \leftarrow [y_k, y_{k-1}, \ldots, y_1]$
  $\beta \leftarrow 1 - \boldsymbol{a}_k^\top \boldsymbol{y}$
  $\alpha_x \leftarrow (b_{k+1} - \boldsymbol{b}_k^\top \breve{\boldsymbol{x}})/\beta$
  $\alpha_y \leftarrow (a_{k+1} - \boldsymbol{a}_k^\top \breve{\boldsymbol{y}})/\beta$
  $\boldsymbol{x} \leftarrow [\boldsymbol{x} - \alpha_x \breve{\boldsymbol{x}}, \alpha_x]^\top$
  $\boldsymbol{y} \leftarrow [\boldsymbol{y} - \alpha_y \breve{\boldsymbol{y}}, \alpha_y]^\top$
return $\boldsymbol{x}$

■

A.7 Functional Analysis

Much of the previous theory on Euclidean vector spaces can be generalized to vector spaces of *functions*. Every element of a (real-valued) *function space* $\mathcal{H}$ is a function from some set $\mathcal{X}$ to $\mathbb{R}$, and elements can be added and scalar multiplied as if they were vectors. In other words, if $f \in \mathcal{H}$ and $g \in \mathcal{H}$, then $\alpha f + \beta g \in \mathcal{H}$ for all $\alpha, \beta \in \mathbb{R}$. On $\mathcal{H}$ we can impose an inner product as a mapping $\langle \cdot, \cdot \rangle$ from $\mathcal{H} \times \mathcal{H}$ to $\mathbb{R}$ that satisfies

FUNCTION SPACE

1. $\langle \alpha f_1 + \beta f_2, g \rangle = \alpha \langle f_1, g \rangle + \beta \langle f_2, g \rangle$;
2. $\langle f, g \rangle = \langle g, f \rangle$;
3. $\langle f, f \rangle \geqslant 0$;
4. $\langle f, f \rangle = 0$ if and only if $f = 0$ (the zero function).

We focus on real-valued function spaces, although the theory for complex-valued function spaces is similar (and sometimes easier), under suitable modifications (e.g., $\langle f, g \rangle = \overline{\langle g, f \rangle}$).

Similar to the linear algebra setting in Section A.2, we say that two elements f and g in $\mathcal{H}$ are *orthogonal* to each other with respect to this inner product if $\langle f, g \rangle = 0$. Given an inner product, we can measure distances between elements of the function space $\mathcal{H}$ using the *norm*

NORM

$$\|f\| := \sqrt{\langle f, f \rangle}.$$

For example, the distance between two functions f_m and f_n is given by $\|f_m - f_n\|$. The space $\mathcal{H}$ is said to be *complete* if every sequence of functions $f_1, f_2, \ldots \in \mathcal{H}$ for which

COMPLETE

$$\|f_m - f_n\| \to 0 \text{ as } m, n \to \infty, \tag{A.28}$$

converges to some $f \in \mathcal{H}$; that is, $\|f - f_n\| \to 0$ as $n \to \infty$. A sequence that satisfies (A.28) is called a *Cauchy sequence*.

CAUCHY SEQUENCE

A complete inner product space is called a *Hilbert space*. The most fundamental Hilbert space of functions is the space L^2. An in-depth introduction to L^2 requires some measure theory [6]. For our purposes, it suffices to assume that $\mathcal{X} \subseteq \mathbb{R}^d$ and that on $\mathcal{X}$ a *measure* μ is defined which assigns to each suitable[4] set A a positive number $\mu(A) \geqslant 0$ (e.g., its volume). In many cases of interest μ is of the form

HILBERT SPACE

MEASURE

$$\mu(A) = \int_A w(\boldsymbol{x})\, \mathrm{d}\boldsymbol{x} \tag{A.29}$$

where $w \geqslant 0$ is a positive function on $\mathcal{X}$ which is called the *density* of μ with respect to the Lebesgue measure (the natural volume measure on $\mathbb{R}^d$). We write $\mu(\mathrm{d}\boldsymbol{x}) = w(\boldsymbol{x})\,\mathrm{d}\boldsymbol{x}$ to indicate that μ has density w. Another important case is where

DENSITY

$$\mu(A) = \sum_{\boldsymbol{x} \in A \cap \mathbb{Z}^d} w(\boldsymbol{x}), \tag{A.30}$$

where $w \geqslant 0$ is again called the density of μ, but now with respect to the counting measure on $\mathbb{Z}^d$ (which counts the points of $\mathbb{Z}^d$). Integrals with respect to measures μ in (A.29) and (A.30) can now be defined as

$$\int f(\boldsymbol{x})\,\mu(\mathrm{d}\boldsymbol{x}) = \int f(\boldsymbol{x})\, w(\boldsymbol{x})\,\mathrm{d}\boldsymbol{x},$$

and

$$\int f(\boldsymbol{x})\,\mu(\mathrm{d}\boldsymbol{x}) = \sum_{\boldsymbol{x}} f(\boldsymbol{x})\, w(\boldsymbol{x}),$$

respectively. We assume for simplicity that μ has the form (A.29). For measures of the form (A.30) (so-called discrete measures), replace integrals by sums in what follows.

Definition A.4: L^2 Space

Let $\mathcal{X}$ be a subset of $\mathbb{R}^d$ with measure $\mu(\mathrm{d}\boldsymbol{x}) = w(\boldsymbol{x})\,\mathrm{d}\boldsymbol{x}$. The Hilbert space $L^2(\mathcal{X}, \mu)$ is the linear space of functions from $\mathcal{X}$ to $\mathbb{R}$ that satisfy

$$\int_{\mathcal{X}} f(\boldsymbol{x})^2\, w(\boldsymbol{x})\,\mathrm{d}\boldsymbol{x} < \infty, \tag{A.31}$$

and with inner product

$$\langle f, g \rangle = \int_{\mathcal{X}} f(\boldsymbol{x})\, g(\boldsymbol{x})\, w(\boldsymbol{x})\,\mathrm{d}\boldsymbol{x}. \tag{A.32}$$

Let $\mathcal{H}$ be a Hilbert space. A set of functions $\{f_i, i \in \mathcal{I}\}$ is called an *orthonormal system* if

ORTHONORMAL SYSTEM

[4]Not all sets have a measure. Suitable sets are Borel sets, which can be thought of as countable unions of rectangles.

1. the norm of every f_i is 1; that is, $\langle f_i, f_i \rangle = 1$ for all $i \in \mathcal{I}$,
2. the $\{f_i\}$ are orthogonal; that is, $\langle f_i, f_j \rangle = 0$ for $i \neq j$.

It follows then that the $\{f_i\}$ are linearly independent; that is, the only linear combination $\sum_j \alpha_j f_j(\boldsymbol{x})$ that is equal to $f_i(\boldsymbol{x})$ for all $\boldsymbol{x}$ is the one where $\alpha_i = 1$ and $\alpha_j = 0$ for $j \neq i$. An orthonormal system $\{f_i\}$ is called an *orthonormal basis* if there is no other $f \in \mathcal{H}$ that is orthogonal to all the $\{f_i, i \in \mathcal{I}\}$ (other than the zero function). Although the general theory allows for uncountable bases, in practice[5] the set $\mathcal{I}$ is taken to be countable.

ORTHONORMAL BASIS

■ **Example A.12 (Trigonometric Orthonormal Basis)** Let $\mathcal{H}$ be the Hilbert space $L^2((0, 2\pi), \mu)$, where $\mu(\mathrm{d}x) = w(x)\,\mathrm{d}x$ and w is the constant function $w(x) = 1, 0 < x < 2\pi$. Alternatively, take $\mathcal{X} = \mathbb{R}$ and w the indicator function on $(0, 2\pi)$. The trigonometric functions

$$g_0(x) = \frac{1}{\sqrt{2\pi}}, \quad g_k(x) = \frac{1}{\sqrt{\pi}}\cos(kx), \quad h_k(x) = \frac{1}{\sqrt{\pi}}\sin(kx), \quad k = 1, 2, \dots$$

form a countable infinite-dimensional orthonormal basis of $\mathcal{H}$. ■

A Hilbert space $\mathcal{H}$ with an orthonormal basis $\{f_1, f_2, \dots\}$ behaves very similarly to the familiar Euclidean vector space. In particular, every element (i.e., function) $f \in \mathcal{H}$ can be written as a unique linear combination of the basis vectors:

$$f = \sum_i \langle f, f_i \rangle f_i, \tag{A.33}$$

exactly as in Theorem A.3. The right-hand side of (A.33) is called a (generalized) *Fourier expansion* of f. Note that such a Fourier expansion does not require a trigonometric basis; any orthonormal basis will do.

FOURIER EXPANSION

■ **Example A.13 (Example A.12 (cont.))** Consider the indicator function $f(x) = \mathbb{1}\{0 < x < \pi\}$. As the trigonometric functions $\{g_k\}$ and $\{h_k\}$ form a basis for $L^2((0, 2\pi), 1\mathrm{d}x)$, we can write

$$f(x) = a_0 \frac{1}{\sqrt{2\pi}} + \sum_{k=1}^{\infty} a_k \frac{1}{\sqrt{\pi}}\cos(kx) + \sum_{k=1}^{\infty} b_k \frac{1}{\sqrt{\pi}}\sin(kx), \tag{A.34}$$

where $a_0 = \int_0^\pi 1/\sqrt{2\pi}\,\mathrm{d}x = \sqrt{\pi/2}$, $a_k = \int_0^\pi \cos(kx)/\sqrt{\pi}\,\mathrm{d}x$ and $b_k = \int_0^\pi \sin(kx)/\sqrt{\pi}\,\mathrm{d}x$, $k = 1, 2, \dots$. This means that $a_k = 0$ for all k, $b_k = 0$ for even k, and $b_k = 2/(k\sqrt{\pi})$ for odd k. Consequently,

$$f(x) = \frac{1}{2} + \frac{2}{\pi}\sum_{k=1}^{\infty} \frac{\sin(kx)}{k}. \tag{A.35}$$

Figure A.7 shows several Fourier *approximations* obtained by truncating the infinite sum in (A.35). ■

[5]The function spaces typically encountered in machine learning and data science are usually *separable spaces*, which allows for the set $\mathcal{I}$ to be considered countable; see, e.g., [106].

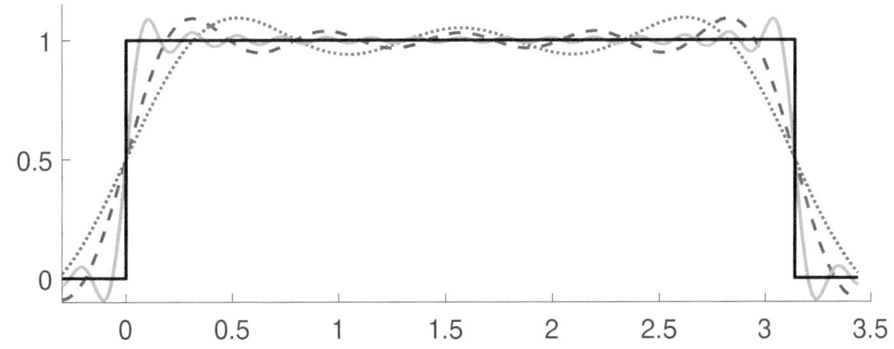

Figure A.7: Fourier approximations of the unit step function f on the interval $(0, \pi)$, truncating the infinite sum in (A.35) to $i = 2$, 4, and 14 terms, giving the dotted blue, dashed red, and solid green curves, respectively.

Starting from any countable basis, we can use the Gram–Schmidt procedure to obtain an orthonormal basis, as illustrated in the following example. ☞ 377

■ **Example A.14 (Legendre Polynomials)** Take the function space $L^2(\mathbb{R}, w(x)\,\mathrm{d}x)$, where $w(x) = \mathbb{1}\{-1 < x < 1\}$. We wish to construct an orthonormal basis of polynomial functions $g_0, g_1, g_2, \ldots$, starting from the collection of monomials: $\iota_0, \iota_1, \iota_2, \ldots$, where $\iota_k : x \mapsto x^k$. Using Gram–Schmidt, the first normalized zero-degree polynomial is $g_0 = \iota_0/\|\iota_0\| = \sqrt{1/2}$. To find g_1 (a polynomial of degree 1), project ι_1 (the identity function) onto the space spanned by g_0. The resulting projection is $p_1 := \langle g_0, \iota_1 \rangle g_0$, written out as

$$p_1(x) = \left(\int_{-1}^{1} x\, g_0(x)\,\mathrm{d}x\right) g_0(x) = \frac{1}{2}\int_{-1}^{1} x\,\mathrm{d}x = 0.$$

Hence, $g_1 = (\iota_1 - p_1)/\|\iota_1 - p_1\|$ is a linear function; that is, of the form $g_1(x) = ax$. The constant a is found by normalization:

$$1 = \|g_1\|^2 = \int_{-1}^{1} g_1^2(x)\,\mathrm{d}x = a^2 \int_{-1}^{1} x^2\,\mathrm{d}x = a^2\frac{2}{3},$$

so that $g_1(x) = \sqrt{3/2}x$. Continuing the Gram–Schmidt procedure, we find $g_2(x) = \sqrt{5/8}(3x^2 - 1)$, $g_3(x) = \sqrt{7/8}(5x^3 - 3x)$ and, in general,

$$g_k(x) = \frac{\sqrt{2k+1}}{2^{k+\frac{1}{2}}k!}\frac{\mathrm{d}^k}{\mathrm{d}x^k}(x^2 - 1)^k, \quad k = 0, 1, 2, \ldots.$$

These are the (normalized) *Legendre polynomials*. The graphs of g_0, g_1, g_2, and g_3 are given in Figure A.8. LEGENDRE POLYNOMIALS

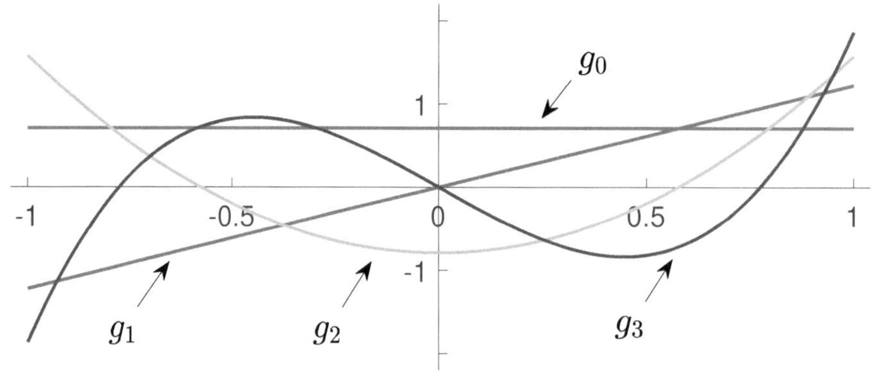

Figure A.8: The first 4 normalized Legendre polynomials.

As the Legendre polynomials form an orthonormal basis of $L^2(\mathbb{R}, \mathbb{1}\{-1 < x < 1\}\,\mathrm{d}x)$, they can be used to approximate arbitrary functions in this space. For example, Figure A.9 shows an approximation using the first 51 Legendre polynomials ($k = 0, 1, \ldots, 50$) of the Fourier expansion of the indicator function on the interval $(-1/2, 1/2)$. These Legendre polynomials form the basis of a 51-dimensional linear subspace onto which the indicator function is orthogonally projected.

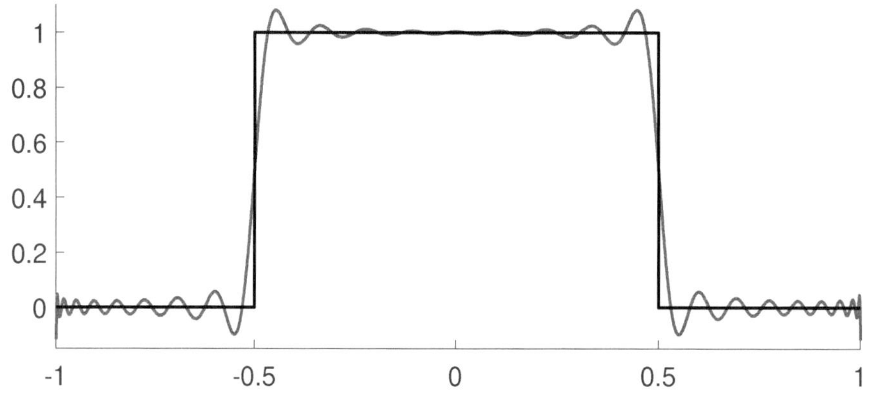

Figure A.9: Approximation of the indicator function on the interval $(-1/2, 1/2)$, using the Legendre polynomials $g_0, g_1, \ldots, g_{50}$.

■

☞ 426

The Legendre polynomials were produced in the following way: We started with an unnormalized probability density on $\mathbb{R}$ — in this case the probability density of the uniform distribution on $(-1, 1)$. We then constructed a sequence of polynomials by applying the Gram–Schmidt procedure to the monomials $1, x, x^2, \ldots$.

ORTHOGONAL POLYNOMIALS

LAGUERRE POLYNOMIALS

By using exactly the same procedure, but with a different probability density, we can produce other such *orthogonal polynomials*. For example, the density of the standard exponential[6] distribution, $w(x) = \mathrm{e}^{-x}, x \geqslant 0$, gives the *Laguerre polynomials*, which are defined

[6]This can be further generalized to the density of a gamma distribution.

by the recurrence

$$(n+1)g_{n+1}(x) = (2n+1-x)g_n(x) - n g_{n-1}(x), \quad n = 1, 2, \dots,$$

with $g_0(x) = 1$ and $g_1(x) = 1 - x$, for $x \geqslant 0$. The *Hermite polynomials* are obtained when using instead the density of the standard normal distribution: $w(x) = \mathrm{e}^{-x^2/2}/\sqrt{2\pi}, x \in \mathbb{R}$. These polynomials satisfy the recursion

HERMITE POLYNOMIALS

$$g_{n+1}(x) = x g_n(x) - \frac{\mathrm{d}g_n(x)}{\mathrm{d}x}, \quad n = 0, 1, \dots,$$

with $g_0(x) = 1$, $x \in \mathbb{R}$. Note that the Hermite polynomials as defined above have not been normalized to have norm 1. To normalize, use the fact that $\|g_n\|^2 = n!$.

We conclude with a number of key results in functional analysis. The first one is the celebrated *Cauchy–Schwarz* inequality.

CAUCHY–SCHWARZ

Theorem A.15: Cauchy–Schwarz

Let $\mathcal{H}$ be a Hilbert space. For every $f, g \in \mathcal{H}$ it holds that

$$|\langle f, g\rangle| \leqslant \|f\|\,\|g\|.$$

Proof: The inequality is trivially true for $g = 0$ (zero function). For $g \neq 0$, we can write $f = \alpha g + h$, where $h \perp g$ and $\alpha = \langle f, g\rangle/\|g\|^2$. Consequently, $\|f\|^2 = |\alpha|^2\,\|g\|^2 + \|h\|^2 \geqslant |\alpha|^2\,\|g\|^2$. The result follows after rearranging this last inequality. □

Let $\mathcal{V}$ and $\mathcal{W}$ be two linear vector spaces (for example, Hilbert spaces) on which norms $\|\cdot\|_{\mathcal{V}}$ and $\|\cdot\|_{\mathcal{W}}$ are defined. Suppose $A : \mathcal{V} \to \mathcal{W}$ is a mapping from $\mathcal{V}$ to $\mathcal{W}$. When $\mathcal{W} = \mathcal{V}$, such a mapping is often called an *operator*; when $\mathcal{W} = \mathbb{R}$ it is called a *functional*. Mapping A is said to be *linear* if $A(\alpha f + \beta g) = \alpha A(f) + \beta A(g)$. In this case we write Af instead of $A(f)$. If there exists $\gamma < \infty$ such that

OPERATOR

FUNCTIONAL

LINEAR MAPPING

$$\|Af\|_{\mathcal{W}} \leqslant \gamma\,\|f\|_{\mathcal{V}}, \quad f \in \mathcal{V}, \tag{A.36}$$

then A is said to be a *bounded mapping*. The smallest γ for which (A.36) holds is called the *norm* of A; denoted by $\|A\|$. A (not necessarily linear) mapping $A : \mathcal{V} \to \mathcal{W}$ is said to be *continuous* at f if for any sequence $f_1, f_2, \dots$ converging to f the sequence $A(f_1), A(f_2), \dots$ converges to $A(f)$. That is, if

BOUNDED MAPPING

NORM

CONTINUOUS MAPPING

$$\forall \varepsilon > 0, \exists \delta > 0 : \forall g \in \mathcal{V}, \|f - g\|_{\mathcal{V}} < \delta \Rightarrow \|A(f) - A(g)\|_{\mathcal{W}} < \varepsilon. \tag{A.37}$$

If the above property holds for every $f \in \mathcal{V}$, then the mapping A itself is called *continuous*.

Theorem A.16: Continuity and Boundedness for Linear Mappings

For a linear mapping, continuity and boundedness are equivalent.

Proof: Let A be linear and bounded. We may assume that A is non-zero (otherwise the statement holds trivially), and that therefore $0 < \|A\| < \infty$. Taking $\delta < \varepsilon/\|A\|$ in (A.37) now ensures that $\|Af - Ag\|_{\mathcal{W}} \leqslant \|A\|\,\|f - g\|_{\mathcal{V}} < \|A\|\,\delta < \varepsilon$. This shows that A is continuous.

Conversely, suppose A is continuous. In particular, it is continuous at $f = 0$ (the zero-element of $\mathcal{V}$). Thus, take $f = 0$ and let ε and δ be as in (A.37). For any $g \neq 0$, let $h = \delta/(2\|g\|_{\mathcal{V}})\, g$. As $\|h\|_{\mathcal{V}} = \delta/2 < \delta$, it follows from (A.37) that

$$\|Ah\|_{\mathcal{W}} = \frac{\delta}{2\|g\|_{\mathcal{V}}} \|Ag\|_{\mathcal{W}} < \varepsilon.$$

Rearranging the last inequality gives $\|Ag\|_{\mathcal{W}} < 2\varepsilon/\delta\|g\|_{\mathcal{V}}$, showing that A is bounded. □

Theorem A.17: Riesz Representation Theorem

Any bounded linear functional ϕ on a Hilbert space $\mathcal{H}$ can be represented as $\phi(h) = \langle h, g \rangle$, for some $g \in \mathcal{H}$ (depending on ϕ).

Proof: Let P be the projection of $\mathcal{H}$ onto the nullspace $\mathcal{N}$ of ϕ; that is, $\mathcal{N} = \{g \in \mathcal{H} : \phi(g) = 0\}$. If ϕ is not the 0-functional, then there exists a $g_0 \neq 0$ with $\phi(g_0) \neq 0$. Let $g_1 = g_0 - Pg_0$. Then $g_1 \perp \mathcal{N}$ and $\phi(g_1) = \phi(g_0)$. Take $g_2 = g_1/\phi(g_1)$. For any $h \in \mathcal{H}$, $f := h - \phi(h)g_2$ lies in $\mathcal{N}$. As $g_2 \perp \mathcal{N}$ it holds that $\langle f, g_2 \rangle = 0$, which is equivalent to $\langle h, g_2 \rangle = \phi(h)\, \|g_2\|^2$. By defining $g = g_2/\|g_2\|^2$ we have found our representation. □

A.8 Fourier Transforms

☞ 387 We will now briefly introduce the Fourier transform. Before doing so, we will extend the concept of L^2 space of real-valued functions as follows.

Definition A.5: L^p Space

Let $\mathcal{X}$ be a subset of $\mathbb{R}^d$ with measure $\mu(\mathrm{d}\boldsymbol{x}) = w(\boldsymbol{x})\,\mathrm{d}\boldsymbol{x}$ and $p \in [1, \infty)$. Then $L^p(\mathcal{X}, \mu)$ is the linear space of functions from $\mathcal{X}$ to $\mathbb{C}$ that satisfy

$$\int_{\mathcal{X}} |f(\boldsymbol{x})|^p\, w(\boldsymbol{x})\,\mathrm{d}\boldsymbol{x} < \infty. \tag{A.38}$$

When $p = 2$, $L^2(\mathcal{X}, \mu)$ is in fact a Hilbert space equipped with inner product

$$\langle f, g \rangle = \int_{\mathcal{X}} f(\boldsymbol{x})\, \overline{g(\boldsymbol{x})}\, w(\boldsymbol{x})\,\mathrm{d}\boldsymbol{x}. \tag{A.39}$$

We are now in a position to define the Fourier transform (with respect to the Lebesgue measure). Note that in the following Definitions A.6 and A.7 we have chosen a particular convention. Equivalent (but not identical) definitions exist that include scaling constants $(2\pi)^d$ or $(2\pi)^{-d}$ and where $-2\pi\boldsymbol{t}$ is replaced with $2\pi\boldsymbol{t}$, $\boldsymbol{t}$, or $-\boldsymbol{t}$.

Definition A.6: (Multivariate) Fourier Transform

The *Fourier transform* $\mathcal{F}[f]$ of a (real- or complex-valued) function $f \in L^1(\mathbb{R}^d)$ is the function $\widetilde{f}$ defined as

FOURIER TRANSFORM

$$\widetilde{f}(\boldsymbol{t}) := \int_{\mathbb{R}^d} \mathrm{e}^{-\mathrm{i}\, 2\pi \boldsymbol{t}^\top \boldsymbol{x}} f(\boldsymbol{x})\, \mathrm{d}\boldsymbol{x}, \quad \boldsymbol{t} \in \mathbb{R}^d.$$

The Fourier transform $\widetilde{f}$ is continuous, uniformly bounded (since $f \in L^1(\mathbb{R}^d)$ implies that $|\widetilde{f}(\boldsymbol{t})| \leqslant \int_{\mathbb{R}^d} |f(\boldsymbol{x})|\, \mathrm{d}\boldsymbol{x} < \infty$), and satisfies $\lim_{\|\boldsymbol{t}\| \to \infty} \widetilde{f}(\boldsymbol{t}) = 0$ (a result known as the *Riemann–Lebesgue lemma*). However, $|\widetilde{f}|$ does not necessarily have a finite integral. A simple example in $\mathbb{R}^1$ is the Fourier transform of $f(x) = \mathbb{1}\{-1/2 < x < 1/2\}$. Then $\widetilde{f}(t) = \sin(\pi t)/(\pi t) = \mathrm{sinc}(\pi t)$, which is not absolutely integrable.

Definition A.7: (Multivariate) Inverse Fourier Transform

The *inverse Fourier transform* $\mathcal{F}^{-1}[\widetilde{f}]$ of a (real- or complex-valued) function $\widetilde{f} \in L^1(\mathbb{R}^d)$ is the function $\breve{f}$ defined as

INVERSE FOURIER TRANSFORM

$$\breve{f}(\boldsymbol{x}) := \int_{\mathbb{R}^d} \mathrm{e}^{\mathrm{i}\, 2\pi \boldsymbol{t}^\top \boldsymbol{x}} \widetilde{f}(\boldsymbol{t})\, \mathrm{d}\boldsymbol{t}, \quad \boldsymbol{x} \in \mathbb{R}^d.$$

As one would hope, it holds that if f and $\mathcal{F}[f]$ are both in $L^1(\mathbb{R}^d)$, then $f = \mathcal{F}^{-1}[\mathcal{F}[f]]$ almost everywhere.

The Fourier transform enjoys many interesting and useful properties, some of which we list below.

1. *Linearity*: For $f, g \in L^1(\mathbb{R}^d)$ and constants $a, b \in \mathbb{R}$,

$$\mathcal{F}[af + bg] = a\mathcal{F}[f] + b\mathcal{F}[g].$$

2. *Space Shifting and Scaling*: Let $\mathbf{A} \in \mathbb{R}^{d \times d}$ be an invertible matrix and $\boldsymbol{b} \in \mathbb{R}^d$ a constant vector. Let $f \in L^1(\mathbb{R}^d)$ and define $h(\boldsymbol{x}) := f(\mathbf{A}\boldsymbol{x} + \boldsymbol{b})$. Then

$$\mathcal{F}[h](\boldsymbol{t}) = \mathrm{e}^{\mathrm{i}\, 2\pi (\mathbf{A}^{-\top} \boldsymbol{t})^\top \boldsymbol{b}} \widetilde{f}(\mathbf{A}^{-\top} \boldsymbol{t}) / |\det(\mathbf{A})|,$$

where $\mathbf{A}^{-\top} := (\mathbf{A}^\top)^{-1} = (\mathbf{A}^{-1})^\top$.

3. *Frequency Shifting and Scaling*: Let $\mathbf{A} \in \mathbb{R}^{d \times d}$ be an invertible matrix and $\boldsymbol{b} \in \mathbb{R}^d$ a constant vector. Let $f \in L^1(\mathbb{R}^d)$ and define

$$h(\boldsymbol{x}) := \mathrm{e}^{-\mathrm{i}\, 2\pi \boldsymbol{b}^\top \mathbf{A}^{-\top} \boldsymbol{x}} f(\mathbf{A}^{-\top} \boldsymbol{x}) / |\det(\mathbf{A})|.$$

Then $\mathcal{F}[h](\boldsymbol{t}) = \widetilde{f}(\mathbf{A}\boldsymbol{t} + \boldsymbol{b})$.

4. *Differentiation*: Let $f \in L^1(\mathbb{R}^d) \cap C^1(\mathbb{R}^d)$ and let $f_k := \partial f / \partial x_k$ be the partial derivative of f with respect to x_k. If $f_k \in L^1(\mathbb{R}^d)$ for $k = 1, \ldots, d$, then

$$\mathcal{F}[f_k](\boldsymbol{t}) = (\mathrm{i}\, 2\pi\, t_k)\, \widetilde{f}(\boldsymbol{t}).$$

5. *Convolution*: Let $f, g \in L^1(\mathbb{R}^d)$ be real or complex valued functions. Their convolution, $f * g$, is defined as

$$(f * g)(\boldsymbol{x}) = \int_{\mathbb{R}^d} f(\boldsymbol{y})\, g(\boldsymbol{x} - \boldsymbol{y})\, \mathrm{d}\boldsymbol{y},$$

and is also in $L^1(\mathbb{R}^d)$. Moreover, the Fourier transform satisfies

$$\mathcal{F}[f * g] = \mathcal{F}[f]\mathcal{F}[g].$$

6. *Duality*: Let f and $\mathcal{F}[f]$ both be in $L^1(\mathbb{R}^d)$. Then $\mathcal{F}[\mathcal{F}[f]](\boldsymbol{t}) = f(-\boldsymbol{t})$.

7. *Product Formula*: Let $f, g \in L^1(\mathbb{R}^d)$ and denote by $\widetilde{f}$, $\widetilde{g}$ their respective Fourier transforms. Then $\widetilde{f}\,g, f\,\widetilde{g} \in L^1(\mathbb{R}^d)$, and

$$\int_{\mathbb{R}^d} \widetilde{f}(\boldsymbol{z})\, g(\boldsymbol{z})\, \mathrm{d}\boldsymbol{z} = \int_{\mathbb{R}^d} f(\boldsymbol{z})\, \widetilde{g}(\boldsymbol{z})\, \mathrm{d}\boldsymbol{z}.$$

There are many additional properties which hold if $f \in L^1(\mathbb{R}^d) \cap L^2(\mathbb{R}^d)$. In particular, if $f, g \in L^1(\mathbb{R}^d) \cap L^2(\mathbb{R}^d)$, then $\widetilde{f}, \widetilde{g} \in L^2(\mathbb{R}^d)$ and $\langle \widetilde{f}, \widetilde{g} \rangle = \langle f, g \rangle$, a result often known as *Parseval's formula*. Putting $g = f$ gives the result often referred to as *Plancherel's theorem*.

The Fourier transform can be extended in several ways, in the first instance to functions in $L^2(\mathbb{R}^d)$ by continuity. A substantial extension of the theory is realized by replacing integration with respect to the Lebesgue measure (i.e., $\int_{\mathbb{R}^d} \cdots \mathrm{d}\boldsymbol{x}$) with integration with respect to a (finite Borel) measure μ (i.e., $\int_{\mathbb{R}^d} \cdots \mu(\mathrm{d}\boldsymbol{x})$). Moreover, there is a close connection between the Fourier transform and characteristic functions arising in probability theory. ☞ 443 Indeed, if $\boldsymbol{X}$ is a random vector with pdf f, then its characteristic function ψ satisfies

$$\psi(\boldsymbol{t}) := \mathbb{E}\, \mathrm{e}^{\mathrm{i}\, \boldsymbol{t}^\top \boldsymbol{X}} = \mathcal{F}[f](-\boldsymbol{t}/(2\pi)).$$

A.8.1 Discrete Fourier Transform

Here, we introduce the (univariate) discrete Fourier transform, which can be viewed as a special case of the Fourier transform introduced in Definition A.6, where $d = 1$, integration is with respect to the counting measure, and $f(x) = 0$ for $x < 0$ and $x > (n - 1)$.

DISCRETE FOURIER TRANSFORM

Definition A.8: Discrete Fourier Transform

The *discrete Fourier transform* (DFT) of a vector $\boldsymbol{x} = [x_0, \dots, x_{n-1}]^\top \in \mathbb{C}^n$ is the vector $\widetilde{\boldsymbol{x}} = [\widetilde{x}_0, \dots, \widetilde{x}_{n-1}]^\top$ whose elements are given by

$$\widetilde{x}_t = \sum_{s=0}^{n-1} \omega^{st}\, x_s, \quad t = 0, \dots, n-1, \tag{A.40}$$

where $\omega = \exp(-\mathrm{i}\, 2\pi/n)$.

In other words, $\widetilde{\boldsymbol{x}}$ is obtained from $\boldsymbol{x}$ via the linear transformation

$$\widetilde{\boldsymbol{x}} = \mathbf{F}\boldsymbol{x},$$

where

$$\mathbf{F} = \begin{bmatrix} 1 & 1 & 1 & \dots & 1 \\ 1 & \omega & \omega^2 & \dots & \omega^{n-1} \\ 1 & \omega^2 & \omega^4 & \dots & \omega^{2(n-1)} \\ \vdots & \vdots & \vdots & \ddots & \vdots \\ 1 & \omega^{n-1} & \omega^{2(n-1)} & \dots & \omega^{(n-1)^2} \end{bmatrix}.$$

The matrix $\mathbf{F}$ is a so-called *Vandermonde matrix*, and is clearly symmetric (i.e., $\mathbf{F} = \mathbf{F}^\top$). Moreover, $\mathbf{F}/\sqrt{n}$ is in fact a unitary matrix and hence its inverse is simply its complex conjugate $\overline{\mathbf{F}}/\sqrt{n}$. Thus, $\mathbf{F}^{-1} = \overline{\mathbf{F}}/n$ and we have that the *inverse discrete Fourier transform* (IDFT) is given by

INVERSE DISCRETE FOURIER TRANSFORM

$$x_t = \frac{1}{n} \sum_{s=0}^{n-1} \omega^{-st}\, \widetilde{x}_s, \quad t = 0, \dots, n-1, \tag{A.41}$$

or in terms of matrices and vectors,

$$\boldsymbol{x} = \overline{\mathbf{F}}\,\widetilde{\boldsymbol{x}}/n.$$

Observe that the IDFT of a vector $\boldsymbol{y}$ is related to the DFT of its complex conjugate $\overline{\boldsymbol{y}}$, since

$$\overline{\mathbf{F}}\,\boldsymbol{y}/n = \overline{\mathbf{F}\,\overline{\boldsymbol{y}}}/n.$$

Consequently, an IDFT can be computed via a DFT.

There is a close connection between circulant matrices $\mathbf{C}$ and the DFT. To make this connection concrete, let $\mathbf{C}$ be the circulant matrix corresponding to the vector $\boldsymbol{c} \in \mathbb{C}^n$ and denote by $\boldsymbol{f}_t$ the t-th column of the discrete Fourier matrix $\mathbf{F}$, $t = 0, 1, \dots, n-1$. Then, the s-th element of $\mathbf{C}\boldsymbol{f}_t$ is

$$\sum_{k=0}^{n-1} c_{(s-k) \bmod n}\, \omega^{tk} = \sum_{y=0}^{n-1} c_y\, \omega^{t(s-y)} = \underbrace{\omega^{ts}}_{s\text{-th element of } \boldsymbol{f}_t} \underbrace{\sum_{y=0}^{n-1} c_y\, \omega^{-ty}}_{\lambda_s}.$$

Hence, the eigenvalues of $\mathbf{C}$ are

$$\lambda_t = \boldsymbol{c}^\top \overline{\boldsymbol{f}}_t, \quad t = 0, 1, \dots, n-1,$$

with corresponding eigenvectors $\boldsymbol{f}_t$. Collecting the eigenvalues into the vector $\boldsymbol{\lambda} = [\lambda_0, \dots, \lambda_{n-1}]^\top = \overline{\mathbf{F}}\boldsymbol{c}$, we therefore have the eigen-decomposition

$$\mathbf{C} = \mathbf{F}\,\mathrm{diag}(\boldsymbol{\lambda})\,\overline{\mathbf{F}}/n.$$

Consequently, one can compute the *circular convolution* of a vector $\boldsymbol{a} = [a_1, \dots, a_n]^\top$ and $\boldsymbol{c} = [c_0, \dots, c_{n-1}]^\top$ by a series of DFTs as follows. Construct the circulant matrix $\mathbf{C}$ corresponding to $\boldsymbol{c}$. Then, the circular convolution of $\boldsymbol{a}$ and $\boldsymbol{c}$ is given by $\boldsymbol{y} = \mathbf{C}\boldsymbol{a}$. Proceed in four steps:

1. Compute $\boldsymbol{z} = \overline{\mathbf{F}}\boldsymbol{a}/n$.

2. Compute $\boldsymbol{\lambda} = \overline{\mathbf{F}}\boldsymbol{c}$.

3. Compute $\boldsymbol{p} = \boldsymbol{z} \odot \boldsymbol{\lambda} = [z_1 \lambda_0, \ldots, z_n \lambda_{n-1}]^\top$.

4. Compute $\boldsymbol{y} = \mathbf{F}\boldsymbol{p}$.

Steps 1 and 2 are (up to constants) in the form of an IDFT, and step 4 is in the form of a DFT. These are computable via the FFT (Section A.8.2) in $O(n \ln n)$ time. Step 3 is a dot product computable in $O(n)$ time. Thus, the circular convolution can be computed with the aid of the FFT in $O(n \ln n)$ time.

☞ 396

One can also efficiently compute the product of an $n \times n$ Toeplitz matrix $\mathbf{T}$ and an $n \times 1$ vector $\boldsymbol{a}$ in $O(n \ln n)$ time by embedding $\mathbf{T}$ into a circulant matrix $\mathbf{C}$ of size $2n \times 2n$. Namely, define

$$\mathbf{C} = \begin{bmatrix} \mathbf{T} & \mathbf{B} \\ \mathbf{B} & \mathbf{T} \end{bmatrix},$$

where

$$\mathbf{B} = \begin{bmatrix} 0 & t_{n-1} & \cdots & t_2 & t_1 \\ t_{-(n-1)} & 0 & t_{n-1} & & t_2 \\ \vdots & t_{-(n-1)} & 0 & \ddots & \vdots \\ t_{-2} & & \ddots & \ddots & t_{n-1} \\ t_{-1} & t_{-2} & \cdots & t_{-(n-1)} & 0 \end{bmatrix}.$$

Then a product of the form $\boldsymbol{y} = \mathbf{T}\boldsymbol{a}$ can be computed in $O(n \ln n)$ time, since we may write

$$\mathbf{C} \begin{bmatrix} \boldsymbol{a} \\ \mathbf{0} \end{bmatrix} = \begin{bmatrix} \mathbf{T} & \mathbf{B} \\ \mathbf{B} & \mathbf{T} \end{bmatrix} \begin{bmatrix} \boldsymbol{a} \\ \mathbf{0} \end{bmatrix} = \begin{bmatrix} \mathbf{T}\boldsymbol{a} \\ \mathbf{B}\boldsymbol{a} \end{bmatrix}.$$

The left-hand side is a product of a $2n \times 2n$ circulant matrix with vector of length $2n$, and so can be computed in $O(n \ln n)$ time via the FFT, as previously discussed.

Conceptually, one can also solve equations of the form $\mathbf{C}\boldsymbol{x} = \boldsymbol{b}$ for a given vector $\boldsymbol{b} \in \mathbb{C}^n$ and circulant matrix $\mathbf{C}$ (corresponding to $\boldsymbol{c} \in \mathbb{C}^n$, assuming all its eigenvalues are non-zero) via the following four steps:

1. Compute $\boldsymbol{z} = \overline{\mathbf{F}}\boldsymbol{b}/n$.

2. Compute $\boldsymbol{\lambda} = \overline{\mathbf{F}}\boldsymbol{c}$.

3. Compute $\boldsymbol{p} = \boldsymbol{z}/\boldsymbol{\lambda} = [z_1/\lambda_0, \ldots, z_n/\lambda_{n-1}]^\top$.

4. Compute $\boldsymbol{x} = \mathbf{F}\boldsymbol{p}$.

Once again, Steps 1 and 2 are (up to constants) in the form of an IDFT, and Step 4 is in the form of a DFT, all of which are computable via the FFT in $O(n \ln n)$ time, and Step 3 is computable in $O(n)$ time, meaning the solution $\boldsymbol{x}$ can be computed using the FFT in $O(n \ln n)$ time.

A.8.2 Fast Fourier Transform

FAST FOURIER TRANSFORM

The *fast Fourier transform* (FFT) is a numerical algorithm for the fast evaluation of (A.40) and (A.41). By using a divide-and-conquer strategy, the algorithm reduces the computational complexity from $O(n^2)$ (for the naïve evaluation of the linear transformation) to $O(n \ln n)$ [60].

The essence of the algorithm lies in the following observation. Suppose $n = r_1 r_2$. Then one can express any index t appearing in (A.40) via a pair (t_0, t_1), with $t = t_1 r_1 + t_0$, where $t_0 \in \{0, 1, \dots, r_1 - 1\}$ and $t_1 \in \{0, 1, \dots, r_2 - 1\}$. Similarly, one can express any index s appearing in (A.40) via a pair (s_0, s_1), with $s = s_1 r_2 + s_0$, where $s_0 \in \{0, 1, \dots, r_2 - 1\}$ and $s_1 \in \{0, 1, \dots, r_1 - 1\}$.

Identifying $\widetilde{x}_t \equiv \widetilde{x}_{t_1,t_0}$ and $x_s \equiv x_{s_1,s_0}$, we may re-express (A.40) as

$$\widetilde{x}_{t_1,t_0} = \sum_{s_0=0}^{r_2-1} \omega^{s_0 t} \sum_{s_1=0}^{r_1-1} \omega^{s_1 r_2 t}\, x_{s_1,s_0}\,, \quad t_0 = 0, 1, \dots, r_1 - 1\,, t_1 = 0, 1, \dots, r_2 - 1. \tag{A.42}$$

Observe that $\omega^{s_1 r_2 t} = \omega^{s_1 r_2 t_0}$ (because $\omega^{r_1 r_2} = 1$), so that the inner sum over s_1 depends only on s_0 and t_0. Define

$$y_{t_0,s_0} := \sum_{s_1=0}^{r_1-1} \omega^{s_1 r_2 t_0}\, x_{s_1,s_0}\,, \quad t_0 = 0, 1, \dots, r_1 - 1\,, s_0 = 0, 1, \dots, r_2 - 1.$$

Computing each y_{t_0,s_0} requires $O(n\, r_1)$ operations. In terms of the $\{y_{t_0,s_0}\}$, (A.42) can be written as

$$\widetilde{x}_{t_1,t_0} = \sum_{s_0=0}^{r_2-1} \omega^{s_0 t}\, y_{t_0,s_0}\,, \quad t_1 = 0, 1, \dots, r_2 - 1\,, t_0 = 0, 1, \dots, r_1 - 1,$$

requiring $O(n\, r_2)$ operations to compute. Thus, calculating the DFT using this two-step procedure requires $O(n\,(r_1 + r_2))$ operations, rather than $O(n^2)$.

Now supposing $n = r_1 r_2 \cdots r_m$, repeated application the above divide-and-conquer idea yields an m-step procedure requiring $O(n\,(r_1 + r_2 + \cdots + r_m))$ operations. In particular, if $r_k = r$ for all $k = 1, 2, \dots, m$, we have that $n = r^m$ and $m = \log_r n$, so that the total number of operations is $O(r\, n\, m) \equiv O(r\, n\, \log_r(n))$. Typically, the *radix* r is a small (not necessarily prime) number, for instance $r = 2$.

Further Reading

A good reference book on matrix computations is Golub and Van Loan [52]. A useful list of many common vector and matrix calculus identities can be found in [95]. Strang's introduction to linear algebra [116] is a classic textbook, and his recent book [117] combines linear algebra with the foundations of deep learning. Fast reliable algorithms for matrices with structure can be found in [64]. Kolmogorov and Fomin's masterpiece on the theory of functions and functional analysis [67] still provides one of the best introductions to the topic. A popular choice for an advanced course in functional analysis is Rudin [106].

APPENDIX B

MULTIVARIATE DIFFERENTIATION AND OPTIMIZATION

The purpose of this appendix is to review various aspects of multivariate differentiation and optimization. We assume the reader is familiar with differentiating a real-valued function.

B.1 Multivariate Differentiation

For a multivariate function f that maps a vector $\boldsymbol{x} = [x_1, \ldots, x_n]^\top$ to a real number $f(\boldsymbol{x})$, the *partial derivative* with respect to x_i, denoted $\frac{\partial f}{\partial x_i}$, is the derivative taken with respect to x_i while all other variables are held constant. We can write all the n partial derivatives neatly using the "scalar/vector" derivative notation: PARTIAL DERIVATIVE

$$\text{scalar/vector:} \qquad \frac{\partial f}{\partial \boldsymbol{x}} := \begin{bmatrix} \frac{\partial f}{\partial x_1} \\ \vdots \\ \frac{\partial f}{\partial x_n} \end{bmatrix}. \tag{B.1}$$

This vector of partial derivatives is known as the *gradient* of f at $\boldsymbol{x}$ and is sometimes written as $\nabla f(\boldsymbol{x})$. GRADIENT

Next, suppose that $\boldsymbol{f}$ is a multivalued (vector-valued) function taking values in $\mathbb{R}^m$, defined by

$$\boldsymbol{x} = \begin{bmatrix} x_1 \\ x_2 \\ \vdots \\ x_n \end{bmatrix} \mapsto \begin{bmatrix} f_1(\boldsymbol{x}) \\ f_2(\boldsymbol{x}) \\ \vdots \\ f_m(\boldsymbol{x}) \end{bmatrix} =: \boldsymbol{f}(\boldsymbol{x}).$$

We can compute each of the partial derivatives $\partial f_i / \partial x_j$ and organize them neatly in a "vector/vector" derivative notation:

$$\text{vector/vector:} \qquad \frac{\partial \boldsymbol{f}}{\partial \boldsymbol{x}} := \begin{bmatrix} \frac{\partial f_1}{\partial x_1} & \frac{\partial f_2}{\partial x_1} & \cdots & \frac{\partial f_m}{\partial x_1} \\ \frac{\partial f_1}{\partial x_2} & \frac{\partial f_2}{\partial x_2} & \cdots & \frac{\partial f_m}{\partial x_2} \\ \vdots & \vdots & \cdots & \vdots \\ \frac{\partial f_1}{\partial x_n} & \frac{\partial f_2}{\partial x_n} & \cdots & \frac{\partial f_m}{\partial x_n} \end{bmatrix}. \tag{B.2}$$

MATRIX OF JACOBI The transpose of this matrix is known as the *matrix of Jacobi* of f at x (sometimes called the *Fréchet derivative* of f at x); that is,

$$\mathbf{J}_f(\boldsymbol{x}) := \left[\frac{\partial \boldsymbol{f}}{\partial \boldsymbol{x}}\right]^\top = \begin{bmatrix} \frac{\partial f_1}{\partial x_1} & \frac{\partial f_1}{\partial x_2} & \cdots & \frac{\partial f_1}{\partial x_n} \\ \frac{\partial f_2}{\partial x_1} & \frac{\partial f_2}{\partial x_2} & \cdots & \frac{\partial f_2}{\partial x_n} \\ \vdots & \vdots & \cdots & \vdots \\ \frac{\partial f_m}{\partial x_1} & \frac{\partial f_m}{\partial x_2} & \cdots & \frac{\partial f_m}{\partial x_n} \end{bmatrix}. \tag{B.3}$$

If we define $\boldsymbol{g}(\boldsymbol{x}) := \nabla f(\boldsymbol{x})$ and take the "vector/vector" derivative of $\boldsymbol{g}$ with respect to $\boldsymbol{x}$, we obtain the matrix of second-order partial derivatives of f:

$$\mathbf{H}_f(\boldsymbol{x}) := \frac{\partial \boldsymbol{g}}{\partial \boldsymbol{x}} = \begin{bmatrix} \frac{\partial^2 f}{\partial^2 x_1} & \frac{\partial^2 f}{\partial x_1 \partial x_2} & \cdots & \frac{\partial^2 f}{\partial x_1 \partial x_m} \\ \frac{\partial^2 f}{\partial x_2 \partial x_1} & \frac{\partial^2 f}{\partial^2 x_2} & \cdots & \frac{\partial^2 f}{\partial x_2 \partial x_m} \\ \vdots & \vdots & \cdots & \vdots \\ \frac{\partial^2 f}{\partial x_m \partial x_1} & \frac{\partial^2 f}{\partial x_m \partial x_2} & \cdots & \frac{\partial^2 f}{\partial^2 x_m} \end{bmatrix}, \tag{B.4}$$

HESSIAN MATRIX which is known as the *Hessian matrix* of f at $\boldsymbol{x}$, also denoted as $\nabla^2 f(\boldsymbol{x})$. If these second-order partial derivatives are *continuous* in a region around $\boldsymbol{x}$, then $\frac{\partial f}{\partial x_i \partial x_j} = \frac{\partial f}{\partial x_j \partial x_i}$ and, hence, the Hessian matrix $\mathbf{H}_f(\boldsymbol{x})$ is *symmetric*.

Finally, note that we can also define a "scalar/matrix" derivative of y with respect to $\mathbf{X} \in \mathbb{R}^{m\times n}$ with (i, j)-th entry x_{ij}:

$$\frac{\partial y}{\partial \mathbf{X}} := \begin{bmatrix} \frac{\partial y}{\partial x_{11}} & \frac{\partial y}{\partial x_{12}} & \cdots & \frac{\partial y}{\partial x_{1n}} \\ \frac{\partial y}{\partial x_{21}} & \frac{\partial y}{\partial x_{22}} & \cdots & \frac{\partial y}{\partial x_{2n}} \\ \vdots & \vdots & \cdots & \vdots \\ \frac{\partial y}{\partial x_{m1}} & \frac{\partial y}{\partial x_{m2}} & \cdots & \frac{\partial y}{\partial x_{mn}} \end{bmatrix}$$

and a "matrix/scalar" derivative:

$$\frac{\partial \mathbf{X}}{\partial y} := \begin{bmatrix} \frac{\partial x_{11}}{\partial y} & \frac{\partial x_{12}}{\partial y} & \cdots & \frac{\partial x_{1n}}{\partial y} \\ \frac{\partial x_{21}}{\partial y} & \frac{\partial x_{22}}{\partial y} & \cdots & \frac{\partial x_{2n}}{\partial y} \\ \vdots & \vdots & \cdots & \vdots \\ \frac{\partial x_{m1}}{\partial y} & \frac{\partial x_{m2}}{\partial y} & \cdots & \frac{\partial x_{mn}}{\partial y} \end{bmatrix}.$$

■ **Example B.1 (Scalar/Matrix Derivative)** Let $y = \boldsymbol{a}^\top \mathbf{X} \boldsymbol{b}$, where $\mathbf{X} \in \mathbb{R}^{m\times n}$, $\boldsymbol{a} \in \mathbb{R}^m$, and $\boldsymbol{b} \in \mathbb{R}^n$. Since y is a scalar, we can write $y = \text{tr}(y) = \text{tr}(\mathbf{X}\boldsymbol{b}\boldsymbol{a}^\top)$, using the cyclic property
☞ 359 of the trace (see Theorem A.1). Defining $\mathbf{C} := \boldsymbol{b}\boldsymbol{a}^\top$, we have

$$y = \sum_{i=1}^{m} [\mathbf{XC}]_{ii} = \sum_{i=1}^{m} \sum_{j=1}^{n} x_{ij} c_{ji},$$

so that $\partial y / \partial x_{ij} = c_{ji}$ or, in matrix form,

$$\frac{\partial y}{\partial \mathbf{X}} = \mathbf{C}^\top = \boldsymbol{a}\boldsymbol{b}^\top.$$

■

■ **Example B.2 (Scalar/Matrix Derivative via the Woodbury Identity)** Let $y = \mathrm{tr}\left(\mathbf{X}^{-1}\mathbf{A}\right)$, where $\mathbf{X}, \mathbf{A} \in \mathbb{R}^{n\times n}$. We now prove that

$$\frac{\partial y}{\partial \mathbf{X}} = -\mathbf{X}^{-\top}\mathbf{A}^{\top}\mathbf{X}^{-\top}.$$

To show this, apply the Woodbury matrix identity to an infinitesimal perturbation, $\mathbf{X} + \varepsilon\mathbf{U}$, of $\mathbf{X}$, and take $\varepsilon \downarrow 0$ to obtain the following: ☞ 373

$$\frac{(\mathbf{X} + \varepsilon\,\mathbf{U})^{-1} - \mathbf{X}^{-1}}{\varepsilon} = -\mathbf{X}^{-1}\,\mathbf{U}\,(\mathbf{I} + \varepsilon\,\mathbf{X}^{-1}\,\mathbf{U})^{-1}\,\mathbf{X}^{-1} \longrightarrow -\mathbf{X}^{-1}\,\mathbf{U}\,\mathbf{X}^{-1}.$$

Therefore, as $\varepsilon \downarrow 0$

$$\frac{\mathrm{tr}\left((\mathbf{X} + \varepsilon\,\mathbf{U})^{-1}\mathbf{A}\right) - \mathrm{tr}\left(\mathbf{X}^{-1}\mathbf{A}\right)}{\varepsilon} \longrightarrow -\mathrm{tr}\left(\mathbf{X}^{-1}\mathbf{U}\,\mathbf{X}^{-1}\mathbf{A}\right) = -\mathrm{tr}\left(\mathbf{U}\,\mathbf{X}^{-1}\mathbf{A}\mathbf{X}^{-1}\right).$$

Now, suppose that $\mathbf{U}$ is an all zero matrix with a one in the (i, j)-th position. We can write,

$$\frac{\partial y}{\partial x_{ij}} = \lim_{\varepsilon\downarrow 0} \frac{\mathrm{tr}\left((\mathbf{X} + \varepsilon\,\mathbf{U})^{-1}\mathbf{A}\right) - \mathrm{tr}\left(\mathbf{X}^{-1}\mathbf{A}\right)}{\varepsilon} = -\mathrm{tr}\left(\mathbf{U}\mathbf{X}^{-1}\mathbf{A}\mathbf{X}^{-1}\right) = -\left[\mathbf{X}^{-1}\mathbf{A}\mathbf{X}^{-1}\right]_{ji}.$$

Therefore, $\frac{\partial y}{\partial \mathbf{X}} = -\left(\mathbf{X}^{-1}\mathbf{A}\mathbf{X}^{-1}\right)^{\top}$. ■

The following two examples specify multivariate derivatives for the important special cases of linear and quadratic functions.

■ **Example B.3 (Gradient of a Linear Function)** Let $\boldsymbol{f}(\boldsymbol{x}) = \mathbf{A}\boldsymbol{x}$ for some $m\times n$ constant matrix $\mathbf{A}$. Then, its vector/vector derivative (B.2) is the matrix

$$\frac{\partial \boldsymbol{f}}{\partial \boldsymbol{x}} = \mathbf{A}^{\top}. \tag{B.5}$$

To see this, let a_{ij} denote the (i, j)-th element of $\mathbf{A}$, so that

$$\boldsymbol{f}(\boldsymbol{x}) = \mathbf{A}\boldsymbol{x} = \begin{bmatrix} \sum_{k=1}^{n} a_{1k}x_k \\ \vdots \\ \sum_{k=1}^{n} a_{mk}x_k \end{bmatrix}.$$

To find the (j, i)-th element of $\frac{\partial \boldsymbol{f}}{\partial \boldsymbol{x}}$, we differentiate the i-th element of $\boldsymbol{f}$ with respect to x_j:

$$\frac{\partial f_i}{\partial x_j} = \frac{\partial}{\partial x_j}\sum_{k=1}^{n} a_{ik}x_k = a_{ij}.$$

In other words, the (i, j)-th element of $\frac{\partial \boldsymbol{f}}{\partial \boldsymbol{x}}$ is a_{ji}, the (i, j)-th element of $\mathbf{A}^{\top}$. ■

■ **Example B.4 (Gradient and Hessian of a Quadratic Function)** Let $f(\boldsymbol{x}) = \boldsymbol{x}^{\top}\mathbf{A}\boldsymbol{x}$ for some $n \times n$ constant matrix $\mathbf{A}$. Then,

$$\nabla f(\boldsymbol{x}) = (\mathbf{A} + \mathbf{A}^{\top})\boldsymbol{x}. \tag{B.6}$$

It follows immediately that if $\mathbf{A}$ is *symmetric*, that is, $\mathbf{A} = \mathbf{A}^\top$, then $\nabla(\boldsymbol{x}^\top \mathbf{A}\boldsymbol{x}) = 2\mathbf{A}\boldsymbol{x}$ and $\nabla^2(\boldsymbol{x}^\top \mathbf{A}\boldsymbol{x}) = 2\mathbf{A}$.

To prove (B.6), first observe that $f(\boldsymbol{x}) = \boldsymbol{x}^\top \mathbf{A}\boldsymbol{x} = \sum_{i=1}^n \sum_{j=1}^n a_{ij}\, x_i x_j$, which is a quadratic form in $\boldsymbol{x}$, is real-valued, with

$$\frac{\partial f}{\partial x_k} = \frac{\partial}{\partial x_k} \sum_{i=1}^n \sum_{j=1}^n a_{ij} x_i x_j = \sum_{j=1}^n a_{kj} x_j + \sum_{i=1}^n a_{ik} x_i.$$

The first term on the right-hand side is equal to the k-th element of $\mathbf{A}\boldsymbol{x}$, whereas the second term equals the k-th element of $\boldsymbol{x}^\top \mathbf{A}$, or equivalently the k-th element of $\mathbf{A}^\top \boldsymbol{x}$. ■

B.1.1 Taylor Expansion

The matrix of Jacobi and the Hessian matrix feature prominently in multidimensional Taylor expansions.

Theorem B.1: Multidimensional Taylor Expansions

Let $\mathcal{X}$ be an open subset of $\mathbb{R}^n$ and let $\boldsymbol{a} \in \mathcal{X}$. If $f : \mathcal{X} \to \mathbb{R}$ is a continuously twice differentiable function with Jacobian matrix $\mathbf{J}_f(\boldsymbol{x})$ and Hessian matrix $\mathbf{H}_f(\boldsymbol{x})$, then for every $\boldsymbol{x} \in \mathcal{X}$ we have the following first- and second-order Taylor expansions:

$$f(\boldsymbol{x}) = f(\boldsymbol{a}) + \mathbf{J}_f(\boldsymbol{a})\,(\boldsymbol{x} - \boldsymbol{a}) + O(\|\boldsymbol{x} - \boldsymbol{a}\|^2) \tag{B.7}$$

and

$$f(\boldsymbol{x}) = f(\boldsymbol{a}) + \mathbf{J}_f(\boldsymbol{a})\,(\boldsymbol{x} - \boldsymbol{a}) + \frac{1}{2}(\boldsymbol{x} - \boldsymbol{a})^\top\, \mathbf{H}_f(\boldsymbol{a})\,(\boldsymbol{x} - \boldsymbol{a}) + O(\|\boldsymbol{x} - \boldsymbol{a}\|^3) \tag{B.8}$$

as $\|\boldsymbol{x} - \boldsymbol{a}\| \to 0$. By dropping the O remainder terms, one obtains the corresponding Taylor approximations.

The result is essentially saying that a smooth enough function behaves locally (in the neighborhood of a point $\boldsymbol{x}$) like a linear and quadratic function. Thus, the gradient or Hessian of an approximating linear or quadratic function is a basic building block of many approximation and optimization algorithms.

■ **Remark B.1 (Version Without Remainder Terms)** An alternative version of Taylor's theorem states that there exists an $\boldsymbol{a}'$ that lies on the line segment between $\boldsymbol{x}$ and $\boldsymbol{a}$ such that (B.7) and (B.8) hold without remainder terms, with $\mathbf{J}_f(\boldsymbol{a})$ in (B.7) replaced by $\mathbf{J}_f(\boldsymbol{a}')$ and $\mathbf{H}_f(\boldsymbol{a})$ in (B.8) replaced by $\mathbf{H}_f(\boldsymbol{a}')$. ■

B.1.2 Chain Rule

Consider the functions $\boldsymbol{f} : \mathbb{R}^k \to \mathbb{R}^m$ and $\boldsymbol{g} : \mathbb{R}^m \to \mathbb{R}^n$. The function $\boldsymbol{x} \mapsto \boldsymbol{g}(\boldsymbol{f}(\boldsymbol{x}))$ is called the *composition* of $\boldsymbol{g}$ and $\boldsymbol{f}$, written as $\boldsymbol{g} \circ \boldsymbol{f}$, and is a function from $\mathbb{R}^k$ to $\mathbb{R}^n$. Suppose $\boldsymbol{y} = \boldsymbol{f}(\boldsymbol{x})$ and $\boldsymbol{z} = \boldsymbol{g}(\boldsymbol{y})$, as in Figure B.1. Let $\mathbf{J}_f(\boldsymbol{x})$ and $\mathbf{J}_g(\boldsymbol{y})$ be the (Fréchet) derivatives of $\boldsymbol{f}$ (at $\boldsymbol{x}$) and $\boldsymbol{g}$ (at $\boldsymbol{y}$), respectively. We may think of $\mathbf{J}_f(\boldsymbol{x})$ as the matrix that describes

COMPOSITION

how, in a neighborhood of $\boldsymbol{x}$, the function f can be approximated by a linear function: $f(\boldsymbol{x}+\boldsymbol{h}) \approx f(\boldsymbol{x}) + \mathbf{J}_f(\boldsymbol{x})\boldsymbol{h}$, and similarly for $\mathbf{J}_g(\boldsymbol{y})$. The well-known *chain rule* of calculus simply states that the derivative of the composition $\boldsymbol{g} \circ \boldsymbol{f}$ is the matrix product of the derivatives of $\boldsymbol{g}$ and $\boldsymbol{f}$; that is, CHAIN RULE

$$\mathbf{J}_{g\circ f}(\boldsymbol{x}) = \mathbf{J}_g(\boldsymbol{y})\,\mathbf{J}_f(\boldsymbol{x}).$$

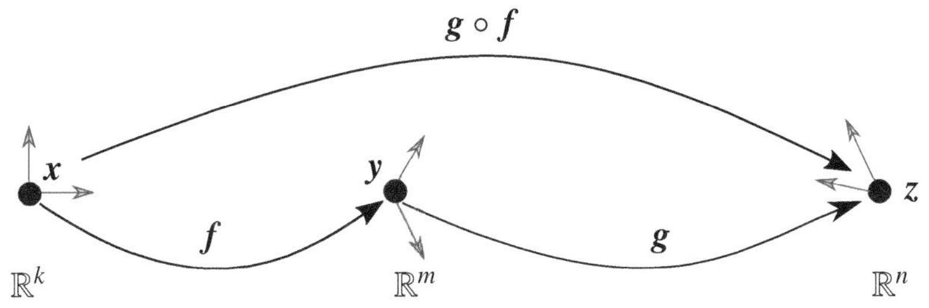

Figure B.1: Function composition. The blue arrows symbolize the linear mappings.

In terms of our vector/vector derivative notation, we have

$$\left[\frac{\partial z}{\partial \boldsymbol{x}}\right]^\top = \left[\frac{\partial z}{\partial \boldsymbol{y}}\right]^\top \left[\frac{\partial \boldsymbol{y}}{\partial \boldsymbol{x}}\right]^\top$$

or, more simply,

$$\frac{\partial z}{\partial \boldsymbol{x}} = \frac{\partial \boldsymbol{y}}{\partial \boldsymbol{x}}\frac{\partial z}{\partial \boldsymbol{y}}. \tag{B.9}$$

In a similar way we can establish a scalar/matrix chain rule. In particular, suppose $\mathbf{X}$ is an $n \times p$ matrix, which is mapped to $\boldsymbol{y} := \mathbf{X}\boldsymbol{\alpha}$ for a fixed p-dimensional vector $\boldsymbol{\alpha}$. In turn, $\boldsymbol{y}$ is mapped to a scalar $z := g(\boldsymbol{y})$ for some function g. Denote the columns of $\mathbf{X}$ by $\boldsymbol{x}_1, \ldots, \boldsymbol{x}_p$. Then,

$$\boldsymbol{y} = \mathbf{X}\boldsymbol{\alpha} = \sum_{j=1}^{p} \alpha_j \boldsymbol{x}_j,$$

and, therefore, $\partial \boldsymbol{y}/\partial \boldsymbol{x}_j = \alpha_j \mathbf{I}_n$. It follows by the chain rule (B.9) that

$$\frac{\partial z}{\partial \boldsymbol{x}_i} = \frac{\partial \boldsymbol{y}}{\partial \boldsymbol{x}_i}\frac{\partial z}{\partial \boldsymbol{y}} = \alpha_i \mathbf{I}_n \frac{\partial z}{\partial \boldsymbol{y}} = \alpha_i \frac{\partial z}{\partial \boldsymbol{y}}.$$

Therefore,

$$\frac{\partial z}{\partial \mathbf{X}} = \left[\tfrac{\partial z}{\partial \boldsymbol{x}_1}, \ldots, \tfrac{\partial z}{\partial \boldsymbol{x}_p}\right] = \left[\alpha_1 \tfrac{\partial z}{\partial \boldsymbol{y}}, \ldots, \alpha_p \tfrac{\partial z}{\partial \boldsymbol{y}}\right] = \frac{\partial z}{\partial \boldsymbol{y}}\,\boldsymbol{\alpha}^\top. \tag{B.10}$$

■ **Example B.5 (Derivative of the Log-Determinant)** Suppose we are given a positive definite matrix $\mathbf{A} \in \mathbb{R}^{p\times p}$ and wish to compute the scalar/matrix derivative $\frac{\partial \ln|\mathbf{A}|}{\partial \mathbf{A}}$. The result is

$$\frac{\partial \ln|\mathbf{A}|}{\partial \mathbf{A}} = \mathbf{A}^{-1}.$$

To see this, we can reason as follows. By Theorem A.8, we can write $\mathbf{A} = \mathbf{Q}\,\mathbf{D}\,\mathbf{Q}^\top$, where ☞ 368

$\mathbf{Q}$ is an orthogonal matrix and $\mathbf{D} = \text{diag}(\lambda_1, \ldots, \lambda_p)$ is the diagonal matrix of eigenvalues of $\mathbf{A}$. The eigenvalues are strictly positive, since $\mathbf{A}$ is positive definite. Denoting the columns of $\mathbf{Q}$ by $(\boldsymbol{q}_i)$, we have

$$\lambda_i = \boldsymbol{q}_i^\top \mathbf{A} \boldsymbol{q}_i = \text{tr}\left(\boldsymbol{q}_i \mathbf{A} \boldsymbol{q}_i^\top\right), \quad i = 1, \ldots, p. \tag{B.11}$$

From the properties of determinants, we have $y := \ln |\mathbf{A}| = \ln |\mathbf{Q}\,\mathbf{D}\,\mathbf{Q}^\top| = \ln(|\mathbf{Q}|\,|\mathbf{D}|\,|\mathbf{Q}^\top|) = \ln |\mathbf{D}| = \sum_{i=1}^p \ln \lambda_i$. We can thus write

$$\frac{\partial \ln |\mathbf{A}|}{\partial \mathbf{A}} = \sum_{i=1}^p \frac{\partial \ln \lambda_i}{\partial \mathbf{A}} = \sum_{i=1}^p \frac{\partial \lambda_i}{\partial \mathbf{A}} \frac{\partial \ln \lambda_i}{\partial \lambda_i} = \sum_{i=1}^p \frac{\partial \lambda_i}{\partial \mathbf{A}} \frac{1}{\lambda_i},$$

where the second equation follows from the chain rule applied to the function composition $\mathbf{A} \mapsto \lambda_i \mapsto y$. From (B.11) and Example B.1 we have $\partial \lambda_i / \partial \mathbf{A} = \boldsymbol{q}_i \boldsymbol{q}_i^\top$. It follows that

$$\frac{\partial y}{\partial \mathbf{A}} = \sum_{i=1}^p \boldsymbol{q}_i \boldsymbol{q}_i^\top \frac{1}{\lambda_i} = \mathbf{Q}\,\mathbf{D}^{-1}\,\mathbf{Q}^\top = \mathbf{A}^{-1}.$$

■

B.2 Optimization Theory

OBJECTIVE FUNCTION

Optimization is concerned with finding minimal or maximal solutions of a real-valued *objective function* f in some set $\mathcal{X}$:

$$\min_{\boldsymbol{x} \in \mathcal{X}} f(\boldsymbol{x}) \quad \text{or} \quad \max_{\boldsymbol{x} \in \mathcal{X}} f(\boldsymbol{x}). \tag{B.12}$$

LOCAL MINIMIZER

GLOBAL MINIMIZER

Since any maximization problem can easily be converted into a minimization problem via the equivalence $\max_{\boldsymbol{x}} f(\boldsymbol{x}) \equiv -\min_{\boldsymbol{x}} -f(\boldsymbol{x})$, we focus only on minimization problems. We use the following terminology. A *local minimizer* of $f(\boldsymbol{x})$ is an element $\boldsymbol{x}^* \in \mathcal{X}$ such that $f(\boldsymbol{x}^*) \leqslant f(\boldsymbol{x})$ for all $\boldsymbol{x}$ in some neighborhood of $\boldsymbol{x}^*$. If $f(\boldsymbol{x}^*) \leqslant f(\boldsymbol{x})$ for all $\boldsymbol{x} \in \mathcal{X}$, then $\boldsymbol{x}^*$ is called a *global minimizer* or *global solution*. The set of global minimizers is denoted by

$$\underset{\boldsymbol{x} \in \mathcal{X}}{\text{argmin}}\, f(\boldsymbol{x}).$$

LOCAL/GLOBAL MINIMUM

The function value $f(\boldsymbol{x}^*)$ corresponding to a local/global minimizer $\boldsymbol{x}^*$ is referred to as the *local/global minimum* of $f(\boldsymbol{x})$.

Optimization problems may be classified by the set $\mathcal{X}$ and the objective function f. If $\mathcal{X}$ is countable, the optimization problem is called *discrete* or *combinatorial*. If instead $\mathcal{X}$ is a nondenumerable set such as $\mathbb{R}^n$ and f takes values in a nondenumerable set, then the problem is said to be *continuous*. Optimization problems that are neither discrete nor continuous are said to be *mixed*.

The search set $\mathcal{X}$ is often defined by means of *constraints*. A standard setting for constrained optimization (minimization) is the following:

$$\begin{aligned} &\min_{\boldsymbol{x} \in \mathcal{Y}} f(\boldsymbol{x}) \\ &\text{subject to: } h_i(\boldsymbol{x}) = 0, \quad i = 1, \ldots, m, \\ &\qquad\qquad\quad g_i(\boldsymbol{x}) \leqslant 0, \quad i = 1, \ldots, k. \end{aligned} \tag{B.13}$$

Here, f is the objective function, and $\{g_i\}$ and $\{h_i\}$ are given functions so that $h_i(\boldsymbol{x}) = 0$ and $g_i(\boldsymbol{x}) \leqslant 0$ represent the *equality* and *inequality* constraints, respectively. The region $\mathcal{X} \subseteq \mathcal{Y}$ where the objective function is defined and where all the constraints are satisfied is called the *feasible region*. An optimization problem without constraints is said to be an *unconstrained* problem. FEASIBLE REGION

For an unconstrained continuous optimization problem, the search space $\mathcal{X}$ is often taken to be (a subset of) $\mathbb{R}^n$, and f is assumed to be a C^k function for sufficiently high k (typically $k = 2$ or 3 suffices); that is, its k-th order derivative is continuous. For a C^1 function the standard approach to minimizing $f(\boldsymbol{x})$ is to solve the equation

$$\nabla f(\boldsymbol{x}) = \boldsymbol{0}, \tag{B.14}$$

where $\nabla f(\boldsymbol{x})$ is the *gradient* of f at $\boldsymbol{x}$. The solutions $\boldsymbol{x}^*$ to (B.14) are called *stationary points*. Stationary points can be local/global minimizers, local/global maximizers, or *saddle points* (which are neither). If, in addition, the function is C^2, the condition

☞ 399

STATIONARY POINTS

SADDLE POINTS

$$\boldsymbol{y}^\top(\nabla^2 f(\boldsymbol{x}^*))\,\boldsymbol{y} > 0 \quad \text{for all } \boldsymbol{y} \neq \boldsymbol{0} \tag{B.15}$$

ensures that the stationary point $\boldsymbol{x}^*$ is a local minimizer of f. The condition (B.15) states that the *Hessian* matrix of f at $\boldsymbol{x}^*$ is *positive definite*. Recall that we write $\mathbf{H} \succ 0$ to indicate that a matrix $\mathbf{H}$ is positive definite.

☞ 400

In Figure B.2 we have a multiextremal objective function on $\mathcal{X} = \mathbb{R}$. There are four stationary points: two are local minimizers, one is a local maximizer, and one is neither a minimizer nor a maximizer, but a saddle point.

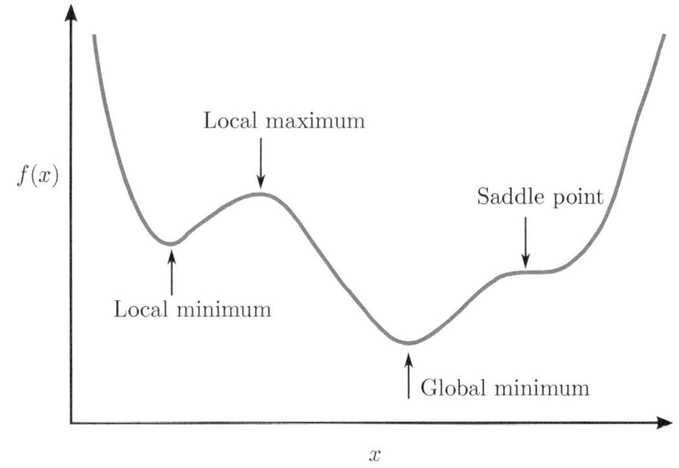

Figure B.2: A multiextremal objective function in one dimension.

B.2.1 Convexity and Optimization

An important class of optimization problems is related to the notion of *convexity*. A set $\mathcal{X}$ is said to be *convex* if for all $\boldsymbol{x}_1, \boldsymbol{x}_2 \in \mathcal{X}$ it holds that $\alpha\,\boldsymbol{x}_1 + (1-\alpha)\,\boldsymbol{x}_2 \in \mathcal{X}$ for all $0 \leqslant \alpha \leqslant 1$.

In addition, the objective function f is a *convex function* provided that for each $\boldsymbol{x}$ in the interior of $\mathcal{X}$ there exists a vector $\boldsymbol{v}$ such that

CONVEX FUNCTION

$$f(\boldsymbol{y}) \geqslant f(\boldsymbol{x}) + (\boldsymbol{y}-\boldsymbol{x})^\top \boldsymbol{v}, \quad \boldsymbol{y} \in \mathcal{X}. \tag{B.16}$$

SUBGRADIENT

☞ 62

The vector $\boldsymbol{v}$ in (B.16) may not be unique and is referred to as a *subgradient* of f.

One of the crucial properties of a convex function f is that *Jensen's inequality* holds (see Exercise 14 in Chapter 2):

$$\mathbb{E}f(\boldsymbol{X}) \geqslant f(\mathbb{E}\boldsymbol{X}),$$

for any random vector $\boldsymbol{X}$.

DIRECTIONAL DERIVATIVE

■ **Example B.6 (Convexity and Directional Derivative)** The *directional derivative* of a multivariate function f at $\boldsymbol{x}$ in the direction $\boldsymbol{d}$ is defined as the right derivative of $g(t) := f(\boldsymbol{x} + t\,\boldsymbol{d})$ at $t = 0$:

$$\lim_{t \downarrow 0} \frac{f(\boldsymbol{x} + t\,\boldsymbol{d}) - f(\boldsymbol{x})}{t} = \lim_{t \uparrow \infty} t\,(f(\boldsymbol{x} + \boldsymbol{d}/t) - f(\boldsymbol{x})).$$

This right derivative may not always exist. However, if f is a convex function, then the directional derivative of f at $\boldsymbol{x}$ in the interior of its domain always exists (in any direction $\boldsymbol{d}$).

To see this, let $t_1 \geqslant t_2 > 0$. By Jensen's inequality we have for any $\boldsymbol{x}$ and $\boldsymbol{y}$ in the interior of the domain:

$$\frac{t_2}{t_1} f(\boldsymbol{y}) + \left(1 - \frac{t_2}{t_1}\right) f(\boldsymbol{x}) \geqslant f\left(\frac{t_2}{t_1}\boldsymbol{y} + \left(1 - \frac{t_2}{t_1}\right)\boldsymbol{x}\right).$$

Making the substitution $\boldsymbol{y} = \boldsymbol{x} + t_1\boldsymbol{d}$ and rearranging the last equation yields:

$$\frac{f(\boldsymbol{x} + t_1\,\boldsymbol{d}) - f(\boldsymbol{x})}{t_1} \geqslant \frac{f(\boldsymbol{x} + t_2\,\boldsymbol{d}) - f(\boldsymbol{x})}{t_2}.$$

In other words, the function $t \mapsto (f(\boldsymbol{x} + t\,\boldsymbol{d}) - f(\boldsymbol{x}))/t$ is increasing for $t > 0$ and therefore the directional derivative satisfies:

$$\lim_{t \downarrow 0} \frac{f(\boldsymbol{x} + t\,\boldsymbol{d}) - f(\boldsymbol{x})}{t} = \inf_{t>0} \frac{f(\boldsymbol{x} + t\,\boldsymbol{d}) - f(\boldsymbol{x})}{t}.$$

Hence, to show existence it is enough to show that $(f(\boldsymbol{x} + t\,\boldsymbol{d}) - f(\boldsymbol{x}))/t$ is bounded from below.

Since $\boldsymbol{x}$ lies in the interior of the domain of f, we can choose t small enough so that $\boldsymbol{x} + t\,\boldsymbol{d}$ also lies in the interior. Therefore, the convexity of f implies that there exists a subgradient vector $\boldsymbol{v}$ such that $f(\boldsymbol{x} + t\,\boldsymbol{d}) \geqslant f(\boldsymbol{x}) + \boldsymbol{v}^\top(t\,\boldsymbol{d})$. In other words,

$$\frac{f(\boldsymbol{x} + t\,\boldsymbol{d}) - f(\boldsymbol{x})}{t} \geqslant \boldsymbol{v}^\top\boldsymbol{d}$$

provides a lower bound for all $t > 0$, and the directional derivative of f at an interior $\boldsymbol{x}$ always exists (in any direction). ■

CONCAVE FUNCTION

A function f satisfying (B.16) with strict inequality is said to be *strictly convex*. It is said to be a (strictly) *concave function* if $-f$ is (strictly) convex. Assuming that $\mathcal{X}$ is an open set, convexity for $f \in C^1$ is equivalent to

$$f(\boldsymbol{y}) \geqslant f(\boldsymbol{x}) + (\boldsymbol{y} - \boldsymbol{x})^\top \nabla f(\boldsymbol{x}) \quad \text{for all } \boldsymbol{x}, \boldsymbol{y} \in \mathcal{X}.$$

Moreover, for $f \in C^2$ strict convexity is equivalent to the Hessian matrix being positive definite for all $\boldsymbol{x} \in \mathcal{X}$, and convexity is equivalent to the Hessian matrix being *positive semidefinite* for all $\boldsymbol{x}$; that is, $\boldsymbol{y}^\top\left(\nabla^2 f(\boldsymbol{x})\right)\boldsymbol{y} \geqslant 0$ for all $\boldsymbol{y}$ and $\boldsymbol{x}$. Recall that we write $\mathbf{H} \succeq 0$ to indicate that a matrix $\mathbf{H}$ is positive semidefinite.

☞ 369

■ **Example B.7 (Convexity and Differentiability)** If f is a continuously differentiable multivariate function, then f is convex if and only if the univariate function

$$g(t) := f(\boldsymbol{x} + t\,\boldsymbol{d}), \quad t \in [0, 1]$$

is a convex function for any $\boldsymbol{x}$ and $\boldsymbol{x} + \boldsymbol{d}$ in the interior of the domain of f. This property provides an alternative definition for convexity of a multivariate and differentiable function.

To see why it is true, first assume that f is convex and $t_1, t_2 \in [0, 1]$. Then, using the subgradient definition of convexity in (B.16), we have $f(\boldsymbol{a}) \geqslant f(\boldsymbol{b}) + (\boldsymbol{a} - \boldsymbol{b})^\top \boldsymbol{v}$ for some subgradient $\boldsymbol{v}$. Substituting with $\boldsymbol{a} = \boldsymbol{x} + t_1\boldsymbol{d}$ and $\boldsymbol{b} = \boldsymbol{x} + t_2\boldsymbol{d}$, we obtain

$$g(t_1) \geqslant g(t_2) + (t_1 - t_2)\,\boldsymbol{v}^\top \boldsymbol{d}$$

for any two points $t_1, t_2 \in [0, 1]$. Therefore, g is convex, because we have identified the existence of a subgradient $\boldsymbol{v}^\top \boldsymbol{d}$ for each t_2.

Conversely, assume that g is convex for $t \in [0, 1]$. Since f is differentiable, then so is g. Then, the convexity of g implies that there is a subgradient v at 0 such that: $g(t) \geqslant g(0) + t\,v$ for all $t \in [0, 1]$. Rearranging,

$$v \geqslant \frac{g(t) - g(0)}{t},$$

and taking the right limit as $t \downarrow 0$ we obtain $v \geqslant g'(0) = \boldsymbol{d}^\top \nabla f(\boldsymbol{x})$. Therefore,

$$g(t) \geqslant g(0) + t\,v \geqslant g(0) + t\,\boldsymbol{d}^\top \nabla f(\boldsymbol{x})$$

and substituting $t = 1$ yields:

$$f(\boldsymbol{x} + \boldsymbol{d}) \geqslant f(\boldsymbol{x}) + \boldsymbol{d}^\top \nabla f(\boldsymbol{x}),$$

so that there exists a subgradient vector, namely $\nabla f(\boldsymbol{x})$, for each $\boldsymbol{x}$. Hence, f is convex by the definition in (B.16). ■

An optimization program of the form (B.13) is said to be a *convex programming problem* if:

CONVEX PROGRAMMING PROBLEM

1. The objective f is a *convex function*.

2. The inequality constraint functions $\{g_i\}$ are convex.

3. The equality constraint functions $\{h_i\}$ are *affine*, that is, of the form $\boldsymbol{a}_i^\top \boldsymbol{x} - b_i$. This is equivalent to both h_i and $-h_i$ being convex for all i.

Table B.1 summarizes some commonly encountered problems, all of which are convex, with the exception of the quadratic programs with $\mathbf{A} \not\succeq 0$.

Table B.1: Some common classes of optimization problems.

Name	$f(\boldsymbol{x})$	Constraints
Linear Program (LP)	$\boldsymbol{c}^\top \boldsymbol{x}$	$\mathbf{A}\boldsymbol{x} = \boldsymbol{b}$ and $\boldsymbol{x} \geqslant 0$
Inequality Form LP	$\boldsymbol{c}^\top \boldsymbol{x}$	$\mathbf{A}\boldsymbol{x} \leqslant \boldsymbol{b}$
Quadratic Program (QP)	$\frac{1}{2}\boldsymbol{x}^\top \mathbf{A}\boldsymbol{x} + \boldsymbol{b}^\top \boldsymbol{x}$	$\mathbf{D}\boldsymbol{x} \leqslant \boldsymbol{d}, \ \ \mathbf{E}\boldsymbol{x} = \boldsymbol{e}$
Convex QP	$\frac{1}{2}\boldsymbol{x}^\top \mathbf{A}\boldsymbol{x} + \boldsymbol{b}^\top \boldsymbol{x}$	$\mathbf{D}\boldsymbol{x} \leqslant \boldsymbol{d}, \ \mathbf{E}\boldsymbol{x} = \boldsymbol{e} \quad (\mathbf{A} \succeq 0)$
Convex Program	$f(\boldsymbol{x})$ convex	$\{g_i(\boldsymbol{x})\}$ convex, $\{h_i(\boldsymbol{x})\}$ of the form $\boldsymbol{a}_i^\top \boldsymbol{x} - b_i$

Recognizing convex optimization problems or those that can be transformed to convex optimization problems can be challenging. However, once formulated as convex optimization problems, these can be efficiently solved using subgradient [112], bundle [57], and cutting-plane methods [59].

B.2.2 Lagrangian Method

The main components of the Lagrangian method are the Lagrange multipliers and the Lagrange function. The method was developed by Lagrange in 1797 for the optimization problem (B.13) with only equality constraints. In 1951 Kuhn and Tucker extended Lagrange's method to inequality constraints. Given an optimization problem (B.13) containing only equality constraints $h_i(\boldsymbol{x}) = 0, \ i = 1, \ldots, m$, the *Lagrange function* is defined as

LAGRANGE FUNCTION

$$\mathcal{L}(\boldsymbol{x}, \boldsymbol{\beta}) = f(\boldsymbol{x}) + \sum_{i=1}^{m} \beta_i \, h_i(\boldsymbol{x}),$$

LAGRANGE MULTIPLIERS

where the coefficients $\{\beta_i\}$ are called the *Lagrange multipliers*. A necessary condition for a point $\boldsymbol{x}^*$ to be a local minimizer of $f(\boldsymbol{x})$ subject to the equality constraints $h_i(\boldsymbol{x}) = 0, \ i = 1, \ldots, m$, is

$$\nabla_{\boldsymbol{x}} \mathcal{L}(\boldsymbol{x}^*, \boldsymbol{\beta}^*) = \mathbf{0},$$
$$\nabla_{\boldsymbol{\beta}} \mathcal{L}(\boldsymbol{x}^*, \boldsymbol{\beta}^*) = \mathbf{0},$$

for some value $\boldsymbol{\beta}^*$. The above conditions are also sufficient if $\mathcal{L}(\boldsymbol{x}, \boldsymbol{\beta}^*)$ is a convex function of $\boldsymbol{x}$.

LAGRANGIAN

Given the original optimization problem (B.13), containing both the equality and inequality constraints, the *generalized Lagrange function*, or *Lagrangian*, is defined as

$$\mathcal{L}(\boldsymbol{x}, \boldsymbol{\alpha}, \boldsymbol{\beta}) = f(\boldsymbol{x}) + \sum_{i=1}^{k} \alpha_i \, g_i(\boldsymbol{x}) + \sum_{i=1}^{m} \beta_i \, h_i(\boldsymbol{x}).$$

Theorem B.2: Karush–Kuhn–Tucker (KKT) Conditions

A necessary condition for a point x^* to be a local minimizer of $f(x)$ in the optimization problem (B.13) is the existence of an α^* and β^* such that

$$\begin{aligned} \nabla_x \mathcal{L}(x^*, \alpha^*, \beta^*) &= \mathbf{0}, \\ \nabla_\beta \mathcal{L}(x^*, \alpha^*, \beta^*) &= \mathbf{0}, \\ g_i(x^*) &\leqslant 0, \quad i = 1, \ldots, k, \\ \alpha_i^* &\geqslant 0, \quad i = 1, \ldots, k, \\ \alpha_i^* \, g_i(x^*) &= 0, \quad i = 1, \ldots, k. \end{aligned}$$

For *convex* programs we have the following important results [18, 43]:

1. Every local solution x^* to a convex programming problem is a global solution and the set of global solutions is convex. If, in addition, the objective function is strictly convex, then any global solution is unique.

2. For a strictly convex programming problem with C^1 objective and constraint functions, the KKT conditions are necessary and sufficient for a unique global solution.

B.2.3 Duality

The aim of duality is to provide an alternative formulation of an optimization problem which is often more computationally efficient or has some theoretical significance (see [43, Page 219]). The original problem (B.13) is referred to as the *primal* problem whereas the reformulated problem, based on Lagrange multipliers, is called the *dual* problem. Duality theory is most relevant to convex optimization problems. It is well known that if the primal optimization problem is (strictly) convex then the dual problem is (strictly) concave and has a (unique) solution from which the (unique) optimal primal solution can be deduced. PRIMAL DUAL

The *Lagrange dual program* (also called the *Wolfe dual*) of the primal program (B.13), is: LAGRANGE DUAL PROGRAM

$$\begin{aligned} \max_{\alpha, \beta} \quad & \mathcal{L}^*(\alpha, \beta) \\ \text{subject to:} \quad & \alpha \geqslant \mathbf{0}, \end{aligned}$$

where $\mathcal{L}^*$ is the *Lagrange dual function*:

$$\mathcal{L}^*(\alpha, \beta) = \inf_{x \in \mathcal{X}} \mathcal{L}(x, \alpha, \beta), \tag{B.17}$$

giving the greatest lower bound (infimum) of $\mathcal{L}(x, \alpha, \beta)$ over all possible $x \in \mathcal{X}$.

It is not difficult to see that if f^* is the minimal value of the primal problem, then $\mathcal{L}^*(\alpha, \beta) \leqslant f^*$ for any $\alpha \geqslant \mathbf{0}$ and any β. This property is called *weak duality*. The Lagrangian dual program thus determines the best lower bound on f^*. If d^* is the optimal value for the dual problem then $d^* \leqslant f^*$. The difference $f^* - d^*$ is called the *duality gap*.

The duality gap is extremely useful for providing lower bounds for the solutions of primal problems that may be impossible to solve directly. It is important to note that for linearly constrained problems, if the primal is infeasible (does not have a solution satisfying the constraints), then the dual is either infeasible or unbounded. Conversely, if the dual is infeasible then the primal has no solution. Of crucial importance is the *strong duality* theorem, which states that for convex programs (B.13) with linear constrained functions h_i and g_i the duality gap is zero, and any $\boldsymbol{x}^*$ and $(\boldsymbol{\alpha}^*, \boldsymbol{\beta}^*)$ satisfying the KKT conditions are (global) solutions to the primal and dual programs, respectively. In particular, this holds for linear and convex quadratic programs (note that not all quadratic programs are convex).

STRONG DUALITY

For a convex primal program with C^1 objective and constraint functions, the Lagrangian dual function (B.17) can be obtained by simply setting the gradient (with respect to $\boldsymbol{x}$) of the Lagrangian $\mathcal{L}(\boldsymbol{x}, \boldsymbol{\alpha}, \boldsymbol{\beta})$ to zero. One can further simplify the dual program by substituting into the Lagrangian the relations between the variables thus obtained.

Further, for a convex primal problem, if there is a *strictly feasible* point $\widetilde{\boldsymbol{x}}$ (that is, a feasible point satisfying all of the inequality constraints with strict inequality), then the duality gap is zero, and strong duality holds. This is known as *Slater's condition* [18, Page 226].

The Lagrange dual problem is an important example of a *saddle-point problem* or *minimax* problem. In such problems the aim is to locate a point $(\boldsymbol{x}^*, \boldsymbol{y}^*) \in \mathcal{X} \times \mathcal{Y}$ that satisfies

$$\sup_{\boldsymbol{y} \in \mathcal{Y}} \inf_{\boldsymbol{x} \in \mathcal{X}} f(\boldsymbol{x}, \boldsymbol{y}) = \inf_{\boldsymbol{x} \in \mathcal{X}} f(\boldsymbol{x}, \boldsymbol{y}^*) = f(\boldsymbol{x}^*, \boldsymbol{y}^*) = \sup_{\boldsymbol{y} \in \mathcal{Y}} f(\boldsymbol{x}^*, \boldsymbol{y}) = \inf_{\boldsymbol{x} \in \mathcal{X}} \sup_{\boldsymbol{y} \in \mathcal{Y}} f(\boldsymbol{x}, \boldsymbol{y}).$$

The equation

$$\sup_{\boldsymbol{y} \in \mathcal{Y}} \inf_{\boldsymbol{x} \in \mathcal{X}} f(\boldsymbol{x}, \boldsymbol{y}) = \inf_{\boldsymbol{x} \in \mathcal{X}} \sup_{\boldsymbol{y} \in \mathcal{Y}} f(\boldsymbol{x}, \boldsymbol{y})$$

MINIMAX

is known as the *minimax* equality. Other problems that fall into this framework are zero-sum games in game theory; see also [24] for a number of combinatorial optimization problems that can be viewed as minimax problems.

B.3 Numerical Root-Finding and Minimization

In order to minimize a C^1 function $f : \mathbb{R}^n \to \mathbb{R}$ one may solve

$$\nabla f(\boldsymbol{x}) = \boldsymbol{0},$$

which gives a stationary point of f. As a consequence, any technique for root-finding can be transformed into an unconstrained optimization method by attempting to locate roots of the gradient. However, as noted in Section B.2, not all stationary points are minima, and so additional information (such as is contained in the Hessian, if f is C^2) needs to be considered in order to establish the type of stationary point.

Alternatively, a root of a continuous function $\boldsymbol{g} : \mathbb{R}^n \to \mathbb{R}^n$ may be found by minimizing the norm of $\boldsymbol{g}(\boldsymbol{x})$ over all $\boldsymbol{x}$; that is, by solving $\min_{\boldsymbol{x}} f(\boldsymbol{x})$, with $f(\boldsymbol{x}) := \|\boldsymbol{g}(\boldsymbol{x})\|_p$, where for $p \geqslant 1$ the *p-norm* of $\boldsymbol{y} = [y_1, \ldots, y_n]^\top$ is defined as

p-NORM

$$\|\boldsymbol{y}\|_p := \left(\sum_{i=1}^{n} |y_i|^p\right)^{1/p}.$$

Hence, any (un)constrained optimization method can be transformed into a technique for locating the roots of a function.

Starting with an initial guess $\boldsymbol{x}_0$, most minimization and root-finding algorithms create a sequence $\boldsymbol{x}_0, \boldsymbol{x}_1, \ldots$ using the iterative updating rule:

$$\boldsymbol{x}_{t+1} = \boldsymbol{x}_t + \alpha_t \, \boldsymbol{d}_t, \quad t = 0, 1, 2, \ldots, \tag{B.18}$$

where $\alpha_t > 0$ is a (typically small) step size, called the *learning rate*, and the vector $\boldsymbol{d}_t$ is the search direction at step t. The iteration (B.18) continues until the sequence $\{\boldsymbol{x}_t\}$ is deemed to have converged to a solution, or a computational budget has been exhausted. The performance of all such iterative methods depends crucially on the quality of the initial guess $\boldsymbol{x}_0$. LEARNING RATE

There are two broad categories of iterative optimization algorithms of the form (B.18):

- Those of *line search* type, where at iteration t we first compute a direction $\boldsymbol{d}_t$ and then determine a reasonable step size α_t along this direction. For example, in the case of minimization, $\alpha_t > 0$ may be chosen to approximately minimize $f(\boldsymbol{x}_t + \alpha \, \boldsymbol{d}_t)$ for fixed $\boldsymbol{x}_t$ and $\boldsymbol{d}_t$. LINE SEARCH

- Those of *trust region* type, where at each iteration t we first determine a suitable step size α_t and then compute an approximately optimal direction $\boldsymbol{d}_t$. TRUST REGION

In the following sections, we review several widely-used root-finding and optimization algorithms of the line search type.

B.3.1 Newton-Like Methods

Suppose we wish to find roots of a function $\boldsymbol{f} : \mathbb{R}^n \to \mathbb{R}^n$. If $\boldsymbol{f}$ is in C^1, we can approximate $\boldsymbol{f}$ around a point $\boldsymbol{x}_t$ as

$$\boldsymbol{f}(\boldsymbol{x}) \approx \boldsymbol{f}(\boldsymbol{x}_t) + \mathbf{J}_f(\boldsymbol{x}_t)(\boldsymbol{x} - \boldsymbol{x}_t),$$

where $\mathbf{J}_f$ is the *matrix of Jacobi* — the matrix of partial derivatives of $\boldsymbol{f}$; see (B.3). When $\mathbf{J}_f(\boldsymbol{x}_t)$ is invertible, this linear approximation has root $\boldsymbol{x}_t - \mathbf{J}_f^{-1}(\boldsymbol{x}_t)\,\boldsymbol{f}(\boldsymbol{x}_t)$. This gives the iterative updating formula (B.18) for finding roots of $\boldsymbol{f}$ with direction $\boldsymbol{d}_t = -\mathbf{J}_f^{-1}(\boldsymbol{x}_t)\,\boldsymbol{f}(\boldsymbol{x}_t)$ and learning rate $\alpha_t = 1$. This is known as *Newton's method* (or the *Newton–Raphson method*) for root-finding. ☞ 400 NEWTON'S METHOD

Instead of a unit learning rate, sometimes it is more effective to use an α_t that satisfies the *Armijo inexact line search* condition: ARMIJO INEXACT LINE SEARCH

$$\|\boldsymbol{f}(\boldsymbol{x}_t + \alpha_t \, \boldsymbol{d}_t)\| < (1 - \varepsilon_1 \, \alpha_t) \, \|\boldsymbol{f}(\boldsymbol{x}_t)\|,$$

where ε_1 is a small heuristically chosen constant, say $\varepsilon_1 = 10^{-4}$. For C^1 functions, such an α_t always exists by continuity and can be computed as in the following algorithm.

Algorithm B.3.1: Newton–Raphson for Finding Roots of $\boldsymbol{f}(\boldsymbol{x}) = \mathbf{0}$

input: An initial guess $\boldsymbol{x}$ and stopping error $\varepsilon > 0$.
output: The approximate root of $\boldsymbol{f}(\boldsymbol{x}) = \mathbf{0}$.

while $\|\boldsymbol{f}(\boldsymbol{x})\| > \varepsilon$ and budget is not exhausted **do**
  Solve the linear system $\mathbf{J}_{\boldsymbol{f}}(\boldsymbol{x})\, \boldsymbol{d} = -\boldsymbol{f}(\boldsymbol{x})$.
  $\alpha \leftarrow 1$
  while $\|\boldsymbol{f}(\boldsymbol{x} + \alpha\, \boldsymbol{d})\| > (1 - 10^{-4}\alpha)\, \|\boldsymbol{f}(\boldsymbol{x})\|$ **do**
    $\alpha \leftarrow \alpha/2$
  $\boldsymbol{x} \leftarrow \boldsymbol{x} + \alpha\, \boldsymbol{d}$
return $\boldsymbol{x}$

We can adapt a root-finding Newton-like method in order to minimize a differentiable function $f : \mathbb{R}^n \to \mathbb{R}$. We simply try to locate a zero of the gradient of f. When f is a C^2 function, the function $\nabla f : \mathbb{R}^n \to \mathbb{R}^n$ is continuous, and so the root of ∇f leads to the search direction

$$\boldsymbol{d}_t = -\mathbf{H}_t^{-1}\, \nabla f(\boldsymbol{x}_t), \tag{B.19}$$

where $\mathbf{H}_t$ is the Hessian matrix at $\boldsymbol{x}_t$ (the matrix of Jacobi of the gradient is the Hessian). When the learning rate α_t is equal to 1, the update $\boldsymbol{x}_t - \mathbf{H}_t^{-1}\, \nabla f(\boldsymbol{x}_t)$ can alternatively be derived by assuming that $f(\boldsymbol{x})$ is approximately quadratic and convex in the neighborhood of $\boldsymbol{x}_t$, that is,

$$f(\boldsymbol{x}) \approx f(\boldsymbol{x}_t) + (\boldsymbol{x} - \boldsymbol{x}_t)^\top \nabla f(\boldsymbol{x}_t) + \frac{1}{2}(\boldsymbol{x} - \boldsymbol{x}_t)^\top \mathbf{H}_t (\boldsymbol{x} - \boldsymbol{x}_t), \tag{B.20}$$

and then minimizing the right-hand side of (B.20) with respect to $\boldsymbol{x}$.

The following algorithm uses an *Armijo inexact line search* for minimization and guards against the possibility that the Hessian may not be positive definite (that is, its Cholesky decomposition does not exist).

☞ 375

Algorithm B.3.2: Newton–Raphson for Minimizing $f(\boldsymbol{x})$

input: An initial guess $\boldsymbol{x}$; stopping error $\varepsilon > 0$; line search parameter $\xi \in (0, 1)$.
output: An approximate minimizer of $f(\boldsymbol{x})$.

$\mathbf{L} \leftarrow \mathbf{I}_n$ (the identity matrix)
while $\|\nabla f(\boldsymbol{x})\| > \varepsilon$ and budget is not exhausted **do**
  Compute the Hessian $\mathbf{H}$ at $\boldsymbol{x}$.
  if $\mathbf{H} \succ \mathbf{0}$ **then** // Cholesky is successful
    Update $\mathbf{L}$ to be the Cholesky factor satisfying $\mathbf{L}\mathbf{L}^\top = \mathbf{H}$.
  else
    Do not update the lower triangular $\mathbf{L}$.
  $\boldsymbol{d} \leftarrow -\mathbf{L}^{-1}\nabla f(\boldsymbol{x})$ (computed by forward substitution)
  $\boldsymbol{d} \leftarrow \mathbf{L}^{-\top}\boldsymbol{d}$ (computed by backward substitution)
  $\alpha \leftarrow 1$
  while $f(\boldsymbol{x} + \alpha\boldsymbol{d}) > f(\boldsymbol{x}) + \alpha\, 10^{-4}\nabla f(\boldsymbol{x})^\top \boldsymbol{d}$ **do**
    $\alpha \leftarrow \alpha \times \xi$
  $\boldsymbol{x} \leftarrow \boldsymbol{x} + \alpha\, \boldsymbol{d}$
return $\boldsymbol{x}$

A downside with all Newton-like methods is that at each step they require the calculation and inversion of an $n \times n$ Hessian matrix, which has computing time of $O(n^3)$, and is thus infeasible for large n. One way to avoid this cost is to use quasi-Newton methods, described next.

B.3.2 Quasi-Newton Methods

The idea behind quasi-Newton methods is to replace the inverse Hessian in (B.19) at iteration t by an $n \times n$ matrix $\mathbf{C}$ satisfying the *secant condition*: SECANT CONDITION

$$\mathbf{C}\boldsymbol{g} = \boldsymbol{\delta}, \tag{B.21}$$

where $\boldsymbol{\delta} \leftarrow \boldsymbol{x}_t - \boldsymbol{x}_{t-1}$ and $\boldsymbol{g} \leftarrow \nabla f(\boldsymbol{x}_t) - \nabla f(\boldsymbol{x}_{t-1})$ are vectors stored in memory at each iteration t. The secant condition is satisfied, for example, by the *Broyden's family* of matrices:

$$\mathbf{A} + \frac{1}{\boldsymbol{u}^\top \boldsymbol{g}}(\boldsymbol{\delta} - \mathbf{A}\,\boldsymbol{g})\,\boldsymbol{u}^\top$$

for some $\boldsymbol{u} \neq \mathbf{0}$ and $\mathbf{A}$. Since there is an infinite number of matrices that satisfy the condition (B.21), we need a way to determine a unique $\mathbf{C}$ at each iteration t such that computing and storing $\mathbf{C}$ from one step to the next is fast and avoids any costly matrix inversion. The following examples illustrate how, starting with an initial guess $\mathbf{C} = \mathbf{I}$ at $t = 0$, such a matrix $\mathbf{C}$ can be efficiently updated from one iteration to the next.

■ **Example B.8 (Low-Rank Hessian Update)** The quadratic model (B.20) can be strengthened by further assuming that $\exp(-f(\boldsymbol{x}))$ is proportional to a probability density that can be approximated in the neighborhood of $\boldsymbol{x}_t$ by the pdf of the $\mathcal{N}(\boldsymbol{x}_{t+1}, \mathbf{H}_t^{-1})$ distribution. This normal approximation allows us to measure the *discrepancy* between two pairs $(\boldsymbol{x}_1, \mathbf{H}_0)$ and $(\boldsymbol{x}_2, \mathbf{H}_1)$ using the Kullback–Leibler divergence between the pdfs of the $\mathcal{N}(\boldsymbol{x}_1, \mathbf{H}_0^{-1})$ and $\mathcal{N}(\boldsymbol{x}_2, \mathbf{H}_1^{-1})$ distributions (see Exercise 4 on page 352): ☞ 42

$$\mathcal{D}(\boldsymbol{x}_1, \mathbf{H}_0^{-1} \,|\, \boldsymbol{x}_2, \mathbf{H}_1^{-1}) := \frac{1}{2}\Big(\mathrm{tr}(\mathbf{H}_1\mathbf{H}_0^{-1}) - \ln|\mathbf{H}_1\mathbf{H}_0^{-1}| + (\boldsymbol{x}_2 - \boldsymbol{x}_1)^\top \mathbf{H}_1 (\boldsymbol{x}_2 - \boldsymbol{x}_1) - n\Big). \tag{B.22}$$

Suppose that the latest approximation to the inverse Hessian is $\mathbf{C}$ and we wish to compute an updated approximation for step t. One approach is to find the symmetric matrix that minimizes its Kullback–Leibler discrepancy from $\mathbf{C}$, as defined above, subject to the constraint (B.21). In other words,

$$\min_{\mathbf{A}} \mathcal{D}(\mathbf{0}, \mathbf{C} \,|\, \mathbf{0}, \mathbf{A})$$
$$\text{subject to: } \mathbf{A}\boldsymbol{g} = \boldsymbol{\delta},\ \mathbf{A} = \mathbf{A}^\top.$$

The solution to this constrained optimization (see Exercise 10 on page 354) yields the Broyden–Fletcher–Goldfarb–Shanno or *BFGS formula* for updating the matrix $\mathbf{C}$ from one iteration to the next: BFGS FORMULA

$$\mathbf{C}_{\text{BFGS}} = \mathbf{C} + \underbrace{\frac{\boldsymbol{g}^\top\boldsymbol{\delta} + \boldsymbol{g}^\top\mathbf{C}\boldsymbol{g}}{(\boldsymbol{g}^\top\boldsymbol{\delta})^2}\,\boldsymbol{\delta}\boldsymbol{\delta}^\top - \frac{1}{\boldsymbol{g}^\top\boldsymbol{\delta}}\big(\boldsymbol{\delta}\boldsymbol{g}^\top\mathbf{C} + (\boldsymbol{\delta}\boldsymbol{g}^\top\mathbf{C})^\top\big)}_{\text{BFGS update}}. \tag{B.23}$$

In a practical implementation, we keep a single copy of $\mathbf{C}$ in memory and apply the BFGS update to it at every iteration. Note that if the current $\mathbf{C}$ is symmetric, then so is the updated matrix. Moreover, the BFGS update is a matrix of rank two.

Since the Kullback–Leibler divergence is not symmetric, it is possible to flip the roles of $\mathbf{H}_0$ and $\mathbf{H}_1$ in (B.22) and instead solve

$$\min_{\mathbf{A}} \mathcal{D}(\mathbf{0}, \mathbf{A} \,|\, \mathbf{0}, \mathbf{C})$$
$$\text{subject to: } \mathbf{A}\boldsymbol{g} = \boldsymbol{\delta}, \ \mathbf{A} = \mathbf{A}^\top.$$

DFP FORMULA

The solution (see Exercise 10 on page 354) gives the Davidon–Fletcher–Powell or *DFP formula* for updating the matrix $\mathbf{C}$ from one iteration to the next:

$$\mathbf{C}_{\text{DFP}} = \mathbf{C} + \underbrace{\frac{\boldsymbol{\delta}\boldsymbol{\delta}^\top}{\boldsymbol{g}^\top\boldsymbol{\delta}} - \frac{\mathbf{C}\boldsymbol{g}\boldsymbol{g}^\top\mathbf{C}}{\boldsymbol{g}^\top\mathbf{C}\boldsymbol{g}}}_{\text{DFP update}}. \tag{B.24}$$

Note that if the curvature condition $\boldsymbol{g}^\top\boldsymbol{\delta} > 0$ holds and the current $\mathbf{C}$ is symmetric positive definite, then so is its update. ■

■ **Example B.9 (Diagonal Hessian Update)** The original BFGS formula requires $\mathcal{O}(n^2)$ storage and computation, which may be unmanageable for large n. One way to circumvent the prohibitive quadratic cost is to only store and update a diagonal Hessian matrix from one iteration to the next. If $\mathbf{C}$ is diagonal, then we may not be able to satisfy the secant condition (B.21) and maintain positive definiteness. Instead the secant condition (B.21) can be relaxed to the set of inequalities $\boldsymbol{g} \geqslant \mathbf{C}^{-1}\boldsymbol{\delta}$, which are related to the definition of a

☞ 405

subgradient for convex functions. We can then find a unique diagonal matrix by minimizing $\mathcal{D}(\boldsymbol{x}_t, \mathbf{C} \,|\, \boldsymbol{x}_{t+1}, \mathbf{A})$ with respect to $\mathbf{A}$ and subject to the constraints that $\mathbf{A}\boldsymbol{g} \geqslant \boldsymbol{\delta}$ and $\mathbf{A}$ is diagonal. The solution (Exercise 15 on page 354) yields the updating formula for a diagonal element c_i of $\mathbf{C}$:

$$c_i \leftarrow \begin{cases} \dfrac{2c_i}{1 + \sqrt{1 + 4c_i u_i^2}}, & \text{if } \dfrac{2c_i}{1 + \sqrt{1 + 4c_i u_i^2}} \geqslant \delta_i/g_i \\ \delta_i/g_i, & \text{otherwise,} \end{cases} \tag{B.25}$$

where $\boldsymbol{u} := \nabla f(\boldsymbol{x}_t)$ and we assume a unit learning rate: $\boldsymbol{x}_{t+1} = \boldsymbol{x}_t - \mathbf{A}\boldsymbol{u}$. ■

■ **Example B.10 (Scalar Hessian Update)** If the identity matrix is used in place of the

STEEPEST DESCENT

Hessian in (B.19), one obtains *steepest descent* or *gradient descent* methods, in which the iteration (B.18) reduces to $\boldsymbol{x}_{t+1} = \boldsymbol{x}_t - \alpha_t \nabla f(\boldsymbol{x}_t)$.

The rationale for the name *steepest descent* is as follows. If we start from any point $\boldsymbol{x}$ and make an infinitesimal move in some direction, then the function value is reduced by the largest magnitude in the (unit norm) direction: $\boldsymbol{u}^* := -\nabla f(\boldsymbol{x})/\|\nabla f(\boldsymbol{x})\|$. This is seen from the following inequality for all unit vectors $\boldsymbol{u}$ (that is, $\|\boldsymbol{u}\| = 1$):

$$\frac{\mathrm{d}}{\mathrm{d}t} f(\boldsymbol{x} + t\,\boldsymbol{u}^*)\Big|_{t=0} \leqslant \frac{\mathrm{d}}{\mathrm{d}t} f(\boldsymbol{x} + t\,\boldsymbol{u})\Big|_{t=0}.$$

Observe that equality is achieved if and only if $\boldsymbol{u} = \boldsymbol{u}^*$. This inequality is an easy con-

☞ 391

sequence of the Cauchy–Schwarz inequality:

$$-\nabla f^\top \boldsymbol{u} \leqslant |\nabla f^\top \boldsymbol{u}| \underbrace{\leqslant}_{\text{Cauchy–Schwartz}} \|\boldsymbol{u}\| \, \|\nabla f\| = \|\nabla f\| = -\nabla f^\top \boldsymbol{u}^*.$$

The steepest descent iteration, $\boldsymbol{x}_{t+1} = \boldsymbol{x}_t - \alpha_t \nabla f(\boldsymbol{x}_t)$, still requires a suitable choice of the learning rate α_t. An alternative way to think about the iteration is to assume that the learning rate is always unity, and that at each iteration we use an inverse Hessian matrix of the form $\alpha_t \mathbf{I}$ for some positive constant α_t. Satisfying the secant condition (B.21) with a matrix of the form $\mathbf{C} = \alpha \mathbf{I}$ is not possible. However, it is possible to choose α so that the secant condition (B.21) is satisfied in the direction of $\boldsymbol{g}$ (or alternatively $\boldsymbol{\delta}$). This gives the *Barzilai–Borwein formulas* for the learning rate at iteration t:

BARZILAI–BORWEIN FORMULAS

$$\alpha_t = \frac{\boldsymbol{g}^\top \boldsymbol{\delta}}{\|\boldsymbol{g}\|^2} \quad \left(\text{or alternatively} \quad \alpha_t = \frac{\|\boldsymbol{\delta}\|^2}{\boldsymbol{\delta}^\top \boldsymbol{g}}\right). \tag{B.26}$$

■

B.3.3 Normal Approximation Method

Let $\varphi_{\mathbf{H}_t^{-1}}(\boldsymbol{x} - \boldsymbol{x}_{t+1})$ denote the pdf of the $\mathcal{N}(\boldsymbol{x}_{t+1}, \mathbf{H}_t^{-1})$ distribution. As we already saw in Example B.8, the quadratic approximation (B.20) of f in the neighborhood of $\boldsymbol{x}_t$ is equivalent (up to a constant) to the minus of the logarithm of the pdf $\varphi_{\mathbf{H}_t^{-1}}(\boldsymbol{x} - \boldsymbol{x}_{t+1})$. In other words, we use $\varphi_{\mathbf{H}_t^{-1}}(\boldsymbol{x} - \boldsymbol{x}_{t+1})$ as a simple model for the density

$$\exp(-f(\boldsymbol{x})) \Big/ \int \exp(-f(\boldsymbol{y}))\, \mathrm{d}\boldsymbol{y}.$$

One consequence of the normal approximation is that for $\boldsymbol{x}$ in the neighborhood of $\boldsymbol{x}_{t+1}$, we can write:

$$-\nabla f(\boldsymbol{x}) \approx \frac{\partial}{\partial \boldsymbol{x}} \ln \varphi_{\mathbf{H}_t^{-1}}(\boldsymbol{x} - \boldsymbol{x}_{t+1}) = -\mathbf{H}_t(\boldsymbol{x} - \boldsymbol{x}_{t+1}).$$

In other words, using the fact that $\mathbf{H}_t^\top = \mathbf{H}_t$,

$$\nabla f(\boldsymbol{x})[\nabla f(\boldsymbol{x})]^\top \approx \mathbf{H}_t(\boldsymbol{x} - \boldsymbol{x}_{t+1})(\boldsymbol{x} - \boldsymbol{x}_{t+1})^\top \mathbf{H}_t,$$

and taking expectations on both sides with respect to $\boldsymbol{X} \sim \mathcal{N}(\boldsymbol{x}_{t+1}, \mathbf{H}_t^{-1})$ gives:

$$\mathbb{E}\, \nabla f(\boldsymbol{X})\, [\nabla f(\boldsymbol{X})]^\top \approx \mathbf{H}_t.$$

This suggests that, given the gradient vectors computed in the past h (where h stands for history) of Newton iterations:

$$\boldsymbol{u}_i := \nabla f(\boldsymbol{x}_i), \quad i = t - (h-1), \ldots, t,$$

the Hessian matrix $\mathbf{H}_t$ can be approximated via the average

$$\frac{1}{h} \sum_{i=t-h+1}^{t} \boldsymbol{u}_i \boldsymbol{u}_i^\top.$$

A shortcoming of this approximation is that, unless h is large enough, the Hessian approximation $\sum_{i=t-h+1}^{t} \boldsymbol{u}_i \boldsymbol{u}_i^\top$ may not be full rank and hence not invertible. To ensure that the

☞ 217

Hessian approximation is invertible, we add a suitable diagonal matrix $\mathbf{A}_0$ to obtain the *regularized* version of the approximation:

$$\mathbf{H}_t \approx \mathbf{A}_0 + \frac{1}{h}\sum_{i=t-h+1}^{t} \boldsymbol{u}_i\boldsymbol{u}_i^\top.$$

With this full-rank approximation of the Hessian, the Newton search direction in (B.19) becomes:

$$\boldsymbol{d}_t = -\left(\mathbf{A}_0 + \frac{1}{h}\sum_{i=t-h+1}^{t} \boldsymbol{u}_i\boldsymbol{u}_i^\top\right)^{-1} \boldsymbol{u}_t. \tag{B.27}$$

☞ 375

Thus, $\boldsymbol{d}_t$ can be computed in $O(h^2\, n)$ time via the Sherman–Morrison Algorithm A.6.1. Further to this, the search direction (B.27) can be efficiently updated to the next one:

$$\boldsymbol{d}_{t+1} = -\left(\mathbf{A}_0 + \frac{1}{h}\sum_{i=t-h+2}^{t+1} \boldsymbol{u}_i\boldsymbol{u}_i^\top\right)^{-1} \boldsymbol{u}_{t+1}$$

in $O(h\, n)$ time, thus avoiding the usual $O(h^2\, n)$ cost (see Exercise 6 on page 353).

B.3.4 Nonlinear Least Squares

☞ 188

Consider the squared-error training loss in nonlinear regression:

$$\ell_\tau(g(\cdot\,|\,\boldsymbol{\beta})) = \frac{1}{n}\sum_{i=1}^{n}(g(\boldsymbol{x}_i\,|\,\boldsymbol{\beta}) - y_i)^2,$$

where $g(\cdot\,|\,\boldsymbol{\beta})$ is a nonlinear prediction function that depends on the parameter $\boldsymbol{\beta}$ (for example, (5.29) shows the nonlinear logistic prediction function). The training loss can be written as $\frac{1}{n}\|\boldsymbol{g}(\tau\,|\,\boldsymbol{\beta}) - \boldsymbol{y}\|^2$, where $\boldsymbol{g}(\tau\,|\,\boldsymbol{\beta}) := [g(\boldsymbol{x}_1\,|\,\boldsymbol{\beta}), \ldots, g(\boldsymbol{x}_n\,|\,\boldsymbol{\beta})]^\top$ is the vector of outputs.

We wish to minimize the training loss in terms of $\boldsymbol{\beta}$. In the Newton-like methods in Section B.3.1, one derives an iterative minimization algorithm that is inspired by a Taylor expansion of $\ell_\tau(g(\cdot\,|\,\boldsymbol{\beta}))$. Instead, given a current guess $\boldsymbol{\beta}_t$, we can consider the Taylor expansion of the nonlinear prediction function $\boldsymbol{g}$:

$$\boldsymbol{g}(\tau\,|\,\boldsymbol{\beta}) \approx \boldsymbol{g}(\tau\,|\,\boldsymbol{\beta}_t) + \mathbf{G}_t(\boldsymbol{\beta} - \boldsymbol{\beta}_t),$$

☞ 400

where $\mathbf{G}_t := \mathbf{J}_g(\boldsymbol{\beta}_t)$ is the matrix of Jacobi of $\boldsymbol{g}(\tau\,|\,\boldsymbol{\beta})$ at $\boldsymbol{\beta}_t$. Denoting the residual $\boldsymbol{e}_t := \boldsymbol{g}(\tau\,|\,\boldsymbol{\beta}_t) - \boldsymbol{y}$ and replacing $\boldsymbol{g}(\tau\,|\,\boldsymbol{\beta})$ with its Taylor approximation in $\ell_\tau(g(\cdot\,|\,\boldsymbol{\beta}))$, we obtain the approximation to the training loss in the neighborhood of $\boldsymbol{\beta}_t$:

$$\ell_\tau(g(\cdot\,|\,\boldsymbol{\beta})) \approx \frac{1}{n}\left\|\mathbf{G}_t(\boldsymbol{\beta} - \boldsymbol{\beta}_t) + \boldsymbol{e}_t\right\|^2.$$

☞ 28

GAUSS–NEWTON

The minimization of the right-hand side is a linear least-squares problem and therefore $\boldsymbol{d}_t := \boldsymbol{\beta} - \boldsymbol{\beta}_t$ satisfies the normal equations: $\mathbf{G}_t^\top\mathbf{G}_t\boldsymbol{d}_t = \mathbf{G}_t^\top(-\boldsymbol{e}_t)$. Assuming that $\mathbf{G}_t^\top\mathbf{G}_t$ is invertible, the normal equations yield the *Gauss–Newton* search direction:

$$\boldsymbol{d}_t = -(\mathbf{G}_t^\top\mathbf{G}_t)^{-1}\mathbf{G}_t^\top\boldsymbol{e}_t.$$

Unlike the search direction (B.19) for Newton-like algorithms, the search direction of a Gauss–Newton algorithm does not require the computation of a Hessian matrix.

Observe that in the Gauss–Newton approach we determine $\boldsymbol{d}_t$ by viewing the search direction as coefficients in a linear regression with feature matrix $\mathbf{G}_t$ and response $-\boldsymbol{e}_t$. This suggests that instead of using a linear regression, we can compute $\boldsymbol{d}_t$ via a *ridge regression* with a suitable choice for the regularization parameter γ:

☞ 217

$$\boldsymbol{d}_t = -(\mathbf{G}_t^\top \mathbf{G}_t + n\gamma \mathbf{I}_p)^{-1} \mathbf{G}_t^\top \boldsymbol{e}_t.$$

If we replace $n\mathbf{I}_p$ with the diagonal matrix $\mathrm{diag}(\mathbf{G}_t^\top \mathbf{G}_t)$, we then obtain the *Levenberg–Marquardt* search direction:

LEVENBERG–MARQUARDT

$$\boldsymbol{d}_t = -(\mathbf{G}_t^\top \mathbf{G}_t + \gamma\, \mathrm{diag}(\mathbf{G}_t^\top \mathbf{G}_t))^{-1} \mathbf{G}_t^\top \boldsymbol{e}_t. \tag{B.28}$$

Recall that the ridge regularization parameter γ has the following effect on the least-squares solution: When it is zero, then the solution $\boldsymbol{d}_t$ coincides with the search direction of the Gauss–Newton method, and when γ tends to infinity, then $\|\boldsymbol{d}_t\|$ tends to zero. Thus, γ controls both the magnitude and direction of vector $\boldsymbol{d}_t$. A simple version of the Levenberg–Marquardt algorithm is the following.

Algorithm B.3.3: Levenberg–Marquardt for Minimizing $\frac{1}{n}\|\boldsymbol{g}(\tau \,|\, \boldsymbol{\beta}) - \boldsymbol{y}\|^2$

input: An initial guess $\boldsymbol{\beta}_0$; stopping error $\varepsilon > 0$; training set τ.
output: An approximate minimizer of $\frac{1}{n}\|\boldsymbol{g}(\tau \,|\, \boldsymbol{\beta}) - \boldsymbol{y}\|^2$.

$t \leftarrow 0$ and $\gamma \leftarrow 0.01$ (or another default value)
while stopping condition is not met **do**
  Compute the search direction $\boldsymbol{d}_t$ via (B.28).
  $\boldsymbol{e}_{t+1} \leftarrow \boldsymbol{g}(\tau \,|\, \boldsymbol{\beta}_t + \boldsymbol{d}_t) - \boldsymbol{y}$
  if $\|\boldsymbol{e}_{t+1}\| < \|\boldsymbol{e}_t\|$ **then**
    $\gamma \leftarrow \gamma/10, \quad \boldsymbol{e}_{t+1} \leftarrow \boldsymbol{e}_t, \quad \boldsymbol{\beta}_{t+1} \leftarrow \boldsymbol{\beta}_t + \boldsymbol{d}_t$
  else
    $\gamma \leftarrow \gamma \times 10$
  $t \leftarrow t + 1$
return $\boldsymbol{\beta}_t$

B.4 Constrained Minimization via Penalty Functions

A constrained optimization problem of the form (B.13) can sometimes be reformulated as a simpler unconstrained problem — for example, the unconstrained set $\mathcal{Y}$ can be transformed to the feasible region $\mathcal{X}$ of the constrained problem via a function $\boldsymbol{\phi} : \mathbb{R}^n \to \mathbb{R}^n$ such that $\mathcal{X} = \boldsymbol{\phi}(\mathcal{Y})$. Then, (B.13) is equivalent to the minimization problem

$$\min_{\boldsymbol{y} \in \mathcal{Y}} f(\boldsymbol{\phi}(\boldsymbol{y})),$$

in the sense that a solution $\boldsymbol{x}^*$ of the original problem is obtained from a transformed solution $\boldsymbol{y}^*$ via $\boldsymbol{x}^* = \boldsymbol{\phi}(\boldsymbol{y}^*)$. Table B.2 lists some examples of possible transformations.

Table B.2: Some transformations to eliminate constraints.

Constrained	Unconstrained
$x > 0$	$\exp(y)$
$x \geqslant 0$	y^2
$a \leqslant x \leqslant b$	$a + (b-a)\sin^2(y)$

Unfortunately, an unconstrained minimization method used in combination with these transformations is rarely effective. Instead, it is more common to use penalty functions.

PENALTY FUNCTIONS

The overarching idea of *penalty functions* is to transform a constrained problem into an unconstrained problem by adding weighted constraint-violation terms to the original objective function, with the premise that the new problem has a solution that is identical or close to the original one.

For example, if there are only equality constraints, then

$$\widetilde{f}(\boldsymbol{x}) := f(\boldsymbol{x}) + \sum_{i=1}^{m} a_i\, |h_i(\boldsymbol{x})|^p$$

for some constants $a_1, \ldots, a_m > 0$ and integer $p \in \{1, 2\}$, gives an *exact penalty function*, in the sense that the minimizer of the penalized function $\widetilde{f}$ is equal to the minimizer of f subject to the m equality constraints $h_1, \ldots, h_m$. With the addition of inequality constraints, one could use

$$\widetilde{f}(\boldsymbol{x}) = f(\boldsymbol{x}) + \sum_{i=1}^{m} a_i\, |h_i(\boldsymbol{x})|^p + \sum_{j=1}^{k} b_j \max\{g_j(\boldsymbol{x}), 0\}$$

for some constants $a_1, \ldots, a_m, b_1, \ldots, b_k > 0$.

☞ 408

ALTERNATING DIRECTION METHOD OF MULTIPLIERS

■ **Example B.11 (Alternating Direction Method of Multipliers)** The Lagrange method is designed to handle convex minimization subject to equality constraints. Nevertheless, some practical algorithms may still use the penalty function approach in combination with the Lagrangian method. An example is the *alternating direction method of multipliers* (ADMM) [17]. The ADMM solves problems of the form:

$$\begin{aligned} \min_{\boldsymbol{x}\in\mathbb{R}^n, \boldsymbol{z}\in\mathbb{R}^m} \quad & f(\boldsymbol{x}) + g(\boldsymbol{z}) \\ \text{subject to:} \quad & \mathbf{A}\boldsymbol{x} + \mathbf{B}\boldsymbol{z} = \boldsymbol{c}, \end{aligned} \tag{B.29}$$

where $\mathbf{A} \in \mathbb{R}^{p\times n}$, $\mathbf{B} \in \mathbb{R}^{p\times m}$, and $\boldsymbol{c} \in \mathbb{R}^p$, and $f : \mathbb{R}^n \to \mathbb{R}$ and $g : \mathbb{R}^m \to \mathbb{R}$ are convex functions. The approach is to form an augmented Lagrangian

$$\mathcal{L}_\varrho(\boldsymbol{x}, \boldsymbol{z}, \boldsymbol{\beta}) := f(\boldsymbol{x}) + g(\boldsymbol{z}) + \boldsymbol{\beta}^\top (\mathbf{A}\boldsymbol{x} + \mathbf{B}\boldsymbol{z} - \boldsymbol{c}) + \frac{\varrho}{2}\, \|\mathbf{A}\boldsymbol{x} + \mathbf{B}\boldsymbol{z} - \boldsymbol{c}\|^2,$$

where $\varrho > 0$ is a penalty parameter, and $\boldsymbol{\beta} \in \mathbb{R}^p$ are dual variables. The ADMM then iterates through updates of the following form:

$$\begin{aligned} \boldsymbol{x}^{(t+1)} &= \underset{\boldsymbol{x}\in\mathbb{R}^n}{\operatorname{argmin}}\, \mathcal{L}_\varrho(\boldsymbol{x}, \boldsymbol{z}^{(t)}, \boldsymbol{\beta}^{(t)}) \\ \boldsymbol{z}^{(t+1)} &= \underset{\boldsymbol{z}\in\mathbb{R}^m}{\operatorname{argmin}}\, \mathcal{L}_\varrho(\boldsymbol{x}^{(t+1)}, \boldsymbol{z}, \boldsymbol{\beta}^{(t)}) \\ \boldsymbol{\beta}^{(t+1)} &= \boldsymbol{\beta}^{(t)} + \varrho\left(\mathbf{A}\boldsymbol{x}^{(t+1)} + \mathbf{B}\boldsymbol{z}^{(t+1)} - \boldsymbol{c}\right). \end{aligned}$$

■

Suppose that (B.13) has inequality constraints only. *Barrier functions* are an important example of penalty functions that can handle inequality constraints. The prototypical example is a *logarithmic barrier function* which gives the unconstrained optimization:

BARRIER FUNCTIONS

$$\widetilde{f}(\boldsymbol{x}) = f(\boldsymbol{x}) - \nu \sum_{j=1}^{k} \ln(-g_j(\boldsymbol{x})), \quad \nu > 0,$$

such that the minimizer of $\widetilde{f}$ tends to the minimizer of f as $\nu \to 0$. Direct minimization of $\widetilde{f}$ via an unconstrained minimization algorithm is frequently too difficult. Instead, it is common to combine the logarithmic barrier function with the Lagrangian method as follows.

The idea is to introduce k nonnegative auxiliary or *slack variables* $s_1, \ldots, s_k$ that satisfy the equalities $g_j(\boldsymbol{x}) + s_j = 0$ for all j. These equalities ensure that the inequality constraints are maintained: $g_j(\boldsymbol{x}) = -s_j \leqslant 0$ for all j. Then, instead of the unconstrained optimization of $\widetilde{f}$, we consider the unconstrained optimization of the Lagrangian:

SLACK VARIABLES

$$\mathcal{L}(\boldsymbol{x}, \boldsymbol{s}, \boldsymbol{\beta}) = f(\boldsymbol{x}) - \nu \sum_{j=1}^{k} \ln s_j + \sum_{j=1}^{k} \beta_j (g_j(\boldsymbol{x}) + s_j), \tag{B.30}$$

where $\nu > 0$ and $\boldsymbol{\beta}$ are the Lagrange multipliers for the equalities $g_j(\boldsymbol{x}) + s_j = 0$, $j = 1, \ldots, k$.

Observe how the logarithmic barrier function keeps the slack variables positive. In addition, while the optimization of $\widetilde{f}$ is over n dimensions (recall that $\boldsymbol{x} \in \mathbb{R}^n$), the optimization of the Lagrangian function $\mathcal{L}$ is over $n + 2k$ dimensions. Despite this enlargement of the search space with the variables $\boldsymbol{s}$ and $\boldsymbol{\beta}$, the optimization of the Lagrangian $\mathcal{L}$ is easier in practice than the direct optimization of $\widetilde{f}$.

■ **Example B.12 (Interior-Point Method for Nonnegativity)** One of the simplest and most common constrained optimization problems can be formulated as the minimization of $f(\boldsymbol{x})$ subject to nonnegative $\boldsymbol{x}$, that is: $\min_{\boldsymbol{x} \geqslant \boldsymbol{0}} f(\boldsymbol{x})$. In this case, the Lagrangian with logarithmic barrier (B.30) is:

$$\mathcal{L}(\boldsymbol{x}, \boldsymbol{s}, \boldsymbol{\beta}) = f(\boldsymbol{x}) - \nu \sum_{k} \ln s_k + \boldsymbol{\beta}^\top (\boldsymbol{s} - \boldsymbol{x}).$$

The KKT conditions in Theorem B.2 are a necessary condition for a minimizer, and yield the nonlinear system for $[\boldsymbol{x}^\top, \boldsymbol{s}^\top, \boldsymbol{\beta}^\top]^\top \in \mathbb{R}^{3n}$:

$$\begin{bmatrix} \nabla f(\boldsymbol{x}) - \boldsymbol{\beta} \\ -\nu/\boldsymbol{s} + \boldsymbol{\beta} \\ \boldsymbol{s} - \boldsymbol{x} \end{bmatrix} = \boldsymbol{0},$$

where $\nu/\boldsymbol{s}$ is a shorthand notation for a column vector with components $\{\nu/s_j\}$. To solve this system, we can use Newton's method for root finding (see, for example, Algorithm B.3.1), which requires a formula for the matrix of Jacobi of $\mathcal{L}$. Here, this $(3n) \times (3n)$ matrix is:

$$\mathbf{J}_{\mathcal{L}}(\boldsymbol{x}, \boldsymbol{s}, \boldsymbol{\beta}) = \begin{bmatrix} \mathbf{H} & \mathbf{O} & -\mathbf{I} \\ \mathbf{O} & \mathbf{D} & \mathbf{I} \\ -\mathbf{I} & \mathbf{I} & \mathbf{O} \end{bmatrix} = \begin{bmatrix} \mathbf{H} & \mathbf{B} \\ \mathbf{B}^\top & \mathbf{E} \end{bmatrix},$$

where $\mathbf{H}$ is the $n \times n$ Hessian of f at $\boldsymbol{x}$; $\mathbf{D} := \mathrm{diag}\,(\nu/(\boldsymbol{s} \odot \boldsymbol{s}))$ is an $n \times n$ diagonal matrix; $\mathbf{B} := [\mathbf{O}, -\mathbf{I}]$ is an $n \times (2n)$ matrix, and[1]

$$\mathbf{E} := \begin{bmatrix} \mathbf{D} & \mathbf{I} \\ \mathbf{I} & \mathbf{O} \end{bmatrix} = \begin{bmatrix} \mathbf{O} & \mathbf{I} \\ \mathbf{I} & -\mathbf{D} \end{bmatrix}^{-1}.$$

Further, we define

$$\mathbf{H}_\nu := (\mathbf{H} - \mathbf{B}\mathbf{E}^{-1}\mathbf{B}^\top)^{-1} = (\mathbf{H} + \mathbf{D})^{-1}.$$

☞ 373 Using this notation and applying the matrix blockwise inversion formula (A.14), we obtain the inverse of the matrix of Jacobi:

$$\begin{bmatrix} \mathbf{H} & \mathbf{B} \\ \mathbf{B}^\top & \mathbf{E} \end{bmatrix}^{-1} = \begin{bmatrix} \mathbf{H}_\nu & -\mathbf{H}_\nu\mathbf{B}\mathbf{E}^{-1} \\ -\mathbf{E}^{-1}\mathbf{B}^\top\mathbf{H}_\nu & \mathbf{E}^{-1} + \mathbf{E}^{-1}\mathbf{B}^\top\mathbf{H}_\nu\mathbf{B}\mathbf{E}^{-1} \end{bmatrix} = \begin{bmatrix} \mathbf{H}_\nu & \mathbf{H}_\nu & -\mathbf{H}_\nu\mathbf{D} \\ \mathbf{H}_\nu & \mathbf{H}_\nu & \mathbf{I} - \mathbf{H}_\nu\mathbf{D} \\ -\mathbf{D}\mathbf{H}_\nu & \mathbf{I} - \mathbf{D}\mathbf{H}_\nu & \mathbf{D}\mathbf{H}_\nu\mathbf{D} - \mathbf{D} \end{bmatrix}.$$

Therefore, the search direction in Newton's root-finding method is given by:

$$-\mathbf{J}_{\mathcal{L}}^{-1} \begin{bmatrix} \nabla f(\boldsymbol{x}) - \boldsymbol{\beta} \\ -\nu/\boldsymbol{s} + \boldsymbol{\beta} \\ \boldsymbol{s} - \boldsymbol{x} \end{bmatrix} = \begin{bmatrix} \mathrm{d}\boldsymbol{x} \\ \mathrm{d}\boldsymbol{x} + \boldsymbol{x} - \boldsymbol{s} \\ \nu/\boldsymbol{s} - \boldsymbol{\beta} - \mathbf{D}(\mathrm{d}\boldsymbol{x} + \boldsymbol{x} - \boldsymbol{s}) \end{bmatrix},$$

where

$$\mathrm{d}\boldsymbol{x} := -(\mathbf{H} + \mathbf{D})^{-1}\left[\nabla f(\boldsymbol{x}) - 2\nu/\boldsymbol{s} + \mathbf{D}\boldsymbol{x}\right],$$

and we have assumed that $\mathbf{H} + \mathbf{D}$ is a positive-definite matrix. If at any step of the iteration the matrix $\mathbf{H} + \mathbf{D}$ fails to be positive-definite, then Newton's root-finding algorithm may fail to converge. Thus, any practical implementation will have to include a fail-safe feature to guard against this possibility.

In summary, for a given penalty parameter $\nu > 0$, we can locate the approximate nonnegative minimizer of f using, for example, the version of the Newton–Raphson root-finding method given in Algorithm B.4.1.

In practice, one needs to choose a sufficiently small value for ν, so that the output $\boldsymbol{x}_\nu$ of Algorithm B.4.1 is a good approximation to $\boldsymbol{x}^* = \mathrm{argmin}_{\boldsymbol{x} \geqslant \mathbf{0}} f(\boldsymbol{x})$. Alternatively, one can create a decreasing sequence of penalty parameters $\nu_1 > \nu_2 > \cdots$ and compute the corresponding solutions $\boldsymbol{x}_{\nu_1}, \boldsymbol{x}_{\nu_2}, \ldots$ of the penalized problems. In the so-called *interior-point method*, a given $\boldsymbol{x}_{\nu_i}$ is used as an initial guess for computing $\boldsymbol{x}_{\nu_{i+1}}$ and so on until the approximation to the minimizer $\boldsymbol{x}^* = \mathrm{argmin}_{\boldsymbol{x} \geqslant \mathbf{0}} f(\boldsymbol{x})$ is deemed accurate.

INTERIOR-POINT METHOD

[1]Here $\mathbf{O}$ is an $n \times n$ matrix of zeros and $\mathbf{I}$ is the $n \times n$ identity matrix.

Algorithm B.4.1: Approximating $\mathbf{x}^* = \operatorname{argmin}_{\mathbf{x} \geqslant \mathbf{0}} f(\mathbf{x})$ with Logarithmic Barrier

input: An initial guess $\mathbf{x}$ and stopping error $\varepsilon > 0$.
output: The approximate nonnegative minimizer $\mathbf{x}_\nu$ of f.

$\mathbf{s} \leftarrow \mathbf{x}, \quad \boldsymbol{\beta} \leftarrow \nu/\mathbf{s}, \quad \mathrm{d}\mathbf{x} \leftarrow \boldsymbol{\beta}$
while $\|\mathrm{d}\mathbf{x}\| > \varepsilon$ and budget is not exhausted **do**
  Compute the gradient $\mathbf{u}$ and the Hessian $\mathbf{H}$ of f at $\mathbf{x}$.
  $\mathbf{s}_1 \leftarrow \nu/\mathbf{s}, \quad \mathbf{s}_2 \leftarrow \mathbf{s}_1/\mathbf{s}, \quad \mathbf{w} \leftarrow 2\mathbf{s}_1 - \mathbf{u} - \mathbf{s}_2 \odot \mathbf{x}$
  if $(\mathbf{H} + \mathrm{diag}(\mathbf{s}_2)) \succ \mathbf{0}$ **then** `// if Cholesky successful`
    Compute the Cholesky factor $\mathbf{L}$ satisfying $\mathbf{L}\mathbf{L}^\top = \mathbf{H} + \mathrm{diag}(\mathbf{s}_2)$.
    $\mathrm{d}\mathbf{x} \leftarrow \mathbf{L}^{-1}\mathbf{w}$ (computed by forward substitution)
    $\mathrm{d}\mathbf{x} \leftarrow \mathbf{L}^{-\top}\mathrm{d}\mathbf{x}$ (computed by backward substitution)
  else
    $\mathrm{d}\mathbf{x} \leftarrow \mathbf{w}/\mathbf{s}_2$ `// if Cholesky fails, do steepest descent`
  $\mathrm{d}\mathbf{s} \leftarrow \mathrm{d}\mathbf{x} + \mathbf{x} - \mathbf{s}, \quad \mathrm{d}\boldsymbol{\beta} \leftarrow \mathbf{s}_1 - \boldsymbol{\beta} - \mathbf{s}_2 \odot \mathrm{d}\mathbf{s}, \quad \alpha \leftarrow 1$
  while $\min_j\{s_j + \alpha\, \mathrm{d}s_j\} < 0$ **do**
    $\alpha \leftarrow \alpha/2$ `// ensure nonnegative slack variables`
  $\mathbf{x} \leftarrow \mathbf{x} + \alpha\, \mathrm{d}\mathbf{x}, \quad \mathbf{s} \leftarrow \mathbf{s} + \alpha\, \mathrm{d}\mathbf{s}, \quad \boldsymbol{\beta} \leftarrow \boldsymbol{\beta} + \alpha\, \mathrm{d}\boldsymbol{\beta}$
return $\mathbf{x}_\nu \leftarrow \mathbf{x}$

■

Further Reading

For an excellent introduction to convex optimization and Lagrangian duality see [18]. A classical text on optimization algorithms and, in particular, on quasi-Newton methods is [43]. For more details on the *alternating direction method of multipliers* see [17].

APPENDIX C

PROBABILITY AND STATISTICS

The purpose of this chapter is to establish the baseline probability and statistics background for this book. We review basic concepts such as the sum and product rules of probability, random variables and their probability distributions, expectations, independence, conditional probability, transformation rules, limit theorems, and Markov chains. The properties of the multivariate normal distribution are discussed in more detail. The main ideas from statistics are also reviewed, including estimation techniques (such as maximum likelihood estimation), confidence intervals, and hypothesis testing.

C.1 Random Experiments and Probability Spaces

The basic notion in probability theory is that of a *random experiment*: an experiment whose outcome cannot be determined in advance. Mathematically, a random experiment is modeled via a triplet $(\Omega, \mathcal{H}, \mathbb{P})$, where: RANDOM EXPERIMENT

- Ω is the set of all possible outcomes of the experiment, called the *sample space*. SAMPLE SPACE
- $\mathcal{H}$ is the collection of all subsets of Ω to which a probability can be assigned; such subsets are called *events*. EVENTS
- $\mathbb{P}$ is a *probability measure*, which assigns to each event A a number $\mathbb{P}[A]$ between 0 and 1, indicating the likelihood that the outcome of the random experiment lies in A. PROBABILITY MEASURE

Any probability measure $\mathbb{P}$ must satisfy the following *Kolmogorov axioms*: KOLMOGOROV AXIOMS

1. $\mathbb{P}[A] \geqslant 0$ for every event A.
2. $\mathbb{P}[\Omega] = 1$.
3. For any sequence $A_1, A_2, \dots$ of events,

$$\mathbb{P}\Big[\bigcup_i A_i\Big] \leqslant \sum_i \mathbb{P}[A_i], \tag{C.1}$$

with strict *equality* whenever the events are *disjoint* (that is, non-overlapping).

SUM RULE

When (C.1) holds as an equality, it is often referred to as the *sum rule* of probability. It simply states that if an event can happen in a number of different but not simultaneous ways, the probability of that event is the sum of the probabilities of the comprising events. If the events are allowed to overlap, then the inequality (C.1) is called the *union bound.*

UNION BOUND

In many applications the sample space is *countable*; that is, $\Omega = \{a_1, a_2, \ldots\}$. In this case the easiest way to specify a probability measure $\mathbb{P}$ is to first assign a number p_i to each *elementary event* $\{a_i\}$, with $\sum_i p_i = 1$, and then to define

ELEMENTARY EVENT

$$\mathbb{P}[A] = \sum_{i: a_i \in A} p_i \quad \text{for all } A \subseteq \Omega.$$

Here the collection of events $\mathcal{H}$ can be taken to be equal to the collection of *all* subsets of Ω. The triple $(\Omega, \mathcal{H}, \mathbb{P})$ is called a *discrete probability space*. This idea is graphically represented in Figure C.1. Each element a_i, represented by a dot, is assigned a weight (that is, probability) p_i, indicated by the size of the dot. The probability of the event A is simply the sum of the weights of all the outcomes in A.

DISCRETE PROBABILITY SPACE

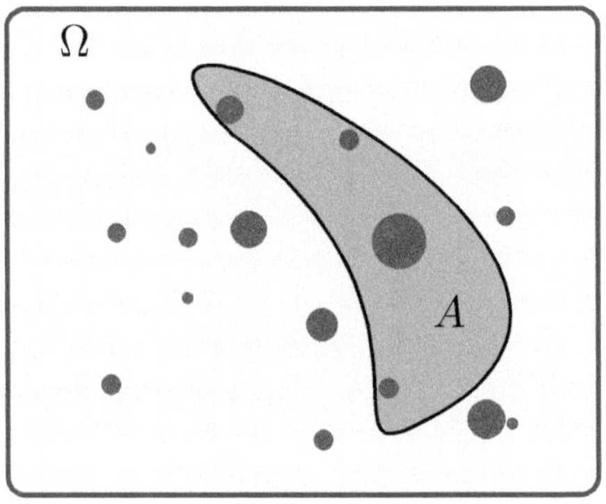

Figure C.1: A discrete probability space.

■ **Remark C.1 (Equilikely Principle)** A special case of a discrete probability space occurs when a random experiment has finitely many outcomes that are all *equally likely*. In this case the probability measure is given by

$$\mathbb{P}[A] = \frac{|A|}{|\Omega|}, \tag{C.2}$$

where $|A|$ denotes the number of outcomes in A and $|\Omega|$ is the total number of outcomes. Thus, the calculation of probabilities reduces to counting numbers of outcomes in events. This is called the *equilikely principle*. ■

EQUILIKELY PRINCIPLE

C.2 Random Variables and Probability Distributions

It is often convenient to describe a random experiment via "random variables", representing numerical measurements of the experiment. Random variables are usually denoted by capital letters from the last part of the alphabet. From a mathematical point of view, a *random variable* X is a function from Ω to $\mathbb{R}$ such that sets of the form $\{a < X \leqslant b\} := \{\omega \in \Omega : a < X(\omega) \leqslant b\}$ are events (and so can be assigned a probability).

RANDOM VARIABLE

All probabilities involving a random variable X can be computed, in principle, from its *cumulative distribution function* (cdf), defined by

CUMULATIVE DISTRIBUTION FUNCTION

$$F(x) = \mathbb{P}[X \leqslant x], \quad x \in \mathbb{R}.$$

For example $\mathbb{P}[a < X \leqslant b] = \mathbb{P}[X \leqslant b] - \mathbb{P}[X \leqslant a] = F(b) - F(a)$. Figure C.2 shows a generic cdf. Note that any cdf is right-continuous, increasing, and lies between 0 and 1.

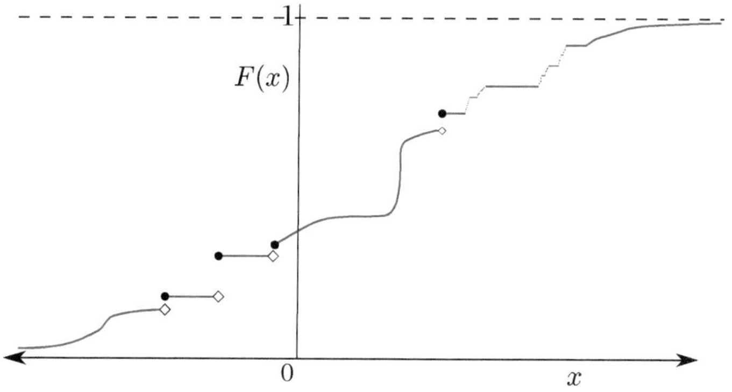

Figure C.2: A cumulative distribution function (cdf).

A cdf F_d is called *discrete* if there exist numbers $x_1, x_2, \ldots$ and probabilities $0 < f(x_i) \leqslant 1$ summing up to 1, such that for all x

DISCRETE CDF

$$F_d(x) = \sum_{x_i \leqslant x} f(x_i). \tag{C.3}$$

Such a cdf is piecewise constant and has jumps of sizes $f(x_1), f(x_2), \ldots$ at points $x_1, x_2, \ldots$, respectively. The function $f(x)$ is called a *probability mass function* or *discrete probability density function* (pdf). It is often easier to use the pdf rather than the cdf, since probabilities can simply be calculated from it via summation:

DISCRETE PDF

$$\mathbb{P}[X \in B] = \sum_{x \in B} f(x),$$

as illustrated in Figure C.3.

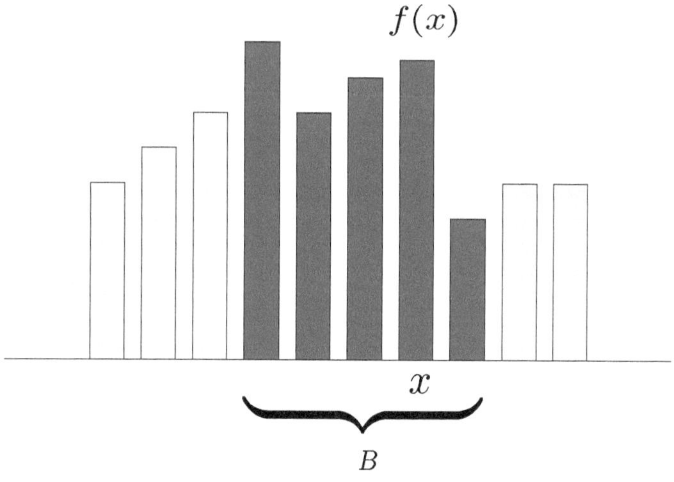

Figure C.3: Discrete probability density function (pdf). The darker area corresponds to the probability $\mathbb{P}[X \in B]$.

CONTINUOUS CDF

A cdf F_c is called *continuous*[1], if there exists a positive function f such that for all x

$$F_c(x) = \int_{-\infty}^{x} f(u)\,\mathrm{d}u. \tag{C.4}$$

PDF

Note that such an F_c is differentiable (and hence continuous) with derivative f. The function f is called the *probability density function (continuous pdf)*. By the fundamental theorem of integration, we have

$$\mathbb{P}[a < X \leqslant b] = F(b) - F(a) = \int_{a}^{b} f(x)\,\mathrm{d}x.$$

Thus, calculating probabilities reduces to integration, as illustrated in Figure C.4.

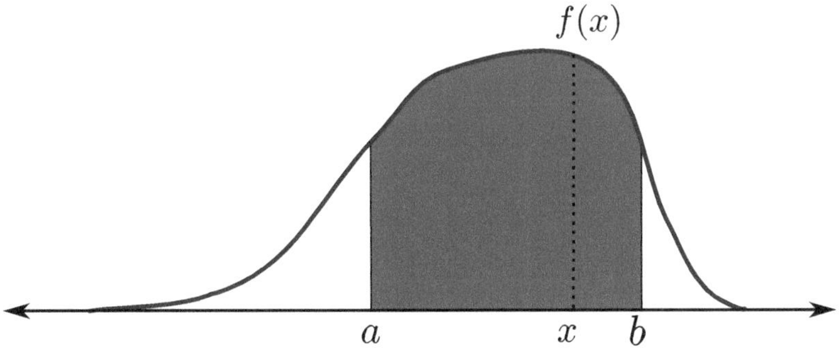

Figure C.4: Continuous probability density function (pdf). The shaded area corresponds to the probability $\mathbb{P}[X \in B]$, with B being here the interval $(a, b]$.

■ **Remark C.2 (Probability Density and Probability Mass)** It is important to note that we deliberately use the *same* name, "pdf", and symbol, f, in both the discrete and the continuous case, rather than distinguish between a probability mass function (pmf) and probability density function (pdf). From a theoretical point of view the pdf plays exactly the same role in the discrete and continuous cases. We use the notation $X \sim$ Dist, $X \sim f$, and $X \sim F$ to indicate that X has distribution Dist, pdf f, and cdf F. ■

Tables C.1 and C.2 list a number of important continuous and discrete distributions. Note that in Table C.1, Γ is the gamma function: $\Gamma(\alpha) = \int_0^\infty \mathrm{e}^{-x} x^{\alpha-1}\,\mathrm{d}x, \quad \alpha > 0$.

[1]In advanced probability, we would say "*absolutely continuous with respect to the Lebesgue measure*".

Table C.1: Commonly used continuous distributions.

Name	Notation	$f(x)$	$x \in$	Parameters
Uniform	$\mathcal{U}[\alpha,\beta]$	$\dfrac{1}{\beta-\alpha}$	$[\alpha,\beta]$	$\alpha<\beta$
Normal	$\mathcal{N}(\mu,\sigma^2)$	$\dfrac{1}{\sigma\sqrt{2\pi}}\,\mathrm{e}^{-\frac{1}{2}\left(\frac{x-\mu}{\sigma}\right)^2}$	$\mathbb{R}$	$\sigma>0,\ \mu\in\mathbb{R}$
Gamma	$\mathsf{Gamma}(\alpha,\lambda)$	$\dfrac{\lambda^\alpha x^{\alpha-1}\mathrm{e}^{-\lambda x}}{\Gamma(\alpha)}$	$\mathbb{R}_+$	$\alpha,\lambda>0$
Inverse Gamma	$\mathsf{InvGamma}(\alpha,\lambda)$	$\dfrac{\lambda^\alpha x^{-\alpha-1}\mathrm{e}^{-\lambda x^{-1}}}{\Gamma(\alpha)}$	$\mathbb{R}_+$	$\alpha,\lambda>0$
Exponential	$\mathsf{Exp}(\lambda)$	$\lambda\,\mathrm{e}^{-\lambda x}$	$\mathbb{R}_+$	$\lambda>0$
Beta	$\mathsf{Beta}(\alpha,\beta)$	$\dfrac{\Gamma(\alpha+\beta)}{\Gamma(\alpha)\Gamma(\beta)}\,x^{\alpha-1}(1-x)^{\beta-1}$	$[0,1]$	$\alpha,\beta>0$
Weibull	$\mathsf{Weib}(\alpha,\lambda)$	$\alpha\lambda\,(\lambda x)^{\alpha-1}\mathrm{e}^{-(\lambda x)^\alpha}$	$\mathbb{R}_+$	$\alpha,\lambda>0$
Pareto	$\mathsf{Pareto}(\alpha,\lambda)$	$\alpha\lambda\,(1+\lambda x)^{-(\alpha+1)}$	$\mathbb{R}_+$	$\alpha,\lambda>0$
Student	t_ν	$\dfrac{\Gamma(\frac{\nu+1}{2})}{\sqrt{\nu\pi}\,\Gamma(\frac{\nu}{2})}\left(1+\dfrac{x^2}{\nu}\right)^{-(\nu+1)/2}$	$\mathbb{R}$	$\nu>0$
F	$\mathsf{F}(m,n)$	$\dfrac{\Gamma(\frac{m+n}{2})\,(m/n)^{m/2}x^{(m-2)/2}}{\Gamma(\frac{m}{2})\,\Gamma(\frac{n}{2})\,[1+(m/n)x]^{(m+n)/2}}$	$\mathbb{R}_+$	$m,n\in\mathbb{N}_+$

The $\mathsf{Gamma}(n/2, 1/2)$ distribution is called the *chi-squared distribution* with n degrees of freedom, denoted χ^2_n. The t_1 distribution is also called the *Cauchy* distribution. χ^2_n DISTRIBUTION

Table C.2: Commonly used discrete distributions.

Name	Notation	$f(x)$	$x \in$	Parameters
Bernoulli	$\mathsf{Ber}(p)$	$p^x(1-p)^{1-x}$	$\{0,1\}$	$0\leqslant p\leqslant 1$
Binomial	$\mathsf{Bin}(n,p)$	$\dbinom{n}{x}p^x(1-p)^{n-x}$	$\{0,1,\ldots,n\}$	$0\leqslant p\leqslant 1,$ $n\in\mathbb{N}$
Discrete uniform	$\mathcal{U}\{1,\ldots,n\}$	$\dfrac{1}{n}$	$\{1,\ldots,n\}$	$n\in\{1,2,\ldots\}$
Geometric	$\mathsf{Geom}(p)$	$p(1-p)^{x-1}$	$\{1,2,\ldots\}$	$0\leqslant p\leqslant 1$
Poisson	$\mathsf{Poi}(\lambda)$	$\mathrm{e}^{-\lambda}\dfrac{\lambda^x}{x!}$	$\mathbb{N}$	$\lambda>0$

C.3 Expectation

It is often useful to consider different kinds of numerical characteristics of a random variable. One such quantity is the expectation, which measures the "average" value of the distribution.

EXPECTATION

The *expectation* (or expected value or mean) of a random variable X with pdf f, denoted by $\mathbb{E}X$ or[2] $\mathbb{E}[X]$ (and sometimes μ), is defined by

$$\mathbb{E}X = \begin{cases} \sum_x x\, f(x) & \text{discrete case,} \\ \int_{-\infty}^{\infty} x\, f(x)\, \mathrm{d}x & \text{continuous case.} \end{cases}$$

If X is a random variable, then a function of X, such as X^2 or $\sin(X)$, is again a random variable. Moreover, the expected value of a function of X is simply a weighted average of the possible values that this function can take. That is, for any real function h

$$\mathbb{E}h(X) = \begin{cases} \sum_x h(x)\, f(x) & \text{discrete case,} \\ \int_{-\infty}^{\infty} h(x)\, f(x)\, \mathrm{d}x & \text{continuous case,} \end{cases}$$

provided that the sum or integral are well-defined.

VARIANCE

The *variance* of a random variable X, denoted by $\mathbb{V}\mathrm{ar}\, X$ (and sometimes σ^2), is defined by

$$\mathbb{V}\mathrm{ar}\, X = \mathbb{E}(X - \mathbb{E}[X])^2 = \mathbb{E}X^2 - (\mathbb{E}X)^2.$$

STANDARD DEVIATION

The square root of the variance is called the *standard deviation*. Table C.3 lists the expectations and variances for some well-known distributions. Both variance and standard deviation measure the spread or dispersion of the distribution. Note, however, that the standard deviation measures the dispersion in the same units as the random variable, unlike the variance, which uses squared units.

Table C.3: Expectations and variances for some well-known distributions.

Dist.	$\mathbb{E}X$	$\mathbb{V}\mathrm{ar}\, X$	Dist.	$\mathbb{E}X$	$\mathbb{V}\mathrm{ar}\, X$
$\mathsf{Bin}(n, p)$	np	$np(1-p)$	$\mathsf{Gamma}(\alpha, \lambda)$	$\dfrac{\alpha}{\lambda}$	$\dfrac{\alpha}{\lambda^2}$
$\mathsf{Geom}(p)$	$\dfrac{1}{p}$	$\dfrac{1-p}{p^2}$	$\mathcal{N}(\mu, \sigma^2)$	μ	σ^2
$\mathsf{Poi}(\lambda)$	λ	λ	$\mathsf{Beta}(\alpha, \beta)$	$\frac{\alpha}{\alpha+\beta}$	$\frac{\alpha\beta}{(\alpha+\beta)^2(1+\alpha+\beta)}$
$\mathcal{U}[\alpha, \beta]$	$\dfrac{\alpha+\beta}{2}$	$\dfrac{(\beta-\alpha)^2}{12}$	$\mathsf{Weib}(\alpha, \lambda)$	$\frac{\Gamma(1/\alpha)}{\alpha\,\lambda}$	$\frac{2\Gamma(2/\alpha)}{\alpha} - \left(\frac{\Gamma(1/\alpha)}{\alpha\,\lambda}\right)^2$
$\mathsf{Exp}(\lambda)$	$\dfrac{1}{\lambda}$	$\dfrac{1}{\lambda^2}$	$\mathsf{F}(m, n)$	$\frac{n}{n-2}\ (n > 2)$	$\frac{2n^2(m+n-2)}{m(n-2)^2(n-4)}\ (n > 4)$
t_ν	$0\ (\nu > 1)$	$\frac{\nu}{\nu-2}\ (\nu > 2)$			

[2]We only use brackets in an expectation if it is unclear with respect to which random variable the expectation is taken.

It is sometimes useful to consider the *moment generating function* of a random variable X. This is the function M defined by

MOMENT GENERATING FUNCTION

$$M(s) = \mathbb{E}\,e^{sX}, \quad s \in \mathbb{R}. \tag{C.5}$$

The moment generating functions of two random variables coincide if and only if the random variables have the same distribution; see also Theorem C.12.

■ **Example C.1 (Moment Generation Function of the Gamma(α, λ) Distribution)** Let $X \sim \mathsf{Gamma}(\alpha, \lambda)$. For $s < \lambda$, the moment generating function of X at s is given by

$$\begin{aligned} M(s) = \mathbb{E}e^{sX} &= \int_0^\infty e^{sx} \frac{e^{-\lambda x} \lambda^\alpha x^{\alpha-1}}{\Gamma(\alpha)} \,dx \\ &= \left(\frac{\lambda}{\lambda - s}\right)^\alpha \int_0^\infty \underbrace{\frac{e^{-(\lambda-s)x} (\lambda - s)^\alpha x^{\alpha-1}}{\Gamma(\alpha)}}_{\text{pdf of Gamma}(\alpha,\lambda-s)} \,dx = \left(\frac{\lambda}{\lambda - s}\right)^\alpha. \end{aligned}$$

For $s \geqslant \lambda$, $M(s) = \infty$. Interestingly, the moment generating function has a much simpler formula than the pdf. ■

C.4 Joint Distributions

Distributions for random vectors and stochastic processes can be specified in much the same way as for random variables. In particular, the distribution of a random vector $\boldsymbol{X} = [X_1, \ldots, X_n]^\top$ is completely determined by specifying the *joint cdf* F, defined by

JOINT CDF

$$F(x_1, \ldots, x_n) = \mathbb{P}[X_1 \leqslant x_1, \ldots, X_n \leqslant x_n], \quad x_i \in \mathbb{R},\ i = 1, \ldots, n.$$

Similarly, the distribution of a *stochastic process*, that is, a collection of random variables $\{X_t, t \in \mathscr{T}\}$, for some index set $\mathscr{T}$, is completely determined by its finite-dimensional distributions; specifically, the distributions of the random vectors $[X_{t_1}, \ldots, X_{t_n}]^\top$ for every choice of n and $t_1, \ldots, t_n$.

STOCHASTIC PROCESS

By analogy to the one-dimensional case, a random vector $\boldsymbol{X} = [X_1, \ldots, X_n]^\top$ taking values in $\mathbb{R}^n$ is said to have a pdf f if, in the continuous case,

$$\mathbb{P}[\boldsymbol{X} \in B] = \int_B f(\boldsymbol{x}) \,d\boldsymbol{x}, \tag{C.6}$$

for all n-dimensional rectangles B. Replace the integral with a sum for the discrete case. The pdf is also called the *joint pdf* of $X_1, \ldots, X_n$. The pdfs of the individual components — called *marginal pdfs* — can be recovered from the joint pdf by "integrating out the other variables". For example, for a continuous random vector $[X, Y]^\top$ with pdf f, the pdf f_X of X is given by

JOINT PDF

MARGINAL PDF

$$f_X(x) = \int f(x, y) \,dy.$$

C.5 Conditioning and Independence

Conditional probabilities and conditional distributions are used to model additional information on a random experiment. Independence is used to model *lack* of such information.

C.5.1 Conditional Probability

Suppose some event $B \subseteq \Omega$ occurs. Given this fact, event A will occur if and only if $A \cap B$ occurs, and the relative chance of A occurring is therefore $\mathbb{P}[A \cap B]/\mathbb{P}[B]$, provided $\mathbb{P}[B] > 0$. This leads to the definition of the *conditional probability* of A given B:

CONDITIONAL PROBABILITY

$$\mathbb{P}[A \mid B] = \frac{\mathbb{P}[A \cap B]}{\mathbb{P}[B]}, \quad \text{if } \mathbb{P}[B] > 0. \tag{C.7}$$

The above definition breaks down if $\mathbb{P}[B] = 0$. Such conditional probabilities must be treated with more care [11].

Three important consequences of the definition of conditional probability are:

PRODUCT RULE

1. *Product rule*: For any sequence of events $A_1, A_2, \ldots, A_n$,

$$\mathbb{P}[A_1 \cdots A_n] = \mathbb{P}[A_1]\,\mathbb{P}[A_2 \mid A_1]\,\mathbb{P}[A_3 \mid A_1 A_2] \cdots \mathbb{P}[A_n \mid A_1 \cdots A_{n-1}], \tag{C.8}$$

   using the abbreviation $A_1 A_2 \cdots A_k := A_1 \cap A_2 \cap \cdots \cap A_k$.

LAW OF TOTAL PROBABILITY

2. *Law of total probability*: If $\{B_i\}$ forms a *partition* of Ω (that is, $B_i \cap B_j = \emptyset, i \neq j$ and $\cup_i B_i = \Omega$), then for any event A

$$\mathbb{P}[A] = \sum_i \mathbb{P}[A \mid B_i]\,\mathbb{P}[B_i]. \tag{C.9}$$

BAYES' RULE

3. *Bayes' rule*: Let $\{B_i\}$ form a partition of Ω. Then, for any event A with $\mathbb{P}[A] > 0$,

$$\mathbb{P}[B_j \mid A] = \frac{\mathbb{P}[A \mid B_j]\,\mathbb{P}[B_j]}{\sum_i \mathbb{P}[A \mid B_i]\,\mathbb{P}[B_i]}. \tag{C.10}$$

C.5.2 Independence

INDEPENDENT EVENTS

Two events A and B are said to be *independent* if the knowledge that B has occurred does not change the probability that A occurs. That is, A, B independent $\Leftrightarrow \mathbb{P}[A \mid B] = \mathbb{P}[A]$. Since $\mathbb{P}[A \mid B]\,\mathbb{P}[B] = \mathbb{P}[A \cap B]$, an alternative definition of independence is

$$A, B \text{ independent} \Leftrightarrow \mathbb{P}[A \cap B] = \mathbb{P}[A]\,\mathbb{P}[B].$$

This definition covers the case where $\mathbb{P}[B] = 0$ and can be extended to arbitrarily many events: events $A_1, A_2, \ldots$ are said to be (mutually) independent if for any k and any choice of distinct indices $i_1, \ldots, i_k$,

$$\mathbb{P}[A_{i_1} \cap A_{i_2} \cap \cdots \cap A_{i_k}] = \mathbb{P}[A_{i_1}]\,\mathbb{P}[A_{i_2}] \cdots \mathbb{P}[A_{i_k}].$$

The concept of independence can also be formulated for random variables. Random variables $X_1, X_2, \ldots$ are said to be *independent* if the events $\{X_{i_1} \leqslant x_{i_1}\}, \ldots, \{X_{i_n} \leqslant x_{i_n}\}$ are independent for all finite choices of n distinct indices $i_1, \ldots, i_n$ and values $x_{i_1}, \ldots, x_{i_n}$. INDEPENDENT RANDOM VARIABLES

An important characterization of independent random variables is the following (for a proof, see [101], for example).

Theorem C.1: Independence Characterization

Random variables $X_1, \ldots, X_n$ with marginal pdfs $f_{X_1}, \ldots, f_{X_n}$ and joint pdf f are independent if and only if

$$f(x_1, \ldots, x_n) = f_{X_1}(x_1) \cdots f_{X_n}(x_n) \quad \text{for all } x_1, \ldots, x_n. \tag{C.11}$$

Many probabilistic models involve random variables $X_1, X_2, \ldots$ that are *independent and identically distributed*, abbreviated as *iid*. We use this abbreviation throughout this book. IID

C.5.3 Expectation and Covariance

Similar to the univariate case, the expected value of a real-valued function h of a random vector $\boldsymbol{X} \sim f$ is a weighted average of all values that $h(\boldsymbol{X})$ can take. Specifically, in the continuous case, $\mathbb{E}h(\boldsymbol{X}) = \int h(\boldsymbol{x})f(\boldsymbol{x})\,\mathrm{d}\boldsymbol{x}$. In the discrete case replace this multidimensional integral with a sum. Using this result, it is not difficult to show that for any collection of dependent or independent random variables $X_1, \ldots, X_n$,

$$\mathbb{E}[a + b_1X_1 + b_2X_2 + \cdots + b_nX_n] = a + b_1\mathbb{E}X_1 + \cdots + b_n\mathbb{E}X_n \tag{C.12}$$

for all constants $a, b_1, \ldots, b_n$. Moreover, for *independent* random variables,

$$\mathbb{E}[X_1X_2 \cdots X_n] = \mathbb{E}X_1\,\mathbb{E}X_2 \cdots \mathbb{E}X_n. \tag{C.13}$$

We leave the proofs as an exercise.

The *covariance* of two random variables X and Y with expectations μ_X and μ_Y, respectively, is defined as COVARIANCE

$$\mathbb{C}\mathrm{ov}(X, Y) = \mathbb{E}[(X - \mu_X)(Y - \mu_Y)].$$

This is a measure of the amount of linear dependency between the variables. Let $\sigma_X^2 = \mathbb{V}\mathrm{ar}\,X$ and $\sigma_Y^2 = \mathbb{V}\mathrm{ar}\,Y$. A scaled version of the covariance is given by the *correlation coefficient*, CORRELATION COEFFICIENT

$$\varrho(X, Y) = \frac{\mathbb{C}\mathrm{ov}(X, Y)}{\sigma_X\,\sigma_Y}.$$

The following properties follow directly from the definitions of variance and covariance.

1. $\mathbb{V}\mathrm{ar}\,X = \mathbb{E}X^2 - \mu_X^2$.
2. $\mathbb{V}\mathrm{ar}[aX + b] = a^2\,\sigma_X^2$.
3. $\mathbb{C}\mathrm{ov}(X, Y) = \mathbb{E}[XY] - \mu_X\,\mu_Y$.
4. $\mathbb{C}\mathrm{ov}(X, Y) = \mathbb{C}\mathrm{ov}(Y, X)$.

5. $-\sigma_X\sigma_Y \leqslant \mathbb{C}\text{ov}(X, Y) \leqslant \sigma_X\sigma_Y$.
6. $\mathbb{C}\text{ov}(aX + bY, Z) = a\,\mathbb{C}\text{ov}(X, Z) + b\,\mathbb{C}\text{ov}(Y, Z)$.
7. $\mathbb{C}\text{ov}(X, X) = \sigma_X^2$.
8. $\mathbb{V}\text{ar}[X + Y] = \sigma_X^2 + \sigma_Y^2 + 2\,\mathbb{C}\text{ov}(X, Y)$.
9. If X and Y are independent, then $\mathbb{C}\text{ov}(X, Y) = 0$.

As a consequence of Properties 2 and 8 we have that for any sequence of *independent* random variables $X_1, \ldots, X_n$ with variances $\sigma_1^2, \ldots, \sigma_n^2$,

$$\mathbb{V}\text{ar}[a_1X_1 + a_2X_2 + \cdots + a_nX_n] = a_1^2\,\sigma_1^2 + a_2^2\,\sigma_2^2 + \cdots + a_n^2\,\sigma_n^2, \qquad \text{(C.14)}$$

for any choice of constants $a_1, \ldots, a_n$.

For random column vectors, such as $\boldsymbol{X} = [X_1, \ldots, X_n]^\top$, it is convenient to write the expectations and covariances in vector and matrix notation. For a random vector $\boldsymbol{X}$ we define its *expectation vector* as the vector of expectations

EXPECTATION VECTOR

$$\boldsymbol{\mu} = [\mu_1, \ldots, \mu_n]^\top = [\mathbb{E}X_1, \ldots, \mathbb{E}X_n]^\top.$$

Similarly, if the expectation of a matrix is the matrix of expectations, then given two random vectors $\boldsymbol{X} \in \mathbb{R}^n$ and $\boldsymbol{Y} \in \mathbb{R}^m$, the $n \times m$ matrix

$$\mathbb{C}\text{ov}(\boldsymbol{X}, \boldsymbol{Y}) = \mathbb{E}[(\boldsymbol{X} - \mathbb{E}\boldsymbol{X})(\boldsymbol{Y} - \mathbb{E}\boldsymbol{Y})^\top] \qquad \text{(C.15)}$$

has (i, j)-th element $\mathbb{C}\text{ov}(X_i, Y_j) = \mathbb{E}[(X_i - \mathbb{E}X_i)(Y_j - \mathbb{E}Y_j)]$. A consequence of this definition is that

$$\mathbb{C}\text{ov}(\mathbf{A}\boldsymbol{X}, \mathbf{B}\boldsymbol{Y}) = \mathbf{A}\mathbb{C}\text{ov}(\boldsymbol{X}, \boldsymbol{Y})\mathbf{B}^\top,$$

where $\mathbf{A}$ and $\mathbf{B}$ are two matrices with n and m columns, respectively.

COVARIANCE MATRIX

The *covariance matrix* of the vector $\boldsymbol{X}$ is defined as the $n \times n$ matrix $\mathbb{C}\text{ov}(\boldsymbol{X}, \boldsymbol{X})$. The covariance matrix is also denoted as $\mathbb{V}\text{ar}(\boldsymbol{X}) = \mathbb{C}\text{ov}(\boldsymbol{X}, \boldsymbol{X})$, in analogy with the scalar identity $\mathbb{V}\text{ar}(X) = \mathbb{C}\text{ov}(X, X)$.

A useful application of the cyclic property of the trace of a matrix (see Theorem A.1) is the following.

☞ 359

Theorem C.2: Expectation of a Quadratic Form

Let $\mathbf{A}$ be an $n \times n$ matrix and $\boldsymbol{X}$ an n-dimensional random vector with expectation vector $\boldsymbol{\mu}$ and covariance matrix $\boldsymbol{\Sigma}$. The random variable $Y := \boldsymbol{X}^\top\mathbf{A}\boldsymbol{X}$ has expectation $\text{tr}(\mathbf{A}\boldsymbol{\Sigma}) + \boldsymbol{\mu}^\top\mathbf{A}\boldsymbol{\mu}$.

Proof: Since Y is a scalar, it is equal to its trace. Now, using the cyclic property: $\mathbb{E}Y = \mathbb{E}\,\text{tr}(Y) = \mathbb{E}\,\text{tr}(\boldsymbol{X}^\top\mathbf{A}\boldsymbol{X}) = \mathbb{E}\,\text{tr}(\mathbf{A}\boldsymbol{X}\boldsymbol{X}^\top) = \text{tr}(\mathbf{A}\,\mathbb{E}[\boldsymbol{X}\boldsymbol{X}^\top]) = \text{tr}(\mathbf{A}(\boldsymbol{\Sigma} + \boldsymbol{\mu}\boldsymbol{\mu}^\top)) = \text{tr}(\mathbf{A}\boldsymbol{\Sigma}) + \text{tr}(\mathbf{A}\boldsymbol{\mu}\boldsymbol{\mu}^\top) = \text{tr}(\mathbf{A}\boldsymbol{\Sigma}) + \boldsymbol{\mu}^\top\mathbf{A}\boldsymbol{\mu}$. □

C.5.4 Conditional Density and Conditional Expectation

Suppose X and Y are both discrete or both continuous, with joint pdf f, and suppose $f_X(x) > 0$. Then, the *conditional pdf* of Y given $X = x$ is given by

CONDITIONAL PDF

$$f_{Y|X}(y \,|\, x) = \frac{f(x, y)}{f_X(x)} \quad \text{for all } y. \tag{C.16}$$

In the discrete case, the formula is a direct translation of (C.7), with $f_{Y|X}(y \,|\, x) = \mathbb{P}[Y = y \,|\, X = x]$. In the continuous case, a similar interpretation in terms of densities can be used; see, for example, [101, Page 221]. The corresponding distribution is called the *conditional distribution* of Y given $X = x$. Note that (C.16) implies that

CONDITIONAL DISTRIBUTION

$$f(x, y) = f_X(x)\, f_{Y\,|\,X}(y \,|\, x).$$

This is useful when the marginal and conditional pdfs are given, rather than the joint one. More generally, for the n-dimensional case we have

$$f(x_1, \ldots, x_n) = f_{X_1}(x_1)\, f_{X_2\,|\,X_1}(x_2 \,|\, x_1) \cdots f_{X_n\,|\,X_1,\ldots,X_{n-1}}(x_n \,|\, x_1, \ldots, x_{n-1}), \tag{C.17}$$

which is in essence a rephrasing of the product rule (C.8) in terms of probability densities. ☞ 430

As a conditional pdf has all the properties of an ordinary pdf, we may define expectations with respect to it. The *conditional expectation* of a random variable Y given $X = x$ is defined as

CONDITIONAL EXPECTATION

$$\mathbb{E}[Y \,|\, X = x] = \begin{cases} \sum_y y\, f_{Y|X}(y \,|\, x) & \text{discrete case,} \\ \int y\, f_{Y|X}(y \,|\, x)\, \mathrm{d}y & \text{continuous case.} \end{cases} \tag{C.18}$$

Note that $\mathbb{E}[Y \,|\, X = x]$ is a function of x. The corresponding random variable is written as $\mathbb{E}[Y \,|\, X]$. A similar formalism can be used when conditioning on a sequence of random variables $X_1, \ldots, X_n$. The conditional expectation has similar properties to the ordinary expectation. Other useful properties (see, for example, [127]) are:

1. *Tower property*: If $\mathbb{E}Y$ exists, then

$$\mathbb{E}\,\mathbb{E}[Y \,|\, X] = \mathbb{E}Y. \tag{C.19}$$

2. *Taking out what is known*: If $\mathbb{E}Y$ exists, then

$$\mathbb{E}\,[XY \,|\, X] = X\mathbb{E}[Y \,|\, X].$$

C.6 Functions of Random Variables

Let $\boldsymbol{x} = [x_1, \ldots, x_n]^\top$ be a column vector in $\mathbb{R}^n$ and $\mathbf{A}$ an $m \times n$ matrix. The mapping $\boldsymbol{x} \mapsto \boldsymbol{z}$, with $\boldsymbol{z} = \mathbf{A}\boldsymbol{x}$, is a linear transformation, as discussed in Section A.1. Now consider a random vector $\boldsymbol{X} = [X_1, \ldots, X_n]^\top$ and let $\boldsymbol{Z} := \mathbf{A}\boldsymbol{X}$. Then $\boldsymbol{Z}$ is a random vector in $\mathbb{R}^m$. The following theorem details how the distribution of $\boldsymbol{Z}$ is related to that of $\boldsymbol{X}$. ☞ 357

Theorem C.3: Linear Transformation

If $\boldsymbol{X}$ has an expectation vector $\boldsymbol{\mu}_X$ and covariance matrix $\boldsymbol{\Sigma}_X$, then the expectation vector of $\boldsymbol{Z}$ is

$$\boldsymbol{\mu}_Z = \mathbf{A}\,\boldsymbol{\mu}_X \tag{C.20}$$

and the covariance matrix of $\boldsymbol{Z}$ is

$$\boldsymbol{\Sigma}_Z = \mathbf{A}\,\boldsymbol{\Sigma}_X\,\mathbf{A}^\top. \tag{C.21}$$

If, in addition, $\mathbf{A}$ is an invertible $n \times n$ matrix and $\boldsymbol{X}$ is a continuous random vector with pdf $f_{\boldsymbol{X}}$, then the pdf of the continuous random vector $\boldsymbol{Z} = \mathbf{A}\boldsymbol{X}$ is given by

$$f_{\boldsymbol{Z}}(\boldsymbol{z}) = \frac{f_{\boldsymbol{X}}(\mathbf{A}^{-1}\boldsymbol{z})}{|\det(\mathbf{A})|}, \quad \boldsymbol{z} \in \mathbb{R}^n, \tag{C.22}$$

where $|\det(\mathbf{A})|$ denotes the absolute value of the determinant of $\mathbf{A}$.

Proof: We have $\boldsymbol{\mu}_Z = \mathbb{E}\boldsymbol{Z} = \mathbb{E}[\mathbf{A}\boldsymbol{X}] = \mathbf{A}\,\mathbb{E}\boldsymbol{X} = \mathbf{A}\boldsymbol{\mu}_X$ and

$$\begin{aligned}\boldsymbol{\Sigma}_Z &= \mathbb{E}[(\boldsymbol{Z}-\boldsymbol{\mu}_Z)(\boldsymbol{Z}-\boldsymbol{\mu}_Z)^\top] = \mathbb{E}[\mathbf{A}(\boldsymbol{X}-\boldsymbol{\mu}_X)(\mathbf{A}(\boldsymbol{X}-\boldsymbol{\mu}_X))^\top]\\ &= \mathbf{A}\,\mathbb{E}[(\boldsymbol{X}-\boldsymbol{\mu}_X)(\boldsymbol{X}-\boldsymbol{\mu}_X)^\top]\mathbf{A}^\top\\ &= \mathbf{A}\,\boldsymbol{\Sigma}_X\,\mathbf{A}^\top.\end{aligned}$$

For $\mathbf{A}$ invertible and $\boldsymbol{X}$ continuous (as opposed to discrete), let $\boldsymbol{z} = \mathbf{A}\boldsymbol{x}$ and $\boldsymbol{x} = \mathbf{A}^{-1}\boldsymbol{z}$. Consider the n-dimensional cube $C = [z_1, z_1 + h] \times \cdots \times [z_n, z_n + h]$. Then,

$$\mathbb{P}[\boldsymbol{Z} \in C] \approx h^n\, f_{\boldsymbol{Z}}(\boldsymbol{z}),$$

by definition of the joint density of $\boldsymbol{Z}$. Let D be the image of C under $\mathbf{A}^{-1}$ — that is, all
☞ 357 points $\boldsymbol{x}$ such that $\mathbf{A}\boldsymbol{x} \in C$. Recall from Section A.1 that any matrix B linearly transforms an n-dimensional rectangle with volume V into an n-dimensional parallelepiped with volume $V\,|\det(B)|$. Thus, in addition to the above expression for $\mathbb{P}[\boldsymbol{Z} \in C]$, we also have

$$\mathbb{P}[\boldsymbol{Z} \in C] = \mathbb{P}[\boldsymbol{X} \in D] \approx h^n |\det(\mathbf{A}^{-1})|\, f_{\boldsymbol{X}}(\boldsymbol{x}) = h^n |\det(\mathbf{A})|^{-1}\, f_{\boldsymbol{X}}(\boldsymbol{x}).$$

Equating these two expressions for $\mathbb{P}[\boldsymbol{Z} \in C]$, dividing both sides by h^n, and letting h go to 0, we obtain (C.22). □

For a generalization of the linear transformation rule (C.22), consider an arbitrary mapping $\boldsymbol{x} \mapsto \boldsymbol{g}(\boldsymbol{x})$, written out:

$$\begin{bmatrix} x_1 \\ x_2 \\ \vdots \\ x_n \end{bmatrix} \mapsto \begin{bmatrix} g_1(\boldsymbol{x}) \\ g_2(\boldsymbol{x}) \\ \vdots \\ g_n(\boldsymbol{x}) \end{bmatrix}.$$

Theorem C.4: Transformation Rule

Let $\boldsymbol{X}$ be an n-dimensional vector of continuous random variables with pdf $f_{\boldsymbol{X}}$. Let $\boldsymbol{Z} = \boldsymbol{g}(\boldsymbol{X})$, where $\boldsymbol{g}$ is an invertible mapping with inverse $\boldsymbol{g}^{-1}$ and *matrix of Jacobi* $\mathbf{J}_{\boldsymbol{g}}$; that is, the matrix of partial derivatives of $\boldsymbol{g}$. Then, at $\boldsymbol{z} = \boldsymbol{g}(\boldsymbol{x})$ the random vector $\boldsymbol{Z}$ has pdf

$$f_{\boldsymbol{Z}}(\boldsymbol{z}) = \frac{f_{\boldsymbol{X}}(\boldsymbol{x})}{|\det(\mathbf{J}_{\boldsymbol{g}}(\boldsymbol{x}))|} = f_{\boldsymbol{X}}(\boldsymbol{g}^{-1}(\boldsymbol{z}))\,|\det(\mathbf{J}_{\boldsymbol{g}^{-1}}(\boldsymbol{z}))|, \quad \boldsymbol{z} \in \mathbb{R}^n. \tag{C.23}$$

Proof: For a fixed $\boldsymbol{x}$, let $\boldsymbol{z} = \boldsymbol{g}(\boldsymbol{x})$; and thus $\boldsymbol{x} = \boldsymbol{g}^{-1}(\boldsymbol{z})$. In the neighborhood of $\boldsymbol{x}$, the function $\boldsymbol{g}$ behaves like a linear function, in the sense that $\boldsymbol{g}(\boldsymbol{x} + \boldsymbol{\delta}) \approx \boldsymbol{g}(\boldsymbol{x}) + \mathbf{J}_{\boldsymbol{g}}(\boldsymbol{x})\,\boldsymbol{\delta}$ for small vectors $\boldsymbol{\delta}$; see also Section B.1. Consequently, an infinitesimally small n-dimensional rectangle at $\boldsymbol{x}$ with volume V is transformed into an infinitesimally small n-dimensional parallelepiped at $\boldsymbol{z}$ with volume $V\,|\det(\mathbf{J}_{\boldsymbol{g}}(\boldsymbol{x}))|$. Now, as in the proof of the linear case, let C be a small cube around $\boldsymbol{z} = \boldsymbol{g}(\boldsymbol{x})$ with volume h^n. Let D be the image of C under $\boldsymbol{g}^{-1}$. Then,

☞ 399

$$h^n f_{\boldsymbol{Z}}(\boldsymbol{z}) \approx \mathbb{P}[\boldsymbol{Z} \in C] \approx h^n |\det(\mathbf{J}_{\boldsymbol{g}^{-1}}(\boldsymbol{z}))|\, f_{\boldsymbol{X}}(\boldsymbol{x}),$$

and since $|\det(\mathbf{J}_{\boldsymbol{g}^{-1}}(\boldsymbol{z}))| = 1/|\det(\mathbf{J}_{\boldsymbol{g}}(\boldsymbol{x}))|$, (C.23) follows as h goes to 0. □

Typically, in coordinate transformations it is $\boldsymbol{g}^{-1}$ that is given — that is, an expression for $\boldsymbol{x}$ as a function of $\boldsymbol{z}$.

■ **Example C.2 (Polar Transform)** Suppose X, Y are independent and have standard normal distribution. The joint pdf is

$$f_{X,Y}(x, y) = \frac{1}{2\pi} \mathrm{e}^{-\frac{1}{2}(x^2+y^2)}, \quad (x, y) \in \mathbb{R}^2.$$

In polar coordinates we have

$$X = R\cos\Theta \qquad \text{and} \qquad Y = R\sin\Theta, \tag{C.24}$$

where $R \geqslant 0$ is the radius and $\Theta \in [0, 2\pi)$ the angle of the point (X, Y). What is the joint pdf of R and Θ? By the radial symmetry of the bivariate normal distribution, we would expect Θ to be uniform on $(0, 2\pi)$. But what is the pdf of R? To work out the joint pdf, consider the inverse transformation $\mathbf{g}^{-1}$, defined by

$$\begin{bmatrix} r \\ \theta \end{bmatrix} \overset{\mathbf{g}^{-1}}{\longmapsto} \begin{bmatrix} r\cos\theta \\ r\sin\theta \end{bmatrix} = \begin{bmatrix} x \\ y \end{bmatrix}.$$

The corresponding matrix of Jacobi is

$$\mathbf{J}_{\mathbf{g}^{-1}}(r, \theta) = \begin{bmatrix} \cos\theta & -r\sin\theta \\ \sin\theta & r\cos\theta \end{bmatrix},$$

which has determinant r. Since $x^2 + y^2 = r^2(\cos^2\theta + \sin^2\theta) = r^2$, it follows by the transformation rule (C.23) that the joint pdf of R and Θ is given by

$$f_{R,\Theta}(r,\theta) = f_{X,Y}(x,y)\, r = \frac{1}{2\pi}\mathrm{e}^{-\frac{1}{2}r^2}\, r, \quad \theta \in (0, 2\pi), \quad r \geqslant 0.$$

By integrating out θ and r, respectively, we find $f_R(r) = r\,\mathrm{e}^{-r^2/2}$ and $f_\Theta(\theta) = 1/(2\pi)$. Since $f_{R,\Theta}$ is the product of f_R and f_Θ, the random variables R and Θ are independent. ■

C.7 Multivariate Normal Distribution

The normal (or Gaussian) distribution — especially its multidimensional version — plays a central role in data science and machine learning. Recall from Table C.1 that a random variable X is said to have a *normal* distribution with parameters μ and σ^2 if its pdf is given by

NORMAL DISTRIBUTION

$$f(x) = \frac{1}{\sigma\sqrt{2\pi}}\,\mathrm{e}^{-\frac{1}{2}\left(\frac{x-\mu}{\sigma}\right)^2}, \quad x \in \mathbb{R}. \tag{C.25}$$

We write $X \sim \mathcal{N}(\mu, \sigma^2)$. The parameters μ and σ^2 are the expectation and variance of the distribution, respectively. If $\mu = 0$ and $\sigma = 1$ then

$$f(x) = \frac{1}{\sqrt{2\pi}}\,\mathrm{e}^{-x^2/2},$$

and the distribution is known as the *standard normal* distribution. The cdf of the standard normal distribution is often denoted by Φ and its pdf by φ. In Figure C.5 the pdf of the $\mathcal{N}(\mu, \sigma^2)$ distribution for various μ and σ^2 is plotted.

STANDARD NORMAL

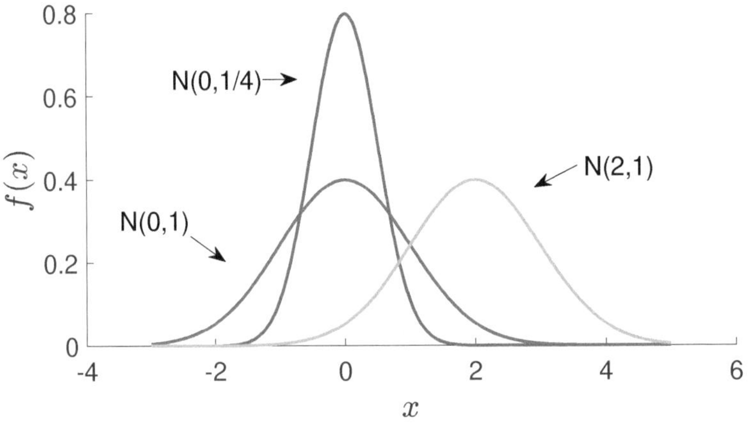

Figure C.5: The pdf of the $\mathcal{N}(\mu, \sigma^2)$ distribution for various μ and σ^2.

We next consider some important properties of the normal distribution.

Theorem C.5: Standardization

Let $X \sim \mathcal{N}(\mu, \sigma^2)$ and define $Z = (X - \mu)/\sigma$. Then Z has a standard normal distribution.

Proof: The cdf of Z is given by

$$\mathbb{P}[Z \leqslant z] = \mathbb{P}[(X-\mu)/\sigma \leqslant z] = \mathbb{P}[X \leqslant \mu + \sigma z]$$
$$= \int_{-\infty}^{\mu+\sigma z} \frac{1}{\sigma\sqrt{2\pi}} \mathrm{e}^{-\frac{1}{2}\left(\frac{x-\mu}{\sigma}\right)^2} \mathrm{d}x = \int_{-\infty}^{z} \frac{1}{\sqrt{2\pi}} \mathrm{e}^{-y^2/2} \mathrm{d}y = \Phi(z),$$

where we make a change of variable $y = (x-\mu)/\sigma$ in the fourth equation. Hence, $Z \sim \mathcal{N}(0,1)$. □

The rescaling procedure in Theorem C.5 is called *standardization*. It follows from Theorem C.5 that any $X \sim \mathcal{N}(\mu, \sigma^2)$ can be written as

STANDARDIZATION

$$X = \mu + \sigma Z, \quad \text{where } Z \sim \mathcal{N}(0,1).$$

In other words, any normal random variable can be viewed as an *affine transformation* — that is, a linear transformation plus a constant — of a standard normal random variable.

AFFINE TRANSFORMATION

We now generalize this to n dimensions. Let $Z_1, \ldots, Z_n$ be independent and standard normal random variables. The joint pdf of $\mathbf{Z} = [Z_1, \ldots, Z_n]^\top$ is given by

$$f_{\mathbf{Z}}(\mathbf{z}) = \prod_{i=1}^{n} \frac{1}{\sqrt{2\pi}} \mathrm{e}^{-\frac{1}{2}z_i^2} = (2\pi)^{-\frac{n}{2}} \mathrm{e}^{-\frac{1}{2}\mathbf{z}^\top \mathbf{z}}, \quad \mathbf{z} \in \mathbb{R}^n. \tag{C.26}$$

We write $\mathbf{Z} \sim \mathcal{N}(\mathbf{0}, \mathbf{I})$, where $\mathbf{I}$ is the identity matrix. Consider the affine transformation

$$\boldsymbol{X} = \boldsymbol{\mu} + \mathbf{B}\,\mathbf{Z} \tag{C.27}$$

for some $m \times n$ matrix $\mathbf{B}$ and m-dimensional vector $\boldsymbol{\mu}$. Note that, by (C.20) and (C.21), $\boldsymbol{X}$ has expectation vector $\boldsymbol{\mu}$ and covariance matrix $\boldsymbol{\Sigma} = \mathbf{B}\mathbf{B}^\top$. We say that $\boldsymbol{X}$ has a *multivariate normal* or *multivariate Gaussian* distribution with mean vector $\boldsymbol{\mu}$ and covariance matrix $\boldsymbol{\Sigma}$. We write $\boldsymbol{X} \sim \mathcal{N}(\boldsymbol{\mu}, \boldsymbol{\Sigma})$.

☞ 434

MULTIVARIATE NORMAL

The following theorem states that any affine combination of independent multivariate normal random variables is again multivariate normal.

Theorem C.6: Affine Transformation of Normal Random Vectors

Let $\boldsymbol{X}_1, \boldsymbol{X}_2, \ldots, \boldsymbol{X}_r$ be independent m_i-dimensional normal random vectors, with $\boldsymbol{X}_i \sim \mathcal{N}(\boldsymbol{\mu}_i, \boldsymbol{\Sigma}_i)$, $i = 1, \ldots, r$. Then, for any $n \times 1$ vector $\boldsymbol{a}$ and $n \times m_i$ matrices $\mathbf{B}_1, \ldots, \mathbf{B}_r$,

$$\boldsymbol{a} + \sum_{i=1}^{r} \mathbf{B}_i \boldsymbol{X}_i \sim \mathcal{N}\left(\boldsymbol{a} + \sum_{i=1}^{r} \mathbf{B}_i \boldsymbol{\mu}_i, \sum_{i=1}^{r} \mathbf{B}_i \boldsymbol{\Sigma}_i \mathbf{B}_i^\top\right). \tag{C.28}$$

Proof: Denote the n-dimensional random vector in the left-hand side of (C.28) by $\boldsymbol{Y}$. By definition, each $\boldsymbol{X}_i$ can be written as $\boldsymbol{\mu}_i + \mathbf{A}_i \boldsymbol{Z}_i$, where the $\{\boldsymbol{Z}_i\}$ are independent (because the $\{\boldsymbol{X}_i\}$ are independent), so that

$$\boldsymbol{Y} = \boldsymbol{a} + \sum_{i=1}^{r} \mathbf{B}_i (\boldsymbol{\mu}_i + \mathbf{A}_i \boldsymbol{Z}_i) = \boldsymbol{a} + \sum_{i=1}^{r} \mathbf{B}_i \boldsymbol{\mu}_i + \sum_{i=1}^{r} \mathbf{B}_i \mathbf{A}_i \boldsymbol{Z}_i,$$

which is an affine combination of independent standard normal random vectors. Hence, $\boldsymbol{Y}$ is multivariate normal. Its expectation vector and covariance matrix can be found easily from Theorem C.3. □

☞ 434

The next theorem shows that the distribution of a subvector of a multivariate normal random vector is again normal.

Theorem C.7: Marginal Distributions of Normal Random Vectors

Let $\boldsymbol{X} \sim \mathcal{N}(\boldsymbol{\mu}, \boldsymbol{\Sigma})$ be an n-dimensional normal random vector. Decompose $\boldsymbol{X}$, $\boldsymbol{\mu}$, and $\boldsymbol{\Sigma}$ as

$$\boldsymbol{X} = \begin{bmatrix} \boldsymbol{X}_p \\ \boldsymbol{X}_q \end{bmatrix}, \quad \boldsymbol{\mu} = \begin{bmatrix} \boldsymbol{\mu}_p \\ \boldsymbol{\mu}_q \end{bmatrix}, \quad \boldsymbol{\Sigma} = \begin{bmatrix} \boldsymbol{\Sigma}_p & \boldsymbol{\Sigma}_r \\ \boldsymbol{\Sigma}_r^\top & \boldsymbol{\Sigma}_q \end{bmatrix}, \tag{C.29}$$

where $\boldsymbol{\Sigma}_p$ is the upper left $p \times p$ corner of $\boldsymbol{\Sigma}$ and $\boldsymbol{\Sigma}_q$ is the lower right $q \times q$ corner of $\boldsymbol{\Sigma}$. Then, $\boldsymbol{X}_p \sim \mathcal{N}(\boldsymbol{\mu}_p, \boldsymbol{\Sigma}_p)$.

Proof: We give a proof assuming that $\boldsymbol{\Sigma}$ is positive definite. Let $\mathbf{B}\mathbf{B}^\top$ be the (lower) Cholesky decomposition of $\boldsymbol{\Sigma}$. We can write

☞ 375

$$\begin{bmatrix} \boldsymbol{X}_p \\ \boldsymbol{X}_q \end{bmatrix} = \begin{bmatrix} \boldsymbol{\mu}_p \\ \boldsymbol{\mu}_q \end{bmatrix} + \underbrace{\begin{bmatrix} \mathbf{B}_p & \mathbf{O} \\ \mathbf{C}_r & \mathbf{C}_q \end{bmatrix}}_{\mathbf{B}} \begin{bmatrix} \boldsymbol{Z}_p \\ \boldsymbol{Z}_q \end{bmatrix}, \tag{C.30}$$

where $\boldsymbol{Z}_p$ and $\boldsymbol{Z}_q$ are independent p- and q-dimensional standard normal random vectors. In particular, $\boldsymbol{X}_p = \boldsymbol{\mu}_p + \mathbf{B}_p\boldsymbol{Z}_p$, which means that $\boldsymbol{X}_p \sim \mathcal{N}(\boldsymbol{\mu}_p, \boldsymbol{\Sigma}_p)$, since $\mathbf{B}_p\mathbf{B}_p^\top = \boldsymbol{\Sigma}_p$. □

By relabeling the elements of $\boldsymbol{X}$ we see that Theorem C.7 implies that *any* subvector of $\boldsymbol{X}$ has a multivariate normal distribution. For example, $\boldsymbol{X}_q \sim \mathcal{N}(\boldsymbol{\mu}_q, \boldsymbol{\Sigma}_q)$.

The following theorem shows that not only the marginal distributions of a normal random vector are normal, but also its *conditional distributions*.

Theorem C.8: Conditional Distributions of Normal Random Vectors

Let $\boldsymbol{X} \sim \mathcal{N}(\boldsymbol{\mu}, \boldsymbol{\Sigma})$ be an n-dimensional normal random vector with $\det(\boldsymbol{\Sigma}) > 0$. If $\boldsymbol{X}$ is decomposed as in (C.29), then

$$\left(\boldsymbol{X}_q \,|\, \boldsymbol{X}_p = \boldsymbol{x}_p\right) \sim \mathcal{N}(\boldsymbol{\mu}_q + \boldsymbol{\Sigma}_r^\top \boldsymbol{\Sigma}_p^{-1}(\boldsymbol{x}_p - \boldsymbol{\mu}_p),\ \boldsymbol{\Sigma}_q - \boldsymbol{\Sigma}_r^\top \boldsymbol{\Sigma}_p^{-1}\boldsymbol{\Sigma}_r). \tag{C.31}$$

As a consequence, $\boldsymbol{X}_p$ and $\boldsymbol{X}_q$ are *independent* if and only if they are *uncorrelated*; that is, if $\boldsymbol{\Sigma}_r = \mathbf{O}$ (zero matrix).

Proof: From (C.30) we see that $\boldsymbol{X}_p = \boldsymbol{\mu}_p + \mathbf{B}_p\boldsymbol{Z}_p$ and $\boldsymbol{X}_q = \boldsymbol{\mu}_q + \mathbf{C}_r\boldsymbol{Z}_p + \mathbf{C}_q\boldsymbol{Z}_q$. Consequently,

$$(\boldsymbol{X}_q \,|\, \boldsymbol{X}_p = \boldsymbol{x}_p) = \boldsymbol{\mu}_q + \mathbf{C}_r\mathbf{B}_p^{-1}(\boldsymbol{x}_p - \boldsymbol{\mu}_p) + \mathbf{C}_q\boldsymbol{Z}_q,$$

where $\boldsymbol{Z}_q$ is a q-dimensional multivariate standard normal random vector. It follows that $\boldsymbol{X}_q$ conditional on $\boldsymbol{X}_p = \boldsymbol{x}_p$ has a $\mathcal{N}(\boldsymbol{\mu}_q + \mathbf{C}_r\mathbf{B}_p^{-1}(\boldsymbol{x}_p - \boldsymbol{\mu}_p),\ \mathbf{C}_q\mathbf{C}_q^\top)$ distribution. The proof of

(C.31) is completed by observing that $\mathbf{\Sigma}_r^\top \mathbf{\Sigma}_p^{-1} = \mathbf{C}_r \mathbf{B}_p^\top (\mathbf{B}_p^\top)^{-1} \mathbf{B}_p^{-1} = \mathbf{C}_r \mathbf{B}_p^{-1}$, and

$$\mathbf{\Sigma}_q - \mathbf{\Sigma}_r^\top \mathbf{\Sigma}_p^{-1} \mathbf{\Sigma}_r = \mathbf{C}_r \mathbf{C}_r^\top + \mathbf{C}_q \mathbf{C}_q^\top - \mathbf{C}_r \mathbf{B}_p^{-1} \underbrace{\mathbf{\Sigma}_r}_{\mathbf{B}_p \mathbf{C}_r^\top} = \mathbf{C}_q \mathbf{C}_q^\top.$$

If $\boldsymbol{X}_p$ and $\boldsymbol{X}_q$ are independent, then they are obviously uncorrelated, as $\mathbf{\Sigma}_r = \mathbb{E}[(\boldsymbol{X}_p - \boldsymbol{\mu}_p)(\boldsymbol{X}_q - \boldsymbol{\mu}_q)^\top] = \mathbb{E}(\boldsymbol{X}_p - \boldsymbol{\mu}_p)\,\mathbb{E}(\boldsymbol{X}_q - \boldsymbol{\mu}_q)^\top = \mathbf{O}$. Conversely, if $\mathbf{\Sigma}_r = \mathbf{O}$, then by (C.31) the conditional distribution of $\boldsymbol{X}_q$ given $\boldsymbol{X}_p$ is the same as the unconditional distribution of $\boldsymbol{X}_q$; that is, $\mathcal{N}(\boldsymbol{\mu}_q, \mathbf{\Sigma}_q)$. In other words, $\boldsymbol{X}_q$ is independent of $\boldsymbol{X}_p$. □

The next few results are about the relationships between the normal, chi-squared, Student, and F distributions, defined in Table C.1. Recall that the chi-squared family of distributions, denoted by χ^2_n, are simply $\mathsf{Gamma}(n/2, 1/2)$ distributions, where the parameter $n \in \{1, 2, 3, \ldots\}$ is called the *degrees of freedom*.

χ^2 DISTRIBUTION

Theorem C.9: Relationship Between Normal and χ^2 Distributions

If $\boldsymbol{X} \sim \mathcal{N}(\boldsymbol{\mu}, \mathbf{\Sigma})$ is an n-dimensional normal random vector with $\det(\mathbf{\Sigma}) > 0$, then

$$(\boldsymbol{X} - \boldsymbol{\mu})^\top \mathbf{\Sigma}^{-1} (\boldsymbol{X} - \boldsymbol{\mu}) \sim \chi^2_n. \tag{C.32}$$

Proof: Let $\mathbf{B}\mathbf{B}^\top$ be the Cholesky decomposition of $\mathbf{\Sigma}$, where $\mathbf{B}$ is invertible. Since $\boldsymbol{X}$ can be written as $\boldsymbol{\mu} + \mathbf{B}\boldsymbol{Z}$, where $\boldsymbol{Z} = [Z_1, \ldots, Z_n]^\top$ is a vector of independent standard normal random variables, we have

$$(\boldsymbol{X} - \boldsymbol{\mu})^\top \mathbf{\Sigma}^{-1} (\boldsymbol{X} - \boldsymbol{\mu}) = (\boldsymbol{X} - \boldsymbol{\mu})^\top (\mathbf{B}\mathbf{B}^\top)^{-1} (\boldsymbol{X} - \boldsymbol{\mu}) = \boldsymbol{Z}^\top \boldsymbol{Z} = \sum_{i=1}^n Z_i^2.$$

Using the independence of $Z_1, \ldots, Z_n$, the moment generating function of $Y = \sum_{i=1}^n Z_i^2$ is given by

☞ 429

$$\mathbb{E}\, \mathrm{e}^{sY} = \mathbb{E}\, \mathrm{e}^{s(Z_1^2 + \cdots + Z_n^2)} = \mathbb{E}\, [\mathrm{e}^{sZ_1^2} \cdots \mathrm{e}^{sZ_n^2}] = \left(\mathbb{E}\, \mathrm{e}^{sZ^2}\right)^n,$$

where $Z \sim \mathcal{N}(0, 1)$. The moment generating function of Z^2 is

$$\mathbb{E}\, \mathrm{e}^{sZ^2} = \int_{-\infty}^{\infty} \mathrm{e}^{sz^2} \frac{1}{\sqrt{2\pi}} \mathrm{e}^{-z^2/2} \mathrm{d}z = \frac{1}{\sqrt{2\pi}} \int_{-\infty}^{\infty} \mathrm{e}^{-\frac{1}{2}(1-2s)z^2} \mathrm{d}z = \frac{1}{\sqrt{1 - 2s}},$$

so that $\mathbb{E}\, \mathrm{e}^{sY} = \left(\frac{1}{2}/(\frac{1}{2} - s)\right)^{\frac{n}{2}}$, $s < \frac{1}{2}$, which is the moment generating function of the $\mathsf{Gamma}(n/2, 1/2)$ distribution; that is, the χ^2_n distribution — see Example C.1. The result now follows from the uniqueness of the moment generating function. □

☞ 429

A consequence of Theorem C.9 is that if $\boldsymbol{X} = [X_1, \ldots, X_n]^\top$ is n-dimensional standard normal, then the squared length $\|\boldsymbol{X}\|^2 = X_1^2 + \cdots + X_n^2$ has a χ^2_n distribution. If instead $X_i \sim \mathcal{N}(\mu_i, 1)$, $i = 1, \ldots$, then $\|\boldsymbol{X}\|^2$ is said to have a *noncentral* χ^2_n *distribution*. This distribution depends on the $\{\mu_i\}$ only through the norm $\|\boldsymbol{\mu}\|$. We write $\|\boldsymbol{X}\|^2 \sim \chi^2_n(\theta)$, where $\theta = \|\boldsymbol{\mu}\|$ is the *noncentrality parameter*.

NONCENTRAL χ^2_n DISTRIBUTION

NONCENTRALITY PARAMETER

Such distributions frequently occur when considering *projections* of multivariate normal random variables, as summarized in the following theorem.

Theorem C.10: Relationship Between Normal and Noncentral χ^2 Distributions

Let $\boldsymbol{X} \sim \mathcal{N}(\boldsymbol{\mu}, \mathbf{I}_n)$ be an n-dimensional normal random vector and let $\mathcal{V}_k \subset \mathcal{V}_m$ be linear subspaces of dimensions k and m, respectively, with $k < m \leqslant n$. Let $\boldsymbol{X}_k$ and $\boldsymbol{X}_m$ be orthogonal projections of $\boldsymbol{X}$ onto $\mathcal{V}_k$ and $\mathcal{V}_m$, and let $\boldsymbol{\mu}_k$ and $\boldsymbol{\mu}_m$ be the corresponding projections of $\boldsymbol{\mu}$. Then, the following holds.

1. The random vectors $\boldsymbol{X}_k$, $\boldsymbol{X}_m - \boldsymbol{X}_k$, and $\boldsymbol{X} - \boldsymbol{X}_m$ are independent.
2. $\|\boldsymbol{X}_k\|^2 \sim \chi^2_k(\|\boldsymbol{\mu}_k\|)$, $\|\boldsymbol{X}_m - \boldsymbol{X}_k\|^2 \sim \chi^2_{m-k}(\|\boldsymbol{\mu}_m - \boldsymbol{\mu}_k\|)$, and $\|\boldsymbol{X} - \boldsymbol{X}_m\|^2 \sim \chi^2_{n-m}(\|\boldsymbol{\mu} - \boldsymbol{\mu}_m\|)$.

Proof: Let $\boldsymbol{v}_1, \ldots, \boldsymbol{v}_n$ be an orthonormal basis of $\mathbb{R}^n$ such that $\boldsymbol{v}_1, \ldots, \boldsymbol{v}_k$ spans $\mathcal{V}_k$ and
☞ 364 $\boldsymbol{v}_1, \ldots, \boldsymbol{v}_m$ spans $\mathcal{V}_m$. By (A.8) we can write the orthogonal projection matrices onto $\mathcal{V}_j$, as $\mathbf{P}_j = \sum_{i=1}^{j} \boldsymbol{v}_i \boldsymbol{v}_i^\top$, $j = k, m, n$, where $\mathcal{V}_n$ is defined as $\mathbb{R}^n$. Note that $\mathbf{P}_n$ is simply the identity matrix. Let $\mathbf{V} := [\boldsymbol{v}_1, \ldots, \boldsymbol{v}_n]$ and define $\boldsymbol{Z} := [Z_1, \ldots, Z_n]^\top = \mathbf{V}^\top \boldsymbol{X}$. Recall from Sec-
☞ 363 tion A.2 that any orthogonal transformation such as $\boldsymbol{z} = \mathbf{V}^\top \boldsymbol{x}$ is *length preserving*; that is, $\|\boldsymbol{z}\| = \|\boldsymbol{x}\|$.

To prove the first statement of the theorem, note that $\mathbf{V}^\top \boldsymbol{X}_j = \mathbf{V}^\top \mathbf{P}_j \boldsymbol{X} = [Z_1, \ldots, Z_j, 0, \ldots, 0]^\top$, $j = k, m$. It follows that $\mathbf{V}^\top(\boldsymbol{X}_m - \boldsymbol{X}_k) = [0, \ldots, 0, Z_{k+1}, \ldots, Z_m, 0, \ldots, 0]^\top$ and $\mathbf{V}^\top(\boldsymbol{X} - \boldsymbol{X}_m) = [0, \ldots, 0, Z_{m+1}, \ldots, Z_n]^\top$. Moreover, being a linear transformation of a normal random vector, $\boldsymbol{Z}$ is also normal, with covariance matrix $\mathbf{V}^\top\mathbf{V} = \mathbf{I}_n$. In particular, the $\{Z_i\}$ are *independent*. This shows that $\boldsymbol{X}_k$, $\boldsymbol{X}_m - \boldsymbol{X}_k$ and $\boldsymbol{X} - \boldsymbol{X}_m$ are independent as well.

Next, observe that $\|\boldsymbol{X}_k\| = \|\mathbf{V}^\top \boldsymbol{X}_k\| = \|\boldsymbol{Z}_k\|$, where $\boldsymbol{Z}_k := [Z_1, \ldots, Z_k]^\top$. The latter vector has independent components with variances 1, and its squared norm has therefore (by definition) a $\chi^2_k(\theta)$ distribution. The noncentrality parameter is $\theta = \|\mathbb{E}\boldsymbol{Z}_k\| = \|\mathbb{E}\boldsymbol{X}_k\| = \|\boldsymbol{\mu}_k\|$, again by the length-preserving property of orthogonal transformations. This shows that $\|\boldsymbol{X}_k\|^2 \sim \chi^2_k(\|\boldsymbol{\mu}_k\|)$. The distributions of $\|\boldsymbol{X}_m - \boldsymbol{X}_k\|^2$ and $\|\boldsymbol{X} - \boldsymbol{X}_m\|^2$ follow by analogy. □

Theorem C.10 is frequently used in the statistical analysis of *normal linear models*; see
☞ 182 Section 5.4. In typical situations $\boldsymbol{\mu}$ lies in the subspace $\mathcal{V}_m$ or even $\mathcal{V}_k$ — in which case $\|\boldsymbol{X}_m - \boldsymbol{X}_k\|^2 \sim \chi^2_{m-k}$ and $\|\boldsymbol{X} - \boldsymbol{X}_m\|^2 \sim \chi^2_{n-m}$, independently. The (scaled) quotient then turns out to have an F distribution — a consequence of the following theorem.

Theorem C.11: Relationship Between χ^2 and F Distributions

Let $U \sim \chi^2_m$ and $V \sim \chi^2_n$ be independent. Then,

$$\frac{U/m}{V/n} \sim \mathsf{F}(m, n).$$

Proof: For notational simplicity, let $c = m/2$ and $d = n/2$. The pdf of $W = U/V$ is given by $f_W(w) = \int_0^\infty f_U(wv)\, v\, f_V(v)\, \mathrm{d}v$. Substituting the pdfs of the corresponding Gamma distributions, we have

$$\begin{aligned} f_W(w) &= \int_0^\infty \frac{(wv)^{c-1}\,\mathrm{e}^{-wv/2}}{\Gamma(c)\,2^c}\, v\, \frac{v^{d-1}\mathrm{e}^{-v/2}}{\Gamma(d)\,2^d}\,\mathrm{d}v = \frac{w^{c-1}}{\Gamma(c)\,\Gamma(d)\,2^{c+d}} \int_0^\infty v^{c+d-1}\,\mathrm{e}^{-(1+w)v/2}\,\mathrm{d}v \\ &= \frac{\Gamma(c+d)}{\Gamma(c)\,\Gamma(d)}\,\frac{w^{c-1}}{(1+w)^{c+d}}, \end{aligned}$$

where the last equality follows from the fact that the integrand is equal to $\Gamma(\alpha)\lambda^{-\alpha}$ times the density of the $\mathsf{Gamma}(\alpha, \lambda)$ distribution with $\alpha = c + d$ and $\lambda = (1 + w)/2$. The density of $Z = \frac{n}{m}\frac{U}{V}$ is given by

$$f_Z(z) = f_W(z\,m/n)\,m/n.$$

The proof is completed by comparing the resulting expression with the pdf of the F distribution given in Table C.1. □ ☞ 427

Corollary C.1 (Relationship Between Normal, χ^2, and t Distributions) Let $Z \sim \mathcal{N}(0, 1)$ and $V \sim \chi^2_n$ be independent. Then,

$$\frac{Z}{\sqrt{V/n}} \sim \mathsf{t}_n.$$

Proof: Let $T = Z/\sqrt{V/n}$. Because $Z^2 \sim \chi^2_1$, we have by Theorem C.11 that $T^2 \sim \mathsf{F}(1, n)$. The result follows now from the symmetry around 0 of the pdf of T and the fact that the square of a t_n random variable has an $\mathsf{F}(1, n)$ distribution. □

C.8 Convergence of Random Variables

Recall that a random variable X is a function from Ω to $\mathbb{R}$. If we have a sequence of random variables $X_1, X_2, \ldots$ (for instance, $X_n(\omega) = X(\omega) + \frac{1}{n}$ for each $\omega \in \Omega$), then one can consider the pointwise convergence:

$$\lim_{n\to\infty} X_n(\omega) = X(\omega), \quad \text{for all } \omega \in \Omega,$$

in which case we say that $X_1, X_2, \ldots$ *converges surely* to X. A more interesting type of convergence uses the probability measure $\mathbb{P}$ associated with X. SURE CONVERGENCE

Definition C.1: Convergence in Probability

The sequence of random variables $X_1, X_2, \ldots$ *converges in probability* to a random variable X if, for all $\varepsilon > 0$,

$$\lim_{n\to\infty} \mathbb{P}\left[|X_n - X| > \varepsilon\right] = 0.$$

We denote the *convergence in probability* as $X_n \xrightarrow{\mathbb{P}} X$. CONVERGENCE IN PROBABILITY

Convergence in probability refers only to the distribution of X_n. Instead, if the sequence $X_1, X_2, \ldots$ is defined on a common probability space, then we can consider the following mode of convergence that uses the joint distribution of the sequence of random variables.

Definition C.2: Almost Sure Convergence

The sequence of random variables $X_1, X_2, \ldots$ *converges almost surely* to a random variable X if for every $\varepsilon > 0$

$$\lim_{n\to\infty} \mathbb{P}\left[\sup_{k\geqslant n} |X_k - X| > \varepsilon\right] = 0.$$

ALMOST SURE CONVERGENCE

We denote the *almost sure convergence* as $X_n \xrightarrow{\text{a.s.}} X$.

Note that in accordance with these definitions $X_n \xrightarrow{\text{a.s.}} 0$ is equivalent to $\sup_{k\geqslant n} |X_k| \xrightarrow{\mathbb{P}} 0$.

■ **Example C.3 (Convergence in Probability Versus Almost Sure Convergence)** Since the event $\{|X_n - X| > \varepsilon\}$ is contained in $\{\sup_{k\geqslant n} |X_k - X| > \varepsilon\}$, we can conclude that almost sure convergence implies convergence in probability. However, the converse is not true in general. For instance, consider the iid sequence $X_1, X_2, \ldots$ with marginal distribution

$$\mathbb{P}[X_n = 1] = 1 - \mathbb{P}[X_n = 0] = 1/n.$$

Clearly, $X_n \xrightarrow{\mathbb{P}} 0$. However, for $\varepsilon < 1$ and any $n = 1, 2, \ldots$ we have,

$$\begin{aligned}
\mathbb{P}\left[\sup_{k\geqslant n} |X_k| \leqslant \varepsilon\right] &= \mathbb{P}[X_n \leqslant \varepsilon, X_{n+1} \leqslant \varepsilon, \ldots] \\
&= \mathbb{P}[X_n \leqslant \varepsilon] \times \mathbb{P}[X_{n+1} \leqslant \varepsilon] \times \cdots \quad \text{(using independence)} \\
&= \lim_{m\to\infty} \prod_{k=n}^{m} \mathbb{P}[X_k \leqslant \varepsilon] = \lim_{m\to\infty} \prod_{k=n}^{m} \left(1 - \frac{1}{k}\right) \\
&= \lim_{m\to\infty} \frac{n-1}{n} \times \frac{n}{n+1} \times \cdots \times \frac{m-1}{m} = 0.
\end{aligned}$$

It follows that $\mathbb{P}[\sup_{k\geqslant n} |X_k - 0| > \varepsilon] = 1$ for any $0 < \varepsilon < 1$ and all $n \geqslant 1$. In other words, it is *not* true that $X_n \xrightarrow{\text{a.s.}} 0$. ■

Another important type of convergence is useful when we are interested in estimating expectations or multidimensional integrals via Monte Carlo methodology.

☞ 67

Definition C.3: Convergence in Distribution

The sequence of random variables $X_1, X_2, \ldots$ is said to *converge in distribution* to a random variable X with distribution function $F_X(x) = \mathbb{P}[X \leqslant x]$ provided that:

$$\lim_{n\to\infty} \mathbb{P}[X_n \leqslant x] = F_X(x) \text{ for all } x \text{ such that } \lim_{a\to x} F_X(a) = F_X(x). \tag{C.33}$$

CONVERGENCE IN DISTRIBUTION

We denote the *convergence in distribution* as either $X_n \xrightarrow{\text{d}} X$, or $X_n \xrightarrow{\text{d}} F_X$.

The generalization to random vectors replaces (C.33) with

$$\lim_{n\to\infty} \mathbb{P}[X_n \in A] = \mathbb{P}[X \in A] \text{ for all } A \subset \mathbb{R}^n \text{ such that } \mathbb{P}[X \in \partial A] = 0, \tag{C.34}$$

where ∂A denotes the boundary of the set A.

A useful tool for demonstrating convergence in distribution is the *characteristic function* ψ_X of a random vector X, defined as the expectation:

CHARACTERISTIC FUNCTION

☞ 225

$$\psi_X(t) := \mathbb{E}\, \mathrm{e}^{\mathrm{i}t^\top X}, \quad t \in \mathbb{R}^n. \tag{C.35}$$

The moment generating function in (C.5) is a special case of the characteristic function evaluated at $t = -\mathrm{i}s$. Note that while the moment generating function of a random variable may not exist, its characteristic function always exists. The characteristic function of a random vector $X \sim f$ is closely related to the Fourier transform of its pdf f.

☞ 392

■ **Example C.4 (Characteristic Function of a Multivariate Gaussian Random Vector)**
The density of the multivariate standard normal distribution is given in (C.26) and thus the characteristic function of $Z \sim \mathcal{N}(\mathbf{0}, \mathbf{I}_n)$ is

$$\begin{aligned}\psi_Z(t) &= \mathbb{E}\, \mathrm{e}^{\mathrm{i}t^\top Z} = (2\pi)^{-n/2} \int_{\mathbb{R}^n} \mathrm{e}^{\mathrm{i}t^\top z - \frac{1}{2}\|z\|^2} \mathrm{d}z \\ &= \mathrm{e}^{-\|t\|^2/2} (2\pi)^{-n/2} \int_{\mathbb{R}^n} \mathrm{e}^{-\frac{1}{2}\|z - \mathrm{i}t^\top\|^2} \mathrm{d}z = \mathrm{e}^{-\|t\|^2/2}, \quad t \in \mathbb{R}^n.\end{aligned}$$

Hence, the characteristic function of the random vector $X = \mu + \mathbf{B}Z$ in (C.27) with multivariate normal distribution $\mathcal{N}(\mu, \Sigma)$ is given by

☞ 437

$$\begin{aligned}\psi_X(t) &= \mathbb{E}\, \mathrm{e}^{\mathrm{i}t^\top X} = \mathbb{E}\, \mathrm{e}^{\mathrm{i}t^\top(\mu + \mathbf{B}Z)} \\ &= \mathrm{e}^{\mathrm{i}t^\top \mu} \mathbb{E}\, \mathrm{e}^{\mathrm{i}(\mathbf{B}^\top t)^\top Z} = \mathrm{e}^{\mathrm{i}t^\top \mu} \psi_Z(\mathbf{B}^\top t) \\ &= \mathrm{e}^{\mathrm{i}t^\top \mu - \|\mathbf{B}^\top t\|^2/2} = \mathrm{e}^{\mathrm{i}t^\top \mu - t^\top \Sigma t/2}.\end{aligned}$$

■

The importance of the characteristic function is mainly derived from the following result, for which a proof can be found, for example, in [11].

Theorem C.12: Characteristic Function

Suppose that $\psi_{X_1}(t), \psi_{X_2}(t), \ldots$ are the characteristic functions of the sequence of random vectors $X_1, X_2, \ldots$ and $\psi_X(t)$ is the characteristic function of X. Then, the following three statements are equivalent:

1. $\lim_{n\to\infty} \psi_{X_n}(t) = \psi_X(t)$ for all $t \in \mathbb{R}^n$.
2. $X_n \xrightarrow{\mathrm{d}} X$.
3. $\lim_{n\to\infty} \mathbb{E}h(X_n) = \mathbb{E}h(X)$ for all bounded continuous functions $h : \mathbb{R}^d \mapsto \mathbb{R}$.

■ **Example C.5 (Convergence in Distribution)** Define the random variables $Y_1, Y_2, \ldots$ as

$$Y_n := \sum_{k=1}^{n} X_k \left(\frac{1}{2}\right)^k, \qquad n = 1, 2, \ldots,$$

where $X_1, X_2, \ldots \stackrel{\text{iid}}{\sim} \mathsf{Ber}(1/2)$. We now show that $Y_n \xrightarrow{\text{d}} \mathcal{U}(0,1)$. First, note that

$$\mathbb{E}\exp(\mathrm{i}tY_n) = \prod_{k=1}^{n} \mathbb{E}\exp(\mathrm{i}tX_k/2^k) = 2^{-n}\prod_{k=1}^{n}(1+\exp(\mathrm{i}t/2^k)).$$

Second, from the collapsing product, $(1-\exp(\mathrm{i}t/2^n))\prod_{k=1}^{n}(1+\exp(\mathrm{i}t/2^k)) = 1-\exp(\mathrm{i}t)$, we have

$$\mathbb{E}\exp(\mathrm{i}tY_n) = (1-\exp(\mathrm{i}t))\frac{1/2^n}{1-\exp(\mathrm{i}t/2^n)}.$$

It follows that $\lim_{n\to\infty}\mathbb{E}\exp(\mathrm{i}tY_n) = (\exp(\mathrm{i}t)-1)/(\mathrm{i}t)$, which we recognize as the characteristic function of the $\mathcal{U}(0,1)$ distribution. ☞ 443 ■

Yet another mode of convergence is the following.

Definition C.4: Convergence in L^p-norm

The sequence of random variables $X_1, X_2, \ldots$ *converges in L^p-norm* to a random variable X if

$$\lim_{n\to\infty}\mathbb{E}|X_n - X|^p = 0, \quad p \geqslant 1.$$

We denote the *convergence in L^p-norm* as $X_n \xrightarrow{L^p} X$.

CONVERGENCE IN L^p-NORM

The case for $p = 2$ corresponds to convergence in mean squared error. The following example illustrates that convergence in L^p-norm is qualitatively different from convergence in distribution.

■ **Example C.6 (Comparison of Modes of Convergence)** Define $X_n := 1 - X$, where X has a uniform distribution on the interval (0,1). Clearly, $X_n \xrightarrow{\text{d}} \mathcal{U}(0,1)$. However, $\mathbb{E}|X_n - X| \longrightarrow \mathbb{E}|1-2X| = 1/2$ and so the sequence does not converge in L^1-norm. In addition, $\mathbb{P}[|X_n - X| > \varepsilon] \longrightarrow 1-\varepsilon \neq 0$ and so X_n does not converge in probability as well.

Thus, in general $X_n \xrightarrow{\text{d}} X$ implies neither $X_n \xrightarrow{\mathbb{P}} X$, nor $X_n \xrightarrow{L^1} X$.

We mention, however, that if $X_n \xrightarrow{\text{d}} c$ for some constant c, then $X_n \xrightarrow{\mathbb{P}} c$ as well. To see this, note that $X_n \xrightarrow{\text{d}} c$ stands for

$$\lim_{n\to\infty}\mathbb{P}[X_n \leqslant x] = \begin{cases} 1, & x > c \\ 0, & x < c \end{cases}.$$

In other words, we can write:

$$\mathbb{P}[|X_n - c| > \varepsilon] \leqslant 1 - \mathbb{P}[X_n \leqslant c+\varepsilon] + \mathbb{P}[X_n \leqslant c-\varepsilon] \longrightarrow 1 - 1 + 0 = 0, \qquad n \to \infty,$$

which shows that $X_n \xrightarrow{\mathbb{P}} c$ by definition. ■

Definition C.5: Complete Convergence

The sequence of random variables $X_1, X_2, \ldots$ is said to *converge completely* to X if for all $\varepsilon > 0$

$$\sum_n \mathbb{P}[|X_n - X| > \varepsilon] < \infty.$$

We denote the *complete convergence* as $X_n \xrightarrow{\text{cpl.}} X$.

COMPLETE CONVERGENCE

■ **Example C.7 (Complete and Almost Sure Convergence)** We show that complete convergence implies almost sure convergence. We can bound the criterion for almost sure convergence as follows:

$$\begin{aligned}
\mathbb{P}[\sup_{k \geqslant n} |X_k - X| > \varepsilon] &= \mathbb{P}[\cup_{k \geqslant n}\{|X_k - X| > \varepsilon\}] \\
&\leqslant \sum_{k \geqslant n} \mathbb{P}[|X_k - X| > \varepsilon] \qquad \text{by union bound in (C.1)} \\
&\leqslant \underbrace{\sum_{k=1}^{\infty} \mathbb{P}[|X_k - X| > \varepsilon]}_{= c < \infty \text{ from } X_n \xrightarrow{\text{cpl.}} X} - \sum_{k=1}^{n-1} \mathbb{P}[|X_k - X| > \varepsilon] \\
&\leqslant c - \sum_{k=1}^{n-1} \mathbb{P}[|X_k - X| > \varepsilon] \longrightarrow c - c = 0, \qquad n \to \infty.
\end{aligned}$$

Hence, by definition $X_n \xrightarrow{\text{a.s.}} X$. ■

The next theorem shows how the different types of convergence are related to each other. For example, in the diagram below, the notation $\overset{p \geqslant q}{\Rightarrow}$ means that L^p-norm convergence implies L^q-norm convergence under the assumption that $p \geqslant q \geqslant 1$.

Theorem C.13: Modes of Convergence

The most general relationships among the various modes of convergence for numerical random variables are shown on the following hierarchical diagram:

$$\begin{array}{ccccc}
\boxed{X_n \xrightarrow{\text{cpl.}} X} & \Rightarrow & \boxed{X_n \xrightarrow{\text{a.s.}} X} & & \\
& & \Downarrow & & \\
& & \boxed{X_n \xrightarrow{\mathbb{P}} X} & \Rightarrow & \boxed{X_n \xrightarrow{\text{d}} X} \\
& & \Uparrow & & \\
\boxed{X_n \xrightarrow{L^p} X} & \overset{p \geqslant q}{\Rightarrow} & \boxed{X_n \xrightarrow{L^q} X} & &
\end{array} .$$

Proof: **1.** First, we show that $X_n \xrightarrow{\mathbb{P}} X \Rightarrow X_n \xrightarrow{\mathrm{d}} X$ using the inequality $\mathbb{P}[A \cap B] \leqslant \mathbb{P}[A]$ for any event B. To this end, consider the distribution function F_X of X:

$$\begin{aligned} F_{X_n}(x) = \mathbb{P}[X_n \leqslant x] &= \mathbb{P}[X_n \leqslant x, |X_n - X| > \varepsilon] + \mathbb{P}[X_n \leqslant x, |X_n - X| \leqslant \varepsilon] \\ &\leqslant \mathbb{P}[|X_n - X| > \varepsilon] + \mathbb{P}[X_n \leqslant x, X \leqslant X_n + \varepsilon] \\ &\leqslant \mathbb{P}[|X_n - X| > \varepsilon] + \mathbb{P}[X \leqslant x + \varepsilon]. \end{aligned}$$

Now, in the arguments above we can switch the roles of X_n and X (there is a symmetry) to deduce the analogous result: $F_X(x) \leqslant \mathbb{P}[|X - X_n| > \varepsilon] + \mathbb{P}[X_n \leqslant x + \varepsilon]$. Therefore, making the switch $x \to x - \varepsilon$ gives $F_X(x - \varepsilon) \leqslant \mathbb{P}[|X - X_n| > \varepsilon] + F_{X_n}(x)$. Putting it all together gives:

$$F_X(x - \varepsilon) - \mathbb{P}[|X - X_n| > \varepsilon] \leqslant F_{X_n}(x) \leqslant \mathbb{P}[|X_n - X| > \varepsilon] + F_X(x + \varepsilon).$$

Taking $n \to \infty$ on both sides yields for any $\varepsilon > 0$:

$$F_X(x - \varepsilon) \leqslant \lim_{n\to\infty} F_{X_n}(x) \leqslant F_X(x + \varepsilon).$$

Since F_X is continuous at x by assumption we can take $\varepsilon \downarrow 0$ to conclude that $\lim_{n\to\infty} F_{X_n}(x) = F_X(x)$.

2. Second, we show that $X_n \xrightarrow{L^p} X \Rightarrow X_n \xrightarrow{L^q} X$ for $p \geqslant q \geqslant 1$. Since the function $f(x) = x^{q/p}$ is concave for $q/p \leqslant 1$, Jensen's inequality yields:

☞ 62

$$(\mathbb{E}|X|^p)^{q/p} = f(\mathbb{E}|X|^p) \geqslant \mathbb{E}f(|X|^p) = \mathbb{E}|X|^q.$$

In other words, $(\mathbb{E}|X_n - X|^q)^{1/q} \leqslant (\mathbb{E}|X_n - X|^p)^{1/p} \longrightarrow 0$, proving the statement of the theorem.

3. Third, we show that $X_n \xrightarrow{L^1} X \Rightarrow X_n \xrightarrow{\mathbb{P}} X$. First note that for any random variable Y, we can write: $\mathbb{E}|Y| \geqslant \mathbb{E}[|Y| \, \mathbb{1}_{\{|Y|>\varepsilon\}}] \geqslant \mathbb{E}[|\varepsilon| \, \mathbb{1}_{\{|Y|>\varepsilon\}}] = \varepsilon \, \mathbb{P}[|Y| > \varepsilon]$. Therefore, we obtain *Chebyshev's inequality*:

CHEBYSHEV'S INEQUALITY

$$\mathbb{P}[|Y| > \varepsilon] \leqslant \frac{\mathbb{E}|Y|}{\varepsilon}. \qquad \text{(C.36)}$$

Using Chebyshev's inequality and $X_n \xrightarrow{L^1} X$, we can write

$$\mathbb{P}[|X_n - X| > \varepsilon] \leqslant \frac{\mathbb{E}|X_n - X|}{\varepsilon} \longrightarrow 0, \quad n \to \infty.$$

Hence, by definition $X_n \xrightarrow{\mathbb{P}} X$.

4. Finally, $X_n \xrightarrow{\text{cpl.}} X \Rightarrow X_n \xrightarrow{\text{a.s.}} X \Rightarrow X_n \xrightarrow{\mathbb{P}} X$ is proved in Examples C.7 and C.3. □

Finally, we will make use of the following theorem.

Theorem C.14: Slutsky

Let $g(\boldsymbol{x}, \boldsymbol{y})$ be a continuous scalar function of vectors $\boldsymbol{x}$ and $\boldsymbol{y}$. Suppose that $\boldsymbol{X}_n \xrightarrow{\mathrm{d}} \boldsymbol{X}$ and $\boldsymbol{Y}_n \xrightarrow{\mathbb{P}} \boldsymbol{c}$ for some finite constant $\boldsymbol{c}$. Then,

$$g(\boldsymbol{X}_n, \boldsymbol{Y}_n) \xrightarrow{\mathrm{d}} g(\boldsymbol{X}, \boldsymbol{c}).$$

Proof: We prove the theorem for scalar X and Y. The proof for random vectors is analogous. First, we show that $\boldsymbol{Z}_n := \begin{bmatrix} X_n \\ Y_n \end{bmatrix} \xrightarrow{\text{d}} \begin{bmatrix} X \\ c \end{bmatrix} =: \boldsymbol{Z}$ using, for example, Theorem C.12. In other words, we wish to show that the characteristic function of the joint distribution of X_n and Y_n converges pointwise as $n \to \infty$: ☞ 443

$$\psi_{X_n,Y_n}(\boldsymbol{t}) = \mathbb{E}\,\mathrm{e}^{\mathrm{i}(t_1 X_n + t_2 Y_n)} \longrightarrow \mathrm{e}^{\mathrm{i}t_2 c}\mathbb{E}\,\mathrm{e}^{\mathrm{i}t_1 X} = \psi_{X,c}(\boldsymbol{t}), \quad \forall \boldsymbol{t} \in \mathbb{R}^2.$$

To show the limit above, consider

$$\begin{aligned}
|\psi_{X_n,Y_n}(\boldsymbol{t}) - \psi_{X,c}(\boldsymbol{t})| &\leqslant |\psi_{X_n,c}(\boldsymbol{t}) - \psi_{X,c}(\boldsymbol{t})| + |\psi_{X_n,Y_n}(\boldsymbol{t}) - \psi_{X_n,c}(\boldsymbol{t})| \\
&= |\mathrm{e}^{\mathrm{i}t_2 c}\,\mathbb{E}\,(\mathrm{e}^{\mathrm{i}t_1 X_n} - \mathrm{e}^{\mathrm{i}t_1 X})| + |\mathbb{E}\,\mathrm{e}^{\mathrm{i}(t_1 X_n + t_2 c)}(\mathrm{e}^{\mathrm{i}t_2(Y_n - c)} - 1)| \\
&\leqslant |\mathrm{e}^{\mathrm{i}t_2 c}| \times |\mathbb{E}(\mathrm{e}^{\mathrm{i}t_1 X_n} - \mathrm{e}^{\mathrm{i}t_1 X})| + \mathbb{E}\,|\mathrm{e}^{\mathrm{i}(t_1 X_n + t_2 c)}| \times |\mathrm{e}^{\mathrm{i}t_2(Y_n - c)} - 1| \\
&\leqslant |\psi_{X_n}(t_1) - \psi_X(t_1)| + \mathbb{E}\,|\mathrm{e}^{\mathrm{i}t_2(Y_n - c)} - 1|.
\end{aligned}$$

Since $X_n \xrightarrow{\text{d}} X$, Theorem C.12 implies that $\psi_{X_n}(t_1) \longrightarrow \psi_X(t_1)$, and the first term $|\psi_{X_n}(t_1) - \psi_X(t_1)|$ goes to zero. For the second term we use the fact that

$$|\mathrm{e}^{\mathrm{i}x} - 1| = \left|\int_0^x \mathrm{i}\,\mathrm{e}^{\mathrm{i}\theta}\,\mathrm{d}\theta\right| \leqslant \left|\int_0^x |\mathrm{i}\,\mathrm{e}^{\mathrm{i}\theta}|\,\mathrm{d}\theta\right| = |x|, \quad x \in \mathbb{R}$$

to obtain the bound:

$$\begin{aligned}
\mathbb{E}|\mathrm{e}^{\mathrm{i}t_2(Y_n - c)} - 1| &= \mathbb{E}|\mathrm{e}^{\mathrm{i}t_2(Y_n - c)} - 1|\mathbb{1}_{\{|Y_n - c| > \varepsilon\}} + \mathbb{E}|\mathrm{e}^{\mathrm{i}t_2(Y_n - c)} - 1|\mathbb{1}_{\{|Y_n - c| \leqslant \varepsilon\}} \\
&\leqslant 2\mathbb{E}\,\mathbb{1}_{\{|Y_n - c| > \varepsilon\}} + \mathbb{E}|t_2(Y_n - c)|\mathbb{1}_{\{|Y_n - c| \leqslant \varepsilon\}} \\
&\leqslant 2\mathbb{P}[|Y_n - c| > \varepsilon] + |t_2|\varepsilon \longrightarrow |t_2|\varepsilon, \quad n \to \infty.
\end{aligned}$$

Since ε is arbitrary, we can let $\varepsilon \downarrow 0$ to conclude that $\lim_{n\to\infty} |\psi_{X_n,Y_n}(\boldsymbol{t}) - \psi_{X,c}(\boldsymbol{t})| = 0$. In other words, $\boldsymbol{Z}_n \xrightarrow{\text{d}} \boldsymbol{Z}$, and by the continuity of g, we have $g(\boldsymbol{Z}_n) \xrightarrow{\text{d}} g(\boldsymbol{Z})$ or $g(\boldsymbol{X}_n, \boldsymbol{Y}_n) \xrightarrow{\text{d}} g(\boldsymbol{X}, \boldsymbol{c})$. □

■ **Example C.8 (Necessity of Slutsky's Condition)** The condition that Y_n converges in probability to a constant cannot be relaxed. For example, suppose that $g(x, y) = x + y$, $X_n \xrightarrow{\text{d}} X \sim \mathcal{N}(0, 1)$ and $Y_n \xrightarrow{\text{d}} Y \sim \mathcal{N}(0, 1)$. Then, our intuition tempts us to *incorrectly* conclude that $X_n + Y_n \xrightarrow{\text{d}} \mathcal{N}(0, 2)$. This intuition is false, because we can have $Y_n = -X_n$ for all n so that $X_n + Y_n = 0$, while both X and Y have the same marginal distribution (in this case standard normal). ■

C.9 Law of Large Numbers and Central Limit Theorem

Two main results in probability are the *law of large numbers* and *the central limit theorem*. Both are limit theorems involving sums of independent random variables. In particular, consider a sequence $X_1, X_2, \ldots$ of iid random variables with finite expectation μ and finite variance σ^2. For each n define $\overline{X}_n := (X_1 + \cdots + X_n)/n$. What can we say about the (random) sequence of averages $\overline{X}_1, \overline{X}_2, \overline{X}_3, \ldots$? By (C.12) and (C.14) we have $\mathbb{E}\,\overline{X}_n = \mu$ and $\mathbb{V}\mathrm{ar}\,\overline{X}_n = \sigma^2/n$. Hence, as n increases, the variance of the (random) average $\overline{X}_n$ goes to 0. This means ☞ 431 that by Definition C.8, the average $\overline{X}_n$ converges to μ in L^2-norm as $n \to \infty$, that is, $\overline{X}_n \xrightarrow{L^2} \mu$.

In fact, to obtain *convergence in probability* the variance need not be finite — it is sufficient to assume that $\mu = \mathbb{E}X < \infty$.

Theorem C.15: Weak Law of Large Numbers

LAW OF LARGE NUMBERS

If $X_1, \ldots, X_n$ are iid with finite expectation μ, then for all $\varepsilon > 0$

$$\lim_{n\to\infty} \mathbb{P}\left[|\overline{X}_n - \mu| > \varepsilon\right] = 0.$$

In other words, $\overline{X}_n \xrightarrow{\mathbb{P}} \mu$.

The theorem has a natural generalization for random vectors. Namely, if $\boldsymbol{\mu} = \mathbb{E}\boldsymbol{X} < \infty$, then $\mathbb{P}\left[\|\overline{\boldsymbol{X}}_n - \boldsymbol{\mu}\| > \varepsilon\right] \to 0$, where $\|\cdot\|$ is the Euclidean norm. We give a proof in the scalar case.

☞ 357

Proof: Let $Z_k := X_k - \mu$ for all k, so that $\mathbb{E}Z = 0$. We thus need to show that $\overline{Z}_n \xrightarrow{\mathbb{P}} 0$. We use the properties of the characteristic function of Z denoted as ψ_Z. Due to the iid assumption, we have

☞ 443

$$\psi_{\overline{Z}_n}(t) = \mathbb{E}\, \mathrm{e}^{\mathrm{i}t\overline{Z}_n} = \mathbb{E}\prod_{i=1}^{n} \mathrm{e}^{\mathrm{i}tZ_i/n} = \prod_{i=1}^{n} \mathbb{E}\,\mathrm{e}^{\mathrm{i}Z_i t/n} = \prod_{i=1}^{n} \psi_Z(t/n) = [\psi_Z(t/n)]^n. \tag{C.37}$$

An application of Taylor's Theorem B.1 in the neighborhood of $t = 0$ yields

$$\psi_Z(t/n) = \psi_Z(0) + o(t/n).$$

Since $\psi_Z(0) = 1$, we have:

$$\psi_{\overline{Z}_n}(t) = [\psi_Z(t/n)]^n = [1 + o(1/n)]^n \longrightarrow 1, \quad n \to \infty.$$

The characteristic function of a random variable that always equals zero is 1. Therefore, Theorem C.12 implies that $\overline{Z}_n \xrightarrow{\mathrm{d}} 0$. However, according to Example C.6, convergence in distribution to a constant implies convergence in probability. Hence, $\overline{Z}_n \xrightarrow{\mathbb{P}} 0$. □

There is also a stronger version of this theorem, as follows.

Theorem C.16: Strong Law of Large Numbers

STRONG LAW OF LARGE NUMBERS

If $X_1, \ldots, X_n$ are iid with expectation μ and $\mathbb{E}X^2 < \infty$, then for all $\varepsilon > 0$

$$\lim_{n\to\infty} \mathbb{P}\left[\sup_{k \geqslant n} |\overline{X}_k - \mu| > \varepsilon\right] = 0.$$

In other words, $\overline{X}_n \xrightarrow{\text{a.s.}} \mu$.

Proof: First, note that any random variable X can be written as the difference of two non-negative random variables: $X = X_+ - X_-$, where $X_+ := \max\{X, 0\}$ and $X_- := -\min\{X, 0\}$. Thus, without loss of generality, we assume that the random variables in the theorem above are nonnegative.

Second, from the sequence $\{\overline{X}_1, \overline{X}_2, \overline{X}_3, \ldots\}$ we can pick up the subsequence $\{\overline{X}_1, \overline{X}_4, \overline{X}_9, \overline{X}_{16}, \ldots\} =: \{\overline{X}_{j^2}\}$. Then, from Chebyshev's inequality (C.36) and the iid condition, we have

$$\sum_{j=1}^{\infty} \mathbb{P}\left[\left|\overline{X}_{j^2} - \mu\right| > \varepsilon\right] \leqslant \frac{\mathbb{V}\mathrm{ar}\, X}{\varepsilon^2} \sum_{j=1}^{\infty} \frac{1}{j^2} < \infty.$$

Therefore, by definition $\overline{X}_{n^2} \xrightarrow{\text{cpl.}} \mu$ and from Theorem C.13 we conclude that $\overline{X}_{n^2} \xrightarrow{\text{a.s.}} \mu$.

Third, for any arbitrary n, we can find a k, say $k = \lfloor \sqrt{n} \rfloor$, so that $k^2 \leqslant n \leqslant (k+1)^2$. For such a k and nonnegative $X_1, X_2, \ldots$, it holds that

$$\frac{k^2}{(k+1)^2}\overline{X}_{k^2} \leqslant \overline{X}_n \leqslant \overline{X}_{(k+1)^2}\frac{(k+1)^2}{k^2}.$$

Since $\overline{X}_{k^2}$ and $\overline{X}_{(k+1)^2}$ converge almost surely to μ as k (and hence n) goes to infinity, we conclude that $\overline{X}_n \xrightarrow{\text{a.s.}} \mu$. □

Note that the condition $\mathbb{E}X^2 < \infty$ in Theorem C.16 can be weakened to $\mathbb{E}|X| < \infty$ and the iid condition on the variables $X_1, \ldots, X_n$ can be relaxed to mere pairwise independence. The corresponding proof, however, is significantly more difficult.

The *Central Limit Theorem* describes the approximate distribution of $\overline{X}_n$, and it applies to both continuous and discrete random variables. Loosely, it states that

CENTRAL LIMIT THEOREM

the average of a large number of iid random variables approximately has a normal distribution.

Specifically, the random variable $\overline{X}_n$ has a distribution that is approximately normal, with expectation μ and variance σ^2/n.

Theorem C.17: Central Limit Theorem

If $X_1, \ldots, X_n$ are iid with finite expectation μ and finite variance σ^2, then for all $x \in \mathbb{R}$,

$$\lim_{n\to\infty} \mathbb{P}\left[\frac{\overline{X}_n - \mu}{\sigma/\sqrt{n}} \leqslant x\right] = \Phi(x),$$

where Φ is the cdf of the standard normal distribution.

Proof: Let $Z_k := (X_k - \mu)/\sigma$ for all k, so that $\mathbb{E}\,Z = 0$ and $\mathbb{E}\,Z^2 = 1$. We thus need to show that $\sqrt{n}\,\overline{Z}_n \xrightarrow{\text{d}} \mathcal{N}(0, 1)$. We again use the properties of the characteristic function. Let ψ_Z be the characteristic function of an iid copy of Z, then due to the iid assumption a similar calculation to the one in (C.37) yields: ☞ 443

$$\psi_{\sqrt{n}\,\overline{Z}_n}(t) = \mathbb{E}\,\mathrm{e}^{\mathrm{i}t\sqrt{n}\,\overline{Z}_n} = [\psi_Z(t/\sqrt{n})]^n.$$

An application of Taylor's Theorem B.1 in the neighborhood of $t = 0$ yields

$$\psi_Z(t/\sqrt{n}) = 1 + \frac{t}{\sqrt{n}}\psi_Z'(0) + \frac{t^2}{2n}\psi_Z''(0) + o(t^2/n).$$

Since $\psi_Z'(0) = \mathbb{E}\, \frac{\mathrm{d}}{\mathrm{d}t}\mathrm{e}^{\mathrm{i}tZ}\big|_{t=0} = \mathrm{i}\,\mathbb{E}Z = 0$ and $\psi_Z''(0) = \mathrm{i}^2\,\mathbb{E}Z^2 = -1$, we have:

$$\psi_{\sqrt{n}\,\overline{Z}_n}(t) = \left[\psi_Z(t/\sqrt{n})\right]^n = \left[1 - \frac{t^2}{2n} + o(1/n)\right]^n \longrightarrow \mathrm{e}^{-t^2/2}, \quad n \to \infty.$$

From Example C.4, we recognize $\mathrm{e}^{-t^2/2}$ as the characteristic function of the standard normal distribution. Thus, from Theorem C.12 we conclude that $\sqrt{n}\,\overline{Z}_n \xrightarrow{\mathrm{d}} \mathcal{N}(0, 1)$. □

Figure C.6 shows the central limit theorem in action. The left part shows the pdfs of $\overline{X}_1, 2\overline{X}_2, \ldots, 4\overline{X}_4$ for the case where the $\{X_i\}$ have a $\mathcal{U}[0, 1]$ distribution. The right part shows the same for the Exp(1) distribution. In both cases, we clearly see convergence to a bell-shaped curve, characteristic of the normal distribution.

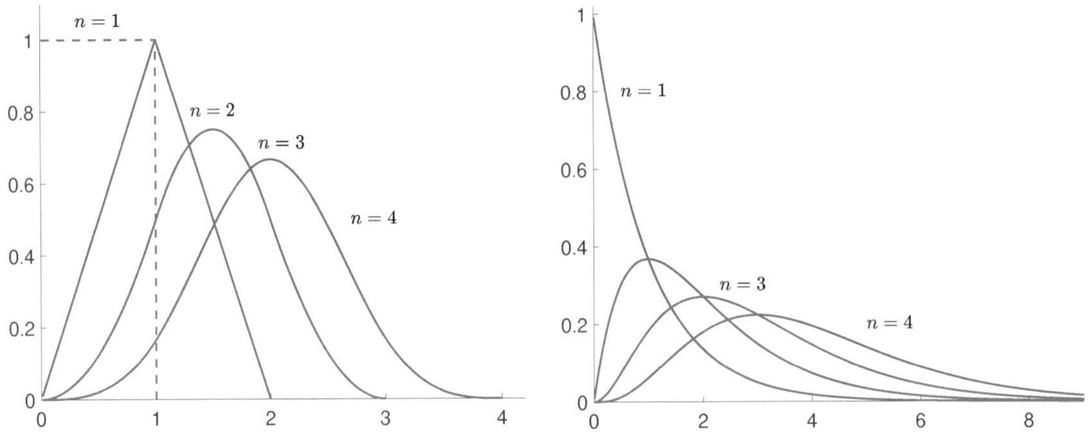

Figure C.6: Illustration of the central limit theorem for (left) the uniform distribution and (right) the exponential distribution.

The multivariate version of the central limit theorem is the basis for many asymptotic (in the size of the training set) results in machine learning and data science.

Theorem C.18: Multivariate Central Limit Theorem

Let $\boldsymbol{X}_1, \ldots, \boldsymbol{X}_n$ be iid random vectors with expectation vector $\boldsymbol{\mu}$ and finite covariance matrix $\boldsymbol{\Sigma}$. Define $\overline{\boldsymbol{X}}_n := (\boldsymbol{X}_1 + \cdots + \boldsymbol{X}_n)/n$. Then,

$$\sqrt{n}\,(\overline{\boldsymbol{X}}_n - \boldsymbol{\mu}) \xrightarrow{\mathrm{d}} \mathcal{N}(\boldsymbol{0}, \boldsymbol{\Sigma}) \quad \text{as } n \to \infty.$$

One application is as follows. Suppose that a parameter of interest, $\boldsymbol{\theta}^*$, is the unique solution of the system of equations $\mathbb{E}\,\boldsymbol{\psi}(\boldsymbol{X}\,|\,\boldsymbol{\theta}^*) = \boldsymbol{0}$, where $\boldsymbol{\psi}$ is a vector-valued (or multi-valued) function and the distribution of $\boldsymbol{X}$ does not depend on $\boldsymbol{\theta}$. An *M-estimator* of $\boldsymbol{\theta}^*$,

M-ESTIMATOR

denoted $\widehat{\boldsymbol{\theta}}_n$, is the solution to the system of equations that results from approximating the expectation with respect to $\boldsymbol{X}$ using an average of n iid copies of $\boldsymbol{X}$:

$$\overline{\boldsymbol{\psi}}_n(\boldsymbol{\theta}) := \frac{1}{n}\sum_{i=1}^{n} \boldsymbol{\psi}(\boldsymbol{X}_i \,|\, \boldsymbol{\theta}).$$

Thus, $\overline{\boldsymbol{\psi}}_n(\widehat{\boldsymbol{\theta}}_n) = \mathbf{0}$.

Theorem C.19: M-estimator

The M-estimator is asymptotically normal as $n \to \infty$:

$$\sqrt{n}\,(\widehat{\boldsymbol{\theta}}_n - \boldsymbol{\theta}^*) \xrightarrow{\text{d}} \mathcal{N}(\mathbf{0}, \mathbf{A}^{-1}\mathbf{B}\mathbf{A}^{-\top}), \qquad \text{(C.38)}$$

where $\mathbf{A} := -\mathbb{E}\frac{\partial \boldsymbol{\psi}}{\partial \boldsymbol{\theta}}(\boldsymbol{X} \,|\, \boldsymbol{\theta}^*)$ and $\mathbf{B} := \mathbb{E}\left[\boldsymbol{\psi}(\boldsymbol{X} \,|\, \boldsymbol{\theta}^*)\, \boldsymbol{\psi}(\boldsymbol{X} \,|\, \boldsymbol{\theta}^*)^\top\right]$ is the covariance matrix of $\boldsymbol{\psi}(\boldsymbol{X} \,|\, \boldsymbol{\theta}^*)$. ☞ 400

Proof: We give a proof under the simplifying assumption[3] that $\widehat{\boldsymbol{\theta}}_n$ is a unique root, that is, for any $\boldsymbol{\theta}$ and ε, there exists a $\delta > 0$ such that $\|\widehat{\boldsymbol{\theta}}_n - \boldsymbol{\theta}\| > \varepsilon$ implies that $\|\overline{\boldsymbol{\psi}}_n(\boldsymbol{\theta})\| > \delta$.

First, we argue that $\widehat{\boldsymbol{\theta}}_n \xrightarrow{\mathbb{P}} \boldsymbol{\theta}^*$; that is, $\mathbb{P}[\|\widehat{\boldsymbol{\theta}}_n - \boldsymbol{\theta}^*\| > \varepsilon] \to 0$. From the multivariate extension of Theorem C.15, we have that

$$\overline{\boldsymbol{\psi}}_n(\boldsymbol{\theta}^*) \xrightarrow{\mathbb{P}} \mathbb{E}\,\overline{\boldsymbol{\psi}}_n(\boldsymbol{\theta}^*) = \mathbb{E}\,\boldsymbol{\psi}(\boldsymbol{X} \,|\, \boldsymbol{\theta}^*) = \mathbf{0}.$$

Therefore, using the uniqueness of $\widehat{\boldsymbol{\theta}}_n$, we can show that $\widehat{\boldsymbol{\theta}}_n \xrightarrow{\mathbb{P}} \boldsymbol{\theta}^*$ via the bound:

$$\mathbb{P}\left[\|\widehat{\boldsymbol{\theta}}_n - \boldsymbol{\theta}^*\| > \varepsilon\right] \leqslant \mathbb{P}\left[\|\overline{\boldsymbol{\psi}}_n(\boldsymbol{\theta}^*)\| > \delta\right] = \mathbb{P}\left[\|\overline{\boldsymbol{\psi}}_n(\boldsymbol{\theta}^*) - \mathbb{E}\,\overline{\boldsymbol{\psi}}_n(\boldsymbol{\theta}^*)\| > \delta\right] \to 0, \quad n \to \infty.$$

Second, we take a Taylor expansion of each component of the vector $\overline{\boldsymbol{\psi}}_n(\widehat{\boldsymbol{\theta}}_n)$ around $\boldsymbol{\theta}^*$ to obtain:

$$\overline{\boldsymbol{\psi}}_n(\widehat{\boldsymbol{\theta}}_n) = \overline{\boldsymbol{\psi}}_n(\boldsymbol{\theta}^*) + \mathbf{J}_n(\boldsymbol{\theta}')(\widehat{\boldsymbol{\theta}}_n - \boldsymbol{\theta}^*),$$

where $\mathbf{J}_n(\boldsymbol{\theta})$ is the Jacobian of $\overline{\boldsymbol{\psi}}_n$ at $\boldsymbol{\theta}$, and $\boldsymbol{\theta}'$ lies on the line segment joining $\widehat{\boldsymbol{\theta}}_n$ and $\boldsymbol{\theta}^*$. Rearrange the last equation and multiply both sides by $\sqrt{n}\,\mathbf{A}^{-1}$ to obtain:

$$-\mathbf{A}^{-1}\mathbf{J}_n(\boldsymbol{\theta}')\sqrt{n}\,(\widehat{\boldsymbol{\theta}}_n - \boldsymbol{\theta}^*) = \mathbf{A}^{-1}\sqrt{n}\,\overline{\boldsymbol{\psi}}_n(\boldsymbol{\theta}^*).$$

By the central limit theorem, $\sqrt{n}\,\overline{\boldsymbol{\psi}}(\boldsymbol{\theta}^*)$ converges in distribution to $\mathcal{N}(\mathbf{0}, \mathbf{B})$. Therefore,

$$-\mathbf{A}^{-1}\mathbf{J}_n(\boldsymbol{\theta}')\sqrt{n}\,(\widehat{\boldsymbol{\theta}}_n - \boldsymbol{\theta}^*) \xrightarrow{\text{d}} \mathcal{N}(\mathbf{0}, \mathbf{A}^{-1}\mathbf{B}\mathbf{A}^{-\top}).$$

Theorem C.15 (the weak law of large numbers) applied to the iid random matrices $\{\frac{\partial}{\partial \boldsymbol{\theta}}\boldsymbol{\psi}(\boldsymbol{X}_i \,|\, \boldsymbol{\theta})\}$ shows that

$$\mathbf{J}_n(\boldsymbol{\theta}) \xrightarrow{\mathbb{P}} \mathbb{E}\,\frac{\partial}{\partial \boldsymbol{\theta}}\boldsymbol{\psi}(\boldsymbol{X} \,|\, \boldsymbol{\theta}).$$

Moreover, since $\widehat{\boldsymbol{\theta}}_n \xrightarrow{\mathbb{P}} \boldsymbol{\theta}^*$ and $\mathbf{J}_n$ is continuous in $\boldsymbol{\theta}$, we have that $\mathbf{J}_n(\boldsymbol{\theta}') \xrightarrow{\mathbb{P}} -\mathbf{A}$. Therefore, by Slutsky's theorem, $-\mathbf{A}^{-1}\mathbf{J}_n(\boldsymbol{\theta}')\sqrt{n}\,(\widehat{\boldsymbol{\theta}}_n - \boldsymbol{\theta}^*) - \sqrt{n}\,(\widehat{\boldsymbol{\theta}}_n - \boldsymbol{\theta}^*) \xrightarrow{\mathbb{P}} \mathbf{0}$. □ ☞ 446

[3]The result holds under far less stringent assumptions.

LAPLACE'S APPROXIMATION

Finally, we mention *Laplace's approximation*, which shows how integrals or expectations behave under the normal distribution with a vanishingly small variance.

Theorem C.20: Laplace's Approximation

Suppose that $\boldsymbol{\theta}_n \to \boldsymbol{\theta}^*$, where $\boldsymbol{\theta}^*$ lies in the interior of the open set $\Theta \subseteq \mathbb{R}^p$ and that $\boldsymbol{\Sigma}_n$ is a $p \times p$ covariance matrix such that $\boldsymbol{\Sigma}_n \to \boldsymbol{\Sigma}^*$. Let $g : \Theta \mapsto \mathbb{R}$ be a continuous function with $g(\boldsymbol{\theta}^*) \neq 0$. Then, as $n \to \infty$,

$$n^{p/2} \int_{\Theta} g(\boldsymbol{\theta})\, \mathrm{e}^{-\frac{n}{2}(\boldsymbol{\theta}-\boldsymbol{\theta}_n)^\top \boldsymbol{\Sigma}_n^{-1}(\boldsymbol{\theta}-\boldsymbol{\theta}_n)}\, \mathrm{d}\boldsymbol{\theta} \to g(\boldsymbol{\theta}^*)\sqrt{|2\pi\, \boldsymbol{\Sigma}^*|}. \tag{C.39}$$

Proof: (Sketch for a bounded domain Θ.) The left-hand side of (C.39) can be written as the expectation with respect to the $\mathcal{N}(\boldsymbol{\theta}_n, \boldsymbol{\Sigma}_n/n)$ distribution:

$$\sqrt{|2\pi\, \boldsymbol{\Sigma}_n|} \int_{\Theta} g(\boldsymbol{\theta}) \frac{\exp\left(-\frac{n}{2}(\boldsymbol{\theta}-\boldsymbol{\theta}_n)^\top \boldsymbol{\Sigma}_n^{-1}(\boldsymbol{\theta}-\boldsymbol{\theta}_n)\right)}{|2\pi\, \boldsymbol{\Sigma}_n/n|^{1/2}}\, \mathrm{d}\boldsymbol{\theta} = \sqrt{|2\pi\, \boldsymbol{\Sigma}_n|}\; \mathbb{E}[g(\boldsymbol{X}_n)\mathbb{1}\{\boldsymbol{X}_n \in \Theta\}],$$

where $\boldsymbol{X}_n \sim \mathcal{N}(\boldsymbol{\theta}_n, \boldsymbol{\Sigma}_n/n)$. Let $\mathbf{Z} \sim \mathcal{N}(\mathbf{0}, \mathbf{I})$. Then, $\boldsymbol{\theta}_n + \boldsymbol{\Sigma}_n^{1/2}\mathbf{Z}/\sqrt{n}$ has the same distribution as $\boldsymbol{X}_n$ and $\left(\boldsymbol{\theta}_n + \boldsymbol{\Sigma}_n^{1/2}\mathbf{Z}/\sqrt{n}\right) \to \boldsymbol{\theta}^*$ as $n \to \infty$. By continuity of $g(\boldsymbol{\theta})\mathbb{1}\{\boldsymbol{\theta} \in \Theta\}$ in the interior of Θ, as $n \to \infty$:[4]

$$\mathbb{E}[g(\boldsymbol{X}_n)\mathbb{1}\{\boldsymbol{X}_n \in \Theta\}] = \mathbb{E}\left[g\left(\boldsymbol{\theta}_n + \frac{\boldsymbol{\Sigma}_n^{1/2}\mathbf{Z}}{\sqrt{n}}\right)\mathbb{1}\left\{\left(\boldsymbol{\theta}_n + \frac{\boldsymbol{\Sigma}_n^{1/2}\mathbf{Z}}{\sqrt{n}}\right) \in \Theta\right\}\right] \longrightarrow g(\boldsymbol{\theta}^*)\mathbb{1}\{\boldsymbol{\theta}^* \in \Theta\}.$$

Since $\boldsymbol{\theta}^*$ lies in the interior of Θ, we have $\mathbb{1}\{\boldsymbol{\theta}^* \in \Theta\} = 1$, completing the proof. □

As an application of Theorem C.20 we can show the following.

Theorem C.21: Approximation of Integrals

Suppose that $r : \boldsymbol{\theta} \mapsto \mathbb{R}$ is twice continuously differentiable with a unique global minimum at $\boldsymbol{\theta}^*$ and $g : \boldsymbol{\theta} \mapsto \mathbb{R}$ is continuous with $g(\boldsymbol{\theta}^*) > 0$. Then, as $n \to \infty$,

$$\ln \int_{\mathbb{R}^p} g(\boldsymbol{\theta})\, \mathrm{e}^{-n\, r(\boldsymbol{\theta})}\, \mathrm{d}\boldsymbol{\theta} \simeq -n\, r(\boldsymbol{\theta}^*) - \frac{p}{2}\ln n. \tag{C.40}$$

More generally, if r_n has a unique global minimum $\boldsymbol{\theta}_n$ and $r_n \to r \Rightarrow \boldsymbol{\theta}_n \to \boldsymbol{\theta}^*$, then

$$\ln \int_{\mathbb{R}^p} g(\boldsymbol{\theta})\, \mathrm{e}^{-n\, r_n(\boldsymbol{\theta})}\, \mathrm{d}\boldsymbol{\theta} \simeq -n\, r(\boldsymbol{\theta}^*) - \frac{p}{2}\ln n.$$

☞ 402

Proof: We only sketch the proof of (C.40). Let $\mathbf{H}(\boldsymbol{\theta})$ be the Hessian matrix of r at $\boldsymbol{\theta}$. By Taylor's theorem we can write

$$r(\boldsymbol{\theta}) - r(\boldsymbol{\theta}^*) = (\boldsymbol{\theta}-\boldsymbol{\theta}^*)^\top \underbrace{\frac{\partial r(\boldsymbol{\theta}^*)}{\partial \boldsymbol{\theta}}}_{=\mathbf{0}} + \frac{1}{2}(\boldsymbol{\theta}-\boldsymbol{\theta}^*)^\top \mathbf{H}(\overline{\boldsymbol{\theta}})(\boldsymbol{\theta}-\boldsymbol{\theta}^*),$$

[4]We can exchange the limit and expectation, as $g(\boldsymbol{\theta})\mathbb{1}\{\boldsymbol{\theta} \in \Theta\} \leqslant \max_{\boldsymbol{\theta}\in\Theta} g(\boldsymbol{\theta})$ and $\int_{\Theta} \max_{\boldsymbol{\theta}\in\Theta} g(\boldsymbol{\theta})\, \mathrm{d}\boldsymbol{\theta} = |\Theta| \max_{\boldsymbol{\theta}\in\Theta} g(\boldsymbol{\theta}) < \infty$.

where $\overline{\boldsymbol{\theta}}$ is a point that lies on the line segment joining $\boldsymbol{\theta}^*$ and $\boldsymbol{\theta}$. Since $\boldsymbol{\theta}^*$ is a unique global minimum, there must be a small enough neighborhood of $\boldsymbol{\theta}^*$, say Θ, such that r is a strictly (also known as strongly) convex function on Θ. In other words, $\mathbf{H}(\boldsymbol{\theta})$ is a positive definite matrix for all $\boldsymbol{\theta} \in \Theta$ and there exists a smallest positive eigenvalue $\lambda_1 > 0$ such that $\boldsymbol{x}^\top \mathbf{H}(\boldsymbol{\theta})\boldsymbol{x} \geqslant \lambda_1 \|\boldsymbol{x}\|^2$ for all $\boldsymbol{x}$. In addition, since the maximum eigenvalue of $\mathbf{H}(\boldsymbol{\theta})$ is a continuous function of $\boldsymbol{\theta} \in \Theta$ and Θ is bounded, there must exist a constant $\lambda_2 > \lambda_1$ such that $\boldsymbol{x}^\top \mathbf{H}(\boldsymbol{\theta})\boldsymbol{x} \leqslant \lambda_2 \|\boldsymbol{x}\|^2$ for all $\boldsymbol{x}$. In other words, denoting $r^* := r(\boldsymbol{\theta}^*)$, we have the bounds: ☞ 405

$$-\frac{\lambda_2}{2}\|\boldsymbol{\theta} - \boldsymbol{\theta}^*\|^2 \leqslant -(r(\boldsymbol{\theta}) - r^*) \leqslant -\frac{\lambda_1}{2}\|\boldsymbol{\theta} - \boldsymbol{\theta}^*\|^2, \quad \boldsymbol{\theta} \in \Theta.$$

Therefore,

$$\mathrm{e}^{-n\,r^*}\int_\Theta g(\boldsymbol{\theta})\,\mathrm{e}^{-\frac{n\lambda_2}{2}\|\boldsymbol{\theta}-\boldsymbol{\theta}^*\|^2}\mathrm{d}\boldsymbol{\theta} \leqslant \int_\Theta g(\boldsymbol{\theta})\,\mathrm{e}^{-n\,r(\boldsymbol{\theta})}\mathrm{d}\boldsymbol{\theta} \leqslant \mathrm{e}^{-n\,r^*}\int_\Theta g(\boldsymbol{\theta})\,\mathrm{e}^{-\frac{n\lambda_1}{2}\|\boldsymbol{\theta}-\boldsymbol{\theta}^*\|^2}\mathrm{d}\boldsymbol{\theta}.$$

An application of Theorem C.20 yields $\int_\Theta g(\boldsymbol{\theta})\,\mathrm{e}^{-n\,r(\boldsymbol{\theta})}\,\mathrm{d}\boldsymbol{\theta} = \mathcal{O}(\mathrm{e}^{-n\,r^*}/n^{p/2})$ and, more importantly,

$$\ln \int_\Theta g(\boldsymbol{\theta})\,\mathrm{e}^{-n\,r(\boldsymbol{\theta})}\,\mathrm{d}\boldsymbol{\theta} \simeq -n\,r^* - \frac{p}{2}\ln n.$$

Thus, the proof will be complete once we show that $\int_{\overline{\Theta}} g(\boldsymbol{\theta})\,\mathrm{e}^{-n\,r(\boldsymbol{\theta})}\,\mathrm{d}\boldsymbol{\theta}$, with $\overline{\Theta} := \mathbb{R}^p \setminus \Theta$, is asymptotically negligible compared to $\int_\Theta g(\boldsymbol{\theta})\,\mathrm{e}^{-n\,r(\boldsymbol{\theta})}\,\mathrm{d}\boldsymbol{\theta}$. Since $\boldsymbol{\theta}^*$ is a global minimum that lies outside any neighborhood of $\overline{\Theta}$, there must exists a constant $c > 0$ such that $r(\boldsymbol{\theta}) - r^* > c$ for all $\boldsymbol{\theta} \in \overline{\Theta}$. Therefore,

$$\begin{aligned}
\int_{\overline{\Theta}} g(\boldsymbol{\theta})\,\mathrm{e}^{-n\,r(\boldsymbol{\theta})}\,\mathrm{d}\boldsymbol{\theta} &= \mathrm{e}^{-(n-1)\,r^*}\int_{\overline{\Theta}} g(\boldsymbol{\theta})\,\mathrm{e}^{-r(\boldsymbol{\theta})}\,\mathrm{e}^{-(n-1)(r(\boldsymbol{\theta})-r^*)}\,\mathrm{d}\boldsymbol{\theta} \\
&\leqslant \mathrm{e}^{-(n-1)\,r^*}\int_{\overline{\Theta}} g(\boldsymbol{\theta})\,\mathrm{e}^{-r(\boldsymbol{\theta})}\,\mathrm{e}^{-(n-1)c}\,\mathrm{d}\boldsymbol{\theta} \\
&\leqslant \mathrm{e}^{-(n-1)(r^*+c)}\int_{\mathbb{R}^p} g(\boldsymbol{\theta})\,\mathrm{e}^{-r(\boldsymbol{\theta})}\,\mathrm{d}\boldsymbol{\theta} = \mathcal{O}(\mathrm{e}^{-n(r^*+c)}).
\end{aligned}$$

The last expression is of order $o(\mathrm{e}^{-n\,r^*}/n^{p/2})$, concluding the proof. □

C.10 Markov Chains

Definition C.6: Markov Chain

A *Markov chain* is a collection $\{X_t, t = 0, 1, 2, \ldots\}$ of random variables (or random vectors) whose futures are conditionally independent of their pasts given their present values. That is, MARKOV CHAIN

$$\mathbb{P}[X_{t+1} \in A \,|\, X_s, s \leqslant t] = \mathbb{P}[X_{t+1} \in A \,|\, X_t] \quad \text{for all } t. \tag{C.41}$$

In other words, the conditional distribution of the future variable X_{t+1}, given the entire past $\{X_s, s \leqslant t\}$, is the same as the conditional distribution of X_{t+1} given only the present X_t. Property (C.41) is called the *Markov property*. MARKOV PROPERTY

The index t in X_t is usually seen as a "time" or "step" parameter. The index set $\{0, 1, 2, \ldots\}$ in the definition above was chosen out of convenience. It can be replaced by any countable index set. We restrict ourselves to *time-homogeneous* Markov chains — Markov chains for which the conditional pdfs $f_{X_{t+1} \mid X_t}(y \mid x)$ do not depend on t; we abbreviate these as $q(y \mid x)$. The $\{q(y \mid x)\}$ are called the *(one-step) transition densities* of the Markov chain. Note that the random variables or vectors $\{X_t\}$ may be *discrete* (e.g., taking values in some set $\{1, \ldots, r\}$) or *continuous* (e.g., taking values in an interval $[0, 1]$ or $\mathbb{R}^d$). In particular, in the *discrete* case, each $q(y \mid x)$ is a probability: $q(y \mid x) = \mathbb{P}[X_{t+1} = y \mid X_t = x]$.

TIME-HOMOGENEOUS

TRANSITION DENSITY

The distribution of X_0 is called the *initial distribution* of the Markov chain. The one-step transition densities and the initial distribution completely specify the distribution of the random vector $[X_0, X_1, \ldots, X_t]^\top$. Namely, we have by the product rule (C.17) and the Markov property that the joint pdf is given by

INITIAL DISTRIBUTION

☞ 433

$$
\begin{aligned}
f_{X_0,\ldots,X_t}(x_0, \ldots, x_t) &= f_{X_0}(x_0)\, f_{X_1 \mid X_0}(x_1 \mid x_0) \cdots f_{X_t \mid X_{t-1},\ldots,X_0}(x_t \mid x_{t-1}, \ldots, x_0) \\
&= f_{X_0}(x_0)\, f_{X_1 \mid X_0}(x_1 \mid x_0) \cdots f_{X_t \mid X_{t-1}}(x_t \mid x_{t-1}) \\
&= f_{X_0}(x_0)\, q(x_1 \mid x_0)\, q(x_2 \mid x_1) \cdots q(x_t \mid x_{t-1}).
\end{aligned}
$$

A Markov chain is said to be *ergodic* if the probability distribution of X_t converges to a fixed distribution as $t \to \infty$. Ergodicity is a property of many Markov chains. Intuitively, the probability of encountering the Markov chain in a state x at a time t far into the future should not depend on the t, provided that the Markov chain can reach every state from any other state — such Markov chains are said to be *irreducible* — and does not "escape" to infinity. Thus, for an ergodic Markov chain the pdf $f_{X_t}(x)$ converges to a fixed *limiting pdf* $f(x)$ as $t \to \infty$, irrespective of the starting state. For the discrete case, $f(x)$ corresponds to the long-run fraction of times that the Markov process visits x.

ERGODIC

LIMITING PDF

Under mild conditions (such as irreducibility) the limiting pdf $f(x)$ can be found by solving the *global balance equations*:

GLOBAL BALANCE EQUATIONS

$$
f(x) = \begin{cases} \sum_y f(y)\, q(x \mid y) & \text{(discrete case)}, \\ \int f(y)\, q(x \mid y)\, \mathrm{d}y & \text{(continuous case)}. \end{cases} \tag{C.42}
$$

For the discrete case the rationale behind this is as follows. Since $f(x)$ is the long-run proportion of time that the Markov chain spends in x, the proportion of transitions *out of* x is $f(x)$. This should be balanced with the proportion of transitions *into* state x, which is $\sum_y f(y)\, q(x \mid y)$.

One is often interested in a stronger type of balance equations. Imagine that we have taken a video of the evolution of the Markov chain, which we may run in forward and reverse time. If we cannot determine whether the video is running forward or backward (we cannot determine any systematic "looping", which would indicate in which direction time is flowing), the chain is said to be time-reversible or simply *reversible*.

REVERSIBLE

Although not every Markov chain is reversible, each ergodic Markov chain, when run backwards, gives another Markov chain — the *reverse Markov chain* — with transition densities $\widetilde{q}(y \mid x) = f(y)\, q(x \mid y)/f(x)$. To see this, first observe that $f(x)$ is the long-run proportion of time spent in x for both the original and reverse Markov chain. Secondly, the "probability flux" from x to y in the reversed chain must be equal to the probability flux from y to x in the original chain, meaning $f(x)\widetilde{q}(y \mid x) = f(y)\, q(x \mid y)$, which yields the

REVERSE MARKOV CHAIN

stated transition probabilities for the reversed chain. In particular, for a *reversible* Markov chain we have

$$f(x)\, q(y \mid x) = f(y)\, q(x \mid y) \quad \text{for all } x, y. \tag{C.43}$$

These are the *detailed (or local) balance equations*. Note that the detailed balance equations imply the global balance equations. Hence, if a Markov chain is irreducible and there exists a pdf such that (C.43) holds, then $f(x)$ must be the limiting pdf. In the discrete state space case an additional condition is that the chain must be *aperiodic*, meaning that the return times to the same state cannot always be a multiple of some integer $\geqslant 2$.

LOCAL BALANCE EQUATIONS

APERIODIC

■ **Example C.9 (Random Walk on a Graph)** Consider a Markov chain that performs a "random walk" on the graph in Figure C.7, at each step jumping from the current vertex (node) to one of the adjacent vertices, with equal probability. Clearly this Markov chain is reversible. It is also irreducible and aperiodic. Let $f(x)$ denote the limiting probability that the chain is in vertex x. By symmetry, $f(1) = f(2) = f(7) = f(8)$, $f(4) = f(5)$ and $f(3) = f(6)$. Moreover, by the detailed balance equations, $f(4)/5 = f(1)/3$, and $f(3)/4 = f(1)/3$. It follows that $f(1) + \cdots + f(8) = 4f(1) + 2 \times 5/3\, f(1) + 2 \times 4/3\, f(1) = 10\, f(1) = 1$, so that $f(1) = 1/10$, $f(3) = 2/15$, and $f(4) = 1/6$.

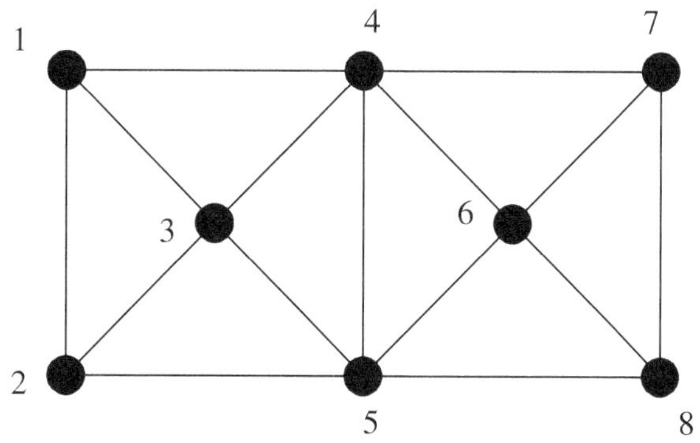

Figure C.7: The random walk on this graph is reversible.

■

C.11 Statistics

Statistics deals with the gathering, summarization, analysis, and interpretation of data. The two main branches of statistics are:

1. *Classical or frequentist statistics*: Here the observed data τ is viewed as the outcome of random data $\mathcal{T}$ described by a probabilistic model — usually the model is specified up to a (multidimensional) parameter; that is, $\mathcal{T} \sim g(\cdot \mid \boldsymbol{\theta})$ for some $\boldsymbol{\theta}$. The statistical inference is then purely concerned with the model and in particular with the parameter $\boldsymbol{\theta}$. For example, on the basis of the data one may wish to

FREQUENTIST STATISTICS

   (a) estimate the parameter,

   (b) perform statistical tests on the parameter, or

(c) validate the model.

BAYESIAN STATISTICS

☞ 47

2. *Bayesian statistics*: In this approach we average over all possible values of the parameter $\boldsymbol{\theta}$ using a user-specified weight function $g(\boldsymbol{\theta})$ and obtain the model $\mathcal{T} \sim \int g(\cdot \mid \boldsymbol{\theta})\, g(\boldsymbol{\theta})\, \mathrm{d}\boldsymbol{\theta}$. For practical computations, this means that we can treat $\boldsymbol{\theta}$ as a random variable with pdf $g(\boldsymbol{\theta})$. Bayes' formula $g(\boldsymbol{\theta} \mid \tau) \propto g(\tau \mid \boldsymbol{\theta})\, g(\boldsymbol{\theta})$ is used to learn $\boldsymbol{\theta}$ based on the observed data τ.

■ **Example C.10 (Iid Sample)** The most fundamental statistical model is where the data $\mathcal{T} = X_1, \ldots, X_n$ is such that the random variables $X_1, \ldots, X_n$ are assumed to be independent and identically distributed:

$$X_1, \ldots, X_n \overset{\text{iid}}{\sim} \mathsf{Dist},$$

according to some known or unknown distribution Dist. An iid sample is often called a *random sample* in the statistics literature. Note that the word "sample" can refer to both a collection of random variables and to a single random variable. It should be clear from the context which meaning is being used.

RANDOM SAMPLE

Often our guess or model for the true distribution is specified up to an unknown parameter $\boldsymbol{\theta}$, with $\boldsymbol{\theta} \in \Theta$. The most common model is:

$$X_1, \ldots, X_n \overset{\text{iid}}{\sim} \mathcal{N}(\mu, \sigma^2),$$

in which case $\boldsymbol{\theta} = (\mu, \sigma^2)$ and $\Theta = \mathbb{R} \times \mathbb{R}_+$. ■

C.12 Estimation

Suppose the model $g(\cdot \mid \boldsymbol{\theta})$ for the data $\mathcal{T}$ is completely specified up to an unknown parameter vector $\boldsymbol{\theta}$. The aim is to estimate $\boldsymbol{\theta}$ on the basis of the observed data τ only (an alternative goal could be to estimate $\boldsymbol{\eta} = \boldsymbol{\psi}(\boldsymbol{\theta})$ for some vector-valued function $\boldsymbol{\psi}$). Specifically, the goal is to find an *estimator* $\boldsymbol{T} = \boldsymbol{T}(\mathcal{T})$ that is close to the unknown $\boldsymbol{\theta}$. The corresponding outcome $\boldsymbol{t} = \boldsymbol{T}(\tau)$ is the *estimate* of $\boldsymbol{\theta}$. The *bias* of an estimator $\boldsymbol{T}$ of $\boldsymbol{\theta}$ is defined as $\mathbb{E}\boldsymbol{T} - \boldsymbol{\theta}$. An estimator $\boldsymbol{T}$ of $\boldsymbol{\theta}$ is said to be *unbiased* if $\mathbb{E}_{\boldsymbol{\theta}}\boldsymbol{T} = \boldsymbol{\theta}$. We often write $\widehat{\boldsymbol{\theta}}$ for both an estimator and estimate of $\boldsymbol{\theta}$. The *mean squared error* (MSE) of a real-valued estimator T is defined as

ESTIMATOR

ESTIMATE

BIAS

MEAN SQUARED ERROR

$$\text{MSE} = \mathbb{E}_\theta (T - \theta)^2.$$

An estimator T_1 is said to be more *efficient* than an estimator T_2 if the MSE of T_1 is smaller than the MSE of T_2. The MSE can be written as the sum

EFFICIENT

$$\text{MSE} = (\mathbb{E}_\theta T - \theta)^2 + \mathbb{V}\text{ar}_\theta T.$$

The first term measures the unbiasedness and the second is the variance of the estimator. In particular, for an unbiased estimator the MSE of an estimator is simply equal to its variance.

For simulation purposes it is often important to include the running time of the estimator in efficiency comparisons. One way to compare two unbiased estimators T_1 and T_2 is to compare their *relative time variance products*,

RELATIVE TIME VARIANCE PRODUCTS

$$\frac{r_i\, \mathbb{V}\text{ar}\, T_i}{(\mathbb{E}\, T_i)^2}, \quad i = 1, 2, \tag{C.44}$$

where r_1 and r_2 are the times required to calculate the estimators T_1 and T_2, respectively. In this scheme, T_1 is considered more efficient than T_2 if its relative time variance product is smaller. We discuss next two systematic approaches for constructing sound estimators.

C.12.1 Method of Moments

Suppose $x_1, \ldots, x_n$ are outcomes from an iid sample $X_1, \ldots, X_n \sim_{\text{iid}} g(x \mid \boldsymbol{\theta})$, where $\boldsymbol{\theta} = [\theta_1, \ldots, \theta_k]^\top$ is unknown. The moments of the sampling distribution can be easily estimated. Namely, if $X \sim g(x \mid \boldsymbol{\theta})$, then the r-th moment of X, that is $\mu_r(\boldsymbol{\theta}) = \mathbb{E}_{\boldsymbol{\theta}} X^r$ (assuming it exists), can be estimated through the *sample r-th moment*: $\frac{1}{n}\sum_{i=1}^n x_i^r$. The *method of moments* involves choosing the estimate $\widehat{\boldsymbol{\theta}}$ of $\boldsymbol{\theta}$ such that each of the first k sample and true moments are matched:

SAMPLE r-TH MOMENT

METHOD OF MOMENTS

$$\frac{1}{n}\sum_{i=1}^n x_i^r = \mu_r(\widehat{\boldsymbol{\theta}}), \quad r = 1, 2, \ldots, k.$$

In general, this set of equations is nonlinear and so its solution often has to be found numerically.

■ **Example C.11 (Sample Mean and Sample Variance)** Suppose the data is given by $\mathcal{T} = \{X_1, \ldots, X_n\}$, where the $\{X_i\}$ form an iid sample from a general distribution with mean μ and variance $\sigma^2 < \infty$. Matching the first two moments gives the set of equations

$$\frac{1}{n}\sum_{i=1}^n x_i = \mu,$$
$$\frac{1}{n}\sum_{i=1}^n x_i^2 = \mu^2 + \sigma^2.$$

The method of moments estimates for μ and σ^2 are therefore the *sample mean*

SAMPLE MEAN

$$\widehat{\mu} = \overline{x} = \frac{1}{n}\sum_{i=1}^n x_i, \tag{C.45}$$

and

$$\widehat{\sigma^2} = \frac{1}{n}\sum_{i=1}^n x_i^2 - (\overline{x})^2 = \frac{1}{n}\sum_{i=1}^n (x_i - \overline{x})^2. \tag{C.46}$$

The corresponding estimator for μ, $\overline{X}$, is unbiased. However, the estimator for σ^2 is biased: $\mathbb{E}\,\widehat{\sigma^2} = \sigma^2(n-1)/n$. An unbiased estimator is the *sample variance*

SAMPLE VARIANCE

$$S^2 = \widehat{\sigma^2}\frac{n}{n-1} = \frac{1}{n-1}\sum_{i=1}^n (X_i - \overline{X})^2.$$

Its square root, $S = \sqrt{S^2}$, is called the *sample standard deviation*. ■

SAMPLE STANDARD DEVIATION

■ **Example C.12 (Sample Covariance Matrix)** The method of moments can also be used to estimate the covariance matrix of a random vector. In particular, let the $\boldsymbol{X}_1, \ldots, \boldsymbol{X}_n$

be iid copies of a d-dimensional random vector $\boldsymbol{X}$ with mean vector $\boldsymbol{\mu}$ and covariance matrix $\boldsymbol{\Sigma}$. We assume $n \geqslant d$. The moment estimator for $\boldsymbol{\mu}$ is, as in the $d = 1$ case, $\overline{\boldsymbol{X}} = (\boldsymbol{X}_1 + \cdots + \boldsymbol{X}_n)/n$. As the covariance matrix can be written (see (C.15)) as

☞ 432

$$\boldsymbol{\Sigma} = \mathbb{E}(\boldsymbol{X} - \boldsymbol{\mu})(\boldsymbol{X} - \boldsymbol{\mu})^\top,$$

the method of moments yields the estimator

$$\widehat{\boldsymbol{\Sigma}} = \frac{1}{n}\sum_{i=1}^{n}(\boldsymbol{X}_i - \overline{\boldsymbol{X}})(\boldsymbol{X}_i - \overline{\boldsymbol{X}})^\top. \quad \text{(C.47)}$$

Similar to the one-dimensional case ($d = 1$), replacing the factor $1/n$ with $1/(n-1)$ gives an unbiased estimator, called the *sample covariance matrix*. ■

SAMPLE COVARIANCE MATRIX

C.12.2 Maximum Likelihood Method

The concept of *likelihood* is central in statistics. It describes in a precise way the information about model parameters that is contained in the observed data.

Let $\mathcal{T}$ be a (random) data object that is modeled as a draw from the pdf $g(\tau \mid \boldsymbol{\theta})$ (discrete or continuous) with parameter vector $\boldsymbol{\theta} \in \Theta$. Let τ be an outcome of $\mathcal{T}$. The function $L(\boldsymbol{\theta} \mid \tau) := g(\tau \mid \boldsymbol{\theta}),\ \boldsymbol{\theta} \in \Theta$, is called the *likelihood function* of $\boldsymbol{\theta}$, based on τ. The (natural) logarithm of the likelihood function is called the *log-likelihood function* and is often denoted by a lower case l.

LIKELIHOOD FUNCTION

LOG-LIKELIHOOD FUNCTION

> Note that $L(\boldsymbol{\theta} \mid \tau)$ and $g(\tau \mid \boldsymbol{\theta})$ have the same formula, but the first is viewed as a function of $\boldsymbol{\theta}$ for fixed τ, where the second is viewed as a function of τ for fixed $\boldsymbol{\theta}$.

The concept of likelihood is particularly useful when $\mathcal{T}$ is modeled as an iid sample $\{X_1, \ldots, X_n\}$ from some pdf $\mathring{g}$. In that case, the likelihood of the data $\tau = \{x_1, \ldots, x_n\}$, as a function of $\boldsymbol{\theta}$, is given by the product

$$L(\boldsymbol{\theta} \mid \tau) = \prod_{i=1}^{n} \mathring{g}(x_i \mid \boldsymbol{\theta}). \quad \text{(C.48)}$$

Let τ be an observation from $\mathcal{T} \sim g(\tau \mid \boldsymbol{\theta})$, and suppose that $g(\tau \mid \boldsymbol{\theta})$ takes its largest value at $\boldsymbol{\theta} = \widehat{\boldsymbol{\theta}}$. In a way this $\widehat{\boldsymbol{\theta}}$ is our best estimate for $\boldsymbol{\theta}$, as it maximizes the probability (density) for the observation τ. It is called the *maximum likelihood estimate* (MLE) of $\boldsymbol{\theta}$. Note that $\widehat{\boldsymbol{\theta}} = \widehat{\boldsymbol{\theta}}(\tau)$ is a function of τ. The corresponding random variable, also denoted $\widehat{\boldsymbol{\theta}}$ is the *maximum likelihood estimator* (also abbreviated as MLE).

MAXIMUM LIKELIHOOD ESTIMATOR

Maximization of $L(\boldsymbol{\theta} \mid \tau)$ as a function of $\boldsymbol{\theta}$ is equivalent (when searching for the maximizer) to maximizing the log-likelihood $l(\boldsymbol{\theta} \mid \tau)$, as the natural logarithm is an increasing function. This is often easier, especially when $\mathcal{T}$ is an iid sample from some sampling distribution. For example, for L of the form (C.48), we have

$$l(\boldsymbol{\theta} \mid \tau) = \sum_{i=1}^{n} \ln \mathring{g}(x_i \mid \boldsymbol{\theta}).$$

If $l(\boldsymbol{\theta} \mid \tau)$ is a differentiable function with respect to $\boldsymbol{\theta}$ and the maximum is attained in the *interior* of Θ, *and there exists a unique maximum point*, then we can find the MLE of $\boldsymbol{\theta}$ by solving the equations

$$\frac{\partial}{\partial \theta_i} l(\boldsymbol{\theta} \mid \tau) = 0, \quad i = 1, \ldots, d.$$

■ **Example C.13 (Bernoulli Random Sample)** Suppose we have data $\tau_n = \{x_1, \ldots, x_n\}$ and assume the model $X_1, \ldots, X_n \sim_{\text{iid}} \mathsf{Ber}(\theta)$. Then, the likelihood function is given by

$$L(\theta \mid \tau) = \prod_{i=1}^{n} \theta^{x_i}(1-\theta)^{1-x_i} = \theta^s (1-\theta)^{n-s}, \quad 0 < \theta < 1, \tag{C.49}$$

where $s := x_1 + \cdots + x_n =: n\overline{x}$. The log-likelihood is $l(\theta) = s \ln \theta + (n-s)\ln(1-\theta)$. Through differentiation with respect to θ, we find the derivative

$$\frac{s}{\theta} - \frac{n-s}{1-\theta} = \frac{s}{\theta(1-\theta)} - \frac{n}{1-\theta}. \tag{C.50}$$

Solving $l'(\theta) = 0$ gives the ML estimate $\widehat{\theta} = \overline{x}$ and ML estimator $\widehat{\theta} = \overline{X}$. ■

C.13 Confidence Intervals

An essential part in any estimation procedure is to provide an assessment of the *accuracy* of the estimate. Indeed, without information on its accuracy the estimate itself would be meaningless. Confidence intervals (also called *interval estimates*) provide a precise way of describing the uncertainty in the estimate. INTERVAL ESTIMATES

Let $X_1, \ldots, X_n$ be random variables with a joint distribution depending on a parameter $\theta \in \Theta$. Let $T_1 < T_2$ be statistics; that is, $T_i = T_i(X_1, \ldots, X_n)$, $i = 1, 2$ are functions of the data, but not of θ.

1. The random interval (T_1, T_2) is called a *stochastic confidence interval* for θ with confidence $1 - \alpha$ if STOCHASTIC CONFIDENCE INTERVAL

$$\mathbb{P}_\theta[T_1 < \theta < T_2] \geqslant 1 - \alpha \quad \text{for all } \theta \in \Theta. \tag{C.51}$$

2. If t_1 and t_2 are the observed values of T_1 and T_2, then the interval (t_1, t_2) is called the *(numerical) confidence interval* for θ with confidence $1 - \alpha$ for every $\theta \in \Theta$. (NUMERICAL) CONFIDENCE INTERVAL

3. If the right-hand side of (C.51) is merely a heuristic estimate or approximation of the true probability, then the resulting interval is called an *approximate confidence interval*.

4. The probability $\mathbb{P}_\theta[T_1 < \theta < T_2]$ is called the *coverage probability*. For a $1 - \alpha$ confidence interval, it must be at least $1 - \alpha$. COVERAGE PROBABILITY

For multidimensional parameters $\boldsymbol{\theta} \in \mathbb{R}^d$ the stochastic confidence interval is replaced with a stochastic *confidence region* $C \subset \mathbb{R}^d$ such that $\mathbb{P}_\theta[\boldsymbol{\theta} \in C] \geqslant 1 - \alpha$ for all $\boldsymbol{\theta}$. CONFIDENCE REGION

■ **Example C.14 (Approximate Confidence Interval for the Mean)** Let $X_1, X_2, \ldots, X_n$ be an iid sample from a distribution with mean μ and variance $\sigma^2 < \infty$ (both assumed to be unknown). By the central limit theorem and the law of large numbers,

☞ 449

$$T = \frac{\overline{X} - \mu}{S/\sqrt{n}} \overset{\text{approx.}}{\sim} \mathcal{N}(0, 1),$$

for large n, where S is the sample standard deviation. Rearranging the approximate equality $\mathbb{P}[|T| \leqslant z_{1-\alpha/2}] \approx 1 - \alpha$, where $z_{1-\alpha/2}$ is the $1 - \alpha/2$ quantile of the standard normal distribution, yields

$$\mathbb{P}\left[\overline{X} - z_{1-\alpha/2}\frac{S}{\sqrt{n}} \leqslant \mu \leqslant \overline{X} + z_{1-\alpha/2}\frac{S}{\sqrt{n}}\right] \approx 1 - \alpha,$$

so that

$$\left(\overline{X} - z_{1-\alpha/2}\frac{S}{\sqrt{n}}, \overline{X} + z_{1-\alpha/2}\frac{S}{\sqrt{n}}\right), \text{ abbreviated as } \overline{X} \pm z_{1-\alpha/2}\frac{S}{\sqrt{n}}, \tag{C.52}$$

is an approximate stochastic $(1 - \alpha)$ confidence interval for μ. ■

Since (C.52) is an asymptotic result only, care should be taken when applying it to cases where the sample size is small or moderate and the sampling distribution is heavily skewed.

C.14 Hypothesis Testing

Suppose the model for the data $\mathcal{T}$ is described by a family of probability distributions that depend on a parameter $\boldsymbol{\theta} \in \Theta$. The aim of *hypothesis testing* is to decide, on the basis of the observed data τ, which of two competing hypotheses holds true; these being the *null hypothesis*, $H_0 : \boldsymbol{\theta} \in \Theta_0$, and the *alternative hypothesis*, $H_1 : \boldsymbol{\theta} \in \Theta_1$.

HYPOTHESIS TESTING

NULL HYPOTHESIS

ALTERNATIVE HYPOTHESIS

In classical statistics the null hypothesis and alternative hypothesis do not play equivalent roles. H_0 contains the "status quo" statement and is only rejected if the observed data are very unlikely to have happened under H_0.

The decision whether to accept or reject H_0 is dependent on the outcome of a *test statistic* $\boldsymbol{T} = \boldsymbol{T}(\mathcal{T})$. For simplicity, we discuss only the one-dimensional case $\boldsymbol{T} \equiv T$. Two (related) types of decision rules are generally used:

TEST STATISTIC

1. **Decision rule 1**: *Reject H_0 if T falls in the critical region.*

   Here the *critical region* is any appropriately chosen region in $\mathbb{R}$. In practice a critical region is one of the following:

   CRITICAL REGION

   - *left one-sided*: $(-\infty, c]$,
   - *right one-sided*: $[c, \infty)$,
   - *two-sided*: $(-\infty, c_1] \cup [c_2, \infty)$.

   For example, for a right one-sided test, H_0 is rejected if the outcome of the test statistic is too large. The endpoints c, c_1, and c_2 of the critical regions are called *critical values*.

   CRITICAL VALUES

2. **Decision rule 2**: *Reject H_0 if the* P-value *is smaller than some* significance level α.

   The *P-value* is the probability that, under H_0, the (random) test statistic takes a value as extreme as or more extreme than the one observed. In particular, if t is the observed outcome of the test statistic T, then

   P-VALUE

   - *left one-sided test*: $P := \mathbb{P}_{H_0}[T \leqslant t]$,
   - *right one-sided*: $P := \mathbb{P}_{H_0}[T \geqslant t]$,
   - *two-sided*: $P := \min\{2\mathbb{P}_{H_0}[T \leqslant t],\ 2\mathbb{P}_{H_0}[T \geqslant t]\}$.

   The smaller the P-value, the greater the strength of the evidence against H_0 provided by the data. As a rule of thumb:

   $$\begin{array}{ll} P < 0.10 & \text{suggestive evidence,} \\ P < 0.05 & \text{reasonable evidence,} \\ P < 0.01 & \text{strong evidence.} \end{array}$$

Whether the first or the second decision rule is used, one can make two types of errors, as depicted in Table C.4.

Table C.4: Type I and II errors in hypothesis testing.

	True statement	
Decision	*H_0 is true*	*H_1 is true*
Accept H_0	Correct	Type II Error
Reject H_0	Type I Error	Correct

The choice of the test statistic and the corresponding critical region involves a multiobjective optimization criterion, whereby both the probabilities of a type I and type II error should, ideally, be chosen as small as possible. Unfortunately, these probabilities compete with each other. For example, if the critical region is made larger (smaller), the probability of a type II error is reduced (increased), but at the same time the probability of a type I error is increased (reduced).

Since the type I error is considered more serious, Neyman and Pearson [93] suggested the following approach: choose the critical region such that the probability of a type II error is as small as possible, while keeping the probability of a type I error below a predetermined small *significance level* α.

SIGNIFICANCE LEVEL

■ **Remark C.3 (Equivalence of Decision Rules)** Note that decision rule 1 and 2 are equivalent in the following sense:

$$\text{Reject } H_0 \text{ if } T \text{ falls in the critical region, at significance level } \alpha.$$
$$\Leftrightarrow$$
$$\text{Reject } H_0 \text{ if the P-value is} \leqslant \text{significance level } \alpha.$$

In other words, the P-value of the test is the smallest level of significance that would lead to the rejection of H_0. ■

In general, a statistical test involves the following steps:

1. Formulate an appropriate statistical model for the data.
2. Give the null (H_0) and alternative (H_1) hypotheses in terms of the parameters of the model.
3. Determine the test statistic (a function of the data only).
4. Determine the (approximate) distribution of the test statistic under H_0.
5. Calculate the outcome of the test statistic.
6. Calculate the P-value *or* the critical region, given a preselected significance level α.
7. Accept or reject H_0.

The actual choice of an appropriate test statistic is akin to selecting a good estimator for the unknown parameter $\boldsymbol{\theta}$. The test statistic should summarize the information about $\boldsymbol{\theta}$ and make it possible to distinguish between the alternative hypotheses.

■ **Example C.15 (Hypothesis Testing)** We are given outcomes $x_1, \ldots, x_m$ and $y_1, \ldots, y_n$ of two simulation studies obtained via independent runs, with $m = 100$ and $n = 50$. The sample means and standard deviations are $\overline{x} = 1.3$, $s_X = 0.1$ and $\overline{y} = 1.5$, $s_Y = 0.3$. Thus, the $\{x_i\}$ are outcomes of iid random variables $\{X_i\}$, the $\{y_i\}$ are outcomes of iid random variables $\{Y_i\}$, and the $\{X_i\}$ and $\{Y_i\}$ are independent. We wish to assess whether the expectations $\mu_X = \mathbb{E}X_i$ and $\mu_Y = \mathbb{E}Y_i$ are the same or not. Going through the 7 steps above, we have:

1. The model is already specified above.
2. $H_0 : \mu_X - \mu_Y = 0$ versus $H_1 : \mu_X - \mu_Y \neq 0$.
3. For similar reasons as in Example C.14, take
$$T = \frac{\overline{X} - \overline{Y}}{\sqrt{S_X^2/m + S_Y^2/n}}.$$
4. By the central limit theorem, the statistic T has, under H_0, approximately a standard normal distribution (assuming the variances are finite).
5. The outcome of T is $t = (\overline{x} - \overline{y})/\sqrt{s_X^2/m + s_Y^2/n} \approx -4.59$.
6. As this is a two-sided test, the P-value is $2\mathbb{P}_{H_0}[T \leqslant -4.59] \approx 4 \cdot 10^{-6}$.
7. Because the P-value is extremely small, there is overwhelming evidence that the two expectations are not the same.

■

Further Reading

Accessible treatises on probability and stochastic processes include [27, 26, 39, 54, 101]. Kallenberg's book [61] provides a complete graduate-level overview of the foundations of modern probability. Details on the convergence of probability measures and limit theorems can be found in [11]. For an accessible introduction to mathematical statistics with simple applications see, for example, [69, 74, 124]. For a more detailed overview of statistical inference, see [10, 25]. A standard reference for classical (frequentist) statistical inference is [78].

APPENDIX D

PYTHON PRIMER

Python has become the programming language of choice for many researchers and practitioners in data science and machine learning. This appendix gives a brief introduction to the language. As the language is under constant development and each year many new packages are being released, we do not pretend to be exhaustive in this introduction. Instead, we hope to provide enough information for novices to get started with this beautiful and carefully thought-out language.

D.1 Getting Started

The main website for Python is

`https://www.python.org/`,

where you will find documentation, a tutorial, beginners' guides, software examples, and so on. It is important to note that there are two incompatible "branches" of Python, called Python 3 and Python 2. Further development of the language will involve only Python 3, and in this appendix (and indeed the rest of the book) we only consider Python 3. As there are many interdependent packages that are frequently used with a Python installation, it is convenient to install a distribution — for instance, the *Anaconda* Python distribution, available from ANACONDA

`https://www.anaconda.com/`.

The Anaconda installer automatically installs the most important packages and also provides a convenient interactive development environment (IDE), called *Spyder*.

> Use the *Anaconda Navigator* to launch *Spyder*, *Jupyter notebook*, install and update packages, or open a command-line terminal.

To get started[1], try out the Python statements in the input boxes that follow. You can either type these statements at the IPython command prompt or run them as (very short)

[1]We assume that you have installed all the necessary files and have launched *Spyder*.

Python programs. The output for these two modes of input can differ slightly. For example, typing a variable name in the console causes its contents to be automatically printed, whereas in a Python program this must be done explicitly by calling the **print** function. Selecting (highlighting) several program lines in *Spyder* and then pressing function key[2] F9 is equivalent to executing these lines one by one in the console.

OBJECT

In Python, data is represented as an *object* or relation between objects (see also Section D.2). Basic data types are numeric types (including integers, booleans, and floats), sequence types (including strings, tuples, and lists), sets, and mappings (currently, dictionaries are the only built-in mapping type).

Strings are sequences of characters, enclosed by single or double quotes. We can print strings via the **print** function.

```
print("Hello World!")
```

```
Hello World!
```

For pretty-printing output, Python strings can be formatted using the **format** function. The bracket syntax {i} provides a placeholder for the i-th variable to be printed, with 0 being the first index. Individual variables can be formatted separately and as desired; formatting syntax is discussed in more detail in Section D.9.

☞ 477

```
print("Name:{1} (height {2} m, age {0})".format(111,"Bilbo",0.84))
```

```
Name:Bilbo (height 0.84 m, age 111)
```

Lists can contain different types of objects, and are created using square brackets as in the following example:

```
x = [1,'string',"another string"]  # Quote type is not important
```

```
[1, 'string', 'another string']
```

MUTABLE

Elements in lists are indexed starting from 0, and are *mutable* (can be changed):

```
x = [1,2]
x[0] = 2  # Note that the first index is 0
x
```

```
[2,2]
```

IMMUTABLE

In contrast, tuples (with round brackets) are *immutable* (cannot be changed). Strings are immutable as well.

```
x = (1,2)
x[0] = 2
```

```
TypeError: 'tuple' object does not support item assignment
```

SLICE

Lists can be accessed via the *slice* notation [start:end]. It is important to note that end is the index of the first element that will *not* be selected, and that the first element has index 0. To gain familiarity with the slice notation, execute each of the following lines.

[2]This may depend on the keyboard and operating system.

```
a = [2, 3, 5, 7, 11, 13, 17, 19, 23]
a[1:4]  # Elements with index from 1 to 3
a[:4]   # All elements with index less than 4
a[3:]   # All elements with index 3 or more
a[-2:]  # The last two elements
```

```
[3, 5, 7]
[2, 3, 5, 7]
[7, 11, 13, 17, 19, 23]
[19, 23]
```

An *operator* is a programming language construct that performs an action on one or more operands. The action of an operator in Python depends on the type of the operand(s). For example, operators such as +, ∗, −, and % that are arithmetic operators when the operands are of a numeric type, can have different meanings for objects of non-numeric type (such as strings). OPERATOR

```
'hello' + 'world'  # String concatenation
```

```
'helloworld'
```

```
'hello' * 2   # String repetition
```

```
'hellohello'
```

```
[1,2] * 2      # List repetition
```

```
[1, 2, 1, 2]
```

```
15 % 4   # Remainder of 15/4
```

```
3
```

Some common Python operators are given in Table D.1. ☞ 469

D.2 Python Objects

As mentioned in the previous section, data in Python is represented by objects or relations between objects. We recall that basic data types included strings and numeric types (such as integers, booleans, and floats).

As Python is an object-oriented programming language, functions are objects too (everything is an object!). Each object has an identity (unique to each object and immutable — that is, cannot be changed — once created), a type (which determines which operations can be applied to the object, and is considered immutable), and a value (which is either mutable or immutable). The unique identity assigned to an object `obj` can be found by calling **`id`**, as in `id(obj)`.

Each object has a list of *attributes*, and each attribute is a reference to another object. The function **`dir`** applied to an object returns the list of attributes. For example, a string object has many useful attributes, as we shall shortly see. Functions are objects with the `__call__` attribute. ATTRIBUTES

A class (see Section D.8) can be thought of as a template for creating a custom type of object.

```
s = "hello"
d = dir(s)
print(d,flush=True)  # Print the list in "flushed" format
```

```
['__add__', '__class__', '__contains__', '__delattr__', '__dir__',
 ... (many left out) ... 'replace', 'rfind',
  'rindex', 'rjust', 'rpartition', 'rsplit', 'rstrip', 'split',
  'splitlines', 'startswith', 'strip', 'swapcase', 'title',
  'translate', 'upper', 'zfill']
```

DOT NOTATION

Any attribute `attr` of an object `obj` can be accessed via the *dot notation*: `obj.attr`. To find more information about any object use the **`help`** function.

```
s = "hello"
help(s.replace)
```

```
replace(...) method of builtins.str instance
    S.replace(old, new[, count]) -> str

    Return a copy of S with all occurrences of substring
    old replaced by new.  If the optional argument count is
    given, only the first count occurrences are replaced.
```

METHOD

This shows that the attribute **`replace`** is in fact a function. An attribute that is a function is called a *method*. We can use the **`replace`** method to create a new string from the old one by changing certain characters.

```
s = 'hello'
s1 = s.replace('e','a')
print(s1)
```

```
hallo
```

In many Python editors, pressing the TAB key, as in `objectname.<TAB>`, will bring up a list of possible attributes via the editor's autocompletion feature.

D.3 Types and Operators

TYPE

Each object has a *type*. Three basic data types in Python are `str` (for string), `int` (for integers), and `float` (for floating point numbers). The function **`type`** returns the type of an object.

```
t1 = type([1,2,3])
t2 = type((1,2,3))
t3 = type({1,2,3})
print(t1,t2,t3)
```

```
<class 'list'> <class 'tuple'> <class 'set'>
```

The *assignment* operator, =, assigns an object to a variable; e.g., `x = 12`. An *expression* is a combination of values, operators, and variables that yields another value or variable.

ASSIGNMENT

> Variable names are case sensitive and can only contain letters, numbers, and underscores. They must start with either a letter or underscore. Note that reserved words such as `True` and `False` are case sensitive as well.

Python is a dynamically typed language, and the type of a variable at a particular point during program execution is determined by its most recent object assignment. That is, the type of a variable does not need to be explicitly declared from the outset (as is the case in C or Java), but instead the type of the variable is determined by the object that is currently assigned to it.

It is important to understand that a variable in Python is a *reference* to an object — think of it as a label on a shoe box. Even though the label is a simple entity, the *contents* of the shoe box (the object to which the variable refers) can be arbitrarily complex. Instead of moving the contents of one shoe box to another, it is much simpler to merely move the label.

REFERENCE

```
x = [1,2]
y = x   # y refers to the same object as x
print(id(x) == id(y))  # check that the object id's are the same
y[0] = 100  # change the contents of the list that y refers to
print(x)
```

```
True
[100,2]
```

```
x = [1,2]
y = x   # y refers to the same object as x
y = [100,2]  # now y refers to a different object
print(id(x) == id(y))
print(x)
```

```
False
[1,2]
```

Table D.1 shows a selection of Python operators for numerical and logical variables.

Table D.1: Common numerical (left) and logical (right) operators.

<table>
<tr><td>+</td><td>addition</td><td>~</td><td>binary NOT</td></tr>
<tr><td>-</td><td>subtraction</td><td>&</td><td>binary AND</td></tr>
<tr><td>*</td><td>multiplication</td><td>^</td><td>binary XOR</td></tr>
<tr><td>**</td><td>power</td><td>|</td><td>binary OR</td></tr>
<tr><td>/</td><td>division</td><td>==</td><td>equal to</td></tr>
<tr><td>//</td><td>integer division</td><td>!=</td><td>not equal to</td></tr>
<tr><td>%</td><td>modulus</td><td></td><td></td></tr>
</table>

Several of the numerical operators can be combined with an assignment operator, as in `x += 1` to mean `x = x + 1`. Operators such as + and * can be defined for other data types as well, where they take on a different meaning. This is called operator *overloading*, an example of which is the use of `<List> * <Integer>` for list repetition as we saw earlier.

D.4 Functions and Methods

FUNCTION

Functions make it easier to divide a complex program into simpler parts. To create a *function*, use the following syntax:

```
def <function name>(<parameter_list>):
    <statements>
```

A function takes a list of input variables that are references to objects. Inside the function, a number of statements are executed which may modify the objects, but not the reference itself. In addition, the function may return an output object (or will return the value `None` if not explicitly instructed to return output). Think again of the shoe box analogy. The input variables of a function are labels of shoe boxes, and the objects to which they refer are the contents of the shoe boxes. The following program highlights some of the subtleties of variables and objects in Python.

> Note that the statements within a function must be indented. This is Python's way to define where a function begins and ends.

```
x = [1,2,3]

def change_list(y):
    y.append(100) # Append an element to the list referenced by y
    y[0]=0        # Modify the first element of the same list
    y = [2,3,4]   # The local y now refers to a different list
            # The list to which y first referred does not change
    return sum(y)

print(change_list(x))
print(x)
```

```
9
[0, 2, 3, 100]
```

Variables that are defined inside a function only have *local scope*; that is, they are recognized only within that function. This allows the same variable name to be used in different functions without creating a conflict. If any variable is used within a function, Python first checks if the variable has local scope. If this is not the case (the variable has not been defined inside the function), then Python searches for that variable outside the function (the global scope). The following program illustrates several important points.

```
from numpy import array, square, sqrt

x = array([1.2,2.3,4.5])

def stat(x):
    n = len(x)         #the length of x
    meanx = sum(x)/n
    stdx = sqrt(sum(square(x - meanx))/n)
    return [meanx,stdx]

print(stat(x))
```

```
[2.6666666666666665, 1.3719410418171119]
```

1. Basic math functions such as sqrt are unknown to the standard Python interpreter and need to be imported. More on this in Section D.5 below.

2. As was already mentioned, indentation is crucial. It shows where the function begins and ends.

3. No semicolons[3] are needed to end lines, but the first line of the function definition (here line 5) must end with a colon (:).

4. Lists are not arrays (vectors of numbers), and vector operations cannot be performed on lists. However, the numpy module is designed specifically with efficient vector/matrix operations in mind. On the second code line, we define x as a vector (ndarray) object. Functions such as square, sum, and sqrt are then applied to such arrays. Note that we used the default Python functions len and sum. More on numpy in Section D.10.

5. Running the program with stat(x) instead of print(stat(x)) in line 11 will not show any output in the console.

> To display the complete list of built-in functions, type (using double underscores) dir(__builtin__) .

D.5 Modules

A Python *module* is a programming construct that is useful for organizing code into manageable parts. To each module with name module_name is associated a Python file module_name.py containing any number of definitions, e.g., of functions, classes, and variables, as well as executable statements. Modules can be imported into other programs using the syntax: import <module_name> as <alias_name>, where <alias_name> is a shorthand name for the module. MODULE

[3]Semicolons can be used to put multiple commands on a single line.

NAMESPACE

When imported into another Python file, the module name is treated as a *namespace*, providing a naming system where each object has its unique name. For example, different modules `mod1` and `mod2` can have different `sum` functions, but they can be distinguished by prefixing the function name with the module name via the dot notation, as in `mod1.sum` and `mod2.sum`. For example, the following code uses the `sqrt` function of the `numpy` module.

```
import numpy as np
np.sqrt(2)
```

```
1.4142135623730951
```

A Python *package* is simply a directory of Python modules; that is, a collection of modules with additional startup information (some of which may be found in its `__path__` attribute). Python's built-in module is called `__builtins__`. Of the great many useful Python modules, Table D.2 gives a few.

Table D.2: A few useful Python modules/packages.

`datetime`	Module for manipulating dates and times.
`matplotlib`	MATLAB™-type plotting package
`numpy`	Fundamental package for scientific computing, including random number generation and linear algebra tools. Defines the ubiquitous `ndarray` class.
`os`	Python interface to the operating system.
`pandas`	Fundamental module for data analysis. Defines the powerful `DataFrame` class.
`pytorch`	Machine learning library that supports GPU computation.
`scipy`	Ecosystem for mathematics, science, and engineering, containing many tools for numerical computing, including those for integration, solving differential equations, and optimization.
`requests`	Library for performing HTTP requests and interfacing with the web.
`seaborn`	Package for statistical data visualization.
`sklearn`	Easy to use machine learning library.
`statsmodels`	Package for the analysis of statistical models.

The `numpy` package contains various subpackages, such as `random`, `linalg`, and `fft`. More details are given in Section D.10.

When using *Spyder*, press `Ctrl+I` in front of any object, to display its help file in a separate window.

As we have already seen, it is also possible to import only specific functions from a module using the syntax: `from <module_name> import <fnc1, fnc2, ...>`.

```
from numpy import sqrt, cos
sqrt(2)
cos(1)
```

```
1.4142135623730951
0.5403023058681397
```

This avoids the tedious prefixing of functions via the (alias) of the module name. However, for large programs it is good practice to always use the prefix/alias name construction, to be able to clearly ascertain precisely which module a function being used belongs to.

D.6 Flow Control

Flow control in Python is similar to that of many programming languages, with conditional statements as well as `while` and `for` loops. The syntax for `if-then-else` flow control is as follows.

```
if <condition1>:
   <statements>
elif <condition2>:
   <statements>
else:
   <statements>
```

Here, `<condition1>` and `<condition2>` are logical conditions that are either `True` or `False`; logical conditions often involve comparison operators (such as `==, >, <=, !=`). In the example above, there is one `elif` part, which allows for an "else if" conditional statement. In general, there can be more than one `elif` part, or it can be omitted. The `else` part can also be omitted. The colons are essential, as are the indentations.

The `while` and `for` loops have the following syntax.

```
while <condition>:
   <statements>

for <variable> in <collection>:
   <statements>
```

Above, `<collection>` is an iterable object (see Section D.7 below). For further control in `for` and `while` loops, one can use a `break` statement to exit the current loop, and the `continue` statement to continue with the next iteration of the loop, while abandoning any remaining statements in the current iteration. Here is an example.

```
import numpy as np
ans = 'y'
while ans != 'n':
  outcome = np.random.randint(1,6+1)
  if outcome == 6:
      print("Hooray a 6!")
      break
  else:
      print("Bad luck, a", outcome)
  ans = input("Again? (y/n) ")
```

D.7 Iteration

Iterating over a sequence of objects, such as used in a for loop, is a common operation. To better understand how iteration works, we consider the following code.

```
s = "Hello"
for c in s:
    print(c,'*', end=' ')
```

```
H * e * l * l * o *
```

ITERABLE

ITERATOR

A string is an example of a Python object that can be iterated. One of the methods of a string object is `__iter__`. Any object that has such a method is called an *iterable*. Calling this method creates an *iterator* — an object that returns the next element in the sequence to be iterated. This is done via the method `__next__`.

```
s = "Hello"
t = s.__iter__()     # t is now an iterator. Same as iter(s)
print(t.__next__() ) # same as next(t)
print(t.__next__() )
print(t.__next__() )
```

```
H
e
l
```

The inbuilt functions **next** and **iter** simply call these corresponding double-underscore functions of an object. When executing a for loop, the sequence/collection over which to iterate must be an iterable. During the execution of the for loop, an iterator is created and the **next** function is executed until there is no next element. An iterator is also an iterable, so can be used in a for loop as well. Lists, tuples, and strings are so-called *sequence* objects and are iterables, where the elements are iterated by their index.

SEQUENCE

RANGE

The most common iterator in Python is the *range* iterator, which allows iteration over a range of indices. Note that **range** returns a range object, not a list.

```
for i in range(4,20):
    print(i, end=' ')
print(range(4,20))
```

```
4 5 6 7 8 9 10 11 12 13 14 15 16 17 18 19
range(4,20)
```

⚠ Similar to Python's slice operator $[i : j]$, the iterator range(i, j) ranges from i to j, *not including* the index j.

SETS

Two other common iterables are sets and dictionaries. Python *sets* are, as in mathematics, unordered collections of unique objects. Sets are defined with curly brackets { }, as opposed to round brackets ( ) for tuples, and square brackets [ ] for lists. Unlike lists, sets do not have duplicate elements. Many of the usual set operations are implemented in Python, including the union `A | B` and intersection `A & B`.

```
A = {3, 2, 2, 4}
B = {4, 3, 1}
C = A & B
for i in A:
    print(i)
print(C)
```

```
2
3
4
{3, 4}
```

A useful way to construct lists is by *list comprehension*; that is, by expressions of the form

LIST COMPREHENSION

`<expression> for <element> in <list> if <condition>`

For sets a similar construction holds. In this way, lists and sets can be defined using very similar syntax as in mathematics. Compare, for example, the mathematical definition of the sets $A := \{3, 2, 4, 2\} = \{2, 3, 4\}$ (no order and no duplication of elements) and $B := \{x^2 : x \in A\}$ with the Python code below.

```
setA = {3, 2, 4, 2}
setB = {x**2 for x in setA}
print(setB)
listA = [3, 2, 4, 2]
listB = [x**2 for x in listA]
print(listB)
```

```
{16, 9, 4}
[9, 4, 16, 4]
```

A *dictionary* is a set-like data structure, containing one or more `key:value` pairs enclosed in curly brackets. The keys are often of the same type, but do not have to be; the same holds for the values. Here is a simple example, storing the ages of Lord of the Rings characters in a dictionary.

DICTIONARY

```
DICT = {'Gimly': 140, 'Frodo':51, 'Aragorn': 88}
for key in DICT:
    print(key, DICT[key])
```

```
Gimly 140
Frodo 51
Aragorn 88
```

D.8 Classes

Recall that objects are of fundamental importance in Python — indeed, data types and functions are all objects. A *class* is an object type, and writing a class definition can be thought of as creating a template for a new type of object. Each class contains a number of attributes, including a number of inbuilt methods. The basic syntax for the creation of a class is:

CLASS

```
class <class_name>:
   def __init__(self):
       <statements>
   <statements>
```

INSTANCE

The main inbuilt method is `__init__`, which creates an *instance* of a class object. For example, `str` is a class object (string class), but `s = str('Hello')` or simply `s = 'Hello'`, creates an instance, `s`, of the `str` class. Instance attributes are created during initialization and their values may be different for different instances. In contrast, the values of class attributes are the same for every instance. The variable `self` in the initialization method refers to the current instance that is being created. Here is a simple example, explaining how attributes are assigned.

```
class shire_person:
   def __init__(self,name): # initialization method
       self.name = name     # instance attribute
       self.age = 0             # instance attribute
   address = 'The Shire'        # class attribute

print(dir(shire_person)[1:5],'...',dir(shire_person)[-2:])
                             # list of class attributes

p1 = shire_person('Sam')   # create an instance
p2 = shire_person('Frodo') # create another instance
print(p1.__dict__)   # list of instance attributes

p2.race = 'Hobbit'   # add another attribute to instance p2
p2.age = 33          # change instance attribute
print(p2.__dict__)

print(getattr(p1,'address'))   # content of p1's class attribute
```

```
['__delattr__', '__dict__', '__dir__', '__doc__'] ...
['__weakref__', 'address']
{'name': 'Sam', 'age': 0}
{'name': 'Frodo', 'age': 33, 'race': 'Hobbit'}
The Shire
```

It is good practice to create all the attributes of the class object in the `__init__` method, but, as seen in the example above, attributes can be created and assigned everywhere, even outside the class definition. More generally, attributes can be added to any object that has a `__dict__`.

An "empty" class can be created via

```
class <class_name>:
   pass
```

INHERITANCE

Python classes can be derived from a parent class by *inheritance*, via the following syntax.

```
class <class_name>(<parent_class_name>):
   <statements>
```

The derived class (initially) inherits all of the attributes of the parent class.

As an example, the class `shire_person` below inherits the attributes `name`, `age`, and `address` from its parent class `person`. This is done using the `super` function, used here to refer to the parent class `person` without naming it explicitly. When creating a new object of type `shire_person`, the `__init__` method of the parent class is invoked, and an additional instance attribute `Shire_address` is created. The `dir` function confirms that `Shire_address` is an attribute only of `shire_person` instances.

```
class person:
   def __init__(self,name):
        self.name = name
        self.age = 0
        self.address= ' '

class shire_person(person):
   def __init__(self,name):
        super().__init__(name)
        self.Shire_address = 'Bag End'

p1 = shire_person("Frodo")
p2 = person("Gandalf")
print(dir(p1)[:1],dir(p1)[-3:] )
print(dir(p2)[:1],dir(p2)[-3:] )
```

```
['Shire_address'] ['address', 'age', 'name']
['__class__'] ['address', 'age', 'name']
```

D.9 Files

To write to or read from a file, a file first needs to be opened. The `open` function in Python creates a file object that is iterable, and thus can be processed in a sequential manner in a `for` or `while` loop. Here is a simple example.

```
fout = open('output.txt','w')
for i in range(0,41):
    if i%10 == 0:
        fout.write('{:3d}\n'.format(i))
fout.close()
```

The first argument of `open` is the name of the file. The second argument specifies if the file is opened for reading (`'r'`), writing (`'w'`), appending (`'a'`), and so on. See `help(open)`. Files are written in text mode by default, but it is also possible to write in binary mode. The above program creates a file `output.txt` with 5 lines, containing the strings 0, 10, . . . , 40. Note that if we had written `fout.write(i)` in the fourth line of the code above, an error message would be produced, as the variable `i` is an integer, and not a string. Recall that the expression `string.format()` is Python's way to specify the format of the output string.

The formatting syntax `{:3d}` indicates that the output should be constrained to a specific width of three characters, each of which is a decimal value. As mentioned in the introduction, bracket syntax {i} provides a placeholder for the i-th variable to be printed, with 0 being the first index. The format for the output is further specified by {i:format}, where format is typically[4] of the form:

[width][.precision][type]
In this specification:

- width specifies the minimum width of output;
- precision specifies the number of digits to be displayed after the decimal point for a floating point values of type f, or the number of digits before *and* after the decimal point for a floating point values of type g;
- type specifies the type of output. The most common types are s for strings, d for integers, b for binary numbers, f for floating point numbers (floats) in fixed-point notation, g for floats in general notation, e for floats in scientific notation.

The following illustrates some behavior of formatting on numbers.

```
'{:5d}'.format(123)
'{:.4e}'.format(1234567890)
'{:.2f}'.format(1234567890)
'{:.2f}'.format(2.718281828)
'{:.3f}'.format(2.718281828)
'{:.3g}'.format(2.718281828)
'{:.3e}'.format(2.718281828)
'{0:3.3f}; {2:.4e};'.format(123.456789,  0.00123456789)
```

```
'  123'
'1.2346e+09'
'1234567890.00'
'2.72'
'2.718'
'2.72'
'2.718e+00'
'123.457; 1.2346e-03;'
```

The following code reads the text file output.txt line by line, and prints the output on the screen. To remove the newline \n character, we have used the strip method for strings, which removes any whitespace from the start and end of a string.

```
fin = open('output.txt','r')
for line in fin:
    line = line.strip()   # strips a newline character
    print(line)
fin.close()
```

```
0
10
20
30
40
```

[4]More formatting options are possible.

When dealing with file input and output it is important to always close files. Files that remain open, e.g., when a program finishes unexpectedly due to a programming error, can cause considerable system problems. For this reason it is recommended to open files via *context management*. The syntax is as follows.

```
with open('output.txt', 'w') as f:
      f.write('Hi there!')
```

Context management ensures that a file is correctly closed even when the program is terminated prematurely. An example is given in the next program, which outputs the most-frequent words in Dicken's *A Tale of Two Cities*, which can be downloaded from the book's GitHub site as `ataleof2cities.txt`.

Note that in the next program, the file `ataleof2cities.txt` must be placed in the current working directory. The current working directory can be determined via `import os` followed by `cwd = os.getcwd()`.

```
numline = 0
DICT = {}
with open('ataleof2cities.txt', encoding="utf8") as fin:
    for line in fin:
        words = line.split()
        for w in words:
           if w not in DICT:
               DICT[w] = 1
           else:
               DICT[w] +=1
        numline += 1

sd = sorted(DICT,key=DICT.get,reverse=True) #sort the dictionary

print("Number of unique words: {}\n".format(len(DICT)))
print("Ten most frequent words:\n")
print("{:8} {}".format("word", "count"))
print(15*'-')
for i in range(0,10):
    print("{:8} {}".format(sd[i], DICT[sd[i]]))
```

```
Number of unique words: 19091

Ten most frequent words:

word     count
---------------
the      7348
and      4679
of       3949
to       3387
a        2768
in       2390
his      1911
was      1672
that     1650
I        1444
```

D.10 NumPy

The package NumPy (module name numpy) provides the building blocks for scientific computing in Python. It contains all the standard mathematical functions, such as **sin**, **cos**, **tan**, etc., as well as efficient functions for random number generation, linear algebra, and statistical computation.

```
import numpy as np   #import the package
x = np.cos(1)
data = [1,2,3,4,5]
y = np.mean(data)
z = np.std(data)
print('cos(1) = {0:1.8f}  mean = {1}  std = {2}'.format(x,y,z))
```

```
cos(1) = 0.54030231  mean = 3.0  std = 1.4142135623730951
```

D.10.1 Creating and Shaping Arrays

The fundamental data type in numpy is the ndarray. This data type allows for fast matrix operations via highly optimized numerical libraries such as LAPACK and BLAS; this in contrast to (nested) lists. As such, numpy is often essential when dealing with large amounts of quantitative data.

ndarray objects can be created in various ways. The following code creates a $2\times3\times2$ array of zeros. Think of it as a 3-dimensional matrix or two stacked 3×2 matrices.

```
A = np.zeros([2,3,2])  # 2 by 3 by 2 array of zeros
print(A)
print(A.shape)   # number of rows and columns
print(type(A))   # A is an ndarray
```

```
[[[ 0.  0.]
  [ 0.  0.]
  [ 0.  0.]]

 [[ 0.  0.]
  [ 0.  0.]
  [ 0.  0.]]]
(2, 3, 2)
<class 'numpy.ndarray'>
```

We will be mostly working with 2D arrays; that is, ndarrays that represent ordinary matrices. We can also use the **range** method and lists to create ndarrays via the **array** method. Note that **arange** is numpy's version of **range**, with the difference that **arange** returns an ndarray object.

```
a = np.array(range(4))    # equivalent to np.arange(4)
b = np.array([0,1,2,3])
C = np.array([[1,2,3],[3,2,1]])
print(a, '\n', b,'\n' , C)
```

```
[0 1 2 3]
[0 1 2 3]
```

```
[[1 2 3]
 [3 2 1]]
```

The dimension of an ndarray can be obtained via its **shape** method, which returns a tuple. Arrays can be reshaped via the **reshape** method. This does not change the current ndarray object. To make the change permanent, a new instance needs to be created.

```
a = np.array(range(9)) #a is an ndarray of shape (9,)
print(a.shape)
A = a.reshape(3,3)   #A is an ndarray of shape (3,3)
print(a)
print(A)
```

```
[0 1 2 3 4 5 6 7 8]
(9,)
[[0, 1, 2]
 [3, 4, 5]
 [6, 7, 8]]
```

One shape dimension for **reshape** can be specified as −1. The dimension is then inferred from the other dimension(s).

The 'T' attribute of an ndarray gives its transpose. Note that the transpose of a "vector" with shape $(n,)$ is the same vector. To distinguish between column and row vectors, reshape such a vector to an $n \times 1$ and $1 \times n$ array, respectively.

```
a = np.arange(3)   #1D array (vector) of shape (3,)
print(a)
print(a.shape)
b = a.reshape(-1,1) # 3x1 array (matrix) of shape (3,1)
print(b)
print(b.T)
A = np.arange(9).reshape(3,3)
print(A.T)
```

```
[0 1 2]
(3,)
[[0]
 [1]
 [2]]
[[0 1 2]]
[[0 3 6]
 [1 4 7]
 [2 5 8]]
```

Two useful methods of joining arrays are **hstack** and **vstack**, where the arrays are joined horizontally and vertically, respectively.

```
A = np.ones((3,3))
B = np.zeros((3,2))
C = np.hstack((A,B))
print(C)
```

```
[[ 1.  1.  1.  0.  0.]
 [ 1.  1.  1.  0.  0.]
 [ 1.  1.  1.  0.  0.]]
```

D.10.2 Slicing

Arrays can be sliced similarly to Python lists. If an array has several dimensions, a slice for each dimension needs to be specified. Recall that Python indexing starts at `'0'` and ends at `'len(obj)-1'`. The following program illustrates various slicing operations.

```
A = np.array(range(9)).reshape(3,3)
print(A)
print(A[0])     # first row
print(A[:,1])   # second column
print(A[0,1])   # element in first row and second column
print(A[0:1,1:2])  # (1,1) ndarray containing A[0,1] = 1
print(A[1:,-1]) # elements in 2nd and 3rd rows, and last column
```

```
[[0 1 2]
 [3 4 5]
 [6 7 8]]
[0 1 2]
[1 4 7]
1
[[1]]
[5 8]
```

Note that `ndarrays` are mutable objects, so that elements can be modified directly, without having to create a new object.

```
A[1:,1] = [0,0] # change two elements in the matrix A above
print(A)
```

```
[[0, 1, 2]
 [3, 0, 5]
 [6, 0, 8]]
```

D.10.3 Array Operations

Basic mathematical operators and functions act *element-wise* on `ndarray` objects.

```
x = np.array([[2,4],[6,8]])
y = np.array([[1,1],[2,2]])
print(x+y)
```

```
[[ 3,  5]
 [ 8, 10]]
```

```
print(np.divide(x,y))   # same as x/y
```

```
[[ 2.  4.]
 [ 3.  4.]]
```

```
print(np.sqrt(x))
```

```
[[1.41421356  2.        ]
 [2.44948974  2.82842712]]
```

In order to compute matrix multiplications and compute inner products of vectors, numpy's **dot** function can be used, either as a method of an `ndarray` instance or as a method of np.

```
print(np.dot(x,y))
```

```
[[10,  10]
 [22,  22]]
```

```
print(x.dot(x))   # same as np.dot(x,x)
```

```
[[28,  40]
 [60,  88]]
```

Since version 3.5 of Python, it is possible to multiply two `ndarrays` using the @ *operator* (which implements the `np.matmul` method). For matrices, this is similar to using the **dot** method. For higher-dimensional arrays the two methods behave differently.

@ OPERATOR

```
print(x @ y)
```

```
[[10 10]
 [22 22]]
```

NumPy allows arithmetic operations on arrays of different shapes (dimensions). Specifically, suppose two arrays have dimensions $(m_1, m_2, \ldots, m_p)$ and $(n_1, n_2, \ldots, n_p)$, respectively. The arrays or shapes are said to be *aligned* if for all $i = 1, \ldots, p$ it holds that

ALIGNED

- $m_i = n_i$, or
- $\min\{m_i, n_i\} = 1$, or
- either m_i or n_i, or both are missing.

For example, shapes $(1, 2, 3)$ and $(4, 2, 1)$ are aligned, as are $(2, ,)$ and $(1, 2, 3)$. However, $(2, 2, 2)$ and $(1, 2, 3)$ are not aligned. NumPy "duplicates" the array elements across the smaller dimension to match the larger dimension. This process is called *broadcasting* and is carried out without actually making copies, thus providing efficient memory use. Below are some examples.

BROADCASTING

```
import numpy as np
A= np.arange(4).reshape(2,2) # (2,2) array

x1 = np.array([40,500])        # (2,) array
x2 = x1.reshape(2,1)         # (2,1) array

print(A + x1) # shapes (2,2) and (2,)
print(A * x2) # shapes (2,2) and (2,1)
```

```
[[ 40 501]
 [ 42 503]]
[[   0   40]
 [1000 1500]]
```

Note that above x1 is duplicated row-wise and x2 column-wise. Broadcasting also applies to the matrix-wise operator @, as illustrated below. Here, the matrix b is duplicated across the third dimension resulting in the two matrix multiplications

$$\begin{bmatrix} 0 & 1 \\ 2 & 3 \end{bmatrix}\begin{bmatrix} 0 & 1 \\ 2 & 3 \end{bmatrix} \quad \text{and} \quad \begin{bmatrix} 4 & 5 \\ 6 & 7 \end{bmatrix}\begin{bmatrix} 0 & 1 \\ 2 & 3 \end{bmatrix}.$$

```
B = np.arange(8).reshape(2,2,2)
b = np.arange(4).reshape(2,2)
print(B@b)
```

```
[[[ 2  3]
  [ 6 11]]

 [[10 19]
  [14 27]]]
```

Functions such as `sum`, `mean`, and `std` can also be executed as methods of an `ndarray` instance. The argument `axis` can be passed to specify along which dimension the function is applied. By default `axis=None`.

```
a = np.array(range(4)).reshape(2,2)
print(a.sum(axis=0)) #summing over rows gives column totals
```

```
[2, 4]
```

D.10.4 Random Numbers

One of the sub-modules in numpy is random. It contains many functions for random variable generation.

```
import numpy as np
np.random.seed(123)  # set the seed for the random number generator
x = np.random.random()      # uniform (0,1)
y = np.random.randint(5,9)  # discrete uniform 5,...,8
z = np.random.randn(4)      # array of four standard normals
print(x,y,'\n',z)
```

```
0.6964691855978616 7
 [ 1.77399501 -0.66475792 -0.07351368  1.81403277]
```

For more information on random variable generation in numpy, see

https://docs.scipy.org/doc/numpy/reference/random/index.html.

D.11 Matplotlib

The main Python graphics library for 2D and 3D plotting is matplotlib, and its subpackage pyplot contains a collection of functions that make plotting in Python similar to that in MATLAB.

D.11.1 Creating a Basic Plot

The code below illustrates various possibilities for creating plots. The style and color of lines and markers can be changed, as well as the font size of the labels. Figure D.1 shows the result.

sqrtplot.py

```
import matplotlib.pyplot as plt
import numpy as np
x = np.arange(0, 10, 0.1)
u = np.arange(0,10)
y = np.sqrt(x)
v = u/3
plt.figure(figsize = [4,2])  # size of plot in inches
plt.plot(x,y, 'g--')         # plot green dashed line
plt.plot(u,v,'r.')           # plot red dots
plt.xlabel('x')
plt.ylabel('y')
plt.tight_layout()
plt.savefig('sqrtplot.pdf',format='pdf')  # saving as pdf
plt.show()                   # both plots will now be drawn
```

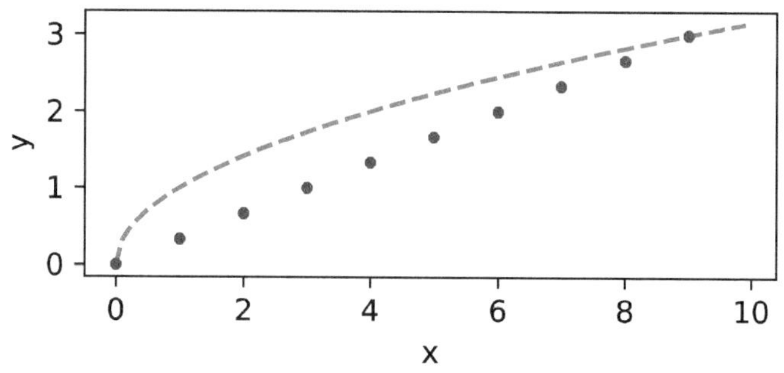

Figure D.1: A simple plot created using pyplot.

The library matplotlib also allows the creation of subplots. The scatterplot and histogram in Figure D.2 have been produced using the code below. When creating a histogram there are several optional arguments that affect the layout of the graph. The number of bins is determined by the parameter bins (the default is 10). Scatterplots also take a number of parameters, such as a string c which determines the color of the dots, and alpha which affects the transparency of the dots.

histscat.py

```
import matplotlib.pyplot as plt
import numpy as np
x = np.random.randn(1000)
u = np.random.randn(100)
v = np.random.randn(100)
plt.subplot(121)            # first subplot
plt.hist(x,bins=25, facecolor='b')
plt.xlabel('X Variable')
plt.ylabel('Counts')
plt.subplot(122)            # second subplot
plt.scatter(u,v,c='b', alpha=0.5)
plt.show()
```

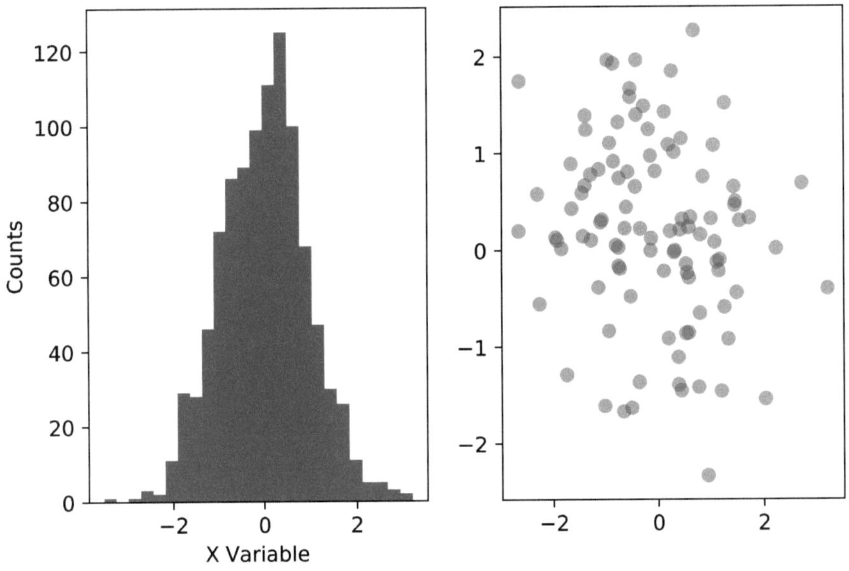

Figure D.2: A histogram and scatterplot.

One can also create three-dimensional plots as illustrated below.

surf3dscat.py

```
import matplotlib.pyplot as plt
import numpy as np
from mpl_toolkits.mplot3d import Axes3D

def npdf(x,y):
    return np.exp(-0.5*(pow(x,2)+pow(y,2)))/np.sqrt(2*np.pi)

x, y  = np.random.randn(100), np.random.randn(100)
z = npdf(x,y)

xgrid, ygrid = np.linspace(-3,3,100), np.linspace(-3,3,100)

Xarray, Yarray = np.meshgrid(xgrid,ygrid)
```

```
Zarray = npdf(Xarray,Yarray)

fig = plt.figure(figsize=plt.figaspect(0.4))
ax1 = fig.add_subplot(121, projection='3d')
ax1.scatter(x,y,z, c='g')
ax1.set_xlabel('$x$')
ax1.set_ylabel('$y$')
ax1.set_zlabel('$f(x,y)$')

ax2 = fig.add_subplot(122, projection='3d')
ax2.plot_surface(Xarray,Yarray,Zarray,cmap='viridis',
                                         edgecolor='none')
ax2.set_xlabel('$x$')
ax2.set_ylabel('$y$')
ax2.set_zlabel('$f(x,y)$')

plt.show()
```

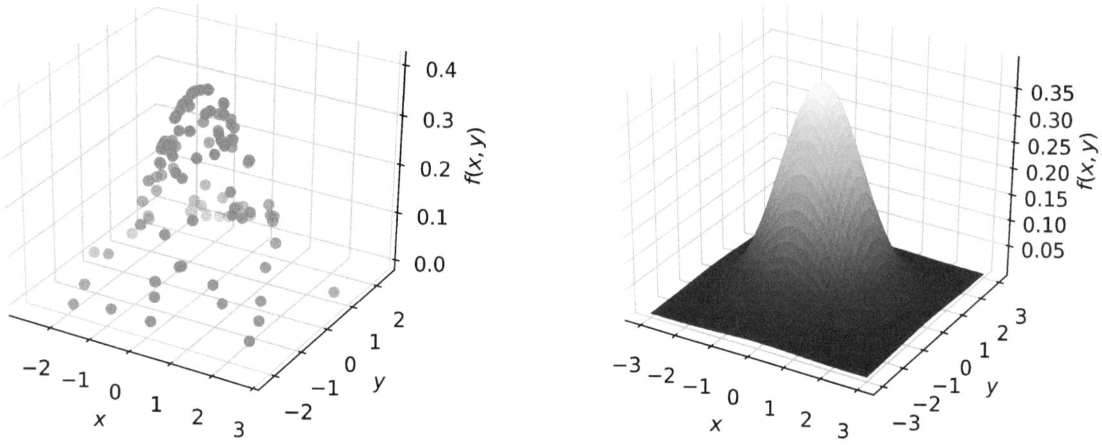

Figure D.3: Three-dimensional scatter- and surface plots.

D.12 Pandas

The Python package Pandas (module name pandas) provides various tools and data structures for data analytics, including the fundamental `DataFrame` class.

> For the code in this section we assume that pandas has been imported via `import pandas as pd`.

D.12.1 Series and DataFrame

The two main data structures in pandas are `Series` and `DataFrame`. A `Series` object can be thought of as a combination of a dictionary and an 1-dimensional `ndarray`. The syntax for creating a Series object is

```
series = pd.Series(<data>, index=['index'])
```

Here, <data> some 1-dimensional data structure, such as a 1-dimensional ndarray, a list, or a dictionary, and index is a list of names of the same length as <data>. When <data> is a dictionary, the index is created from the keys of the dictionary. When <data> is an ndarray and index is omitted, the default index will be [0, ..., len(data)-1].

```
DICT = {'one':1, 'two':2, 'three':3, 'four':4}
print(pd.Series(DICT))
```

```
one      1
two      2
three    3
four     4
dtype: int64
```

```
years =  ['2000','2001','2002']
cost = [2.34, 2.89, 3.01]
print(pd.Series(cost,index = years, name = 'MySeries')) #name it
```

```
2000    2.34
2001    2.89
2002    3.01
Name: MySeries, dtype: float64
```

The most commonly-used data structure in pandas is the two-dimensional DataFrame, which can be thought of as pandas' implementation of a spreadsheet or as a dictionary in which each "key" of the dictionary corresponds to a column name and the dictionary "value" is the data in that column. To create a DataFrame one can use the pandas DataFrame method, which has three main arguments: data, index (row labels), and columns (column labels).

```
DataFrame(<data>, index=['<row_name>'], columns=['<column_name>'])
```

If the index is not specified, the default index is [0, ..., len(data)-1]. Data can also be read directly from a CSV or Excel file, as is done in Section 1.1. ☞ 1 If a dictionary is used to create the data frame (as below), the dictionary keys are used as the column names.

```
DICT = {'numbers':[1,2,3,4], 'squared':[1,4,9,16] }
df = pd.DataFrame(DICT, index = list('abcd'))
print(df)
```

```
   numbers  squared
a        1        1
b        2        4
c        3        9
d        4       16
```

D.12.2 Manipulating Data Frames

Often data encoded in DataFrame or Series objects need to be extracted, altered, or combined. Getting, setting, and deleting columns works in a similar manner as for dictionaries. The following code illustrates various operations.

```
ages = [6,3,5,6,5,8,0,3]
d={'Gender':['M', 'F']*4, 'Age': ages}
df1 = pd.DataFrame(d)
df1.at[0,'Age']= 60  # change an element
df1.at[1,'Gender'] = 'Female' # change another element
df2 = df1.drop('Age',1)  # drop a column
df3 = df2.copy();     # create a separate copy of df2
df3['Age'] = ages   # add the original column
dfcomb = pd.concat([df1,df2,df3],axis=1)  # combine the three dfs
print(dfcomb)
```

```
   Gender  Age  Gender  Gender  Age
0       M   60       M       M    6
1  Female    3  Female  Female    3
2       M    5       M       M    5
3       F    6       F       F    6
4       M    5       M       M    5
5       F    8       F       F    8
6       M    0       M       M    0
7       F    3       F       F    3
```

Note that the above DataFrame object has two Age columns. The expression dfcomb['Age'] will return a DataFrame with both these columns.

Table D.3: Useful pandas methods for data manipulation.

agg	Aggregate the data using one or more functions.
apply	Apply a function to a column or row.
astype	Change the data type of a variable.
concat	Concatenate data objects.
replace	Find and replace values.
read_csv	Read a CSV file into a DataFrame.
sort_values	Sort by values along rows or columns.
stack	Stack a DataFrame.
to_excel	Write a DataFrame to an Excel file.

It is important to correctly specify the data type of a variable before embarking on data summarization and visualization tasks, as Python may treat different types of objects in dissimilar ways. Common data types for entries in a DataFrame object are float, category, datetime, bool, and int. A generic object type is object.

```
d={'Gender':['M', 'F', 'F']*4, 'Age': [6,3,5,6,5,8,0,3,6,6,7,7]}
df=pd.DataFrame(d)
print(df.dtypes)
df['Gender'] = df['Gender'].astype('category')  #change the type
print(df.dtypes)
```

```
Gender     object
Age         int64
dtype: object
Gender    category
Age            int64
dtype: object
```

D.12.3 Extracting Information

☞ 1 Extracting statistical information from a DataFrame object is facilitated by a large collection of methods (functions) in pandas. Table D.4 gives a selection of data inspection methods. See Chapter 1 for their practical use. The code below provides several examples of useful methods. The **apply** method allows one to apply general functions to columns or rows of a DataFrame. These operations do not change the data. The **loc** method allows for accessing elements (or ranges) in a data frame and acts similar to the slicing operation for lists and arrays, *with the difference that the "stop" value is included*, as illustrated in the code below.

```
import numpy as np
import pandas as pd
ages = [6,3,5,6,5,8,0,3]
np.random.seed(123)
df = pd.DataFrame(np.random.randn(3,4), index = list('abc'),
                  columns = list('ABCD'))
print(df)
df1 = df.loc["b":"c","B":"C"]     # create a partial data frame
print(df1)
meanA = df['A'].mean()           # mean of 'A' column
print('mean of column A = {}'.format(meanA))
expA = df['A'].apply(np.exp)  # exp of all elements in 'A' column
print(expA)
```

```
          A         B         C         D
a -1.085631  0.997345  0.282978 -1.506295
b -0.578600  1.651437 -2.426679 -0.428913
c  1.265936 -0.866740 -0.678886 -0.094709
          B         C
b  1.651437 -2.426679
c -0.866740 -0.678886
mean of column A = -0.13276486552118785
a    0.337689
b    0.560683
c    3.546412
Name: A, dtype: float64
```

The **groupby** method of a DataFrame object is useful for summarizing and displaying the data in manipulated ways. It groups data according to one or more specified columns, such that methods such as **count** and **mean** can be applied to the grouped data.

Table D.4: Useful pandas methods for data inspection.

`columns`	Column names.
`count`	Counts number of non-NA cells.
`crosstab`	Cross-tabulate two or more categories.
`describe`	Summary statistics.
`dtypes`	Data types for each column.
`head`	Display the top rows of a DataFrame.
`groupby`	Group data by column(s).
`info`	Display information about the DataFrame.
`loc`	Access a group or rows or columns.
`mean`	Column/row mean.
`plot`	Plot of columns.
`std`	Column/row standard deviation.
`sum`	Returns column/row sum.
`tail`	Display the bottom rows of a DataFrame.
`value_counts`	Counts of different non-null values.
`var`	Variance.

```
df = pd.DataFrame({'W':['a','a','b','a','a','b'],
          'X':np.random.rand(6),
          'Y':['c','d','d','d','c','c'], 'Z':np.random.rand(6)})
print(df)
```

```
   W         X  Y         Z
0  a  0.993329  c  0.641084
1  a  0.925746  d  0.428412
2  b  0.266772  d  0.460665
3  a  0.201974  d  0.261879
4  a  0.529505  c  0.503112
5  b  0.006231  c  0.849683
```

```
print(df.groupby('W').mean())
```

```
          X         Z
W
a  0.662639  0.458622
b  0.136502  0.655174
```

```
print(df.groupby(['W', 'Y']).mean())
```

```
            X         Z
W Y
a c  0.761417  0.572098
  d  0.563860  0.345145
b c  0.006231  0.849683
  d  0.266772  0.460665
```

To allow for multiple functions to be calculated at once, the `agg` method can be used. It can take a list, dictionary, or string of functions.

```
print(df.groupby('W').agg([sum,np.mean]))
```

```
          X                     Z
        sum      mean        sum      mean
W
a  2.650555  0.662639  1.834487  0.458622
b  0.273003  0.136502  1.310348  0.655174
```

D.12.4 Plotting

The `plot` method of a DataFrame makes plots of a DataFrame using Matplotlib. Different types of plot can be accessed via the `kind = 'str'` construction, where `str` is one of `line` (default), `bar`, `hist`, `box`, `kde`, and several more. Finer control, such as modifying the font, is obtained by using `matplotlib` directly. The following code produces the line and box plots in Figure D.4.

```
import numpy as np
import pandas as pd
import matplotlib
df = pd.DataFrame({'normal':np.random.randn(100),
         'Uniform':np.random.uniform(0,1,100)})
font = {'family' : 'serif', 'size'   : 14} #set font
matplotlib.rc('font', **font)   # change font
df.plot()   # line plot (default)
df.plot(kind = 'box')  # box plot
matplotlib.pyplot.show()  #render plots
```

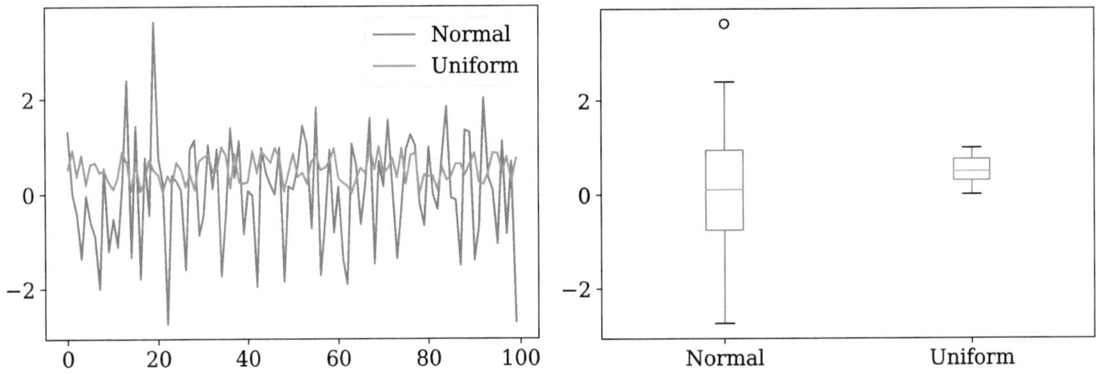

Figure D.4: A line and box plot using the `plot` method of `DataFrame`.

D.13 Scikit-learn

Scikit-learn is an open-source machine learning and data science library for Python. The library includes a range of algorithms relating to the chapters in this book. It is widely used due to its simplicity and its breadth. The module name is `sklearn`. Below is a brief introduction into modeling the data with `sklearn`. The full documentation can be found at https://scikit-learn.org/.

D.13.1 Partitioning the Data

Randomly partitioning the data in order to test the model may be achieved easily with sklearn's function train_test_split. For example, suppose that the training data is described by the matrix X of explanatory variables and the vector y of responses. Then the following code splits the data set into training and testing sets, with the testing set being half of the total set.

```
from sklearn.model_selection import train_test_split
X_train, X_test, y_train, y_test = train_test_split(X, y,
                                                    test_size = 0.5)
```

As an example, the following code generates a synthetic data set and splits it into equally-sized training and test sets.

syndat.py

```
import numpy as np
import matplotlib.pyplot as plt
from sklearn.model_selection import train_test_split

np.random.seed(1234)

X=np.pi*(2*np.random.random(size=(400,2))-1)
y=(np.cos(X[:,0])*np.sin(X[:,1])>=0)

X_train , X_test , y_train , y_test = train_test_split(X, y,
    test_size=0.5)


fig = plt.figure()
ax = fig.add_subplot(111)
ax.scatter(X_train[y_train==0,0],X_train[y_train==0,1], c='g',
           marker='o',alpha=0.5)
ax.scatter(X_train[y_train==1,0],X_train[y_train==1,1], c='b',
           marker='o',alpha=0.5)
ax.scatter(X_test[y_test==0,0],X_test[y_test==0,1], c='g',
           marker='s',alpha=0.5)
ax.scatter(X_test[y_test==1,0],X_test[y_test==1,1], c='b',
           marker='s',alpha=0.5)

plt.savefig('sklearntraintest.pdf',format='pdf')
plt.show()
```

D.13.2 Standardization

In some instances it may be necessary to standardize the data. This may be done in sklearn with scaling methods such as MinMaxScaler or StandardScaler. Scaling may improve the convergence of gradient-based estimators and is useful when visualizing data on vastly different scales. For example, suppose that X is our explanatory data (e.g., stored as a numpy array), and we wish to standardize such that each value lies between 0 and 1.

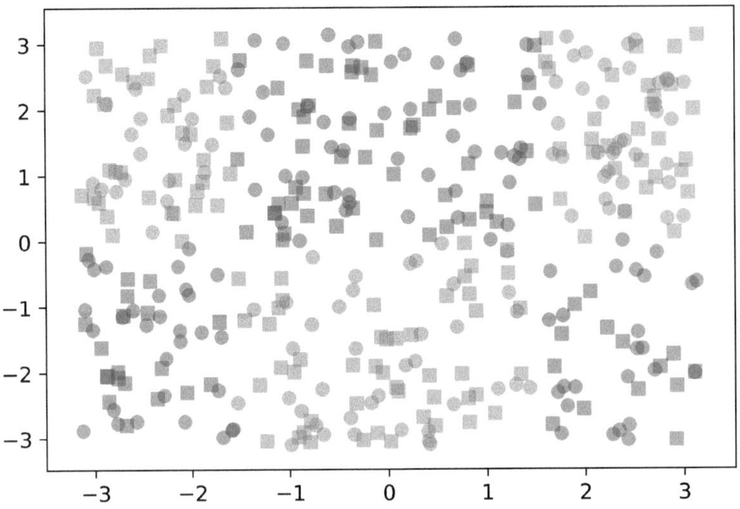

Figure D.5: Example training (circles) and test (squares) set for two class classification. Explanatory variables are the (x, y) coordinates, classes are zero (green) or one (blue).

```
from sklearn import preprocessing
min_max_scaler = preprocessing.MinMaxScaler(feature_range=(0, 1))
x_scaled = min_max_scaler.fit_transform(X)
# equivalent to:
x_scaled = (X - X.min(axis=0)) / (X.max(axis=0) - X.min(axis=0))
```

D.13.3 Fitting and Prediction

Once the data has been partitioned and standardized if necessary, the data may be fitted to a statistical model, e.g., a classification or regression model. For example, continuing with our data from above, the following fits a model to the data and predicts the responses for the test set.

```
from sklearn.someSubpackage import someClassifier
clf = someClassifier()        # choose appropriate classifier
clf.fit(X_train, y_train)     # fit the data
y_prediction = clf.predict(X_test)  # predict
```

Specific classifiers for logistic regression, naïve Bayes, linear and quadratic discriminant analysis, K-nearest neighbors, and support vector machines are given in Section 7.8.

☞ 279

D.13.4 Testing the Model

Once the model has made its prediction we may test its effectiveness, using relevant metrics. For example, for classification we may wish to produce the confusion matrix for the test data. The following code does this for the data shown in Figure D.5, using a support vector machine classifier.

```
from sklearn import svm
clf = svm.SVC(kernel = 'rbf')
clf.fit(X_train , y_train)
y_prediction = clf.predict(X_test)

from sklearn.metrics import confusion_matrix
print(confusion_matrix(y_test , y_prediction))
```

```
[[102  12]
 [  1  85]]
```

D.14 System Calls, URL Access, and Speed-Up

Operating system commands (whether in Windows, MacOS, or Linux) for creating directories, copying or removing files, or executing programs from the system shell can be issued from within Python by using the package os. Another useful package is requests which enables direct downloads of files and webpages from URLs. The following Python script uses both. It also illustrates a simple example of exception handling in Python.

misc.py

```
import os
import requests
for c in "123456":
  try:                       # if it does not yet exist
     os.mkdir("MyDir"+ c)    # make a directory
  except:                    # otherwise
      pass                   # do nothing

uname = "https://github.com/DSML-book/Programs/tree/master/
   Appendices/Python Primer/"
fname = "ataleof2cities.txt"
r = requests.get(uname + fname)
print(r.text)
open('MyDir1/ato2c.txt', 'wb').write(r.content) #write to a file
                                      # bytes mode is important here
```

The package numba can significantly speed up calculations via smart compilation. First run the following code.

jitex.py

```
import timeit
import numpy as np
from numba import jit
n = 10**8

#@jit
def myfun(s,n):
   for i in range(1,n):
```

```
        s = s+ 1/i
    return s

start = timeit.time.clock()
print("Euler's constant is approximately {:9.8f}".format(
                                   myfun(0,n) - np.log(n)))
end = timeit.time.clock()
print("elapsed time: {:3.2f} seconds".format(end-start))
```

```
Euler's constant is approximately 0.57721566
elapsed time: 5.72 seconds
```

Now remove the # character before the @ character in the code above, in order to activate the "just in time" compiler. This gives a 15-fold speedup:

```
Euler's constant is approximately 0.57721566
elapsed time: 0.39 seconds
```

Further Reading

To learn Python, we recommend [82] and [110]. However, as Python is constantly evolving, the most up-to-date references will be available from the Internet.

BIBLIOGRAPHY

[1] S. C. Ahalt, A. K. Krishnamurthy, P. Chen, and D. E. Melton. Competitive learning algorithms for vector quantization. *Neural Networks*, 3:277–290, 1990.

[2] H. Akaike. A new look at the statistical model identification. *IEEE Transactions on Automatic Control*, 19(6):716–723, 1974.

[3] N. Aronszajn. Theory of reproducing kernels. *Transactions of the American Mathematical Society*, 68:337–404, 1950.

[4] D. Arthur and S. Vassilvitskii. K-means++: The advantages of careful seeding. In *Proceedings of the Eighteenth Annual ACM-SIAM Symposium on Discrete Algorithms*, pages 1027–1035, Philadelphia, 2007. Society for Industrial and Applied Mathematics.

[5] S. Asmussen and P. W. Glynn. *Stochastic Simulation: Algorithms and Analysis*. Springer, New York, 2007.

[6] R. G. Bartle. *The Elements of Integration and Lebesgue Measure*. John Wiley & Sons, Hoboken, 1995.

[7] D. Bates and D. Watts. *Nonlinear Regression Analysis and Its Applications*. John Wiley & Sons, Hoboken, 1988.

[8] J. O. Berger. *Statistical Decision Theory and Bayesian Analysis*. Springer, New York, second edition, 1985.

[9] J. Bezdek. *Pattern Recognition with Fuzzy Objective Function Algorithms*. Plenum Press, New York, 1981.

[10] P. J. Bickel and K. A. Doksum. *Mathematical Statistics*, volume I. Pearson Prentice Hall, Upper Saddle River, second edition, 2007.

[11] P. Billingsley. *Probability and Measure*. John Wiley & Sons, New York, third edition, 1995.

[12] C. M. Bishop. *Pattern Recognition and Machine Learning*. Springer, New York, 2006.

[13] P. T. Boggs and R. H. Byrd. Adaptive, limited-memory BFGS algorithms for unconstrained optimization. *SIAM Journal on Optimization*, 29(2):1282–1299, 2019.

[14] Z. I. Botev, J. F. Grotowski, and D. P. Kroese. Kernel density estimation via diffusion. *Annals of Statistics*, 38(5):2916–2957, 2010.

[15] Z. I. Botev and D. P. Kroese. Global likelihood optimization via the cross-entropy method, with an application to mixture models. In R. G. Ingalls, M. D. Rossetti, J. S. Smith, and B. A. Peters, editors, *Proceedings of the 2004 Winter Simulation Conference*, pages 529–535, Washington, DC, December 2004.

[16] Z. I. Botev, D. P. Kroese, R. Y. Rubinstein, and P. L'Ecuyer. The cross-entropy method for optimization. In V. Govindaraju and C.R. Rao, editors, *Machine Learning: Theory and Applications*, volume 31 of *Handbook of Statistics*, pages 35–59. Elsevier, 2013.

[17] S. Boyd, N. Parikh, E. Chu, B. Peleato, and J. Eckstein. Distributed optimization and statistical learning via the alternating direction method of multipliers. *Foundations and Trends in Machine Learning*, 3:1–122, 2010.

[18] S. Boyd and L. Vandenberghe. *Convex Optimization*. Cambridge University Press, Cambridge, 2004. Seventh printing with corrections, 2009.

[19] R. A. Boyles. On the convergence of the EM algorithm. *Journal of the Royal Statistical Society, Series B*, 45(1):47–50, 1983.

[20] L. Breiman. *Classification and Regression Trees*. CRC Press, Boca Raton, 1987.

[21] L. Breiman. Bagging predictors. *Machine Learning*, 24(2):123–140, 1996.

[22] L. Breiman. Heuristics of instability and stabilization in model selection. *Annals of Statistics*, 24(6):2350–2383, 12 1996.

[23] L. Breiman. Random forests. *Machine Learning*, 45(1):5–32, 2001.

[24] F. Cao, D.-Z. Du, B. Gao, P.-J. Wan, and P. M. Pardalos. Minimax problems in combinatorial optimization. In D.-Z. Du and P. M. Pardalos, editors, *Minimax and Applications*, pages 269–292. Kluwer, Dordrecht, 1995.

[25] G. Casella and R. L. Berger. *Statistical Inference*. Duxbury Press, Pacific Grove, second edition, 2001.

[26] K. L. Chung. *A Course in Probability Theory*. Academic Press, New York, second edition, 1974.

[27] E. Cinlar. *Introduction to Stochastic Processes.* Prentice Hall, Englewood Cliffs, 1975.

[28] T. M. Cover and J. A. Thomas. *Elements of Information Theory.* John Wiley & Sons, New York, 1991.

[29] J. W. Daniel, W. B. Gragg, L. Kaufman, and G. W. Stewart. Reorthogonalization and stable algorithms for updating the Gram-Schmidt QR factorization. *Mathematics of Computation*, 30(136):772–795, 1976.

[30] P.-T. de Boer, D. P. Kroese, S. Mannor, and R. Y. Rubinstein. A tutorial on the cross-entropy method. *Annals of Operations Research*, 134(1):19–67, 2005.

[31] A. P. Dempster, N. M. Laird, and D. B. Rubin. Maximum likelihood from incomplete data via the EM algorithm. *Journal of the Royal Statistical Society*, 39(1):1 – 38, 1977.

[32] L. Devroye. *Non-Uniform Random Variate Generation.* Springer, New York, 1986.

[33] N. R. Draper and H. Smith. *Applied Regression Analysis.* John Wiley & Sons, New York, third edition, 1998.

[34] Q. Duan and D. P. Kroese. Splitting for optimization. *Computers & Operations Research*, 73:119–131, 2016.

[35] R. O. Duda, P. E. Hart, and D. G. Stork. *Pattern Classification.* John Wiley & Sons, New York, 2001.

[36] B. Efron and T. J. Hastie. *Computer Age Statistical Inference: Algorithms, Evidence, and Data Science.* Cambridge University Press, Cambridge, 2016.

[37] B. Efron and R. Tibshirani. *An Introduction to the Bootstrap.* Chapman & Hall, New York, 1994.

[38] T. Fawcett. An introduction to ROC analysis. *Pattern Recognition Letters*, 27(8):861–874, June 2006.

[39] W. Feller. *An Introduction to Probability Theory and Its Applications*, volume I. John Wiley & Sons, Hoboken, second edition, 1970.

[40] J. C. Ferreira and V. A. Menegatto. Eigenvalues of integral operators defined by smooth positive definite kernels. *Integral Equations and Operator Theory*, 64:61–81, 2009.

[41] N. I. Fisher and P. K. Sen, editors. *The Collected Works of Wassily Hoeffding.* Springer, New York, 1994.

[42] G. S. Fishman. *Monte Carlo: Concepts, Algorithms and Applications.* Springer, New York, 1996.

[43] R. Fletcher. *Practical Methods of Optimization.* John Wiley & Sons, New York, 1987.

[44] Y. Freund and R. E. Schapire. A decision-theoretic generalization of on-line learning and an application to boosting. *J. Comput. Syst. Sci.*, 55(1):119–139, 1997.

[45] J. H. Friedman. Greedy function approximation: A gradient boosting machine. *Annals of Statistics*, 29:1189–1232, 2000.

[46] A. Gelman. *Bayesian Data Analysis*. Chapman & Hall, New York, second edition, 2004.

[47] A. Gelman and J. Hall. *Data Analysis Using Regression and Multilevel/Hierarchical Models*. Cambridge University Press, Cambridge, 2006.

[48] S. Geman and D. Geman. Stochastic relaxation, Gibbs distribution and the Bayesian restoration of images. *IEEE Transactions on Pattern Analysis and Machine Intelligence*, 6(6):721–741, 1984.

[49] J. E. Gentle. *Random Number Generation and Monte Carlo Methods*. Springer, New York, second edition, 2003.

[50] W. R. Gilks, S. Richardson, and D. J. Spiegelhalter. *Markov Chain Monte Carlo in Practice*. Chapman & Hall, New York, 1996.

[51] P. Glasserman. *Monte Carlo Methods in Financial Engineering*. Springer, New York, 2004.

[52] G. H. Golub and C. F. Van Loan. *Matrix Computations*. Johns Hopkins University Press, Baltimore, fourth edition, 2013.

[53] I. Goodfellow, Y. Bengio, and A. Courville. *Deep Learning*. MIT Press, Cambridge, 2016.

[54] G. R. Grimmett and D. R. Stirzaker. *Probability and Random Processes*. Oxford University Press, third edition, 2001.

[55] T. J. Hastie, R. J. Tibshirani, and J. H. Friedman. *The Elements of Statistical Learning: Data mining, Inference, and Prediction*. Springer, New York, 2009.

[56] T. J. Hastie, R. J. Tibshirani, and M. Wainwright. *Statistical Learning with Sparsity: The Lasso and Generalizations*. CRC Press, Boca Raton, 2015.

[57] J.-B. Hiriart-Urruty and C. Lemarèchal. *Fundamentals of Convex Analysis*. Springer, New York, 2001.

[58] W. Hock and K. Schittkowski. *Test Examples for Nonlinear Programming Codes*. Springer, New York, 1981.

[59] J. E. Kelley, Jr. The cutting-plane method for solving convex programs. *Journal of the Society for Industrial and Applied Mathematics*, 8(4):703–712, 1960.

[60] A. K. Jain. *Fundamentals of Digital Image Processing*. Prentice Hall, Englewood Cliffs, 1989.

[61] O. Kallenberg. *Foundations of Modern Probability*. Springer, New York, second edition, 2002.

[62] A. Karalic. Linear regression in regression tree leaves. In *Proceedings of ECAI-92*, pages 440–441, Hoboken, 1992. John Wiley & Sons.

[63] C. Kaynak. Methods of combining multiple classifiers and their applications to handwritten digit recognition. Master's thesis, Institute of Graduate Studies in Science and Engineering, Bogazici University, 1995.

[64] T. Keilath and A. H. Sayed, editors. *Fast Reliable Algorithms for Matrices with Structure*. SIAM, Pennsylvania, 1999.

[65] C. Nussbaumer Knaflic. *Storytelling with Data: A Data Visualization Guide for Business Professionals*. John Wiley & Sons, Hoboken, 2015.

[66] D. Koller and N. Friedman. *Probabilistic Graphical Models: Principles and Techniques - Adaptive Computation and Machine Learning*. The MIT Press, Cambridge, 2009.

[67] A. N. Kolmogorov and S. V. Fomin. *Elements of the Theory of Functions and Functional Analysis*. Dover Publications, Mineola, 1999.

[68] D. P. Kroese, T. Brereton, T. Taimre, and Z. I. Botev. Why the Monte Carlo method is so important today. *Wiley Interdisciplinary Reviews: Computational Statistics*, 6(6):386–392, 2014.

[69] D. P. Kroese and J. C. C. Chan. *Statistical Modeling and Computation*. Springer, 2014.

[70] D. P. Kroese, S. Porotsky, and R. Y. Rubinstein. The cross-entropy method for continuous multi-extremal optimization. *Methodology and Computing in Applied Probability*, 8(3):383–407, 2006.

[71] D. P. Kroese, T. Taimre, and Z. I. Botev. *Handbook of Monte Carlo Methods*. John Wiley & Sons, New York, 2011.

[72] H. J. Kushner and G. G. Yin. *Stochastic Approximation and Recursive Algorithms and Applications*. Springer, New York, second edition, 2003.

[73] P. Lafaye de Micheaux, R. Drouilhet, and B. Liquet. *The R Software: Fundamentals of Programming and Statistical Analysis*. Springer, New York, 2014.

[74] R. J. Larsen and M. L. Marx. *An Introduction to Mathematical Statistics and Its Applications*. Prentice Hall, New York, third edition, 2001.

[75] A. M. Law and W. D. Kelton. *Simulation Modeling and Analysis*. McGraw-Hill, New York, third edition, 2000.

[76] P. L'Ecuyer. A unified view of IPA, SF, and LR gradient estimation techniques. *Management Science*, 36:1364–1383, 1990.

[77] P. L'Ecuyer. Good parameters and implementations for combined multiple recursive random number generators. *Operations Research*, 47(1):159 – 164, 1999.

[78] E. L. Lehmann and G. Casella. *Theory of Point Estimation*. Springer, New York, second edition, 1998.

[79] T. G. Lewis and W. H. Payne. Generalized feedback shift register pseudorandom number algorithm. *Journal of the ACM*, 20(3):456–468, 1973.

[80] R. J. A. Little and D. B. Rubin. *Statistical Analysis with Missing Data*. John Wiley & Sons, Hoboken, second edition, 2002.

[81] D. C. Liu and J. Nocedal. On the limited memory BFGS method for large scale optimization. *Mathematical Programming*, 45(1-3):503–528, 1989.

[82] M. Lutz. *Learning Python*. O'Reilly, fifth edition, 2013.

[83] M. Matsumoto and T. Nishimura. Mersenne twister: A 623-dimensionally equidistributed uniform pseudo-random number generator. *ACM Transactions on Modeling and Computer Simulation*, 8(1):3–30, 1998.

[84] W. McKinney. *Python for Data Analysis*. O'Reilly Media, Inc., second edition, 2017.

[85] G. J. McLachlan and T. Krishnan. *The EM Algorithm and Extensions*. John Wiley & Sons, Hoboken, second edition, 2008.

[86] G. J. McLachlan and D. Peel. *Finite Mixture Models*. John Wiley & Sons, New York, 2000.

[87] N. Metropolis, A. W. Rosenbluth, M. N. Rosenbluth, A. H. Teller, and E. Teller. Equations of state calculations by fast computing machines. *Journal of Chemical Physics*, 21(6):1087–1092, 1953.

[88] C. A. Micchelli, Y. Xu, and H. Zhang. Universal kernels. *Journal of Machine Learning Research*, 7:2651–2667, 2006.

[89] Z. Michalewicz. *Genetic Algorithms + Data Structures = Evolution Programs*. Springer, New York, third edition, 1996.

[90] J. F. Monahan. *Numerical Methods of Statistics*. Cambridge University Press, London, 2010.

[91] T. A. Mroz. The sensitivity of an empirical model of married women's hours of work to economic and statistical assumptions. *Econometrica*, 55(4):765–799, 1987.

[92] K. P. Murphy. *Machine Learning: A Probabilistic Perspective*. The MIT Press, Cambridge, 2012.

[93] J. Neyman and E. Pearson. On the problem of the most efficient tests of statistical hypotheses. *Philosophical Transactions of the Royal Society of London, Series A*, 231:289–337, 1933.

[94] M. A. Nielsen. *Neural Networks and Deep Learning*, volume 25. Determination Press, 2015.

[95] K. B. Petersen and M. S. Pedersen. The Matrix Cookbook. *Technical University of Denmark*, 2008.

[96] J. R. Quinlan. Learning with continuous classes. In A. Adams and L. Sterling, editors, *Proceedings AI'92*, pages 343–348, Singapore, 1992. World Scientific.

[97] C. E. Rasmussen and C. K. I. Williams. *Gaussian Processes for Machine Learning*. MIT Press, Cambridge, 2006.

[98] B. D. Ripley. *Stochastic Simulation*. John Wiley & Sons, New York, 1987.

[99] C. P. Robert and G. Casella. *Monte Carlo Statistical Methods*. Springer, New York, second edition, 2004.

[100] S. M. Ross. *Simulation*. Academic Press, New York, third edition, 2002.

[101] S. M. Ross. *A First Course in Probability*. Prentice Hall, Englewood Cliffs, seventh edition, 2005.

[102] R. Y. Rubinstein. The cross-entropy method for combinatorial and continuous optimization. *Methodology and Computing in Applied Probability*, 2:127–190, 1999.

[103] R. Y. Rubinstein and D. P. Kroese. *The Cross-Entropy Method: A Unified Approach to Combinatorial Optimization, Monte-Carlo Simulation and Machine Learning*. Springer, New York, 2004.

[104] R. Y. Rubinstein and D. P. Kroese. *Simulation and the Monte Carlo Method*. John Wiley & Sons, New York, third edition, 2017.

[105] S. Ruder. An overview of gradient descent optimization algorithms. *arXiv: 1609.04747*, 2016.

[106] W. Rudin. *Functional Analysis*. McGraw–Hill, Singapore, second edition, 1991.

[107] D. Salomon. *Data Compression: The Complete Reference*. Springer, New York, 2000.

[108] G. A. F. Seber and A. J. Lee. *Linear Regression Analysis*. John Wiley & Sons, Hoboken, second edition, 2003.

[109] S. Shalev-Shwartz and S. Ben-David. *Understanding Machine Learning: From Theory to Algorithms*. Cambridge University Press, Cambridge, 2014.

[110] Z. A. Shaw. *Learning Python 3 the Hard Way*. Addison–Wesley, Boston, 2017.

[111] Y. Shen, S. Kiatsupaibul, Z. B. Zabinsky, and R. L. Smith. An analytically derived cooling schedule for simulated annealing. *Journal of Global Optimization*, 38(2):333–365, 2007.

[112] N. Z. Shor. *Minimization Methods for Non-differentiable Functions.* Springer, Berlin, 1985.

[113] B. W. Silverman. *Density Estimation for Statistics and Data Analysis.* Chapman & Hall, New York, 1986.

[114] J. S. Simonoff. *Smoothing Methods in Statistics.* Springer, New York, 2012.

[115] I. Steinwart and A. Christmann. *Support Vector Machines.* Springer, New York, 2008.

[116] G. Strang. *Introduction to Linear Algebra.* Wellesley–Cambridge Press, Cambridge, fifth edition, 2016.

[117] G. Strang. *Linear Algebra and Learning from Data.* Wellesley–Cambridge Press, Cambridge, 2019.

[118] W. N. Street, W. H. Wolberg, and O. L. Mangasarian. Nuclear feature extraction for breast tumor diagnosis. In *IS&T/SPIE 1993 International Symposium on Electronic Imaging: Science and Technology, San Jose, CA*, pages 861–870, 1993.

[119] V. M. Tikhomirov. On the representation of continuous functions of several variables as superpositions of continuous functions of one variable and addition. In *Selected Works of A. N. Kolmogorov*, pages 383–387. Springer, Berlin, 1991.

[120] S. van Buuren. *Flexible Imputation of Missing Data.* CRC Press, Boca Raton, second edition, 2018.

[121] V. N. Vapnik. *The Nature of Statistical Learning Theory.* Springer, New York, 1995.

[122] V. N. Vapnik and A. Ya. Chervonenkis. On the uniform convergence of relative frequencies of events to their probabilities. *Theory of Probability and Its Applications*, 16(2):264–280, 1971.

[123] G. Wahba. *Spline Models for Observational Data.* SIAM, Philadelphia, 1990.

[124] L. Wasserman. *All of Statistics: A Concise Course in Statistical Inference.* Springer, 2010.

[125] A. Webb. *Statistical Pattern Recognition.* Arnold, London, 1999.

[126] H. Wendland. *Scattered Data Approximation.* Cambridge University Press, Cambridge, 2005.

[127] D. Williams. *Probability with Martingales.* Cambridge University Press, Cambridge, 1991.

[128] C. F. J. Wu. On the convergence properties of the EM algorithm. *The Annals of Statistics*, 11(1):95–103, 1983.

INDEX

A

acceptance probability, 78–80, 97
acceptance–rejection method, 73, 78
accuracy (classification–), 256
activation function, 204, 327
AdaBoost, 319–322
AdaGrad, 341
Adam method, 341, 348
adjoint operation, 363
affine transformation, 407, 437
agglomerative clustering, 147
Akaike information criterion, 126, 176, 177
algebraic multiplicity, 365
aligned arrays (Python), 483
almost sure convergence, 442
alternating direction method of multipliers, 220, 418
alternative hypothesis, 460
anaconda (Python), 465
analysis of variance (ANOVA), 183, 184, 195, 208
annealing schedule, 97
approximation error, 32–34, 184
approximation–estimation tradeoff, 32, 41, 325
Armijo inexact line search, 411
assignment operator (Python), 469
attributes (Python), 467
auxiliary variable methods, 128
axioms of Kolmogorov, 423

B

back-propagation, 333
backward elimination, 201
backward substitution, 372
bagged estimator, 308
bagging, 307, 309, 312
balance equations (Markov chains), 78, 79, 454
bandwidth, 131, 134, 226
barplot, 9
barrier function, 419
Barzilai–Borwein formulas, 336, 415
basis
  of a vector space, 357
  orthogonal –, 363
Bayes
  empirical, 242
  error rate, 254
  factor, 57
  naïve –, 260
  optimal decision rule, 260
Bayes' rule, 47, 48, 430, 456
Bayesian information criterion, 54
Bayesian statistics, 47, 49, 456
Bernoulli distribution, 427, 459
Bessel distribution, 164, 227

beta distribution, 52, 427
bias of an estimator, 456
bias vector (deep learning), 328
bias–variance tradeoff, 35, 307
binomial distribution, 427
Boltzmann distribution, 96
bootstrap aggregation, *see* bagging
bootstrap method, 88, 308
bounded mapping, 391
boxplot, 10, 14
broadcasting (Python), 483
Broyden's family, 413
Broyden–Fletcher–Goldfarb–Shanno (BFGS) updating, 269, 340, 413
burn-in period, 78

C

categorical variable, 3, 177, 178, 191, 192, 253, 301
Cauchy sequence, 246, 387
Cauchy–Schwarz inequality, 223, 247, 391, 414
central difference estimator, 106
central limit theorem, 449, 460
  multivariate, 450
centroid, 144
chain rule for differentiation, 403
characteristic function, 225, 227, 247, 394, 443
characteristic polynomial, 365
Chebyshev's inequality, 446
chi-squared distribution, 438, 441
Cholesky decomposition, 70, 154, 248, 266, 375
circulant matrix, 383, 395
class (Python), 475
classification, 20, 253–288
  hierarchical, 258
  multilabel, 258
classifier, 21, 253
coefficient of determination, 181, 195
  adjusted, 181
coefficient profiles, 221
combinatorial optimization, 404
comma separated values (CSV), 2
common random numbers, 106, 119
complete Hilbert space, 224, 386
complete vector space, 216
complete convergence, 445
complete-data
  likelihood, 128
  log-likelihood, 138
composition of functions, 402
concave function, 406, 409
conditional
  distribution, 433
  expectation, 433
  pdf, 74, 433
  probability, 430
confidence interval, 85, 89, 94, 186, 459
  Bayesian, 51
  bootstrap, 89
confidence region, 459
confusion matrix, 255, 256
constrained optimization, 405
context management (Python), 479
continuous mapping, 391
continuous optimization, 404
control variable, 92
convergence
  almost sure, 442
  in L^p norm, 444
  in distribution, 442
  in probability, 441
  sure, 441
convex
  function, 62, 220, 405
  program, 407–410
  set, 42, 405
convolution, 382, 394
convolution neural network, 331
Cook's distance, 212
cooling factor, 97
correlation coefficient, 71, 431
cost-complexity
  measure, 305
  pruning, 305
countable sample space, 424
covariance, 431
  matrix, 45, 70, 432–434, 437, 438
  properties, 431
coverage probability, 459

credible
  interval, 51
  region, 51
critical
  region, 460
  value, 460
cross tabulate, 7
cross-entropy
  method, 100, 110
  risk, 53, 122, 125
    in-sample, 176
  training loss, 123
cross-validation, 37, 38
  leave-one-out, 40, 173
  linear model, 174
crude Monte Carlo, 85
cubic spline, 237
cumulative distribution function (cdf), 72, 425
  joint, 429
cycle, 81

D

Davidon–Fletcher–Powell updating, 354, 414
decision tree, 290
deep learning, 332
degrees of freedom, 181
dendrogram, 147
density, 387
dependent variable, 168
derivatives
  multidimensional, 400
  partial, 399
design matrix, 179
detailed balance equations, 455
determinant of a matrix, 359
diagonal matrix, 359
diagonalizable, 366
dictionary (Python), 475
digamma function, 127, 162
dimension, 357
direct sum, 217
directional derivative, 406
discrete
  distribution, 425
  Fourier transform, 394
  optimization, 404
  probability space, 424
  sample space, 424
  uniform distribution, 427
discriminant analysis, 261
distribution
  Bernoulli, 427
  Bessel, 227
  beta, 52, 427
  binomial, 427
  chi-squared, 438, 441
  discrete, 425
  discrete uniform, 427
  exponential, 427
  extreme value, 114
  F, 441
  gamma, 427
  Gaussian, *see* normal
  geometric, 427
  inverse-gamma, 50, 83
  joint, 429
  multivariate normal, 45, 437
  noncentral χ^2, 439
  normal, 44, 427, 436
  Pareto, 427
  Poisson, 427
  probability, 424, 429
  Student's t, 441
  uniform, 427
  Weibull, 427
divisive clustering, 147
dot notation (Python), 468
dual optimization problem, 409–410

E

early stopping, 49, 251
efficiency
  of estimators, 456
  of acceptance–rejection, 72
eigen-decomposition, 366
eigenvalue, 365
eigenvector, 365
elementary event, 424
elite sample, 100
empirical

Bayes, 242
cdf, 11, 76
distribution, 131
entropy impurity, 294
epoch (deep learning), 351
equilikely principle, 424
ergodic Markov chain, 454
error of the first and second kind, 461
estimate, 456
estimator, 456
bias of, 456
control variable, 92
efficiency of, 456
unbiased, 456
Euclidean norm, 362
evaluation functional, 223, 246
event, 423
elementary, 424
independent, 430
exact match ratio, 258
exchangeable variables, 40
expectation, 428
conditional, 433
properties, 431, 433
vector, 45, 432, 434, 437
expectation–maximization (EM) algorithm, 128, 137, 209
expected generalization risk, 24
expected optimism, 36
explanatory variable, 22, 168
exponential distribution, 427
extreme value distribution, 114

F

factor, 3, 178
false negative, 256
false positive, 256
fast Fourier transform, 396
F_β score, 257
F distribution, 183, 197, 426, 441
feasible region, 405
feature, 1, 20
importance, 313
map, 189, 216, 225, 231, 244, 276
feed-forward network, 328
feedback shift register, 69
finite difference method, 107, 113
finite-dimensional distributions, 429
Fisher information matrix, 124
Fisher's scoring method, 127
folds (cross-validation), 38
forward selection, 200
forward substitution, 372
Fourier expansion, 388
Fourier transform, 393
discrete, 394
frequentist statistics, 455
full rank matrix, 28
function (Python), 470
function space, 386
function, C^k, 405
functional, 391
functions of random variables, 433

G

gamma
distribution, 427
function, 426
Gauss–Markov inequality, 59
Gauss–Newton search direction, 416
Gaussian distribution, *see* normal distribution
Gaussian kernel, 226
Gaussian kernel density estimate, 131
Gaussian process, 71, 239
Gaussian rule of thumb, 134
generalization risk, 23, 86
generalized inverse-gamma distribution, 163
generalized linear model, 204
geometric cooling, 97
geometric distribution, 427
geometric multiplicity, 365
Gibbs pdf, 97
Gibbs sampler, 81, 83, 84
random, 82
random order, 82
reversible, 82
Gini impurity, 294
global balance equations, 454
global minimizer, 404
gradient, 399, 405

boosting, 318
descent, 414
Gram matrix, 218, 222, 272
Gram–Schmidt procedure, 377

H

Hamming distance, 142
Hermite polynomials, 391
Hermitian matrix, 364, 367
Hessian matrix, 124, 400, 405, 406
hidden layer, 327
hierarchical classification, 258
Hilbert matrix, 33
inverse, 60
Hilbert space, 215, 387
isomorphism, 247
hinge loss, 271
histogram, 10
Hoeffding's inequality, 62
homotopy paths, 221
hyperparameters, 50, 241
hypothesis testing, 460

I

immutable (Python), 466
importance sampling, 93–96
improper prior, 50, 83
in-sample risk, 35
incremental effects, 179
independence
of event, 430
of random variables, 431
independence sampler, 79
independent and identically distributed (iid), 431, 448, 456
indicator, 11
indicator feature, 178
indicator loss, 253
infinitesimal perturbation analysis, 113
information matrix equality, 124
inheritance (Python), 476
initial distribution (Markov chain), 454
inner product, 362
instance (Python), 476
integration
Monte Carlo, 86
interaction, 179, 193
interior-point method, 420
interval estimate, *see* confidence interval
inverse
discrete Fourier transform, 395
Fourier transform, 393
matrix, 372
inverse-gamma distribution, 50, 83
inverse-transform method, 72
irreducible risk, 32
iterable (Python), 474
iterative reweighted least squares, 213, 351
iterator (Python), 474

J

Jacobi
matrix of, 411, 435
Jensen's inequality, 62
joint
cdf, 429
pdf, 429
jointly normal, *see* multivariate normal
jointly normal distribution, *see* multivariate normal distribution

K

Karush–Kuhn–Tucker (KKT) conditions, 409, 410
kernel density estimation, 131, 135, 226, 331
kernel trick, 232
Kiefer–Wolfowitz algorithm, 107
K-nearest neighbors method, 270
Kolmogorov axioms, 423
Kullback–Leibler divergence, 42, 100, 128, 352

L

Lagrange
dual program, 409
function, 408
method, 408–409
multiplier, 408
Lagrangian, 408, 418
penalty, 418
Laguerre polynomials, 390
Lance–Williams update, 149

Laplace's approximation, 452
lasso (regression), 220
latent variable methods, *see* auxiliary variable methods
law of large numbers, 67, 448, 460
law of total probability, 430
learner, 22, 168
learning rate, 336, 411
least-squares
  iterative reweighted, 213
  nonlinear, 190, 337, 416
  ordinary, 27, 46, 171, 191, 211, 380
  regularized, 172, 236, 378
leave-one-out cross-validation, 40, 173
left pseudo-inverse, 362
left-eigenvector, 367
Legendre polynomials, 389
length preserving transformation, 363
length of a vector, 362
level set, 103
Levenberg–Marquardt search direction, 417
leverage, 173
Levinson–Durbin, 71, 384
likelihood, 42, 48, 123, 458
  complete-data, 128
  log-, 136, 458
  optimization, 137
  ratio, 93
limited memory BFGS, 338
limiting pdf, 454
limiting pdf (Markov chain), 454
line search, 411
linear
  discriminant function, 262
  kernel, 225, 273
  mapping, 391
  model, 43, 211
  program, 408
  subspace, 364
  transformation, 358, 433
linearly independent, 357
link function, 204
linkage, 148
  matrix, 150
list comprehension (Python), 475
local balance equations, *see* detailed balance equations
local minimizer, 404
local/global minimum, 404
log-likelihood, 458
log-odds ratio, 268
logarithmic efficiency, 117
logistic distribution, 204
logistic regression, 204
long-run average reward, 89
loss function, 20
loss matrix, 255

M

M-estimator, 450
Manhattan distance, 142
marginal distribution, 429, 438
Markov chain, 74, 78, 80, 83, 453
  ergodic, 454
  reversible, 454
  simulation of, 75
Markov chain Monte Carlo, 78
Markov property, 74, 453
Matérn kernel, 227
matplotlib (Python), 485–487
matrix, 358
  blockwise inverse, 372
  covariance, 70, 438
  determinant, 359
  diagonal —, 359
  inverse, 359
  of Jacobi, 400, 411, 416, 435
  pseudo-inverse, 362
  sparse, 381
  Toeplitz, 381
  trace, 359
  transpose, 359
matrix multiplication (Python), 483
max-cut problem, 151
maximum a posteriori, 52
maximum distance, 142
maximum likelihood estimation, 42, 46, 100, 127, 136, 137, 458
mean integrated squared error, 133
mean squared error, 32, 88, 456
measure, 387

Mersenne twister, 69
method (Python), 468
method of moments, 457
Metropolis–Hastings algorithm, 78, 81
minibatch, 337
minimax
  equality, 410
  problem, 410
minimization, 413
minimizer, 404
minimum
  global, 404
  local, 404
misclassification error, 255
misclassification impurity, 294
mixture density, 135
model, 40
  evidence, 54
  linear, 211
  matrix, 43, 170, 174
  multiple linear regression, 169
  normal linear, 174, 182, 183, 440
  regression, 191
  response surface, 189
  simple linear regression, 187
modified Bessel function of the second kind, 163, 227
module (Python), 471
modulo 2 generators, 69
modulus, 69
moment
  generating function, 429, 438
  sample-, 457
momentum method, 342
Monte Carlo
  integration, 86
  sampling, 68–85
  simulation, 67
Moore–Penrose pseudo-inverse, 362
multi-logit, 268
multi-output linear regression, 213
  nonlinear, 330
multilabel classification, 258
multiple linear regression, 169
multiple-recursive generator, 69
multiplier
  Lagrange, 408
multivariate
  central limit theorem, 450
  normal distribution, 44–46, 437
mutable (Python), 466

N

naïve Bayes, 260
namespace (Python), 472
nested models, 58, 180
network architecture, 331
network depth, 329
network width, 329
neural networks, 325
Newton's method, 127, 205, 213, 338, 411
  — for root-finding, 411
  quasi —, 338
Neyman–Pearson approach, 461
noisy optimization, 105
nominal distribution, 93
noncentral χ^2 distribution, 439
norm, 386, 391
normal distribution, 45, 427, 436, 437
normal equations, 28
normal linear model, 46, 174, 182, 183, 440
normal matrix, 367
normal method (bootstrap), 89
normal model, 44
  Bayesian, 49, 50, 83
normal updating (cross-entropy), 101
null hypothesis, 460
null space, 365

O

object (Python), 466
objective function, 404, 405, 409, 417
Occam's razor, 173
operator, 391
operator (Python), 467
optimal decision boundary, 272
optimization
  combinatorial, 404
  constrained, 405
  continuous, 404
  unconstrained, 405

ordinary least-squares, 27
orthogonal
  basis, 363
  complement, 364
  matrix, 363, 384
  polynomial, 390
  projection, 364
  vector, 362
orthonormal, 363
  basis, 388
  system, 387
out-of-bag, 309
overfitting, 23, 35, 141, 172, 216, 237, 291, 295, 302, 316
overloading (Python), 470

P

p-norm, 220, 410
P-value, 195, 461
pandas (Python), 2, 487–492
Pareto distribution, 427
Parseval's formula, 394
partial derivative, 399
partition, 430
peaks function, 233
Pearson's height data, 207
penalty function, 417, 421
  exact, 418
percentile, 7
percentile method (bootstrap), 89, 91
permutation matrix, 370
Plancherel's theorem, 394
PLU decomposition, 370
pointwise squared bias, 35
pointwise variance, 35
Poisson distribution, 427
polynomial kernel, 230
polynomial regression model, 26
positive definite
  matrix, 405
positive semidefinite
  function, 223
  matrix, 369, 406
posterior
  pdf, 48
  predictive density, 49
precision, 257
predicted residual, 173
  — sum of squares (PRESS), 173
prediction function, 20
prediction interval, 186
predictive mean, 240
predictor, 168
primal optimization problem, 409
principal axes, 154
principal component analysis (PCA), 153, 155
principal components, 154
prior
  improper, 83
  pdf, 48
  predictive density, 49
  uninformative, 49
probability
  density function (pdf), 426
  density function (pdf), joint, 429
  distribution, 424, 429
  mass function, 426
  measure, 423
  space, 424
product rule, 74, 430, 454
projected subgradient method, 106
projection matrix, 27, 173, 211, 267, 364, 440
projection pursuit, 351
proposal (MCMC), 78
pseudo-inverse, 28, 211, 362, 380
Pythagoras' theorem, 180, 181, 183, 232, 363

Q

quadratic discriminant function, 262
quadratic program, 408
qualitative variable, 3
quantile, 51, 85
quantile–quantile plot, 199
quantitative variable, 3
quartile, 7
quasi-Newton method, 338, 413
quasi-random point set, 233
quotient rule for differentiation, 160

R

radial basis function (rbf) kernel, 226, 278
random
  experiment, 423
  number generator, 68
  numbers (Python), 484
  sample
    see iid sample, 456
  variable, 424
  vector, 429, 433
    covariance of, 432
    expectation of, 432
  walk sampler, 80
range (Python), 474
rank, 28, 358
rarity parameter (cross-entropy), 100
ratio estimator, 89
read_csv (Python), 2
recall, 257
reference (Python), 469
regional prediction functions, 290
regression, 20, 167
  function, 21
  line, 169
  model, 191
  simple linear, 181
regularization, 216, 217
  paths, 221
regularization parameter, 217
regularizer, 217
relative error (estimated), 85
relative time variance product, 456
renewal reward process, 89
representational capacity, 325
representer of evaluation, 223
reproducing kernel Hilbert space (RKHS), 223
reproducing property, 223
resampling, 76, 88
residual squared error, 171
residual sum of squares, 171
residuals, 171, 173
response surface model, 189
response variable, 20, 168
reverse Markov chain, 454
reversibility, 454
reversible Gibbs sampler, 82
ridge regression, 216, 217
Riemann–Lebesgue lemma, 393
right pseudo-inverse, 362
risk, 20, 167
Robbins–Monro algorithm, 106
root finding, 410
R^2, *see* coefficient of determination

S

saddle point, 405
  problem, 410
sample
  mean, 7, 85, 457
  median, 7
  quantile, 7
  range, 8
  space, 423
    countable, 424
    discrete, 424
  standard deviation, 8, 457
  variance, 8, 89, 457
saturation, 334
Savage–Dickey density ratio, 58
scale-mixture, 164
scatterplot, 13
scikit-learn (Python), 492–495
score function, 42, 123
  method, 113
secant condition, 413
semi-simple matrix, 366
sequence object (Python), 474
set (Python), 474
shear operation, 361
Sherman–Morrison
  formula, 174, 248, 373
  recursion, 374, 375, 416
significance level, 461
simple linear regression, 169, 187
simulated annealing, 96, 97
sinc kernel, 226
singular value, 379, 380
singular value decomposition, 154, 378
slack variable, 419
Slater's condition, 410

slice (Python), 3, 466
smoothing parameter, 100
softmax function, 269, 330
source vectors, 143
sparse matrix, 381
specificity, 257
spectral representation, 379
sphere the data, 266
splitting for continuous optimization, 103
splitting rule, 291
squared-error loss, 167
standard basis, 358
standard deviation, 428
    sample-, 457
standard error (estimated), 85
standard normal distribution, 436
standardization, 437
stationary point, 405
statistical (estimation) error, 32, 95
statistical test
    one-sided –, 460
    two-sided –, 460
statistics
    Bayesian, 456
    frequentist, 455
steepest descent, 332, 414
step-size parameter γ, 316
stochastic approximation, 106, 337
stochastic confidence interval, 459
stochastic counterpart, 107
stochastic gradient descent, 337, 351
stochastic process, 429
strict feasibility, 410
strong duality, 410
Student's t distribution, 183, 426, 441
    multivariate, 162, 164, 227
studentized residual, 212
stumps, 321
subgradient, 406
subgradient method, 106
sum rule, 424
supervised learning, 22
support vectors, 273
Sylvester equation, 381
systematic Gibbs sampler, 82

T

tables
    counts, 6
    frequency, 6
    margins, 7
target distribution, 78
Taylor's theorem
    multidimensional, 402
test
    loss, 24
    sample, 24
    statistic, 460
theta KDE, 134
time-homogeneous, 454
Tobit regression, 209
Toeplitz matrix, 381
total sum of squares, 181
tower property of expectation, 433
trace of a matrix, 359
training loss, 23
training set, 21
transformation
    of random variables, 433, 435
    rule, 95, 434
transition
    density, 74, 454
    graph, 75
transpose of a matrix, 358, 359
tree branch, 303
true negative, 256
true positive, 256
trust region, 411
type (Python), 468
type I and type II errors, 461

U

unbiased, 59
unbiased estimator, 456
unconstrained optimization, 405
uniform distribution, 427
union bound, 424
unitary matrix, 364
universal approximation property, 227
unsupervised learning, 22

V

validation set, 25, 305

Vandermonde matrix, 29, 395
Vapnik–Chernovenkis bound, 62
variance, 428, 432
    properties, 431
    sample, 89, 457
    sample-, 457
vector quantization, 143
vector space, 357
    basis, 357
    dimension, 357
Voronoi tessellation, 143

W

weak derivative, 113
weak duality, 409
weak learners, 315
Weibull distribution, 427
weight matrix (deep learning), 328
Wolfe dual program, 409
Woodbury identity, 249, 353, 373, 401